W0018777

ANDROGEN EXCESS DISORDERS IN WOMEN

CONTEMPORARY ENDOCRINOLOGY

P. Michael Conn, SERIES EDITOR

ANDROGEN EXCESS DISORDERS IN WOMEN

POLYCYSTIC OVARY SYNDROME AND OTHER DISORDERS

SECOND EDITION

Edited by

RICARDO AZZIZ, MD, MPH, MBA

Center for Androgen Related Disorders, Cedars-Sinai Medical Center;
and Department of Obstetrics and Gynecology, and Department of Medicine,
The David Geffen School of Medicine at UCLA, Los Angeles, CA

JOHN E. NESTLER, MD

Department of Internal Medicine, Virginia Commonwealth University,
Richmond, VA

DIDIER DEWAILLY, MD

Department of Endocrine Gynaecology and Reproductive Medicine,
Lille University Hospital, Lille, France

HUMANA PRESS
TOTOWA, NEW JERSEY

© 2006 Humana Press Inc.
999 Riverview Drive, Suite 208
Totowa, New Jersey 07512
humanapress.com
All rights reserved. No part of this book may be reproduced, stored in a retrieval system, or transmitted in any form
or by any means, electronic, mechanical, photocopying, microfilming, recording, or otherwise without written
permission from the Publisher.
All papers, comments, opinions, conclusions, or recommendations are those of the author(s), and do not necessarily
reflect the views of the publisher.

Due diligence has been taken by the publishers, editors, and authors of this book to assure the accuracy of the
information published and to describe generally accepted practices. The contributors herein have carefully
checked to ensure that the drug selections and dosages set forth in this text are accurate and in accord with the
standards accepted at the time of publication. Notwithstanding, as new research, changes in government
regulations, and knowledge from clinical experience relating to drug therapy and drug reactions constantly
occurs, the reader is advised to check the product information provided by the manufacturer of each drug for any
change in dosages or for additional warnings and contraindications. This is of utmost importance when the
recommended drug herein is a new or infrequently used drug. It is the responsibility of the treating physician to
determine dosages and treatment strategies for individual patients. Further it is the responsibility of the health
care provider to ascertain the Food and Drug Administration status of each drug or device used in their clinical
practice. The publisher, editors, and authors are not responsible for errors or omissions or for any consequences
from the application of the information presented in this book and make no warranty, express or implied, with
respect to the contents in this publication.

This publication is printed on acid-free paper. ∞
ANSI Z39.48-1984 (American Standards Institute) Permanence of Paper for Printed Library Materials.

Production Editor: Robin B. Weisberg

Cover design by Patricia F. Cleary

For additional copies, pricing for bulk purchases, and/or information about other Humana titles, contact Humana at
the above address or at any of the following numbers: Tel.: 973-256-1699; Fax: 973-256-8341; E-mail:
orders@humanapr.com; or visit our Website: www.humanapress.com

Photocopy Authorization Policy:

Photocopy Authorization Policy: Authorization to photocopy items for internal or personal use, or the internal or
personal use of specific clients is granted by Humana Press, provided that the base fee of US $30.00 per copy is
paid directly to the Copyright Clearance Center (CCC), 222 Rosewood Dr., Danvers MA 01923. For those
organizations that have been granted a photocopy license from the CCC, a separate system of payment has been
arranged and is acceptable to the Humana Press. The fee code for users of the Transactional Reporting Service is 1-
58829-663-6/06 $30.00.

Printed in the United States of America. 10 9 8 7 6 5 4 3 2 1

eISBN 1-59745-179-7

Library of Congress Cataloging-in-Publication Data

Androgen excess disorders in women : polycystic ovary syndrome and other disorders / edited
by Ricardo Azziz, John E. Nestler, Didier Dewailly.-- 2nd ed.
       p. ; cm. -- (Contemporary endocrinology)
 Includes bibliographical references and index.
 ISBN 1-58829-663-6 (alk. paper)
 1. Adrenal glands--Diseases. 2. Androgens--Metabolism. 3. Androgens--Physiological effect.
4. Polycystic ovary syndrome.
 [DNLM: 1. Adrenocortical Hyperfunction. 2. Hyperandrogenism. 3. Androgens.
4. Polycystic Ovary Syndrome. WK 770 A573 2006] I. Azziz, Ricardo. II. Nestler, John E.
III. Dewailly, Didier. IV. Series: Contemporary endocrinology (Totowa, N.J.)
 RG207.5.A53 2006
 618.1--dc22
                                                                          2006003453

DEDICATION

To Cindy, Ashlee, Jonathon, and Mallory, for their loving and patient
support and encouragement.

—*RA*

To my brother, Gregory, whose love, encouragement, and sense of humor
were always a source of comfort.

—*JEN*

To my family and friends, with love and fondness.

—*DD*

PREFACE

The field of androgen excess disorders has advanced substantially since the original publication of this book. The Androgen Excess Society (AES) was founded to bring together investigators in the field. A better understanding of the screening, progression, and molecular genetics of nonclassic adrenal hyperplasia (NCAH) has improved the clinical care and diagnostic accuracy of these patients. New criteria for the diagnosis of the polycystic ovary syndrome (PCOS) were proposed in Rotterdam, criteria that have resulted in controversy and, hopefully, initiation of new studies. The association of insulin resistance with PCOS has been strengthened, and the role of metformin in treating the infertility of the PCOS has been validated. Risks for diabetes and, more controversially, cardiovascular disease in women with PCOS have received substantial investigation. Our understanding of the epidemiology and economic impact of these disorders has expanded, emphasizing their critical importance. These are but a few highlights of how the terrain has changed in a relatively brief period of time.

In keeping with these advances, the title of this book has been revised to reflect the growing importance of PCOS as the most prevalent androgen excess disorder in women, and arguably, as the one that might have the most serious adverse consequences for general health. There are fewer chapters to provide a more focused elucidation of the area. Several chapters were penned by new (and young) authors who are conducting cutting-edge research in the field. The time line for publication of the book was intentionally kept tight, so that the information would be fresh and current. Although the editors have grown grayer, and not necessarily wiser, we nonetheless hope that this second edition of *Androgen Excess Disorders in Women: Polycystic Ovary Syndrome and Other Disorders* will prove to be a helpful guide and resource for all clinicians and researchers in the field of androgen excess.

Ricardo Azziz, MD, MPH, MBA
John E. Nestler, MD
Didier Dewailly, MD

ACKNOWLEDGMENTS

A text of this size and depth cannot be created by only a dedicated few and the authors must, out of fairness, acknowledge the support and assistance of many. We would like to thank our staff who helped in the preparation and formatting of the many drafts involved, particularly acknowledging the expert administrative assistance of Lois Dollar and April Moore. We also greatly appreciate the support, patience, and guidance offered to us by the editorial staff at Humana Press, notably Richard Lansing, Executive Director for Continuing Medical Education and New Project Development; Patrick Marton, Acquisitons Editor; and Robin B. Weisberg, Director of Editorial Services. Our young collaborators, and past and present fellows, deserve special thanks for their generous and enthusiastic intellectual contributions; a notable example is Dr. Daniela Jakubowicz who has played a significant role in furthering our understanding of the therapeutic value of metformin in the polycystic ovary syndrome. We are also deeply indebted to the many authors who gave freely of their time and effort, and who endured our frequent requests for revisions under a tight and unforgiving timeline. Finally, we acknowledge with profound gratitude our many patients who suffer from androgen excess disorders, and who trusted us to care for them, teaching us so much about these disorders and about living gracefully despite their disability.

Ricardo Azziz, MD, MPH, MBA
John E. Nestler, MD
Didier Dewailly, MD

CONTENTS

IV. The Polycystic Ovary Syndrome

CONTRIBUTORS

DAVID H. ABBOTT, PhD • *National Primate Research Center and Department of Obstetrics and Gynecology, University of Wisconsin, Madison, WI*

YVES ARDAENS, MD • *Department of Radiology, Hopital Jeanne de Flandre, Centre Hospitalier et Universitaire de Lille, Lille, France*

RICARDO AZZIZ, MD, MPH, MBA • *Department of Obstetrics and Gynecology, Cedars-Sinai Medical Center; Department of Obstetrics and Gynecology, and Department of Medicine, The David Geffen School of Medicine at University of California, Los Angeles, CA*

HANADI BA-AKDAH, MD • *Department of Obstetrics and Gynecology, McGill University, Royal Victoria Hospital, Montreal, Quebec, Canada*

JEAN-PATRICE BAILLARGEON, MD, MSC • *Division of Endocrinology, Department of Medicine, Université de Sherbrooke, Sherbrooke, QC, Canada*

ADAM H. BALEN, MD • *Department of Reproductive Medicine, The General Infirmary, Leeds, UK*

NICOLE BERGER, MD • *Laboratoire d Anatomie et de Cytologie pathologiqus, Hopital Lyon Sud, Pienne-Benite, France*

AUDE BRAC DE LA PERRIÉRE, MD • *Fédération d Endocrinologie, Pole Est des Hopitaux de Lyon, Bron, France*

GLENN D. BRAUNSTEIN, MD • *Department of Medicine, Cedars-Sinai Medical Center, Department of Medicine, The David Geffen School of Medicine at University of California, Los Angeles, CA*

GRANT D. BRINKWORTH, PhD • *CSIRO Human Nutrition, Adelaide, Australia*

WILLIAM BUCKETT, MD • *Department of Obstetrics and Gynecology, McGill University, Royal Victoria Hospital, Montreal, Quebec, Canada*

ENRICO CARMINA, MD • *Department of Clinical Medicine, University of Palermo, Palermo, Italy*

R. JEFFREY CHANG, MD • *Department of Reproductive Medicine, University of California, San Diego School of Medicine, La Jolla, CA*

KAI I. CHEANG, PharmD • *Department of Pharmacy, Virginia Commonwealth University School of Pharmacy, Richmond, VA*

THEODORE P. CIARALDI, PhD • *Department of Medicine, University of California, San Diego, VA San Diego Healthcare System, San Diego, CA*

HENRI DÉCHAUD, MD • *Laboratoire de Radioanalyse, Pole Est des Hopitaux de Lyon, Hopital Neuro-Cardio, Lyon, France*

EZGI DEMIRTAS, MD • *Department of Obstetrics and Gynecology, McGill University, Royal Victoria Hospital, Montreal, Quebec, Canada*

DIDIER DEWAILLEY, MD • *Service d'Endocrinologie et Diabetologie, Department of Endocrine Gynaecology and Reproductive Medicine, Hopital Jeanne de Flandre, Centre Hospitalier et Universitaire de Lille, Lille, France*

EVANTHIA DIAMANTI-KANDARAKIS, MD • *Department of Medicine, Athens University Medical School, Athens, Greece*

DANIEL A. DUMESIC, MD • *National Primate Research Center, University of Wisconsin, Madison, WI and Reproductive Medicine & Infertility Associates, Woodbury, MN*

ANDREA DUNAIF, MD • *Division of Endocrinology, Metabolism and Molecular Medicine, The Feinberg School of Medicine, Northwestern University, Chicago, IL*

DAVID A. EHRMANN, MD • *Section of Endocrinology, The University of Chicago, Chicago, IL*

SIGRID ELSENBRUCH, PhD • *Institute of Medical Psychology, University Hospital of Essen Medical School, Essen, Germany*

PAULINA A. ESSAH, MD • *Division of Endocrinology and Metabolism, Department of Internal Medicine, Virginia Commonwealth University, Richmond, VA*

WALTER FUTTERWEIT, MD, FACP • *Division of Endocrinology, Mount Sinai School of Medicine, New York, NY*

ALESSANDRA GAMBINERI, MD • *Department of Internal Medicine, S. Orsola-Malpighi, Bologna, Italy*

DAVID H. GELLER, MD, PhD • *Ahmanson Department of Pediatrics Cedars-Sinai Medical Center; Department of Pediatrics, The David Geffen School of Medicine at UCLA, Los Angeles, CA*

MARK O. GOODZARZI, MD, PhD • *Division of Endocrinology, Diabetes, and Metabolism, Cedars-Sinai Medical Center; Department of Medicine, The David Geffen School of Medicine at University of California, Los Angeles, CA*

SUSANNE HAHN, MD • *Division of Endocrinology, Department of Medicine, University Hospital of Essen Medical School, Essen, Germany*

PAUL HARDIMAN, MBBS, MD, FRCOG • *University Department of Obstetrics and Gynaecology, Royal Free and University College of Medical School, London, UK*

MARIE-FRANCE HIVERT, MD • *Division of Endocrinology, Department of Internal Medicine, Université de Sherbrooke, Sherbrooke, QC, Canada*

T. KETA HODGSON, BSN, RN • *Department of Medicine, Cedars-Sinai Medical Center, Los Angeles, CA*

KATHLEEN M. HOEGER, MD • *Department of Obstetrics and Gynecology, University of Rochester Medical Center, Rochester, NY*

HANANEL HOLZER, MD • *Department of Obstetrics and Gynecology, McGill University, Royal Victoria Hospital, Montreal, Quebec, Canada*

ONNO E. JANSSEN, MD • *Division of Endocrinology, Department of Medicine, University Hospital of Essen Medical School, Essen, Germany*

MARY C. JOHNSON, BS • *Department of Obstetrics and Gynecology, Penn State College of Medicine, Penn State University College of Medicine, M.S. Hershey Medical Center, Hershey, PA*

SOPHIE JONARD, MD • *Department of Endocrine Gynaecology and Reproductive Medicine, Hopital Jeanne de Flandre, Centre Hospitalier et Universitaire de Lille, Lille, France*

RICHARD S. LEGRO, MD • *Department of Obstetrics and Gynecology, Penn State College of Medicine, Hershey, PA*

JON E. LEVINE, MD • *Department of Neurobiology and Physiology, Northwestern University, Evanston, IL*

DENIS A. MAGOFFIN, PhD • *Department of Obstetrics and Gynecology, Cedars-Sinai Medical Center, Los Angeles, CA*

HOWARD D. MCCLAMROCK, MD • *Division of Reproductive Endocrinology and Infertility, Department of Obstetrics and Gynecology, and Reproductive Sciences, University of Maryland School of Medicine, Baltimore, MD*

RINKU V. MEHTA, MD • *Department of Reproductive Medicine, University of California, San Diego School of Medicine, La Jolla, CA*

WALTER L. MILLER, MD • *Division of Endocrinology, UCSF Children's Hospital, Division of Endocrinology, Department of Pediatrics, University of California San Francisco, San Francisco, CA*

PASCALE MIRAKIAN, MD • *Fédération d Endocrinologie, Pole Est des Hopitaux de Lyon, Bron, France*

PAOLO MOGHETTI, MD • *Division of Endocrinology and Metabolism, Department of Biomedical and Surgical Sciences, University of Verona, Verona, Italy*

DAVID E. MOLLER, MD • *Department of Molecular Endocrinology, Merck Research Laboratories, Rahway, NJ*

LISA J. MORAN, BND • *Research Center for Reproductive Health and CSIRO Human Nutrition, Adelaide, Australia*

BARBARA NERAUD, MD • *Service d'Endocrinologie et Diabetologie, Department of Endocrine Gynaecology and Reproductive Medicine, Hopital Jeanne de Flandre, Centre Hospitalier et Universitaire de Lille, Lille, France*

JOHN E. NESTLER, MD • *Division of Endocrinology and Metabolism, Department of Internal Medicine, Virginia Commonwealth University, Richmond, VA*

MANNY NOAKES, PhD • *Diet & Lifestyle Interventions, CSIRO Human Nutrition, Adelaide, Australia*

ROBERT J. NORMAN, PhD • *Department of Obstetrics and Gynecology, Research Centre for Reproductive Health, University of Adelaide, Adelaide, Australia*

FRANCESCO ORIO, MD, PhD • *Department of Molecular and Clinical Endocrinology and Oncology, University "Federico II," Naples, Italy*

VASANTHA PADMANABHAN, MD • *Departments of Pediatrics, Obstetrics and Gynecology, and Molecular and Integrative Physiology, University of Michigan Reproductive Sciences Program, Ann Arbor, MI*

STEFANO PALOMBA, MD • *Department of Obstetrics and Gynecology, University "Magna Graecia" of Catanzaro, Catanzaro, Italy*

C. RICHARD PARKER, JR., PhD • *Department of Obstetrics and Gynecology, University of Alabama at Birmingham, Birmingham, AL*

RENATO PASQUALI, MD • *Department of Internal Medicine, S. Orsola-Malpighi, Bologna, Italy*

JEAN LOUIS PEIX, MD • *Département de Chirurgie, Hopital Lyon Sud, Pienne-Benite, France*

INGRID PLOTTON, MD • *Fédération d Endocrinologie, Pole Est des Hopitaux de Lyon, Bron, France*

MICHEL PUGEAT, MD • *Fédération d Endocrinologie, Pole Est des Hopitaux de Lyon, Bron, France*

GÉRALD RAVEROT, MD • *Fédération d Endocrinologie, Pole Est des Hopitaux de Lyon, Bron, France*

ANNE-CÉLINE REYSS, MD • *Department of Endocrine Gynaecology and Reproductive Medicine, Hopital Jeanne de Flandre, Centre Hospitalier et Universitaire de Lille, Lille, France*

YANN ROBERT, MD • *Department of Radiology, Hopital Jeanne de Flandre, Centre Hospitalier et Universitaire de Lille, Lille, France*

MITCHELL ROSEN, MD • *Division of Reproductive Endocrinology & Infertility, Department of Obstetrics and Gynecology, UCSF Center for Reproductive Health, San Francisco, CA*

FRANK Z. STANCZYK, PhD • *Departments of Obstetrics and Gynecology, University of Southern California, Keck School of Medicine, Women & Children's Hospital, Los Angeles, CA*

SEANG LIN TAN, MBBS, FRCOG, FRCSC, MMed (O&G), MBA • *Department of Obstetrics and Gynecology, McGill University, Royal Victoria Hospital, Montreal, Quebec, Canada*

ANTONIO VIDAL-PUIG, MD, PhD • *Metabolic Medicine, Addenbrooke's Hospital, Cambridge, UK*

ABDULKADIR WAGLEY, MBchB, MRCOG • *Department of Obstetrics and Gynecology, Royal Free & University College Medical School, London, UK*

BULENT O. YILDIZ, MD • *Department of Medicine, Hacettepe University School of Medicine, and Department of Internal Medicine, Endocrinology & Metabolism Unit Hacettepe, Ankara, Turkey*

I Introduction

A Brief History of Androgen Excess

Ricardo Azziz

SUMMARY

Male-like hair growth and masculinization of women and the ambiguity of genders has fascinated mankind for millennia, frequently appearing in mythology and the arts. The earliest reports of androgen excess, beginning 400 years BC, focused on the appearance of male-like hair growth and features in women, often accompanied by menstrual cessation. The first etiologies identified as a cause of androgenization in the female were adrenal disorders, primarily adrenocortical neoplasms, but also eventually adrenal hyperplasia. The first report of a patient with nonclassic adrenal hyperplasia (NCAH) was made in 1957. The Achard–Thiers syndrome, which was originally reported in 1921 and was felt to primarily affect postmenopausal women, included the development of diabetes mellitus, hirsutism, and menstrual irregularity or amenorrhea in conjunction with adrenocortical disease. Androgen production by the ovary was not recognized until the early 1900s, with the first case of a patient with glucose intolerance, hirsutism, and ovarian pathology reported by Tuffier in 1914. As early as the mid-18th century, the presence of sclerocystic or multicystic ovaries was recognized, although this pathology was felt to be primarily associated with pelvic pain and/or menorrhagia. It was not until the seminal report of Drs. Stein and Leventhal of 1935 that the association of polycystic ovaries and amenorrhea, and possibly obesity and/or hirsutism, was noted. Subsequent investigations have elucidated the ovarian source of the androgens the gonadotropic abnormalities, the insulin resistance, and the high prevalence of the disorder, currently known as the polycystic ovary syndrome (PCOS). This syndrome was initially treated by ovarian wedge resection, but subsequent ovulatory therapies, including clomiphene citrate, menopausal gonadotropins, and most recently insulin sensitizers, have replaced this surgery as the treatment of choice for fertility improvement in PCOS. Notwithstanding, laparoscopic ovarian drilling retains a place in our current therapeutic armamentarium for these patients.

Key Words: History; hirsutism; adrenal hyperplasia; polycystic ovary syndrome; androgen-secreting neoplasms; Stein and Leventhal; Achard and Thiers.

1. INTRODUCTION

The earliest reports of androgen excess, beginning around 400 BC, focused on the appearance of male-like hair growth and features in women, often accompanied by menstrual cessation. Two other pathologies were then recognized, beginning in the late 17th and early 18th centuries, including adrenal pathologies and sclerocystic ovaries. These seemingly separate and disparate observations only begin to converge in the 20th century into the disorders we recognize today. The burgeoning research interest in androgen excess disorders can be indirectly determined from a survey of PubMed, the database managed by the National Center for Biotechnology Information, for articles published in scientific journals relating to the most common androgen excess disorders (Fig. 1). For example, in 1980 there were 69 articles published referencing "polycystic ovary syndrome" or "Stein–Leventhal syndrome"; in 1990 this number had more than doubled to 169, and in the year 2000 more than 241 articles were published on the subject.

From: *Contemporary Endocrinology: Androgen Excess Disorders in Women:*
Polycystic Ovary Syndrome and Other Disorders, Second Edition
Edited by: R. Azziz et al. © Humana Press Inc., Totowa, NJ

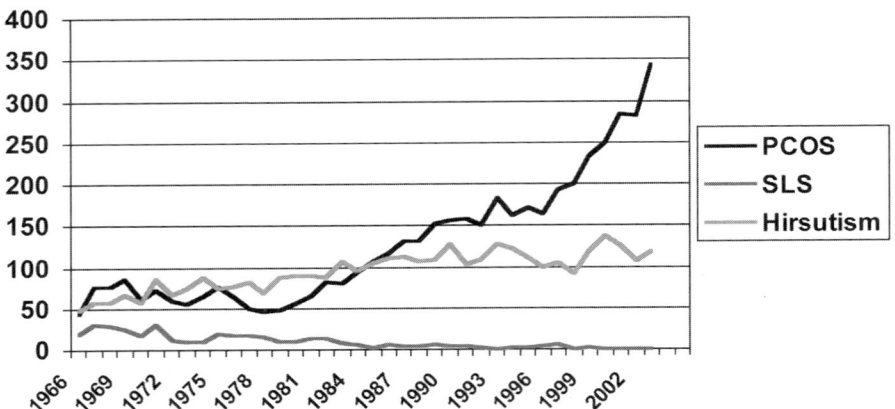

Fig. 1. Number of publications on polycystic ovaries and polycystic ovary syndrome (PCOS), Stein–Leventhal Syndrome (SLS), and hirsutism from 1966 to 2003 in PubMed.

In this chapter we briefly review the development of the study and treatment of androgen excess disorders with a historical perspective. This will not only illuminate how medical knowledge develops, but may assist us in placing this disorder, and in particular its definition, in perspective. In general, I will limit myself to studies published before 1980.

2. BACKGROUND

2.1. Hyperandrogenism in Mythology and the Arts

Male-like hair growth and masculinization of women, and the ambiguity of genders, has fascinated mankind for millennia. Hapi (or Hapy), god of the Nile, was described as an obese bearded man with the breasts of a woman *(1)*. Pharaoh Maatkare Hatshepsut of Egypt (1473–1458 BC), one of only two female pharaohs, was reported to be hirsute, although it would appear that she actually wore a false beard and dressed like a man primarily to emphasize her role as pharaoh and ruler. The Venus Barbata, the bearded cross-dressing Venus, was called on to repel unwanted husbands and suitors. A fresco in Pompei (painted sometime before the 97 AD eruption of Mount Vesuvius) depicts the toilet of Hermaphrodite with what appears to be a bearded woman assisting in the event *(2)*, although some investigators believe that it is a man dressed in eastern style (Dr. John Clarke, Professor of Art History, University of Texas at Austin, personal communication). The early medieval legend of St. Willegefortis (i.e., *virgo forte*, or strong virgin), or Uncumber, is described as growing a beard to ward off an unwanted suitor in order to remain chaste for God (and for which her father rewarded her by having her crucified) *(3)*. The fable about a bearded female pope, who later bore the name of Johanna (Joan), was first noted in the middle of the thirteenth century *(4)*.

Artists of all times have been fascinated by the hirsute woman. The most famous of these portraits is that of Magdalena Ventura de Los Abruzos, "La Mujer Barbuda" (1631, currently in the Hospital de Tavera, Toledo, Spain), by the Italian painter José (Jusepe) de Ribera. This portrait depicts Magdalena, who had arrived in Naples at 52 years of age from Acumulo in the region of Los Abruzos, alongside her second husband and a nursing child in arms. Apparently, after three miscarriages at the age of 37, she began to experience increasing hair. The Duke of Alcalá, then Viceroy of Naples, was so taken by her case that he commissioned Ribera to paint the woman and have her history recorded in the painting. Other important portraits include that of Brigida del Rio "La Barbuda de Peñaranda" (1590, currently in the museum El Prado), by the Spanish painter Juan Sánchez Cotán (ca. 1560–1627); Eugenia Martinez Vallejo "La Monstrua' Vestida," by Spanish painter J. Carreño De Miranda (1614–1685, currently in the museum El Prado); the portrait of Rosina Margerita Mullerin or Rosine-

Fig. 2. Portrait of Rosina Margerita Mullerin (or Rosine-Marguerite Müller), daughter of a servant to the court of the Saxonian elector Johann-Georg III, who died in 1732 (chalk, unknown artist, currently in the Kupferstich-Kabinett, Dresden, reproduced with permission).

Marguerite Müller (Fig. 2); Margret Halsebner of Basel or "The Woman with Two Beards" (or "Portrait of Old Woman With a Beard"), by Willem Key (~1515–1568; stolen in 1972 from the Suermondt Museum, Aachen; copy made by Anthonis Mor van Dashorst, ~1517–1577, currently in the Alte Pinakothek, Munich) (Fig. 3); and that of Helene Antonia of Liege (Belgium), depicted in an engraving by Johannes Loselius (~late 16th century).

Shakespeare referred to hirsute women in Macbeth (c.1607) when he wrote about three sisters:

> You should be women and yet
> Your beards forbid me
> To interpret that you are so

In "Don Quixote" (1615) Cervantes tells the tale of the bearded Countess Trifaldi (La Dolorosa, or the Distressed One), where the giant Malabruno, in revenge for the death of his first cousin, the Queen Maguncia, casts a spell on the Countess Trifaldi and her friends, making their faces hairy and bearded. In response, the lady Dolorosa laments *(5)*: ". . . where, I ask, can a duenna [female chaperon] with a beard go to? What father or mother will feel pity for her? Who will help her? For, if even

Fig. 3. "Old Woman with a Beard," by Anthonis Mor van Dashorst (c.1517–1577; currently in the Alte Pinakothek, Munich, reproduced with permission), copy of a portrait of Margret Halsebner of Basel, or "The Woman with Two Beards," by Willem Key (c.1515–1568; stolen in 1972 from the Suermondt Museum, Aachen).

when she has a smooth skin, and a face tortured by a thousand kinds of washes and cosmetics, she can hardly get anybody to love her, what will she do when she shows a countenance turned into a thicket?"

Hyperandrogenized women have frequently been held as a curiosity for all to view. Caufield recounts the story told by one D. George Sagari describing the case of the 22-year-old Augustina Barbara, daughter of Balthazer Ursler (or Ulster), whose whole body and face was covered by yellowish hair, including "a thick beard that reached her girdle"; he noted that her husband had married her ". . . merely to make a shew of her, for which purpose he traveled into various countries . . ." *(6)*. The famous Julia Pastrana, the "Nondescript" or Bearded and Hairy Lady, a 23-year-old of Mexican origin, attracted throngs of gawkers during her tour of Britain in 1857 *(7)*. However, it is likely that neither of these cases represented women solely with hyperandrogenism, as they appeared to suffer far more extensive hair growth and, at least in the case of Pastrana, facial distortion *(8)*.

Between 1840 and 1940, freak museums, circus sideshows, and carnivals were popular entertainment and frequently exhibited famous bearded ladies such as Annie Jones, Lady Olga, Clementine Delait, Madame Devere, and Princess Gracie *(9,10)*. By the early part of the century, entertaining and extensive descriptions of a multitude of hirsute or bearded women and their history, true or invented, had been published *(11,12)*. Even now, our fascination with bearded women continues. A contemporary news article presents the biography of Jennifer Miller, circus artist, performance artist, juggler and clown, and founder and director of Circus Amok *(13)*. Ms. Miller, by her mid-20s, had grown a full beard, although she was not sure what made it grow. Once, she said, "a doctor told me I had high progesterone."

2.2. Hyperandrogenism in Ancient Medicine

As early as the fifth century BC, Hippocrates (~460–~370 BC) described two separate cases of women who become generally hairy with beards and whose bodies assumed a masculine appearance; both became amenorrheic and died at an early age *(14)*:

> *Phaethufa* in *Abdera*, the wife of *Pytheus*, who had a child formerly when she was very young, upon her husband's being banfh'd, iffs'd her *Menses* a long time; and her joints grew afterwards painful and red. Upon this her body became *manly*, and *hairy* all over; a beard thrust out, and her voice became rough. Every thing was try'd by us that was likely to bring down her *Menses*, but to no purpose; and not long after she dy'd.

> The same thing happn'd in *Thasus* to *Namufias* the wife of *Gorgippus*. All physicians that I talk'd with were of the opinion, that the only hope left was in her *Menses* coming down again as they ought: But this cou'd never be brought about, tho' we try'd every thing; and she dy'd not long after.

Maimonides (1135–1204) noted that *(15)*:

> Just as there are men whose nature resembles the nature of women, and their skin is fair and soft, similar to the body of a woman, so too there are women whose skin is dry and hard, and whose nature resembles the nature of a man. However, if any woman's nature tends to be transformed to the nature of a man, this does not arise from medications, but is causes by heavy menstrual activity. (*Fin Liber Comm. Epidemirum VI; 8*)

The famous French surgeon and obstetrician Amboise Paré (1510–1590), in his "Concerning the Generation of Man" *(16)*, described affected women, noting:

> Many women, when their flowers or tearmes be stopped, degenerate after a manner into a certaine manly nature, whence they are called *Viragines*, that is to say stout, or manly women; therefore their voice is loud and bigge, like unto a mans, and they become bearded.

Notably, these early physicians clearly noted the association of menstrual irregularity and the development of hirsutism and masculine features. Clinical interest in the hirsutism continued in the ensuing centuries, as evidenced by the proliferation of portraits (*see* Figs. 2–4) recording the appearance and lives of these unfortunate women. Of note, during the late 19th century and early part of the 20th century, hirsutism in women was often ascribed to a number of conditions, including mental disorders *(12,17,18)* or hermaphroditsim *(19–21)*.

2.3. Adrenocortical Disorders

The first case describing androgenization of a female related to an adrenal disorder, most likely an adrenocortical carcinoma, appears to be that reported by Henry Sampson in 1697 *(22)*:

> *Hannah Taylor* was born in *Crouched Fryars June* 12 1682. She was till three Years old very sickly, lean and not able to go alone; but about *Bartholomewtide*, 1685. she began to grow strong and fat, which increased till the time of her Death: She was also a very forward Child of Understanding, had her *Pubes* grown thick and long, as also Hair under her Arm-pits, and Downyness upon her Chin, unusual with those of her Sex, except in some aged Persons. . . . She had a Face as big and broad as any fat grown Woman of 20 Years . . . the left Kidney (where was the seat of her Misery) exceeding large, and double the bigness of that on the right side. . . . The Testicles* were large, but smooth and white, without protuberances or shew of eggs.

The case appears to be also the earliest illustration of the effect that adrenal androgens may have on the ovaries. In 1905, Bulloch and Sequiera reported on an 11-year-old girl with almost identical

---

*Following Galen, the ovaries were termed female testis or testicles.

findings to that described by Sampson, which they diagnosed as having suffered from a left suprarenal carcinoma *(23)*. In addition to reviewing all similar prior cases, they make the then-novel observation that the "cortex of the suprarenal gland is probably connected in some way with the growth of the body and the development of puberty and sexual maturity."

De Crecchio in 1865 described the case of Giuseppe Marzo, who lived like a man and upon his death was found to have female reproductive organs and massively enlarged suprarenal (adrenal) capsules, essentially being the first report of virilizing congenital adrenal hyperplasia *(24)*. The Danish pathologist Johannes Andreas Grib Fibiger, recipient of the 1926 Nobel Prize for Physiology or Medicine, went on to describe a similar case *(25)*, followed by Debre and Semelaigne *(26)* and von Gierke *(27)*, such that the adrenogenital or adrenal virilizing syndrome became known as the Fibiger–Debré–von Gierke syndrome, a term that was in use until the early 1960s. Because previous treatment with various analogs of androgens (including 17-ethyl-testosterone, 17-vinyl-testosterone, 17-methylandrostenediol, and 17-methylandrostanediol) in an effort to block the effect of excess androgens in these patients had been unsuccessful, Lawson Wilkins and colleagues went on to describe the first successful treatment of a patient with congenital adrenal hyperplasia using intramuscular injections of cortisone crystals in aqueous solutions administered every 6 hours for 15 days in 1950 *(28)*, which was confirmed the following year *(29,30)*. Subsequent analysis of urinary steroids suggested that the likely biosynthetic defect was 21-hydroxylase deficiency *(31,32)*. Cloning of the active 21-hydroxylase gene (CYP21B or CYP21) and an associated pseudogene (CYP21B or CYP21P) confirmed the genetic etiology of the syndrome as a single gene mutation *(33)*.

The first report of a patient with nonclassic (also called nonclassical, late-onset, adult-onset, attenuated, acquired, cryptic, mild, partial, or postmenarchial) adrenal hyperplasia was made by Decourt et al in 1957 *(34)*; the following year these investigators reported on six cases and suggested that the disorder was the result of defective 21-hydroxylation *(35)*. The diagnosis of this disorder by adrenocorticotropic hormone stimulation test was suggested in 1979 by various investigators *(36–38)*. However, not until the Val 281Leu mutation was reported in nine patients with similar human leukocyte antigen (HLA) haplotypes (i.e., HLA-B14, DR1), establishing the molecular genetic basis of the disorder *(39)*, was NCAH confirmed as its own distinct entity.

2.4. Insulin Resistance, Glucose Intolerance, and Hyperandrogenism

At a session of the Société Medicale des Hôspitaux de Paris on July 19, 1921, Professor Emile Charles Achard, with the assistance of Dr. Joseph Thiers, presented the case of a 71-year-old woman who presented with poor health, incontinence, facial hair, and a history of glycosuria *(40)*. Although this is considered to be the first report documenting an association between glucose intolerance and hyperandrogenism, we should note that a number of other investigators had previously made this observation (*see* refs. *18* and *40*). Second, the patient described was not glycosuric at the time of her exam and remained so despite the ingestion of 100 g of glucose. Finally, the investigators noted on autopsy normal ovaries, evidence of chronic pancreatic, and hyperplastic and pigmented adrenals and ascribed the cause of the disorder, probably correctly, to the adrenal. The Achard–Thiers syndrome, as the disorder later became known, was felt to primarily affect postmenopausal women and included the development of diabetes mellitus, hirsutism, menstrual irregularity or amenorrhea, accompanied by suprarenal (adrenocortical) disease *(41–43)*. Hence, this syndrome does not refer to patients with insulin resistance and ovarian hyperandrogenism, but primarily to patients with adrenocortical pathology, notably carcinoma, glucose intolerance, and virilization. The hypercortisolism present may account for the high prevalence of psychiatric abnormalities observed in hirsute and amenorrheic women diagnosed at the time *(18)*.

In fact, the first case of a patient with glucose intolerance (i.e., glycosuria), hirsutism, and ovarian pathology appears to have been reported by Tuffier in 1914 *(44)*. This physician described a woman with virilization, glycosuria, bilateral adrenal hypertrophy, and a right ovarian mass the size of a walnut, possibly a luteoma; the glycosuria resolved after removal of the ovarian tumor. In 1947

Kierland reported on three patients with acanthosis nigricans who also presented with amenorrhea, hirsutism, obesity, and, in the two who were explored, normal adrenals *(45)*. Subsequent investigators continued to report on the association of metabolic abnormalities, hirsutism, and menstrual dysfunction, acanthosis nigricans, and frequently ovarian stromal hyperthecosis, in the absence of adrenocortical pathology.

In 1976 Kahn and colleagues described six patients with acanthosis nigricans and variable degrees of glucose intolerance, hyperinsulinemia, and marked resistance to exogenous insulin *(46)*. After the study, they divided these women into two unique clinical syndromes: type A, a syndrome in younger females with signs of virilization or accelerated growth, in whom the receptor defect may be primary, and type B, a syndrome in older females with signs of an immunological disease, in whom circulating antibodies to the insulin receptor are found. Women with the type A insulin resistance syndrome were later referred to as suffering from the hyperandrogenism, insulin resistance, and acanthosis nigricans (HAIR-AN) syndrome, a term still used today *(47)*.

2.5. The Stein–Leventhal Syndrome

As described earlier, Sampson, in his description of the autopsy findings of Hannah Taylor, observed that, in addition to the suprarenal tumor, the ovaries were large, smooth, white, and without protuberances, like eggs *(22)*. In 1721 Antonio Vallisneri (1661–1730) described the case of a young married infertile peasant woman, moderately obese, who had two larger than normal ovaries that were smooth and shiny *(48)*:

> But let us move on to the ovaries of the woman. On the day of February 3, a young married farmwoman who was infertile fell from a tree and died soon thereafter. I wanted to see the reason for this sterility, since the husband was also young and vigorous, and she was moderately plump, of good color, and well built.

> I found the two ovaries to be larger than ordinary: the left one was larger than a dove's egg, with a slightly dark color and more or less round, the right one was a little smaller, more white, and considerably crushed; and both of them were a tiny bit tuberculous, smooth, and shiny, almost as if from a paint that had been spread. Once the first tunica was removed, which was very fibrous and almost completely fleshy on the left, a blister was uncovered that was the size of a hazelnut with membranes that were very swollen and dense, covered and packed full of a material the color of soot, cloudy, nauseating, rather dense, but considerably liquid. On the left side, there were another two similar blisters that appeared to be only half as big which were also filled with the same material, as well as many others of various sizes although all of them were smaller, all dirty, stained, and likewise blackish. A healthy lymphatic blister does not appear this way, as it is usually pure and filled with transparent lymph.

> Between these, the usual fine membranes, fibers, and blood vessels could be distinguished, but they were nearly empty, and nothing else.

In 1844 Chereau *(49)* and Rokitansky *(50)* both described sclerocystic ovaries, which subsequent observers referred to as microcystic, cystic degeneration, polymicrocystic, cystic oophoritis, or hydrops folliculi, among other terms. Bulius and Kretschmar provided an early description of hyperthecosis *(51)*. Von Kahlden in 1902 *(52)* and Fogue and Massabuau in 1910 *(53–55)* published excellent in-depth reviews on the pathology and known clinical implications of these ovaries.

Multicystic ovaries were initially observed to be associated with menorrhagia, pelvic pain, and emotional disturbances. In 1872 Battey reported performing a bilateral oophorectomy in a 30-year-old patient who had intense pain during her menstrual periods, which had resulted in morphine addiction and epileptiform convulsions, among other complaints *(56)*. Although generally uninterested in the ovarian pathology of the hundreds of ovaries he subsequently removed as cure for similar ailments, Battey noted that while most ovaries were normal in appearance, some demonstrated cystic degeneration or sclerosis *(57)*. Encouraging this practice, Lawson Tait in 1879 affirmed the need for castration for the treatment of symptomatic cystic degeneration of the ovaries *(58)*. However, more conservative procedures, such as partial resection, were soon proposed *(59)*. Even then, critical voices

Fig. 4. (Left) Irving Freiler Stein (1887–1976); (right) Michael Leo Leventhal (1901–1971). (Reproduced with permission from: Harold Speert, Obstetric and Gynecologic Milestones. Parthenon Publishing Group, 1996.)

still abounded. John A. McGlinn, reading before the Obstetrical Society of Philadelphia on November 14, 1915, decried the use of ovarian resection in cases of microcystic disease of the ovary, and instead suggested simply puncturing "those cysts which are upon the surface" *(60)*.

Despite the proposed aggressive treatment of cystic degeneration of the ovaries, the etiology underlying this abnormality remained unclear. Fogue and Massabuau summarize the concepts of the time noting three potential mechanisms: inflammation, congestion, and dystrophy *(54)*. The inflammation theory proposed that the microcystic ovary was the result of an infection of either internal or external provenance. The congestion theory suggested that the lesion was the result of pressure, partial torsion, or other interruption in circulatory flow to the ovary. Finally, the dystrophy theory proposed that the abnormalities were caused by modifications or abnormalities in the nutrition of the ovary.

In 1935, in a seminal report, Stein and Leventhal brought together the disparate observations of menstrual dysfunction, microcystic or sclerocystic ovaries, and virilism or hirsutism *(61)*. Operating at Michael Reese Hospital in Chicago, these gynecologists (Fig. 4) reported on seven women in whom amenorrhea was associated with the presence of bilateral polycystic and enlarged ovaries; the heterogeneity of the syndrome was evident even at the initial description, with three of the women being obese and five hirsute. These investigators referred to the ovarian changes as "polycystic" rather than the terms used previously. Wedge resection was performed in all these patients, resulting in two pregnancies (both in the same patient, Case 1) and regular cycles in the remainder.

The etiology of the Stein–Leventhal syndrome remained unclear. Stein and Leventhal in their original report suggested that "the ovarian change in bilateral cystic ovaries is most probably a result of some hormonal stimulation" and that this stimulation was most likely the result of anterior pituitary secretions. They also proposed that "mechanical crowding of the [ovarian] cortex by cysts interferes with the progress of the normal Graafian follicles to the surface of the ovary." Although originally described as a distinct masculinization syndrome *(62)*, Culiner and Shippel suggested that theca luteinization (i.e., hyperthecosis) was an important ovarian mechanism in the Stein–Leventhal syndrome *(63)*.

A hypothalamic–pituitary–ovarian axis abnormality was initially hypothesized by Stein and Leventhal *(61)*, citing the works of investigators that had reported polycystic-looking ovaries after the injection of anterior pituitary extracts to patients undergoing surgery for uterine fibroids *(64)*. A number of investigators observed excess interstitial cell-stimulating hormone (a.k.a. luteinizing hormone) activity in the urine of women with the Stein–Leventhal syndrome, determined using a bioassay (i.e., the ovarian response of immature female rats or the prostatic response of hypophysectomized male rats to urinary extracts) *(65–68)*. These findings were later confirmed by plasma levels determined by radioimmunoassay *(69)*.

Prior observations in patients suffering from adrenal hyperplasia or adrenal neoplasms had suggested that the adrenal cortex was the only source of androgens in women. However, grafting experiments in mice and rats demonstrated the ovaries were able to restore androgenicity in castrate animals *(70,71)*, and Plate went on to postulate that some androgenic substance secreted by the ovaries in patients with the Stein–Leventhal syndrome was responsible for the symptoms observed *(72)*. Nevertheless, an adrenal etiology for the androgens in Stein–Leventhal syndrome was still considered relevant by some investigators, who used cortisone therapy in an attempt to improve ovulatory function in these women *(73,74)*. The basis for this recommendation lay primarily in the supposition that women with postpubertal hirsutism, oligomenorrhea, and infertility, and in particular those with elevated levels of urinary 17-ketosteroids, suffered from a form of adrenal hyperplasia *(75)*. It is notable that regardless of source, significant confusion still reigned regarding the role of androgens in this hyperandrogenic syndrome, highlighted by the proposal by Netter and Lambert in 1953 to treat sclerocystic ovaries with exogenous testosterone *(76)*.

In an effort to more fully understand the pathophysiology of this elusive disorder, Goldzieher and collaborators produced a thorough and in-depth series of studies of the clinical, histological, and biochemical features of the disorder in patients whose pathology was confirmed by surgical resection *(77–81)*. Concerning an adrenal component for PCOS, biochemical studies were inconclusive, leading Goldzieher to note that "[o]n the basis of so much work and so little yield in the attempt to distinguish ovarian from adrenal factors, one might well begin to wonder if there is not indeed an adrenocortical components in some cases of polycystic ovarian disease" *(82)*. These investigators also built on the work of others to identify the complete series of steps from pregnenolone to estrogen in normal and polycystic ovary tissue, confirming the production of androgens by normal ovarian tissue and the excessive production of the same in patients with polycystic ovaries *(79)*. Abnormally high production of urinary 17-ketosteroids had been used as the primary sign of excess androgen production in affected women. However, following the description of a method for measuring testosterone in plasma in 1961 *(83)*, increased circulating levels of this androgen in patients with polycystic ovaries and/or hirsutism was demonstrated shortly thereafter *(84,85)*.

Recommended treatment of women with polycystic ovaries consisted primarily of bilateral ovarian wedge resection *(86)*, with Stein claiming high rates of success in the treatment of infertility *(87)*. The Stein–Leventhal syndrome was subsequently defined as patients with secondary amenorrhea, hirsutism, sterility, hypoplasia of the uterus, and bilaterally enlarged polycystic or sclerocystic ovaries *(88)*, a definition primarily used to identify women who would appear to benefit from ovarian wedge resection. The syndrome was deemed rare, however, with Stein himself collecting only 90 cases in almost 30 years of practice *(89)*. The advent of clomiphene citrate *(90)* and the observation that ovarian wedge resection was associated with significant periovarian and peritubal adhesion formation *(91)* began to relegate this procedure to women who were resistant to clomiphene ovulation induction. And with the introduction of ovulation induction with menopausal gonadotropins *(92–94)* and laparoscopic ovarian cautery *(95)* for the treatment of clomiphene-resistant patients in the early to mid-1960s, the use of bilateral ovarian wedge resection for the treatment of the Stein–Leventhal syndrome was extinguished.

The heritable nature of PCOS was recognized in the late 1960s *(96,97)*, leading to our current search for the responsible gene variants. More recently, Burghen and colleagues demonstrated that patients with PCOS were more hyperinsulinemic than weight-matched controls, suggesting that these women were insulin resistant *(98)*, a finding rapidly confirmed by others *(99,100)*. It is noteworthy that Burghen and colleagues postulated finding hyperinsulinemia in patients with polycystic ovaries not because of the previous observations of the bearded diabetic woman or the observations made in women with acanthosis nigricans, but because of prior evidence suggesting that androgen administration altered carbohydrate metabolism. Shortly thereafter, the stimulatory effect of insulin in normal and polycystic ovary theca cells was demonstrated *(101–103)*, although it was not until the advent of insulin-sensitizing agents that reliable proof was obtained that insulin played an active role in the hyperandrogenism of patients with PCOS.

The utility of a polycystic ovarian morphology in the diagnosis of PCOS decreased considerably following the decline in the use of bilateral ovarian wedge resection for its treatment. However, the advent of ultrasonography once again revived the use of ovarian morphology as a sign of PCOS *(104,105)*. Additional details of the history of the discovery of PCOS have been published *(82,106,107)*.

3. CONCLUSIONS

Androgenization of women has captivated humankind for millennia, with early recognition of the relationship between menstrual dysfunction and the development of hirsutism and other virilizing features. Most early patients described appeared to suffer from ovarian or adrenal neoplasms, such that the hyperandrogenic symptoms were generally marked. Not until the early 20th century were lesser degrees of hyperandrogenism recognized as meriting medical evaluation, and only in the past century have significant strides been made in elucidating the etiology and pathophysiology underlying these disorders. Initially the adrenal cortex was recognized as a potential cause of androgen excess, with androgen-secreting neoplasms and frequently concomitant cushingoid features and later adrenal hyperplasia identified as a cause. Many of these women were also found to develop glucose intolerance or diabetes, a disorder known as "diabetes of the bearded woman" or the Achard–Thiers syndrome. NCAH was initially recognized in 1957, but it was not until three decades later that the molecular etiology could be established.

The presence of microcystic, sclerocystic, or cystic degeneration of the ovaries was recognized as early as the mid-19th century, although this pathology was primarily associated with pelvic pain, dysmenorrhea, and menorrhagia. Initially, treatment consisted of castration, although this was soon followed by the more conservative bilateral cuneiform or wedge resection. Recognition that this ovarian pathology could also be associated with amenorrhea, infertility, and hirsutism was not made until the report by Stein and Leventhal in 1935. The ovarian wedge procedure was used extensively to treat these women, although recognition that it could result in significant adhesion formation followed by the introduction of clomiphene citrate and then menotropins and laparoscopic electrocautery led to the demise of this surgical procedure for the treatment of polycystic ovary-associated amenorrhea and infertility. Our understanding of the steroidogenic, gonadotropic, heritable, and metabolic features of PCOS has increased in the past 50 years. It is hoped that progress in understanding the pathophysiology, and genetic and molecular basis, of this common and pervasive disorder will continue to increase exponentially.

KEY POINTS

- Multiple instances of hyperandrogenism, often significant and in association with amenorrhea, have been recorded, with the first cases attributed to Hippocrates in approx 500 BC.
- The first case of what appears to have been an adrenocortical carcinoma with cushingoid and virilizing features was reported by Sampson in 1697.
- Achard and Thiers first described the syndrome that eventually would bear their name in 1921; affected patients were generally postmenopausal and developed diabetes mellitus, hirsutism and/or virilization,

menstrual irregularity or amenorrhea, and suprarenal (adrenocortical) pathology.

- De Crecchio described the first patient with virilizing adrenal hyperplasia in 1865.
- Decourt and colleagues described the first patient with what would be later known as 21-hydroxylase-deficient NCAH in 1957.
- Large, smooth, and possibly polycystic-appearing ovaries were described initially by Vallisneri in 1752; the pathology of the sclerocystic or microcystic ovaries was then detailed separately in 1844 by Chereau and Rokitansky.
- In 1935, Stein and Leventhal described seven patients with bilateral polycystic ovaries and amenorrhea, reporting for the first time the disparate observations of menstrual dysfunction, microcystic or sclerocystic ovaries, and virilism or hirsutism.

ACKNOWLEDGMENTS

Supported in part by an endowment from the Helping Hand of Los Angeles and by grants RO1-HD29364 and K24-D01346 from the National Institutes of Health.

REFERENCES

1. Funk & Wagnalls Standard Dictionary of Folklore, Mythology, and Legend, Vol. 2. New York: Funk & Wagnalls Co., 1950:479.
2. Raehs A. Zur Ikonographie des Hermaphroditen. Frankfurt am Main: Peter Lang, 1990:53.
3. Lacey JH. Anorexia nervosa and a bearded female saint. Br Med J (Clin Res Ed) 1982;285:1816–1817.
4. Kirsch JP (transcribed by Marie Jutras). Popess Joan. In: The Catholic Encyclopedia, Volume VIII. Robert Appleton Company, 1910. (online edition, 2003, K. Knight; http://www.newadvent.org/cathen/08407a.htm).
5. Miguel de Cervantes Saavedra. Don Quijote de la Mancha. Segunda Parte, Ch. XXXIX, 1615 (trans. Ormsby, J. 1885;see http://www.donquixote.com/partwochap23.html).
6. Caufield J. Augustina Barbara Vanbecke, in Portraits, Memoirs, and Characters of Remarkable Persons, from the Reign of Edward the Third to the Revolution. Kirby, London 1813, pp. 168–169.
7. Laurence JZ. A short account of the bearded and hairy female. Lancet 1857;(July 7):48.
8. Browne J, Messenger S. Victorian spectacle: Julia Pastrana, the bearded and hairy female. Endeavour 2003;27:155–159.
9. Nohain J, Caradec F. La vie Exemplaire de la Femme a Barbe. Paris: La Jeune Parque, 1969.
10. Drimmer F. Very Special People. New York: Bantam Books, 1976.
11. Le Doulle A-F, Houssay F. Les Velus. Vigot Freres. 1912.
12. Berillon E. Les femmes a barbe. Rev Hypnotism Psychol Physiol 1905;4–11, 37–46, 67–74, 97–108, 132–142, 161–170, 195–203, 267–275, 290–297, 322–328, 356–362.
13. Smith D. Step right up! See the bearded person! New York Times 1995;(June 9):1.
14. Hippocrates. Epidemics Book VI, Aphorisms 55 & 56. In: Francis Clifton, ed. Hippocrates, Upon Air, Water and Situation; Uupon Epidemical Diseases; and Upon Prognosticks, in Acute Cases Especially. London: J. Watts Pub., 1734:172.
15. Rosner F, Munter S. The Medical Aphorism of Moses Maimonides, Vol. II. New York: Yeshiva University Press, 1971;34.
16. Paré, A. The causes of the suppression of the courses or menstrual fluxe. Chap. LI, Lib. 24. In: Johnson T, trans. The Workes of that famous Chirurgion Ambrole Parey: Translated out of Latine and compared with the French. London: Th. Cotes and R. Young, 1634:947.
17. Harris-Liston L. Coton Hill lunatic asylum—cases of bearded women. Br Med J 1894;(June 2):1190–1191.
18. Laignel-Lavastine M. Femmes à barbe et endocrinopsychiatrie. Paris Méd 1921;41:325–333.
19. Chowne WD. Remarkable case of hirsute growth in a female. Lancet 1852;1:421–422.
20. Chowne WD. Remarkable case of hirsute growth in a female (cont.). Lancet 1852;1:514–516.
21. Chowne WD. Remarkable case of hirsute growth in a female (cont.). Lancet 1852;2:51–53.
22. Sampson H. A Relation of one Hannah Taylor, a very Extraordinary Child of about Six Years of Age, who in Face, etc., was as large as a full grown Woman; and of what appeared on the Dissection of her Body. Philos Trans R Soc 1697;19:80–82.
23. Bullock W, Siquiera JH. The relation of the suprarenal capsules to the sexual organs. Trans Pathol Soc London 1905;56:189–208.
24. de Crecchio L. Sopra un Caso di Apparenze Virili in una Donna. Napoli: Morgagni, 1865:151–183.
25. Fibiger J. Beiträge zur Kenntniss des weiblichen Scheinzwittertums. Virchows Arch Pathol Anat Physiol Klin Med (Berl) 1905;181:1–51.
26. Debré R, Sémélaigne G. Hypertrophie considerable des capsules surrénales chez un nourrison mort a 10 mois sans avoir augmenté de poids depuis sa naissance. Bull Soc Pediatr (Paris) 1925; 23:270–271.

27. von Gierke E. Über Interrenalismus und interrenale Intoxikation. Verhandlung Dtsch Gesellschaft Pathol 1928;23:449–456.
28. Wilkins L, Lewis RA, Klein R, Rosemberg E. The suppression of androgen secretion in a case of congenital adrenal hyperplasia. Preliminary report. Bull Johns Hopkins Hosp 1950;86:249–252.
29. Wilkins L, Lewis RA, Klein R, et al. Treatment of congenital adrenal hyperplasia with cortisone. J Clin Endocrinol Metab 1951;11:1–25.
30. Bartter FC, Albright F, Forbes AP, et al. The effects of adrenocorticotropic hormone and cortisone in the adrenogenital syndrome associated with congenital adrenal hyperplasia: an attempt to explain and correct its disordered hormonal pattern. J Clin Invest 1951;30:237–251.
31. Eberlein WR, Bongiovanni AM. Partial characterization of urinary adrenocortical steroids in adrenal hyperplasia. J Clin Invest 1955;34:1337–1343.
32. Jailer JW, Gold JI, Vande Wiele R, Lieberman S. 17Alpha-hydroxyprogesterone and 21-desoxyhydrocortisone; their metabolism and possible role in congenital adrenal virilism. J Clin Invest 1955; 34:1639–1646.
33. White PC, Grossberger D, Onufer BJ, et al. Two genes encoding steroid 21-hydroxylase are located near the genes encoding the fourth component of complement in man. Proc Natl Acad Sci USA 1985;82:1089–1093.
34. Decourt J, Jayle MF, Baulieu E. Virilisme cliniquement tardif avec excretion de pregnanetriol et insuffisance de production de cortisol. Ann Endocrinol (Paris) 1957;18:416–422.
35. Jayle MF, Weinmann SH, Baulieu EE, et al. Post-pubertaire discret par deficience de l'hydroxylation en C_{21}. Acta Endocrinol 1958;29:512–524.
36. Gourmelen M, Pham-Huu-Trung MT, Bredon MG, Girard F. 17-Hydroxyprogesterone in the cosyntropin test: results in normal and hirsute women and in mild congenital adrenal hyperplasia. Acta Endocrinol (Copenh) 1979;90:481–489.
37. Rosenwaks Z, Lee PA, Jones GS, et al. An attenuated form of congenital virilizing adrenal hyperplasia. J Clin Endocrinol Metab 1979;49:335–339.
38. New MI, Lorenzen F, Pang S, et al. "Acquired" adrenal hyperplasia with 21-hydroxylase deficiency is not the same genetic disorder as congenital adrenal hyperplasia. J Clin Endocrinol Metab 1979;48:356–359.
39. Speiser PW, New MI, White PC. Molecular genetic analysis of nonclassic steroid 21-hydroxylase deficiency associated with HLA-B14,DR1. N Engl J Med 1988;319:19–23.
40. Achard EC, Thiers J. Le virilisme pilaire et son association à l'insuffisance glycotique (diabète des femmes à barbe). Bull Acad Natl Med (Paris) 1921;86:51–56.
41. Brown W Hurst. A case of pluriglandular syndrome: "diabetes of bearded women." Lancet 1928;2:1022–1033.
42. Shepardson HC, Shapiro E. The diabetes of bearded women (suprarenal tumor, diabetes, and hirsutism): a clinical correlation of the suprarenal cortex in carbohydrate metabolism. Endocrinol 1939;24:237–252.
43. Jeffcoate W, Kong M-F. Diabéte des femmes á barbe; a classic paper reread. Lancet 2000;356:1183–1185.
44. Tuffier B. Le virilisme surrénal. Bull Acad Natl Med (Paris) 1914; 71:726–731.
45. Kierland RR. Acanthosis nigricans: An analysis of data in twenty-two cases and a study of its frequency in necropsy material. J Invest Dermatol 1947;9:299–305.
46. Kahn CR, Flier JS, Bar RS, et al. The syndromes of insulin resistance and acanthosis nigricans. Insulin-receptor disorders in man. N Engl J Med 1976;294:739–745.
47. Barbieri RL, Ryan KJ. Hyperandrogenism, insulin resistance, and acanthosis nigricans syndrome: a common endocrinopathy with distinct pathophysiologic features. Am J Obstet Gynecol 1983;147:90–101.
48. Vallisneri A. Istoria della Generazione dell'Uomo, e degli Animali, se sia da' vermicelli spermatici, o dalle uova. Venezia, Appresso Gio. Gabbriel Hertz, 1721:146.
49. Chereau, Achilles. Mémoires pour Servir à l'Etude des Maladies des Ovaires. Paris: Fortin, Masson & Cie, 1844.
50. Rokitansky C. A Manual of Pathological Anatomy—Vol II. Philadelphia: Blanchard & Lea, 1855:246 (trans. by Edward Sieveking from original German 1844 ed.).
51. Bulius G, Kretschmar C. Angiodystrophia. Stuttgart: F. Enke. Verlag von Ferdinand Enke, 1897.
52. von Kahlden C. Uber die kleincystische Degeneration der Ovarien und ihre Beziehungen zu den sogenannten Hydrops folliculi. In: Ziegler E, ed. Beiträge zur pathologischen Anatomie und zur allgemeinen Pathologie. Jena, Germany: Verlag von Gustav Fischer, 1902:1–102.
53. Forgue E, Massabuau G. L'ovaire a petits kystes. Rev Gynecol Chirurg Abdom 1910;14:97–152.
54. Forgue E, Massabuau G. L'ovaire a petits kystes (cont.). Rev Gynecol Chirurg Abdom 1910;14:209–284.
55. Forgue E, Massabuau G. L'ovaire a petits kystes (cont.). Rev Gynecol Chirurg Abdom 1910;14:306–332.
56. Battey R. Normal ovariotomy. Atlanta Med Surg J 1872–1873;10:321–329.
57. Thiery M. Battey's operation: an exercise in surgical frustration. Eur J Obstet Gynecol Reprod Biol 1998;81:243–246.
58. Tait L. Removal of normal ovaries. Br Med J 1879;813:284.
59. Martin A. Ergebnisse der Ovarien und Tubenresektion. Verhandl Dtsch Ges Gynak 1891;4:242–257.
60. McGlinn JA. The end results of resection of the ovaries for microcystic disease. Am J Obstet Dis Women Child 1916;73:435–439.
61. Stein IF, Leventhal ML. Amenorrhea associated with bilateral polycystic ovaries. Am J Obstet Gynecol 1935;29:181–191.

62. Geist SH, Gains JA. Diffuse luteinization of the ovaries associated with the masculinization syndrome. Am J Obstet Gynecol 1942;43:975–983.
63. Culiner A, Shippel S. Virilism and theca cell hyperplasia of the ovary syndrome. J Obstet Gynaecol Br Comm 1949;14:631–653.
64. Geist SH. Reaction of the mature human ovary to antiutrin-s. Am J Obstet Gynecol 1933;26:588–592.
65. Keettel WC, Bradbury JT, Stoddard PJ. Observations on the polycystic ovary syndrome. Am J Obstet Gynecol 1957;73:954–965.
66. McArthur JW, Ingersoll FM, Worcester J. The urinary excretion of interstitial-cell and follicle-stimulating hormone activity by women with diseases of the reproductive system. J Clin Endocrinol Metab 1958;18:1202–1215.
67. Ingersoll FM, McArthur JW. Longitudinal studies of gonadotropin excretion in the Stein-Leventhal syndrome. Am J Obstet Gynecol 1959;11:795–805.
68. Taymor Ml, Barnard R. Luteinizing hormone excretion in the polycystic ovary syndrome. Fertil Steril 1962;13:501–512.
69. Yen SS, Vela P, Rankin J. Inappropriate secretion of follicle-stimulating hormone and luteinizing hormone in polycystic ovarian disease. J Clin Endocrinol Metab 1970;30:435–442.
70. Hill HT. Ovaries secrete male hormones: I. Restoration of the castrate type of seminal vesicle and prostate glands to normal by grafts of ovaries in mice. Endocrinology 1937; 21:495–502.
71. Deanesly R. The androgenic activity of ovarian grafts in castrated male rats. Proc R Soc Lond B Biol Sci 1938;126:122–135.
72. Plate WP. Hirsutism in ovarian hyperthecosis. Acta Endicrinol (Kbh) 1951;8:17–32.
73. Jones GE, Howard JE, Langford H. The use of cortisone in follicular phase disturbances. Fertil Steril 1953;4:49–62.
74. Greenblatt RB. Cortisone in treatment of the hirsute woman. Am J Obstet Gynecol 1953;66:700–710.
75. Jones HW, Jones GES. The gynecological aspects of adrenal hyperplasia and allied disorders. Am J Obstet Gynecol 1954;68:1330–1365.
76. Netter MA, Lambert A. Therapeutique medicale de l'ovarite sclero-kystique. C R Soc Fr Gyncol 1954;24:78–81.
77. Goldzieher JW, Green JA. The polycystic ovary. I. Clinical and histologic features. J Clin Endocrinol Metab 1962;22:325–338.
78. Goldzieher JW, Axelrod LR. The polycystic ovary. II. Urinary steroid excretion. J Clin Endocrinol Metab 1962;22:425–430.
79. Axelrod LR, Goldzieher JW. The polycystic ovary. III. Steroid biosynthesis in normal and polycystic ovarian tissue. J Clin Endocrinol Metab 1962;22:431–440.
80. Green JA, Goldzieher JW. The polycystic ovary. IV. Light and electron microscope studies. Am J Obstet Gynecol 1965;91:173–181.
81. Axelrod LR, Goldzieher JW. The polycystic ovary. V. Alternate pathways of steroid aromatization in normal, pregnancy and polycystic ovaries. J Clin Endocrinol Metab 1965;25:1275–1278.
82. Goldzieher JW, Axelrod LR. Clinical and biochemical features of polycystic ovarian disease. Fertil Steril 1963;14:631–653.
83. Finkelstein M, Forchielli E, Dorfman RI. Estimation of testosterone in human plasma. J Clin Endocrinol Metab 1961;21:98–101.
84. Forchielli E, Orcini G, Nightingale MS, et al. Testosterone in human plasma. Anal Biochem 1963;5:416–421.
85. Dignam WJ, Pion RJ, Lamb EJ, Simmer HH. Plasma androgens in women. II Patients with polycystic ovaries and hirsutism. Acta Endocrinol (Copenh) 1964;45:254–271.
86. Stein IF, Cohen MR. Surgical treatment of bilateral polycystic ovaries—amenorrhea and sterility. Am J Obstet Gynecol 1939;38:465–480.
87. Stein IF Sr. Duration of fertility following ovarian wedge resection—Stein-Leventhal syndrome. West J Surg Obstet Gynecol 1964;72:237–242.
88. Stein IF Sr. The Stein-Leventhal syndrome; a curable form of sterility. N Engl J Med 1958;259:420–423.
89. Stein IF Sr. Duration of fertility following ovarian wedge resection—Stein-Leventhal syndrome. West J Surg Obstet Gynecol 1964;72:237–242.
90. Greenblatt RB. Chemical induction of ovulation. Fertil Steril 1961;12:402–404.
91. Kistner RW. Peri-tubal and peri-ovarian adhesions subsequent to wedge resection of the ovaries. Fertil Steril 1969;20:35–41.
92. Gemzell CA, Diczfalusy E, Tillinger G. Clinical effect of human pituitary follicle stimulating hormone. J Clin Endocrinol Metab 1958;18:138–148.
93. Lunenfeld B, Menzi A, Volet B. Clinical effects of human postmenopausal gonadotropins. Acta Endocrinol (Kbh) 1960;51(Suppl):587.
94. Bettendorf G. Human hypophyseal gonadotropin in hypophysectomized women. Int J Fertil 1963;8:799–807.
95. Palmer R, de Brux J. [Histological, biochemical and therapeutic results obtained in women with Stein-Leventhal ovaries diagnosed by celioscopy]. Bull Fed Soc Gynecol Obstet Lang Fr 1967;19:405–412.
96. Cooper HE, Spellacy WN, Prem KA, Cohen WD. Hereditary factors in the Stein-Leventhal syndrome. Am J Obstet Gynecol 1968;100:371–387.

97. Moncada Lorenzo E. Familial study of hirsutism. J Clin Endocrinol Metab 1970;31:556–564.
98. Burghen GA, Givens JR, Kitabchi AE. Correlation of hyperandrogenism with hyperinsulinism in polycystic ovarian disease. J Clin Endocrinol Metab 1980;50:113–116.
99. Pasquali R, Venturoli S, Paradisi R, et al. Insulin and C-peptide levels in obese patients with polycystic ovaries. Horm Metab Res 1982;14:284–287.
100. Chang RJ, Nakamura RM, Judd HL, Kaplan SA. Insulin resistance in nonobese patients with polycystic ovarian disease. J Clin Endocrinol Metab 1983;57:356–359.
101. Barbieri RL, Makris A, Ryan KJ. Effects of insulin on steroidogenesis in cultured porcine ovarian theca. Fertil Steril 1983;40:237–241.
102. Barbieri RL, Makris A, Ryan KJ. Insulin stimulates androgen accumulation in incubations of human ovarian stroma and theca. Obstet Gynecol 1984;64(Suppl):73S–80S.
103. Barbieri RL, Makris A, Randall RW, et al. Insulin stimulates androgen accumulation in incubations of ovarian stroma obtained from women with hyperandrogenism. J Clin Endocrinol Metab 1986;62:904–910.
104. Swanson M, Sauerbrei EE, Cooperberg PL. Medical implications of ultrasonically detected polycystic ovaries. J Clin Ultrasound. 1981;9:219–222.
105. Parisi L, Tramonti M, Casciano S, Zurli A, Gazzarrini O. The role of ultrasound in the study of polycystic ovarian disease. J Clin Ultrasound. 1982;10:167–172.
106. Givens JT, Wild RA. Historical overview of the polycystic ovary. In: Dunaif A, Givens JR, Haseltine FP, Merriam GR, eds. Polycystic Ovary Syndrome. Boston: Blackwell Scientific Publications, 1992:3–18.
107. Goldzieher JW. Historical perspectives. In: Chang RJ, Heindel JJ, Dunaif A, eds. Polycystic Ovary Syndrome. New York: Marcel Dekker, 2002:1–14.

II   Physiology of Androgens

2

Ovarian and Adrenal Androgen Biosynthesis and Metabolism

Walter L. Miller, David H. Geller, and Mitchell Rosen

SUMMARY

The pathways of adrenal and ovarian steroid biosynthesis use the same enzymes for the initial steps of steroidogenesis but express different enzymes that convert steroid precursors to the final active products. Both the adrenal and ovary produce dehydroepiandrosterone (DHEA), the principal precursor of androgens and estrogens. The key enzyme in DHEA production is P450c17, which catalyzes both 17α-hydroxylation and 17,20-lyase activities. The 17,20-lyase activity of human P450c17 strongly favors 17-hydroxypregnenolone rather than 17-hydroxyprogesterone (17-OHP) as a substrate, producing abundant DHEA, so that most human androgens and estrogens derive from DHEA. Understanding the biochemistry of P450c17 is central to understanding the hyperandrogenism of polycystic ovary syndrome (PCOS). Rare genetic disorders of steroidogenesis provide human genetic knockout experiments of nature, yielding important information about the biosynthesis and physiological roles of steroids.

Key Words: Androgens; steroidogenesis; ovary; adrenal; 17,20-lyase; 17-hydroxylase; CYP21; P450c17; StAR; 3β-hydroxysteroid dehydrogenase; 17β-hydroxysteroid dehydrogenase; 5α-reductase; 11β-hydroxylase.

1. INTRODUCTION

The pathways of steroidogenesis employ a relatively small number of steroidogenic enzymes, but variations in their tissue specificity of expression and in the availability of substrates and cofactors result in the widely varying patterns of steroid production in each steroidogenic tissue (1). Although no cell type expresses all the steroidogenic enzymes, their interrelationships can be seen in the idealized integrated pathway shown in Fig. 1. Cholesterol is the precursor for all steroid hormones. The human adrenal and ovary can synthesize cholesterol *de novo* from acetate, but most cholesterol is provided by plasma low-density lipoproteins (LDLs) derived from dietary cholesterol. The presence of adequate LDL suppresses 3-hydroxy-3-methylglutaryl coenzyme A (HMG-CoA) reductase, the rate-limiting step in cholesterol synthesis. HMG-CoA reductase, as well as LDL receptor number and uptake of LDL cholesterol, are stimulated by adrenocorticotropic hormone (ACTH) in the adrenal and by follicle-stimulating hormone (FSH) and luteinizing hormone (LH) in the ovary. Steroidogenic cells take up LDL cholesterol esters by receptor-mediated endoctytosis to be either stored or immediately converted to free cholesterol for use as substrate in steroidogenesis. Storage of cholesterol esters in lipid droplets is under the control of two opposing enzymes, cholesterol esterase (cholesterol ester hydrolase) and cholesterol synthetase. LH and ACTH stimulate esterase to increase the availability of free cholesterol for steroidogenesis while inhibiting synthetase.

From: *Contemporary Endocrinology: Androgen Excess Disorders in Women: Polycystic Ovary Syndrome and Other Disorders, Second Edition*
Edited by: R. Azziz et al. © Humana Press Inc., Totowa, NJ

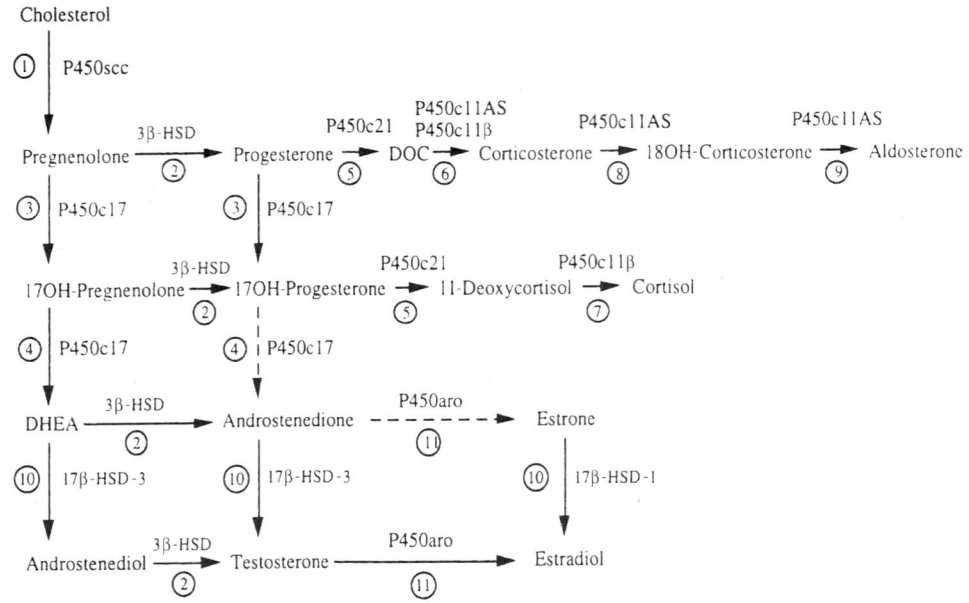

Fig. 1. Integrated view of human steroidogenesis showing adrenal and gonadal pathways. Reaction 1: P450scc converts cholesterol to pregnenolone. Reaction 2: 3β-Hydroxysteroid dehydrogenase (3β-HSD) converts Δ^5 steroids (pregnenolone, 17α-hydroxypregnenolone, dehydroepiandrosterone [DHEA], androstenediol) to the corresponding Δ^4 steroids (progesterone, 17α-hydroxyprogesterone, androstenedione, testosterone). Reaction 3: P450c17 catalyzes the 17α-hydroxylation of pregnenolone and progesterone. Reaction 4: The 17,20-lyase activity of P450c17 converts 17α-hydroxypregnenolone to DHEA; the conversion of 17α-hydroxyprogesterone to androstenedione occurs in cattle and rodents, but human P450c17 cannot catalyze this reaction efficiently. Reaction 5: P450c21 catalyzes the 21-hydroxylation of progesterone and 17α-hydroxyprogesterone. Reaction 6: Deoxycorticosterone (DOC) can be converted to corticosterone by either P450c11AS (in the adrenal zona glomerulosa) or P450c11β (in the adrenal zona fasciculata). Reaction 7: P450c11β converts 11-deoxycortisol to cortisol. Reactions 8 and 9: P450c11AS catalyzes 18 hydroxylase (reaction 8) and 18 methyl oxidase activities (reaction 9) to produce aldosterone in the adrenal zona glomerulosa. Reaction 10: Two isozymes of 17β-hydroxysteroid dehydrogenase (17β-HSD) activate sex steroids: 17β-HSD1 produces estradiol and 17β-HSD3 produces androgens. In peripheral tissues 17β-HSD5 has similar activity to 17β-HSD3, and 17β-HSD2 and 4 catalyze the "reverse" reactions to inactivate sex steroids. Reaction 11: P450aro aromatizes C19 androgenic steroids to C18 estrogens.

2. BACKGROUND

2.1. The Steroidogenic Enzymes

2.1.1. Early Steps: From Cholesterol to DHEA

2.1.1.1. THE STEROIDOGENIC ACUTE REGULATORY PROTEIN

Chronic regulation of steroidogenesis by LH or ACTH occurs at the level of gene transcription *(1)*, whereas more acute regulation leading to steroid secretion following an LH surge is controlled by cholesterol access to the rate-limiting enzyme P450scc *(2)*. This acute regulation is mediated by the steroidogenic acute regulatory protein (StAR), which facilitates the movement of cholesterol into the mitochondrion, where it becomes the substrate for the cholesterol side-chain cleavage enzyme, P450scc, the first steroid biosynthetic enzyme. StAR was first identified as short-lived 30- and 37-kDa phosphoproteins rapidly synthesized by steroidogenic cells in response to trophic hormone stimulation *(2)*.

The central role of StAR was proven by two observations. First, robust steroid hormone synthesis follows co-transfection of StAR and the cholesterol side-chain cleavage system into nonsteroidogenic COS-1 cells *(3,4)*. Second, patients with mutations of StAR have congenital lipoid adrenal hyperplasia, in which all adrenal and gonadal steroidogenesis is disrupted *(4,5)*. Thus, StAR is needed for the rapid flux of cholesterol from the outer to inner mitochondrial membrane to facilitate acute synthesis of aldosterone following angiotensin II stimulation, of cortisol following ACTH stimulation, and of gonadal sex steroids following LH stimulation.

StAR acts exclusively at the outer mitochondrial membrane, but its mechanism of action is not fully understood. Deletion of up to 62 of its N-terminal residues prevents StAR from entering the mitochondrion, yet it remains fully active *(6)*. When attached to the outer mitochondrial membrane, StAR becomes constitutively active, but it is inactive when localized to the mitochondrial intramembranous space, and protein import studies show that its level of activity is directly related to the time it spends on the outer membrane *(7)*. Conformational changes, apparently induced by the acidified outer mitochondrial membrane, are essential for StAR's activity *(8–10)*. StAR is ultimately targeted to the mitochondrial matrix, where it is degraded more rapidly than other mitochondrial proteins. Thus, StAR is an unusual, perhaps unique protein that exerts its action in a cellular compartment other than that to which it is ultimately targeted.

However, some steroidogenesis, notably that in the placenta, is independent of StAR. Non-steroidogenic COS-1 cells transfected with the cholesterol side chain cleavage system convert cholesterol to pregnenolone at 14% of the rate achieved by co-transfection with the cholesterol side-chain cleavage system plus StAR, establishing the presence of StAR-independent steroidogenesis *(4,5)*. The carboxyl half of a protein termed MLN64 is structurally related to StAR, exhibits StAR activity in vitro, and is cleaved from full-length MLN64 in the placenta, suggesting that it may play a role in placental steroidogenesis *(11,12)*. The mitochondrial peripheral benzodiazepine receptor also plays a role in movement of cholesterol into mitochondria, but its precise role is less well understood *(13)*.

2.1.1.2. CYTOCHROME P450

Steroidogenic enzymes fall into two broad categories: the cytochrome P450 enzymes and the hydroxysteroid dehydrogenases *(1)*. Cytochrome P450 includes a large group of enzymes containing about 500 amino acids and a single heme group; their name derives from the characteristic absorption peak at 450 nm. There are two classes of P450 enzymes. Type I enzymes are found in mitochondria and include P450scc and the two isozymes of P450c11. Type II enzymes are found in the endoplasmic reticulum and include the steroidogenic enzymes P450c17, P450c21, and P450aro. The human genome project has identified 57 P450 genes: 7 encode type I enzymes, all of which play key roles in sterol biosynthesis, and 50 encode type II enzymes. Of these 50 type II enzymes, about 20 participate in the biosynthesis of steroids, sterols, fatty acids, and eicosanoids, about 15 principally metabolize xenobiotic agents and drugs, and about 15 are "orphan" enzymes whose functions and activities remain unclear.

2.1.1.3. P450SCC

Conversion of cholesterol to pregnenolone by mitochondrial P450scc is the initial, rate-limiting, and hormonally regulated step in steroid hormone biosynthesis *(1)*. P450scc catalyzes three sequential chemical reactions: 20α-hydroxylation, 22-hydroxylation, and scission of the cholesterol side chain to yield pregnenolone and isocaproic acid. P450scc, encoded by a single gene on chromosome 15 *(14)*, possesses a single active site in contact with the hydrophobic lipid bilayer of the inner mitochondrial membrane. Deletion of the gene for P450scc in rabbits eliminates all steroidogenesis *(15)*, indicating that all steroid hormone biosynthesis is initiated through the action of this one enzyme. Haploinsufficiency of P450scc as a result of *de novo* heterozygous mutation causes a late-onset form of congenital lipoid adrenal hyperplasia *(16,17)*.

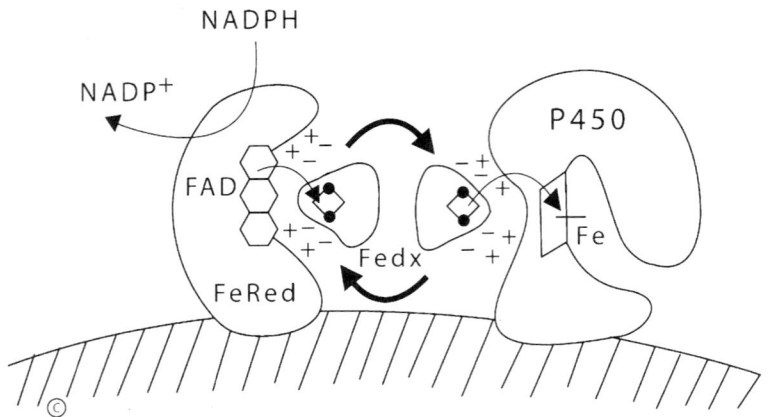

Fig. 2. Electron transfer by mitochondrial (type I) P450 enzymes. Nicotinamide adenine dinucleotide phosphate (NADPH) donates a pair of electrons to ferredoxin reductase, which is bound to the inner mitochondrial membrane. The flavin adenine dinucleotide (FAD) moiety of ferredoxin reductase passes the electrons to the iron/sulfur center of ferredoxin, depicted by a ball-and-stick diagram. Ferredoxin then dissociates from ferredoxin reductase, diffuses through the mitochondrial matrix, and interacts with the redox-partner binding-site of a type I P450. The electrons from the Fe_2S_2 center of ferredoxin then travel through an ill-described protein conduit in the P450 to reach the heme ring of the P450, which mediates catalysis.

Electrons are transported to P450scc by ferredoxin reductase and ferredoxin. All mitochondrial P450 enzymes function as the terminal oxidase in an electron transport chain *(1,18)* (Fig. 2). Electrons from reduced nicotinamide adenine-dinucleotide phosphate (NADPH) are accepted by a flavoprotein (ferredoxin reductase), loosely associated with the inner mitochondrial membrane *(19)*. Ferredoxin reductase then transfers these electrons to an iron/sulfur protein (ferredoxin) located either in the mitochondrial matrix or in the inner mitochondrial membrane *(20)*. In the final transfer, electrons are donated by ferredoxin to P450scc. No genetic disorders of these electron-transfer proteins have been described.

2.1.1.4. 3β-Hydroxysteroid Dehydrogenase/ Δ^5->Δ^4 Isomerase

Once cholesterol has been converted to pregnenolone, it may undergo 17α-hydroxylation by P450c17 to yield 17-hydroxypregnenolone, or it may be converted to progesterone, the first biologically important steroid hormone in the pathway (Fig. 1). A single 42-kDa microsomal enzyme, 3β-hydroxysteroid dehydrogenase (3β-HSD), performs both the conversion of a hydroxyl to a ketone group and the subsequent isomerization of the double bond from the B ring (Δ^5 steroids) to the A ring (Δ^4 steroids) *(21–23)*. A single enzyme therefore converts pregnenolone to progesterone, 17α-hydroxypregnenolone to 17α-hydroxyprogesterone (17-OHP), dehydroepiandrosterone (DHEA) to androstenedione, as well as androstenediol to testosterone, all with the same catalytic efficiency. Therefore, this enzyme is essential for the synthesis of both sex steroids and corticoids. Characteristic of members of the hydroxysteroid dehydrogenase family, there are two isozymes for 3β-HSD, encoded by separate genes with nearly identical nucleotide sequence, but different tissue expression. The type II enzyme is found in the adrenals and gonads, and the type I enzyme, encoded by a closely linked gene with identical intron/exon organization, is found in placenta, breast, and other extraglandular tissues, such as skin. Mutations have been identified only in the 3β-HSD-II gene; mutations in the 3β-HSD-I gene would presumably prevent adequate placental production of progesterone, thereby precipitating spontaneous abortion.

2.1.1.5. P450c17

Both pregnenolone and progesterone may undergo 17α-hydroxylation to 17α-hydroxypregnenolone and 17-OHP, respectively. The 17α-hydroxypregnenolone may also undergo scission of its C17,20 carbon bond to yield DHEA (Fig. 1). However, a small fraction of the 17-OHP is converted to androstenedione, although the human P450c17 enzyme catalyzes the Δ^4 substrate reaction at only 3% of the Δ^5 substrate rate (i.e.,17α-hydroxypregnenolone to DHEA) *(24)*. A single enzyme, P450c17, mediates all four reactions and also acts as a 16α-hydroxylase. Whereas P450scc functions as the quantitative regulator in determining the amount of steroid hormone production, P450c17 serves as the qualitative regulator of steroidogenesis, apportioning cholesterol metabolites to the three principal classes of steroid hormone. If neither activity of P450c17 is present (e.g., in the adrenal zona glomerulosa), pregnenolone is converted to mineralocorticoids; if 17α-hydroxylase activity is present in the absence of 17,20-lyase activity (e.g., in the adrenal zona fasciculata), pregnenolone is converted to the glucocorticoid cortisol. If both activities are present (e.g., in the gonads and the zona reticularis), pregnenolone is converted to precursors of sex steroids.

17α-Hydroxylase and 17,20-lyase were once thought to be separate enzymes. The adrenals of prepubertal, preadrenarchal children synthesize ample cortisol but negligible DHEA, indicating the presence of 17α-hydroxylase activity but not 17,20-lyase activity. During adrenarche the adrenal begins to produce DHEA and other C-19 steroids, suggesting that 17,20-lyase activity is turned on. Furthermore, some patients with apparently normal 17α-hydroxylase activity yet nearly absent 17,20-lyase activity have been described. However, both 17α-hydroxylase and 17,20-lyase activities reside in a single protein *(25)*. P450c17 is encoded by a single gene residing on chromosome 10q24.3 that bears structural relation to the gene for P450c21 *(26,27)*.

Thus, the distinction between 17α-hydroxylase and 17,20-lyase is functional, not genetic or structural. Human P450c17 catalyzes the 17α-hydroxylation of Δ^5 pregnenolone and Δ^4 progesterone with equal efficiency, but catalyzes the 17,20-lyase conversion of 17-OHP to Δ^4 androstenedione very poorly *(24)*. Thus, most sex steroid synthesis proceeds through DHEA, and little proceeds through 17-OHP. This is evidenced by the large amounts of DHEA produced by both fetal and adult adrenal glands. Moreover, the 17α-hydroxylase reaction is 20- to 25-fold more efficient (higher V_{max}/Km) than the 17,20-lyase reaction *(24)*. The major factor regulating the 17,20-lyase reaction, and thus the production of all androgens and estrogens, is electron transfer.

All microsomal P450 enzymes (including P450c17, P450c21, and aromatase) receive electrons from a membrane-bound flavoprotein, P450 oxidoreductase (POR) *(18)*. POR is an 82-kDa membrane-associated flavoprotein that has a bilobed structure *(28)*. A pair of electrons from NADPH is accepted by the flavin adenine dinucleotide (FAD) moiety in one lobe; this elicits flexion of a hinge region permitting the FAD moiety to move close to the flavin mononucleotide (FMN) moiety in the second lobe. The electrons jump from the FAD to the FMN, and then the protein "unflexes" on its hinge, permitting the FMN moiety to interact with the redox-partner binding site of the P450. The FMN of POR then gives up the electrons, which migrate through multiple paths to the heme group of the P450, where they mediate catalysis (Fig. 3). The 17,20-lyase activity of P450c17 can be increased by increasing the molar ratio of POR to P450c17 or by factors that increase the affinity of POR for P450c17.

Two posttranslational mechanisms, the presence of cytochrome b_5 (24) and the serine phosphorylation of P450c17 *(29)*, facilitate the interaction of P450c17 with POR to optimize electron transfer *(30)*. The adrenal zona reticularis, the site of production of adrenal androgen precursors, contains abundant cytochrome b_5, whereas the other adrenal zones have virtually none *(31)*. Thus, cytochrome b_5 appears to play a major role in human adrenal androgen synthesis, but its potential role in the ovary has not been explored. Increasing the ratio of either P450 oxidoreductase or cytochrome b_5 to P450c17 in vitro or in vivo favors the 17,20-lyase activity and P450c17 mutations that interfere with electron receipt from POR cause isolated 17,20-lyase deficiency *(32,33)*.

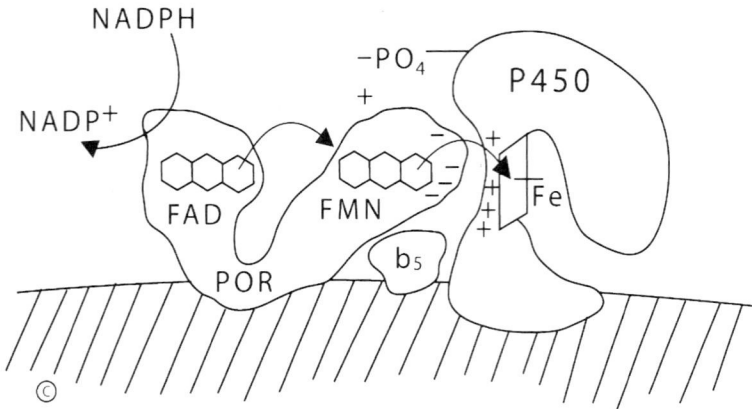

Fig. 3. Electron transfer by microsomal (type II) P450 enzymes. Nicotinamide adenine dinucleotide phosphate (NADPH) donates two electrons to the flavin adenine dinucleotide (FAD) moiety of P450 oxidoreductase (POR), bound to the endoplasmic reticulum. Electron receipt elicits a conformational change, permitting the FAD and flavin mononucleotide (FMN) moieties to come close together, so that the electrons pass from the FAD to the FMN. The protein then returns to its original orientation, and the FMN domain interacts with the redox-partner binding site of the P450, donating electrons to the heme group to achieve catalysis. The interaction of POR and the P450 is coordinated by negatively charged acidic residues on the surface of the FMN domain of POR and positively charged basic residues in the redox-partner binding site of the P450. In the case of human P450c17, this interaction is facilitated by the allosteric action of cytochrome b_5 and by the serine phosphorylation of P450c17.

There are three forms of cytochrome b_5, encoded by two genes. The gene on chromosome 18q23 undergoes alternative splicing to yield two proteins: the widely expressed 134AA form bound to the endoplasmic reticulum (which interacts with P450c17) and a soluble 98AA form found in erythropoietic tissues. A second gene on chromosome 16q22.1 encodes a form of cytochrome b_5 termed OMb_3 associated with the outer mitochondrial membrane. OMb_3 can support the 17,20-lyase activity of P450c17 in vitro, but it is not clear whether it serves this function in vivo *(18,30)*.

Although the kinase responsible for the serine/threonine phosphorylation of P450c17 has not yet been identified, it is clear that the balance achieved between this kinase and a counteracting phosphatase must regulate the level of P450c17 phosphorylation. Protein phosphatase 2A (PP2A), but not the closely related PP4 or PP6, will specifically dephosphorylate P450c17, and the level of PP2A activity in adrenal cells is regulated by a phosphoprotein termed SET *(34)*. SET is of interest because it acts as a transcriptional regulator of P450c17 gene transcription in mouse Leydig MA-10 cells, although a similar action with the human gene has not yet been demonstrated. The acquisition of 17,20-lyase activity through serine phosphorylation may provide a direct link to the insulin resistance of PCOS *(35)*, as girls who have increased 17,20-lyase activity and premature exaggerated adrenarche tend to develop PCOS as adults.

Although ablation of POR in mice results in embryonic lethality, numerous patients have been described as having POR mutations *(36–40)*. Most of these patients came to medical attention because they had a severe skeletal malformation disorder termed Antley–Bixler syndrome in addition to having disordered adrenal and gonadal steroidogenesis with a pattern suggesting combined deficiencies of P450c17, P450c21, and P450aro. Careful analysis of the enzymology of all known mutants shows that all affect P450c17, with some affecting the 17,20-lyase activity to a greater degree than 17α-hydroxylase activity *(40)*. The altered steroidogenesis in these patients leads to undervirilization of males and partial virilization of females. One of the initial patients reported was an adult woman who presented with PCOS *(36)*; whether this is a rare event or indicates that POR may be more commonly associated with some forms of PCOS is not yet known.

2.1.2. Adrenal-Specific Enzymes

2.1.2.1. P450c21

Progesterone and 17-OHP may be hydroxylated at their C21 position to produce deoxycorticosterone (DOC) and 11-deoxycortisol, respectively (Fig. 1). The nature of the 21α-hydroxylating step has been of intense interest because more than 90% of cases of congenital adrenal hyperplasia (CAH) are caused by defects in P450c21. The severe form of this common genetic disease can be fatal. Compromised cortisol and aldosterone synthesis can lead to hyponatremia, hyperkalemia, hypotension, and circulatory collapse, leading to death within the first month of life if not treated appropriately. Decreased cortisol production *in utero* results in a compensatory increase in ACTH, with consequent overstimulation of the adrenal steroid biosynthesis. 17-OHP levels rise in the face of impaired 21-hydroxylation and the limited ability of P450c17 to convert 17-OHP to androstenedione. However, continued ACTH-stimulated flux of cholesterol results in accumulation of 17-hydroxypregnenolone, which is then converted to DHEA, androstenedione, and ultimately testosterone, resulting in severe prenatal virilization of female fetuses *(41)*. Mild, "nonclassical" variants of CAH are a common cause of adrenal (but not gonadal) hyperandrogenism.

The gene for P450c21 lies within the major histocompatibility locus on chromosome 6p21, and mutations in P450c21 are linked to specific human leukocyte antigen types. The genetics of this locus are complicated by its very high rate of genetic recombination. As a result, there can be one, two, three, or four copies of the P450c21 gene, but only one is functional, and its mutations arise from recombination with the nearby inactive pseudogene *(41)*. Extra-adrenal 21α-hydroxylase activity, observed in a variety of fetal and adult tissues remains poorly characterized, but is not mediated by the same P450c21 enzyme that is found in the adrenal glands *(42)*.

2.1.2.2. P450c11β AND P450c11AS

The final steps in the mineralocorticoid and glucocorticoid pathways are catalyzed by P450c11β and P450c11AS *(43,44)* (Fig. 1). These two isozymes share 93% amino acid sequence identity and are encoded by tandemly duplicated genes located on chromosome 8q21-22. Both forms of P450c11 reside on the inner mitochondrial membrane and, like P450scc, utilize ferredoxin reductase and ferredoxin to receive electrons from NADPH. P450c11β is abundantly expressed in the zona fasciculata, where it catalyzes the classic 11β-hydroxylase reactions that convert 11-deoxycortisol to cortisol and DOC to corticosterone. The less abundant P450c11AS (aldosterone synthase) is uniquely expressed in the zona glomerulosa, where it catalyzes the sequential steps 11β-hydroxylase, 18-hydroxylase, and 18-methyloxidase necessary to produce aldosterone.

2.1.2.3. STEROID SULFOTRANSFERASE AND SULFATASE

Steroid sulfates may be synthesized directly from cholesterol sulfate or by sulfating Δ^5 steroids by the main sulfotransferases, SULT2A1 and SULT2B1, and the phenolic sulfotransferase SULT1E1. The addition of a sulfate group prevents the activation of Δ^5 steroids to Δ^4 steroids by 3β-HSD. SULT2A1, which is expressed primarily the adrenals and to a lesser degree the liver and kidney, sulfates DHEA *(45)*. SULT2A1 is not expressed in the ovary *(45)*, whereas SULT2B1 and SULT1E1 are widely expressed *(46)*. SULT2B1 and SULT1E1 sulfate the other 3β-hydroxysteroids and estrogens, respectively. Steroid sulfates may be hydrolyzed to their native state by steroid sulfatase. Steroid sulfatase, encoded on chromosome Xp22.3, desulfates most 3β-hydroxysteroids and is expressed in most tissues, including the ovary. The principal role of steroid sulfatase is in disposing of excess steroid sulfates. When steroid sulfatase activity is absent in X-linked icthyosis, steroid sulfates accumulate in the stratum corneum of the skin. However, this condition has little impact on adrenal or gonadal steroidogenesis.

2.1.3. Enzymes Involved in Sex Steroid Synthesis

2.1.3.1. 17β-HYDROXYSTEROID DEHYDROGENASES

The 17β-hydroxysteroid dehydrogenases (17β-HSDs) interconvert 17-ketosteroids with the corresponding 17-hydroxysteroids, thereby controlling the synthesis and metabolism of sex steroids. Although these enzymes can catalyze both oxidative and reductive reactions in vitro, the presence of substantial excess of a suitable cofactor in vivo (or in the absence of a preferred cofactor) will typically direct each enzyme in the oxidative or reductive direction (47–49). The reductive enzymes utilize NAD(P)H as the cofactor and the oxidative enzymes utilize NAD(P)$^+$. Five principal human isoforms have been identified, some of which are preferential oxidases, whereas others are principally reductases. These isoforms share only 20–30% sequence identity, and yet their secondary and tertiary structures are remarkably similar.

Type I 17β-HSD (17β-HSD-I) is the estrogenic form expressed in the placenta, where it produces estriol, and in the ovary, where it catalyzes the last step in estradiol synthesis (50,51). 17β-HSD-I, encoded by a gene on chromosome 17q21 near the BRCA locus, is also expressed in endometrium, breast, testis, adipose tissue, skin, liver, and prostate. 17β-HSD-I is a homodimer that uses NADPH as its cofactor and converts estrone to estradiol. The crystallographic structure of 17β-HSD-I shows that the active site binds steroids that have a planar A-ring and are missing the C-19 methyl group, whereas steroids with a C-19 methyl group or a nonplanar A-ring cannot be bound (52). Human ovarian granulosa cells, but not theca cells, express 17β-HSD-I (51).

17β-HSD-II oxidizes estradiol to estrone and testosterone to Δ^4-androstenedione with equal efficiency using NAD$^+$ as a cofactor (53,54). This enzyme can also oxidize C-20 substrates (20α-HSD activity), converting 20α-hydroxyprogesterone to progesterone (54). 17β-HSD-II, encoded on chromosome 16q24, shares only 20% amino acid sequence identity with 17β-HSD-I. 17β-HSD-II is found in the placenta, breast, liver, small intestine, prostate, secretory endometrium, kidney, and ovary. 17β-HSD-II is expressed in the endothelial cells of the placental intravillous vessels, consistent with its role in defending the fetal circulation from transplacental passage of maternal estradiol or testosterone (55). In the endometrium, 17β-HSD-II is expressed in the secretory phase and maintains predominance of progestational activity by oxidizing estradiol and converting 20α-hydroxyprogesterone back to progesterone.

17β-HSD-III is an androgenic enzyme that uses NADPH as a cofactor and reduces androstenedione to testosterone and DHEA to androstenediol. The 17β-HSD-III gene on chromosome 9q22 is expressed primarily in the testis and adipose tissue, but is not expressed in the ovary. This is the only form of 17β-HSD for which a deficiency state is known, causing the classic syndrome of male pseudohermaphroditism, often termed 17-ketosteroid reductase deficiency (56).

17β-HSD-IV, originally identified as a NAD$^+$-dependent oxidase with activities similar to 17β-HSD-II, is a widely distributed peroxisomal protein that acts primarily as a 2-enoyl-CoA hydratase and 3-hydroxyacyl-CoA dehydrogenase (47,48). The gene for 17β-HSD-IV on chromosome 5q2 is ubiquitously expressed. Its role in sex-steroid metabolism is not established, but in vitro it oxidizes the C-18 and C-19 substrates using NAD$^+$ as a cofactor.

17β-HSD-V, initially cloned as a 3β-hydroxysteroid dehydrogenase, primarily catalyzes the conversion of Δ^4 androstenedione to testosterone (57) but also has 20β-HSD activity. 17β-HSD-V is expressed in most tissues, including liver, kidney, blood vessels, and testis, prostate, adrenal, bone, and ovary (57). Whereas the other 17β-HSDs are members of the short-chain dehydrogenase family, 17β-HSD-V is a member of the aldoketoreductase family. The peripheral conversion of Δ^4 androstenedione to testosterone by 17β-HSD-V in target tissues is apparently responsible for the "weak androgen" action of androstenedione. It is also the only form of 17β-HSD expressed in the ovarian theca and corpus luteum, suggesting that 17β-HSD-V is the enzyme responsible for the last step in testosterone production in the ovary.

2.1.3.2. 5α-Reductase

Testosterone can be converted to the more potent androgen dihydrotestosterone (DHT) by 5α-reductase in target tissues, especially the skin. There are two isoforms of 5α-reductase, which share about 50% homology. Each is a membrane-associated enzyme that utilizes NADPH as a cofactor to reduce the 4-to-5 double bond in ring A. Both enzymes are expressed in several tissues, including the urogenital tract, skin, gastrointestinal and respiratory tracts, liver, brain, and ovary. The type II enzyme is critical for the development of normal male external genitalia, but an essential role is not established in women. By contrast, type I may have a role in fecundity and parturition *(58)*.

2.1.3.3. P450aro: Aromatase

Estrogens are produced from androgens by a complex series of reactions catalyzed by a single aromatase enzyme P450aro *(59)*. This microsomal enzyme is encoded by a single gene on chromosome 15q21.1; this gene is unusual in that it uses several different alternative transcriptional start sites and first exons driven by different upstream promoter sequences, permitting the same protein to be expressed under different control in different cell types *(59)*. P450aro in peripheral tissues, especially fat, can convert substantial portions of circulating androstenedione and testosterone in women to estrone and estradiol. The placenta expresses large amounts of aromatase, protecting the fetus from maternal androgens and permitting disposal of fetal C-19 steroids *(60)*. Although the placenta produces huge amounts of estriol from fetally produced DHEA, estriol is not needed for normal pregnancy, as shown by the normal development, labor, and parturition of fetuses that have genetic lesions that prevent estriol production *(61)*.

2.2. Circulating Sex Steroids

Dehydroepiandrosterone sulfate (DHEAS) is the most abundant steroid in the circulation of adults of reproductive age *(62)*. DHEA, DHEAS, and androstenedione are produced almost exclusively by the adrenal zona reticularis. The adrenal does not express 17β-HSD-III; the minimal adrenal production of testosterone is probably a result of 17β-HSD-V *(57)*. Adrenal C-19 steroids do not bind significantly to the androgen receptor; hence these steroids are primarily precursors that are converted to active androgens or estrogens by isozymes of 17β-HSD in target tissues. The normal adrenal does not express aromatase and therefore does not synthesize estrogens, but adrenocorticol carcinomas may produce estrogens.

Metabolism of steroidal precursors by skin and fat, which express aromatase and 17β-HSD-I, -III and -V *(63)*, produces most of the circulating testosterone. The relative expression of these enzymes determines how androstenedione will be metabolized. These tissues also express 3β-HSD-I and steroid sulfatase, converting DHEAS to androstenedione. Most of the DHT produced in target tissues acts in an autocrine or paracrine fashion. Sebaceous glands and hair follicles express 5α-reductase-I and -II respectively, converting testosterone to DHT at this important site of action *(31)*.

2.3. Cell Biology of Sex Steroid Synthesis

2.3.1. In the Ovary

Ovarian theca and granulosa cells in individual follicular units synthesize both androgens and estrogens. The steroidogenic enzymes expressed in the theca or granulosa cells vary with the menstrual cycle *(64)*. In the follicular phase, theca cells express StAR, P450scc, 3β-HSD-II, and P450c17 to produce androstenedione, some of which is converted to testosterone by 17β-HSD-V. Most thecal steroids diffuse to the granulosa cell, but some are secreted into the circulation. Rising thecal androstenedione production during the follicular phase plus increased expression of 17β-HSD-I and P450aro in the granulosa cells results in abundant estradiol production by the preovulatory follicle. Theca cells continue to provide androstenedione to the granulosa cell during the luteal phase, while the granulosa cell undergoes granulosa-lutein transformation and expresses StAR, P450scc, and 3β-HSD-II, but not P450c17 *(65)*. This permits the granulosa-lutein cell to produce progesterone while converting C-19 steroids from the theca to estrogens.

Numerous follicles grow simultaneously in the reproductive ovary. During the recruitment and growth phase, the granulosa cells proliferate, differentiate, and become steroidogenic. During the growth phase, most follicles undergo atresia, while some remain viable. The "antral" follicular stage is characterized by antrum formation and acquisition of differentiated steroidogenic cells *(66)*. LH induces theca cells to synthesize androgens before FSH induces the granulosa cells to aromatize them. Only the granulosa cells surrounding the follicle express aromatase and produce estradiol. During the 2 weeks before ovulation (the follicular phase), LH increases the theca cell expression of the LH receptor, StAR, P450scc, 3β-HSD-II, and P450c17, whereas FSH increases granulosa cell expression of aromatase and 17β-HSD-I. The majority of ovarian C-19 steroids are produced from DHEA. While rodents and other species produce androstenedione directly from 17-OHP, the 17,20-lyase activity of human P450c17 strongly favors 17α-hydroxypregnenolone as a substrate and utilizes 17-OHP inefficiently *(24)*.

Intraovarian peptides, including members of the insulin-like growth factor, transforming growth factor, and epidermal growth factor families, modulate both folliculogenesis and steroidogenesis. Germ cell differentiating factor-9, expressed by the oocyte throughout folliculogenesis, appears to promote granulosa cell differentiation, stimulate theca cells, and inhibit luteal cell formation *(67)*. Insulin-like growth factors appear to enhance responses to FSH. Both insulin-like growth factor (IGF)-1 and IGF-2 can increase granulosa cell proliferation and estradiol secretion, but only IGF-2 mRNA is expressed in the ovary, particularly in the granulosa cells *(68)*. Women with IGF-1 deficiency can be induced to ovulate by stimulation with gonadotropins *(69)*, consistent with the view that IGF-1 is not required for folliculogenesis.

Granulosa cells also produce inhibin, an α/β heterodimer that exists in two forms differing in the β-subunit. Serum concentrations of inhibin A and B vary with the menstrual cycle: inhibin A rises in follicular fluid with increasing follicular size, beginning just before ovulation and peaking during the luteal phase, and inhibin B increases in granulosa cells following stimulation with FSH *(70)*. Serum concentrations of inhibin B correlate with the volume of granulosa cells, serving as an index for the size of the growing cohort of follicles (ovarian "reserve"). Inhibin B peaks in the early follicular phase and inhibits pituitary production of FSH during the later follicular phase. The antral follicle grows 1–2 mm in diameter daily during the antral phase of follicular development in response to gonadotropins *(66)*, reaching a diameter of about 20 mm, primarily representing accumulated follicular fluid. The theca interna continues to differentiate into interstitial cells, generating more androstenedione. The granulosa cell layers differentiate into discrete zones: the membrana layer subsequently acquires LH receptors in response to FSH, whereas the cumulus layer constituitively expresses LH receptors. One dominant follicle destined for ovulation develops into a mature graafian follicle.

Following ovulation, LH induces granulosa membrana cells to differentiate into granulosa lutein cells and the remaining thecal interstitial cells to differentiate into theca lutein cells, thus forming the corpus luteum. LH also induces the granulosa lutein cells to produce vascular endothelial growth factor, which is required for neovascularization of the corpus luteum that penetrates the basement membrane to provide the LDLs necessary for steroidogenesis. After ovulation, the luteal cells accumulate more LH receptors, allowing basal LH levels to maintain the corpus luteum. Early in pregnancy, human chorionic gonadotropin stimulates these LH receptors, inducing progesterone synthesis by the corpus luteum during the first trimester. In contrast to the preovulatory follicle, the granulosa-lutein cells and the corpus luteum of pregnancy express large amounts of P450scc and 3β-HSD, permitting the synthesis of progesterone.

2.3.2. In the Adrenal

The regulation of adrenal androgen production is poorly understood *(35)*. The fetal adrenal produces large amounts of DHEAS, but these serve no essential role, because fetuses that cannot pro-

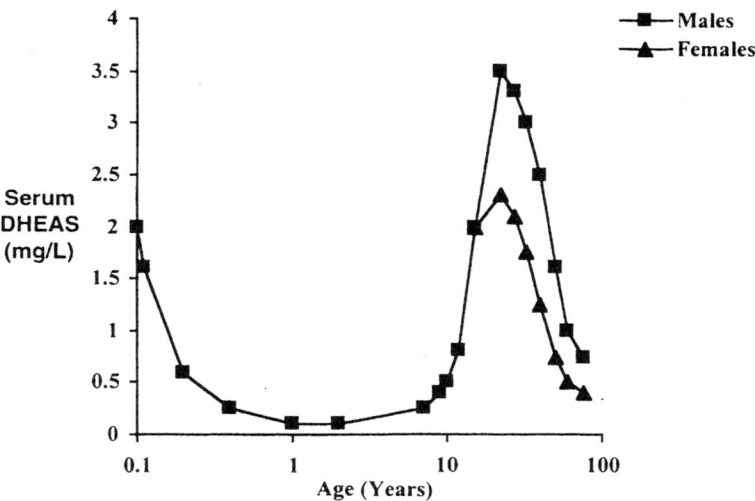

Fig. 4. Concentrations of dehydroepiandrosterone sulfate (DHEAS) as a function of age. Note that the *x*-axis is on a logarithmic scale.

duce these steroids (e.g., those having no P450c17) develop normally. Following birth, the fetal zone of the adrenal involutes and the production of DHEAS falls to very low levels. The adrenal zona reticularis becomes morphologically identifiable after about 3 years of age, but production of DHEAS does not begin until the onset of adrenarche at about 7–8 years of age.

Adrenarche is defined by this rise in serum DHEAS and is the basis of so-called pubarche—the onset of small amounts of pubic hair before the onset of puberty. Adrenarche typically precedes true puberty (i.e., the activation of the hypothalamic/pituitary/gonadal axis) by about 2 years and is completely independent of puberty, gonadotropins, and the gonads themselves. The stimulus to adrenarche is unknown. Adrenarche occurs only in human beings, chimpanzees, gorillas, and possibly orangutans, and hence is rather difficult to study *(71)*. Some have hypothesized a peptide stimulator of the zona reticularis (analogous to ACTH), but no solid evidence for an adrenal androgen-stimulating hormone has appeared.

Serum concentrations of DHEAS continue to rise after puberty, reaching maximal levels at 25–30 years, and then begin to decline ("adrenopause") (Fig. 4) *(62)*. Despite the 100-fold change in DHEAS concentrations during adrenarche, concentrations of ACTH and cortisol do not change. Thus most contemporary studies of adrenarche focus on intra-adrenal events *(35)*. The onset of adrenarche is accompanied by a decrease in 3β-HSD and an increase in P450c17 and cytochrome b_5 in the zona reticularis *(72)*, favoring the production of DHEA. Serine phosphorylation of P450c17 also favors DHEA production, but it is not known whether the phosphorylation of P450c17 changes during adrenarche.

A developmentally programmed trigger, possibly IGF-1, may induce cellular proliferation in the zona reticularis while promoting synthesis of the steroidogenic machinery *(35,73)*. At the same time, increased synthesis of cytochrome b_5 promotes the 17,20-lyase activity of P450c17, which, together with decreased 3β-HSD expression, favors DHEA production. Serine phosphorylation of P450c17 may also be developmentally programmed, and also facilitates 17,20-lyase activity *(73)*. Unlike the neighboring fasciculata, the reticularis expresses sulfotransferase preferentially over steroid sulfatase, ensuring augmented DHEAS production.

Premature exaggerated adrenarche has been linked to insulin resistance, and girls with premature exaggerated adrenarche appear to be at greater risk of developing PCOS as adults *(73–76)*. PCOS is characterized by hirsutism, virilism, hyperandrogenism, menstrual irregularities, chronic anovulation, obesity, insulin resistance, acanthosis nigricans, high concentrations of LH, and ovarian cysts *(77,78)*. Hyperandrogenism and insulin resistance appear to be primary lesions, and the other findings are secondary events. The hyperandrogenism in women with PCOS is of both ovarian and adrenal origin *(77,78)*. The adrenal hyperandrogenism of PCOS resembles an exaggerated form of adrenarche, and girls with premature adrenarche are more likely to develop PCOS *(74,75)*. A gain-of-function disorder in the pathway leading to the serine/threonine phosphorylation of P450c17 could account for such increases in both adrenal and ovarian androgen secretion and an earlier age of adrenarche *(29,34)*, but such mutations have not yet been reported.

The hyperinsulinism and insulin resistance of PCOS is at the level of insulin receptor signal transduction *(78)*. Serine phosphorylation of the β chain of the insulin receptor interferes with the tyrosine phosphorylation of the receptor that normally follows binding of insulin *(79–81)*. Furthermore, some PCOS women appear to have insulin receptors in their fibroblasts that are hyperphosphorylated *(78)*. Thus, a gain-of-function mutation in a serine-threonine kinase or its signal transduction pathway might increase the serine hyperphosphorylation of both P450c17 and the β chain of the insulin receptor, thus accounting for both the hyperandrogenism and insulin resistance of PCOS with a single molecular lesion *(29,34,35,73)*.

3. CONCLUSION

The pathways of adrenal and ovarian steroid biosynthesis use the same enzymes for the initial steps of steroidogenesis, but express different enzymes that convert steroid precursors to the final active products. Both the adrenal and ovary produce DHEA as the key precursor of androgens and estrogens. The key enzyme in DHEA production is P450c17, which catalyzes both 17α-hydroxylation and 17,20-lyase activities. The 17,20-lyase activity of human P450c17 strongly favors the Δ^5 pathway, so that most human androgens and estrogens derive from DHEA. Consequently, understanding the biochemistry and regulation of P450c17, and in particular its 17,20-lyase activity, is central to understanding the hyperandrogenism of PCOS.

4. FUTURE AVENUES OF INVESTIGATION

While the pathways of steroidogenesis have been described in textbooks for more than 40 years, much remains to be learned. First, the transcriptional mechanisms leading to cell-type-specific, developmentally programmed, and hormonally regulated expression of each steroidogenic enzyme and cofactor will need to be delineated in detail before the underlying genetic control of steroidogenesis can be understood. Second, the enzymology of each biosynthetic reaction, and especially the posttranslational mechanisms regulating these, require further investigation. Third, the identity, nature, and activities of steroid modifying enzymes in target tissues such as skin, uterus, breast, fat, and muscle require further investigation. Fourth, steroid synthesis in the brain and its potential role in reproductive and other behaviors are only beginning to be explored. Finally, the factors governing the hyperandrogenic states commonly grouped under the diagnosis of PCOS remain to be elucidated.

KEY POINTS

- Steroidogenic enzymes fall into two broad categories: cytochromes P450 and hydroxysteroid dehydrogenases.
- Each P450 enzyme is encoded by a single gene but has multiple activities.
- Each hydroxysteroid dehydrogenase is encoded by multiple genes, but all have similar activities.
- The adrenal and ovary share the same enzymes catalyzing early steps in steroidogenesis, but possess different enzymes leading to the different final products.
- P450c17, which catalyzes 17α-hydroxylase and 17,20-lyase activities, is a key factor in androgen and estrogen synthesis.
- The 17,20-lyase activity of human P450c17 strongly favors the Δ^5 pathway, so that most androgen and estrogen synthesis proceeds through DHEA.

- The 17,20-lyase activity of human P450c17 is regulated posttranslationally by serine phosphorylation of P450c17 and by the allosteric action of cytochrome b_5.
- Serine phosphorylation may provide an important mechanistic link between the hyperandrogenism and insulin resistance of PCOS.

REFERENCES

1. Miller WL. Molecular biology of steroid hormone synthesis. Endocr Rev 1988;9:295–318.
2. Stocco DM, Clark BJ. Regulation of the acute production of steroids in steroidogenic cells. Endocr Rev 1996;17:221–244.
3. Clark BJ, Wells J, King SR, Stocco DM. The purification, cloning and expression of a novel luteinizing hormone-induced mitochondrial protein in MA-10 mouse Leydig tumor cells: Characterization of the steroidogenic acute regulatory protein (StAR). J Biol Chem 1994;269:28314–28322.
4. Lin D, Sugawara T, Strauss JF III, et al. Role of steroidogenic acute regulatory protein in adrenal and gonadal steroidogenesis. Science 1995;267:1828–1831.
5. Bose HS, Sugawara T, Strauss JF III, Miller WL. The pathophysiology and genetics of congenital lipoid adrenal hyperplasia. N Engl J Med 1996;335:1870–1878.
6. Arakane F, Sugawara T, Nishino H, et al. Steroidogenic acute regulatory protein (StAR) retains activity in the absence of its mitochondrial targeting sequence: Implications for the mechanism of StAR action. Proc Natl Acad Sci USA 1996;93:13731–13736.
7. Bose HS, Lingappa VR, Miller WL. Rapid regulation of steroidogenesis by mitochondrial protein import. Nature 200;417:87–91.
8. Bose HS, Whittal RM, Baldwin MA, Miller WL. The active form of the steroidogenic acute regulatory protein, StAR, appears to be a molten globule. Proc Natl Acad Sci USA 1999;96:7250–7253.
9. Yaworsky DC, Baker BY, Bose HS, et al. pH-dependent interaction of the carboxyl-terminal helix of steroidogenic acute regulatory protein with synthetic membranes. J Biol Chem 2005;280:2045–2054.
10. Baker BY, Yaworsky DC, Miller WL. A pH-dependent molten globule transition is required for activity of the steroidogenic acute regulatory protein (StAR). J Biol Chem 2006;280:4753–4760.
11. Watari H, Arakane F, Moog-Lutz C, et al. MLN64 contains a domain with homology to the steroidogenic acute regulatory protein (StAR) that stimulates steriodogenesis. Proc Natl Acad Sci USA 1997;94:8462–8467.
12. Bose HS, Whittal RM, Huang MC, Baldwin MA, Miller WL. N-218 MLN64, a protein with StAR-like steroidogenic activity is folded and cleaved similarly to StAR. Biochemistry 2000;39:11722–11731.
13. Papadopoulos V. Peripheral-type benzodiazepine/diazepam binding inhibitor receptor: Biological role in steroidogenic cell function. Endocr Rev 1993;14:222–240.
14. Chung B, Matteson KJ, Voutilainen R, Mohandas TK, Miller WL. Human cholesterol side-chain cleavage enzyme, P450scc: cDNA cloning, assignment of the gene to chromosome 15, and expression in the placenta. Proc Natl Acad Sci USA 1986;83:8962–8966.
15. Yang X, Iwamoto K, Wang M, Artwohl J, Mason JI, Pang S. Inherited congenital adrenal hyperplasia in the rabbit is caused by a deletion in the gene encoding cytochrome P450 cholesterol side-chain cleavage enzyme. Endocrinology 1993;132:1977–1982.
16. Tajima T, Fujieda K, Kouda N, Nakae J, Miller WL. Heterozygous mutation in the cholesterol side chain cleavage enzyme (P450scc) gene in a patient with 46,XY sex reversal and adrenal insufficiency. J Clin Endocrinol Metab 2001;86:3820–3825.
17. Katsumata N, Ohtake M, Hojo T, et al. Compound heterozygous mutations in the cholesterol side-chain cleavage enzyme gene (CYP11A) cause congenital adrenal insufficiency in humans. J Clin Endocrinol Metab 2002;87:3808–3813.
18. Miller WL. Regulation of steroidogenesis by electron transfer. Endocrinology 2005;146:2544–2550.
19. Solish SB, Picado-Leonard J, Morel Y, et al. Human adrenodoxin reductase: Two mRNAs encoded by a single gene of chromosome 17cen→q25 are expressed in steroidogenic tissues. Proc Natl Acad Sci USA 1988;71:7104–7108.
20. Picado-Leonard J, Voutilainen R, Kao L, Chung B, Strauss JF III, Miller WL. Human adrenodoxin: Cloning of three cDNAs and cycloheximide enhancement in JEG-3 cells. J Biol Chem 1988;263:3240–3244.
21. Luu-The V, Lachance Y, Labrie C, et al. Full length cDNA structure and deduced amino acid sequence of human 3β-hydroxy-5-ene steroid dehydrogenase. Mol Endocrinol 1989;3:1310–1312.
22. Lorence MC, Murry BA, Trant JM, Mason JI. Human 3β-hydroxysteroid dehydrogenase/$\Delta^5 \rightarrow \Delta^4$ isomerase from placenta: Expression in nonsteroidogenic cells of a protein that catalyzes the dehydrogenation/isomerization of C21 and C19 steroids. Endocrinology 1990;126:2493–2498.
23. Morel Y, Mebarki F, Rheaume E, Sanchez R, Forest MG, Simard J. Structure-function relationships of 3β-hydroxysteroid dehydrogenase: contribution made by the molecular genetics of 3β-hydroxysteroid dehydrogenase deficiency. Steroids 1979;62:176–184.
24. Auchus RJ, Lee TC, Miller WL. Cytochrome b_5 augments the 17,20-lyase activity of human P450c17 without direct electron transfer. J Biol Chem 1998;273:3158–3165.

25. Nakajin S, Hall PF. Microsomal cytochrome P-450 from neonatal pig testis. Purification and properties of A C21 steroid side-chain cleavage system (17α-hydroxylase-C17,20-lyase). J Biol Chem 1981;256:3871–3876.

26. Chung B, Picado-Leonard J, Haniu M, et al.Cytochrome P450c17 (steroid 17α-hydroxylase/17,20-lyase): cloning of human adrenal and testis cDNAs indicates the same gene is expressed in both tissues. Proc Natl Acad Sci USA 1987;84:407–411.

27. Picado-Leonard J, Miller WL. Cloning and sequence of the human gene encoding P450c17 (steroid 17α-hydroxylase/17,20-lyase): Similarity to the gene for P450c21. DNA 1987;6:439–448.

28. Wang M, Roberts DL, Paschke R, Shea TM, Masters BSS, Kim JP. Three-dimensional structure of NADPH-cytochrome P450 reductase: Prototype for FMN-and FAD-containing enzymes. Proc Natl Acad Sci USA 1997;94:8411–8416.

29. Zhang L, Rodriguez H, Ohno S, Miller WL. Serine phosphorylation of human P450c17 increases 17,20-lyase activity: Implications for adrenarche and for the polycystic ovary syndrome. Proc Natl Acad Sci USA 1995;92:10619–10623.

30. Pandey AV, Miller WL. Regulation of 17,20-lyase activity by cytochrome b_5 and by serine phosphorylation of P450c17. J Biol Chem 2005;280:13265–13271.

31. Suzuki T, Sasano H, Tamura M, et al.Temporal and spatial localization of steroidogenic enzymes in premenopausal human ovaries: in situ hybridization and immunohistochemical study. Mol Cell Endocrinol 1993;97:135–143.

32. Geller DH, Auchus RJ, Mendonça BB, Miller WL. The genetic and functional basis of isolated 17,20-lyase deficiency. Nature Genet 1997;17:201–203.

33. Geller DH, Auchus RJ, Miller WL. P450c17 mutations R347H and R358Q selectively disrupt 17,20-lyase activity by disrupting interactions with P450 oxidoreductase and cytochrome b_5. Mol Endocrinol 1999;13:167–175.

34. Pandey AV, Mellon SH, Miller WL. Protein phosphatase 2A and phosphoprotein SET regulate androgen production by P450c17. J Biol Chem 2003;278:2837–2844.

35. Miller WL. The molecular basis of adrenarche: A hypothesis. Acta Pediatr 1999;88 (Suppl 433):60–66.

36. Flück CE, Tajima T, Pandey AV, et al. Mutant P450 oxidoreductase causes disordered steroidogenesis with and without Antley-Bixler syndrome. Nat Genet 2004;36:228–230.

37. Arlt W, Walker EA, Draper N, et al. Congenital adrenal hyperplasia caused by mutant P450 oxidoreductase and human androgen synthesis: analytical study. Lancet 2004;363:2128–2135.

38. Adachi M, Tachibana K, Asakura Y, Yamamoto T, Hanaki K, Oka A. Compound heterozygous mutations of cytochrome P450 oxidoreductase gene (POR) in two patients with Antley-Bixler syndrome. Am J Med Genet 2004;128A:333–339.

39. Fukami M, Horikawa R, Nagai T, et al. POR (P450 oxidoreductase) mutations and Antley-Bixler syndrome with abnormal genitalia and/or impaired steroidogenesis: molecular and clinical studies in 10 patients. J Clin Endocrinol Metab 2005;90:414–426.

40. Huang N, Pandey AV, Agrawal V, et al. Diversity and function of mutations in P450 oxidoreductase in patients with Antley-Bixler syndrome and disordered steroidogenesis. Am J Hum Genet 2005;76:729–749.

41. Morel Y, Miller WL. Clinical and molecular genetics of congenital adrenal hyperplasia due to 21-hydroxylase deficiency. Adv Hum Genet 1991;20:1–68.

42. Mellon SH, Miller WL. Extra-adrenal steroid 21-hydroxylation is not mediated by P450c21. J Clin Invest 1989;84:1497–1502.

43. White PC, Curnow KM, Pascoe L. Disorders of steroid 11β-hydroxylase isozymes. Endocr Rev 1994;15:421–438.

44. Fardella CE, Miller WL. Molecular biology of mineralocorticoid metabolism. Annu Rev Nutr 1996;16:443–470.

45. Luu-The V, Dufort I, Paquet N, Reimnitz G, Labrie F. Structural characterization and expression of the human dehydroepiandrosterone sulfotransferase gene. DNA Cell Biol 1995;14:511–518.

46. Miki Y, Nakata T, Suzuki T, et al. Systemic distribution of steroid sulfatase and estrogen sulfotransferase in human adult and fetal tissues. J Clin Endocrinol Metab 2002;87:5760–5768.

47. Peltoketo H, Luu-The V, Simard J, Adamski J. 17β-Hydroxysteroid dehydrogenase (HSD)/17-ketosteroid reductase (KSR) family: nomenclature and main characteristics of the 17 HSD/KSR enzymes. J Mol Endocrinol 1999;23:1–11.

48. Mindnich R. Möller G, Adamski J. The role of 17β-hydroxysteroid dehydrogenases. Mol Cell Endocrinol 2004;218:7–20.

49. Agarwal AK, Auchus RJ. Cellular redox state regulates hydroxysteroid dehydrogenase activity and intracellular hormone potency. Endocrinology 2005;146:2531–2538.

50. Peltoketo H, Isomaa V, Maenlavsta O, Vihko R. Complete amino acid sequence of human placental 17β-hydroxysteroid dehydrogenase deduced from cDNA. FEBS Lett 1988;239:73–77.

51. Tremblay Y, Ringler GE, Morel Y, et al. Regulation of the gene for estrogenic 17-ketosteroid reductase lying on chromosome 17cen→q25. J Biol Chem 1989;264:20458–20462.

52. Sawicki MW, Erman M, Puranen T, Vihko P, Ghosh D. Structure of the ternary complex of human 17β-hydroxysteroid dehydrogenase type I with 3-hydroxyestra-1,3,5,7-tetraen-17-one (equilin) and $NADP^+$. Proc Natl Acad Sci USA 1999;96:840–845.

53. Wu L, Einstein M, Geissler WM, Chan HK, Elliston KO, Andersson S. Expression cloning and characterization of human 17 beta-hydroxysteroid dehydrogenase type 2, a microsomal enzyme possessing 20 alpha-hydroxysteroid dehydrogenase activity. J Biol Chem 1993;268:12964–12969.

54. Lu ML, Huang YW, Lin SX. Purification, reconstitution, and steady-state kinetics of the trans-membrane 17β-hydroxysteroid dehydrogenase 2. J Biol Chem 2002;277:22123–22130.
55. Takeyama J, Sasano H, Suzuki T, Iinuma K, Nagura H, Andersson S. 17β-Hydroxysteroid dehydrogenase types 1 and 2 in human placenta: An immunohistochemical study with correlation to placental development. J Clin Endocrinol Metab 1998;83:3710–3715.
56. Geissler WM, David DL, Wu L, et al. Male pseudohermaphroditism caused by mutations of testicular 17β-hydroxysteroid dehydrogenase 3. Nat Genet 1994;7:34–39.
57. Dufort I, Rheault P, Huang XF, Soucy P, Luu-The V. Characteristics of a highly labile human type 5 17β-hydroxysteroid dehydrogenase. Endocrinology 1999;140:568–574.
58. Mahendroo MS, Russell DW. Male and female isoenzymes of steroid 5α-reductase. Rev Reprod 1999;4:179–183.
59. Simpson ER, Mahendroo MS, Means GD, et al. Aromatase cytochrome P450, the enzyme responsible for estrogen biosynthesis. Endocr Rev 1994;15:342–355.
60. Grumbach MM, Auchus RJ. Estrogen: consequences and implications of human mutations in synthesis and action. J Clin Endocrinol Metab 1999;84:4677–4694.
61. Miller WL. Steroid hormone biosynthesis and actions in the materno-feto-placental unit. Clinics Perinatol 1998;25:799–817.
62. Orentreich N, Brind JL, Rizer RL, Vogelman JH. Age changes and sex differences in serum dehydroepiandrosterone sulfate concentrations throughout adulthood. J Clin Endocrinol Metab 1984;59:551–555.
63. Corbould AM, Judd SJ, Rodgers RJ. Expression of types 1, 2, and 3 17β-hydroxysteroid dehydrogenase in subcutaneous abdominal and intra-abdominal adipose tissue of women. J Clin Endocrinol Metab 1998;83:187–194.
64. Thiboutot D, Bayne E, Thorne J, et al. Immunolocalization of 5α-reductase isozymes in acne lesions and normal skin. Arch Dermatol 2000;136:1125–1129.
65. Voutilainen R, Tapanainen J, Chung BC, Matteson KJ, Miller WL. Hormonal regulation of P450scc (20,22-desmolase) and P450c17 (17α-hydroxylase/17,20-lyase) in cultured human granulosa cells. J Clin Endocrinol Metab 1986;63:202–207.
66. Gougeon A. Regulation of ovarian follicular development in primates: facts and hypotheses. Endocr Rev 1996;17:121–155.
67. McNatty KP, Moore LG, Hudson NL, et al. The oocyte and its role in regulating ovulation rate: a new paradigm in reproductive biology. Reproduction 2004;128:379–386.
68. Voutilainen R, Miller WL. Coordinate tropic hormone regulation of mRNAs for insulin-like growth factor II and the cholesterol side-chain cleavage enzyme, P450scc, in human steroidogenic tissues. Proc Natl Acad Sci USA 1987;84:1590–1594.
69. Dor J, Ben-Shlomo I, Lunenfeld B, et al.Insulin-like growth factor-I (IGF-I) may not be essential for ovarian follicular development: evidence from IGF-I deficiency. J Clin Endocrinol Metab 1992;74:539–542.
70. Welt CK, Smith ZA, Pauler DK, Hall JE. Differential regulation of inhibin A and inhibin B by luteinizing hormone, follicle-stimulating hormone, and stage of follicle development. J Clin Endocrinol Metab 2001;86:2531–2537.
71. Arlt W, Martens JWM, Song M, Wang JT, Auchus RJ, Miller WL. Molecular evolution of adrenarche: structural and functional analysis of P450c17 from four primate species. Endocrinology 2002;143:4665–4672.
72. Suzuki T, Sasano H, Takeyama J, et al. Developmental changes in steroidogenic enzymes in human postnatal adrenal cortex: immunohistochemical studies. Clin Endocrinol 2000;53:739–747.
73. Auchus RJ, Geller DH, Lee TC, Miller WL. The regulation of human P450c17 activity: relationship to premature adrenarche, insulin resistance and the polycystic ovary syndrome. Trends Endocrinol Metab 1998;9:47–50.
74. Ibañez L, Potau N, Virdis R, et al. Postpubertal outcome in girls diagnosed of premature pubarche during childhood: Increased frequency of functional ovaian hyperandrogenism. J Clin Endocrinol Metab 1993;76:1599–1603.
75. Oppenheimer E, Linder B, DiMartino-Nardi J. Decreased insulin sensitivity in prepubertal girls with premature adrenarche and acanthosis nigricans. J Clin Endocrinol Metab 1995;80:614–618.
76. Ibañez L, Potau N, Zampolli M, et al. Hyperinsulinemia in post-pubertal girls with a history of premature pubarche and functional ovarian hyperandrogenism. J Clin Endocrinol Metab 1996;81:1237–1243.
77. Ehrmann DA, Barnes RB, Rosenfield RL. Polycystic ovary syndrome as a form of functional ovarian hyperandrogenism due to dysregulation of androgen secretion. Endocr Rev 1995;16:322–353.
78. Dunaif A. Insulin resistance and the polycystic ovary syndrome: mechanisms and implications for pathogenesis. Endocr Rev 1997;18:774–800.
79. Bollag G, Roth R, Beaudoin J, Mochley-Rosen D, Koshland D Jr. Protein kinase C directly phosphorylates the insulin receptor in vitro and reduces its protein-tyrosine kinase activity. Proc Natl Acad Sci USA 1986;83:5822–5824.
80. Stadtmauer L, Rosen OM. Increasing the cAMP content of IM-9 cells alters the phosphorylation state and protein kinase activity of the insulin receptor. J Biol Chem 1986;261:3402–3407.
81. Takayama S, White MF, Kahn CR. Phorbol ester-induced serine phosphorylation of the insulin receptor decreases its tyrosine kinase activity. J Biol Chem 1988;263:3440–3447.

3

Androgens Throughout the Life of Women

C. Richard Parker, Jr.

SUMMARY

Androgens consist of 19 carbon steroids that are synthesized in endocrine tissues from cholesterol as substrate or via conversion from other androgens or precursor steroids in the periphery, including liver, gonads, and adrenals. The commonly studied androgens and androgen metabolites in the human include dehydroepiandrosterone (DHEA) and its metabolite dehydroepiandrosterone sulfate (DHEAS), androstenedione (A4), and testosterone and its 5α-reduced metabolite, the potent androgen 5α-dihydrotestosterone (DHT). Circulating levels of androgens are equally low in male and female children prior to adrenarche. During adrenarche, plasma levels of DHEA and DHEAS achieve adult levels at an earlier stage of development than do those of testosterone. Production of androgens in the human female occurs to varying extents as a function of age and physiological status. DHEA and DHEAS are primarily products of the adrenal zona reticularis, whereas A4 and testosterone are synthesized in both the ovary and adrenal; substantial quantities of testosterone arise from peripheral conversion from A4. Modest changes in circulating levels of testosterone occur during the ovarian cycle of women during the reproductive years, with the highest levels seen at mid-cycle; plasma levels of DHEA, DHEAS, and A4 show little change during the ovarian cycle. The suppression that occurs with oral contraceptive use is associated with reductions in plasma levels of androgens, including those produced by the adrenal. Circulating levels of testosterone and A4 increase during pregnancy, probably as a result of human chorionic gonadotropin stimulation of the ovary and increased protein binding. There is also increased production of adrenal androgens in pregnancy through unknown mechanisms, but because of the increased conversion of DHEAS to estrogen in the placenta, plasma levels of this steroid decline during pregnancy. Although the ovary is an important source of androgens in women, there seems to be little impact of menopause on circulating levels of A4 and testosterone. Striking reductions in plasma DHEA and DHEAS occur during aging in women, and these changes appear independent of ovarian status. The best evidence to date is suggestive of a selective deficiency in the Δ^5 steroid pathway in the zona reticularis during aging, and this may be because of a loss of cells in this zone. Finally, estrogen treatment of postmenopausal women has not been shown to have consistent effects on the androgenic milieu of women.

Key Words: Androgens; adrenarche; menopause; adrenopause; adrenal.

1. INTRODUCTION

Androgens consist of 19 carbon steroids that are synthesized in endocrine tissues from cholesterol as the substrate. The cholesterol employed for steroid synthesis can be derived from that produced locally within the steroidogenic cell or may be derived from lipoprotein-associated cholesterol that is imported into the cell from the plasma. The commonly studied androgens and androgen metabolites in the human, shown in Fig. 1 as they are related via the synthetic pathway, are dehydroepiandrosterone (DHEA), dehydroepiandrosterone sulfate (DHEAS), androstenedione (A4), testosterone, and 5α-dihydrotestosterone

From: *Contemporary Endocrinology: Androgen Excess Disorders in Women:*
Polycystic Ovary Syndrome and Other Disorders, Second Edition
Edited by: R. Azziz et al. © Humana Press Inc., Totowa, NJ

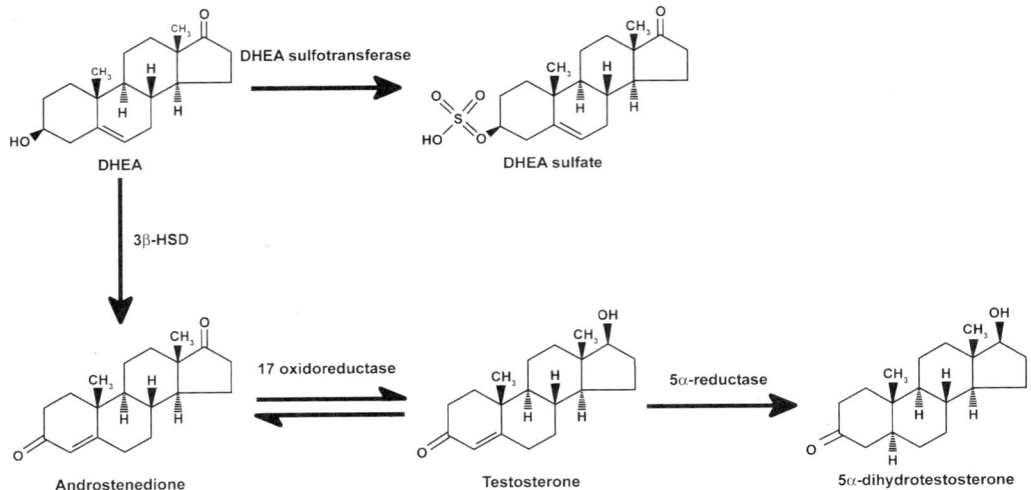

Fig. 1. Structures of commonly studied androgens. The structures and steroidogenic enzymes that are involved in the conversion of the indicated steroids from dehydroepiandrosterone (DHEA) are shown. HSD, hydroxysteroid dehydrogenase.

(DHT), a peripheral metabolite of testosterone. Several other C19 steroids also circulate in the human, such as androsterone and androstenediol and their conjugates, as well as 11-hydroxyandrostenedione, among others. These are less often studied and will not be discussed further, since a consideration of their mechanisms of production and metabolism does not appear to offer any additional insights to the general regulation of androgen levels in the human above those mentioned earlier.

We have found that one feature of androgen-producing cells that distinguishes them from steroidogenic cells that are not usually associated with the capacity for androgen formation is the co-localization of cytochrome P450c17 (CYP17) and cytochrome b_5 *(1)*. The steroidogenic cells in the human that co-express these two factors are the testicular Leydig cell, the ovarian theca cell, the theca-luteal cell of the corpus luteum, the zona reticularis cell of the adrenal, and the fetal zone cell of the fetal adrenal cortex *(1)*. The formation of C19 steroids from their immediate precursors, such as pregnenolone and progesterone (21-carbon-containing steroids), requires 17α-hydroxylation followed by cleavage of the bond between carbons 17 and 20 of the steroid side chain. CYP17 accomplishes both of these steroidogenic steps, with the 17,20-lyase activity of CYP17 promoted by cytochrome b_5. In addition, cytochrome P450 oxido-reductase augments both 17α-hydroxylation and 17,20-lyase activities of CYP17 *(2)*.

DHEA and DHEAS are mainly derived from the zona reticularis, which also expresses a hydroxysteroid sulfotransferase that can sulfurylate pregnenolone and DHEA *(3,4)*. Because of the preference of 17-hydroxypregnenolone over 17α-hydroxyprogesterone (17-OHP) in the human as substrate for the 17,20-lyase reaction, little A4 or testosterone is actually derived from 17-OHP. Rather, they are mainly formed from DHEA via the action of 3β-hydroxysteroid dehydrogenase (3β-HSD) (Fig. 1). The predominant steroidogenic pathway in the adrenal zona fasciculata, which expresses CYP17 and 3β-HSD but lacks appreciable quantities of cytochrome b_5, proceeds mainly toward the synthesis of cortisol rather than C19 steroids. The adrenal contribution to circulating levels of A4 and testosterone, however, likely also involves their synthesis in the zona fasciculata, because of the virtual absence of 3β-HSD in the zona reticularis *(5)*. In this chapter we discuss androgen production throughout the life span of women.

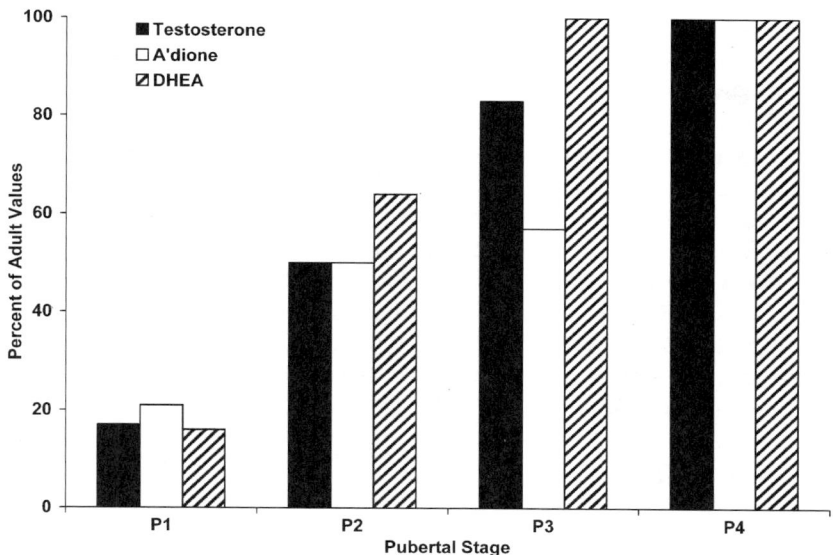

Fig. 2. Androgens throughout puberty. The concentrations of testosterone, androstenedione, and dehydroepiandrosterone (DHEA) are shown as a percentage of adult levels in adolescent girls as the transition through the various Tanner stages for pubic hair. (Data from ref. *6*.)

2. BACKGROUND

2.1. Adrenal and Gonadal Maturation in Adolescence

Circulating levels of androgens are similar in male and female children prior to adrenarche. The pubertal changes in androgens in females plotted as a function of Tanner pubic hair stage are demonstrated in Fig. 2 *(6)*. As can be seen, plasma levels of DHEA achieve adult levels at an earlier stage of development than do those of testosterone. The changes in circulating levels of A4 are intermediate. The pattern of increase in DHEAS levels (not shown) is similar to those of DHEA. A similar discordance in the chronological evolution of DHEA/DHEAS compared to that of testosterone also occurs during adolescence in both girls and boys. The striking increases in plasma levels of DHEA and DHEAS in children coincide with the appearance of the zona reticularis in the adrenal and its progressive broadening during adolescence *(7)*.

The absence of 3β-HSD in the zona reticularis, coupled with the presence of CYP17, cytochrome b_5, and DHEA sulfotransferase in this zone *(1,4,5,8,9)*, is permissive to the increasing production of DHEA and DHEAS during adrenarche. The immunohistochemical evidence for functional changes in the capacity for adrenal androgen production during development is consistent with enzymatic changes, as noted by Schiebinger et al. *(10)*, who observed a fourfold increase in 17,20-lyase and 17α-hydroxylase activities but no alteration in 3β-HSD in the postpubertal adrenal compared to that of adolescents.

The mechanism for the growth and development of the zona reticularis during the transition from childhood to adulthood has yet to be determined. The increased production of adrenal androgens is not attributable to any increases in circulating levels of adrenocorticotropic hormone (ACTH) and is not accompanied by alterations in cortisol levels. Although the existence of an adrenal androgen-stimulating hormone has been postulated, no factor has yet been identified that satisfies the expectations for such a substance. It is likely that acute and chronic regulation of steroid secretion in the zona reticularis is dependent on pituitary ACTH, and the functional phenotype (downregulation of 3β-HSD, but enhanced expression of CYP17, cytochrome b_5, and hydroxysteroid sulfotransferase) of these cells probably is influenced by other, as yet unidentified, factors. Alternatively, there is evidence for central nervous system and pituitary regulation of gonadal maturation during puberty, which is initially mediated by the occurrence of pulsatile luteinizing hormone (LH) secretion during sleep in late adolescence.

Although they are usually temporally linked, the onsets of adrenal and gonadal maturation are not functionally interdependent. For example, adrenarche occurs normally in girls with ovarian dysgenesis (e.g., Turner's syndrome), precocious adrenarche can occur without premature gonadarche, and gonadal development at puberty occurs normally in the presence of adrenal insufficiency (e.g., Addison's disease). It seems clear that nutritional status and hormones involved in growth regulation during childhood play important roles in modulating both ovarian and adrenal maturation. Factors involved in bone maturation may also be linked to adrenal androgens in the peripubertal period. Among similar-aged premenarcheal girls chosen for low or high dietary calcium intake, those with low dietary calcium had significantly reduced levels of DHEA and DHEAS compared to the high-calcium-intake group; however, no differences in testosterone or estradiol levels were noted between the two groups of girls *(11)*.

Recently, the concept of intrauterine programming has been extended to include possible fetal influences on androgen production in the peripubertal period. For example, Ibanez et al. *(12)* found evidence for precocity, ovarian androgen excess, and hyperinsulinism among some girls who had experienced restricted fetal growth. Other investigators have also found evidence for differences in the androgenic milieu of low-birthweight infants during childhood and beyond. For example, Korhonen et al. *(13)* reported significantly increased levels of DHEAS and androstenedione at age 7 among previously growth-restricted infants compared to normal infants. Szathmari et al. *(14)* reported increased levels of DHEA, DHEAS, and A4 during young adulthood in formerly low-birthweight girls; no such effects were seen among formerly low-birthweight males, however. Future studies of these relationships are obviously required.

2.2. Androgen Regulation in Adult Women

The adrenals and ovaries contribute variably to the circulating levels of androgens in adult women *(15)*. Generally, it is thought that A4 is derived in roughly equal amounts from the ovary and the adrenal, while testosterone is derived approximately 25% from the adrenal, 25% from the ovary, and 50% from the peripheral conversion of A4. DHT is produced in peripheral tissues from testosterone and circulates at levels about one-third to one-half that of testosterone. DHEA and DHEAS are almost exclusively of adrenal origin. DHEA is secreted in a pulsatile manner and demonstrates a diurnal rhythm similar to that of cortisol in young women *(16)*. Because of the low metabolic clearance rate (MCR) of DHEAS, however, there are only minor changes in its concentration throughout the day. Androstenedione and testosterone levels also exhibit a diurnal rhythm *(15)*, although less variable than that of DHEA and cortisol.

In the ovary, androgens are synthesized in the theca cells, which have this synthetic capacity by virtue of their expression of CYP17 and cytochrome b_5; follicular granulosa cells do not have such factors *(1)* and therefore are dependant on the theca cell as a source of substrate for estrogen formation. There are no substantial changes in the circulating levels of DHEA, DHEAS, or A4 throughout the ovarian cycle. On the other hand, Abraham *(17)* described the occurrence of a slight mid-cycle peak of testosterone and also noted higher levels of this androgen in the luteal phase than during the early to midfollicular phase of the menstrual cycle. Such findings have generally been observed in other subsequent studies of ovulatory women. The concentrations of A4 and testosterone are higher in ovarian vein blood than in the periphery and are usually highest in blood draining a mature ovarian follicle or functional corpus luteum (Fig. 3) *(18)*. The slight increase in the circulating levels of testosterone and A4 during the luteal phase are probably a result of the fact that CYP17 and cytochrome b_5 also are expressed in the theca-luteal cells of the ovary *(1)*.

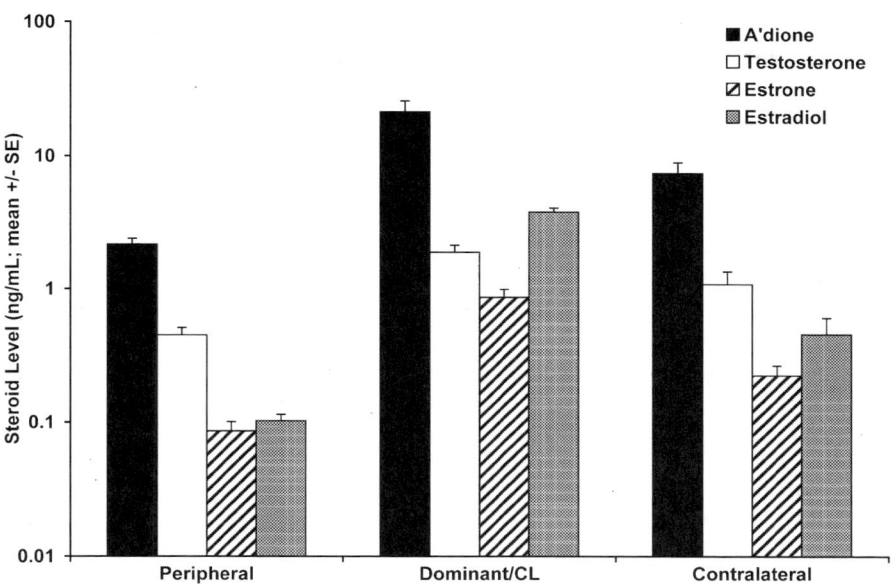

Fig. 3. Steroid concentrations in peripheral and ovarian venous plasma of premenopausal women. The concentrations of androstenedione (A4), testosterone, estrone, and estradiol in peripheral blood, blood draining the ovary having either a dominant follicle and/or an active corpus luteum, and in blood draining the contralateral ovary are shown for women having gynecological surgery. CL, corpus luteum. (Data from ref. *18*.)

Ovarian suppression in adult women because of the ingestion of oral contraceptive steroids is often accompanied by reductions in circulating levels of androgens, including those who are treated with low-estrogen preparations *(19,20)* . Such observations have also been made among women who were treated with high levels of estrogens *(21,22)*. While the mechanism for oral-contraceptive-induced reductions in ovarian androgens seems clear, the mechanisms for the reductions in DHEAS levels noted in each of the above studies are not readily apparent. Androgen levels are strikingly decreased among women with hypopituitarism, particularly among those displaying combined hypoadrenalism and hypogonadism *(23)*. Among both reproductive and postmenopausal aged women with hypopituitarism, those treated with estrogen replacement therapy had androgen levels that were similar to those not treated.

2.3. Pregnancy/Fetal Development

In pregnancy, circulating levels of testosterone, A4, and DHT are all increased relative to the levels found in nonpregnant women *(24–26)*. The mechanisms for increased levels of testosterone, A4, and DHT are varied. Increased circulating levels of testosterone and A4 are evident as early as a few days after ovulation in a conception cycle, whereas there are no early changes in DHEAS, suggesting that only ovarian androgen production is augmented in early pregnancy *(27)*. Late in gestation, the increase in testosterone and DHT levels is the result of substantial increases in sex-hormone-binding globulin (SHBG) and a resultant decrease in the MCRs of these androgens. On the other hand, the production rates for DHEA and DHEAS are increased in late pregnancy. Nevertheless, circulating levels of DHEAS at term are 50% or less than those in nonpregnant women. The reduction in circulating levels of DHEAS is progressive during gestation and is considered to result largely from ever-increasing rates of uptake and utilization in the placenta for estrogen formation *(28)*. Shortly after delivery, circulating androgen levels in women return to levels seen in the nonpregnant state. Among infants delivered at or near term, umbilical cord concentrations of A4 and testosterone are significantly lower than maternal levels, and there are no differences in maternal or umbilical cord blood concentrations related to the gender of the infant. Between 10 and 20 weeks gestation, however, testosterone levels in the fetal compartment are strikingly higher in pregnancies carrying a male fetus than are those with a female fetus; this difference arises from fetal testicular responses to the high levels of gonadotropins produced at this time *(29)*.

Interestingly, the maternal concentrations of DHEAS, A4, and testosterone have recently been shown to decrease with increasing maternal age in women in late gestation *(30)*. The cause and physiological significance of this observation is unclear at present. Racial differences have also been recently reported with respect to the endocrine milieu of pregnancy: African American women had significantly increased levels of total, free, and bioavailable testosterone compared to Caucasian women in both the first and third trimesters of pregnancy *(31)*. These increases could be related to greater levels of SHBG among African American women, particularly at term. Further studies of the effects of race and ethnicity on the endocrine milieu of pregnancy seem warranted, particularly in view of the widely recognized disparities in pregnancy outcomes among the races.

2.4. Perimenopause

Johnston and colleagues *(32)* evaluated cross-sectional differences in the endocrine milieu of women transitioning into the menopause. Significant correlations between bone mass and concentrations of estrogens and testosterone were seen, and the investigators concluded that vertebral bone loss might begin before menses cease. Other investigators *(33)* have also noted in cross-sectional studies a significant positive correlation between circulating levels of free testosterone, DHEAS, and bone mass in postmenopausal women. In this instance, there were differences noted in the relationship according to bone type (cortical vs trabecular).

Endocrine differences as a function of menopausal transitional status are also of interest. Serum estradiol levels are strikingly reduced among women in late perimenopause (with menstrual irregularity and a follicle-stimulating hormone [FSH] > 40 mIU/mL), compared to women in the early perimenopause (FSH < 40 mIU/mL with minimal menstrual irregularity) *(32)* (Fig. 4). Alternatively, serum levels of testosterone were indistinguishable between women in these groups. A further clear decline in serum estradiol levels was noted in menopausal women having no menses during the prior year compared to perimenopausal women; testosterone levels were only slightly lower compared to those in women in early perimenopause. Among women 12–55 months postmenopause, serum testosterone levels were similar to those of women just entering menopause. The reductions in circulating estradiol levels were attributable to striking reductions in rates of production; the stability of testosterone levels was associated with a fairly consistent production rate among all the groups of women studied. In no instance was there evidence for alterations in MCR of steroids among these women; therefore, circulating levels of androgens and estrogens in the perimenopause and perhaps for long periods thereafter reflect glandular secretion and peripheral conversion from precursors.

As part of a 7-year longitudinal surveillance of women undergoing menopause, Rannevik and colleagues *(34)* reported detailed analyses of hormonal parameters during the 6 months prior to cessation of menses and the first 6 months of postmenopause. None of these subjects were using hormone replacement therapy. Whereas FSH and LH levels increased 78% and 57% during this time period, respectively, estradiol levels declined 67%. They also noted slight reductions in the serum levels of A4 (16%) and testosterone (18%) that were, nevertheless, statistically significant during this interval of striking change in the functions of the hypothalamic–pituitary–ovarian axis. The serum levels of testosterone and A4 were fairly stable for the 3 years prior to menopause, with a tendency to decline progressively thereafter. Alternatively, serum levels of DHEA and DHEAS generally tended to decline progressively over the entire 7-year period of study. These results are compatible with the view that DHEA and DHEAS are primarily of adrenal origin and that age-associated declines are more related to age *per se* than abrupt changes in ovarian function during the perimenopausal period.

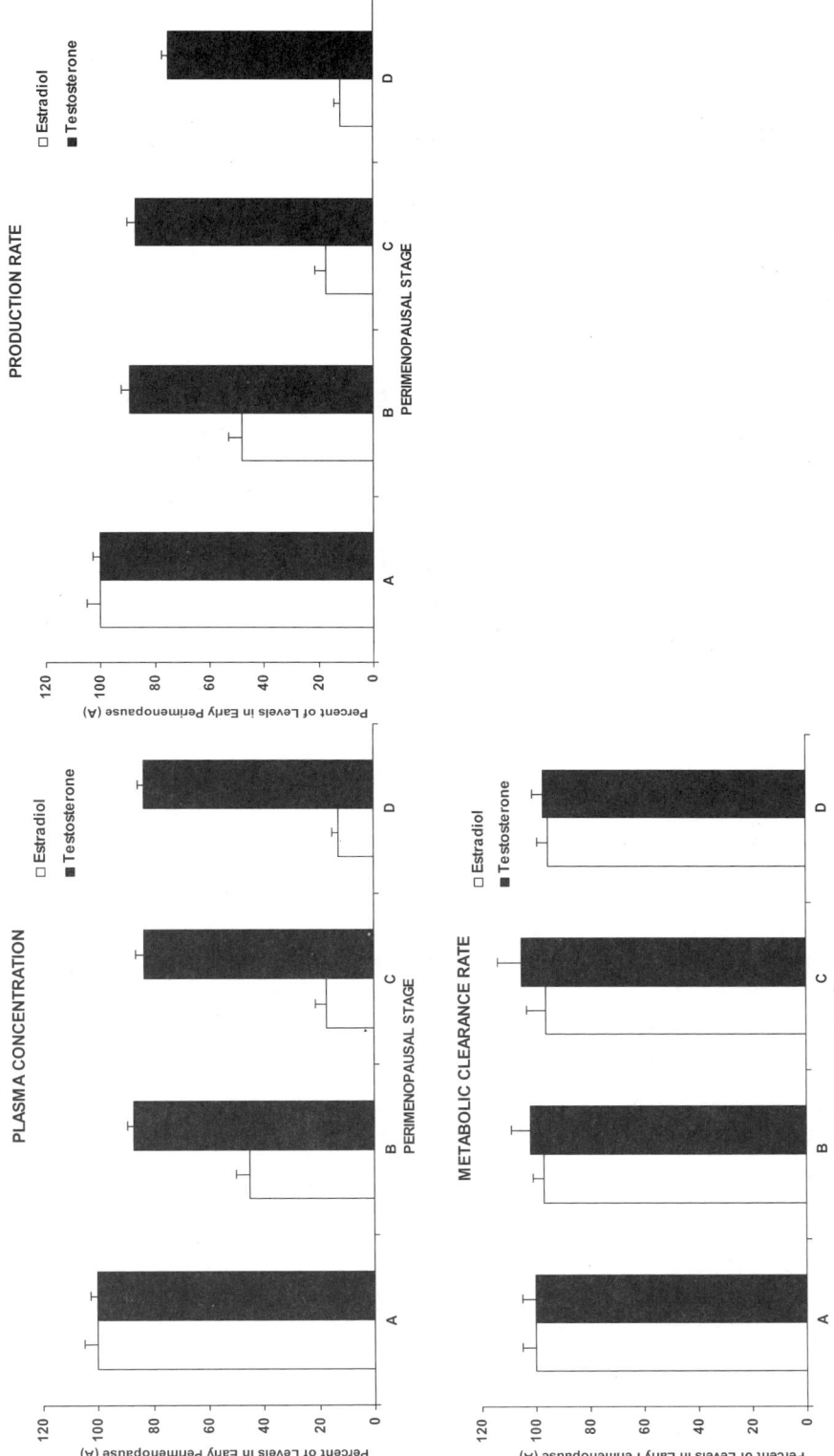

Fig. 4. Estradiol and testosterone parameters during the menopausal transition. The concentrations of estradiol and testosterone, their blood production rates, and their metabolic clearance rates are shown for women at various indicated phases of perimenopause. FSH, follicle-stimulating hormone. (Data from ref. 32.)

41

Burger and associates *(35)* also conducted longitudinal studies of women through the menopausal transition. As found by others, they noted little, if any, change in total testosterone levels leading up to, during, and for several years after menopause. However, reductions in SHBG levels were observed in concert with the reduction in estradiol observed leading up to the menopause and thereafter. The net effect of declining SHBG levels in the face of unwavering levels of testosterone is to increase the free testosterone fraction by 80% over the interval beginning 4 years prior to menopause and ending 2 years after the menopause. If free testosterone levels are actually increased during the perimenopausal period, this could have physiological impacts on androgen-dependent systems, such as libido, in these women.

Although the utilization of hormone replacement therapy (estrogen with or without a progestin) in postmenopausal women is in a period of flux because of concerns relating to potential adverse side effects, the impact of such agents is still of academic if not practical importance to our understanding of the regulation of androgen production and metabolism in women. Several investigators have compared women treated with various hormone replacement regimens to those who were not treated. Abraham and Maroulis *(36)* observed increased levels of DHEA and DHEAS among postmenopausal women treated with estrogen compared to untreated women; no impact on testosterone, DHT, or A4 was noted in this study. In contrast, Casson and associates *(37)* found that DHEAS and testosterone levels declined significantly in women during 12 weeks of treatment with an oral micronized estradiol. Just as SHBG levels have been found to increase in premenopausal women being treated with estrogen (e.g., oral contraceptives) or experiencing natural significant increases in estrogenicity (e.g., pregnancy), treatment of postmenopausal women with estrogens (±progestogen) also leads to significantly increased levels of SHBG. Among such women, Gower and Nyman *(38)* found no changes in total testosterone, albeit a 50% reduction in calculated free testosterone. When the group of subjects was considered as a whole, there was a positive correlation between free testosterone levels and total lean and leg lean mass; such findings could indicate that there is a risk that oral estrogen use could accelerate muscle loss already occurring during aging in women.

2.5. Androgens in Aging Women

The role of the ovary in androgen production in postmenopausal women has been the topic of many studies, often with conflicting results. Sluijhmer et al. *(39)* found that treatment of postmenopausal women with a gonadotropin-releasing hormone (GnRH) agonist reduced peripheral levels of testosterone, which declined further after ovariectomy, whereas neither A4 nor DHEAS levels were affected by either maneuver. Nevertheless, significant ovarian/peripheral venous gradients for A4 and testosterone were noted in GnRH agonist-treated and control subjects in this study. Others have also noted ovarian/peripheral gradients for A4 and testosterone. Vermeulen *(15)* reported that testosterone levels were lower in perimenopausal ovariectomized women compared to similarly aged women with intact ovaries; there were no differences in circulating levels of A4, DHEA, or DHEAS. In a more comprehensive study, Laughlin and associates *(40)* reported that among postmenopausal women 50–89 years of age, total testosterone and bioavailable testosterone levels were significantly lower among those with bilateral oophorectomy compared to intact women; A4 levels were unaffected by oophorectomy. These relationships were consistent, regardless of age or years postmenopause or postoophorectomy.

Davidson and colleagues *(41)* also observed a reduction in testosterone but not A4 levels in ovariectomized women. They reported in their cross-sectional study of women ages 18–75 years that testosterone, DHEAS, and A4 decline with age and that the decline is greatest among younger women. Couzinet and colleagues *(42)*, however, concluded that the postmenopausal ovary played a very minor role in contributing to the circulating androgenic milieu. They based their conclusions on the findings that testosterone, bioavailable testosterone, and A4 were similar among intact and ovariectomized women, whereas the levels of all of these were extremely low or undetectable among postmenopausal women with adrenal insufficiency (with or without their ovaries). They also noted

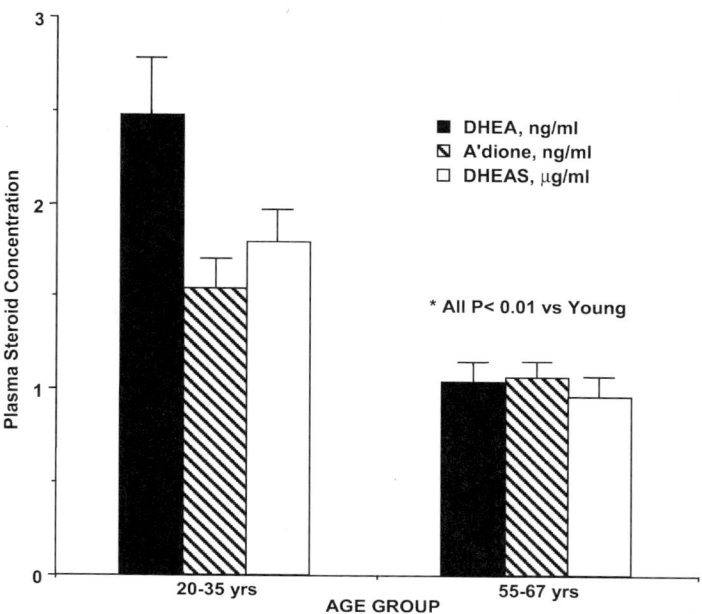

Fig. 5. Effect of aging on androgen levels in women. The concentrations in peripheral plasma of dehydroepiandrosterone (DHEA), androstenedione, and DHEA sulfate (DHEAS) in healthy postmenopausal women, none of whom were being treated with hormone replacement therapy, are compared to those in healthy, normally ovulating young women. (Data from ref. *43*.)

that the levels of testosterone, A4, and DHEAS all declined 80% or more after thorough adrenal suppression (1 mg dexamethasone twice daily for 4 days) among normal postmenopausal women, but did not respond to human chorionic gonadotropin (hCG) stimulation. Finally, they found immunoreactivity for steroidogenic enzymes in only 1 of 7 frozen ovarian specimens and only 3 of 10 paraffin-embedded ovarian samples tested from postmenopausal women. We recently studied the responses of androgens to overnight adrenal suppression in premenopausal and postmenopausal women and found that the extent of the reduction in A4 levels was greatest in postmenopausal women; no differences in suppression of DHEA or cortisol were noted as a function of age or menopausal status *(43)*. These findings suggest that the adrenal may contribute more to the circulating pool of A4 in postmenopausal women than in young ovulatory women.

Age-associated reductions in circulating levels of androgens primarily considered to be of adrenal origin, such as DHEA and DHEAS, as well as A4 have been noted in several cross-sectional studies. As shown in Fig. 5, plasma levels of DHEA, DHEAS, and A4 in young ovulatory women are higher than those in postmenopausal women not exposed to any hormone replacement regimens. In general, cross-sectional studies have concluded that the highest levels of these steroids are found in women (and men) during their 20s and that a decrease in circulating levels occurs fairly steadily thereafter through the sixth or seventh decade of life, at which time the decline begins to plateau. Some evidence for racial differences in the endocrine milieu of women during aging has been noted, although this topic of inquiry has not been thoroughly addressed as yet. Manson et al. *(44)* found significantly lower levels of DHEAS, but not testosterone, among African American women approaching menopause compared to Caucasian women, and this relationship persisted over several months of observation.

Based on studies that employed dynamic adrenal testing strategies, it appears that the Δ^4 steroid pathway in the adrenal is not altered substantially with aging, whereas the Δ^5 pathway and 17,20-lyase activity in the adrenal are impaired to varying degrees *(16,43,45)*. For example, Liu and colleagues *(16)* found that there was a striking decrease in the pulse amplitude of DHEA secretion in postmenopausal women compared to that of young women; no alteration was noted for cortisol secretory patterns. They also found evidence for reduced adrenal secretion of DHEA and DHEAS, but not cortisol, in response to corticotropin-releasing hormone infusions. Vermeulen and associates *(45)* tested the adrenal responses to an acute bolus injection of ACTH and found that the responses of Δ^5 steroids, such as DHEA and 17-hydroxypregnenolone, were reduced in postmenopausal women compared to those in young women; Δ^4 steroids, such as A4 and cortisol, were not impaired with aging. Similar findings were observed when comparing young and old men, suggesting that the age-associated defect in adrenal androgen production in women is not a gender-specific phenomenon *(45)*.

We sought to determine if the impairment in adrenal DHEA/DHEAS production during aging in women was a consequence of reduced sensitivity of the adrenal to ACTH and/or because of a reduction in total secretory capacity in response to ACTH *(43)*. We found, using graded infusions of varying doses of ACTH followed by a standard 250 μg bolus in women who had undergone an overnight dexamethasone suppression prior to the infusion, that the minimal required ACTH dose to elicit a statistically significant rise in DHEA levels was similar in young and postmenopausal women, as were the minimal doses of ACTH required to activate A4 and cortisol in both age groups. On the other hand, we found that the maximal increment in DHEA levels over baseline in the postmenopausal women was significantly reduced compared to that of young women; however, no impairments in the maximal output of A4 or cortisol were noted.

Based on the above-mentioned studies, and in view of the functional zonation of the adrenal cortex, the steroidogenic defect that occurs in aging appears to be localized primarily to the zona reticularis. The potential changes in the zona reticularis that could lead to reduced C19 steroid production in aging are numerous. For example, the reductions noted during in vivo studies could result from a selective decrease in zona reticularis cells (but not cells of other cortical zones) in the expression of one or many of the genes that encode steroidogenic enzymes, factors involved with cholesterol production and import/transport, the ACTH receptor, or various elements of signal transduction. To date, there is no evidence to support any of the above-mentioned possibilities.

Another possibility is that there could be a selective loss of zona reticularis cells with age through a variety of mechanisms. We have obtained some evidence for such a phenomenon in aging humans. Based on image analysis studies of adrenals obtained from adults who died suddenly after traumatic injury, we found that the width of the zona reticularis was significantly reduced in elderly adults than in young adult men and women *(46,47)*; the thickness of the zona fasciculata/zona glomerulosa was not subnormal in aging adults. We also have recently reported that the width of the cell population that contains cytochrome b_5 (presumably synonymous with the zona reticularis) is also significantly reduced in the adrenal cortex of aging humans *(47)*. Our additional evaluations have suggested that a decrease in the thickness of the zona reticularis during aging is not because of cell shrinkage, but rather to a reduction in the number of zona reticularis cells *(48)*. The potential mechanisms for such cell loss are numerous, including increased rates of apoptosis, among others. Confirmation of these morphological findings should provide important directions to future studies of adrenal androgen deficiency in aging.

3. CONCLUSIONS

Androgens consist of 19 carbon steroids that are synthesized in endocrine tissues from cholesterol as substrate or via conversion from other androgens or precursor steroids in the periphery, including liver, gonads, and adrenal. The commonly studied androgens and androgen metabolites in the human include DHEA and its metabolite DHEAS, A4, and testosterone and its 5α-reduced metabolite, the potent androgen DHT. Circulating levels of androgens are similar in male and female children prior to adrenarche.

During adrenarche, plasma levels of DHEA and DHEAS achieve adult levels at an earlier stage of development than do those of testosterone. Although they are usually temporally linked, the onsets of adrenal and gonadal maturation are not functionally interdependent. In adult women, the adrenals and ovaries contribute variably to the circulating levels of androgens, although in general A4 is derived in roughly equal amounts from the ovary and the adrenal, whereas testosterone is derived approximately 25% from the adrenal, 25% from the ovary, and 50% from the peripheral conversion of A4. DHT is produced in peripheral tissues from testosterone and circulates at levels of about one-third to one-half that of testosterone. DHEA and DHEAS are almost exclusively of adrenal origin. DHEA is secreted in a pulsatile manner and demonstrates a diurnal rhythm that is similar to that of cortisol in young women.

In pregnancy, circulating levels of testosterone, A4, and DHT are all increased relative to those in nonpregnant women, although circulating levels of DHEAS at term are 50% or less than those in nonpregnant women. Interestingly, the maternal concentrations of DHEAS, A4, and testosterone have recently been shown to decrease with increasing maternal age in women in late gestation. The serum levels of testosterone and A4 are fairly stable for about 3 years prior to menopause, with a tendency to decline progressively thereafter. Alternatively, the age-associated declines in DHEA and DHEAS, consistent with the fact that these steroids are primarily of adrenal origin, are more related to age *per se* than to abrupt changes in ovarian function during the perimenopausal period. The steroidogenic defect that occurs in aging appears to be localized primarily to the zona reticularis of the adrenal. Although the role of the ovary in androgen production in postmenopausal women has been the topic of many studies, it still remains ill-defined. Finally, estrogen treatment of postmenopausal women has not been shown to have consistent effects on the androgenic milieu of women.

4. FUTURE AVENUES OF INVESTIGATION

A number of issues remain to be addressed and better understood, including (1) the role of androgens in the quality of life of adult pre- and postmenopausal women, including its role in determining well-being, mood, and libido; (2) the mechanism(s) underlying adrenarche and the adrenopause; and (3) the role and regulation of ovarian androgen production in the postmenopause.

KEY POINTS

- Production of androgens in the human female occurs to varying extents as a function of age and physiological status.
- DHEA and DHEAS are primarily products of the adrenal zona reticularis, whereas A4 and testosterone are synthesized in both the ovary and adrenal; substantial quantities of testosterone arise from peripheral conversion from A4.
- Circulating levels of androgens are quite low in early childhood, and an increase in plasma concentrations of DHEA and DHEAS heralds the onset of adrenarche, which appears to be linked to a thickening of the zona reticularis. Increases of A4 and testosterone occur slightly later and are probably reflective of maturation of steroid production in the ovary and adrenal.
- Adrenal and ovarian maturation are not functionally linked.
- Modest changes in circulating levels of testosterone occur during the ovarian cycle of women during the reproductive years, with the highest levels seen at mid-cycle; plasma levels of DHEA, DHEAS, and A4 show little change during the ovarian cycle.
- Ovarian suppression that occurs with oral contraceptive use is associated with reductions in plasma levels of androgens, including those produced primarily in the adrenal; the mechanism for reduced adrenal androgen levels in such circumstances is unclear.
- Circulating levels of testosterone and A4 increase during pregnancy, probably as a result of hCG stimulation of the ovary and increased protein binding. There is also an increased production of adrenal androgens in pregnancy through unknown mechanisms, but because of the ever-increasing conversion of DHEAS to estrogen in the placenta, plasma levels of this steroid decline during pregnancy.
- Although the ovary is an important source of androgens in women, there seems to be little impact of menopause on circulating levels of A4 and testosterone.
- Striking reductions in plasma DHEA and DHEAS occur during aging in women, and these changes appear independent of ovarian status.

- The best evidence to date is suggestive of a selective deficiency in the [5] steroid pathway in the zona reticularis during aging, and this may be because of a loss of cells in this zone.
- Estrogen treatment of postmenopausal women has not been shown to have consistent effects on the androgenic milieu of women.

ACKNOWLEDGMENTS

The author thanks Barbara Staton and Jo Taylor for their important contributions during the generation of this chapter.

REFERENCES

1. Dharia SP, Slane A, Jian M, et al. Co-localization of P450c17 and cytochrome b5 in androgen synthesizing tissues of the human. Biol Reprod 2004;71:83–88.
2. Miller WL, Auchus RJ, Geller DH. The regulation of 17,20 lyase activity. Steroids 1997;62:133–142.
3. Cameron EHD, Jones T, Cones D, Anderson ABM, Griffiths K. Further studies on the relationship between C19- and C21-steroid synthesis in the human adrenal gland. J Endocrinol 1969;45:215–230.
4. Kennerson AR, McDonald DA, Adams JB. Dehydroepiandrosterone sulfotransferase localization in human adrenal glands: a light and electron microscopic study. J Clin Endocrinol Metab 1983;56:786–790.
5. Endoh A, Kristiansen SB, Casson PR, Buster JE, Hornsby PJ. The zona reticularis is the site of biosynthesis of dehydroepiandrosterone and dehydroepiandrosterone sulfate in the adult human adrenal cortex resulting from its low expression of 3 beta-hydroxysteroid dehydrogenase. J Clin Endocrinol Metab 1996;81:3558–3565.
6. Dean HJ, Winter HSD. Abnormalities of pubertal development. In: Collu R, Ducharme JR, Guyda HJ, ed. Pediatric Endocrinology, 2nd ed. New York: Raven Press, Ltd, 1989:331–366.
7. Dhom G. The prepuberal and puberal growth of the adrenal (adrenarche). Beitr Path Bd 1973;150:357–377.
8. Gell JS, Carr BR, Sasano H, et al. Adrenarche results from development of a 3β-hydroxysteroid dehydrogenase-deficient adrenal reticularis. J Clin Endocrinol Metab 1998;83:3695–3701.
9. Suzuki T, Sasano H, Takeyama J, et al. Developmental changes in steroidogenic enzymes in human postnatal adrenal cortex: immunohistochemical studies. Clin Endocrinol 2000;53:739–747.
10. Schiebinger RJ, Albertson BD, Cassorla FG, et al. The developmental changes in plasma adrenal androgens during infancy and adrenarche are associated with changing activities of adrenal microsomal 17-hydroxylase and 17,20-desmolase. J Clin Invest 1981;67:1177–1182.
11. Bongofiglio D, Garofalo C, Catalano S, et al. Low calcium intake is associated with decreased androgens and reduced bone age in premenarcheal girls in the last pubertal stages. J Bone Miner Metab 2004;22:64–70.
12. Ibanez L, Potau N, Francois I, de Zegher F. Precocious pubarche, hyperinsulinism, and ovarian hyperandrogenism in girls: relation to reduced fetal growth. J Clin Endocrinol Metab 1998;84:3945–3949.
13. Korhonen P, Hyodynmaa E, Lenko H-L, Tammela O. Growth and adrenal androgen status at 7 years in very low birth weight survivors with and without brochopulmonary dysplasia. Arch Dis Child 2004;89:320–324.
14. Szathmari M, Vasarhelyi B, Tulassay T. Effect of low birth weight on adrenal steroids and carbohydrate metabolism in early adulthood. Horm Res 2001;55:172–178.
15. Vermeulen A. The hormonal activity of the postmenopausal ovary. J Clin Endocrinol Metab1976;42:247–253.
16. Liu CH, Laughlin GA, Fischer UG, Yen SSC. Marked attenuation of ultradian and circadian rhythms of dehydroepiandrosterone in postmenopausal women: evidence for a reduced 17,20-desmolase enzymatic activity. J Clin Endocrinol Metab 1990;71:900–906.
17. Abraham GE. Ovarian and adrenal contribution to peripheral androgens during the menstrual cycle. J Clin Endocrinol Metab 1974;39:340–346.
18. Aiman J, Forney JP, Parker CR Jr. Androgen and estrogen secretion by normal and neoplastic ovaries in premenopausal women. Obstet Gynecol 1986;68:327–332.
19. Fern M, Rose DP, Fern EB. Effect of oral contraceptives on plasma androgenic steroids and their precursors. Obstet Gynecol 1978;51:541–544.
20. Madden JD, Milewich L, Parker CR Jr, et al. The effect of oral contraceptive treatment on the serum concentration of dehydroisoandrosterone sulfate. Am J Obstet Gynecol 1978;132:380–384.
21. Coenen CMH, Thomas CMG, Borm GF, Hollanders JMG, Rolland R. Changes in androgens during treatment with four low-dose contraceptives. Contraception 1996;53:171–176.
22. Thorneycroft IH, Stanczyk FZ, Bradshaw KD, et al. Effect of low-dose oral contraceptives on androgenic markers and acne. Contraception 1999;60:255–262.
23. Miller KK, Sesmilo G, Schiller A, et al. Androgen deficiency in women with hypopituitarism. J Clin Endocrinol Metab 2001;86:561–567.

24. Mizuno M, Lobotsky J, Lloyd CW, Kobayashi T, Murasawa Y. Plasma androstenedione and testosterone during pregnancy and in the newborn. J Clin Endocr 1968;28:1133–1142.
25. Rivarola M, Forest MG, Migeon CJ. Testosterone, androstenedione, and dehydroeipandrosterone in plasma during pregnancy and at delivery: concentration and protein binding. J Clin Endocr 1968;28:34–40.
26. Tulchinsky D. Adrenal androgens in pregnancy. In: Genazzani AR, Thijssen JHH, Siiteri PK, eds. Adrenal Androgens. New York: Raven Press, 1980:189–198.
27. Castracane VD, Stewart DR, Gimpel T, Overstreet JW, Lasley BL. Maternal serum androgens in human pregnancy: early increases within the cycle of conception. Human Reproduction 1998;13:460–464.
28. Siiteri PK, MacDonald PC. Placental estrogen biosynthesis during human pregnancy. J Clin Endocrinol Metab 1966;26:751–761.
29. Reyes FI, Boroditsky RS, Winter JS, Faiman C. Studies on human sexual development. II. Fetal and maternal serum gonadotropin and sex steroid concentrations. J Clin Endocrinol Metab 1974;38:612–617.
30. Carlsen SM, Jacobsen G, Bjerve KS. Androgen levels in pregnant women decrease with increasing maternal age. Scand J Clin Lab Invest 2003;63:23–26.
31. Zhang Y, Graubard BI, Klebanoff MA, et al. Maternal hormone levels among populations at high and low risk of testicular germ cell cancer. Brit J Cancer 2005;92:1787–1793.
32. Johnston CC, Hui SL, Witt RM, et al. Early menopausal changes in bone mass and sex steroids. J Clin Endocrinol Metab 1985;61:905–911.
33. Tok EC, Ertunc D, Oz U, et al. The effect of circulating androgens on bone mineral density in postmenopausal women. Maturitas 2004;48:235–242.
34. Rannevik G, Carlstrom K, Jeppsson S, Bjerre B, L Svanberg. A prospective long-term study in women from pre-menopause to post-menopause: Changing profiles of gonadotrophins, oestrogens and androgens. Maturitas 1986;8:297–307.
35. Burger HG, Dudley EC, Cui J, Dennerstein L, Hopper JL. A prospective longitudinal study of serum testosterone, dehydroepiandrosterone sulfate and sex hormone-binding globulin levels through the menopause transition. J Clin Endocrinol Metab 2000;85:2832–2838.
36. Abraham GE, Maroulis GB. Effect of exogenous estrogen on serum pregnenolone, cortisol, and androgens in postmenopausal women. Obstet Gynecol 1975;45:271–274.
37. Casson PR, Elkind-Hirsh KE, Buster JE, et al. Effect of postmenopausal estrogen replacement on circulating androgens. Obstet Gynecol 1997;90:995–998.
38. Gower BA, Nyman L. Associations among oral estrogen use, free testosterone concentration, and lean body mass among postmenopausal women. J Clin Endocrinol Metab 2000;85:4476–4480.
39. Sluijmer AV, Heineman MJ, De Jong FH, Evers JLH. Endocrine activity of the postmenopausal ovary: The effects of pituitary down-regulation and oophorectomy. J Clin Endocrinol Metab 1995;80:2163–2167.
40. Laughlin GA, Barrett-Connor E, Kritz-Silverstein D, von Muhlen D. Hysterectomy, oophorectomy, and endogenous sex hormone levels in older women: The Rancho Bernardo study. J Clin Endocrinol Metab 2000;85:645–651.
41. Davidson SL, Bell R, Donath S, Montalto JG, Davis SR. Androgen levels in adult females: changes with age, menopause and oophorectomy. J Clin Endocrinol Metab 2005;90:3847–3853.
42. Couzinet B, Meduri G, Lecce MG, et al. The postmenopausal ovary is not a major androgen-producing gland. J Clin Endocrinol Metab 2001;86:5060–5066.
43. Parker CR Jr, Slayden SM, Azziz R, et al. Effects of aging on adrenal function in the human: responsiveness and sensitivity of adrenal androgens and cortisol to adrenocorticotropin in premenopausal and postmenopausal women. J Clin Endocrinol Metab 2000;85:48–54.
44. Manson JM, Sammel MD, Freeman EW, Grisso JA. Racial differences in sex hormone levels in women approaching the transition to menopause. Fert Steril 2001;75:297–304.
45. Vermeulen A, Deslypere JP, Schelfhout W, Verdonk L, Rubens R. Adrenocortical function in old age: response to acute adrenocorticotropin stimulation. J Clin Endocrinol Metab 1982;54:187–191.
46. Parker CR Jr., Mixon RL, Brissie RM, Grizzle WE. Aging alters zonation in the adrenal cortex of men. J Clin Endocrinol Metab 1997;82:3898–3901.
47. Dharia S, Slane A, Jian M, et al. Effects of aging on cytochrome b5 expression in the human adrenal gland. J Clin Endocrinol Metab 2005;90:4357–4361.
48. Staton BA, Mixon RL, Dharia S, Brissie RM, Parker CR Jr. Is reduced cell size the mechanism for shrinkage of the adrenal zona reticularis in aging? Endocr Res 2004;30:529–534.

4

Physiological Effects of Androgens in Women

T. Keta Hodgson and Glenn D. Braunstein

SUMMARY

Androgens are important sex steroid hormones for women as well as men. Previously, the focus of androgen research in women was almost exclusively on issues of excess. Only recently have we become aware that androgens play an important role in many aspects of the health and well-being of women. From sexual orientation to brain development to bone health, androgens are closely intertwined with estrogens, and only by understanding the former can we have a full understanding of women's health and their medical care issues. This chapter focuses on the normal physiology of androgens in women—what we know and what we have yet to learn.

Key Words: Androgens; DHEA; intracrinology; pilosebaceous unit; sex steroid hormones; sexual differentiation; SHBG; testosterone.

1. INTRODUCTION

Androgens are the most abundant of the sex hormones in women; while their concentrations are measured in nanomoles, estrogens are measured in mere picomoles *(1)*. Many of the major biological events of a woman's life—adrenarche, menarche, sexuality, fertility, parturition, lactation, and menopause—are mediated in part by sex hormones. Yet we know relatively little about the effects of androgens in women. Much of what we do know is deduced from our understanding of conditions of androgen excess or insufficiency in men and women, from mutations involving the genes for the androgen receptor and for key enzymes in the steroid sex hormone pathways, and from extrapolation of the results of in vitro studies and experiments in other animal species *(2)*. Although sex hormones are gender-typical, they are not gender-limited *(3)* —males and females use the same pathways, hormones, and enzymes for synthesis and metabolism. Where there is sexual dimorphism, the different response may result from differences in (a) concentration, (b) duration of exposure, (c) tissue-specific receptivity or sensitivity, and (d) the presence or absence of other modulating hormones, growth or inhibiting factors, or enzymes.

2. BACKGROUND

2.1. Androgen Physiology

The sex hormones dehydroepiandrostenedione (DHEA) and its sulfated form, DHEAS, androstenedione (A4), testosterone (T), dihydrotestosterone (DHT), estrone (E1), and estradiol (E2) are produced in the adrenal glands, the gonads, and in numerous peripheral sites. Although adrenal androgen secretion increases in response to adrenocorticotropic hormone (ACTH), androgens do not influence ACTH secretion, and ACTH plays a primarily permissive role in adrenal androgen physiology. In fact, a specific regulator of adrenal androgen secretion has been proposed, but, so far,

From: *Contemporary Endocrinology: Androgen Excess Disorders in Women:*
Polycystic Ovary Syndrome and Other Disorders, Second Edition
Edited by: R. Azziz et al. © Humana Press Inc., Totowa, NJ

has eluded isolation *(4)*. Ovarian androgen secretion increases following stimulation of the theca cells by luteinizing hormone (LH). However, unlike the situation in men in which testosterone inhibits the secretion of LH, either directly or through aromatization to estradiol, there is no known feedback regulatory loop controlling androgen secretion in women, at least not at the levels normally observed.

Most of the circulating androgens are bound to sex hormone-binding globulin (SHBG) or albumin. SHBG has a high affinity for the biologically active sex hormones DHT, T, A4, E2, and E1, whereas DHEA and DHEAS exhibit little or no binding *(5)*. The role of SHBG in the regulation of sex hormone action has yet to be fully elucidated. It is thought to regulate the concentration of circulating steroid hormones, act as a reservoir for ready-made hormones, and serve as a partner for nongenomic steroid action. Between 0.5 and 7.5% of the androgens in women exist in the free or unbound state and are available to freely act upon cells *(6)*. Unlike SHBG, albumin has a low affinity for sex hormones; therefore, the albumin-bound steroids may be readily available to the tissues. The free and albumin-bound fractions of sex steroid hormones together are termed *bioavailable*.

Although serum measurements of the sex steroid hormones are generally relied upon to diagnose androgen excess or deficiency, most of the androgens produced in women are made in peripheral tissues, which contain the enzymes to convert DHEAS to DHEA (which can then be transformed into A4 and then T) or 5α-reductase (where T may be further converted to the potent androgen DHT). Alternatively, in selected tissues such as liver, skin, fat, muscle, kidney and bone, T can undergo aromatization to E2, which in turn may be converted to E1 in those tissues that contain 17β-hydroxysteroid dehydrogenase. These locally produced hormones can act on neighboring cells in a paracrine fashion or on the cell of origin in an intracrine manner. The extent of the intracrine process can be appreciated by measurements of androgen metabolites, such as androsterone glucuronide, androstane-3α,17β-diolglucuronide, androstane-3β,17β-diolglucuronide, and androsterone sulfate *(7)*.

2.2. Mechanisms of Action

The classic model of sex steroid hormone action is a nucleus-based, ligand/receptor-mediated event that acts on DNA to bring about transcription and translation (i.e., a genomic effect). This complex pathway is well suited for routine maintenance of sex hormone action, such as protein synthesis. However, its relative slowness (minutes to hours) means that it is unable to respond to rapid shifts in physiological demands. Although not the first to describe the effect, Tchernitchin was the first to use the term *nongenomic* to describe a hormonal action sequence that could not otherwise happen if it were regulated by classic genomic processes *(8)*. Nongenomic effects differ from genomic effects in at least three ways: (a) they are more rapid (from seconds to minutes), (b) they do not rely on an operational nucleus and can be shown to take place in nonnucleated cells, and (c) they are not sensitive to the effects of inhibitors of transcription or translation *(9)*. It has taken the better part of three decades to begin to understand the nongenomic actions of sex steroids, yet many controversies (well reviewed by Lösel and colleagues *[10]*) remain to be settled.

It is possible that some of the nongenomic actions of androgens are the result of direct action by the hormone on cell membrane fluidity through their interaction with the membrane phospholipids. Other evidence suggests that nongenomic actions are possibly mediated by receptors— either cell surface receptors that generate signals across the plasma membrane or androgen receptors without associated transcription action *(9,11)*. Triggers for nongenomic reactions include calcium ions, which act as a second messenger to stimulate transmembrane cascades or stimulation of the mitogen-activated protein kinase or cyclic adenosine monophosphate (cAMP) pathways *(11)*.

SHBG has been shown to have a role in the nongenomic action of sex hormones. SHBG receptors have been found on the cell surface of the testes, prostate, breast, and liver, and represent either a G protein-coupled receptor or a closely related one. SHBG devoid of steroids binds to this receptor, and sex steroids then interact with the SHBG–receptor complex, resulting in the generation of cAMP, either directly or by stimulating an influx of calcium into the cell *(11)*.

Megalin is a member of the family of low-density lipoprotein receptors, and there is evidence that it may serve as a helper molecule in nongenomic steroid hormone action *(12)*. It is known that certain tissues require more steroid hormones than can be supplied by the classic pathway. Not only is the stimulatory process too slow, but the classic paradigm does not allow for the local storage of steroid hormones. Tissues needing large quantities of steroid hormone—and in which abundant megalin receptors have been found—are the breast, uterus, prostate, and epididymis. Lipophilic molecules such as megalin may help these cells store large amounts of sex hormones *(13)*.

Although the mechanisms for the nongenomic effects of androgens remain controversial, their cellular effects have been clearly demonstrated. The principal androgen within follicular cells of the ovary, A4, can cause a rapid, dose-related increase in calcium ion concentrations in granulosa-lutein cells of humans and in ovarian granulosa cells of pigs. No such effect has been noted for T. Likewise, osteoblasts of male, but not female, rats experience a rapid increase in calcium ion concentration in the presence of T. The nongenomic effects of progesterone, E2, T, and A4 have been studied in vitro and in vivo in granulosa, endometrial, and Sertoli cells as well as in the oocytes and spermatozoa of various animal models, including humans *(9)*. However, it is yet to be determined whether the nongenomic pathways for androgens have any physiologically important effect in humans.

2.3. Overview of Androgen Action Throughout the Female Life Span

2.3.1. Prenatal

2.3.1.1. DIFFERENTIATION OF INTERNAL GENITAL DUCTS AND EXTERNAL GENITALIA

Primordial germ cells formed in the dorsal endoderm of the yolk sac migrate into the undifferentiated, bipotential, embryonic gonads at the urogenital ridge—a process that is well underway by 6 weeks of gestation. In a male fetus with a normal Y chromosome, the sex-determining region Y gene (SRY), SOX9, and other genes initiate the differentiation of the gonad into testes at 6–7 weeks. The Leydig cells, which appear by 8–9 weeks, contain functional LH/human chorionic gonadotropin (hCG) receptors. They are stimulated initially by hCG secreted by placental trophoblasts, then later by fetal pituitary LH. The result is T production. Between 9 and 14 weeks of gestation, T stimulates the wolffian duct structures to differentiate into the epididymis, vas deferens, and seminal vesicles. The normal fetal testes also secretes anti-müllerian hormone (AMH), which results in the local regression of the müllerian duct between 8 and 12 weeks of gestation. The external genitalia remain undifferentiated until about 8 weeks. The urogenital tubercle, urogenital sinus, and labioscrotal tissue contain type 2 5α-reductase, which converts T into DHT. The DHT effects the fusion of the labioscrotal folds to form a scrotum and penile urethra, stimulates the growth of the genital tubercle to form a glans penis, and enhances prostate differentiation and growth *(14)*. That DHT is required for the normal formation of the male external genitalia is most clearly demonstrated by patients with the autosomal recessive 5α-reductase, type 2 deficiency (pseudovaginal perineal-scrotal hypospadias). These genetic males, whose testes secrete ample quantities of T, are unable to form normal amounts of DHT. At birth, affected individuals have ambiguous genitalia, with a phallus that resembles a clitoris, bifid, empty scrotum, and a urogenital sinus that contains the urethra and a blind vaginal pouch opening into the perineum. Wolffian duct structures are present, having developed under the influence of T, while the prostate is underdeveloped, and the müllerian duct-derived structures are absent *(15)*.

The female fetus begins her process of sexual differentiation when the bipotential embryonic gonad begins to develop oocytes—between 11 and 12 weeks of gestation. In the absence of gonadal T production, the wolffian duct structures regress, and, lacking AMH, the müllerian ducts differentiate into the fallopian tubes, the uterus, and the upper third of the vagina. Because T levels are low, the 5α-reductase, type 2 at the genital ridge lacks the substrate to form DHT. Therefore, there is no "zipping up" of the labioscrotal folds, resulting in formation of open labia, a perineal vaginal orifice, and a perineal urethra. The regression of the wolffian duct structures, the differentiation of the müllerian duct structures, and the maintenance of open labioscrotal folds along normal female lines does not require a gonad, nor does it require the fetus to be a genetic female. XX or XY fetuses with dysgenetic or streak gonads also exhibit development of the internal and external genitalia along female lines *(14)*. XY fetuses with complete androgen insensitivity resulting from mutations in the androgen receptor gene (complete testicular feminization) have functional testes that produce T and AMH. However, as a result of the inability of T or DHT to act and the presence of AMH, the fetus does not develop wolffian or müllerian duct structures, respectively, and the external genitalia develop along female lines *(15,16)*.

Although the prenatal female internal and external genital development appears to be a passive process, the tissues of the female fetus are capable of responding to T. This is well demonstrated by patients with congenital adrenal hyperplasia (CAH) who have mutations in the *CYP21* gene that cause a deficiency of 21-hydroxylase. This enzyme deficiency results in reduced cortisol production, which leads to an elevation of ACTH and, in turn, increases the conversion of cholesterol to pregnenolone in the adrenals. The buildup of cortisol precursors results in increased quantities of 17-hydroxyprogesterone, A4, and T. The high T levels result in fusion of the labioscrotal folds and clitoral enlargement. The responsiveness of the external genitalia to T is time dependent; labioscrotal fusion only takes place when the fetus is exposed to high T and DHT before 12 weeks gestation. After 12 weeks, high T levels only result in clitoral enlargement. Of interest, the wolffian ducts in affected females regress normally, suggesting that the T levels achieved are not sufficient to stimulate their growth and differentiation *(17)*.

2.3.1.2. BRAIN IMPRINTING

The concept that the sex steroid milieu of the brain during gestation and shortly after birth can permanently influence the reproductive cycles and sexual behavior emanates from studies in nonhuman mammals, especially rats *(2,18)*. During species-specific critical periods, androgen administration to females masculinizes the brain, leading to anovulatory infertility and loss of sexual responsivity to males. Administration of androgens to female rhesus monkeys during gestation also leads to masculinized sexual behavior, play, and grooming. It is much more difficult to determine what role, if any, prenatal androgen exposure has on human behavior. Sex differences in childhood play behaviors, including playmate preferences and types of activities and objects chosen for play, can be detected by 12 months of age in normal children. Girls born to women who were given androgenic progestins during pregnancy demonstrated increased male-typical play behavior. In addition, a longitudinal study of the relationship between endogenous maternal serum T levels and the sex-type behavior of the offspring at 3.5 years of age failed to show a correlation in males, but did in female children; the higher the maternal T during pregnancy, the more likely the girls were to exhibit male-type play behavior. Females with the 21-hydroxylase deficiency form of CAH exhibit more male-typical activities during childhood, adolescence, and adulthood than do control females. Spatial orientation, visualization, and targeting are higher in women with CAH than controls, more closely resembling the male pattern. Finally, and in contrast to their unaffected sisters, females with CAH have a greater likelihood of being sexually attracted to women *(19,20)*. Other evidence that prenatal androgen exposure is important for subsequent masculine behavior comes from studies of males with complete androgen insensitivity, who exhibit female-type behavior in regards to interests, gender identity, and sexual orientation *(21,22)*.

2.3.2. Postnatal, Prepubertal Period

In the first 4 months of life, the neonate experiences brief increases in the levels of sex hormones. The male neonate experiences surges in LH and T, in part because the maternal-placental estrogens are no longer suppressing the hypothalamic–pituitary–testicular axis. The hormone levels peak at 1–2 months and then decline, and by age 6 months they are at a nadir, remaining low until adrenarche. In females, the postnatal hormone activity is less dramatic than in the males, but more complex.

There is a modest increase in LH, which stimulates ovarian E2 secretion for 2–4 months, whereas the follicle-stimulating hormone (FSH) rise, which peaks at 3–6 months, brings about maturation of the ovarian follicles. FSH then steadily decreases but is measurable up to 2 years of age. All neonates experience a precipitous decrease in circulating DHEA secondary to the disappearance of the adrenal fetal zone. By age 1, the two-zone fetal adrenal gland has been replaced by an adrenal with three zones (granulosa, fasciculate, and reticularis), which initially produce little DHEA *(23)*.

The physiological roles, if any, of androgens in the neonatal or preadrenarchal female are unknown. Unlike female rats administered androgens in the postnatal period, high androgen exposure during this time in females with CAH does not suppress subsequent menstrual cyclicity in gonadotropins or sex steroids *(24)*. As noted above, females with CAH may exhibit an increased rate of attraction to females, and this may be secondary to the prenatal or postnatal exposure to high androgen levels, as well as other factors such as psychological issues arising from their ambiguous genitalia and their social milieu.

It is also unknown whether physiological levels of androgens are important for bone health in childhood *(25,26)*. Girls with untreated CAH show rapid growth of the long bones with early epiphyseal maturation, resulting in the individual being tall for her age during childhood and shorter as an adult than would have been predicted based on parental height. In this instance, it is possible that the elevated androgens were serving as substrate for estrogen production in the bone and that it is this estrogen, and not the androgens, that stimulates bone growth. Evidence for this comes from a female with aromatase deficiency, a defect that prevents the conversion of T to E2. This patient was virilized but was shorter than expected and had a delay in bone age despite marked elevations in androgens *(27)*.

2.3.3. Adrenarche

The adrenal sex hormones remain suppressed following the regression of the fetal adrenal zone. The zona reticularis forms at about 3 years, but DHEA and DHEAS secretion from this region does not begin to rise until about age 6 years, which marks the biochemical onset of adrenarche. By about age 8 years, there is sufficient peripheral conversion of the adrenal androgens into T and DHT to stimulate axillary and pubic (ambisexual) hair growth. Androgen stimulation of the apocrine glands in these areas may result in a change in body odor. The child also may experience a transient growth spurt at this time *(28)*.

2.3.4. Puberty

The onset of puberty is heralded by an increase in gonadotropin-releasing hormone secretion, which initially induces a nocturnal rise in LH and then pulsatile increases in both LH and FSH throughout the day and night. The gonadotropins stimulate gonadal growth and development, and sex steroid hormone production in both sexes. The effects of increased pubertal androgen secretion in males is dramatic and easily observed: growth of the penis, increased pubic and axillary hair, development of body and facial hair, growth of the larynx with deepening of the voice, skeletal muscle enlargement and increase in muscle strength, growth spurt, increase in bone mass, increased erythrocytosis, increased malodorous perspiration, acne, and an increase in libido and aggressive behavior *(29)*.

Pubertal girls also experience a rise in T, which follows a diurnal variation with higher levels in the morning than evening. When menstrual cycles are established, there is a midcycle rise of T and A4 that is concordant with the rise in E2 and follows the peak in LH secretion. One of the physiological roles of T in females is to serve as the major prohormone for the production of E2, a hormone essential for secondary sexual development and the pubertal growth spurt. Other physiological actions of androgens during normal female puberty are more difficult to discern, although they likely have a role in the increased activity of sudoriferous glands, a further increase in the amount and thickness of ambisexual hair (axilla, pubis), and enhanced sebum production with the likelihood of concurrent acne.

2.3.5. Adulthood

Peak levels of androgens and androgen precursors in women are achieved between ages 20 and 30 years, followed by a steady decline. Between the ages of 21 and 40 years, there is approximately a 50% drop in T, DHEA, and DHEAS levels *(30,31)*. There is a further decline of close to 25% between the ages of 42 and 50 years *(32)*. Several longitudinal studies have demonstrated that there is no significant change in T levels during the menopausal transition, and, in fact, free T may actually increase owing to the reduction in SHBG levels that occurs following the profound decline in estrogen production by the ovaries *(32,33)*. That the theca cells in the postmenopausal ovary are steroidogenically active in androgen production has been demonstrated by studies carried out in postmenopausal women undergoing bilateral oophorectomy, in which serum T and A4 levels were reduced by 50% *(34)*. DHEA, DHEAS, T, A4, and DHT continue to decrease between ages 60 years and 70 years, but at a slower rate than prior to age 50. Thus, adrenal and ovarian secretion of androgens decrease over time with age.

The physiological role of androgens in women during adulthood is unclear. Women with hyperandrogenic disorders may develop hirsutism, acne, deepening of the voice, androgenic alopecia, clitoromegaly, malodorous perspiration, increased muscle mass, and aggressive behavior—essentially confirming the fact that female androgen target tissues can respond in a manner similar to those in men when exposed to a high enough concentration of androgens for a sufficient duration. Low free T levels have been found in women with hypopituitarism, adrenal insufficiency, oophorectomy, or premature ovarian failure and after institution of oral estrogen therapy in menopausal women *(6)*. Based upon observations in such women, a consensus panel developed the clinical construct of the female androgen insufficiency syndrome: low libido with a global decrease in sexual desire, fantasy, or arousability; persistent, unexplained fatigue; decreased sense of well-being; blunted motivation; flattened mood; thinning or loss of pubic hair; decreased lean body mass; and osteopenia or osteoporosis *(35)*. Thus, it is likely that androgens play at least a permissive role in female sexuality, mood, and body composition in adulthood, although unassailable direct support for this concept is currently lacking.

2.4. Androgen Effects on Specific Target Tissues in Women

The sites of sex hormone production are also, in most cases, the sites of tissue-specific action. Following is a brief review of these sites and their functional aspects.

2.4.1. Brain

Functional differences in some aspects of brain function exist between males and females, and studies in rodents and nonhuman primates indicate that sex steroid hormones are important contributors to these differences. Hormonal effects are categorized as organizational or activational *(2)*. Organizational refers to the hard wiring by which a male or female phenotype is actualized. Activational refers to those effects that are malleable and respond to current conditions. Each of these effects may be structural, or functional, or both *(2,36)*. Androgen receptors are distributed throughout the brain and are generally within close proximity to estrogen receptors *(37)*. High concentrations of the receptors are present in the preoptic area of the hypothalamus, with smaller numbers located in the amygdala, hippocampus, and cerebral cortex. Areas of the brain also contain 5α-reductase and aromatase and are thus able to convert T to DHT or E2, which may mediate some of the effects of T. In addition, as well demonstrated in rat studies, myelinating glial cells in the central and peripheral nervous system can synthesize DHEA directly from cholesterol as well as from precursors of extraneural origin *(38)*. At present it is unknown whether local androgen synthesis in the human brain is physiologically relevant. Small trials of DHEA given orally to patients have not shown a clear effect on cognitive function, mood, sense of well-being, perimenopausal symptoms, or memory *(10,39)*. These negative studies may indicate a lack of physiological importance of DHEA in brain function, an inadequate achievement of brain DHEA concentrations following oral administration, or that the locally produced neurosteroids have exerted their maximal physiological effects and that the exogenous DHEA can exert no additional effect.

As previously noted, the prenatal exposure of female rats and monkeys to androgens during the critical prenatal and early postnatal periods results in masculinized behavior and, in the rats, inhibits the cyclicity of the hypothalamic–pituitary–ovarian axis. CAH represents the clinical correlate in humans, and affected females do demonstrate more male-typical activities than controls *(2)*. In contrast, males with complete androgen insensitivity have female-typical behavior, suggesting that in the absence of androgen action, the default behavior is female. It is clear that androgens can have an activational effect on behavior in women. In some studies, women who fulfill the criteria for the female androgen insufficiency syndrome note an improvement in their overall sense of well-being when given exogenous T, increasing their serum T concentrations into the physiological or slightly supraphysiological range for women in their reproductive age group *(40–43)*. In addition, an increase in the scores for aggression has been noted in women receiving T injections that result in supraphysiological levels *(41,44)*. These studies suggest that androgens may be a modulator of mood and behavior in women.

The relationship between androgens and libido and sexual function in women has been an active area of investigation. In reproductive-age women, there is a correlation between the midcycle increase in T and libido, the ability to become sexually aroused, and frequency of sexual activity *(45–49)*. Within the normal range for women, those that have the highest T levels across the menstrual cycle have less depression and more sexual gratification than do the women with the lowest levels of T *(46)*. Some, but not all, cross-sectional populations studies have also shown correlations between serum T concentrations and desire, arousal, responsiveness, and frequency of sexual activity *(6)*. Although a recent cross-sectional study failed to show a relationship between T levels and self-reported sexual function in women, it did note that there was a significant association between reduced sexual desire, arousal, and responsiveness, and low DHEAS levels *(50)*. Since DHEAS serves as a prohormone for tissue production of A4, T, and DHT, it is likely that the tissue concentrations of these more potent androgens were also low in these women. Decreased sexual function has been noted in women with hyperandrogenic disorders given antiandrogens, such as cyproterone acetate *(51)*. Finally, the administration of T to women with hypoactive sexual desire disorder results in an improvement in libido and sexual function. The levels of free T achieved in these studies were within the high physiological range to slightly supraphysiological range for premenopausal women. Of interest, there is a positive, though not robust, correlation between the change in T levels and improvement in sexual desire and other parameters of sexual function in women *(6,52–55)*. Thus, the relationship between physiological levels of androgens and female sexual desire is well established and most likely reflects an action of T on the brain *(56)*.

2.4.2. Bone

Androgen receptors are present on osteoblasts and, to a lesser extent, on osteoclasts and osteocytes. Osteoblasts from young bone have greater expression of androgen receptors than in older bone, and the receptors are more expressed in osteoblasts from cortical bone than from cancellous bone *(57)*. In vitro, androgens stimulate osteoblast proliferation, enhance their differentiation, and prevent osteoblast apoptosis, while stimulating osteoclast apoptosis *(25,57)*. Estrogen receptors and aromatase are also present in bone. Thus, it is difficult to dissect out the physiological role that androgens play directly vs indirectly through conversion to estrogen in bone development, growth, and health.

Endogenous serum T levels are correlated with bone mineral density in adolescent females and premenopausal women; in pre- and perimenopausal women, there is also an inverse correlation between bone density and SHBG levels. Conflicting data exist regarding the relationship between endogenous T levels and bone density in postmenopausal women *(57)*. Women with hyperandrogenism resulting from polycystic ovary syndrome (PCOS) have increased bone mass, and the bone mineral density correlates with the T and A4 levels *(25,57)*. Patients with complete androgen insufficiency syndrome exhibit a reduction in bone mineral density of the spine, but not of the hips, compared to normal females *(58)*. Administration of androgens along with estrogens to women with postmenopausal osteoporosis increases bone mass in the spine and hips to a greater extent than with estrogen alone *(25,56,59)*. That this is not merely a result of aromatization of the T to E2 in the bone arises from two observations in women. First, women who receive norethindrone acetate or nandrolone decanoate, compounds related to T, but which are not aromatized to estrogen, exhibit significant increases in bone mineral density *(60,61)*. Second, in a study that compared the effects of esterified estrogens alone or with methyltestosterone on serum and urine markers of bone formation and resorption, estrogen or estrogen plus methyltestosterone both decreased the markers of osteoclastic bone resorption, while only the estrogen plus methyltestosterone combination increased the levels of markers of osteoblastic bone formation *(62)*.

Androgens stimulate growth plate closure during puberty, but this action is mediated through aromatization to estrogens and binding to the estrogen receptor. Men and women with aromatase deficiency have an absent pubertal growth spurt and delayed epiphyseal closure, as do men with a mutation in the estrogen receptor-α gene *(25,27)*.

2.4.3. Breast

Normal mammary epithelial cells contain androgen receptors in addition to estrogen and progesterone receptors. To date there is a paucity of in vitro studies of the effect of androgens on normal human breast cells. In breast cancer cell lines, androgens antagonize the proliferation induced by estrogens and increase cellular apoptosis. Studies in animals also have generally shown that T and DHEA inhibit mammary carcinoma development. Dimitrakakis and coworkers examined the effects of estrogen alone, estrogen plus progesterone, estrogen plus T, and vehicle control in oophorectomized rhesus monkeys. These investigators observed that the addition of T significantly decreased mammary epithelial proliferation, altered the ratio of type of estrogen receptor (α or β) expressed, and reduced mammary epithelial estrogen receptor signaling *(63)*. The administration of the androgen receptor antagonist flutamide to intact, cycling female monkeys led to an increase in mammary epithelial cell proliferation, indicating that physiological levels of T in female monkeys protect the breast from the proliferative effects of estrogen *(63)*.

Few well-controlled clinical studies in women have examined the relationship between endogenous androgen levels and breast health. Some, but not all, studies have found that women with higher total T levels have an increased risk of breast cancer. In contrast, women with hyperandrogenemia resulting from PCOS do not appear to have an increased risk of breast cancer *(64)*. A recent retrospective, observational study in women receiving hormone therapy found that women who were given a combination of estrogen, progestin, and T had a lower rate of breast cancer development than historical control women who received only estrogen and progestin and was actually similar to that found in women who had never used hormones *(65)*. The limited data available at this time support the notion that endogenous T has a protective role in the breast, counterbalancing the effects of estrogens and progesterone.

2.4.4.Pilosebaceous Unit

Skin is an important site for peripheral conversion of androgens; at least 50% of a woman's androgen production occurs here as DHEAS is converted to DHEA to A4, A4 to T, and T to DHT. Hair is produced by the pilosebaceous unit (PSU), which extends outward from the hypodermis. The PSU is composed of a hair and a sebaceous gland component, which can differentiate into a terminal hair follicle or to a sebaceous follicle with a fine, vellus hair and an active sebaceous gland. Prior to puberty, the hair in the androgen-sensitive areas of the body is primarily vellus with small sebaceous glands. During puberty, the increased androgens stimulate vellus hairs to develop into terminal hairs. The growth of hair in sexual areas (i.e., hair that is terminal primarily in males, such as that of the upper lip, chin, chest, abdomen, back, thighs, or upper arms) depends upon the sensitivity of the PSU

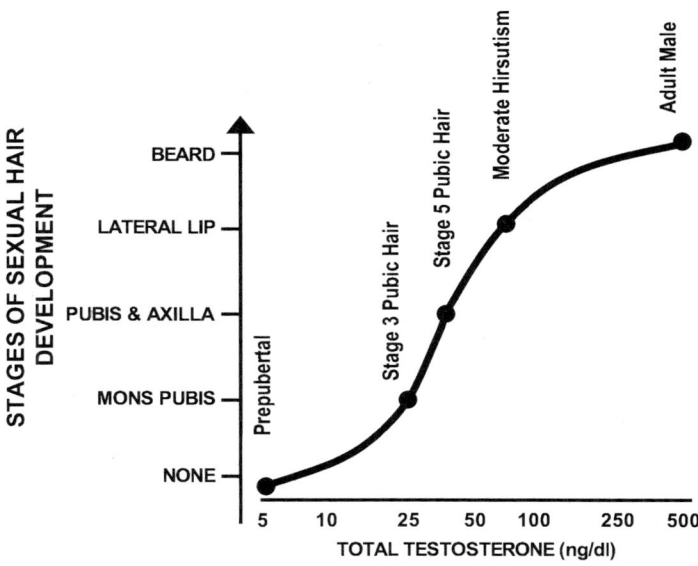

Fig. 1. Relationship of stages of sexual hair development to T circulating levels (log scale): (**A**) prepubertal; (**B**) stage 3 pubic hair; (**C**) stage 5 pubic hair (adult female); (**D**) moderate hirsutism; (**E**) adult male. (From ref. *72*.)

to androgens, the level of androgens they are exposed to, and the duration of exposure. Figure 1 illustrates the relationship between plasma T levels and the degree of sexual hair development, demonstrating that the sensitivity of hair follicles to T is greatest in the pubic region and less in the beard area *(66)*. Androgen receptors present in the dermal papilla, sebaceous, and eccrine sweat epithelium have the highest concentration in the genital skin, followed by the pubic area, and then the nonsexual areas (e.g., lateral aspects of the scalp, eyebrows, etc.). The concentration of 5α-reductase in the skin follows the same pattern, and the fact that men with the 5α-reductase deficiency syndrome have sparse body and facial hair growth suggests that local production of DHT from T is important for hair growth in some areas of the body. Hirsutism—growth of excessive hair in the androgen responsive hair follicles in sexual areas of the body—can occur in women whose PSUs are exquisitely sensitive to normal serum androgen levels or may occur when androgen production increases above normal.

The sebaceous glands also are sensitive to androgen action. Sebum production increases with increasing concentrations of T, as shown in Fig. 2 *(66)*. Androgens stimulate the prepubertal vellus follicles to form large sebaceous glands in the acne-prone areas of the body *(67)*. The presence of acne in women correlates best with free T concentration. Again, there appears to be a large degree of variability in the sensitivity of the sebaceous glands to the effect of androgens. The sebocytes contain both 3β-hydroxysteroid dehydrogenase and 5α-reductase, allowing DHEA to serve as a prohormone for the local production of DHT *(67)*. Acne initially arises when there is overproduction of sebum by the sebaceous glands. The result is a noninflammatory process that produces simple blackheads and whiteheads. Acne becomes pathological when the lining of the sebaceous follicle is overstimulated by androgens. The cells slough off faster than the follicle can expel them, and they begin to adhere to one another. This accumulation blocks egress from the follicle, and a comodone (pimple) develops. If the condition continues, the follicle may become infected with *Propionibacterium acnes* (*P. acnes*). If the lesions (papules, pustules, nodules, and cysts) become rampant, the condition is termed acne vulgaris, a potentially devastating condition that can leave lifelong physical and psychological scars.

Pattern alopecia is thinning or loss of hair that accelerates with age in genetically susceptible men and women and is the result of miniaturization of scalp hairs (i.e., conversion of terminal hairs to vellus hairs) *(67)*. Although the two patterns of balding are labeled male pattern (temporo-occipital)

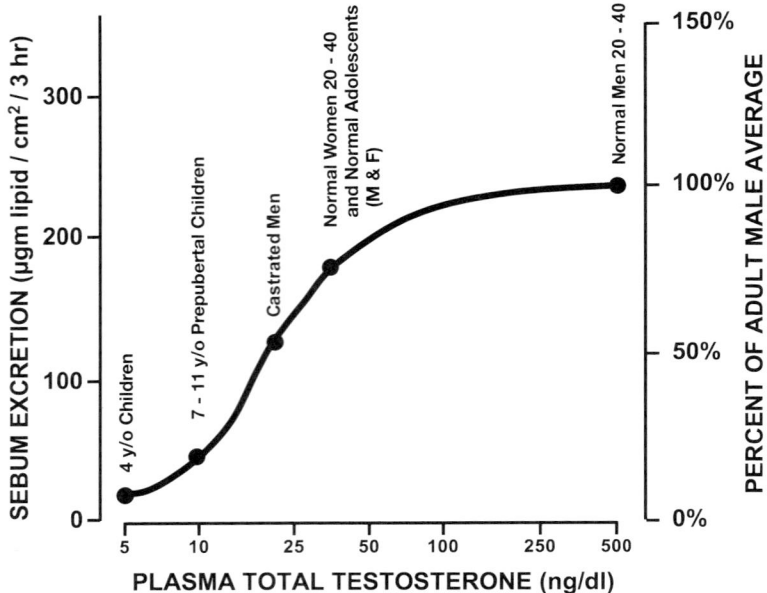

Fig. 2. Relationship of stages of sebum output and T (log scale): **(A)** 4-year-old children; **(B)** 7- to 11-year-old prepubertal children; **(C)** castrated men; **(D)** normal adult women, 20–40 years as well as average sebum level of normal 15- to 19-year-old males and females; **(E)** normal adult men, aged 20–40. (From ref. *72*.)

and female pattern (crown of scalp or diffuse), women can have male-pattern alopecia and men the female type. DHT appears to be the most important androgen involved in the process in males, because the 5α-reductase, type 2 isoenzyme is located in the hair follicles in the scalp. Patients with the 5α-reductase deficiency syndrome lack this isoenzyme and do not develop male pattern baldness. Also, men with male pattern baldness may respond to inhibitors of the 5α-reductase type 2 enzyme activity, such as finasteride. However, women with female pattern alopecia usually have normal androgen levels, and they do not generally respond to finasteride, suggesting that this pattern is not androgen sensitive *(68)*. Women with severe hyperandrogenism more often display a male pattern of hair thinning, which may respond to antiandrogens.

2.4.5. Cardiovascular System

There is a high degree of interest in the role of sex hormones in cardiovascular disease (CVD), the leading cause of death in both men and women. A few studies have examined the relationship between endogenous T and A4 levels and the presence of CVD in women, with conflicting results *(69)*. The largest prospective cohort study to date did not observe a correlation *(70)*. Conflicting data also exist concerning the risk of heart disease in patients with hyperandrogenism. These studies have examined women with PCOS and have amply demonstrated an adverse risk profile (metabolic syndrome) for CVD. However, it is likely that the insulin resistance associated with the syndrome is more relevant to the pathogenesis of CVD than are the elevated androgen levels. For example, exogenous administration of T to women or to female-to-male transsexuals has not been associated with an increased risk of CVD *(69)*. Thus, at present, there is no evidence for either a physiological or a pathological effect of androgens on the cardiovascular system.

3. CONCLUSIONS

Table 1 summarizes the areas where physiological functions of androgens in women have been demonstrated through a collection of observational and investigational studies, as well as the areas

Table 1
Physiological Actions of Androgens in Females

Demonstrated

- Prohormone for E2 production
- Axillary and pubic hair growth in adrenarche and puberty
- Stimulation of sebum production in adrenarche and puberty
- Bone health
- Sexuality

Probable but not conclusively demonstrated

- Imprinting of behavior in prenatal or early postnatal period
- Influence on mood and behavior
- Influence on some aspects of cognition
- Antagonize physiological concentrations of E2 in some tissues

where data supporting a role exist, but are not yet proven. It is clear that DHEA, DHEAS, A4, and T serve as prohormones for the production of estrogens in the adrenals, ovaries, and a variety of peripheral tissues. Therefore, the physiological effects of estrogens in women indirectly reflect a role for androgens, for if there were no androgens, there would be no estrogens. The rise in androgens during adrenarche and puberty directly stimulates axillary and pubic hair growth, as well as increased sebum production. In addition to serving as an important prohormone for estrogen synthesis in the bone, androgens directly influence osteoblastic and osteoclastic function. The importance for androgens in female sexual desire has been well demonstrated; studies of postmenopausal women with low sexual desire and low T levels demonstrate an increase in desire and sexual activity when their T levels are increased to within the reference range for premenopausal women. Areas in which there are suggestive but inconclusive data concern the physiological roles on the organizational and activational aspects of brain function, including the issues surrounding prenatal imprinting of behavior and postnatal influence on mood and behavior. The roles of androgens in the breast and the cardiovascular system remain to be better determined.

4. FUTURE AVENUES OF INVESTIGATION

Several areas concerning the normal physiological effects of androgens in women warrant additional study. These include:

1. Determining the role of androgens and estrogens in the sexual dimorphism of brain structure and function.
2. Understanding the steroid-specific effects in relation to the enzymatic conversion of the sex steroids in tissues, i.e., discerning which hormone is responsible for which effect.
3. Developing sensitive and specific assays for measurement of androgens and their metabolites in women, as well as methods to accurately access the tissue production and intracrine action of androgens.
4. Determining the physiological contribution of androgens to body composition, muscle mass, and strength.
5. Ascertaining if androgens contribute to the development of the metabolic syndrome in hyperandrogenic women or if the increased cardiovascular risk is a result of other factors such as hyperinsulinism.
6. Understanding the relationship of the immune system and autoimmune diseases to androgens. Although not discussed in the text, an important aspect of sexual dimorphism is that females are much more likely to develop autoimmune diseases than men. Further investigation is required to understand the role of physiological concentrations of androgens in the normal function of the immune system and possible protection against the development of autoimmune disorders (*see* ref. *71*).

KEY POINTS

- DHEA, DHEAS, A4, and T are prohormones for the production of estrogens in the adrenals, ovaries, and peripheral tissues.
- Androgens directly stimulate the pilosebaceous unit, stimulating axillary and pubic hair growth and increasing sebum production at puberty.

- Androgens influence osteoblastic and osteoclastic function in the bone.
- Androgens influence female sexual desire.
- Androgens may have a physiological role in the organizational and activational aspects of brain function.

REFERENCES

1. Burger HG. Androgen production in women. Fertil Steril 2002;77(Suppl 4):S3–S5.
2. Cohen-Bendahan CC, van de Beek C, Berenbaum SA. Prenatal sex hormone effects on child and adult sex-typed behavior: methods and findings. Neurosci Biobehav Rev 2005;29(2):353–384.
3. Rhen T, Cidlowski JA. Steroid hormone action. In: Strauss J III, Barbieri R, eds. Yen and Jaffe's Reproductive Endocrinology, 5th ed. Philadelphia: Elsevier Saunders, 2004:155–174.
4. Parker LN, Odell WD. Evidence for existence of cortical androgen-stimulating hormone. Am J Physiol 1979;236(6):E616–620.
5. Rivera-Woll LM, Papalia M, Davis SR, Burger HG. Androgen insufficiency in women: diagnostic and therapeutic implications. Hum Reprod Update 2004;10(5):421–432.
6. Cameron DR, Braunstein GD. Androgen replacement therapy in women. Fertil Steril 2004;82(2):273–289.
7. Labrie F, Luu-The V, Labrie C, et al. Endocrine and intracrine sources of androgens in women: inhibition of breast cancer and other roles of androgens and their precursor dehydroepiandrosterone. Endocr Rev 2003;24(2):152–182.
8. Tchernitchin A. The role of eosinophil receptors in the non-genomic response to oestrogens in the uterus. J Steroid Biochem 1979;11(1B):417–424.
9. Revelli A, Massobrio M, Tesarik J. Nongenomic actions of steroid hormones in reproductive tissues. Endocr Rev 1998;19(1):3–17.
10. Lösel RM, Falkenstein E, Feuring M, et al. Nongenomic steroid action: controversies, questions, and answers. Physiol Rev 2003;83(3):965–1016.
11. Heinlein CA, Chang C. The roles of androgen receptors and androgen-binding proteins in nongenomic androgen actions. Mol Endocrinol 2002;16(10):2181–2187.
12. Willnow TE, Nykjaer A, Herz J. Lipoprotein receptors: new roles for ancient proteins. Nat Cell Biol 1999;1(6):E157–E162.
13. Nykjaer A, Willnow TE. The low-density lipoprotein receptor gene family: a cellular Swiss army knife? Trends Cell Biol 2002;12(6):273–280.
14. Lee MM. Molecular genetic control of sex differentiation. In: Pescovitz OH, Eugster EA, eds. Pediatric Endocrinology: Mechanisms, Manifestations, and Management. Philadelphia: Lippincott Williams & Wilkins; 2004:231–242.
15. Sultan C, Lumbroso S, Paris F, et al. Disorders of androgen action. Semin Reprod Med 2002;20(3):217–228.
16. Gottlieb B, Pinsky L, Beitel LK, Trifiro M. Androgen insensitivity. Am J Med Genet 1999;89(4):210–217.
17. New MI. Clinical and endocrinological aspects of 21-hydroxylase deficiency. Ann NY Acad Sci 1985;458:1–27.
18. Kelly SJ, Ostrowski NL, Wilson MA. Gender differences in brain and behavior: hormonal and neural bases. Pharmacol Biochem Behav 1999;64(4):655–664.
19. Hines M, Brook C, Conway GS. Androgen and psychosexual development: core gender identity, sexual orientation and recalled childhood gender role behavior in women and men with congenital adrenal hyperplasia (CAH). J Sex Res 2004;41(1):75–81.
20. Berenbaum SA, Bailey JM. Effects on gender identity of prenatal androgens and genital appearance: evidence from girls with congenital adrenal hyperplasia. J Clin Endocrinol Metab 2003;88(3):1102–1106.
21. Hines M, Ahmed SF, Hughes IA. Psychological outcomes and gender-related development in complete androgen insensitivity syndrome. Arch Sex Behav 2003;32(2):93–101.
22. Wisniewski AB, Migeon CJ, Meyer-Bahlburg HF, et al. Complete androgen insensitivity syndrome: long-term medical, surgical, and psychosexual outcome. J Clin Endocrinol Metab 2000;85(8):2664–2669.
23. Quigley CA. Editorial: The postnatal gonadotropin and sex steroid surge-insights from the androgen insensitivity syndrome. J Clin Endocrinol Metab 2002;87(1):24–28.
24. Reiter EO, Grumbach MM, Kaplan SL, Conte FA. The response of pituitary gonadotropes to synthetic LRF in children with glucocorticoid-treated congenital adrenal hyperplasia: lack of effect of intrauterine and neonatal androgen excess. J Clin Endocrinol Metab 1975;40(2):318–325.
25. Vanderschueren D, Vandenput L, Boonen S, Lindberg MK, Bouillon R, Ohlsson C. Androgens and bone. Endocr Rev 2004;25(3):389–425.
26. Frank GR. Role of estrogen and androgen in pubertal skeletal physiology. Med Pediatr Oncol 2003;41(3):217–221.
27. Conte FA, Grumbach MM, Ito Y, Fisher CR, Simpson ER. A syndrome of female pseudohermaphrodism, hypergonadotropic hypogonadism, and multicystic ovaries associated with missense mutations in the gene encoding aromatase (P450arom). J Clin Endocrinol Metab 1994;78(6):1287–1292.
28. Auchus RJ, Rainey WE. Adrenarche—physiology, biochemistry and human disease. Clin Endocrinol 2004;60(3):288–296.
29. Hiort O. Androgens and puberty. Best Pract Res Clin Endocrinol Metab 2002;16(1):31–41.

30. Zumoff B, Strain GW, Miller LK, Rosner W. Twenty-four-hour mean plasma testosterone concentration declines with age in normal premenopausal women. J Clin Endocrinol Metab 1995;80(4):1429–1430.

31. Labrie F, Belanger A, Cusan L, Gomez JL, Candas B. Marked decline in serum concentrations of adrenal C19 sex steroid precursors and conjugated androgen metabolites during aging. J Clin Endocrinol Metab 1997;82(8):2396–2402.

32. Lasley BL, Santoro N, Randolf JF, et al. The relationship of circulating dehydroepiandrosterone, testosterone, and estradiol to stages of the menopausal transition and ethnicity. J Clin Endocrinol Metab 2002;87(8):3760–3767.

33. Burger HG, Dudley EC, Cui J, Dennerstein L, Hopper JL. A prospective longitudinal study of serum testosterone, dehydroepiandrosterone sulfate, and sex hormone-binding globulin levels through the menopause transition. J Clin Endocrinol Metab 2000;85(8):2832–2838.

34. Hughes CL, Jr., Wall LL, Creasman WT. Reproductive hormone levels in gynecologic oncology patients undergoing surgical castration after spontaneous menopause. Gynecol Oncol 1991;40(1):42–45.

35. Bachmann G, Bancroft J, Braunstein G, et al. Female androgen insufficiency: the Princeton consensus statement on definition, classification, and assessment. Fertil Steril 2002;77(4):660–665.

36. Cooke BM, Woolley CS. Gonadal hormone modulation of dendrites in the mammalian CNS. J Neurobiol 2005;64(1):34–46.

37. Patchev VK, Schroeder J, Goetz F, Rohde W, Patchev AV. Neurotropic action of androgens: principles, mechanisms and novel targets. Exp Gerontol 2004;39(11–12):1651–1660.

38. Baulieu EE. Neurosteroids: a novel function of the brain. Psychoneuroendocrinology 1998;23(8):963–987.

39. Cameron DR, Braunstein GD. The use of dehydroepiandrosterone therapy in clinical practice. Treat Endocrinol 2005;4(2):95–114.

40. Greenblatt RB, Barfield WE, Garner JF, Calk GL, Harrod JP Jr. Evaluation of an estrogen, androgen, estrogen-androgen combination, and a placebo in the treatment of the menopause. J Clin Endocrinol Metab 1950;10(12):1547–1558.

41. Sherwin BB, Gelfand MM, Brender W. Androgen enhances sexual motivation in females: a prospective, crossover study of sex steroid administration in the surgical menopause. Psychosom Med 1985;47(4):339–351.

42. Goldstat R, Briganti E, Tran J, Wolfe R, Davis SR. Transdermal testosterone therapy improves well-being, mood, and sexual function in premenopausal women. Menopause 2003;10(5):390–398.

43. Shifren JL, Braunstein GD, Simon JA, et al. Transdermal testosterone treatment in women with impaired sexual function after oophorectomy. N Engl J Med 2000;343(10):682–688.

44. Sherwin BB, Gelfand MM. Differential symptom response to parenteral estrogen and/or androgen administration in the surgical menopause. Am J Obstet Gynecol 1985;151(2):153–160.

45. Morris NM, Udry JR, Khan-Dawood F, Dawood MY. Marital sex frequency and midcycle female testosterone. Arch Sex Behav 1987;16(1):27–37.

46. Van Goozen SH, Wiegant VM, Endert E, Helmond FA, Van de Poll NE. Psychoendocrinological assessment of the menstrual cycle: the relationship between hormones, sexuality, and mood. Arch Sex Behav 1997;26(4):359–382.

47. Riley A, Riley E. Controlled studies on women presenting with sexual drive disorder: I. Endocrine status. J Sex Marital Ther 2000;26(3):269–283.

48. Persky H, Dreisbach L, Miller WR, et al. The relation of plasma androgen levels to sexual behaviors and attitudes of women. Psychosom Med 1982;44(4):305–319.

49. Floter A, Nathorst-Boos J, Carlstrom K, von Schoultz B. Addition of testosterone to estrogen replacement therapy in oophorectomized women: effects on sexuality and well-being. Climacteric 2002;5(4):357–365.

50. Davis SR, Davison SL, Donath S, Bell RJ. Circulating androgen levels and self-reported sexual function in women. JAMA 2005;294(1):91–96.

51. Appelt H, Strauss B. Effects of antiandrogen treatment on the sexuality of women with hyperandrogenism. Psychother Psychosom 1984;42(1–4):177–181.

52. Buster JE, Kingsberg SA, Aguirre O, et al. Testosterone patch for low sexual desire in surgically menopausal women: a randomized trial. Obstet Gynecol 2005;105(5 Pt 1):944–952.

53. Braunstein GD, Sundwall DA, Katz M, et al. Safety and efficacy of a testosterone patch for the treatment of hypoactive sexual desire disorder in surgically menopausal women: a randomized, placebo-controlled trial. Arch Intern Med 2005;165(14):1582–1589.

54. Simon J, Braunstein G, Nachtigall L, et al. Testosterone patch increases sexual activity and desire in surgically menopausal women with hypoactive sexual desire disorder. J Clin Endocrinol Metab 2005;90(9)5226–5233.

55. Lobo RA, Rosen RC, Yang HM, Block B, Van Der Hoop RG. Comparative effects of oral esterified estrogens with and without methyltestosterone on endocrine profiles and dimensions of sexual function in postmenopausal women with hypoactive sexual desire. Fertil Steril 2003;79(6):1341–1352.

56. Davis SR, Tran J. Testosterone influences libido and well being in women. Trends Endocrinol Metab 2001;12(1):33–37.

57. Notelovitz M. Androgen effects on bone and muscle. Fertil Steril 2002;77(Suppl 4):S34–S41.

58. Marcus R, Leary D, Schneider DL, Shane E, Favus M, Quigley CA. The contribution of testosterone to skeletal development and maintenance: lessons from the androgen insensitivity syndrome. J Clin Endocrinol Metab 2000;85(3):1032–1037.

59. Kearns AE, Khosla S. Potential anabolic effects of androgens on bone. Mayo Clin Proc 2004;79(4 Suppl):S14–S18.

60. Speroff L, Rowan J, Symons J, Genant H, Wilborn W. The comparative effect on bone density, endometrium, and lipids of continuous hormones as replacement therapy (CHART study). A randomized controlled trial. JAMA 1996;276(17):1397–1403.

61. Erdtsieck RJ, Pols HA, van Kuijk C, et al. Course of bone mass during and after hormonal replacement therapy with and without addition of nandrolone decanoate. J Bone Miner Res 1994;9(2):277–283.

62. Raisz LG, Wiita B, Artis A, et al. Comparison of the effects of estrogen alone and estrogen plus androgen on biochemical markers of bone formation and resorption in postmenopausal women. J Clin Endocrinol Metab 1996;81(1):37–43.

63. Dimitrakakis C, Zhou J, Wang J, et al. A physiologic role for testosterone in limiting estrogenic stimulation of the breast. Menopause 2003;10(4):292–298.

64. Somboonporn W, Davis SR. Testosterone effects on the breast: implications for testosterone therapy for women. Endocr Rev 2004;25(3):374–388.

65. Dimitrakakis C, Jones RA, Liu A, Bondy CA. Breast cancer incidence in postmenopausal women using testosterone in addition to usual hormone therapy. Menopause 2004;11(5):531–535.

66. Rosenfield RL, Deplewski D. Role of androgens in the developmental biology of the pilosebaceous unit. Am J Med 1995;98(1A):80S–88S.

67. Deplewski D, Rosenfield RL. Role of hormones in pilosebaceous unit development. Endocr Rev 2000 2000;21(4):363–392.

68. Kaufman KD. Androgens and alopecia. Mol Cell Endocrinol 2002;198(1–2):89–95.

69. Eckardstein Av, Wu FCW. Testosterone and atherosclerosis. Growth Hormone IGF Res 2003;13(Suppl 1):S72–S84.

70. Barrett-Connor E, Goodman-Gruen D. Prospective study of endogenous sex hormones and fatal cardiovascular disease in postmenopausal women. BMJ 1995;311(7014):1193–1196.

71. Olsen NJ, Kovacs WJ. Gonadal steroids and immunity. Endocr Rev 1996;17(4):369–384.

72. Rosenfield RL. Pilosebaceous physiology in relation to hirsutism and acne. Clin Endocrinol Metab 1986;15(2):341–362.

5

Androgen Measurements

Methods, Interpretation, and Limitations

Frank Z. Stanczyk

SUMMARY

Androgens are measured primarily in serum by direct and indirect (with one or two purification steps) immunoassay methods. Direct immunoassays are seldom thoroughly validated and often lack sensitivity and specificity. Free testosterone measurements also have their limitations, but the equilibrium dialysis assay or calculated method using an algorithm is considered reliable. There is growing use of assays that combine mass spectrometry with either gas or liquid chromatography for quantifying androgens. Liquid chromatography–tandem mass spectrometry is touted to become the "gold standard" for all steroid hormone measurements.

Key Words: Androgens; radioimmunoassays; immunoassays; free testosterone; mass spectrometry.

1. INTRODUCTION

The initial methods used to measure steroid hormones included bioassays and different chemical methods. These assays were restricted to quantifying conjugated steroids in urine and lacked sensitivity. Subsequently, development of the radioimmunoassay (RIA) method made it possible to measure steroid hormones routinely in serum or plasma with high sensitivity (i.e., in the picogram and low nanogram range instead of the microgram and milligram range). The first RIA method, developed in 1959 by Yallow and Berson *(1,2)*, was for insulin. Ten years later, Abraham *(3)* reported the development of the first steroid RIA, which was for estradiol (E_2). The immediate impact of the RIA method allowed measurement of an immensely wide range of compounds of clinical and biological importance and opened new horizons in endocrinology.

The purpose of the present chapter is to discuss the RIA method developed by Abraham *(3)* and its application to the measurement of other steroid hormones, including androgens, as well as its modification to less time-consuming direct immunoassays. The advantages and disadvantages of both the conventional and direct immunoassays will be pointed out. Thereafter, assay methods used to measure free testosterone will be evaluated. In addition to immunoassay methods, use of mass spectrometry assay methods for quantifying androgens will be discussed, as well as the potential for these assays to become the gold standard for steroid hormone measurements. Finally, advantages and disadvantages of biological fluids, specifically serum, plasma, urine, and saliva, will be compared.

2. BACKGROUND

2.1. Radioimmunoassay Method

The general principle of the RIA method, using E_2 as an example, involves competition between E_2 and radioactive E_2—both in excess—for a limited amount of antibody against E_2 (Fig. 1). The antibody-bound radioactive fraction is separated from the unbound radioactive fraction and used to

From: *Contemporary Endocrinology: Androgen Excess Disorders in Women: Polycystic Ovary Syndrome and Other Disorders, Second Edition*
Edited by: R. Azziz et al. © Humana Press Inc., Totowa, NJ

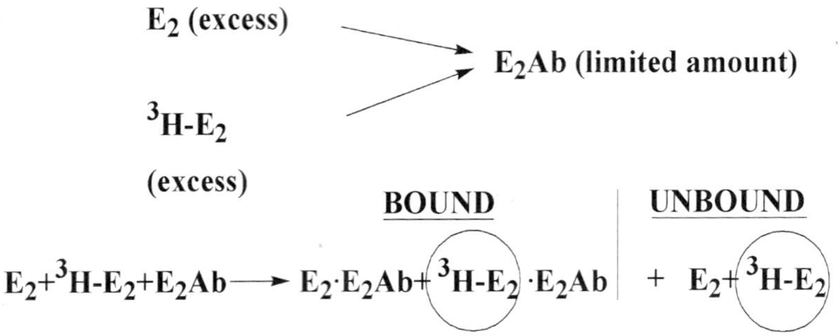

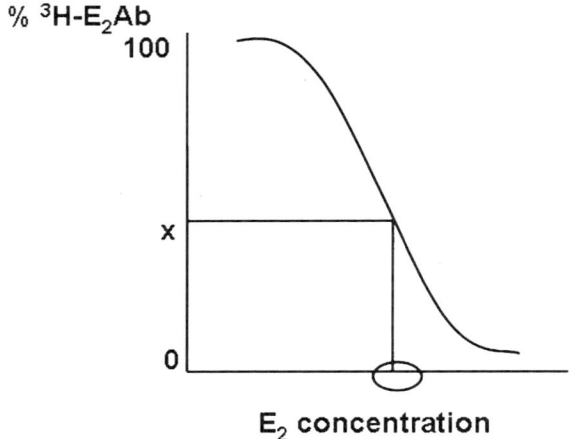

Fig. 1. General principle of the radioimmunoassay (RIA) method, using estradiol (E_2) as an example. The method involves competition between E_2 and radioactive E_2 (^{3}H-E_2), both in excess, for a limited amount of antibody against E_2 (E_2Ab) Antibody-bound and unbound fractions are separated, and the bound fraction is used for quantification.

Fig. 2. The radioactivity in the antibody-bound fraction is quantified for different estradiol (E_2) concentrations in preparing the E_2 standard curve. The E_2 concentration in a sample is determined from the corresponding radioactive antibody-bound E_2 fraction extrapolated off the standard curve.

prepare an E_2 standard curve and quantify E_2 in a sample. The E_2 concentration in a sample is determined from the corresponding radioactive antibody-bound E_2, extrapolated off the E_2 standard curve (Fig. 2).

From a practical standpoint, the procedure for the E_2 RIA described by Abraham *(3,4)* involved purification of E_2 in serum samples by organic solvent extraction and chromatography prior to its quantification by RIA. The purification step was necessary because E_2 is converted to many different unconjugated and conjugated (sulfates and glucuronides) metabolites that may potentially interfere with the specificity of the assay.

Quantification of E_2 involved preparation of different concentrations of the E_2 standard for the standard curve and was followed by addition of tritiated E_2 and E_2 antibody to the standards and purified E_2 from the samples. After an incubation period, antibody-bound E_2 was separated from unbound E_2. The bound radioactive fraction was counted and was used to extrapolate the E_2 concentration off the standard curve. The E_2 RIA developed by Abraham was shown to be sensitive, specific, accurate, and precise.

Subsequently, the extraction/chromatographic RIA method described for E_2 was applied successfully to many other steroid hormones. This methodology has remained essentially the same to the present time, except for use of an iodinated instead of tritiated marker, which has increased assay sensitivity.

Advantages of RIA methods with purification steps, which will be referred to as conventional RIAs, include the following: (a) steroid binding proteins, such as sex hormone-binding globulin (SHBG), are denatured by the organic solvent in the extraction step, thereby releasing steroids such as testosterone and E_2; (b) metabolites are removed prior to quantification of the steroid by RIA; (c) relatively large serum aliquots can be used for the assay, allowing the analyte to be "read" on a more accurate part of the standard curve; (d) multiple steroid hormones can be measured in the same sample aliquot after separation of the steroids by column chromatography; and (e) the assay is highly reliable when properly validated.

Steroid RIA methods with purification steps have been used in numerous studies that have enriched the field of endocrinology with new knowledge, and their use in diagnostic testing has provided physicians with valuable information for diagnosing and treating countless number of patients. However, the conventional steroid RIA also has disadvantages. It is cumbersome, time-consuming, and relatively costly. It usually takes 2 days to measure a single steroid hormone in about 40 samples.

2.2. Direct Immunoassays

In the late 1970s, direct RIA methods were developed to quantify steroid hormones (5). They differed from the conventional RIA methods, primarily by not including any purification steps. Subsequently, the radioactive label used for RIAs was replaced with a chemiluminescent, fluorescent, or enzymatic label to allow immunoassay to be carried out in an automated analyzer. This allowed for a tremendously increased throughput of samples to be analyzed in clinical laboratories. Direct steroid immunoassays are convenient, simple, rapid, and relatively inexpensive. In contrast to conventional immunoassays, it takes only several hours to quantify a steroid hormone in about 40 samples. Direct steroid immunoassays also have the following major disadvantages: (a) steroid measurements are often overestimated due to lack of specificity of the antibody; (b) matrix differences between the serum sample and solutions of the standard (for standard curve) may exist; (c) testosterone and E_2 may not be released efficiently from SHBG; and (d) they generally lack the sensitivity to measure low levels of steroids efficiently.

The disadvantages just described for direct steroid hormone immunoassays are evident in one of our studies evaluating direct E_2 and testosterone immunoassay kits (6). In the study, we evaluated eight commercial direct E_2 immunoassay kits, which were used to perform assays either on an analyzer or manually. Three of the kits were for RIAs, three were for enzyme immunoassays, and two were for chemiluminescent immunoassays. A ninth kit, which required a purification (organic solvent extraction) step prior to RIA, was also evaluated. The resulting E_2 values obtained with all of the kits were compared to those obtained with our conventional E_2 RIA. Forty female samples containing low, medium, and high levels of E_2 were analyzed. We determined intraclass correlation coefficients and validity, which reflects assay accuracy, for low, high, and all E_2 measurements obtained with the different kits. Overall, the RIA with the preceding purification step performed the best. All of the direct assays had either poor correlations and/or failed validity (Table 1).

In the study just described, we also evaluated four different direct testosterone immunoassays kits and used our conventional testosterone RIA as the standard for comparison. Three of the kits were for RIAs, and one was for chemiluminescent immunoassay. Testosterone was analyzed in 10 premenopausal, 10 postmenopausal, and 10 male samples. The results showed that the assays performed generally well for the male samples, but gave either poor intraclass correlations and/or failed validity for the female samples (Table 2).

On the basis of the results from our study, we concluded the following: (a) wide differences were observed in the levels of each hormone measured with kits from different manufacturers; (b) the E_2

Table 1
**Intraclass Correlation Coefficients and Validity for Low, High,
and All Estradiol (E$_2$) Measurements in Female Serum**

Assay method[a]	E$_2$ <183 pmol/L	E$_2$ >183 pmol/L	All data
EX-RIA	0.61	0.98	0.99
RIA 1	0.54	0.63*	0.81*
RIA 2	0.41	0.09	0.38
RIA 3	0.74	0.16	0.38
EIA 1	0.73	0.22*	0.53*
EIA 2	−0.15*	0.86*	0.94*
EIA 3	0.27	0.59	0.82
CIA 1	0.54	0.46*	0.74*
CIA 2	0.27*	0.83*	0.92*

[a]The measurements were carried out with an E$_2$ radioimmunoassay kit requiring a prior purification step (EX-RIA) and eight different direct E$_2$ immunoassay kits that included radioactive (RIA), chemiluminescent (CIA), or enzyme (EIA) markers. Comparisons were made to E$_2$ values determined by conventional RIA.
*$p < 0.05$ (failed test for validity).

Table 2
**Intraclass Correlation Coefficients and Validity for Testosterone Measurements
in Premenopausal, Postmenopausal, and Male Serum Samples**

Assay method[a]	Premenopausal	Postmenopausal	Male
RIA 1	0.71*	0.88*	0.87
RIA 2	0.52*	0.64*	0.84
RIA 3	0.20	0.62	0.93*
CIA 1	0.70*	0.95*	0.98

[a]The measurements were carried out with four different direct testosterone immunoassay kits that included radioactive (RIA) or chemiluminescent (CIA) markers. Comparisons were made to testosterone values determined by conventional RIA.
*$p < 0.05$ (failed test for validity).

RIA kit that required a prior purification step gave values that were similar to our conventional E$_2$ RIA; and (c) testosterone levels in both premenopausal and postmenopausal samples were not measured reliably.

The findings pertaining to testosterone in our study are consistent with those recently reported by Taieb and coworkers *(7)*, who measured serum testosterone levels in women, men, and children using 10 different direct testosterone immunoassay kits and by isotope dilution gas chromatography–mass spectrometry (GC-MS). Taieb and coworkers *(7)* concluded that the direct assays were generally acceptable for quantifying testosterone in male samples but not in samples from women or children. In an accompanying editorial on the study by Taieb et al. *(7)*, the editors concluded that "guessing appears to be nearly as good as most commercially available immunoassays and clearly superior to some" *(8)*.

Androgen levels in premenopausal women, and to a lesser extent in postmenopausal women, are well documented. Most of our knowledge about the androgen levels is derived from studies in which reliable conventional RIAs have been used to quantify the androgens. However, it is important to realize that even though direct immunoassays are generally not reliable for quantifying female testosterone levels, these assays can be used to measure serum levels of androgens that have relatively higher concentrations than testosterone, e.g., dehydroepiandrosterone (DHEA), DHEA sulfate (DHEAS), and androstenedione. The important point is that such assays should be thoroughly validated with respect to sensitivity, precision, specificity, and accuracy in the laboratory where the assays are being performed.

2.3. Measuring Free and Bioavailable Testosterone

In blood, testosterone is present predominantly in a protein-bound form and only a very small portion is free. In premenopausal women, approximately 66 and 30% of total testosterone is bound to SHBG and albumin, respectively, and the free fraction generally comprises less than 2% of the total *(9)*. Testosterone is bound with high affinity ($K_a = 1.7 \times 10^9 \, M^{-1}$) and low capacity to SHBG, and it is bound with low affinity ($K_a = 1 \times 10^4 \, M^{-1}$ to $1 \times 10^5 \, M^{-1}$) but high capacity to albumin *(9,10)*.

For many years it was accepted that only the free fraction of testosterone in the circulation can be taken up by tissues and that the protein-bound testosterone complex is inactive. However, some investigators observed that the fraction of testosterone bound to albumin dissociates rapidly and is taken up by tissues in a manner similar to that of the free steroid *(11–13)*. Testosterone bound to the large pool of albumin, together with the small amount of the free steroid, likely forms the circulating pool of bioavailable (non-SHBG-bound) testosterone. This fraction of testosterone enters cells, where it may undergo metabolism or bind to the androgen receptor and exert biological activity.

Commonly used methods for measuring free testosterone involve the addition of a small amount of ^{3}H-testosterone to serum or plasma and, after a suitable incubation period, separation of the protein (SHBG and albumin)-bound fractions from the free fraction of testosterone by means of a membrane (e.g., equilibrium dialysis) or filter (e.g., centrifugal ultrafiltration). These barriers retain the protein-bound fractions but allow free testosterone to pass through. The percentage of tritiated free testosterone is then calculated on the basis of the total ^{3}H-testosterone added. Recovery of free components through a barrier is sometimes monitored using a small-labeled molecule such as ^{14}C-glucose.

Several technical limitations exist in the assay methods used to measure free testosterone. The equilibrium dialysis method is influenced by dilution of the serum sample. The centrifugal ultrafiltration method is subject to adsorption of testosterone to the filter. Both the dialysis and ultrafiltration methods can be affected by impurities of tritiated testosterone not bound by SHBG or albumin; these impurities may increase the percentage of free testosterone. Also, the use of too large an amount of ^{3}H-testosterone in the assays may increase the concentration of total testosterone and possibly disturb the equilibrium of endogenous testosterone. Despite its limitations, the equilibrium dialysis assay is considered the gold standard method for quantifying free testosterone.

Two methods used to determine the percentage of bioavailable testosterone in serum include centrifugal ultrafiltration with heat-treated serum and ammonium sulfate precipitation. In the centrifugal ultrafiltration method the percentage of albumin-bound testosterone is determined after SHBG is inactivated by heating the serum sample to 60°C for 1 hour. After the temperature of the sample returns to 37°C, the testosterone dissociated from SHBG is reequilibrated in the serum, and the testosterone fraction bound to albumin can be determined by ultrafiltration. The fraction of testosterone bound to albumin along with the free testosterone fraction determined before heating the sample comprises the total bioavailable testosterone fraction. A much simpler method to determine bioavailable testosterone involves addition of a small amount of ^{3}H-testosterone to serum and, after a suitable incubation period, precipitation of the globulins (including the SHBG-testosterone complex) with saturated ammonium sulfate, centrifugation, counting the tritium in the supernatant, and calculating the percentage of the total ^{3}H-testosterone that is not SHBG bound.

Technical difficulties are also encountered in the measurement of bioavailable testosterone. When this fraction is measured by use of a barrier method after inactivation of SHBG, the same technical problems exist as described for the measurement of free testosterone. The most frequently encountered sources of error in the ammonium sulfate precipitation assay are a result of the use of impure tritiated testosterone, insufficient counting time of the small amount of radiolabeled testosterone, and incomplete precipitation of globulins. The deficiencies in both assays are often the cause of poor intraassay and interassay reproducibility.

Using the methods just described, the concentration of free or bioavailable testosterone is usually calculated from the percentage of free or bioavailable testosterone multiplied by the total testosterone concentration, which is quantified separately by an immunoassay method. Free testosterone concentrations are sometimes measured directly in the dialysate following equilibrium dialysis. However, a highly sensitive conventional RIA is essential to measure the very low testosterone levels.

Because the assays described above for quantifying free or bioavailable testosterone are time-consuming and expensive, they are available in a limited number of reference laboratories. The most widely used method for measurement of free testosterone in clinical laboratories is direct RIA. In general, this assay uses a ^{125}I-labeled testosterone analog that has very low affinity for SHBG and albumin and competes with free testosterone for binding sites on an immobilized specific testosterone antibody. Although this approach provides a simple and rapid test for quantifying free testosterone, it has been pointed out that the assay method has several deficiencies; these include low antibody affinity, major biasing effects resulting from dilution of serum samples, significant binding of the analog to serum proteins, and lack of parallelism between measurements of serially diluted serum samples and free testosterone *(14)*. For these reasons the reliability of the assay that utilizes the analog-based free testosterone RIA kit has been questioned *(15,16)*.

One study *(17)* showed that plasma free testosterone levels in samples from normal women and patients with polycystic ovarian syndrome were approximately three to four times higher when measured by use of a commercial analog-based RIA kit compared with measurements using the equilibrium dialysis method. The results obtained with the latter method were comparable to published data. Nevertheless, good correlations between the results of the two methods were obtained. The investigators concluded that the free testosterone values measured by use of the kit had a mean bias of –76%, thereby making comparison with published data difficult. The higher levels of free testosterone measured by direct RIA may result from the fact that the antibody in the RIA system has a greater affinity for testosterone weakly bound to albumin than albumin does. This may allow the antibody to strip some of the testosterone that is bound to albumin. In a subsequent study *(18)* it was shown that the direct testosterone RIA had unacceptably high systematic bias and random variability and did not correlate well with equilibrium dialysis. In a letter to the editor by Rosner *(19)* about the direct free testosterone analog RIA, he concluded: "the literature of science ought not to use a method so grossly inaccurate when better ones exist." In addition, Rosner *(19)* suggested that the "journal might choose to return manuscripts that use it without further evaluation to discourage its use."

Some laboratories and investigators have measured total testosterone and SHBG and have used the testosterone:SHBG ratio, referred to as the free androgen index (FAI), as an estimate of free testosterone. The validity of the FAI as an accurate reflection of free testosterone has been questioned. In one small study in men *(15)*, the FAI was shown to be unreliable, based on its comparison to free testosterone quantified by equilibrium dialysis; the ratio of FAI to free testosterone determined by dialysis was 0.12–0.26. In another small study *(20)*, a high correlation coefficient (0.858) was found between the FAI and free testosterone levels determined by centrifugal ultrafiltration in serum samples from women, whereas in male samples the correlation was only 0.435. In a more recent study *(18)* in women, a good correlation was found between FAI and equilibrium dialysis. However, the authors of that study pointed out that the FAI can be altered by changes in either testosterone or SHBG and that using this quotient alone can be misleading. Therefore, use of the FAI is limited.

Both free and bioavailable testosterone can also be calculated by use of an algorithm that requires the concentrations of total testosterone, SHBG, and albumin, as well as the binding constants of testosterone to SHBG and albumin obtained from published equations *(15)*. Calculated free testosterone levels in men and women were found to be nearly identical with corresponding values measured by equilibrium dialysis *(18,21,22)*.

It is important to realize that when indirect methods, such as equilibrium dialysis or centrifugal ultrafiltration, are used to determine the percentage of free testosterone, the accuracy of the total testosterone concentration is very important. This percentage is multiplied by the total testosterone concentration to obtain the free testosterone concentration. Thus, direct immunoassay methods should not be used to quantify total testosterone levels in female samples. RIAs with preceding organic solvent extraction, and chromatography steps will provide reliable values. Similarly, if the FAI or algorithm is used to calculate free testosterone, the accuracy of both total testosterone and SHBG values is essential. Although the concentration of albumin is also required in the algorithm method, an average normal albumin value can be used without any significant change in the calculated free testosterone concentration.

Differences in SHBG concentrations obtained with different commercially available SHBG kits have been reported *(23)*. In one study *(18)*, an approximate twofold greater absolute value was found using an immunoradiometric assay (IRMA) compared to RIA, and better accuracy was found with IRMA. The IRMA method was calibrated against a dihydrotestosterone (DHT)-binding capacity IRMA, which is considered to provide SHBG values that reflect more physiologically relevant SHBG concentrations in blood. Thus, it seems reasonable to use SHBG assay methods that correlate well with assay methods based on testosterone- or DHT-binding capacity.

2.4. Monitoring of DHEA and Androstenedione Supplementation

During the past several years there has been increasing use of the androgens DHEA and androstenedione by men and women as supplements to enhance athletic performance, cognitive function, mood, and/or libido. A primary reason for the increased use of these androgens is their classification as a food instead of a drug. Before 1994, DHEA and androstenedione were available by prescription only. However, in 1994 the Dietary Supplement Health and Education Act classified these androgens as foods instead of drugs.

Although DHEA and androstenedione are usually sold over the counter in tablets containing 25 mg of the steroid, their use in doses of 200 mg or higher has been advertised for enhancing athletic performance and building muscle mass. Because DHEA and androstenedione are readily converted to potent androgens such as testosterone and DHT, as well as to the active estrogens E_2 and estrone, their long-term use at high doses may lead to adverse effects. In women, increased androgens may lead to a hyperandrogenic state with clinical manifestations of hirsutism, acne, and/or alopecia, whereas elevated estrogens may stimulate hormonally sensitive tissues, such as the endometrium and breast, leading to hyperplasia and possibly cancer. Elevated androgens and/or estrogens may adversely affect reproductive function and normal physiological body processes. Because potentially harmful circulating levels of potent androgens and estrogens are formed when DHEA and androstenedione are administered in high doses, it is very important that serum levels of these steroids be monitored to ensure that they are not abnormally elevated.

2.5. Mass Spectrometry Assays

In addition to immunoassays, another major advance in assay methodology for quantifying steroid hormones also occurred in the 1970s—GC-MS. This method combines the resolving power of GC with the high sensitivity and specificity of the mass spectrometer. Separation of steroids by GC requires that they be first derivatized to increase their volatility, selectivity, and detectability. The mass spectrometer functions as a unique detector that provides structural information on individual solutes as they elute from the GC column.

MS can also be combined with liquid chromatography (LC-MS), which has high resolving power (Fig. 3). LC has the advantage of not requiring derivatization of compounds for their separation. In recent years there has been increasing use of LC–tandem spectrometry, usually referred to as LC-tandem MS (LC-MS/MS), for measuring steroid hormones. Tandem MS consists of two mass spectrometers in series connected by a chamber (collision cell). After chromatography, the sample is processed in the first mass spectrometer to obtain the precursor ion, which is then fragmented in the collision cell into product ions. The mass of the product ions is then determined in the detector of the second mass spectrometer. This method has high specificity, sensitivity, and throughput.

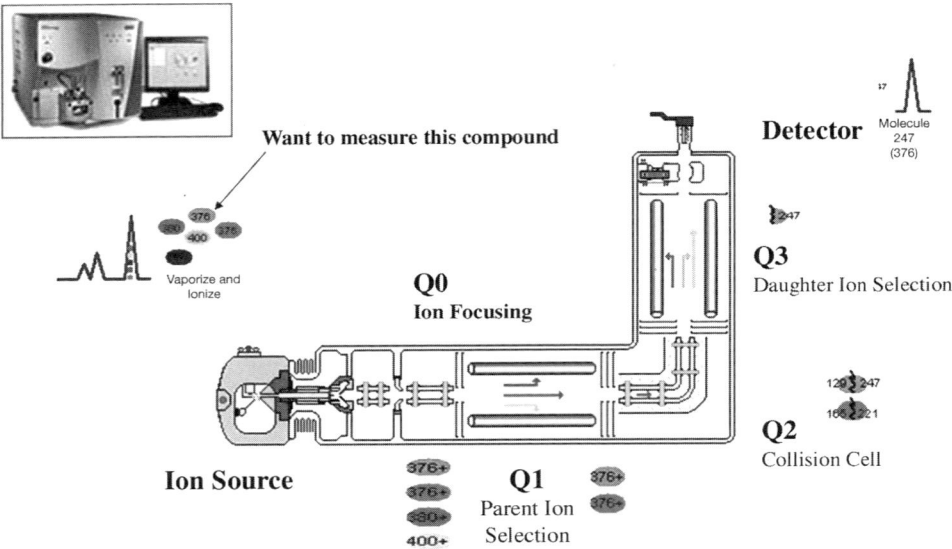

Fig. 3. Depiction of liquid chromatography–mass spectrometry (LC-MS). This method combines the resolving power of LC with the high sensitivity and specificity of the mass spectrometer (Courtesy of Nigel Clarke, PhD).

As with immunoassay methods, MS assays also have their advantages and disadvantages. Advantages of MS assays are they produce highly accurate results when properly validated, and they have the capability of high throughput of samples. The disadvantages include costly instrumentation; need for a highly trained technician, and the underestimation of results caused by incomplete derivatization of compounds.

There has been considerable discussion about the standardization of steroid hormone assays, and LC-MS/MS assays have been proposed as the new gold standard for these measurements. However, it is a misconception that any steroid hormone quantified by a MS method gives a gold standard result. It is important to realize that interlaboratory differences exist in MS assays because of the lack of standardization of assay reagents and procedures as well as instrumentation. Assay conditions must first be standardized between laboratories to achieve gold standard results.

2.6. Use of Serum/Plasma, Urine, or Saliva for Quantifying Androgens: Advantages and Disadvantages

2.6.1. Serum or Plasma

Use of serum or plasma is, overall, convenient for the patient, clinician, and laboratory. It is also appropriate for rapid and repeated sample analysis (e.g., dynamic testing). However, serum or plasma collection is invasive and subject to episodic, diurnal, and cyclic variability of the hormone. In addition, it is representative only of the concentration that existed at the time of blood sampling.

2.6.2. Urine

In contrast to serum or plasma, collection of urine is noninvasive and usually represents an approximate proportion of the steroid secreted during the period of collection. Also, urine contains metabolites in high concentrations, which does not require sensitive assays for measurement of metabolites. However, urine collection has several disadvantages. It is inconvenient because a 24-hour collection is usually recommended, requires creatinine determination to monitor completeness of collection, may require special interpretation if renal function is altered, and contains steroids that are predominantly in a conjugated (inactive) form.

Prior to the development of RIA methods, androgens were measured in urine primarily by colorimetric determination of 17-ketosteroids, which are present in conjugated form. However, urinary 17-ketosteroids are primarily metabolites of adrenal androgen precursors, and they therefore mainly test adrenal, not ovarian, androgen biosynthesis. Subsequently, an RIA was developed to measure urinary testosterone in its glucuronidated form. However, this assay is cumbersome, time-consuming, and not practical for routine diagnostic testing.

2.6.3. Saliva

Collection of saliva is noninvasive, simple, nonstressful, and allows collection at frequent intervals. However, use of saliva for measurement of steroid hormones has some major disadvantages. Interpretation of results may be complicated by contamination with blood and/or steroid metabolism by salivary glands. Also, steroid levels are generally only 2–3% of corresponding levels in serum/plasma, often requiring highly sensitive assays. In addition, assays are usually performed only by highly specialized laboratories.

3. CONCLUSION

Conventional RIAs are highly reliable for measuring androgens. Direct immunoassays should not be used to measure testosterone levels in women. The equilibrium dialysis assay and calculation using an algorithm are reliable methods for determining free testosterone concentrations. LC-MS/MS assays are considered to be the gold standard, but interlaboratory comparisons using standardized reagents, instrumentation, and procedures are essential. Overall, serum or plasma samples are more convenient for the patient, clinician, and laboratory, than are urine or saliva samples.

4. FUTURE AVENUES OF INVESTIGATION

As stated earlier, the LC-MS/MS assay method for quantifying steroid hormones in serum has the capability of achieving not only high assay sensitivity and specificity, but also high throughput of samples. It is thought that it will become the gold standard for steroid hormone measurements. However, before any assay can be considered a gold standard, it is first essential to standardize among laboratories the reagents, supplies, instrumentation, and conditions used for the assay. This will require considerable effort by participating laboratories, but it is essential to avoid interlaboratory differences in quantifying a particular analyte by LC-MS/MS.

Presently, the overall cost of instrumentation, a highly trained technician, supplies, and reagents for measuring steroid hormones by LC-MS/MS is generally still prohibitive for small clinical laboratories. However, major advances in mass spectrometry instrumentation have been made in a relatively short period of time in recent years. Therefore, it is not unreasonable to expect less expensive MS instruments with lower accompanying costs that can achieve high assay sensitivity and specificity to become available in the near future. This would allow smaller clinical laboratories to obtain highly accurate measurements of steroid hormones with a high throughput of samples. In addition, those laboratories would be able to participate in the standardization of steroid hormones assays.

KEY POINTS

- Well-validated RIAs that include preceding organic solvent extraction and chromatography steps are highly reliable for measuring androgens.
- Direct immunoassays should not be used to quantify testosterone levels in women.
- Well-validated LC-MS/MS assays can give highly accurate results with high sample throughput.
- The LC-MS/MS assay method is touted as the future gold standard for steroid hormone measurements, although this remains to be determined.
- Interlaboratory comparisons of steroid hormone measurements using LC-MS/MS are essential before this assay method becomes the gold standard.
- The equilibrium dialysis assay and calculation method using mass action equations are both reliable for determining free testosterone concentrations.

- Serum or plasma samples for quantifying androgens are overall more convenient for the patient, clinician, and laboratory than urine or saliva samples.

REFERENCES

1. Yalow RS, Berson SA. Assay of plasma insulin in human subjects by immunological methods. Nature 1959;184:1648.
2. Yalow RS, Berson SA. Immunoassay of endogenous plasma insulin in man. J Clin Invest 1960;38:1157–1175.
3. Abraham GE. Sold-phase radioimmunoassay of estradiol-17β. J Clin Endocrinol Metab 1969;29:866
4. Abraham GE, Odell WD. Solid-phase radioimmunoassay of serum estradiol-17β: a semi-automated approach. In: Peron FG, Caldwell BV, eds. Immunologic Methods in Steroid Determination. New York: Appleton-Century-Crofts, 1970:87–112.
5. Chard T. An introduction of radioimmunoassay and related technique. In: Van der Vliet PC, ed. Laboratory Techniques in Biochemistry and Molecular Biology. Amsterdam: Elsevier Science, 1995.
6. Stanczyk FZ, Cho MM, Endres DB, et al. Limitations of direct estradiol and testosterone immunoassay kits. Steroids 2003;68:1173–1178.
7. Taieb J, Mathian B, Millot F, et al. Testosterone measured by 10 immunoassays and by isotope-dilution gas chromatography - mass spectrometry in sera from 116 men, women, and children. Clin Chem 2003;49:1381–1395.
8. Fitzgerald RL, Herold DA. Ciba Corning ACS:180 direct total testosterone assay can be used on female sera. Clin Chem 1997;43:1466–1467.
9. Westphal U. Steroid-Protein Interactions. Berlin: Springer-Verlag, 1986.
10. Westphal U. Steroid-Protein Interactions II. Berlin: Springer-Verlag, 1971.
11. Manni A, Pardrige WM, Cefalus W, et al. Bioavailability of albumin-bound testosterone. J Clin Endocrinol Metab 1985;61:705.
12. Pardridge W, Landaw EM. Tracer kinetics model of blood brain barrier transport of plasma protein-bound ligands: Empiric testing of free hormone hypothesis. J Clin Invest 1984;74:74.
13. Pardridge WM. Transport of protein bound hormones into tissues in vivo. Endocr Rev 1981;2:103.
14. Ekins R. Hirsutism: Free and bound testosterone. Am Clin Biochem 1990;27:91.
15. Vermeulen A, Verdonck L, Kaufman JM. A critical evaluation of simple methods for the estimation of free testosterone in serum. J Clin Endocrinol Metab 1999;84:3666–3672.
16. Winters SJ, Kelley DE, Goodpaster B. The analog free testosterone assay: are the results in man clinically useful? Clin Chem [Erratum (1999) 45:444] 1998;44:2178–2182.
17. Cheng RN, Reed MJ, James VHT. Plasma free testosterone: Equilibrium dialysis vs direct radioimmunoassay. Clin Chem 1986;32:1411.
18. Miller KK, Rosner W, Lee H, et al. Measurement of free testosterone in normal women and women with androgen deficiency: comparison of methods. J Clin Endocrinol Metab 2004;89:525–533.
19. Rosner W. An extraordinary inaccurate assay for free testosterone is still with us [lett]. J Clin Endocrinol Metab 2001;86:2903.
20. Kapoor P, Luttrell B, Williams D. The free androgen index is not valid for adult males. J Steroid Biochem Mol Biol 1993;45:325–326.
21. Morley JE, Patrick P, Perry III HM. Evaluation of assays available to measure free testosterone. Metabolism 2002;51:554–559.
22. Emadi-Konjin P, Bain J, Bromberg IL. Evaluation of an algorithm for calculation of serum "bioavailable" testosterone (BAT). Clin Biochem 2003;36:591–596.
23. Bukowski C, Grigg MA, Longcope C. Sex-hormone-binding globulin concentrations: differences among commercially available methods. Clin Chem 2000;46:1415–1416.

III Uncommon Causes of Androgen Excess

6

Androgen-Secreting Adrenal and Ovarian Neoplasms

Michel Pugeat, Gérald Raverot, Ingrid Plotton, Aude Brac de la Perrière, Pascale Mirakian, Henri Déchaud, Nicole Berger, and Jean Louis Peix

SUMMARY

Androgen-secreting neoplasms (ASNs) are generally associated with rapidly progressive symptoms of hyperandrogenism, which result in various degrees of virilization. A plasma concentration of testosterone of more than 200 ng/dL (8.7 nmol/L) (or two to three times the upper normal range) with a normal dehydroepiandrosterone sulfate (DHEAS) level is highly suggestive of an ovarian ASN. The value of low dexamethasone suppression test is associated with high sensitivity but limited specificity in differential diagnosis of hyperandrogenism. Suppression of testosterone levels by administration of a progestogen or gonadotropin-releasing hormone agonist will not discriminate an ovarian ASN from hyperthecosis, but will strongly orientate the diagnosis to the ovarian origin of androgen excess. Ovarian and adrenal venous catheterization and sampling should be reserved for patients in whom the presence of a small ovarian tumor cannot be excluded on imaging studies and restrictive to expert unit.

The prognosis for ovarian ASNs is generally good, although some Sertoli and granulosa cell tumors can be aggressive and malignant, requiring surgery and chemotherapy. In postmenopausal women hysterectomy and bilateral salpingo-oophorectomy is the preferred treatment because of the high incidence of associated endometrial lesions. In young women in whom fertility is an issue, a unilateral salpingo-oophorectomy must be performed. In contrast, the prognosis of adrenocortical carcinoma is poor, but early surgical treatment can be life-saving. Mitotane (o,p'-DDD) is the recommended treatment for adrenal carcinoma in patients who cannot be cured by surgery.

Key Words: Androgen-secreting neoplasms; adrenocortical carcinoma; Sertoli–Leydig cell tumors; virilization; hirsutism.

1. INTRODUCTION

In women, acne and excess hair growth are generally associated with excessive androgen production *(1)*. Their prevalence is estimated to be 5–15% within the female population, with a likely genetic variability *(2)*. Although hirsutism and acne are not considered diseases *(3)*, the social prejudice of pilosebaceous male pattern is a frequent reason for a medical consultation. In practical medicine, consultation provides the opportunity to identify androgen disorders that require treatment and prevention, such as the polycystic ovary syndrome (PCOS) or 21-hydroxylase-deficient nonclassic adrenal hyperplasia (NCAH). The clinician should not misdiagnose the rare androgen-secreting neoplasm (ASN), which is potentially malignant and requires specific treatment. This chapter will review our current knowledge and propose a paradigm for laboratory investigation to identify adrenal and ovarian ASNs.

From: *Contemporary Endocrinology: Androgen Excess Disorders in Women:*
Polycystic Ovary Syndrome and Other Disorders, Second Edition
Edited by: R. Azziz et al. © Humana Press Inc., Totowa, NJ

2. BACKGROUND

2.1. Pathogenesis

2.1.1. Adrenal ASNs

The prevalence of adrenocortical carcinomas approximates two new cases per million of population per year (*see* ref. *4*). In about 60% of cases *(5–7)* adrenocortical carcinoma produces hormones independently or in combination. The incidence of Cushing's syndrome or symptoms of excessive androgens is estimated to be approximately 30% and approximately 10%, respectively, and the incidence of combined excessive androgen and cortisol secretion is approximately 20%.

Hormone-secreting benign adrenal adenomas are rare; equally rare are hormonally silent or hormone-secreting adrenocortical carcinomas. Diffuse or nodular adrenocortical hyperplasia can be associated with genetic deficiency of cortisol synthesis or a part of the Carney's complex. The differentiation from benign to malignant adrenocortical neoplasms seems to occur via a multistep process that has been described within a tumor from one patient *(8)*. Local recurrence or distant metastasis cannot be predicted solely on the microscopic appearance of the tumor. However, histological classification and tumor–node–metastasis staging *(9)* and Weiss criteria *(10)* are currently used to suspect the malignancy of steroid-secreting adrenal tumor. A few cases of adrenocortical tumor have been described in patients with 21-hydroxylase deficiency *(11,12)*, but causality is unlikely. Indeed, adrenocorticotropic hormone (ACTH)-induced diffuse or nodular hyperplasia is polyclonal, in contrast to the monoclonal identity of most adrenal adenomas and carcinomas *(13,14)*.

2.1.2. Ovarian ASNs

Ovarian cancer is the sixth most common cancer among women. The annual incidence rates differ according to geographic area, with a high rate in Scandinavia (15/100,000) and a low rate in Japan (3/100,000). Ovarian cancer may derive from coelomic epithelium, stromal (e.g., granulosa) cells, and germ cells, the latter almost always arising in children and young adults during the first decades of life. The 2003 World Health Organization (WHO) histological classification of sex cord-stromal, Sertoli-stromal, and steroid cell tumors is depicted in Table 1. This classification encompasses most ASNs that are well differentiated and benign, although some intermediate or poorly differentiated tumors can be malignant and aggressive.

The vast majority of epithelial ovarian cancers are sporadic, whereas 5–10% are estimated to be inherited. The breast–ovarian cancer syndrome is the most common inherited type and has been linked to germline mutations in the *BRCA1* tumor-suppressor gene *(15)*. No heritability has been reported for ovarian ASN. Interestingly, one study reported that a heterozygous inactivating mutation (F591S) located in the sixth transmembrane domain of the gene encoding follicle-stimulating hormone (FSH) receptor was found in 9 of 13 patients with sex cord tumors (69%) and 2 of 3 ovarian small cell carcinomas, but not in a control population *(16)*. In contrast, no evidence for a role for activating mutations or polymorphisms of FSH receptors was subsequently reported in 15 granulosa cell tumors *(17)*. This discrepancy suggested that the F591S mutation, which was observed in young patients, might be a distinct subgroup of granulosa cell tumors *(17)*.

Significant concern has been focused on the risk of epithelial ovarian cancer after treatment of infertility *(18,19)*. Several cases of ovarian cancer were reported in infertile women receiving infertility drugs. In one study, granulosa cell tumors were the most common ovarian tumor in infertile women receiving clomiphene citrate and gonadotropin *(20)*. It has been proposed that gonadotropins might be associated with the development of ovarian tumors. One hypothesis is that after ovulation, the mitotic activity required to repair the ovarian epithelium increases the likelihood of genetic abnormalities, such as mutations of the *p53* gene, which leads to malignant transformation *(21)*. However, women with increasing parity have been found to have a strong protection against epithelial ovarian cancer (*see* ref. *22*). Therefore, this "repetitious ovulatory activity" or "gonadotropins" hypothesis in relation to ovarian carcinogenesis remains to be elucidated.

Table 1
World Health Organization 2003 Histological Classification
Including Most Ovarian Androgen-Secreting Tumors

Sex cord stromal tumors	Sertoli stromal cell tumor
Granulosa stromal cell tumors	Sertoli–Leydig cell tumor group
Granulosa cell tumor group	(androblastomas)
Adult granulosa cell tumor	Well differentiated
Juvenile granulosa cell tumor	Of intermediate differentiation
	Variant with heterologous element
Thecoma-fibroma group	Poorly differentiated (sarcomoid)
Thecoma, not otherwise specified	Variant with heterologous element
Typical	Stromal Leydig cell tumor
Luteinized	Retiform
Fibroma	Variant with heterologous element
Cellular fibroma	Sex cord stromal tumors of mixed or
Fibrosarcoma	unclassified type
Stromal tumor with minor	Sex cord tumor with annular tubules
sex cord elements	Gynandroblastoma (specific components)
Sclerosing stromal tumor	Sex cord stromal tumor, unclassified
Signet-ring stromal tumor	Steroid cell tumors
Unclassified (fibrothecoma)	Stromal luteoma
	Leydig cell tumor group
	Hilus cell tumor
	Leydig cell tumor, nonhilar type
	Leydig cell tumor, not otherwise specified
	Steroid cell tumor, not otherwise specified
	Well differentiated
	Malignant

Adapted from ref. *14a.*

2.2. Diagnosis

2.2.1. History and Symptoms

Virilization is defined as the development of male secondary sex characteristics, including enlargement of the clitoris (transverse diameter >10 mm), deepening of the voice, marked growth of facial and body hair, acne and seborrhea, and male-pattern balding. However, although hirsutism with menstrual irregularity is generally associated with PCOS or less frequently with NCAH *(23)*, in less than 1% isolated hirsutism may be the presenting symptom of an ASN. Therefore, this diagnosis must be ruled out in any patient with apparently simple hirsutism *(3)*.

There are some specific clinical expressions of ASNs, such as age of presentation, progression of symptoms, and size of the tumor. In prepubertal females, because of their relatively large volume, ovarian tumors are often easily palpable and can be identified by computed tomography (CT) scanning and pelvic ultrasonography. Isosexual precocious puberty in girls, defined by early development of the secondary sexual characteristics before the age of 8 years, with breast development and pubic and axillary hair growth followed by menarche, has been described in both ovarian and adrenal ASNs. However, heterosexual precocious puberty with appearance of sexual hair not accompanied by breast development as a result of predominant androgen effects is highly suggestive of an ASN. In adult premenopausal women, rapidly progressive symptoms of virilization are highly suggestive of tumoral hyperandrogenism.

Secondary amenorrhea or extreme oligomenorrhea is reported in most but not all premenopausal patients with an ASN. These symptoms may be masked by the use of oral contraceptives, which may also decrease androgen secretion by the ovarian tumor. After the discontinuation of oral contraceptives, secondary amenorrhea accompanied by a rapid virilization is therefore one possible clinical finding.

In postmenopausal women, a long history of clinical hyperandrogenism cannot exclude the presence of an ovarian ASN. In most cases facial hirsutism with frequent shaving, male-pattern baldness, increased muscular development, and some deepening of the voice can be associated with a slowly growing Leydig (or hilus) cell tumor. Clitoromegaly, with an enlarged phallus diameter (>10 mm), is usual but sometimes difficult to identify. In contrast, a history of non-insulin-dependent diabetes and of infertility, with or without irregular menstrual cycles, late menopause, abdominal obesity, and acanthosis nigricans, is more suggestive of hyperthecosis and either PCOS or the Hyperandrogenic-insulin resistant-acanthosis nigricans (HAIR-AN) syndrome.

Adrenal androgen-secreting adenomas have been reported in both pre and postmenopausal women *(24)*. Some are human chorionic gonadotropin (hCG) or gonadotropin dependent. In contrast, androgen-secreting adrenocarcinomas are gonadotropin independent. They can secrete androgens exclusively *(25)* or in association to cortisol.

Symptoms of hypercortisolism (Cushing's syndrome) indicate an adrenal ASN in a patient with virilization, but can be observed in rare steroid cell tumors of adrenocortical type *(26)*. Combined symptoms of Cushing's syndrome and virilization are highly indicative for the diagnosis of secreting adrenal carcinoma. Cushing's syndrome is usually recognizable, with proximal muscle wasting, striae, and thin skin, although the presence of central obesity, glucose intolerance, and hypertension causes some confusion with PCOS.

2.2.2. Laboratory Investigation

2.2.2.1. Basal Measurement of Androgens

Screening for ASNs is generally achieved by the measurement of basal testosterone and dehydroepiandrosterone sulfate (DHEAS) plasma concentrations. In most ovarian ASNs, testosterone level is more than 200 ng/dL and DHEAS is within the normal range for age. Unfortunately, these cutoff values cannot guarantee the diagnosis in all cases *(27,28)*. Indeed, 20% of ovarian ASNs have total testosterone levels of below150 ng/dL *(15)*, although conversely testosterone levels greater than 150 ng/dL are extremely uncommon in patients with PCOS, the most common cause of hyperandrogenism in premenopausal women *(23)*. Moreover, while a normal DHEAS level is unlikely to be associated with an adrenal ASN *(29)*, a high DHEAS level can be observed in premenopausal patients with functional hyperandrogenism (30) and in a few ovarian steroid cell (lipid cell) or Sertoli–Leydig cell tumors *(31,32)*.

The basal androgen levels in one premenopausal and five postmenopausal patients presenting with virilization are depicted in Table 2 *(31)*. This limited experience and review of the literature suggests the following:

1. Small ovarian tumors can demonstrate episodic secretion of androgens; therefore, repeated androgen measurements for diagnosis may be required, notably in postmenopausal women presenting symptoms of virilization.
2. Postmenopausal women have the highest incidence of ovarian ASNs. In normal postmenopausal women, the plasma concentrations of androgens are considerably lower than in normal premenopausal women, and therefore androgen concentrations must be interpreted accordingly *(31)*.
3. Circulating levels of precursors of testosterone, Δ^4-androstenedione, and 17α-hydroxyprogesterone (17-OHP), as well as estradiol should be measured, because some ovarian secreting tumors may predominantly secrete testosterone precursors or estradiol, being either androgenic, estrogenic, or even both.
4. The 17-OHP concentrations were increased in all patients in our study, with the exception of one woman, who presented with a Leydig (hilar) cell tumor (patient 2, Table 2). An increased 17-OHP plasma concentration may also be suggestive of 21-hydroxylase deficiency, and therefore an ACTH test should be performed. On rare occasions, 21-hydroxylase deficiency has been reported in patients with Leydig cell or lipid ovarian tumor *(11,12,33)*. We found one patient with virilization and hyperthecosis (patient 5, Table 2) who had a 17-OHP concentration as high as 6289 ng/dL 60 minutes after intravenous ACTH. The diagnosis of homozygous 21-hydroxylase-deficient NCAH was confirmed subsequently by genetic molecular analysis.

Table 2
Basal Plasma Concentrations of Testosterone (T), Δ4-Androstenedione (A4),
17α-Hydroxyprogesterone (17-OHP), and Dehydroepiandrosterone Sulfate
(DHEAS) in Six Women Presenting Symptoms of Virilization

Case	Age (yr)	T (ng/dL)	A4 (ng/dL)	17-OHP (ng/dL)	DHEAS (μg/dL)	Final diagnosis
1	77	706[a]	340[a]	350[a]	88	Granulosa cell tumor
2	66	310[a]	420[a]	53	202	Leydig (hilar) cell tumor
3	68	300[a]	215[a]	157[a]	202	Bilateral hyperthecosis
4	62	234[a]	233[a]	350[a]	90	Bilateral hyperthecosis
5	62	235[a]	132[a]	534[a]	39	Bilateral hyperthecosis + 21-hydroxylase deficiency (nonclassic)
Range for normal, Postmenopausal		16–32	31–75	12–103	35–210	
6	24	655[a]	512[a]	663[a]	273	Sertoli–Leydig cell tumor
Range for normal, premenopausal		12–43	45–230	16–85	60-295	

[a]Higher than normal range for age.

2.2.2.2. GONADOTROPIN LEVELS

Gonadotropin measurements are not required for the screening of patients with virilization. They are usually within the normal range, whether pre- or postmenopausal *(31)*. To date, no evidence of gonadotropin suppression by the excess androgens of functioning ASNs has been reported.

2.2.2.3. SUPPRESSION TESTS OF ANDROGEN SECRETION

Several suppression tests have been proposed for the evaluation of patients with suspected ASNs:

1. Dexamethasone-suppression test: This is the classic method for diagnosing adrenal ASNs. It has been claimed that an adrenal ASN is unlikely if the DHEAS level is within the normal range after dexamethasone administration *(29)*. The value of a 48-hour low-dose (2 mg) dexamethasone-suppression test has been evaluated in the differential diagnosis of hyperandrogenism in 211 hyperandrogenic women *(34)*. Testosterone suppression (>40% reduction or normalization) was associated with 100% sensitivity and 88% specificity in distinguishing patients with ovarian and adrenal ASNs from patients with nontumorous hyperandrogenism.

2. Gonadotropin-suppression test: This test can be achieved by the administration of a progestogen or long-acting gonadotropin-releasing hormone agonist (GnRHa). Progestogens can suppress androgens in patients with virilizing ovarian tumors, an effect probably mediated through gonadotropin suppression. GnRHa administration has also been reported to decrease plasma gonadotropin and androgen levels in patients with hyperthecosis or Leydig cell ovarian tumors. However, androgen suppression by a GnRHa will not discriminate hyperthecosis from ovarian secreting tumors, as shown in Table 3, although it will serve to indicate that the ovaries are the source of the excessive androgen secretion in a virilized patient *(31)*. These findings would need further investigation. Indeed, ovarian ASNs which are poorly differentiated or have luteinizing hormone receptor mutations may have androgen secretion independent of gonadotropins.

2.2.2.4. NONSTEROID MARKERS OF OVARIAN SECRETING TUMORS

hCG-producing ovarian tumors are of the germ cell type and do not secrete androgens, although some may be associated with minor symptoms of hyperandrogenism and slightly increased androgen levels.

Table 3
Total Testosterone (T), Androstenedione (A4,) and 17α-Hydroxyprogesterone (17-OHP)
Levels in Five Women With Ovarian Androgen-Secreting Tumors, at Baseline,
After Administration of a Long-Acting Gonadotropin-Releasing Hormone
Agonist (GnRHa),[a] and Following Surgery

	T (ng/dL)			A4 (ng/dL)			17-OHP (ng/dL)		
	Basal	GnRHa	Postsurg	Basal	GnRHa	Postsurg	Basal	GnRHa	Postsurg
1	706	3	3	340	16	5	350	18	10
2	310	5	7	420	23	31	53	19	77
3	300	24	19	215	94	77	157	ND	ND
4	234	26	36	233	98	189	103	92	49
5	655	13	15	512	86	50	430	112	16

[a]D-Trp-6-GnRH, 3.75 mg im, monthly.
ND, not determined.

α-Fetoprotein (AFP) is generally expressed by germ cell ovarian tumors containing yolk sac elements. It has been reported that some Sertoli–Leydig cell tumors also express AFP, with immunostaining within Leydig and/or Sertoli cells or within heterologous elements (hepatoid cells) *(35)*. Of well-documented AFP-secreting Sertoli-Leydig cell tumors, 20–30% have a malignant prognosis. Thus, the measurement of AFP may be useful in detecting recurrence and/or metastasis in these cases.

Inhibin serum concentration was found to be high in three postmenopausal women with granulosa cell tumors. Furthermore, among 209 patients, inhibin was high in 92% of the mucinous, 17% of clear-cell, 15% of undifferentiated, and 19% of all the other ovarian carcinomas *(36)*. The measurement of serum inhibin levels is a useful diagnostic aid when monitoring for recurrence following surgery. The assays are applicable for postmenopausal women, when inhibin is physiologically low, and when studies are underway to assess its use in premenopausal women *(37)*.

The anti-Müllerian hormone (AMH) serum concentration, also known as müllerian-inhibiting substance was found to be very high in one patient with an ovarian sex cord tumor with annular tubules. The serum concentration of AMH correlated with the recurrence and/or the metastasis of the tumor throughout the patient's evolution *(38)*. AMH has been shown to be a specific marker of Sertoli and granulosa cell origin in ovarian tumors *(39)*. A highly sensitive AMH assay has been proposed for the monitoring of patients with granulosa cell tumor *(40)*.

2.2.3. Imaging Techniques

Adrenocortical tumors can be visualized by various radiological techniques, although ultrasound can only delineate an adrenal mass larger than 2 cm in diameter. Because of the adipose tissue that surrounds them, the adrenal glands are easily visible by CT scan, which can detect adrenal nodules smaller than 5 mm *(4)*. Magnetic resonance imaging (MRI) can provide additional information regarding the invasion of an adrenocortical carcinoma into blood vessels. However, it remains to be further documented whether MRI can distinguish between malignant tumors and nonfunctioning adenomas by comparing the ratio of the signal intensity of each type of adrenal mass to that of liver.

Androgen-secreting ovarian neoplasms can be visualized by transabdominal pelvic ultrasonography when tumor size is larger than 6–8 cm in diameter, although differentiation from normal ovarian structure is inaccurate. Transvaginal ultrasonography is able to detect neoplasms 1–3 cm in size and allows reproducible measurement of the ovaries, and thus it is more useful than pelvic ultrasonography in locating ovarian tumors. However, because stromal hyperthecosis mimics an ASN sonographically, caution is advisable. CT scanning has limited value in the diagnosis of ovarian

ASNs. In contrast, iodomethyl-norcholesterol scanning has been found to be useful in identifying ovarian tumors larger than 2 cm in size *(26)*.

2.2.4. Selective Venous Catheterization

Selective venous catheterization is an invasive technique that is not easy to perform and that carries a risk of intracorporeal hemorrhage and thrombosis. Its accuracy is operator-dependent and varies greatly according to the anatomical variability of the venous system *(41)*. However, in skillful hands, ovarian and adrenal vein catheterization is helpful in the preoperative assessment of virilized women to locate the source of excessive androgen production in patients suspected of having an ASN *(27,32,41)*.

2.2.5. Exploratory Laparotomy

Exploratory laparotomy is an invasive method of diagnosing ovarian ASNs, during which tumors can be located by elementary palpation. However, because ASNs can be associated with the development of typical polycystic-like ovaries *(42)*, direct ultrasonography of both ovaries during laparotomy will help identify the tumor *(31)*. Alternatively intraoperative measurement of testosterone in serum samples taken from each ovarian vein has also been useful in identifying small Leydig cell tumors *(43)*.

An ovarian ASN should be treated by ovariectomy. In addition, hysterectomy in postmenopausal women may be indicated when an endometrial lesion (e.g., endometrial hyperplasia/carcinoma) is suspected *(44)*. In our experience, three of four postmenopausal patients with an ASN or hyperthecosis had associated endometrial pathology, with subclinical adrenocarcinoma in one case *(31)*.

3. CONCLUSIONS/SYNOPSIS

Androgen-secreting neoplasms are generally associated with distinct clinical features and presentations, and are associated with rapidly progressive symptoms of hyperandrogenism that generally result in various degrees of virilization. The patient's history and clinical presentation are strong predictors for ASNs. A plasma concentration of testosterone greater than 200 ng/dL (8.7 nmol/L) (or two to three times the upper normal range) with a normal DHEAS level is highly suggestive of an ovarian ASN. A combined increased testosterone of greater than 200 ng/dL (8.7 nmol/L) with an elevated DHEAS level of more than 600 μg/dL (16.3 μmol/L) is highly suggestive of an adrenal ASN. Suppression and stimulation testing has a high degree of sensitivity, albeit low specificity, for the diagnosis of ASNs and is generally of limited value in the diagnosis of these neoplasms. Ovarian and adrenal venous catheterization and sampling should be reserved for patients in whom the presence of a small ovarian tumor cannot be excluded on imaging studies and restrictive to expert unit. The prognosis of ovarian ASNs is generally good, although some Sertoli and granulosa cell tumors can be aggressive and malignant, requiring surgery and chemotherapy. Alternatively, the prognosis of adrenocortical ASNs, namely carcinomas, is poor, although early surgical treatment can be lifesaving.

4. FUTURE AVENUES OF INVESTIGATION

Our understanding of the pathogenesis of ASNs has undergone major advances during the past decade. The identification of molecular defects in the hereditary syndrome responsible for adrenocortical tumors has guided the search for a candidate gene mutation in sporadic tumors *(46,47)*. Much remains to be done to identify relevant molecular alteration that will open new avenues for treatment.

Human ovarian ASNs are a potential model for studying theca cell steroidogenesis. From an autonomous Sertoli-Leydig cell tumor, a human ovarian theca-like cell culture model has been developed, found to be appropriate for the study of the molecular mechanisms regulating steroidogenesis *(48)*. A pluripotential model for human adrenocortical studies has been developed, the NCI-H295R cell line *(49)*, which is widely used for understanding the mechanism(s) of multidrug resistance of steroid-secreting adrenocortical carcinoma and adrenocortical physiology in general *(50)*.

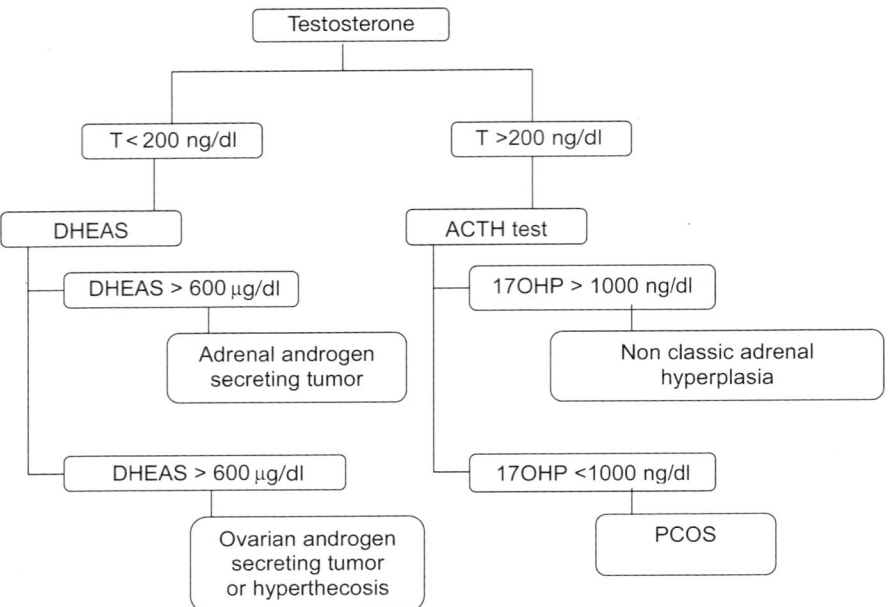

Fig. 1. Scheme of the laboratory investigation for identifying adrenal or ovarian androgen-secreting neoplasms in hirsute patients. T, testosterone; 17-OHP, 17α-hydroxyprogesterone; DHEAS, dehydroepiandrosterone sulfate; ACTH, adrenocorticotropic hormone.

KEY POINTS

- Regardless of age of presentation, ASNs are generally associated with rapidly progressive symptoms of hyperandrogenism, which result in various degrees of virilization; secondary amenorrhea is common in premenopausal women.
- A plasma testosterone concentration of more than 200 ng/dL (or two to three times the upper normal range) with a normal DHEAS level is highly suggestive of an ovarian ASN. Combined increased testosterone greater than 200 ng/dL (8.7 nmol/L) with DHEAS greater than 600 μg/dL (16.3 μmol/L) is highly suggestive of an adrenal ASN (*see* Fig. 1).
- The value of the low dexamethasone suppression test in differential diagnosis of hyperandrogenism is associated with high sensitivity but limited specificity. Suppression of testosterone levels by administration of a progestagen or GnRHa will not distinguish an ovarian ASN from hyperthecosis, but will strongly orientate the diagnosis to the ovarian origin of tumoral testosterone level. Ovarian and adrenal venous catheterization and sampling should be reserved for patients in whom the presence of a small ovarian tumor cannot be excluded on imaging studies and restrictive to expert unit.
- The prognosis of adrenocortical carcinoma is poor, although surgical treatment can be life-saving. Mitotane (*o,p'*-DDD) is the only adrenal-specific agent available for treatment of adrenal carcinoma in patients who cannot be cured by surgery *(45)*.
- In postmenopausal women suspected of having an ovarian ASN, an abdominal hysterectomy and bilateral salpingo-oophorectomy is the preferred treatment because of the high incidence of associated endometrial lesions. In young women in whom fertility is an issue, only a unilateral salpingo-oophorectomy need be performed.
- Ovarian ASNs generally have a good prognosis, although some Sertoli and granulosa cell tumors can be aggressive, with malignant tumors requiring appropriate surgery and chemotherapy.

ACKNOWLEDGMENTS

We are particularly indebted to Iain McGill for his help in revising the manuscript and to Myriam Richard for expert typing. This research was funded by Université Claude Bernard—UER Lyon-Nord, and Ipsen—France Laboratory.

REFERENCES

1. Ferriman D, Gallwey JD. Clinical assessment of body hair growth in women. J Clin Endocrinol Metab 1961;21:1440–1447.
2. Lorenzo EM. Familial study of hirsutism. J Clin Endocrinol Metab 1970;31:57–73.
3. Rittmaster RS Hirsutism Lancet 1997;349:191–195.
4. Latronico AC, Chrousos GP. Extensive personal experience: Adrenocortical tumors. J Clin Endocrinol Metab 1997;82:1317–1324.
5. Bertagna C, Orth DN. Clinical and laboratory findings and results of therapy in 58 patients with adrenocortical tumors admitted to a single medical center (1951 to 1978). Am J Med 1981;71:855–875.
6. Luton J-P, Cerdas S, Billaud L, et al. Clinical features of adrenocortical carcinoma, prognosis factor, and the effect of mitotane therapy. N Engl J Med 1990;322:1195–1201.
7. Nader S, Hickey RC, Sellin RV, Samaan NA. Adreno cortical carcinoma: A study of 77 cases. Cancer 1983;52:707–711.
8. Bernard MH, Sidhu S, Berger N, et al. A case report in favor of a multistep adrenocortical tumorigenesis. J Clin Endocrinol Metab 2003;88:998–1001.
9. DeLellis RA, Lloyd RV, Heitz PU, Eng C, eds. World Health Organization. Classification of Tumours. Pathology and Genetics of Endocrine Organs. IARC Press. Lyon 2004.
10. Weiss L. Comparative histologic study of 43 metastasizing and non metastasizing adrenocortical tumors. Am J Surg Pathol 1984;8:163–169.
11. Seters AP van, Aalderen W, Moolenaar AJ, Gorsiro MCB, Roon F, Backer ET. Adrenocortical tumour in untreated congenital adrenocortical hyperplasia associated with adequate ACTH suppressibility. Clin Endocrinol 1981;14:325–334.
12. Pang S, Becker D, Cotelingam J, Foley TP, Drash AL. Adrenal tumor in a patient with congenital adrenal hyperplasia due to 21-hydroxylase deficiency. Pediatrics 1981;68:242–246.
13. Beuschlein F, Reincke M, Karle M, et al. Clonal composition of human adrenocortical neoplasms. Cancer Res 1994;54:4927–4932.
14. Gicquel C, Leblond-Francillard M, Bertagna X, et al. Clonal analysis of human adrenocarcinomas and secreting adenomas. Clin Endocrinol (Oxf) 1994;40:465–477.
14a. Tavassoli FA, Devilee P. World Health Organization. Classification of Tumours. Pathology and Genetics of the Breast and Female Genital Organs. Lyon: IARC Press, 2003.
15. Miki Y, Swensen J, Shattuck-Eidens D, et al. A strong candidate for the breast ovarian cancer susceptibility gene BRAC1. Science 1994;266:66–71.
16. Kotlar TJ, Young RH, Albanese C, Crowley WF, Scully RE, Jameson JL. A mutation in the follicle-stimulating hormone receptor occurs frequently in human ovarian sex cord tumors. J Clin Endocrinol Metab 1997;82:1020–1026.
17. Fuller PJ, Verity K, Shen Y, Mamers P, Jobling T, Burger HG. No evidence of a role for mutations or polymorphisms in the follicle-stimulating hormone receptot in ovarian granulosa cell tumors. J Clin Endocrinol Metab 1998;83:274–279.
18. Rossing MA, Daling JR, Weiss NS, Moore D, Self S. Ovarian tumors in a cohort of infertile women. N Engl J Med 1994;331:771–776.
19. Kristensen GB, Trpoé C. Epithelial ovarian carcinoma. Lancet 1997;349:113–117.
20. Willimsen W, Kruitwagen R, Bastaliaans B, et al. Ovaran stimulation and granulosa-cell tumour. Lancet 1993;341:986–988.
21. Godwin AK, Testa JR, Hamilton TC. The biology of ovarian cancer development. Cancer 1993;71:530–536.
22. Riman T, Persson I, Nilsson S. Hormonal aspects of epithelial ovarian cancer: review of epidemiological evidence. Clin Endocrinol 1998;49:695–707.
23. Azziz R, Sanchez LA, Knochenhauer ES, et al. Andogen exces in women: experiencewith over 1000 consecutive patients. J Clin Endocrinol 2004; 89:453–462.
24. Danilowicz K, Albiger N, Vanegas M, Gomez RM, Cross G, Bruno OD. Androgen-secreting adrenal adenomas. Obstet Gynecol 2002;100:1099–1102.
25. Moreno S, Montoya G, Armstrong J, et al. Profile and outcome of pure androgen-secreting adrenal tumors in women: experience of 21 cases. Surgery 2004 136;1192–1198.
26. Young RH, Scully RE. Ovarian steroid cell tumors associated with Cushing's syndrome: A report of three cases. Int J Gynaecol Pathol 1987;6:40–48.
27. Meldrum DR, Abraham GE. Peripheral and ovarian venous concentrations of various steroid hormones in virilizing ovarian tumors. Obstet Gynecol 1979;53:36–43.
28. Waggoner W, Boots LR, Azziz R. Total testosterone and DHEAS levels as predictors of androgen-secreting neoplasms: a population study. Gynecol Endocrinol 1999;13:394–400.
29. Derksen J, Nagesser SK, Meinders AE, Haak HR, van de Velde CJH. Identification of virilizing adrenal tumors in hirsute women. N Engl J Med 1994;331:968–973.
30. Pugeat M, Nicolas MH, Craves J, et al. Androgens in polycystic ovary syndrome. Ann NY Acad Sci 1993;687:124–135.
31. Pascale Miras-Mirakian, Pugeat M, Robert M, et al. Androgen suppressive effect of GnRH agonist in ovarian hyperthecosis and virilizing tumours. Clin Endocrinol 1994;41:571–576.

32. Surrey ES, de Ziegler D, Gambone JC, Judd HL. Preoperative localization of androgen-secreting tumors: clinical, endocrinologic, and radiologic evaluation of ten patients. Am J Obstet Gynecol 1998;158:1313–1322.
33. Rosenfield RM, Cohen RM, Talemran A. Lipid cell tumor of the ovary in reference to adult onset congenital adrenal hyperplasic and polycystic ovary syndrome. J Reprod Med 1987;32:363–366.
34. Kaltsa GA, Isidori AM, Kola BP, et al. The value of the low-dose dexamethasone suppression test in the differential diagnosis of hyperandrogenism in women. J Clin Endocrinol Metab 2003;88:2634–2643.
35. Hammad A, Jasnosz KM, Olson PR. Expression of alpha-fetoprotein by ovarian Sertoli-Leydig cell tumors. Case report and review of the literature. Arch Pathol Lab Med 1995;119:1075–1079.
36. Healy DL, Burger HG, Mamers P, et al. Elevated serum inhibin concentrations in postmenopausal women with ovarian tumors. N Engl J Med 1993;329:1539–1542.
37. Robertson DM, Burger HG, Fuller PJ. Inhibin/activin and ovarian cancer. Endocrine-Related Cancer 2004;11:35–49.
38. Gustafson ML, Lee MM, Scully RE, et al. Müllerian inhibiting substance as a marker for ovarian sex-cord tumor. N Engl J Med 1992;326:466–471.
39. Rey R, Sabourin JC, Venara M, et al. Anti-Mullerian hormone is a specific marker of sertoli-and granulosa-cell origin in gonadal tumors. Hum Pathol 2000;10:1202–1203.
40. Long WQ, Ranchin V, Pautier P, et al. Detection of minimal levels of serum anti-Müllerian hormone during follow-up of patients with ovarian granulosa tumor by means of a highly sensitive enzyme-linked immunosorbent assay. J Clin Endocrinol Metab 2000;85:540–544.
41. Kaltas GA, Mukherje JJ, Kola B, et al. Is ovarian and adrenal venous catheterization and sampling in the investigation of hyperandrogenic women? Clin Endocrinol (Oxf) 2003;59:34–43.
42. Kaltas GA, Isidori A, Besser GM, Grossman AB. Secondary forms of polycystic ovary syndrome. Trends Endocrinol Metab 2004;15:204–210.
43. Regnier C, Bennet A, Malet D, et al. Intraoperative testosterone assay for virilizing ovarian tumor topographic assessment : report of a Leydig cell tumor of the ovary in a premenopausal woman with an adrenal incidentaloma. J Clin Endocrinol Metab 2002;87:3074–3077.
44. Roddick JW Jr, Green RR. Relation of ovarian stromal hyperplasia to endometrial carcinoma. Am J Obstet Gynecol 1957;73:843–852.
45. Haner S, Fassnacht M. Mitotane for adrenal carcinoma treatment. Curr Opin Invest Drugs 2005;6:386–394.
46. Gicquel C, Bertagna X, Le Bouc YL. Recent advances in the pathogenesis of adrenocortical tumors. Eur J Endocrinol 1995;133:133–144.
47. Libé R, Bertherat J. Molecular genetics of adrenocortical tumors, from familial to sporadic diseases. Eur J Endocrinol 2005;153:1–12.
48. Rainey WE, Sawetawan C, McCarthy JL, et al. Human ovarian tumor cells: a potential model for thecal cell steroidogenesis. J Clin Endocrinol Metab 1996;81:257–263.
49. Rayney W, Bird I, Mason J. NCI-H295R cell line: a pluripotent model for human adrenocortical studies. Mol Cell Endocrinol 1994;100:45–50.
50. Beltran S, de Ravel R, Emptoz-Bonneton A, et al. Evaluation of the effects of a C-21 derivative of progesterone to enhance cytotoxicity of chemotherapeutic agents in a human adrenocortical carcinoma NCI H295R cell line. (Poster P1-577) 86th Annual Meeting of the Endocrine Society. New Orleans, June 16–19, 2004.

7

Cushing's Syndrome, Acromegaly, and Androgen Excess

Anne-Céline Reyss and Didier Dewailly

SUMMARY

In textbooks, hirsutism is often mentioned as a classical cutaneous manifestation of Cushing's syndrome and acromegaly. However, in our experience, hirsutism is seldom the presenting symptom that leads to the diagnosis of these two diseases. In fact, the prevalence of Cushing's syndrome and acromegaly in large series of patient presenting with hirsutism is very low, ranging from 0.2 to 3% and 0 to 0.003%, respectively. However, these life-threatening diseases should not go undiagnosed. Therefore, they need to be excluded in any hyperandrogenic woman by clinical evaluation and, when necessary, by appropriate work-up.

Key Words: Hirsutism; androgen; Cushing's syndrome; acromegaly.

1. INTRODUCTION

In many textbooks of endocrinology, hirsutism is often mentioned as a classical cutaneous manifestation of Cushing's syndrome (CS) and acromegaly. However, in our experience, hirsutism is seldom the symptom leading to the diagnosis of these two diseases.

2. BACKGROUND

2.1. Cushing's Syndrome and Androgen Excess

Cushing's syndrome is a generic term that refers to the clinical state resulting from prolonged and inappropriate exposure to cortisol excess. Symptoms associated with hypercortisolism include central obesity, facial plethora, hypertension, thin skin, bruising, purplish skin striae, lethargy, and proximal muscle wasting. In women, hirsutism, acne, and menstrual irregularities can be associated with these symptoms.

2.1.1. Etiology and Epidemiology

To begin with, it must be remembered that the most common cause of the Cushing phenotype is the use of exogenous glucocorticoids. This being excluded, the etiology of endogenous CS can be divided into two groups: (1) adrenocorticotropic hormone (ACTH)-dependent, resulting from an increased pituitary or ectopic secretion of ACTH, and (2) ACTH-independent, resulting from an autonomous cortisol hypersecretion by the adrenals. ACTH-dependent CS accounts for 85% of noniatrogenic cases; in 80% of these cases, the cause is autonomous pituitary ACTH secretion by a corticotroph adenoma, and the disorder is referred to as Cushing's disease (1).

The incidence of endogenous CS is about two to four new cases per million persons annually, with a female-to-male ratio of 9:1 (2). The estimated incidence of Cushing's disease is 2.4 per million persons per year (3). In different large series of patients presenting with hirsutism, the prevalence of Cushing's syndrome and Cushing's disease ranged from 0.2 to 3% and 0 to 1%, respectively (Table 1) (4–8).

From: *Contemporary Endocrinology: Androgen Excess Disorders in Women:*
Polycystic Ovary Syndrome and Other Disorders, Second Edition
Edited by: R. Azziz et al. © Humana Press Inc., Totowa, NJ

Table 1
Prevalence of Cushing's Syndrome in Five Studies of Hyperandrogenic Women

No. of patients	Cushing's syndrome (%)	Ref.
1000	0.2	4
350	0.3	5
100	1	6
120	0.4	7
150	2.1	8

Table 2
Frequency of Hirsutism and Menstrual Disorders in Five Series of Adults and Two of Children with Cushing's Syndrome

No. of patients	Hirsutism (%)	Menstrual disorders (%)	Ref.
33	73	86	9
100	74	35	10
50	84	72	11
31	64	69	12
70	81	84	13
59	78	78	14
12	58	20	15

2.1.2. Frequency of Hirsutism in Patients With Cushing's Syndrome

Hirsutism is commonly present in CS, ranging from 58 to 84% of patients. The frequency of menstrual irregularity is similar (Table 2) (9–15). Therefore, when the two symptoms exist, the patient's presentation can mimic polycystic ovary syndrome (PCOS) with obesity. In fact, CS seldom presents with hirsutism alone, but sometimes hirsutism may be the complaint that takes the patient to the doctor.

2.1.3. Differentiating Cushing's Syndrome From Other Hyperandrogenic States

In most cases, hirsutism associated with CS is mild. It must be differentiated from the excessive hair growth induced by glucocorticoid excess, which differs from hirsutism by involving different parts of the body, such as the forehead, forearms, or lower legs. This is termed *hypertrichosis*, and such hair is typically relatively fine (vellus) (16).

Hyperandrogenism is present in most ACTH-dependent CS (Cushing's disease or ectopic ACTH secretion) because ACTH stimulates adrenal androgen (17). In some cases these patients may erroneously be considered as having a "standard" PCOS with metabolic syndrome. Indeed, menstrual irregularities (e.g., oligomenorrhea) and oligo-ovulation or anovulation are often associated with hyperandrogenism in CS, presumably because of coexisting PCOS, as suggested by Kaltsas et al., who found that 6 of their 13 patients with CS had an ovarian morphology suggestive of polycystic ovaries on pelvic ultrasound (18). As in congenital adrenal hyperplasia (see Chapter 8), this is probably a morphological reaction to high-circulating adrenal androgen levels.

Even though they are very rare, adrenal carcinomas should be excluded in any woman who develops hirsutism or virilization (male-pattern baldness, clitoromegaly and deepening of voice) in a short time period, particularly if they are accompanied by features of CS (see Chapter 6). In this situation, amenorrhea is frequent, secondary to functional hypogonadism hypogonadotropic because of the very high androgen levels (19).

Finally, prepubertal girls with CS may present with heterosexual precocious puberty (20).

2.1.4. Diagnosis

When CS is suspected on clinical grounds, 24-hour urine-free cortisol (UFC) measure or an overnight low-dose (1 mg) dexamethasone suppression test should be performed. Collection of urine for estimation of cortisol is a noninvasive procedure and is widely used as a screening test for the diagnosis of Cushing's disease *(2)*. The 24-hour UFC measurement provides a good estimate of the integrated 24-hour cortisol secretion. The upper normal range in most assays is between 220 and 330 nmol/24 hours (80–120 µg/24 hours). The 24-hour UFC measurement was shown to have a diagnostic sensitivity of about 100%, and assuming complete collection has been performed, there are virtually no false-negative results. On the other hand, false-positive results are more common because a raised 24-hour UFC level has been documented in some circumstances such as depression, chronic active alcoholism, and glucocorticoid resistance *(2)*.

The overnight 1-mg dexamethasone suppression test is also a simple screening procedure for hypercorticism. It consists of oral administration of 1 mg of dexamethasone at 2400 hours and measurement of plasma cortisol at 0800 hours in the next morning sample. The 0800-hour value must be less than 5 µg/100 mL to exclude a CS *(2)*. The false-negative rate is less than 3%, but the incidence of false-positives is high—about 20–30%, similar to that of the 24-hour UFC.

2.2. Acromegaly and Androgen Excess

Acromegaly is a chronic multisystemic disease with a prevalence of about 30–40 cases per million persons *(21)*. It is most often associated with anterior pituitary tumor. Hirsutism and menstrual irregularity are commonly associated with symptoms of acromegaly in women.

2.2.1. Epidemiology

In a prospective study including 350 patients presenting with hirsutism or androgenic alopecia, only one case of acromegaly was found *(5)*. In other studies describing patients with clinical hirsutism, no case of acromegaly was reported *(4,6–8)*.

The frequency of hisutism in acromegaly varies from 0% *(22)* to 50% *(23)*. In a recent study *(24)*, hirsutism was noted in 26 of 47 acromegalic women (55.5%). Other symptoms suggestive of androgen excess were less frequent, such as acne (17%) and male pattern alopecia (6.4%). Of 47 women with acromegaly, only 19% presented with normal menstrual patterns.

Amenorrhea or oligomenorrhea has been reported as the principal symptom at presentation of some patients with acromegaly *(25,26)*. However, it should not be viewed as a sign of hyperandrogenism. In the majority of cases, it is owing to a complete or partial gonadotropin deficiency with hypoestrogenism, accompanying large tumors with or without hyperprolactinemia.

However, other patients may present with a PCOS-like picture, including hyperandrogenism, menstrual irregularity, increased free androgen levels, luteinizing hormone (LH) hyperesponsiveness to gonadotropin-releasing hormone (GnRH) stimulation, and insulin resistance *(24)*. Whether these patients truly have PCOS is unclear.

2.2.2. Association of Acromegaly and PCOS: Pathophysiological Hypothesis

Kaltsas et al. observed that many acromegalic women had a low sex hormone-binding globulin (SHBG) level, in inverse relationship to growth hormone (GH) levels *(24)*. This may be explained by a negative effect on SHBG synthesis by the hepatocytes, either directly by GH or indirectly by insulin-like growth factor (IGF)-1, or through the hyperinsulinism induced by the GH excess *(24)*. Indeed, acromegaly is well-known to be associated with insulin resistance, glucose intolerance (29– 45% cases), and diabetes (10–20%) because of GH oversecretion *(27)*. As in other such situations (*see* Chapter 24), hyperinsulinism may stimulate ovarian testosterone secretion through insulin and/or IGF-1 receptors in the presence of LH *(28)*, thus inducing ovarian functional hyperandrogenism.

GH on its own might also be involved in the pathophysiology of PCOS, as suggested by many studies. GH receptors have been characterized in human ovaries, especially on granulosa cells, and are physiologically operant *(29,30)*. In 10 women with PCOS, the somatostatin analog octreotide

(100 µg twice daily) given subcutaneously for 7 days induced a significant decrease of serum test-osterone, androgen, and estradiol levels *(1)*. However, mechanisms other than the inhibition in GH secretion by octreotide might be argued. The authors have observed a significant decrease in the basal level, peak amplitude, and LH response to GnRH. Furthermore, by lowering insulin secretion, octreotide may have an additional suppressive effect on androgen secretion.

Lastly, the presence of IGF-1 receptors has been documented in human ovaries, especially in theca-interstitial cells. In cultured rat theca cells, IGF-1 potentiates the LH-induced androgen synthesis in a dose-dependent manner *(32)*. Such an effect can be hypothesized to be present in acromegaly, where IGF-1 serum levels are frankly elevated.

2.2.3. *Diagnosis*

The biochemical evaluation of patients with suspected acromegaly includes the measurement of both basal serum GH and IGF-1 levels as well as the measurement of GH during an oral glucose test tolerance test (OGTT). A single GH sample has low specificity and sensitivity. For the evaluation of GH secretion, multiple samples during both day and nighttime must be measured. There is some consensus that a mean integrated 24-hour GH level below 2.5 µg/L excludes active acromegaly *(33)*. For OGTT testing, after an overnight fast, blood samples are taken at baseline and at 30, 60, 90, and 120 minutes after 75 g of glucose orally administered. The nadir GH level should be less than1 µg/L *(34)*.

The IGF-1 level is invariably high in acromegaly, although the cut-off value depends on the age and gender of the patient *(35)*.

3. CONCLUSIONS AND FUTURE AVENUES OF INVESTIGATION

PCOS associated with CS or acromegaly is rare, but represents a unique occasion to study the relationship between PCOS and adrenal androgens or the GH–IGF system, respectively. More attention should be paid in the future to the investigation of such patients, as well as to their follow-up, in order to verify that PCOS subsides with the treatment of the disease with which it is associated.

KEY POINTS

- Hirsutism is seldom the presenting symptom that leads to the diagnosis of CS and acromegaly. However, these life-threatening diseases should not be missed.
- Symptoms associated with hypercortisolism include central obesity, facial plethora, hypertension, thin skin, bruising, purplish skin striae, lethargy, and proximal muscle wasting.
- In mild cases of CS, the patient's presentation can mimic PCOS. Ovarian morphology suggestive of poly-cystic ovaries can be found on pelvic ultrasound.
- When CS is suspected on clinical grounds, a 24-hour urine free cortisol measurement or an overnight low-dose (1 mg) dexamethasone suppression test should be performed.
- Patients with acromegaly may present with a PCOS-like picture, with hyperandrogenism, menstrual irregularity, increased free androgen levels, LH hyperresponsiveness to GnRH stimulation, and insulin resistance.
- For optimal evaluation of GH secretion, multiple samples during the day and nightime must be measured. There is some consensus that a mean integrated 24-hour GH level below 2.5 µg/L excludes active acromegaly.

REFERENCES

1. Trainer P, Newell-Price J, Besser M. Cushing syndrome. In: Wass JAH, Shalet SM, eds. Endocrinology and Diabetes. New York: Oxford University Press 2002:817–836.
2. Newell-Price J, Trainer P, Besser M, Grossman A. The diagnosis and differential diagnosis of Cushing's syndrome and pseudo-Cushing's states. Endoc Rev 1998;19:647–672.
3. Extabe J, Vasquez JA. Morbidity and mortality in Cushing's disease: an epidemiological approach. Clin Endocrinol 1994;40:479–484.
4. Azziz R, Sanchez A, Knochenhauer ES, et al. Androgen excess in women: experience with over 1000 consecutive patients. J Clin Endocrinol Metab 2004;89:453–462.

5. O'Driscoll JB, Mamtora H, Higginson J, et al. A prospective study of the prevalence of clear-cut endocrine disorders and polycystic ovaries in 350 patients presenting with hirsutism or androgenic alopecia. Clin Endocrinol (Oxf) 1994;41:231–236.

6. Barbieri RL. Hyperandrogenic disorders. Clin Obst Gynecol 1990;133:640–654.

7. Moran C, Tapia Mdel C, Hernandez E, et al. Etiological review of hirsutism in 250 patients. Arch Med Res 1994;25:311–314.

8. Zargar AH, Wani AI, Masoodi SR. Epidemiologic and etiologic aspects of hirsutism in Kasmiri women in the indian subcontinent. Fertil Steril 2002;77:674–678.

9. Plotz C, Knowlton A, Ragan C. The natural history of Cushing's syndrome. Am J Med 1952;13:597–614.

10. Sprague RG, Randall RV, Salassa RM, et al. Cushing syndrome: a progressive and often fatal disease. A review of 100 cases seen between July 1945 and July 1954. Arch Inter Med 1956;98:389–398.

11. Soffer L, Iannaccone A, Gabrilove J. Cushing's syndrome: a study of fifty patients. Arch Inter Med 1961;300:215–219.

12. Urbanic RC, George JM. Cushing's disease—18 years' experience. Medicine (Baltimore)1981;60:4–24.

13. Ross EJ, Linch DC. Cushing's syndrome-killing disease: discriminatory value of signs and symptoms aiding early diagnosis. Lancet 1982;2:646–649.

14. Magagiakou MA, Mastorakos G, Odlfield EH, et al. Cushing's syndrome in children and adolescents; presentation diagnosis, and therapy. N Engl J Med 1994;331:629–636.

15. Weber A, Trainer PJ, Grossman AB, et al. Investigation, management and therapeutic outcome in 12 cases of childhood and adolescent Cushing's syndrome. Clin Endocrinol (Oxf) 1995;43:19–28.

16. Hatch R, Rosenfield RL, Kim MH, Tredway D. Hirsutism: implications, etiology and management 1981;140:815–830.

17. Cunningham SK, McKenna TJ. Dissociation of adrenal androgen and cortisol secretion in Cushing's syndrome. Clin Endocrinol (Oxf) 1994;41:795–800.

18. Kaltsas GA, Korbonit M, Isidori AM, et al. How common are polycystic ovaries and the polycystic ovarian syndrome in women with Cushing's syndrome? Clin Endocrinol (Oxf) 2000;53:493–500.

19. Derksen J, Nagesser SK, Meinders AE, et al. Identification of virilising adrenal tumor in hirsute women. N Engl J Med 1994;331:968–973.

20. Auchus RJ. Rainey WE. Adrenarche-physiologie, biochemistry and human disease. Clin Endocrinol 2004;60:288–296.

21. Holdaway IM, Rajassorya C. Epidemiology of acromegaly. Pituitary 1999;2:29–41.

22. Melmed S, Ho K, Klibanski A, Reichlin S, Thorner M. Recent advances in pathogenesis, diagnosis, and management of acromegaly. J Clin Endocrinol Metab 1995;80:3395–3401.

23. Thibouto DM. Dermatological manifestations of endocrine disorders. J Clin Endocrinol Metab 1995;80:3082–3087.

24. Kaltsas GA, Mukherjee PJ, Jenkins PJ, et al. Menstrual irregularity in women with acromegaly. J Clin Endocrinol Metabol 1999;84:2731–2735.

25. Jadresic A, Banks LM, Child DF, et al. The acromegaly syndrome. Q J Med 1982;202:189–204.

26. Nabarro JDN. Acromegaly. Clin Endocrinol 1987;26:481–512.

27. Melmed S. Acromegaly. N Engl J Med 1990;322:966–997.

28. Poretski L. On the paradox of insulin-induced hyperandrogenism in insulin-resistant states. Endocrine Rev 1991;12:3–10

29. Lazone A, Di Simone N, Castelanni R, Fulghesu AM, Caruso A, Mancuso S. Human growth hormone enhances proges-terone production by human luteal cells in vitro: evidence of a synergistic effect with human chorionic gonadotropin. Fertil Steril 1990;57:92–98.

30. Mason HD, Martkaninen H, Beard RW, Anyroku V, Francks S. Direct gonadotrophic effect of growth hormone on oestradiol production by human granulosa cells. J Endocrinol 1990;126:R1–R4.

31. Prelevic GM, Wueburger MI, Balint-Peric L, Nesic JS. Inhibitory effect of sandostatin on secretion of luteinizing hormone and ovarian steroids in polycystic ovary syndrome. Lancet 1990;336:900–903.

32. Cara JF, Rosenfeild R. Insulin-like growth factor I and insulin potentiate luteinizing hormone-induced androgen syn-thesis by rat ovarian thecal-interstitial cells. Endocrinology 1988;123:733–742.

33. Giustina A, Barkan A, Casanueva FF, et al. Criteria for cure of acromegaly : a consensus statement. J Clin Endocrinol Metab 2000;8:526–529.

34. Freda Pu, Reyes CM, Nuruzzaman AT, et al. Basal and glucose-suppresses levels less than 1 µg/L in newly diagnosed acromegaly. Pituitary 2003;6:175–180.

35. Ferone D, Resmini E, Bocca L, et al. Current diagnostic guidelines for biochemical diagnosis of acromegaly. Minerva Endocrinol 2004;29:207–223.

8

The Nonclassic Adrenal Hyperplasias

Didier Dewailly and Ricardo Azziz

SUMMARY

Nonclassic adrenal hyperplasia (NCAH) resulting from 11β-hydroxylase or 3β-hydroxysteroid dehydrogenase is very rare and may not even exist in women presenting with hyperandrogenism in adulthood. Consequently, screening for these disorders in hyperandrogenic patients is not generally necessary. Alternatively, 21-hydroxylase (21-OH)-deficient NCAH has been increasingly recognized in adolescent or adult hyperandrogenic patients. It is now widely accepted that neither clinical presentation nor androgen plasma levels can be used for the screening or diagnosis of 21-OH-deficient NCAH in hyperandrogenic women, especially those presenting with a polycystic ovary syndrome (PCOS)-like phenotype. Therefore, the measurement of a follicular morning level of serum 17α-hydroxyprogesterone (17-OHP) should be included in the initial investigation of all hyperandrogenic women, including those with premature pubarche. A basal screening level of 17-OHP of more than 2–4 ng/mL mandates an acute adrenocorticotropic hormone (ACTH) stimulation test to confirm the diagnosis. A post-ACTH stimulation 17-OHP level of 10–12 ng/mL is consistent with the diagnosis of NCAH. The diagnosis of 21-OH-deficient NCAH has important implications for preconception counseling and potentially for the prevention of adrenal insufficiency during illness or surgery.

Key Words: Adrenal; adrenal hyperplasia; hyperandrogenism;17-hydroxyprogesterone; ACTH test; androgens; androgen excess; polycystic ovary syndrome; 21-hydroxylase; 3β-hydroxysteroid dehydrogenase; 11β-hydroxylase.

1. INTRODUCTION

Since its initial description by Jayle et al. *(1)*, adrenal enzymatic deficiencies causing hyperandrogenic symptoms some time after birth have been called postpubertal, attenuated, mild, acquired, or late-onset adrenal hyperplasia. However, the term *nonclassic adrenal hyperplasia* (NCAH) is currently preferred, in contrast to the classic forms of congenital adrenal hyperplasia (CAH), in which clinical features are apparent at or immediately after birth. Presumably, the variability of the clinical and biological phenotypes is less linked to genetic polymorphism than to yet poorly documented extra-adrenal factors that may alter the clinical expression of the disease. Although deficiencies in 11β-hydroxylase (11-OH) and 3β-hydroxysteroid dehydrogenase (3β-HSD) may result in the disorder, defects in 21-hydroxylase (21-OH) account for more than 95% of patients with NCAH and probably all NCAH observed in adult women. In fact, 21-OH-deficient NCAH is one of the most common homozygous recessive disorders, affecting one of every 1000–2000 individuals *(2)*. In view of this, we will primarily review 21-OH-deficient NCAH, although we will also briefly summarize our scant knowledge regarding 11-OH- and 3β-HSD-deficient NCAH.

From: *Contemporary Endocrinology: Androgen Excess Disorders in Women:*
Polycystic Ovary Syndrome and Other Disorders, Second Edition
Edited by: R. Azziz et al. © Humana Press Inc., Totowa, NJ

2. BACKGROUND

2.1. 3β-HSD-Deficient NCAH

NCAH resulting from 3β-HSD deficiency has been purportedly diagnosed in 1–15% of children with premature pubarche *(3,4)* and a variable frequency of females with hirsutism and menstrual disorders with pubertal or postpubertal onset *(5–7)*. In these early studies, patients were presumed to suffer from 3β-HSD-deficient NCAH if they demonstrated a pregnenolone (PREG), 17-hydroxy-pregnenolone (17-HPREG), dehydroepiandrosterone (DHEA), and/or androstenediol (ADIOL) peak value, or a PREG:P4, 17-HPREG:17-HP, DHEA:A4, or ADIOL:testosterone ratio above the 90th or 95th percentile of normal during acute adrenocorticotropic hormone (ACTH)-(1-24) stimulation testing *(5,7–9)*. However, when compared to those used for diagnosing 21-OH-deficient NCAH (*see* Subsection 2.3.4.3.), these criteria appear to be exceedingly lax, particularly since it is well documented that the adrenal cortex is overactive in a significant proportion of individuals with polycystic ovary syndrome (PCOS) *(10)*.

Only the study of individuals with a confirmed mutation of the 3β-HSD gene can yield an accurate assessment of the endocrinological features and diagnostic criteria of the disorder. Two types of 3β-HSD genes, type I and II, have been reported (*HSD3B1* and *HSD3B2*, respectively). Although both encode for gonadal enzymes, the *HSD3B1* primarily determines extra-adrenal enzymatic activity and the *HSD3B2* encodes for the intra-adrenal 3β-HSD protein. Consequently, 3β-HSD-deficiency CAH (and presumably NCAH) results from mutations of *HSD3B2*. However, several studies of hirsute females and children with premature pubarche presumed to have 3β-HSD-deficient NCAH by the previously published hormonal criteria have been unable to confirm the diagnosis upon molecular analysis of *HSD3B2 (11–14)*. In fact, Pang and colleagues have nicely demonstrated that the majority of patients with an exaggerated 17-HPREG response to ACTH stimulation primarily suffer from PCOS *(15)*.

To date, patients with molecularly proven defects of *HSD3B2* (and no mutation of *HSD3B1*) have had stimulated 17-HPREG values that were at least 20–50 standard deviations greater than age-matched controls (i.e., in adult women >90 ng/mL), and all have presented in childhood *(14)*. Excluding children diagnosed at birth or those with ambiguous genitalia, these studies suggested that 3β-HSD-deficient NCAH could be present in children with premature pubarche when the 17-HPREG level was 95.8 ng/mL or more (294 nmol/L), 54 standard deviations (SD) or more above Tanner II pubic hair stage-matched control mean level; and in adults when the 17-HPREG levels were greater than 94.2 ng/mL (289 nmol/L), equivalent to or greater than 21 SD above the normal mean level *(14)*.

Consequently, the number of patients with true, genetically verifiable, 3β-HSD-deficient NCAH appears to be extremely small, if at all. In a prospective study of 86 consecutive patients with either hirsutism and/or hyperandrogenic oligomenorrhea, none demonstrated an ACTH-stimulated 17-HPREG value greater than threefold the upper normal limit (~15 ng/mL), nowhere close to the values found in patients with genetically confirmed type II 3β-HSD gene defects *(8)*. In fact, as yet no patient with genetically proven 3β-HSD-deficient NCAH presenting in adulthood has been observed. In view of the limited number of patients with genetically confirmed 3β-HSD-deficient NCAH diagnosed to date, it is difficult, if not impossible, to make any kind of accurate statement regarding their clinical features, pathophysiology, or diagnostic scheme at this time.

2.2. 11β-Hydroxylase-Deficient NCAH

Investigators have suggested that some hyperandrogenic women suffer from 11-OH-deficient NCAH *(16–19)*. However, it should be stressed that the same diagnostic uncertainties noted for 3β-HSD-deficient NCAH are found in the study of 11-OH (i.e., P450c11)-deficient NCAH. P450c11 is encoded by *CYP11B1*, a gene located on chromosome 8q21-22 in tandem with the *CYP11B2* gene, which encodes for aldosterone synthase *(20)*. Deficiency of 11-OH results from mutations in *CYP11B1*, and patients with both alleles encoding inactive enzymes have 11-OH-deficient CAH.

To confirm the diagnosis of 11-OH-deficient NCAH and potentially develop diagnostic criteria, we genotyped five patients with presumed 11-OH-deficient NCAH *(21)*. Three were children (two females and one male) who presented with advanced bone age, accelerated growth, acne, and precocious adrenarche. Two of these were found to have mutations of both *CYP11B1* alleles, defects that affected enzymatic expression in vitro. One of these two individuals had an 11-deoxycortisol level after ACTH stimulation of 41 ng/mL (upper normal limit of controls was <8 ng/mL), the other had a urinary tetrahydro-11-deoxycortisol level of 1690 µg/24 hours (upper normal limit of controls was less than 50 µg/24 hours). No mutations of the coding regions or intron/exon boundaries of the *CYP11B1* genes in the third child could be detected. In addition, we studied two adult hyperandrogenic women with presumed 11-OH-deficient NCAH, neither of which had mutations of their *CYP11B1* genes *(21)*. These last two patients had ACTH-stimulated 11-deoxycortisol levels of 22 and 24 ng/mL and were the only women with 11-deoxycortisol values threefold greater than the normal limit discovered during a study of 260 consecutive hyperandrogenic patients *(19)*.

In conclusion, 11-OH-deficient NCAH is an extremely rare cause of adult hyperandrogenism, if it exists at all. Patients with 11-OH-deficient NCAH appear to have 11-deoxycortisol values (either after ACTH stimulation or as urinary metabolite) that are at least fivefold the upper normal limit, although the number of patients studied is insufficient to arrive at any firm conclusion.

2.3. 21-Hydroxylase-Deficient NCAH

2.3.1. Pathophysiology of 21-OH-Deficient NCAH

The pathophysiology of 21-OH-deficient NCAH must be considered from both a genetic and an endocrinological perspective.

2.3.1.1. GENETIC PATHOPHYSIOLOGY OF 21-OH-DEFICIENT NCAH

Cytochrome P450c21 is responsible for adrenal 21-OH activity, catalyzing the conversion of 17-hydroxyprogesterone (17-OHP) to 11-deoxycortisol, and progesterone (P4) to deoxycorticosterone *(20)*. The gene encoding for P450c21 *(CYP21)* exists in tandem with a pseudogene *(CYP21P)*. Each gene has 10 exons and 9 introns, although *CYP21P* has 88 mutations that render it nontranscribable. These genes are located in tandem with the human leukocyte antigen (HLA) locus on chromosome 6 next to the gene for the fourth component of complement (C4) and a gene for the extracellular matrix protein called tenascin-x (X) *(22)*. The genes for CYP21, C4, and X are all duplicated, such that there are C4A and C4B, CYP21 and CYP21P, and XA and XB pairs. Both C4 genes are actively transcribed, whereas CYP21P and XA are inactive.

To date, approx 100 different *CYP21* mutations have been reported *(23–31)*, mostly point mutations, alhough small deletions and complete gene deletions and small insertions have also been described. Approximately 17 mutations account for about 95% of all of affected alleles (Table 1). The majority are primarily derived from the intergenic recombination of DNA sequences between the *CYP21* gene and the highly homologous *CYP21P* (a process termed *gene conversion*), while the remaining are spontaneous mutations. Occasionally, patients may demonstrate multiple mutations of their *CYP21* alleles.

Initially CAH and NCAH were not thought to share genetic defects because the severity of the disorders is so different. However, with the genotyping of increasing numbers of affected patients it has become clear that most individuals with NCAH and CAH are "compound heterozygotes" carrying different genetic mutations on each *CYP21* allele (Fig. 1). In fact, approximately two-thirds of NCAH patients are compound heterozygotes *(10)*, carrying a gene defect encoding for a mutation resulting in severe defects in P450c21 function (<2% wild-type activity) on one allele, whereas the other carries a mutation that determines a mild defect in the enzyme (i.e., 20–50% of wild-type activity) (Table 1).

Table 1
Common Mutations of CYP21 Resulting in 21-Hydroxylase (21-OH)-Deficient Nonclassic Adrenal Hyperplasia (NCAH) and Salt-Wasting (SW-CAH) and Simple Virilizing (SV-CAH) Classic Adrenal Hyperplasia

Mutation	Location on CYP21	% of wild-type 21-OH activity	Clinical presentation if homozygous
Mild[a]			
Pro30Leu	Exon 1	30–60	NCAH
Val281Leu	Exon 7	50	NCAH
Arg339His	Exon 8	20–50	NCAH
Arg441Trp	Exon 10	?	NCAH
Pro453-Ser	Exon 10	20–50	NCAH
Severe[a]			
30 kb deletion	Entire gene	0	SW-CAH
8 b deletion (nt 7-7-714)	Exon 3	0	SW-CAH
Ile172Asn	Exon 4	3–7	SV-CAH
I236N, V237G, M239L (mutation cluster)	Exon 6	0	SW-CAH
Gly291Ser	Exon 7	0	SW-CAH
T insertion, codon 306 (nt 1757)	Exon 7	0	SW-CAH
Gly318Stop	Exon 8	0	SW-CAH
Arg356Trp	Exon 8	2	SV-CAH, SW-CAH
Trp405Stop	Exon 9	0	SW-CAH
GG483C	Exon 10	0	SW-CAH
Arg484Pro	Exon 10	0	SW-CAH
Variable[a,b]			
nt 656A > G	Intron 2	0–5	NCAH, SW-CAH, SV-CAH

[a]Refers to the severity of the defective enzymatic activity of P45021 determined by the mutation.

[b]The variations in phenotype with Intron 2 mutations are thought to be caused by the variable region around nucleotide 656, which affects the splicing of the intron, although the exact mechanism has yet to be elucidated. (Adapted from refs. *22–31.*)

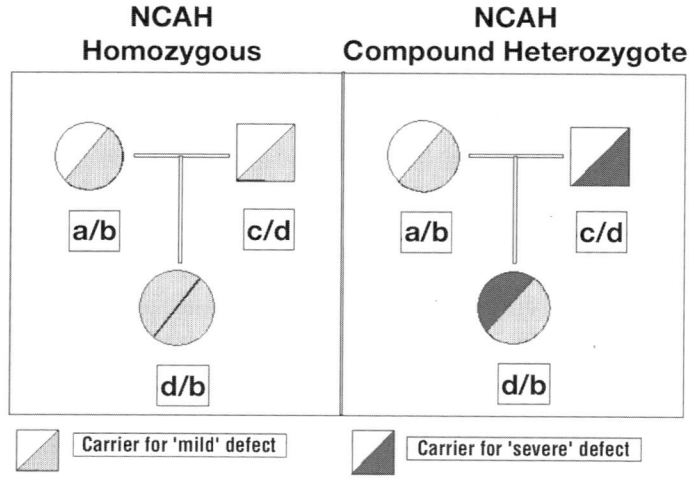

Fig. 1. Allelic variants in 21-hydroxylase-deficient nonclassic adrenal hyperplasia (NCAH). 'Mild' and 'severe' refer to the severity of the defect in the enzymatic activity P45021 determined by defects of *CYP21* that, when present in homozygous form, result in NCAH and CAH, respectively.

2.3.1.2. ENDOCRINE PATHOPHYSIOLOGY OF 21-OH-DEFICIENT NCAH

Patients with NCAH are both hyperandrogenic (primarily with an excessive production of androstenedione [A4]) and hyperprogestogenic (i.e., elevated 17-OHP and P4), although the mechanisms underlying the excess hormonal secretion in these patients are complex and varied *(32)*. Most frequently, the signs and symptoms of NCAH generally begin peripubertally. The physiological alterations of adrenocortical steroidogenesis occurring at this time (increased 17,20-lyase and decreased 3β-HSD activities), which result in the increased output of adrenal androgens known as adrenarche, serve to exaggerate the tendency of patients with NCAH to develop adrenal hyperandrogenism.

Mechanisms underlying the excess androgen and progestogen production in NCAH include alterations in adrenal enzymatic dynamics as well as ovarian hyperandrogenism. Alternatively, although extra-adrenal 21-OH activity, mediated by non-P450c21 enzymes *(33)*, is observed in humans, its role in the phenotypic presentation of women with defects of P450c21 function remains unclear *(34)*.

2.3.1.3. ADRENAL DYSFUNCTION IN NCAH

Classically, adrenal hyperplasia and the associated androgen and progesterone excess have been proposed to arise as the result of chronic ACTH oversecretion resulting from an incipient defect in the secretion of cortisol. However, and in contrast to CAH *(35–38)*, the mutations producing 21-OH-deficient NCAH do not generally impair cortisol production. Thus, the majority of NCAH patients do not demonstrate either overt cortisol insufficiency or ACTH hypersecretion, whether basally or after acute corticotropin-releasing hormone (CRH) stimulation *(38–41)*. Essentially, NCAH is essentially a hyperandrogenic disorder, without overt abnormality in the glucocorticoid and mineralocorticoid pathways. Although an overt abnormality of the hypothalamic–pituitary–adrenal axis is not obvious in NCAH, some patients may have a subtle degree of cortisol deficiency and/or ACTH hypersecretion. For example, adrenocortical adenomas or hyperplasia can be detected radiographically in up to 40% of patients with NCAH *(42,43)*, suggesting that in these patients a state of chronic ACTH hypersecretion may be present. Furthermore, a sluggish cortisol response to prolonged ACTH stimulation has been found in some patients *(44)*. Thus, subtle oversecretion of ACTH may be a factor in the androgen excess and hyperprogesteronemia of some patients with NCAH.

Another mechanism underlying the excess 17-OHP and P4 secretion of NCAH is the intrinsic change in P450c21 kinetics. As we previously demonstrated mathematically *(32)*, the diminished kinetic efficiency of P450c21 is sufficient to cause an increase in the amount of precursor hormone present within the adrenal cortex without requiring an increase in its production rate (i.e., without requiring increased ACTH-stimulated production of 17-HPREG). The increased concentration of 17-OHP enables P450c21 to produce a relatively normal amount of product (i.e., 11-deoxycortsol and cortisol), effectively compensating for its diminished function. This effect, together with the fact that the Δ^4 17,20-lyase pathway activity in humans is minimally active *(45,46)*, serves to explain the extremely high 17-OHP and P4 levels found in NCAH patients in the absence of obvious ACTH oversecretion. In fact, this may explain why some patients continue to demonstrate a chronic exaggeration in P4 and 17-HP levels despite adequate glucocorticoid replacement doses *(47)*.

Nonetheless, how the highly exaggerated 17-OHP production results in an exaggerated production of A4 despite the limited Δ^4 17,20-lyase activity observed in humans remains unclear. Possible mechanisms include (a) intra-adrenal levels of 17-OHP that are sufficient to produce increased amounts of A4 despite the decreased Δ^4 17,20-lyase activity, (b) increased conversion of 17-OHP to A4 by extra-adrenal enzymes, and (c) a concomitant accumulation of 17-HPREG, resulting in increased DHEA and subsequently A4 production. In fact, the viability of this latter mechanism is supported by our findings that the adrenal response to ACTH was generally exaggerated in patients with 21-OH-deficient NCAH, including elevated levels of 17-HPREG *(48)*.

2.3.1.4. OVARIAN DYSFUNCTION IN NCAH

Androgen excess in NCAH may also result from associated ovarian hyperandrogenism. Many NCAH patients have polycystic-appearing ovaries *(49)* and mild luteinizing hormone (LH) hypersecretion *(50)*. Furthermore, gonadotropin-releasing hormone analog suppression results in a reduction in the circulating hyperandrogenemia in these patients *(51)*. Thus, women with NCAH may present with PCOS-like features, which may aggravate their hyperandrogenism, and needs to be considered in formulating their treatment plans.

2.3.2. Phenotypic Aspects of NCAH: Clinical and Biological Features

2.3.2.1. IN CHILDREN

Before the age of 7–8 years, NCAH may mimic idiopathic premature pubarche or, more rarely, a virilizing ovarian or adrenal tumor, especially when there are symptoms of precocious pseudopuberty, such as accelerated height velocity and clitoromegaly. Initially, the prevalence of NCAH as a cause of premature pubarche was debated, suggested to involve 0% to more than 20% of these patients, mainly because of differences in diagnostic criteria. More recently it has been estimated that 6–8% of children with premature pubarche suffer from NCAH *(52)*, such that screening for this disorder is justified in these children. Recently, NCAH has been detected in newborns undergoing routine screening for 21-OH deficiency. Mild-to-moderate elevations in 17-OHP serum levels were observed in infants who had no symptoms other than mild clitoromegaly in some affected girls *(53)*.

2.3.2.2. IN ADOLESCENT AND ADULT WOMEN

NCAH is encountered with relatively high frequency (1–10%) among adolescent and adult patients with hyperandrogenism. This incidence may vary from one geographic area to another, as reported for classical 21-OH deficiency, because of ethnic and racial clusters *(54–56)*. It appears lowest among Anglo-Saxon individuals, intermediate among Mediterranean or Latin populations (arising in Portugal, Spain, Italy, France, and Greece), and highest among Ashkenazi Jews and some Middle Eastern populations. Additional populational studies are needed to clearly establish the ethnic and racial variations in prevalence.

Numerous studies have demonstrated that adolescent or adult women with NCAH are seldom more virilized than women with ovarian causes of hyperandrogenism and that no symptom of hyperandrogenism is specific to NCAH. Furthermore, NCAH may also be detected in individuals presenting with only mild symptoms, such as women with eumenorrhea and acne *(57)*. Therefore, the clinical presentation cannot be used for the diagnosis of NCAH. Clitoromegaly, male habitus, and temporal baldness are infrequent findings, unless the patient suffers from undiagnosed simple virilizing CAH. Recent data from a multicenter study including 220 patients suggested that NCAH is a progressive disorder, with the prevalence of hirsutism increased with the patient's age *(58)*.

In addition, it has been reported that NCAH is sometimes detected in women having no symptom of hyperandrogenism, generally either by chance or through family studies, and that they have the same biochemical profile as symptomatic patients with NCAH. This asymptomatic form of the disorder was termed *cryptic* NCAH *(59)*. However, not all of these patients are actually asymptomatic when investigated more closely (e.g., not complaining of hirsutism when present does not constitute an absence of symptoms) *(59)*, and further longitudinal studies of these individuals are needed.

2.3.3. Relationship Between Genotype and Phenotype

In CAH, relatively good correlation exists between the genotype (i.e., the defects in *CYP21*) and the phenotype, although this holds true mainly for homozygotes. As noted previously, approximately 75% of NCAH patients are compound heterozygotes, carrying a mild and a severe mutation *(30)*. When compared to individuals with the mild/severe genotype, patients with the mild/mild genotype had lower rates of hirsutism, oligomenorrhea, and acne (Fig. 2). Also, they tended to have lower serum 17-OHP levels, basally and in response to ACTH(1-24) stimulation, although the differences did not achieve statistical significance *(30)*. These results generally agree with those of a prior study

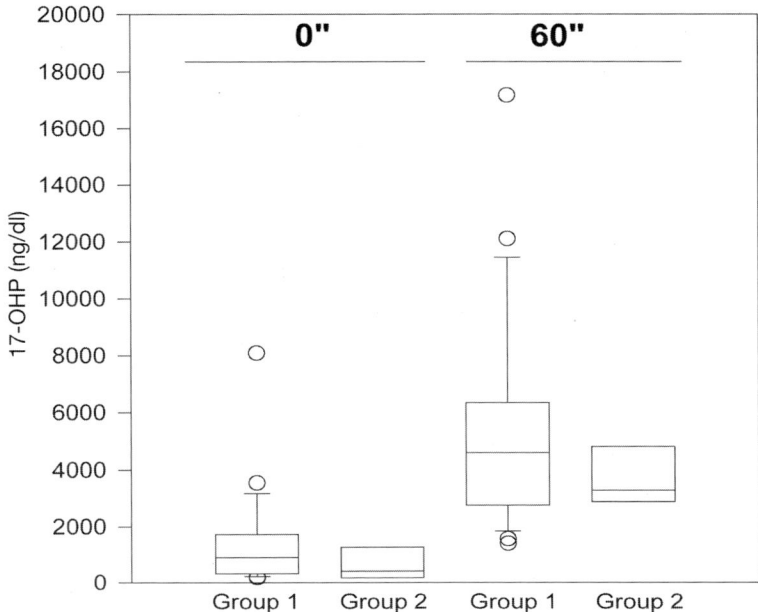

Fig. 2. Serum 17-hydroxyprogesterone (17-OHP in ng/dL) is shown for basal (0 min) and adrenocorticotropic hormone (ACTH)-stimulated (60 min) values in groups 1 and 2. The y-axis is log scale. The shaded boxes represent 25–75%. Crossbars within the boxes indicate the median; the open circles are data points that lie outside the 5–95%. Group 1 carries a mild/severe CYP21 genotype, and group 2 a mild/mild genotype. (From ref. *30*.)

in which the distinction between homozygotes and compound heterozygotes was based on HLA linkage associations *(60)*. Alternatively, other investigators reported that the phenotype in NCAH could not be accurately predicted from the genotype alone *(61)*.

The weak relationship between genotype and phenotype of NCAH indicates that the genetic defects of CYP21 alone do not fully explain the variability in the clinical and biological expression of NCAH and that other genetic and/or acquired abnormalities may play a role. For example, the frequency of metabolic derangement in NCAH, potentially playing a role in the phenotypic expression of the disorder, is still poorly understood. However, one study noted a reduction in insulin sensitivity in these patients, possibly attributed to their hyperandrogenemia *(62)*. More recently, the homeostasis model assessment was used to assess insulin sensitivity in 18 patients with NCAH and 26 healthy weight-matched control women *(63)*. The results suggested diminished insulin sensitivity in the former, in whom, interestingly, the fasting insulin serum level correlated significantly and positively with the circulating levels of 17-OHP. It is therefore possible that the clinical expression of hyperandrogenism in some NCAH women might be amplified or triggered by the presence of concurrent hyperinsulinism, possibly secondary to insulin resistance, as noted in PCOS.

2.3.4. Diagnosis

2.3.4.1. Screening With a Basal 17-OHP Level

It is possible to effectively screen for NCAH using a basal 17-OHP measurement. However, in NCAH the elevation in basal plasma 17-OHP levels is much less than in CAH. Therefore, several diagnostic pitfalls must be avoided. First, blood sampling preferably should be performed early in the morning, as this will minimize the false-negative results that may occur following the circadian decrease in 17-OHP levels paralleling those of cortisol and ACTH (Fig. 3). Second, patients must be investigated in the early follicular phase, which will avoid the false-positive results as a result of 17-OHP

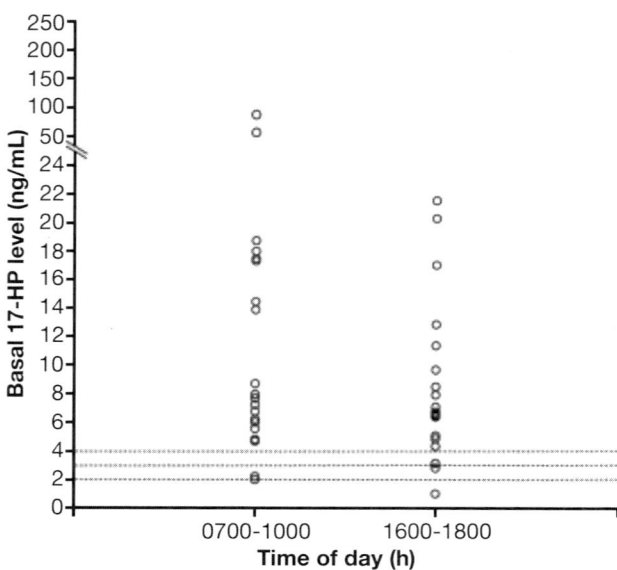

Fig. 3. The basal 17-hydroxyprogesterone (17-HP) level was determined in 20 patients with 21-hydroxy-lase-deficient nonclassic adrenal hyperplasia (NCAH) in the morning (7–10 AM) and in the afternoon (4–5:30 PM). All 20 samples obtained in the morning had 17-HP levels of more than 2 ng/mL, and 18 of 20 had levels of greater than 3 or greater than 4 ng/mL. Of samples obtained in the afternoon, 19, 18, and 17 had levels of more than 2, more than 3, and more than 4 ng/mL, respectively (dashed horizontal lines). (From ref. *64.*)

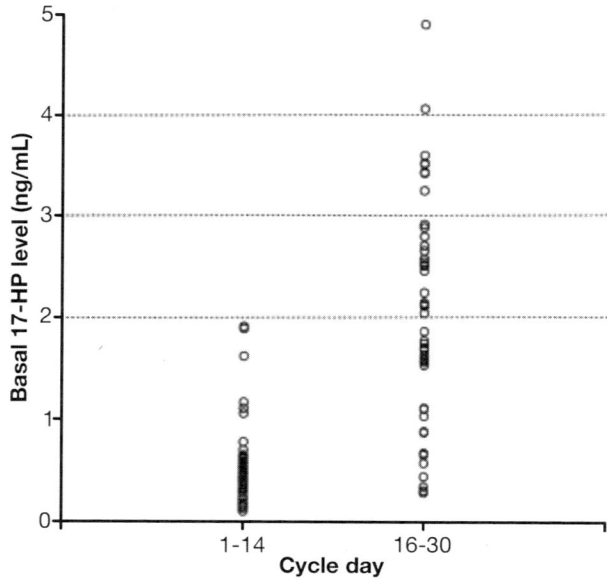

Fig. 4. The variability of the basal 17-hydroxyprogesterone (17-HP) level with the menstrual cycle was deter-mined in eight healthy control women. Blood was sampled every other day throughout the cycle, with an average of 13.4 samples per subject. Considering the first day of menstrual flow as day 1 of the menstrual cycle, 56 samples were obtained in the follicular phase of the menstrual cycle (i.e., cycle days 1–14) and 51 in the luteal phase (i.e., cycle days 16–30). Of the samples obtained on or before cycle day 14, none demonstrated a 17-HP level of at least 2 ng/mL. Of those obtained on or after cycle day 16, 24, 9, and 2 samples had 17-HP levels of more than 2, more than 3, and more than 4 ng/mL, respectively (dashed horizontal lines). (From ref. *64.*)

production by the corpus luteum (Fig. 4). Third, patients with NCAH should obviously not be pre-treated with corticosteroids before screening. Fourth, one should keep in mind that ovarian hyperandrogenism, such as in PCOS, is also associated with mild elevation in 17-OHP *(65)*.

In a study including 284 consecutive patients with hyperandrogenism *(64)*, a follicular unsuppressed morning 17-OHP level >4 ng/mL (12.0 nmol/L) had maximum specificity (100%) and good sensitivity (90%) and a positive predictive value (PPV) of 40% for the diagnosis of NCAH. However, if not all subjects can be studied in the early morning, we prefer to use a cutoff value of 2 ng/mL, which has a lower PPV (18.8%) but a virtually a 100% sensitivity regardless of the time of the day obtained *(64)*. A basal 17-OHP level of less than 2 ng/mL (6.0 nmol/L) effectively rules out NCAH *(66)*.

In our experience, and that of others, false-positive results (i.e., elevated levels) may be observed in patients with androgen-secreting tumors *(67;* R. Azziz, unpublished observation; D. Dewailly, unpublished observation). The risk of a false-negative is more worrisome. Up to 10% of NCAH patients may demonstrate basal 17-OHP levels that overlap with those of controls *(64)*. Striving to sample patients in the early morning hours, benefiting from the peak in adrenocortical secretion can minimize this rate of false-negatives. The use of plasma 21-deoxycortisol (21-DOF) has been proposed as a more specific marker for NCAH *(68)*. This steroid is secreted exclusively by the adrenals because it arises from the 11-hydroxylation of 17-OHP. However, the limited availability of the 21-DOF assay has restricted its use to the detection of heterozygotes, whose post-ACTH 17-OHP level often overlaps control values. Overall, a follicular phase morning level of basal 17-OHP should be included in the first screening investigation of all hyperandrogenic women, including those with premature pubarche *(69)*.

2.3.4.2. CIRCULATING ANDROGEN PLASMA LEVELS

Circulating androgen levels are not generally helpful for the diagnosis of NCAH. Individual values of plasma testosterone, A4, or dehydroepiandrosterone sulfate completely overlap with those of non-NCAH hyperandrogenic patients *(70)*. Recently, the diagnostic value of serum 11β-hydroxyandrostenedione and Δ^5-androstenediol was found to be no better for detecting NCAH *(71)*. Consequently, androgen levels are not of value in screening for NCAH patients.

2.3.4.3. ACTH STIMULATION TEST FOR THE DIAGNOSIS OF NCAH

For many investigators the ACTH test is mandatory to ascertain the diagnosis of NCAH. However, there is no evidence that its sensitivity in detecting 21-OH deficiency is higher than the use of a basal 17-OHP determination, providing the above-mentioned caveats are understood. Alternatively, the acute response to ACTH stimulation remains the standard for clinically diagnosing patients with NCAH. Therefore, the ACTH test should not be used routinely to screen patients, but should be used primarily when the morning follicular unsuppressed 17-OHP levels exceed at least 2 ng/mL (6 nmol/L) to confirm the diagnosis *(64)*.

From the study of NCAH patients whose disorder has been genetically confirmed, either through family studies or via molecular analysis, it is clear that the diagnosis of NCAH should be considered in those patients in whom the poststimulation 17-OHP level exceeds 10–12 ng/mL (30–36 nmol/L). Individuals whose stimulated 17-OHP exceeds the upper normal limit (usually 3–4 ng/mL) but do reach this threshold often include heterozygotes for *CYP21* mutations *(72)* or patients with functional ovarian and/or adrenal hyperandrogenism *(65)*. Sometimes a heterozygote will demonstrate a post-stimulation 17-OHP level as high as 15 ng/mL *(72)*, but only rarely do their basal 17-OHP levels exceed 2 ng/mL, at least in our experience *(73)*.

2.3.5. *Potential Complications of NCAH*

2.3.5.1. ADRENAL INSUFFICIENCY

In contrast to CAH patients, the majority of NCAH patients do not generally demonstrate cortisol insufficiency; normal cortisol and ACTH levels at baseline and following CRH stimulation are usually observed *(38–41)*. However, a deficient cortisol response following prolonged ACTH stimulation (72 hours) has been reported in one study *(44)*, suggesting that some patients with NCAH may demonstrate relative hypoadrenalism in the face of prolonged stress (e.g., illness, surgery, trauma). Therefore, it is advisable that patients with NCAH carry medical alert bracelets denoting the fact that they may require supplemental glucocorticoids in the event of an accident or severe illness.

2.3.5.2. ADRENAL HYPERPLASIA AND ADENOMAS

The incidence of adrenal hyperplasia and adrenal adenomas in NCAH is unclear. In small series, estimated retrospectively, it appears that 40–50% of these women will demonstrate radiological evidence of adrenal abnormalities 45% *(42,43)*. If it is assumed that adrenal size is a good cumulative index of ACTH secretion, the mildness and low frequency of hyperplasia emphasizes the relative normalcy of the hypothalamic–pituitary–adrenal axis in NCAH. Because adrenal adenomas are a frequent finding in NCAH, the finding of an adrenal incidentaloma should prompt the search for this disorder. Alternatively, the finding of an adenoma in a patient with NCAH mandates caution before proceeding to surgical intervention, and expectant management with regular radiological follow-up is recommended.

2.3.5.3. DEVELOPMENT OF PCOS-LIKE FEATURES

The occurrence of PCOS-like features in NCAH has been mentioned in several studies, but its exact prevalence is disputed, mainly because there is no consensual definition of PCOS. As an example, our previously reported rate was 36% using hormonal data *(49)*, whereas it is now 53% using sonographic criteria for PCOS (i.e., increased ovarian area >5.5 cm^2) and 47% by combining this criterion with either abnormal menstrual cycles or elevated LH levels *(74)*. Conversely, the prevalence of a pattern of inappropriate secretion of gonadotropins (i.e., exaggerated ratio of LH to follicle-stimulating hormone) was less frequent among our patients *(49)* than in other reports *(50)*.

Such an association between NCAH and PCOS-like features is in keeping with clinical data concerning the more severe forms of 21-OH deficiency *(75)*. Likewise, experimental data indicate that the chronic impact of excessive extra-ovarian androgens may alter ovarian follicular maturation and induce morphological features of PCOS, for example, in female-to-male transsexuals *(76)*. Alternatively, the excess androgens or progesterone produced by the adrenal in NCAH may act on the hypothalamic–pituitary–ovarian axis. For example, despite good adrenal control of hyperandrogenism by glucocorticoids in CAH women, gonadotropin abnormalities persisted, suggesting the effect of prenatal masculinization on neuroendocrine function *(77)*. However, this mechanism may not be pertinent to NCAH, because hyperandrogenism appears later and is milder.

2.3.5.4. INFERTILITY

Although anovulation is the main cause of infertility in NCAH patients, hyperprogesteronemia can also result in unfavorable cervical mucus and, more importantly, a persistently atrophic or decidualized endometrium. In addition, it is possible that NCAH women suffer from an increased miscarriage rate (~30%) if untreated *(78)*, presumably because of the deleterious effect of chronically elevated androgens on the oocytes and/or the presence of PCOS. Nevertheless, it should be remembered that approximately one-half of women with NCAH do not have menstrual disturbances, and many conceive spontaneously *(79)*.

2.3.5.5. THE RISK OF VIRILIZED FEMALE NEWBORNS AND PRECONCEPTION TREATMENT IN WOMEN WITH NCAH

There are two theoretical possibilities for such a risk. First, the mother's hyperandrogenism could virilize a female fetus. So far, such an event has not been reported in NCAH mothers, consistent with the fact that the fetus is protected against mild to moderate endogenous maternal hyperandrogenemia by extensive placental aromatase activity. Therefore, glucocorticoid suppression solely for this reason is not required during pregnancy.

Second, there is the possibility that a female fetus will be born with CAH. This possibility depends on the parents' genotype. If the NCAH-affected mother is a homozygote with two mild alleles (*see* Table 1), there is no risk that her daughter will have CAH, although she may have NCAH. Alternatively, if the mother is a compound heterozygote carrying one mild and one severe CYP21 allele, the risk to the child will depend on whether the father is a heterozygote or not and, if this is the case, whether he bears a mild or severe allele. Genetic counseling should take into account the lower frequency of severe CYP21 mutations among Ashkenazi Jewish NCAH patients (~25%) compared to other ethnic groups (~40%) *(80)*. This suggests that for NCAH patients of Ashkenazi Jewish origin, there is approximately a 12% likelihood—compared with about a 20% for non-Jews—of transmitting a classic mutation to the offspring. About 1.7% (1 in 60) of the general population carries a classic *CYP21* mutation *(2)*; again there is a 50% chance of transmitting this mutation. Multiplying these probabilities, there is about a 0.1% (~1 in 1000) chance of an Ashkenazi Jewish NCAH patient having a child affected with severe CAH and about a 0.17% (~1 in 600) chance for the non-Jewish NCAH patient.

Theoretically, before conceiving, a patient with NCAH and her partner should be genotyped to determine their risk for having a child with CAH, with the caveat that not all the defects resulting in 21-OH deficiency are detected in by current screening. Only in cases of paternal heterozygozity for CAH should antepartum maternal glucocorticoid suppression be recommended, according to the schedule that has been proposed when both parents are heterozygotes and have given birth to a first child with CAH *(81)*.

In a multinational multicenter database *(78)* we reported on the outcome of 206 pregnancies among 107 women with 21-OH-deficient NCAH. Of the 206 pregnancies, 145 occurred prior to the mother's diagnosis of NCAH and 61 following the diagnosis of NCAH. Among pregnancies occurring prior to the maternal diagnosis of NCAH, 21.4% ended in a spontaneous miscarriage; alternatively, only 4.9% of pregnancies occurring after the diagnosis of NCAH was established were lost ($p < 0.007$). Of the 156 live births, 2.6% of were diagnosed with classic CAH; 15.4% of children (13 females and 11 males) have been diagnosed with NCAH so far. We concluded that the risk of NCAH women giving birth a child affected with CAH is around 2.6% (95% CI 0.7–6.4%), and for NCAH it is at least 15.4% (95% CI 9.4–21.3%). These data also suggest that preconception diagnosis of women with 21-OH-deficient NCAH appears to improve their reproductive outcome

2.3.6. Therapeutic Implications

2.3.6.1. Treatment of Hyperandrogenism

As in CAH, the standard treatment of NCAH is glucocorticoid therapy. It aims mainly to reduce adrenal hyperandrogenism, while the necessity for cortisol replacement is less evident. We generally use dexamethasone (DXM), rather than less potent ACTH-inhibiting compounds such as hydrocortisone or prednisone, because it can to be used once daily and maximizes compliance. Generally, very low doses of DXM are needed (0.25 daily to every other day), and practitioners should seek to normalize the morning A4 plasma levels. The fact that much of the intra-adrenal abnormalities in NCAH may be ACTH-independent may explain why 17-OHP serum levels are frequently not completely normalized by glucocorticoid treatment despite adequate ACTH suppression *(47)*. Chronic glucocorticoid suppression may also be recommended in those NCAH patients whose adrenals are hyperplastic adenomatous on computed tomography to prevent any further growth *(43)*.

Despite this standard, many patients with NCAH do not tolerate glucocorticoids well, as they do not require cortisol replacement, complaining of weight gain, mood changes, and insomnia. Furthermore, glucocorticoids alone may not be sufficient to improve their hyperandrogenism, particularly if they have developed PCOS-like features and ovarian hyperandrogenism. This appears to be more important in NCAH women who are first diagnosed when older than 20 years *(82).* These patients will benefit from the use of an oral contraceptive with or even without glucocorticoids. In NCAH patients with hirsutism, the addition of an antiandrogen should be considered *(83)*.

2.3.6.2. Treatment of Adrenal Insufficiency and Its Consequences

Although we have never observed a case of acute adrenal insufficiency, or symptoms of chronic hypocortisolism, in patients with NCAH, we cannot exclude this possibility. Therefore, we caution all patients with NCAH about the theoretical risk of adrenal insufficiency, describe the presenting symptoms, and ask them to carry a medical alert bracelet and hydrocortisone tablets for emergency use. So far, very few, if any, of our patients had to resort to transient hydrocortisone treatment. Prophylactic glucocorticoid supplementation should be given to NCAH patients undergoing surgery or during significant illnesses or trauma.

2.3.6.3. Treatment of Subfertility

Glucocorticoid suppression, as described above, normalizes ovulatory function and improves fertility and pregnancy outcome in many patients with NCAH. Alternatively, particularly the older patient with NCAH may require additional ovulation-induction agents (e.g., clomiphene citrate, menotropins) to achieve successful ovulation *(79)*. Because many patients with NCAH also demonstrate polycystic ovarian morphology, the same precautions against overstimulation as is taken for the treatment of patients with PCOS should be followed in these women. Therefore, in all women with NCAH desiring pregnancy, we recommend that the ovaries be evaluated ultrasonographically. In addition, persistent hyperprogesteronemia may result in persistently poor cervical mucus, overcome by the use of intrauterine inseminations. Alternatively, a persistently atrophic or decidualized endometrium may require a significant increase in the dose of glucocorticoids being used or even the use of a surrogate gestational carrier.

3. CONCLUSIONS

Although NCAH is now a well-characterized genetic disorder, it is clear that it remains undiagnosed in many adolescent and adult women suffering from hyperandrogenism. The diagnostic accuracy would be improved by the greater use of a screening basal 17-OHP measurement in the initial investigation of all hyperandrogenic women, including those with premature pubarche. The diagnosis of NCAH has important implications for preconception counseling and potentially for the prevention of adrenal insufficiency during illness or surgery.

4. FUTURE AVENUES OF INVESTIGATION

A number of areas remain to be better understood, including:

1. How the highly exaggerated 17-OHP production results in results in an exaggerated production of A4, despite the limited ^{4} 17,20-lyase activity observed in humans.
2. What the prevalence of insulin resistance/hyperinsulinism in NCAH is and whether it influences the phenotypic variability of the disorder.
3. What the long-term outcome of NCAH patients is into the menopause and older age.
4. What the optimum therapeutic plan is and whether the age of diagnosis alters this, through the use of prospective, randomized protocols.
5. What the actual prevalence of the disorder is, including its ethnic and racial variation, possibly through the use of large-scale neonatal screening studies.

KEY POINTS

- NCAH resulting from 11β-hydroxylase or 3β-hydroxysteroid dehydrogenase is very rare and may not exist in women presenting with hyperandrogenism in adulthood. Consequently, screening for these disorders in hyperandrogenic patients is not necessary.
- Neither clinical presentation nor androgen plasma levels can be used for the screening or diagnosis of NCAH in hyperandrogenic women.
- The measurement of a follicular phase morning level of serum 17-OHP should be included in the initial investigation of all hyperandrogenic women.

- A basal screening level of 17-OHP of more than 2–4 ng/mL mandates an acute ACTH stimulation test to confirm the diagnosis.
- A post-ACTH stimulation 17-OHP level of 10–12 ng/mL is consistent with the diagnosis of NCAH.
- Up to 50% of women with NCAH will develop radiologically evident adrenal hyperplasia or adenomas, and these should be followed expectantly.
- Many patients with NCAH develop PCOS-like features, including ovulatory dysfunction; in contrast, few patients develop adrenal insufficiency.
- Althoughglucocorticoids may improve clinical features in many women with NCAH, particularly if younger, many patients also benefit from the use of ovulation induction agents, oral contraceptives or antiandrogens as needed.
- Preconception treatment with glucocorticoids in NCAH may improve reproductive outcome; and all couples should undergo genotyping, if possible, to establish the exact risk of giving birth to a child affected with classic adrenal hyperplasia.

REFERENCES

1. Jayle MF, Weinmann SH, Baulieu EE, Villin Y. Virilisme post-pubertaire discret par déficience de l'hydroxylation en C21. Acta Endocrinol (Copenh) 1958;29:513–524.
2. Speiser PW, Dupont B, Rubinstein P, Piazza A, Kastelan A, New MI. High frequency of nonclassical steroid 21-hydroxylase deficiency. Am J Hum Genet 1985;37:650.
3. Temeck JW, Pang S, Nelson C, New MI. Genetic defects of steroidogenesis in premature pubarche. J Clin Endocrinol Metab 1987;64:609.
4. Hawkins LA, Chasalow FI, Blethen SL. The role of adrenocorticotropin testing in evaluating girls with premature adrenarche and hirsutism/oligomenorrhea. J Clin Endocrinol Metab 1992;74:248.
5. Pang S, Lerner AJ, Stoner E, et al. Late-onset adrenal steroid 3β-hydroxysteroid dehydrogenase deficiency. I. A cause of hirsutism in pubertal and postpubertal women. J Clin Endocrinol Metab 1985;60:428.
6. Schram P, Zerah M, Mani P, Jewelewics R, Jaffe S, New MI. Nonclassical 3β-hydroxysteroid dehydrogenase deficiency: A review of our experience with 25 female patients. Fertil Steril 1992;58:129.
7. Morán C, Tena G, Herrera J, Bermúdez JA., Zárate A. Heterogeneity of late-onset adrenal 3β-ol-hydroxysteroid dehydrogenase deficiency in patients with hirsutism and polycystic ovaries. Arch Med Res 1994;25:315.
8. Azziz R, Bradley EL Jr, Potter HD, Boots LR. 3β-Hydroxysteroid dehydrogenase deficiency in hyperandrogenism. Am J Obstet Gynecol 1993;168:889.
9. Mathieson J, Couzinet B, Wekstein-Noel S, Nahoul K, Turpin G, Schiason G. The incidence of late-onset congenital adrenal hyperplasia resulting from 3β-hydroxysteroid dehydrogenase deficiency among hirsute women. Clin Endocrinol 1992;36:383.
10. Azziz R, Black V, Hines GA, Fox LM, Boots LR. Adrenal androgen excess in the polycystic ovary syndrome: sensitivity and responsivity of the hypothalamic-pituitary-adrenal axis. J Clin Endocrinol Metab 1998;83(7):2317–2323.
11. Zerah M, Rheaume E, Mani P, et al. No evidence of mutations in the genes for type I and type II 3β-hydroxysteroid dehydrogenase (3β-HSD) in nonclassical 3ß-HSD deficiency. J Clin Endocrinol Metab 1994;79:1811.
12. Chang YT, Zhang L, Alkaddour HS, et al. Absence of molecular defect in the type II 3β-hydroxysteroid dehydrogenase (3β-HSD) gene in premature pubarche children and hirsute female patients with moderately decreased adrenal 3β-HSD activity. Pediatr Res 1995;37:820.
13. Sakkal-Alkaddour H, Zhang L, Yang X, et al. Studies of 3β-hydroxysteroid dehydrogenase genes in infants and children manifesting premature pubarche and increased ACTH stimulation Δ5 steroid levels. J Clin Endocrinol Metab 1996;81:3961.
14. Lutfallah C, Wang W, Mason JI, et al. Newly proposed hormonal criteria via genotypic proof for type II 3beta-hydroxysteroid dehydrogenase deficiency. J Clin Endocrinol Metab 2002;87(6):2611–2622.
15. Pang S, Carbunaru G, Haider A, et al. The hormonal phenotype of nonclassic 3 beta-hydroxysteroid dehydrogenase (HSD3B) deficiency in hyperandrogenic females is associated with insulin-resistant polycystic ovary syndrome and is not a variant of inherited HSD3B2 deficiency. J Clin Endocrinol Metab 2004;89(2):783–794.
16. Zachmann M, Tassinari D, Prader A. Clinical and biochemical variability of congenital adrenal hyperplasia resulting from 11B-hydroxylase deficiency. A study of 25 patients. J Clin Endocrinol Metab 1983;56:222.
17. Cathelineau G, Brerault JL, Fiet J, Julien R, Dreux C, Canivet J. Adrenocortical 11 beta-hydroxylation defect in adult women with postmenarchial onset of symptoms. J Clin Endocrinol Metab 1980;51:287.
18. Carmina E, Malizia G, Janni A. Prevalence of late-onset 21-hydroxylase deficiency in hirsutism of western sicily. J Endocrinol Invest 1987;10:75.
19. Azziz R, Boots LR, Parker CR Jr, Bradley E Jr, Zacur HA. 11β-Hydroxylase deficiency in hyperandrogenism. Fertil Steril 1991;55:733.

20. Miller WL. Molecular biology of steroid hormone synthesis. Endo Rev 1988;9:295.
21. Joehrer K, Geley S, Strasser-Wozak EMC, et al. CYP11B1 mutations causing nonclassic adrenal hyperplasia resulting from 11β-hydroxylase deficiency. Hum Mol Genet 1997;6:1829.
22. Miller WL. Genetics, diagnosis, and management of 21-hydroxylase deficiency. J Clin Endocrinol Metab 1994;78:241.
23. Morel Y, Miller WL. Clinical and molecular genetics of congenital adrenal hyperplasia resulting from 21-hydroxylase deficiency. Adv Hum Genet 1991;20:1.
24. Chiou SH, Hu MC, Chung B. A missense mutation of Ile172 to Asn or Arg 356 to Trp causes steriod 21-hydroxylase deficiency. J Biol Chem 1990;256:3549.
25. Higashi Y, Hiromasa T, Tanae A, et al. Effects of individual mutations in the P-450c21 pseudogene on P-450c21 activity and their distrobution in patient genomes of congenital steroid 21-hydroxylase deficiency. J Biochem 1991;109:638.
26. Wu DA, Chung B. Mutations of P450c21 (steroid 21-hydroxylase) at Cys428, Val281, or Ser268 result in complete, partial, or no loss of enzymatic activity. J Clin Invest 1991;88:519.
27. Mornet E, Crete P, Kuttenn F, et al. Distribution of deletions and seven point mutations on CYP21B genes in three clinical forms of steroid 21-hydroxylase deficiency. Am J Hum Genet 1991;48:79.
28. Speiser PW, Dupont J, Zhu D, et al. Disease expression and molecular genotype in congenital adrenal hyperplasia resulting from 21-hydroxylase deficiency. J Clin Invest 1992;90:584.
29. Owerbach D, Sherman L, Ballard AL, Azziz R. Pro 453 to Ser mutation in CYP21 is associated with nonclassical steroid 21-hydroxylase deficiency. Mol Endocrinol 1992;6:1211.
30. Bobba A, Iolascon A, Giannattasio S, et al. A multicenter study of women with nonclassical congenital adrenal hyperplasia: relationship between genotype and phenotype. Mol Genet Metab 2000;71:527–534.
31. Krawczak M, Cooper DN. The human gene mutation database. Trends Genet 2003;13:121–122.
32. Azziz R, Slayden SM. Mechanisms of steroid excess in 21-hydroxylase deficient non-classic adrenal hyperplasia. J Soc Gynecol Invest 1996;3:297.
33. Mellon SH., Miller WL. Extra-adrenal steroid 21-hydroxylase is not mediated by P450c21. J Clin Invest 1989;84:1497.
34. Speiser PW, Agdere L, Veshiba H, White PC, New MI. Aldosterone synthesis in patients with salt-wasting congenital adrenal hyperplasia (21-hydroxylase deficiency) and complete absence of adrenal 21-hydroxylase (P450c21). N Engl J Med 1991;3221:145.
35. Sydnor KL, Kelley VC, Raile RB, Ely RS, Sayers G. Blood adrenocorticotrophin in children with congenital adrenal hyperplasia. Proc Soc Exp Biol Med 1953;82:695.
36. Cacciari E, Cicognani A, Pirazzoli P, et al. GH, ACTH, TSH, LH, and FSH reserve in prepubertal girls with congenital adrenal hyperplasia. J Clin Endocrinol Metab 1976;43:1146.
37. Richards GE, Grumbach MM, Kaplan SL, Conte FA. The effect of long acting glucocorticoids on menstrual abnormalities in patients with virilizing congenital adrenal hyperplasia. J Clin Endocrinol Metab 1978;47:1208.
38. Moreira AC, Elias LLK. Pituitary-adrenal responses to corticotropin-releasing hormone in different degrees of adrenal 21-hydroxylase deficiency. J Clin Endocrinol Metab 1992;74:198.
39. Feuillan P, Pang S, Schürmeyer T, Avgerinos PC, Chrousos GP. The hypothalamic-pituitary-adrenal axis in partial (late-onset) 21-hydroxylase deficiency. J Clin Endocrinol Metab 1988;67:154.
40. Carmina E, Lobo R. Pituitary-adrenal responses to corticotropin-releasing factor in late onset 21-hydroxylase deficiency. Fertil Steril 1990;54:79.
41. Ghizzoni L, Bernasconi S, Virdis R, et al. Dynamics of 24-hour pulsatile cortisol, 17-hydroxyprogesterone, and androstenedione release in prepubertal patients with nonclassic 21-hydroxylase deficiency and normal prepubertal children. Metabolism 1994;43:372.
42. Azziz R, Kenney PH. Magnetic resonance imaging of the adrenal gland in women with late-onset adrenal hyperplasia. Fertil Steril 1991;56:142–144.
43. Jaresch S, Kornely E, Kley HK, Schlaghecke R. Adrenal incidentaloma and patients with homozygous or heterozygous congenital adrenal hyperplasia. J Clin Endocrinol Metab 1992;74:685–689.
44. Kater CE, Biglieri EG, Wajchenberg B. Effects of continued adrenal corticotropin stimulation on the mineralocorticoid hormones in classical and nonclassical symbol virilizing types of 21-hydroxylase deficiency. J Clin Endocrinol Metab 1985;60:l057.
45. Waterman MR, Mason JI, Zuber MX, et al. Expression of P-450 enzyme activities in heterologous cells by transfection. Arch Toxicol 1989;13:155.
46. Lin D, Harikrishna JA, Moore CCD, Jones KL, Miller WL. Missense mutation Ser[106]→Pro causes 17α-hydroxylase deficiency. J Biol Chem 266:15992.
47. Sanchez LA, Moran C, Reyna R, Ochoa T, Boots LR, Azziz R. Adrenal progestogen and androgen production in 21-hydroxylase-deficient nonclassic adrenal hyperplasia is partially independent of adrenocorticotropic hormone stimulation. Fertil Steril 2002;77(4):750–753.
48. Huerta R, Dewailly D, Decanter C, Knochenhauer ES, Boots LR, Azziz R. Adrenocortical hyperresponsivity to adrenocorticotropic hormone: a mechanism favoring the normal production of cortisol in 21-hydroxylase-deficient nonclassic adrenal hyperplasia. Fertil Steril 2000;74(2):329–334.

49. Dewailly D, Vantyghem-Haudiquet MC, Sainsard C, et al. Clinical and biological phenotypes in late-onset 21-hydroxylase deficiency. J Clin Endocrinol Metab 1986;63:418.
50. Levine LS, Dupont B, Lorenzen F, et al. Cryptic 21-hydroxylase deficiency in families of patients with classical congenital adrenal hyperplasia. J Clin Endocrinol Metab 1980;51:1316.
51. Carmina E, Longo RA. Ovarian suppression reduces clinical and endocrine expression of late-onset congenital adrenal hyperplasia due to 21-hydroxlase deficiency. Fertil Steril 1994;62:738–743.
52. Forrest MG. Adrenal diseases and steroids. Curr Opinion Pediatr 1990;2:775–785.
53. Therrell BL Jr, Berenbaum SA, Manter-Kapanke V, et al. Results of screening 1.9 million Texas newborns for 21-hydroxylase-deficient congenital adrenal hyperplasia. Pediatrics 1998;101:583–590.
54. New MI, Speiser PW. Genetics of adrenal steroid 21-hydroxylase deficiency. Endocr Rev 1986;7:331–349.
55. Carmina E, Rosato F, Janni A, Rizzo M, Longo RA. Relative prevalence of different androgen excess disorders in 950 women referred because of clinical hyperandrogenism. J Clin Endocrinol Metab 2005, in press.
56. Azziz R, Sanchez LA, Knochenhauer ES, et al. Androgen excess in women: experience with over 1000 consecutive patients. J Clin Endocrinol Metab 2004;89:453–462.
57. Lucky AW, Rosenfield RL, McGuire J, Rudy S, Helke J. Adrenal androgen hyperresponsiveness to adrenocorticotropin in women with acne and/or hirsutism: adrenal enzyme defects and exaggerated adrenarche. J Clin Endocrinol Metab 1986;62:840–848.
58. Moran C, Azziz R, Carmina E, et al. 21-hydroxylase deficient non-classic adrenal hyperplasia is a progressive disorder: a multicenter study. Am J Obstet Gynecol 2000;183:1468–1474.
59. Levine LS, Dupont BO, Lorenzen F, et al. Cryptic 21-hydroxylase deficiency in families of patients with classical congenital adrenal hyperplasia. J Clin Endocrinol Metab 1980;51:1316–1324.
60. Speiser PW, New MI. Genotype and hormonal phenotype in nonclassical 21-hydroxylase deficiency. J Clin Endocrinol Metab 1987;64:86–91.
61. Deneux C, Tardy V, Dib A, et al. Phenotype-genotype correlation in 56 women with nonclassical congenital adrenal hyperplasia resulting from 21-hydroxylase deficiency. J Clin Endocrinol Metab 2001;86:207–213.
62. Speiser PW, Serrat J, New MI, Gertner JM. Insulin insensitivity in adrenal hyperplasia resulting from nonclassical steroid 21-hydroxylase deficiency. J Clin Endocrinol Metab 1992;75:1421–1424.
63. Saygili F, Oge A, Yilmaz C. Hyperinsulinemia and insulin insensitivity in women with nonclassical congenital adrenal hyperplasia resulting from 21-hydroxylase deficiency: the relationship between serum leptin levels and chronic hyperinsulinemia. Horm Res 2005;63:270–274.
64. Azziz R, Hincapie LA, Knochenhauer ES, Dewailly D, Fox L, Boots LR. Screening for 21-hydroxylase deficient non-classic adrenal hyperplasia among hyperandrogenic women: a prospective study. Fertil Steril 1999;72:915–925.
65. Azziz R, Rafi A, Smith BR, Bradley EL, Zacur HA. On the origin of the elevated 17-hydroxyprogesterone levels after adrenal stimulation in hyperandrogenism. J Clin Endocrinol Metab 1990.70:431–436.
66. Azziz R, Zacur HA. 21-Hydroxylase deficiency in female hyperandrogenism: Screening and Diagnosis. J Clin Endocrinol Metab 1989;69:577.
67. Katz E, Scherzer WJ, Mansfield RJ, Adashi EY. Effect of systemic hyperandrogenism on the adrenal response to adrenocorticotropin hormone. Fertil Steril 1994;61(3):567–569.
68. Fiet J, Gueux B, Gourmelen M, et al. Comparison of basal and adrenocorticotropin-stimulated plasma 21-desoxycortisol and 17-hydroxyprogesterone values as biological markers of late-onset adrenal hyperplasia. J Clin Endocrinol Metab 1988;66:659–667.
69. Solyom J, Gacs G, Keszei K, et al. Detection of late-onset adrenal hyperplasia in girls with peripubertal virilization. Acta Endocrinol 1987;115:413–418.
70. Kuttenn F, Couillin P, Girard F, et al. Late-onset adrenal hyperplasia in hirsutism. N Engl J Med 1985;313:224–231.
71. Huerta R, Dewailly D, Decanter C, Knochenhauer ES, Boots LR, Azziz R. 11-Beta-hydroxyandrostenedione and delta5-androstenediol as markers of adrenal androgen production in 21-hydroxylase deficient nonclassic adrenal hyperplasia. Fertil Steril 1999;72:996–1000.
72. New MI, Lorenzen F, Lerner AJ, et al. Genotyping steroid 21-hydroxylase deficiency: hormonal reference data. J Clin Endocrinol Metab 1983;57(2):320–326.
73. Knochenhauer ES, Cortet-Rudelli C, Cunningham RD, Boots LR, Dewailly D, Azziz R. Heterozygote for 21-hydroxylase (21-OH) deficient classic adrenal hyperplasia (CAH) women are not at risk for hyperandrogenism. J Clin Endocrinol Metab 1997;82:479–485.
74. Dewailly D, Robert Y, Helin I, et al. Ovarian stromal hypertrophy in hyperandrogenic women. Clin Endocrinol 1994;14:557–556.
75. Hague WM, Adams J, Rodda C, et al. The prevalence of polycystic ovaries in patients with congenital adrenal hyperplasia and their close relatives. Clin Endocrinol 1990;33:501–510.
76. Pache TD, Schadha S, Gooren LJG, et al. Ovarian morphology in long-term androgen-treated female to male transsexuals. A human model for the study of polycystic ovarian syndrome? Histopathology 1991;19:445–452.
77. Barnes RB, Rosenfield RL, Ehrmann DA, et al. Ovarian hyperandrogenism as a result of congenital adrenal virilizing disorders: evidence for perinatal masculinization of neuroendocrine function in women. J Clin Endocrinol Metab 79:1328–1333.

78. Moran C, Azziz R, Carmina E, et al. Reproductive outcome of women with 21-hydroxylase deficient non-classic adrenal hyperplasia: a multicenter study (in preparation).
79. Feldman S, Billaud L, Thalabard JC, et al. Fertility in women with late-onset adrenal hyperplasia resulting from 21-hydroxylase deficiency. J Clin Endocrinol Metab 1992;74:635–639.
80. Speiser PW, Krochenhauer ES, Dewailly D, Frazetti R, Marcondes JA, Azziz R. A multicenter study of women with nonclassical congenital adrenal hyperplasia: relationship between genotype and phenotype. Mol Genet Metab 2000;71:527–534.
81. Forest MG, Betuel H, David M. Prenatal treatment in congenital adrenal hyperplasia resulting from 21-hydroxylase deficiency : Update 88 of the French Multicentric Study. Endocr Res 1989;15:277–301.
82. Azziz R, Markham S, Huth J, Smith B, Zacur HA. Long-term follow-up of hyperandrogenic women with adrenal 21-hydroxylase deficiency. American Fertility Society, No. P-144, 1988.
83. Spritzer P, Billaud L, Thalabard JC, et al. Cyproterone acetate versus hydrocortisone treatment in late-onset adrenal hyperplasia. J Clin Endocrinol Metab 1990;70:642–646.

9

Pregnancy-Related Androgen Excess

Howard D. McClamrock

SUMMARY

Pregnancy-related androgen excess and virilization almost always arise from conditions occurring during pregnancy. It is often stressful and challenging for the clinician because of the need for a timely diagnosis in order to estimate the likelihood of virilization of an unborn female fetus. Androgen excess in pregnancy is relatively rare; the two most common causes are gestational luteoma and hyperreactio luteinalis (theca-lutein cysts of the ovaries). Increases in androgen production in pregnant women may cause hirsutism and virilization of the mother and at times virilization of a female fetus. The risk to the fetus depends on a number of factors, including the onset of increased maternal androgen production, the severity of the increase, and the condition causing the increase. Luteoma should be suspected in women with unilateral or bilateral (47%) solid ovarian masses, whereas Krukenberg tumor remains a possibility, especially in those with bilateral lesions. The natural course of the luteoma is regression and disappearance of the lesion after delivery, and it appears that elevated cord levels of testosterone and/or androstenedione are necessary for virilization of a female fetus based on a few reported cases. Hyperreactio luteinalis should be suspected in pregnancies complicated by androgen excess with bilateral cystic ovaries noted on exam or ultrasound. This condition does not appear to put the female fetus at risk for virilization even in situations in which circulating maternal androgens are high. As with luteoma, these cysts normally regress after delivery and should not require surgical therapy. Consideration should always be given to exogenous hormone therapy in pregnancies complicated by androgen excess. Unilateral solid ovarian lesions in pregnancies complicated by androgen excess increase the risk of malignancy. Current management is mostly limited to evaluating the risk to the fetus rather than implementing intervention that may alter the course of the pregnancy.

Key Words: Gestational hyperandrogenism; androgen excess in pregnancy; maternal/fetal masculinization and virilization; hyperreactio luteinalis (theca-lutein cysts); luteoma pregnancy complication.

1. INTRODUCTION

Pregnancy-related androgen excess and virilization are almost always the result of conditions arising during pregnancy. Hyperandrogenism usually results in infertility through anovulation, even when hirsutism or virilization is not clinically evident. Not surprisingly, clinically evident cases of pregnancy-related androgen excess are rare, making the condition interesting from a theoretical point but also challenging and urgent from a clinical point. The clinician may be called on to estimate the likelihood of virilization of an unborn female fetus while being faced with a lack of available knowledge to aid in this estimate. Immediate biochemical corroboration is desirable to assist the clinician in this diagnostic dilemma.

From: *Contemporary Endocrinology: Androgen Excess Disorders in Women:*
Polycystic Ovary Syndrome and Other Disorders, Second Edition
Edited by: R. Azziz et al. © Humana Press Inc., Totowa, NJ

Pregnancy-related androgen excess is a rare condition. The two most common causes are gestational luteomas and hyperreactio luteinalis (theca-lutein cysts of the ovary). Increases in androgen production in pregnant women may cause hirsutism and virilization of the mother and virilization of a female fetus. The risk of virilization to the female fetus depends on the time of onset of the increased maternal androgen production, the severity of the increase, and unknown factors.

This chapter will summarize the current available information regarding pregnancy-related androgen disorders and provide guidelines for diagnosis and therapeutic options.

2. BACKGROUND

2.1. Androgen Environment in Normal Pregnancy

Before discussing the physiology of pregnancy-related androgen excess, it is best to first review androgen production, transport, and metabolism in normal pregnant women. Pregnancy is associated with an increase in the circulating serum concentrations of some but not all androgens (1–6). Total testosterone elevations are noticeable in the first trimester of pregnancy, and further increases are observed through term (1,2,6). Increases in testosterone can be measured in normal pregnancies as early as 15 days after the luteinizing hormone surge, implying luteal phase ovarian production (7). The percentage of free testosterone remains fairly constant throughout pregnancy, ranging from 0.2 to 0.8%—slightly less than the fraction that is unbound in the nonpregnant state (1,8). Normal pregnancy testosterone levels vary somewhat in published reports, presumably because of assay variation. However, most studies report normal levels of total testosterone ranging from 100 to 150 ng/dL (3–6,8).

Much of the testosterone increases are thought to be a result of enhanced hepatic synthesis of sex hormone-binding globulin (SHBG) following estrogenic stimulation in pregnancy (5,6,9–11). In contrast to the total levels, the circulating levels of free testosterone are only slightly increased in the first and second trimesters, but increase approximately twofold in the third trimester (1). Elevations in serum SHBG concentrations do not explain the late increase in serum free testosterone concentrations seen in third trimester pregnancies. Although the cause of the increased testosterone is unknown, this rise can only be explained by an increase in testosterone production

One possibility is human chorionic gonadotropin (hCG) stimulation of the theca-interstitial cells of the ovary leading to testosterone synthesis. However, serum hCG concentrations reach a peak at the end of the first trimester and decline as serum free testosterone concentrations increase toward the end of pregnancy. The corpus luteum may also be a source of testosterone production, and this has been documented in subhuman primates (12). A third possibility is that testosterone may come from the maternal adrenal cortex, which would be in parallel with the increase in maternal cortisol secretion seen in late pregnancy.

Circulating levels of androstenedione have also been shown to be higher in the latter part of pregnancy (4). Dehydroepiandrosterone sulfate (DHEAS) concentrations decrease steadily in maternal circulation as pregnancy advances, falling to 50–30% of the levels found in nonpregnant women despite n increase in DHEAS production, which may be of fetal origin (11,13). The metabolic clearance rate of DHEAS increases sharply in the early second trimester and is associated with the greatest decrease in circulating levels (11,13). It has also been shown that the levels of 3α-17β-androstanediol glucuronide (3α-dio-G) are elevated in maternal serum. 3α-dio-G is a product of dihydrotestosterone (DHT) in hair follicles of the skin, and it is unclear whether the increase in level is a result of increased activity of 5α-reductase and other androgen-metabolizing enzymes or increased substrate availability.

2.2. Gestational Protection Against Maternal and Fetal Virilization

Pregnancy is associated with clear-cut increases in total and free androgens, yet most women and their infants are not virilized. Several mechanisms have been postulated to account for this phenomenon (14).

2.2.1. Increases in Serum SHBG Concentration

As noted earlier, serum-free androgens increase relatively little and only late in pregnancy. Earlier elevations in circulating androgen concentrations are for the most part increases in SHBG-bound androgens, with relatively little free androgen available for action at the target tissues, especially during the time of fetal development.

2.2.2. Competition for Androgen–Receptor Binding by Placental Progesterone

Progesterone production increases dramatically during pregnancy (approximately 10-fold) and may approach 250 mg per day at term. Androgen receptors are thought to have a low affinity for progesterone; nevertheless, such a massive increase in progesterone production may lead to inhibition of testosterone binding to the androgen receptor.

2.2.3. Competition for Androgen Activation in Target Tissues by Placental Progesterone

5α-Reductase acts to convert testosterone to the more biologically potent DHT in target tissues. There is a weak affinity of progesterone for the 5α-reductase enzyme, but excessive amounts of progesterone may be sufficient to inhibit the conversion of testosterone to DHT.

2.2.4. Aromatization of Androgens by the Placenta

It has been demonstrated that the placenta has a massive ability to convert androgens to estrogens *(15)*. It is therefore likely that the placenta affords a protective mechanism to both the mother and the female fetus from virilization by metabolizing androgens to estrogens. In one report *(16)*, the maternal concentration of testosterone was 15,000 ng/dL, while the cord level was 252 ng/dL, or 1.7% of the maternal level. The estradiol level in the cord serum was elevated compared to that of the mother, suggesting that testosterone was converted to estradiol as part of a protective mechanism against the passage of testosterone from mother to fetus. Other evidence consistent with the placental protective mechanism hypothesis includes the observation that female infants in pregnancies where there is a deficiency of placental aromatase are virilized *(17,18)*. In one such infant, cord serum levels of testosterone, DHT, and androstenedione were markedly elevated compared with normal infants, whereas estrone, estradiol, and estriol concentrations were somewhat lower than expected.

The findings just discussed suggest that aromatization diminishes the exposure of the fetus to androgens but does not fully explain the lack of fetal virilization. In the case mentioned earlier with a cord serum testosterone concentration of 252 ng/dL, the female infant was not virilized despite androgen levels that would normally be sufficient for virilization. There have also been reports of pregnancies complicated by luteomas of the ovaries in which high serum DHT concentrations were associated with normal female infants *(19–21)*. DHT cannot be aromatized to estrogens, which suggests that female fetuses are also likely protected against virilization in normal pregnancies by some form of decreased androgen action.

2.3. Gestational Hyperandrogenism

Most reported cases of virilization in pregnancy have been the result of ovarian pathology arising during pregnancy, such as luteomas or hyperreactio luteinalis (gestational ovarian theca-lutein cysts), or rarely from iatrogenic insult. The incidence of hyperandrogenism during pregnancy is even lower than the incidence of ovarian tumors, which may be associated with this condition. The true incidence is unclear because many of these tumors go unnoticed. In the case of luteoma, most reported cases were discovered incidentally at the time of laparotomy in asymptomatic pregnant patients. A significant number may go entirely unrecognized because most luteomas regress in the postpartum period. Such circumstances are thought to decrease the reported incidence of luteomas.

Hyperreactio luteinalis is another condition that may be associated with pregnancy-related androgen excess. Multiple pregnancies and pregnancies associated with hydatidiform mole, choriocarcinoma, diabetes, and isoimmunization have an increased frequency of hyperreactio luteinalis. Mothers with preexisting hirsutism such as those with polycystic ovary syndrome (PCOS) are more likely to have hyperreactio luteinalis as the source of their androgen excess. Trophoblastic disease is associated with 10–20% risk of gross theca-lutein cysts. As with luteoma, this estimate may be inaccurate because it is based on cases in which tumors were demonstrated radiologically, at laparotomy, or on examination. Hyperreactio luteinalis may also at times occur in ovaries of normal size or in ovaries that are only slightly enlarged, and careful histological examination of ovarian tissue is necessary for the unequivocal diagnosis.

Other neoplasms associated with hyperandrogenism in pregnancy are even less common. Sertoli–Leydig cell tumors (arrhenoblastomas) have been reported in 18 pregnant women *(22–24)*. The rarity of these slow-growing tumors likely reflects the effects of hyperandrogenism and its association with anovulation, although there have been a few cases reported in which the tumor was thought to have been present at the time of conception. Virilization is common (69–87%) in both pregnant and nonpregnant women *(22)*, but the risk of malignancy in Sertoli–Leydig cell tumor cells is higher in pregnancy than in a nonpregnant state (40–50% vs 12–22%) *(22)*. The tumors have also been associated with substantial maternal (31%) and perinatal (50%) mortality.

Krukenberg tumors are primary tumors of the gastrointestinal tract metastatic to the ovary that in some way activate the surrounding stroma. They are solid tumors, generally occur bilaterally (80%), and have occurred in only a few cases in pregnancy. Most of the women had hirsutism or virilization, and half had virilized infants.

There have been sporadic reports of other ovarian tumors associated with hyperandrogenism in pregnancy, including Brenner, Leydig cell, granulosa cell, and Y-cell tumors; mucinous and serous cystadenocarcinomas; and dermoid cysts. The true incidence of these tumors in pregnancy is difficult to validate because of the small numbers. Maternal adrenal tumors (three adenomas and one adrenocortical carcinoma) have also been associated with virilization of female infants *(25)*. There have been numerous reports of Cushing's syndrome in pregnancy, but there has been no mention of fetal masculinization or virilization *(26)*. Most of the mothers were not masculinized, and most of the pregnancies did not progress to term. PCOS has been associated with maternal virilization in pregnancy, although very few cases have been described despite the relatively common occurrence of PCOS *(27)*. One case reported associated fetal virilization, but the PCOS was not confirmed histologically.

Circulating androgen levels in pregnant patients with PCOS have been compared to normally pregnant women *(28)*. At 10–16 weeks of gestation, levels of androstenedione, testosterone, DHEAS, and the free androgen index tended to be higher in the PCOS group. These differences became significant when the studies were repeated at 22–28 weeks. Insulin levels 2 hours after an oral glucose tolerance test were also significantly higher in PCOS than in normally pregnant women at 10–16 and 22–28 weeks gestation. The same group had previously demonstrated that the ovarian volumes of PCOS patients were increased after delivery, suggesting that the ovaries were persistently stimulated during pregnancy *(29)*. It has been suggested that increases in insulin levels may stimulate androgen production, but that the fetal-placental unit is functioning normally with its protection mechanism in place. The gestational administration of progestins, androgens, and even estrogens has been associated with several cases of masculinization of female fetus *(14,30,31)*. A virilized female fetus has also been associated with placental aromatase deficiency *(17)*. A recent report describes a pregnancy associated with severe virilization in the second trimester that ended in an uneventful delivery of a normal male child. Maternal androgens were elevated, but fetal levels were normal and there was normal placental aromatase activity. Ultrasound revealed no evidence of ovarian lesions. There was dramatic resolution of both the androgen levels and the virilization postnatally. This case was thought to be associated with ovarian androgen elevations not associated with an ovarian mass *(32)*.

2.4. Etiologies of Gestational Hyperandrogenism

2.4.1. Luteoma

Luteoma was first described as a pathological entity by Sternberg in 1963 *(33)*. Luteomas are benign solid ovarian lesions (not true tumors) made up of hyperplastic masses of large lutein cells.

They range in size from 1 to 25 cm in diameter, with most being from 6 to 10 cm. Garcia-Bunuel reviewed 20 cases, finding 9 (45%) to be bilateral *(34,35)*. This incidence of bilaterality is confirmed (47%) when reviewing all the cases reported in the literature. Luteomas regress and disappear after delivery, and many may be unnoticed because they secrete little androgen or have minimal hormonal effects. Some subclinical luteomas are discovered incidentally at the time of cesarean section or other surgery for unrelated reasons. There is some evidence that the incidence is increased in the African American population *(36)*. Increased androgen secretion (e.g., urinary 17-keto steroids) was suggested by earlier reports of virilization associated with luteomas. Elevated levels of testosterone, androstenedione, and DHT have been demonstrated in pregnant women with luteomas regardless of the virilizing effect of the tumor *(19,20,35,37,38)*. Maternal hirsutism or virilization has been seen in approximately 30–35% of reported cases of pregnancy luteoma *(36,37)*, suggesting that either serum androgen concentrations are often not very elevated or that androgen action is blunted. Increases in SHBG concentrations that occur normally during pregnancy may be important to decrease the amount of free testosterone available at the tissue level. High levels of circulating androgens may not necessarily imply maternal virilization. This was convincingly demonstrated in a study in which two hyperandrogenic but nonvirilizing patients were found to have luteomas at the time of laparotomy. Circulating testosterone levels were 2830 ng/dL in one patient and 2650 ng/dL in the other, with normal pregnancy levels ranging from 50 to 120 ng/dL *(19)*. Approximately 75% of female infants born to virilized mothers with luteomas are virilized, whereas in the absence of maternal virilization, fetuses are not virilized. Exposure of male infants to excess androgens from luteomas has not been associated with physical signs of hyperandrogenism.

In estimating the risk of virilization to a fetus *in utero*, several factors merit consideration, including the sex of the fetus, the duration of exposure to the androgens, and most importantly the state of pregnancy in which the exposure took place *(20,39)*. Cord serum androgen concentrations have been measured in only a few cases of infants whose mothers had marked hyperandrogenism as a result of luteomas in pregnancy (Table 1). Normal cord serum androgen concentrations were found in three infants—a male and two nonvirilized females *(19,40)* —whereas elevated cord serum testosterone levels were found in a virilized female infant *(20)* and a premature nonvirilized male *(41)*. Based on these limited data, it appears that the risk of virilization of a female infant is correlated with serum testosterone concentrations in the fetus, but not the mother. Therefore, maternal hyperandrogenemia is a necessary but not sufficient cause of virilization of the female infant, suggesting that other factors such as placental androgen degradation may play a role.

2.4.2. Hyperreactio Luteinalis

The ovaries in hyperreactio luteinalis (i.e., gestational ovarian theca-lutein cysts) generally contain multiple follicular cysts beneath a focal, hyalinized, thickened cortex. Pregnancies associated with multiple or isoimmunized gestations, trophoblastic disease or molar pregnancies, and diabetes mellitus are thought to have an increased risk of gestational theca-lutein cysts. The highest incidence is thought to be associated with trophoblastic disease, where the risk ranges from 10 to 22%; however, as with luteoma, the incidence is likely higher than reported. The risk of maternal hirsutism or virilization in pregnancies complicated by hyperreactio luteinalis is thought to be approximately 30%, a rate that is similar to that for luteomas *(14,42–44)*. Many if not all virilized women have shown high serum testosterone and androstenedione concentrations, whereas cord serum testosterone concentrations from newborns of virilized mothers are either normal or increased *(3,14,16,42,43)*. Despite elevations in cord testosterone levels in some infants associated with maternal virilization, none of the female infants reported have been virilized *(37)*. One possible exception was not biopsy confirmed *(45)*.

The reasons why female infants in pregnancies complicated by theca-lutein cysts remain protected from virilization are not clear. Given that elevated cord blood androgens have been observed, it would seem that placental aromatization of androgens did not entirely protect the fetus from expo-

Table 1
Androgen Levels in Cord Blood of Infants of Mothers With Pregnancy Luteomas

Reference	T (ng/dL)	DHT (ng/dL)	A (ng/dL)	DHEA (ng/dL)	DHEAS (μg/dL)	Maternal hirsutism or virilization	Fetal virilization	Sex
42	461	ND	648	ND	ND	Present	Absent	Male
19								
Case 1	26	Not detected	27	ND	ND	Absent	Absent	Female
Case 2	35	5	110	ND	ND	Absent	Absent	Male
20	109	48	2400	240	162	Present	Present	Female
41	35	—	—	—	—	Present	Absent	Female
Normal cord blood[a]	13–57	4–40	93–126	203–503	98			

T, testosterone; DHT, dihydrotestosterone; A, androstenedione; DHEA, dehydroepiandrosterone; DHEAS, dehydroepiandrosterone sulfate; ND, not determined.
[a]Normal cord blood levels obtained from refs. *4–6, 14,* and *46.* (Adapted from ref. *14.*)

sure to elevated androgens. In one case elevated cord blood androgens were associated with a 17β-estradiol level of 33, 000 pg/mL (about —five to seven times the normal value). The high estrogen effect may maintain elevated SHBG levels, thereby lowering free androgen concentrations. Therefore, current data would suggest that neither maternal virilization nor fetal hyperandrogenemia is related to fetal virilization in this disorder.

2.4.3. Other Ovarian Lesions

Hyperandrogenism in pregnancy has been associated with numerous ovarian tumors, including Sertoli–Leydig cell tumors, Krukenberg tumors, Brenner cell tumors, lipoid cell tumors, and sclerosing stromal cell tumors. Sertoli–Leydig cell tumors are rare in pregnancy, reflecting the effects of hyperandrogenism on ovulation, but there have been a few cases of the tumor thought to be present at the time of conception. The risk of malignancy in Sertoli–Leydig cell tumors is thought to be higher in pregnancy (40–50% vs 12–22%) than in the nonpregnant state. Maternal virilization has been seen in approximately 69–87% of the tumors, a rate that is similar to that found in nonpregnant women. In the few cases with surviving female infants, one virilized infant was born to a virilized mother and two other infants were born to mothers with nonvirilizing tumors. Therefore, the risk of fetal virilization is thought to be high in virilized mothers with Sertoli–Leydig cell tumors.

Krukenberg tumors are not primary ovarian tumors but are metastatic lesions within the ovary from primary tumors of the gastrointestinal tract. These are solid tumors that are bilateral approximately 80% of the time. In the few cases reported in the literature, the risk of maternal and female fetal virilization was thought to be approximately 80–100% *(37,46)*. There have been a few isolated reports of other virilizing ovarian tumors and also a few reports of virilizing adrenal tumors in pregnancy *(25)*. The small number of these cases makes it difficult to comment on their incidence and risk of virilization.

PCOS with no associated ovarian tumor has been associated with virilization in pregnancy, although the number of cases is small given the relatively common occurrence of the condition *(27)*.

2.4.4. Placental Aromatase Deficiency

Placental aromatase deficiency may be associated with irreversible virilization in female infants, but maternal masculinization would be expected to regress in the postpartum period.

2.4.5. Iatrogenic Causes

Masculinization of female infants has been reported secondary to the gestational administration of synthetic androgens, progestins, and diethylstilbestrol *(30,31,47,48)*. Similar outcomes may be seen with placental aromatase deficiency in which the conversion of androgen to estrogen is decreased, leading to androgen accumulation *(17,18)*. In the cases reporting masculinized female infants, the medications were generally begun in the first trimester. Not all of these mothers demonstrated overt signs of virilization, and these cases may be exceptions where masculinization of the fetus is unaccompanied by maternal virilization. The exact dose of the virilizing drug needed to cause masculinization of the female fetus remains unknown. There have been no reports of virilization of female infants born to women who have continued oral contraceptive pills after conception *(49)*. External female genital development occurs between 7 and 12 weeks of gestation, and it is at this time that androgen exposure may lead to partial or complete labial fusion and clitoral hypertrophy. After 12 weeks of gestation, clitoral hypertrophy remains a risk, but labial fusion is not thought to occur. Male fetuses do not appear to be affected.

Interestingly, exposure to exogenous androgens during pregnancy does not invariably result in masculinization of the female fetus. Large doses of methylandrostenediol were given to a pregnant patient beginning in the first trimester for treatment of breast cancer with no evidence of virilization of the patient or the female infant *(50)*. It is difficult at this time to advise pregnant patients exposed to androgens because of the lack of data in this area.

Table 2
Clinical Correlations of Gestation-Associated Tumors

Clinical parameter	Luteoma	Theca-lutein cyst	Sertoli–Leydig cell	Krukenberg tumor
Incidence (%)	Unknown	Unknown	Rare	Rare
Bilaterally (%)	~47	~96	~5	~80
Maternal virilization (%)	~36	~30	69-87	100
Female fetal virilization (%)	~75	0	High if mother is virilized	80–100
Malignancy (%)	0	0	~44	100

Adapted from ref. *14.*

2.5. Diagnostic Evaluation

Virilization in pregnancy deserves investigation because of the possibility that a masculinizing lesion may be malignant and the condition may cause hirsutism and virilization in women and virilization of the female fetus (Table 2). Sertoli–Leydig cell tumors, granulosa theca cell tumors, Krukenberg tumors, ovarian mucinous cystadenocarcinomas, and adrenocortical carcinomas are examples of malignant tumors that may be associated with gestational hyperandrogenism. Cushing's syndrome and placental aromatase deficiency, although rare in pregnancy, may need to be considered.

2.5.1. History and Physical Examination

Androgen excess in pregnancy is considered when a pregnant women presents with a rapid onset of masculinization. The signs and symptoms may include hirsutism, temporal balding, acne, clitoromegaly, and deepening of the voice. Some conditions such as luteomas, hyperreactio luteinalis, and other ovarian and adrenal tumors may be associated with palpable pelvic or abdominal masses. Special attention should be paid to the timing of the onset of virilization and to any history of drug exposure. The potential risk of fetal virilization and the likely causes of hyperandrogenism may be estimated by considering the time of onset of maternal symptoms. The fetal risk is related to the duration of exposure to the androgens, the sex of the fetus, and, most important, the stage of pregnancy during which the exposure takes place *(20,39)*. Exposure to androgens early in pregnancy (7–12 weeks) may cause labioscrotal fusion and clitoromegaly, which may irreversible and require surgical correction if the child is to be raised as a female. Androgen exposure after 12 weeks of gestation generally does not cause labial fusion, but clitoral hypertrophy may be apparent. Chronic hirsutism (as seen with PCOS) is more likely to be associated with the clinical findings of theca-lutein cysts, although the gestational recurrence of virilizing tumors has been reported. Theca-lutein cysts are also associated with multiple gestation, fetal hydrops, and trophoblastic disease, all of which should be excluded.

2.5.2. Sonography

Ultrasound may help differentiate cystic from solid tumors, unilateral from bilateral lesions, and ovarian from adrenal pathology. This information may be helpful in that theca-lutein cysts are almost always bilateral, whereas luteomas are bilateral in approximately half of cases. Other ovarian tumors are usually unilateral, with the exception of the Krukenberg tumor, which is more often bilateral. Therefore, a unilateral mass, particularly if solid, increases the likelihood of malignancy. Currently, the ability of ultrasonography to predict the fetal sex is not reliable until after the most crucial time for androgen exposure in the developing fetus, which is at the time of genital differentiation.

2.5.3. Diagnostic Laparoscopy

There is little information on the use of laparoscopy in the diagnosis of gestational hyperandrogenism, but it stands to reason that laparoscopy with ovarian biopsy may potentially be useful if virilization is observed in the first trimester. Laparoscopic decompression of large theca-lutein cysts has been reported *(51)*.

2.5.4. Maternal Hormonal Determinations and Umbilical Cord Blood Sampling

Maternal serum androgen determinations are of questionable benefit because there are elevations in androgens in normal pregnancy and even extremely elevated levels are not always associated with fetal virilization. Furthermore, if maternal virilization is caused by exogenous hormone administration, circulating levels may be low because the offending hormone is often a synthetic steroid that may not be measured in assays for testosterone or other androgens. High circulating maternal androgen levels appear necessary for fetal virilization, but virilization may be absent even in the face of high circulating maternal or cord blood androgen levels. Percutaneous umbilical blood sampling for testosterone, androstenedione, and karyotype could theoretically provide prognostic information in women who have conditions that would be associated with fetal virilization if the procedure could be done early enough in gestation. Currently, however, this procedure can not be reliably and safely performed prior to the period of differentiation of the fetal genitalia, the time at which the damage is done.

2.6. Prognosis and Treatment

The prognosis and therapeutic approach to women with gestational hyperandrogenism is determined in large part by the natural history of the underlying cause.

2.6.1. Luteomas

Luteomas should be suspected in women with bilateral solid ovarian masses, but a Krukenberg tumor is still a possibility. With luteoma the mother can expect regression and disappearance of the lesion after delivery. It appears that elevated cord levels of testosterone and or androstenedione are necessary for virilization of a female fetus in luteoma-complicated pregnancies, based on the few reported cases. Therefore, in the face of normal cord androgens, the risk of virilization of the female infant would appear to be much lower, and there may perhaps be no risk at all. If percutaneous umbilical blood sampling revealed a normal female XX karyotype and elevated cord androgens, the mother could be informed of the potential high risk for fetal virilization (*see* Table 2). However, because this information currently cannot be obtained early enough in gestation for antepartum therapeutic interventions, later percutaneous umbilical blood sampling is not recommended.

2.6.2. Hyperreactio Luteinalis

In pregnancies complicated by theca-lutein cysts, the fetus does not appear to be at risk for virilization even if maternal serum androgen concentrations are high. As with luteomas, these cysts regress after delivery and should not require surgical therapy.

2.6.3. Other Tumors

Unilateral solid tumors in pregnancy associated with virilization have an increased risk (approaching 50%) of malignancy compared to nonvirilizing tumors.

2.6.4. Exogenous Hormone Therapy

The risk of maternal and fetal virilization from exogenous hormone therapy remains unknown. The fetal risk would likely depend on the time of exposure, as described earlier, with early exposure being associated with the risk of labioscrotal fusion. There is no evidence that measurements of androgens and progestins in the cord blood are helpful in determining the risk to the fetus, and maternal hirsutism or virilization should regress after cessation of the hormone administration.

3. CONCLUSIONS

Luteomas or hyperreactio luteinalis (theca-lutein cysts) represent the most frequent causes of gestational hyperandrogenism. Luteomas are solid tumors occurring bilaterally in approximately 47% of the cases. Mothers with luteomas can expect regression and disappearance of the lesion as well as improvement of the symptoms in the postpartum period. Luteomas are associated with maternal masculinization in approximately 36% of cases, and the risk of virilization of the female fetus in cases associated with maternal masculinization is approximately 75%. The risk of virilization of the female fetus appears to be based on the onset of the maternal affliction.

Hyperreactio luteinalis (theca-lutein cysts) are cystic, almost always bilateral, and regress during the postpartum period. There is an approximately 30% risk of maternal hirsutism or virilization associated with theca-lutein cysts, but there is no association with fetal masculinization or virilization, even in the face of elevated maternal and cord androgens. This condition is relatively increased compared to other androgenizing lesions in mothers with long-term hirsutism such as PCOS.

Malignancy should be suspected in unilateral solid ovarian lesions in pregnancy, but bilateral tumors do not exclude malignancy. Krukenberg tumors are malignant tumors that are bilateral in approximately 80% of cases. In general, ovarian tumors occurring during pregnancy have a 6.7% risk of malignancy. If virilization accompanies a unilateral ovarian gestational lesion, however, the malignancy rate increases to approximately 50%.

Little is known about hormonal exposure/iatrogenic etiology and the risk of maternal androgenization and fetal virilization. As with other conditions, the risk to the fetus is thought to be related to the time of exposure, and maternal masculinization would be expected to regress after cessation of hormone administration.

Antepartum preventative therapy is not yet available despite our ability to recognize the clinical manifestations of pregnancy-related androgen excess and the technology available to evaluate fetal androgen exposure. Therefore, we are limited to evaluating the risk to the fetus rather than practicing intervention that may alter the course of the pregnancy.

4. FUTURE AVENUES OF INVESTIGATION

Future efforts should be aimed at developing appropriate screening tests that would allow physicians to assess the risk for developing gestational hyperandrogenism early in pregnancy. Along with earlier diagnosis modalities, it may be possible to develop therapies that would decrease androgen exposure, thereby limiting fetal exposure to androgens during the critical period of sex determination.

KEY POINTS

- Several protective mechanisms exist to protect the fetus from maternal androgen excess.
- Luteomas and theca-lutein cysts represent the most frequent causes of gestational hyperandrogenism:
  o Luteomas are solid unilateral or bilateral lesions. They regress after pregnancy but are associated with virilization of the female infant.
  o Hyperreactio luteinalis (theca-lutein cysts) are usually bilateral and are relatively increased in mothers with preexisting hirsutism, multiple pregnancy, gestational trophoblastic disease, diabetes, and isoimmunization. Regression of the lesions is seen in the postpartum state, and fetal virilization is not seen.
- Gestational hyperandrogenism associated with a unilateral solid lesion is associated with an increased risk for malignancy.
- While diagnostic studies may give information about the risk of fetal virilization, these studies are generally unavailable until after the time of sexual differentiation.
- Future therapeutic modalities should be aimed at earlier diagnosis and treatment to limit androgen exposure.

REFERENCES

1. Bammann BL, Coulan CB, Jiang N-S. Total and free testosterone during pregnancy. Am J Obstet Gynecol 1980;137:293.
2. Berger NG, Tepke JT, Woodrull JD. Markedly elevated serum testosterone in pregnancy without fetal virilization. Obstet Gynecol 1984;63:260.
3. Dawood MT, Saxena BB. Testosterone and dihydrotestosterone in maternal and cord blood and in amniotic fluid. Am J Obstet Gynecol 1977;129;37.
4. Mizuno M, Lobotsky J, Lloyd CW, Kobayashi T, Murasawa Y. Plasma androstenedione and testosterone during pregnancy and in the newborn. J Clin Endocrinol Metab 1968;28:1133.
5. Rivarola MA, Forest MG, Migeon CJ. Testosterone, androstenedione and dehydroepiandrosterone in plasma during pregnancy and at delivery: Concentration and protein binding. J Clin Endocrinol Metab 1968;28:34.
6. Saez JM, Forest MG, Morera A, Bertrand J. Metabolic clearance rate and blood production rate of testosterone and dihydrotestosterone in normal subjects, during pregnancy, in hyperthyroidism. J Clin Invest 1972;51:1226.
7. Castracane VD, Stewart DR, Gimpel T, et al. Maternal serum and androgens in human pregnancy: early increases with the cycle of conception. Human Reprod 1998;13:460.
8. Anderson DC, Lasley BL, Fisher RA, Shepherd JH, Newlam L, Hendrickx AG. Transplacental gradients of sex-hormone-binding globulin in human and simian pregnancy. J Clin Endocrinol Metab 1976;5:657.

9. Vermeulen A, Verdonch L, an Der Straeten M, Orie N. Capacity of the testosterone-binding globulin in human plasma and influence of specific binding of testosterone on its metabolic clearance rates. J Clin Endocrinol Metab 1969;29:1470.

10. Pearlman WH, Crepy O, Murphy M. Testosterone-binding levels in the serum of women during normal menstrual cycle, pregnancy and the post-partum period. J Clin Endocrinol Metab 1967;27:1012.

11. Tulchinsky D, Ryan KJ. Maternal-Fetal Endocrinology. Philadephia: W.B. Saunders, 1980.

12. Treloar OL, Wolf RC, Meyer RK. The corpus luteum of the rhesus monkey during late pregnancy. Endocrinology 1972;91:665.

13. Milewich L, Gomez-Sanchez C, Madden JD, et al. Dehydroisoandrosterone sulfate in peripheral blood premenopausal, pregnant and postmenopausal women and men. J Steroid Biochem 1978;9:1159.

14. McClamrock HD, Adashi EY. Gestational hyperandrogenism. In: Adashi EY, Rock JA, Rosenwaks Z, eds. Reproductive Endocrinology, Surgery, and Technology: A Textbook in Reproductive Medicine. Philadelphia: Lipincott-Raven Publishers, 1966:1599.

15. Edman CD, Toofanian A, MacDonald PC, Gant NF. Placental clearance rate of maternal plasma androstenedione through placental estradiol formation: an indirect method of assessing uteroplacental blood flow. Am J Obstet Gynecol 1981;141:1029.

16. Hensleigh PA, Carter RP, Grotjan HE Jr. Fetal protection against masculinization with hyperreactio luteinalis and virilization. J Clin Endocinol Metab 1975;40:816.

17. Shozu M, Akasofu K, Harada T, Kubota Y. A new cause of female pseudohermaphroditism: placental aromatase deficiency. J Clin Endocrinol Metab 1991;72:560.

18. Conte FA, Gumbach MM, Ito Y, et al. A syndrome of female pseudohermaphroditism, hypergonadotropic hypogonadism, and multicystic ovaries associated with missense mutations in the gene encoding aromatase (P450arom). J Clin Endocrinol Metab 1994;78:1287.

19. Nagamani M, Gomez LG, Garza J. In vivo steroid studies in luteoma of pregnancy. Obstet Gynecol 1982;59:105S.

20. Verkauf BS, Reiter EO, Hernandez L, Burns SA. Virilization of mother and fetus associated with luteoma of pregnancy: a case report with endocrinologic studies. Am J Obstet Gynecol 1977;129:274.

21. Jewelewicz R, Perkins RP, Dyrenfurth I, Vande Wiele RL. Luteomas of pregnancy: a cause for maternal virilization. Am J Obstet Gynecol 1971;109(1):24.

22. Galle PC, McCool JA, Elsner CW. Arrhenoblastoma during pregnancy. Obstet Gynecol 1978;51:359.

23. Barkan A, Cassorla F, Loriaux DL, Marshall JC. Pregnancy in a patient with virilizing arrhenoblastoma. Am J Obstet Gynecol 1984;149:901.

24. Widschwendter M, Meduri G, Loosfelt H, et al. Fulminant recurrence of a Sertoli-Leydig cell tumour during pregnancy. Br J Obstet Gynaecol 1999;106:284.

25. Kirk JMW, Perry LA, Shand WR, Kirby RS, Besser GM, Savage MO. Female pseudohermaphroditism due to a maternal adrenocortical tumor. J Clin Endocrinol Metab 1990;70:1280.

26. Buescher MA, McClamrock HD, Adashi EY. Cushing syndrome in pregnancy. Obstet Gynecol 1992;79:130.

27. Ben-Chetrit A, Greenblatt EM. Recurrent maternal virilization during pregnancy associated with polycystic ovary syndrome: a case report and review of the literature. Hum Reprod 1995;10:3057.

28. Sir-Petermann T, Maliqueo M, Angel B, Laura HE, Perez-Bravo F, Recabarren SE. Maternal serum androgens in pregnant women with polycystic ovary syndrome: possible implications in prenatal androgenization. Hum Reprod 2002;17:2573.

29. Sir-Petermann T, Devoto L, Maliqueo M, Peirano P, Recabarren SE, Wildt L. Resumption of ovarian function during lactational amenorrhea in breastfeeding women with polycystic ovary syndrome: endocrine aspects. Hum Reprod 2001;16:1603.

30. Wilkins L, Jones Jr HW, Holman GH, Stempfel RS. Masculinization of the female fetus associated with administration of oral and intramuscular progestins during gestation: Non-adrenal female pseudohermaphrodism. J Clin Endocrinol Metab 1958;18:559.

31. Grumbach MM, Ducharme JR, Moloshok RE. On the fetal masculinizing action of certain oral progestins. J Clin Endocrinol Metab 1959;19:1369.

32. Holt HB, Medbak S, Kirk D, et al. Recurrent severe hyperandrogenism during pregnancy: a case report. J Clin Pathol 2005;58(4):439.

33. Sternberg WH: Non-functioning ovarian neoplasma. In: Grady HG, Smith DE, eds. The Ovary. International Academy of Pathologists, monogram 3. Baltimore: Williams & Wilkins, 1963:289.

34. Garcia-Bunuel R, Berek JS, Woodruff JD. Luteomas of pregnancy. Obstet Gynecol 1975;45:407.

35. Garcia-Bunuel R, Berek JS, Woodruff JD. Tumors, including endometriosis: luteomas of pregnancy. Obstet Gynecol Surv 1975;45:635.

36. Joshi R, Dunaif A. Ovarian disorders of pregnancy. Endocrinol Metab Clin North Am 1995;24:153.

37. McClamrock HD, Adashi EY. Gestational hyperandrogenism. Fertil Steril 1992;57:257.

38. Wolff E, Glasser M, Gordon GG, et al. Virilization luteoma of pregnancy. Report of a case with measurements of testosterone and testosterone binding in plasma. Am J Med 1973;54:229.

39. Grumbach MM, Ducharme JR. The effects of androgens on fetal sexual development: Androgen-induced female pseudohermaphroditism. Fertil Steril 1960;11:157.
40. Illingworth PJ, Johnstone FD, Steel J, Seth J. Luteoma of pregnancy: masculinization of a female fetus preventd by placental aromatization. Br J Obstet Gynaecol 1992;99(12):1019.
41. Baxi L, Holub D, Hembree W. Bilateral luteomas of pregnancy in a patient with diabetes. Am J Obstet Gynecol 1988;159:454.
42. Bradshaw KD, Santos-Ramos R, Rawlins SC, et al. Endocrine studies in pregnancy complicated by ovarian theca lutein cysts and hyperreactio luteinalis. Obstet Gynecol 1986;67:66S.
43. Muechler EK, Fichter J, Zongrone J. Human chorionic gonadotropin, estradiol, testosterone changes in two pregnancies with hyperreactio luteinalis. Am J Obstet Gynecol 1987;157:1126.
44. Wajda KJ, Lucas JG, Marsh WL. Hyperreactio luteinalis. Arch Pathol Lab Med 1989;113;921.
45. Bilowus M, Abbassi V, Gibbons MD. Female pseudohermaphroditism is a neonate born to a mother with polycystic ovary disease. J Urol 1986;136:1098.
46. Ritter DB, McGill FM, Greston WM. Krukenberg tumor part II: identification during pregnancy. Female Patient 1999;24:19.
47. Duck SC, Katayama DP. Danazol may cause female pseudohermaphroditism. Fertil Steril 1981;35:230.
48. Bongiovanni AM, DiGeorge AM, Grumbach MM. Masculinization of the female infant associated with estrogenic therapy alone during gestation: four causes. J Clin Endocrinol Metab 1959;19:1004.
49. Bracken MB. Oral contraception and congenital malformations in offspring: a review and meta-analysis of the prospective studies. Obstet Gynecol 1990;76;552.
50. Vandekerckhove D. A Propus de l'administration de methylandrostendiol pendant la grossesse. Ann Endocrinol (Paris) 1954;15;513.
51. Montz FJ, Schlaerth JB, Morrow CP. The natural history of theca-lutein cysts. Obstet Gynecol 1988;72:247.

10

Drug-Induced Hyperandrogenism

Barbara Neraud and Didier Dewailly

SUMMARY

Many drugs can induce hyperandrogenic symptoms, such as hirsutism, acne, seborrheic dermatitis, male-pattern baldness, menstrual irregularities, clitoral hypertrophy, or deepening of voice. Anabolic-androgenic steroids, synthetic progestins, and antiepileptics are the most frequently implicated drugs in hyperandrogenism. The mechanisms of androgen excess are various: (1) intrinsic androgenic activity of the drug, (2) interaction with sex hormone-binding globulin, or (3) functional alterations of the hypothalamic–pituitary–ovary axis. Nevertheless, the physiology of these drugs remains generally unclear. Such is the case for valproate, which seems to induce a form of polycystic ovary syndrome (PCOS) by increasing ovarian androgen synthesis. It should be remembered that it is always necessary to rule out other causes of androgen excess, especially an ovarian or adrenal tumor or PCOS, before attributing the symptom(s) to a drug.

Key Words: Hyperandrogenism; hypertrichosis; alopecia; anabolic-androgenic steroids; doping; synthetic progestins; valproic acid.

1. INTRODUCTION

Administration of drugs can induce hyperandrogenic symptoms, including hirsutism, acne, seborrheic dermatitis, male-pattern baldness, menstrual irregularities, clitoral hypertrophy, or deepening of voice. The principal mechanisms resulting in hyperandrogenism are either an increase in circulating androgen levels or an intrinsic androgenic activity of the drug. The drugs most commonly responsible for the development of virilizing effects include anabolic steroids, progestins, and antiepileptics. In addition, the use or abuse of androgens such as testosterone, androstenodione, or dehydroepiandrosterone (DHEA) for improving athletic performance or treating menopausal- or androgen deficiency- (e.g., hyposexuality) related symptoms may result in the development of frank hyperandrogenism.

2. BACKGROUND

2.1. Mechanisms of Drug-Induced Androgen Excess

In addition to the direct contribution of androgens (testosterone, androstenodione, DHEA) use/abuse to increased circulating androgen levels, several mechanisms, which may be associated and operate simultaneously, can explain clinical and biological drug-induced androgen excess. Nevertheless, the physiological mechanisms of this effect sometimes remain unclear.

From: *Contemporary Endocrinology: Androgen Excess Disorders in Women:*
Polycystic Ovary Syndrome and Other Disorders, Second Edition
Edited by: R. Azziz et al. © Humana Press Inc., Totowa, NJ

Table 1
Selected Androgenic Anabolic Steroids

Androisoxazole	Methyltestosterone
Androstanolone	Metribolone
Bolandiol	Mibolerone
Bolasterone	Nandrolone
Boldenone	Norboletone
Bolenol	Norclostebol
Bolmantalate	Norethandrolone
Chlordrolone	Oxabolone
Cloxostestosterone	Oxandrolone
Dehydrochlormethyltestosterone	Oxymesterone
Drostanolone	Oxymetholone
Ethylestrenol	Penmesterol
Fluoxymesterone	Propetandrol
Formebolone	Quinbolone
Furazabol	Silandrone
Mebolazine	Stanozolol
Mesabolone	Clostebol
Mestanolone	Stenbolone
Mesterolone	Testoslactone
Metandienone	Testosterone
Metenolone	Tibolone
Methandriol	Tiomesterone
Methylandrostandiol	Trenbolone
Testolactone	Zeranol

2.1.1. Intrinsic Androgenic Activity of the Drug

Anabolic-androgenic steroids (AASs) (Table 1) and progestins are often synthetic derivatives of testosterone. As such, they may bind to the androgen receptor and exert direct androgenic action. Because the androgen receptor mediates both the anabolic and androgenic actions, no anabolic steroid is devoid of androgenic properties.

2.1.2. Interactions with Sex Hormone-Binding Globulin

Progestins and AASs decrease sex hormone-binding globulin (SHBG) plasma concentration, and thereby increase free testosterone level, by suppressing the hepatic synthesis of SHBG in a dose-dependent manner (1). They may also bind to SHBG, thereby displacing testosterone and increasing the plasma level of free active testosterone. In addition, drugs that induce obesity and hyperinsulinism, such as valproate, indirectly decrease the plasma concentration of SHBG through the negative effect of insulin on SHBG hepatic synthesis. Therefore, an increased concentration of free testosterone explains, at least in part, the high prevalence of hirsutism in women taking these drugs.

2.1.3. Functional Alterations of the Hypothalamic–Pituitary–Ovarian Axis

Some investigators have suggested that valproate and other antiepileptic drugs may interfere with the hypothalamic control of the hypothalamic–pituitary–ovarian (HPO) axis by altering luteinizing hormone secretion through mechanisms involving γ-aminobutyric acid-ergic neurotransmission. Nevertheless, others have suggested that epileptic or psychotic problems themselves may induce these abnormalities.

2.1.4. Increased Endogenous Androgens

By inducing excessive adrenal androgen secretion, chronic treatment with adrenocorticotropic hormone (Cortrosyn® or Synacten®) or metyrapone may cause mild hyperandrogenism in women through peripheral transformation of adrenal androgens to testosterone and dihydrotestosterone. The increase in serum prolactin concentration may in part explain the hirsutism induced by phenothiazines. Indeed, prolactin is thought to increase the concentration of adrenal androgens, notably the metabolite dehydroepiandrosterone sulfate, and to decrease the concentration of SHBG, thereby raising the concentration of free testosterone.

2.2. Anabolic-Androgenic Steroids

AASs are obtained after several chemical modifications of testosterone, which increase the therapeutic index (i.e., the anabolic:androgenic activity ratio): (1) alkylation at the 17α position (which decreases strongly the hepatic metabolism of the hormone), (2) reduction of C-1,2 binding (which decreases the androgenic activity), halogenation at the C-4 and C-9 positions, (3) esterification of the 17β-hydroxyl group (which prolongs their duration of action and increases their solubility in the fatty vehicles used for injection), and (4) modifications in the ring structure of the steroid, demethylation at C-19 position (which increases the anabolic action). Most agents contain a combination of these structural changes (Table 1). Oral, parenteral, transdermal, or nasal administration of AASs is possible.

2.2.1. AASs in Sport and Fitness

For the past 50 years, AASs have been used by a wide variety of competitive athletes (bodybuilders, weightlifters, field athletes, swimmers, runners, etc.) with the hope of improving their training, endurance, and performance. A significant number of recreational athletes also seem to be using these drugs. And more and more nonathletes are also abusing AASs with the hope of improving their physical appearance.

The frequency and severity of androgenic effects of AASs depend on the dose, the method of use, the duration of administration, individual differences in susceptibility, and the properties of the molecule (the anabolic:androgenic activity ratio varies between 30:1 and 100:1). Athletes often take more than one steroid at the same time and use dosages that sometimes exceed physiological replacement levels by 10–100 times or even more. The early side effects in women are acne, oily hair and skin, seborrheic dermatitis, hirsutism, and menstrual irregularities. These clinical abnormalities can subside if AAS use is discontinued. With prolonged treatment deepening of the voice, shrinkage of the breasts, male-pattern baldness, and hypertrophy of the clitoris develop and are generally irreversible after discontinuation of AAS use. In adolescents, premature halting of growth has been described (2). The detection of AAS abuse requires specific procedures, which allow for the measurement of the derivative itself.

2.2.2. Other Uses of AASs

With the exception of their use in male hypogonadism treatment, anabolic steroids have been used in a variety of clinical situations with the hope that the beneficial actions would outweigh the risk of androgenic side effects (3). Although androgens increase the production of erythropoietin in the kidney, their efficiency in the treatment of bone marrow failure resulting from aplastic anemia or myelofibrosis or anemia as a result of renal failure is still uncertain.

Danazol is a steroid molecule designed for the treatment of endometriosis. It is more infrequently used because of its androgenic side effects (facial acne and acne of the back, hirsutism, hypertrophy of the clitoris, hair loss, deepening of the voice, weight gain, alterations in lipid metabolism) (4). These manifestations appear early, are dose-dependent, and decrease with discontinuation of treatment. Gonadotropin-releasing hormone analogs, which are equally effective in treating endometriosis, are now more widely used (5,6).

The use of AAS, such as stanozolol and danazol, has been proven successful for the treatment of lipodermatosclerosis, cryofibrogenemia, and secondary Raynaud's phenomenon, probably via fibrinolytic properties. Oral AAS alkylated in the C-17 position increase the level of the serum inhibitor of the activated first component of complement, which is deficient in hereditary angioneurotic edema. Low doses are required, and thus, no serious adverse effects are generally seen (7).

Defective secretion of androgens has been suggested as a component of growth failure in patients with Turner's syndrome. Oxandrolone in combination with growth hormone has been shown to result in growth acceleration and to improve final adult height in these patients without serious side effects *(8)*.

2.3. Synthetic Progestins

Progestins contained in oral second-generation contraceptive agents are 19-nortestosterone derivatives with attenuated androgenic properties. Norsteroids are classified in two main groups: the estranes group, which includes norethisterone (the main active molecule), norethisterone acetate, norgestrienone, lynestrenone, and ethynodiol diacetate, and the gonanes group, which includes norgestrel, levonorgestrel, and 3-cetodesogestrel.

The androgenic activity of progestins is based on two mechanisms. First, the strong similarities in structure between norsteroids and testosterone explain their ability to bind the androgen receptor, inducing directly androgenic effects. Second, progestins lower SHBG and bind to it, thus displacing testosterone and increasing levels of free active testosterone. The androgenic effects are dose-dependent and are variable according to the characteristics of each synthetic progestin.

Few women are affected by hyperandrogenism when using current oral contraceptive pills (OCPs) because of their ethynilestradiol content, which attenuates the androgenic effects of the progestins. In fact, most OCPs are able to reduce circulating free testosterone levels and ameliorate acne and hirsutism. The most frequent symptoms are acne and weight gain, whereas hirsutism is quite uncommon. The exception is progestin-only contraceptives, which tend to cause a higher prevalence of androgenic side effects. Thus, it is essential to exclude an endogenous source of androgens in case of hyperandrogenic manifestations occurring when taking an OCP. New third-generation OCPs containing 19-norpregnane progestins (e.g., gestodene, desogestrel, and norgestimate) have fewer androgenic side effects and are increasingly being used.

2.4. Antiepileptic Drugs

Epidemiological studies have reported that reproductive endocrine disorders, in particular polycystic ovary syndrome (PCOS) and hyperandrogenemia, are more common in women with epilepsy than in normal women. These disorders have been attributed to epilepsy for a long time. More recently it has been suggested that antiepileptic drugs themselves, particularly valproic acid or valproate (VPA), may be directly involved. Valproic acid is a short-chained fatty acid that is commonly used to treat epilepsy (both partial and generalized seizures), bipolar disorders, and migraines. Long-term treatment of women with VPA has been reported to be associated with reproductive disturbances. Moreover, an increased prevalence of clinical and biological hyperandrogenism, hyperinsulinism, and ultrasound criteria of PCOS has been observed in women taking prolonged VPA treatment, but the direct responsibility of VPA is still debated *(9,10)*.

Animal studies examining the effect of long-term VPA treatment have yielded variable results. In the rat, prolonged VPA treatment results in the formation of ovarian follicular cysts and a reduced number of corporea lutea; however, it decreases androgen and progesterone biosynthesis. In isolated porcine follicles, studies showed that overall androgen production is decreased in response to VPA. However, an increase in the testosterone:estrogen ratio resulting from a decrease in granulosa cell aromatase activity was observed. Finally, therapeutic exposure to VPA for a long period (12–15 months) does not reproduce the hormonal or morphological abnormalities of PCOS in nonepileptic, normally cycling rhesus monkeys *(11)*. The discrepancy between these animal and in vitro studies and human data remains so far unexplained.

A number of physiopathological mechanisms have been proposed to account for the reproductive effects of antiepileptic drugs.

2.4.1. Effect on Weight and/or Insulin Resistance

First, significant weight gain is commonly observed with VPA treatment, and it can be associated with the development of metabolic abnormalities, especially insulin resistance. So far the mechanism

Table 2
Drugs or Drug Classes Inducing Generalized Hypertrichosis

Acetazolamide
Calcium channel blockers
Cyclosporin
Diazoxide
Erythropoietin
Hydantoin
Glucocorticoids
Minoxidil
Penicillamine
Psoralens
Retinoic acid
Streptomycin

by which VPA induces such metabolic abnormalities is unknown. However, in a number of cases, reproductive disorders have been observed in patients using VPA in the absence of excessive body weight gain *(12)*. Thus, it seems that the weight gain caused by therapeutic exposure to VPA by itself cannot always explain the reproductive abnormalities observed.

2.4.2. Direct Effect on Ovarian Androgen Production

In vitro experiments have noted a direct effect of VPA on androgen biosynthesis. Indeed, human ovarian theca cells isolated from follicles of both normally cycling and PCOS women have the capacity to respond to a therapeutically relevant concentration of VPA with an increase in androgen biosynthesis. This finding suggests that VPA treatment could independently induce PCOS symptoms in the absence of genetic predisposition. The observation that PCOS theca cells respond more strongly to VPA than normal cells suggests that VPA treatment can further increase androgen synthesis in patients with PCOS. The intracellular mechanisms by which VPA induces a PCOS-like phenotype in these normal ovarian cells have been examined, and an increase in CYP17 and CYP11A, responsible for the expression of production of DHEA sulfate and androstenedione, has been observed. VPA-dependent changes in steroid biosynthesis may result from a modification of chromatin structure, enhancing transcription of genes encoding steroidogenic enzymes *(13)*.

According to these data, it appears that VPA could increase ovarian androgen biosynthesis directly, rather than through an alteration in hypothalamic pituitary function. Nevertheless, the occurrence of PCOS in women treated by VPA probably also depends on additional factors such as frequency and type of seizures, duration of epilepsy, age at onset of seizures, ethnic and genetic predispositions to PCOS, or environmental characteristics.

2.5. Androgen-Independent Drug-Induced Hair Growth or Loss

Drug-induced hypertrichosis must be differentiated from hirsutism because it is independent of androgen stimulation. It describes the growth of terminal and/or vellus hair on areas of the body where the hair is usually short. It depends on the drug dosage and is reversible after drug withdrawal. A large number of drugs may induce this phenomenon, through several mechanisms (Table 2) *(14)*.

The hair growth-promoting action of cyclosporine is partially attributable to its growth-promoting influence on hair epithelial cells caused by a decrease in protein kinase C. It stimulates telogen follicles to reenter anagen. Hirsutism and hypertrichosis could occur simultaneously with cyclosporine treatment. Benoxaprofen induces an abnormal prolongation of anagen growth. Antihypertensive agents that affect potassium channels also affect hair growth. Erythropoietin decreases the inhibiting effect of cortisol on hair follicular activity by lowering its level. Minoxidil is a treatment for male-pattern baldness: when directly applied to the balding scalp, it induces androgen-independent hair growth. Diazoxide and psoralens also often induce hypertrichosis *(14)*. Phenytoin therapy is known to induce a generalized increase in hair growth.

Table 3
Drugs or Drug Classes Inducing Hair Loss

Pro-androgen action
Oral contraceptive pill, danazol, testosterone, anabolic steroids

Antithyroid action
Propylthiouracils, carbimazole, amiodarone, lithium

Prothyroid action
Thyroxine

Lichenoid cicatricial alopecia
Chloroquine, mepacrine, proguanil

Telogen effluvium
Albendazol, allopurinol, amphetamine, boric acid,
bromocriptine, captopril, cimetidine, colchicine, coumarins,
enalapril, glibenclamide, gold, heparin, heparinoïds,
hypolipidemic agents (clofibrate, triparanol), indandiones,
interferons, levodopa, methyldopa, methysergide,
metoprolol,penicillamine, phenytoin, propranolol, retinol,
sulphasalazine

Anagen effluvium
Radiation, cyclophosphamide, doxorubicin, colchicine (high
dose), thallium, mercury, arsenic, cantharadin, azathioprine,
methotrexate

Alternatively, drugs can induce alopecia through loss of either anagen or telogen hairs (Table 3). Anagen hair loss is usually dramatic, inducing an abrupt cessation of mitotic activity in rapidly dividing hair matrix cells. Telogen effluvium (shedding of telogen or dead hairs) induces premature rest of the follicle.

3. CONCLUSION AND FUTURE AVENUES OF INVESTIGATION

The diagnosis of drug-induced androgen excess is often evident except in the case of illegal administration of anabolic androgenic steroids. However, other etiologies of hyperandrogenism should not be ignored, particularly an androgen-secreting tumor or PCOS (which may also be induced or worsened by certain drugs). Our knowledge of the physiological mechanisms underlying drug-induced hyperandrogenism continues to develop, particularly concerning VPA. Indeed, this common antiepileptic drug has been suggested to induce a PCOS-like phenotype. The role of the weight gain often associated with VPA treatment in the development of reproductive disturbances is still unclear. Further investigations are required to elucidate the mechanism underlying the interaction between VPA and ovarian theca cells. These studies have the potential for improving our understanding of the mechanisms underlying ovarian hyperandrogenism in PCOS, because this molecule induces alteration of ovarian steroidogenic enzyme activity that mimic those observed in spontaneously ocurring PCOS (*see* Chapter 18). Investigative efforts should also be turned to designing newer antiepileptic drugs devoid of hyperandrogenic and reproductive side effects.

KEY POINTS

• Intrinsic androgenic activity of the drug, interactions with SHBG, or alterations of the HPO axis are the main mechanisms of drug-induced hyperandrogenism. Nevertheless, for some drugs, the mechanism(s) of hyperandrogenism remain unclear.

- The androgenic side effects of AASs depend on the dose, duration of administration, properties of the molecule (i.e., the anabolic:androgenic activity ratio), and individual differences in susceptibility; with intensive and prolonged use, side effects may become irreversible, even with discontinuation.
- Unlike oral second-generation contraceptive agents, new third-generation OCPs contain progestins with few androgenic effects.
- Valproic acid appears to stimulate increased ovarian androgen biosynthesis, which appears to be independent of the metabolic changes induced by this antiepileptic drug.
- In the case of severe symptoms in a woman taking a drug known to induce hyperandrogenism, others etiologies should be considered, particularly an androgen-secreting ovarian or adrenal neoplasm or PCOS.

REFERENCES

1. Song S, Chen J, Lu C, et al. Effects of different doses of norethisterone on ovarian function, serum sex hormone binding globulin and high density lipoprotein-cholesterol. Contraception 1993;47:527–537.
2. Rogol A, Yesalis CE. Anabolic-androgenic steroids and the adolescent. Pediatr Ann 1992;21:175–188.
3. Wilson JD, Griffin JE. Progress in endocrinology and metabolism: the use and misuse of androgens. Metabolism 1980;29:1278–1295.
4. Barbieri RL, Evans S, Kistner RW. Danazol in the treatment of endometriosis: analysis of 100 cases with a 4 year follow-up. Fertil Steril 1982;37:737–746.
5. Cirkel U, Ochs H, Schneider HPG. A randomized, comparative trial of triptorelin depot (D-Trp-LHRH) and danazol in the treatment of endometriosis. Eur J Obstet Gynecol 1995;59:61–69.
6. Wright S, Valdes CT, Dunn RC, Franklin RR. Short-term Lupron or danazol therapy for pelvic endometriosis. Fertil Steril 1995;63:504–507.
7. Elfman T, Falanga V. Stanazolol as a novel therapeutic agent in dermatology. J Am Acad Dermatol 1995;33:254–258.
8. Nilsson KO, Albertsson-Wikland K, Alm J, et al. Improved final height in girls with Turner's syndrome treated with growth hormone and oxandrolone. J Clin Endocrinol Metab 1996;81:635–640.
9. Isojarvi JI, Laatikainen TJ, Pakarinen AJ, Juntunen KT, Myllyla VV. Polycystic ovaries and hyperandrogenism in women taking valproate for epilepsy. N Engl J Med 1993;329:1383–1388.
10. Verrotti A, Greco R, Latini G, Chiarelli F. Endocrine and metabolic changes in epileptic patients receiving valproic acid. J Pediatr Endocr Metab 2005;18:423–430.
11. Ferin M, Morrell M, Xiao E, et al. Endocrine and metabolic responses to long-term monotherapy with the antiepileptic drug valproate in the normally cycling rhesus monkey. J Clin Endocrinol Metab 2003;88:2908–2915.
12. Bilo L, Meo R, Valentino R, Di Carlo C, Striano S, Nappi C. Characterization of reproductive endocrine disorders in women with epilepsy. J Clin Endocrinol Metab 2001;86:2950–2956.
13. Nelson-DeGrave V, Wickenheisser J, Cockrell J, et al. Valproate potentiates androgen biosynthesis in human ovarian theca cells. Endocrinology 2003;145:799–808.
14. Australasian Hair and Wool Research Society's website: http://www.alopecia.com

11

Severe Insulin-Resistance Hyperandrogenic Syndromes

David E. Moller, Antonio Vidal-Puig, and Ricardo Azziz

SUMMARY

Insulin resistance is associated with a wide variety of markedly heterogeneous clinical disorders, either inherited or acquired, that may result in acanthosis nigricans, ovarian hyperandrogenism, and ovulatory dysfunction. These include the type A (primarily affecting lean women and resulting from defects of the insulin receptor), type B (resulting from an autoimmune process affecting the insulin receptor), and type C (a variant of type A, the hyperandrogenic, insulin-resistant, and acanthosis nigricans [HAIR-AN] syndrome) insulin resistance syndromes. More rare syndromes include leprechaunism, the Rabson–Mendenhall syndrome, and a heterogeneous group of lipodystrophic syndromes. HAIR-AN is generally present in obese women who do not demonstrate defects of the insulin receptor, although they may exhibit postreceptor defects. Although the type A and B insulin resistance syndromes are rare causes of ovarian hyperandrogenism, type C may affect up to 3% of women with androgen excess. These patients may be treated as other women with ovarian hyperandrogenism, namely using lifestyle modification, oral contraceptives, insulin sensitizers, and antiandrogens, although a few patients may require further suppression of their hypothalamic–pituitary–ovarian axis with a long-acting gonadotropin-releasing hormone analog. It is likely that HAIR-AN patients are at greater risk for metabolic dysfunction and diabetes mellitus, although long-term studies are lacking. Finally, the distinction between HAIR-AN and the polycystic ovary syndrome (PCOS) is becoming less clear as more patients with PCOS are being recognized as potentially having postreceptor abnormalities in insulin signaling. Better characterization awaits more detailed phenotyping and genotyping studies.

Key Words: Insulin resistance; insulin; androgen excess; type A insulin resistance; type B insulin resistance; type C insulin resistance; HAIR-AN syndrome; acanthosis nigricans; Rabson–Mendenhall syndrome; lipodystrophy.

1. INTRODUCTION

Insulin resistance is defined as a subnormal biological response to insulin and, in the presence of normal pancreatic function, is associated with hyperinsulinemia and an exaggerated insulin secretion (i.e., hyperinsulinism). Insulin resistance is associated with a wide variety of markedly heterogeneous clinical disorders, either inherited or acquired. These may range from mild to moderate forms observed in women with uncomplicated polycystic ovary syndrome (PCOS) and simple obesity to more severe disorders including various forms of lipodystrophy, leprechaunism, and the hyperandrogenic, insulin-resistant, and acanthosis nigricans (HAIR-AN), or type C insulin resistance, syndrome (Table 1). In many of these individuals, ovarian hyperandrogenism secondary to the prevalent hyperinsulinism is present. In general, the duration or degree of insulin resistance is often correlated with the severity of ovarian dysfunction.

From: *Contemporary Endocrinology: Androgen Excess Disorders in Women:
Polycystic Ovary Syndrome and Other Disorders, Second Edition*
Edited by: R. Azziz et al. © Humana Press Inc., Totowa, NJ

Table 1
**Classification of Insulin Resistance Syndromes Associated
With Ovarian Hyperandrogenism Dysfunction**

Syndromes of severe insulin resistance:
Congenital-Hereditary
Type A syndrome[a]
Type C (HAIR-AN) syndrome ?
Leprechaunism[a]
Rabson–Mendenhall syndrome?
Congenital lipodystrophies
Acquired
Type C (HAIR-AN) syndrome?
Type B syndrome[b]
Acquired lipodystrophies
States of mild–moderate insulin resistance:
"Typical" PCOS
Obesity

[a]May be associated with insulin receptor mutations.
[b]Caused by autoantibodies directed against the insulin receptor.
HAIR-AN, hyperandrogenic, insulin-resistant, and acanthosis nigricans
syndrome; PCOS, polycystic ovary syndrome.

These women usually demonstrate acanthosis nigricans, which consists of thickened hyperpigmented areas of skin, occurring most commonly in the crural areas of the skin, such as the axilla, nape of the neck, and other skinfold locations (Fig. 1) *(1)*. Histopathologically, the epidermis demonstrates irregular hyperplasia with prominent rete ridges, epidermal hyperpigmentation, moderate ortho/parakeratotic hyperkeratosis, and no significant inflammatory infiltrate (Fig. 2). This cutaneous disorder is frequently associated with moderate to severe degrees of insulin resistance *(2)*.

Here we will describe the clinical spectrum, putative molecular mechanisms, prevalence, and diagnosis of the severe insulin-resistance syndromes associated with ovarian hyperandrogenism. It is important to note that although some syndromes (e.g., leprechaunism) represent discrete clinical disorders, others (e.g., type A and HAIR-AN) cannot be easily distinguished from one each other. These terms may therefore refer to patients within a spectrum of overlapping clinical features and molecular defects. However, these disorders are considered to result from primary abnormalities of insulin action and not from defects in ovarian steroidogenesis.

2. BACKGROUND

2.1. Clinical Spectrum of Insulin Resistance Hyperandrogenic Syndromes

2.1.1. Type A Insulin Resistance Syndrome

This syndrome was first described in 1976 in three lean adolescent females with severe insulin resistance and acanthosis nigricans *(3)*. The phenotype of the type A insulin resistance syndrome (Online Mendelian Inheritance in Man catalog number 147670; *see* http://www.ncbi.nlm.nih.gov/omim/) includes extreme insulin resistance, acanthosis nigricans, polycystic ovaries, and severe levels of hyperandrogenism in patients who are usually not obese. In general, this syndrome is the result of mutations in the insulin receptor, although this is not a consistent finding *(4)*. One recent study reported a patient who appeared to have the type A syndrome and who carried a novel heterozygous missense mutation in the lamin A/C (*LMNA*) gene, predicting a G602S amino acid substitution in lamin A *(5)*. Lamins are the main components of the nuclear lamina, a filamentous network located between the inner nuclear membrane and chromatin that plays a fundamental role in nuclear organi-

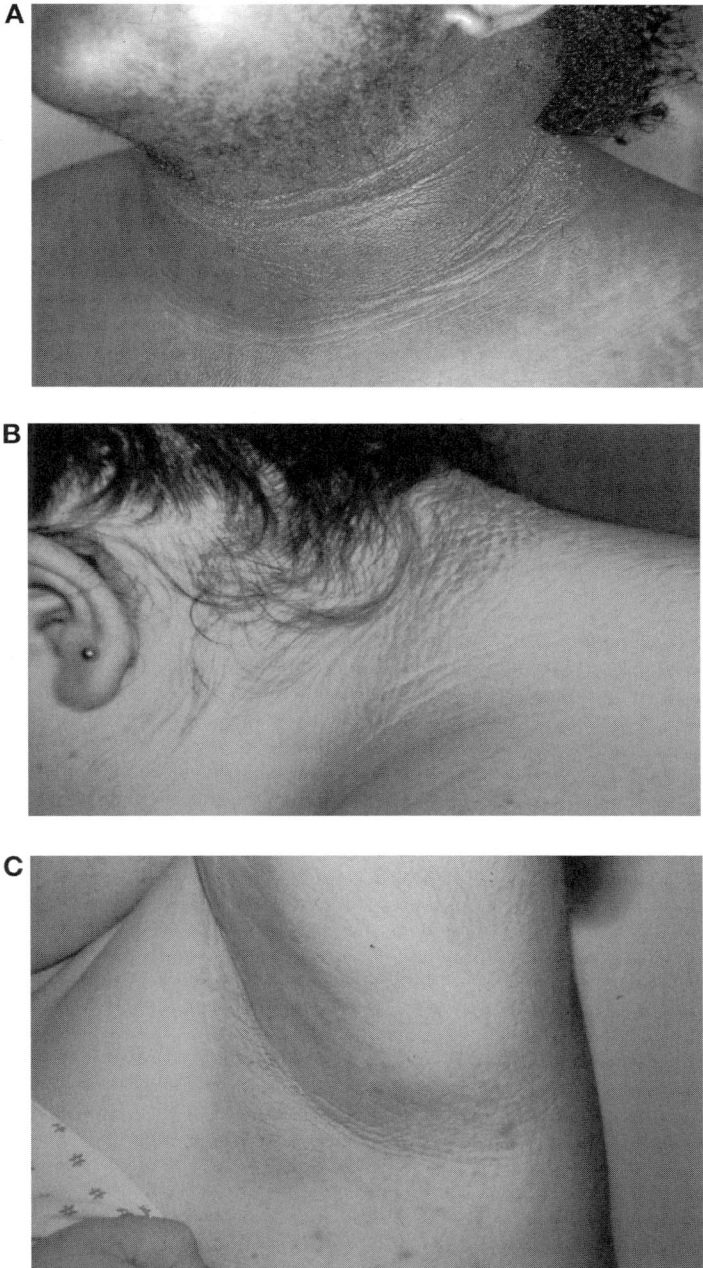

Fig. 1. Acanthosis nigrican. Nape of the neck in a black (**A**) and white (**B**) woman and (**C**) in the axilla of the white woman. The patches of cutaneous hyperpigmentation in these locations are indicative of acanthosis nigricans.

zation in all differentiated cells. Mutations in lamins A and C, which are alternatively spliced products of the lamin A/C (*LMNA*) gene, are responsible for several genetic diseases, including familial partial lipodystrophy of the Dunnigan type (FPLD). Women with FPLD often have irregular menses and hyperandrogenemia as well as severe insulin resistance, dyslipidemia, and atherosclerotic vascular disease, similar to patients with the type A syndrome.

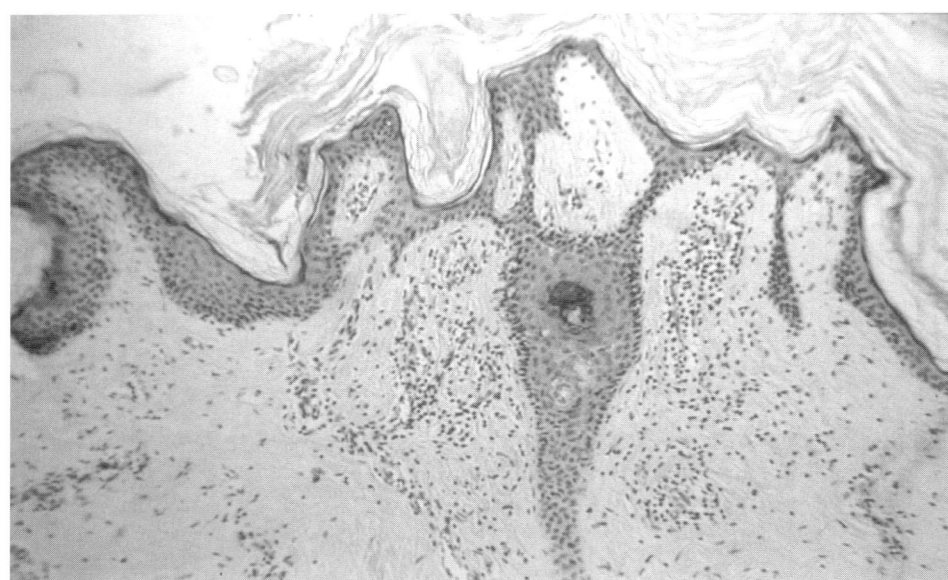

Fig. 2. Acanthosis nigricans—microscopic. The photomicrograph demonstrates the typical histological features in the affected epidermis, which demonstrates irregular hyperplasia with prominent rete ridges, epidermal hyperpigmentation, and moderate ortho/parakeratotic hyperkeratosis, with no significant inflammatory infiltrate.

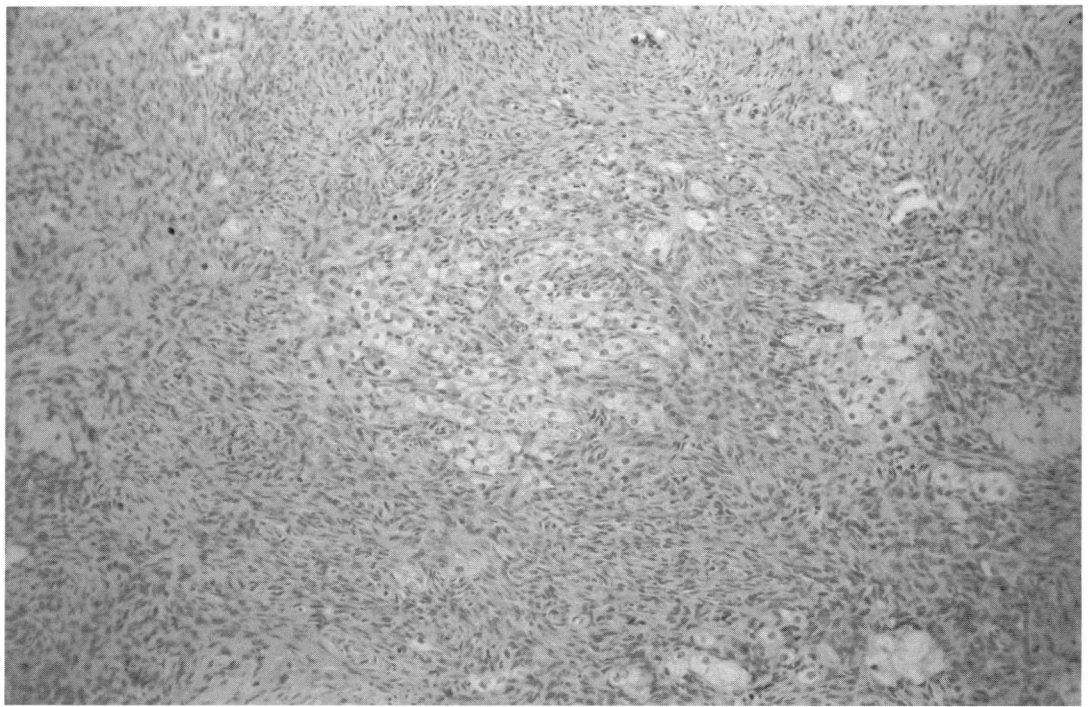

Fig. 3. Stromal hyperthecosis of the ovary of a patient with the hyperandrogenic, insulin-resistant, and acanthosis nigricans (HAIR-AN) syndrome. Note the isolated island of luteinized theca cells independent of a follicle.

Patients with the type A syndrome reportedly have substantial degrees of ovarian hyperandrogenism, including ovarian hyperthecosis (Fig. 3). Fasting insulin levels in these patients may be modestly elevated (20–50 μU/mL) but are more typically higher (over 500 μU/mL), with postprandial levels ranging from 500 to 3000 μU/mL. A number of atypical clinical findings have also been reported in patients who appear to share other features of the type A syndrome. These include hypertrichosis, muscle cramps *(7)*, and short stature *(8)*. In addition, a rare variant syndrome referred to as "pseudoacromegaly" has been reported, characterized by accelerated growth, acral hypertrophy, and other clinical features of acromegaly that occur in the context of extreme insulin resistance and the features of the type A syndrome *(9,10)*. Importantly, there is no biochemical evidence to suggest that excessive growth hormone or insulin-like growth factor-1 secretion is present, unlike "typical" cases of acromegaly.

In contrast to the type B syndrome (*see* next section), which is acquired but may present with a similar phenotype, the type A syndrome seems to be primarily inherited. Both dominant and recessive modes of transmission have been observed, and male family members with acanthosis nigricans and evidence of severe insulin resistance have also been described *(11,12)*. In general, type A insulin resistance probably accounts for less than 1% of women presenting with hyperandrogenism and insulin resistance.

2.1.2. Type B Insulin Resistance Syndrome

Type B insulin resistance syndrome is used to describe patients with a clinical phenotype that (in some cases) is similar to that of type A syndrome, but where the underlying cause is a distinct acquired autoimmune disorder *(4,13)*. The molecular defect resulting in type B insulin resistance is generally the development of autoantibodies directed against the insulin receptor *(4,13)*. As in many other autoimmune disorders, patients are predominantly young to middle-aged women (mean age 44 years, with a reported range of 15–69 years). The majority of patients are females of African or Caribbean origin.

Most patients have features of other autoimmune diseases, and approximately 30–50% meet established criteria for the diagnosis of systemic lupus erythematosus *(14,15)*. It should be noted that the majority of patients do not meet the criteria for a distinct autoimmune disease but may have one or more features of autoimmunity including alopecia, nephritis, anemia, leukopenia, elevated sedimentation rate, and positive antinuclear antibodies *(14)*.

Patients with the type B syndrome typically have marked insulin resistance that is often directly proportional to the titer of anti-insulin receptor antibodies. In some cases overt hyperglycemia, which is nearly impossible to control with insulin, has been observed *(4,14)*. These patients may require treatment with immunosuppressive agents *(16–19)*. Alternatively, insulin receptor autoantibodies may also exert a partial agonist (insulin-mimetic) activity. This accounts for the fact that certain patients present with fasting hypoglycemia and explains the clinical paradox of alternating hyperglycemia and hypoglycemia observed in others with this syndrome *(14)*.

This syndrome is also associated with ovarian hyperandrogenism when premenopausal females are affected, and, in addition to amenorrhea and hirsutism, some may develop virilization *(4,14,22,23)*. The extent to which these overt features of ovarian hyperandrogenism occur may depend on the degree and duration of insulin resistance.

Overall, the type B insulin resistance syndrome is an extremely rare cause of ovarian hyperandrogenism. Definitive diagnosis depends upon the demonstration of circulating antibodies that either block insulin binding or interact with the insulin receptor (immunoprecipitation) via specialized in vitro tests *(14)*.

2.1.3. Type C Insulin Resistance Syndrome

Unlike type A (primarily lean with insulin receptor defects) and type B (an autoimmune process) insulin resistance syndromes, the type C (or HAIR-AN) syndrome is characterized by the presence of marked acanthosis, hyperandrogenism, insulin resistance, obesity, and the absence of insulin receptor defects *(22)*. In general, the type C syndrome is considered a variant of the type A disorder with less severe insulin resistance. The first description of this disorder appears to be that of Brown and Winkelmann, who in 1968 described 90 cases of acanthosis nigricans, 2 of which were women with clinical and histopathological signs of the Stein–Leventhal syndrome, both having marked amenorrhea and hirsutism, with one being very obese, whereas the other was not obese (and this latter patient had already developed diabetes) *(23)*. Consequently, it appears that these investigators may be describing one patient with type C and the other with type A insulin resistance. Barbieri and Ryan went on to codify the features of the HAIR-AN syndrome as it was understood in the early 1980s *(24)*. These investigators suggested that the syndrome was more common than previously reported, potentially affecting 1–3% of hyperandrogenemic women.

In addition to generally extreme degrees of hyperandrogenism, often accompanied by normal to low luteinizing hormone levels, these patients generally have post-glucose-challenge insulin levels greater than 300–500 µU/mL. We should note that although some of these patients may have markedly elevated levels of fasting insulin (> 80 µU/mL), the fasting insulin level might not be the most sensitive marker of hyperinsulinemia. As is observed in patients with PCOS, determination of the insulin response to an oral glucose tolerance test may be the most sensitive method of detecting severe hyperinsulinemia *(25)*. Accurate measurement of insulin is critical to the assessment of insulin resistance and hyperinsulinism. However, the plasma insulin values obtained with current assay methodology may vary greatly from one laboratory to the next *(26)*, limiting the sensitivity of the basal insulin values for the detection of the HAIR-AN (or for that matter type A or B) syndrome.

The ovaries of these patients often demonstrate stromal hyperthecosis *(27,28)*. The degree of hyperandrogenemia is such that it may raise concerns that the patient actually suffers from an androgen-secreting ovarian neoplasm, which can be alleviated by the finding of very high insulin levels following a glucose challenge and symmetrical ovarian enlargement on ultrasound.

In patients with type C syndrome, the degree of obesity appears to play a significant role in determining the degree of insulin resistance *(22,29)*. These patients also appear to have increased basal rates of hepatic glucose production and peripheral insulin resistance, with a right shift in the insulin dose–response curve determined by the euglycemic insulin clamp *(30)*. These women generally demonstrate normal insulin receptor *(32)* and peroxisome proliferator-activated receptor-γ *(33)* genes, although a few patients have been suggested to have insulin receptor defects *(8)*. However, insulin binding to monocytes and fibroblasts appears to be normal *(27,34)*, suggesting the presence of postreceptor defects of insulin signaling.

The presence of ill-defined circulating "factors" or cellular components that serve as inhibitors of insulin receptor function have been suggested as an etiological factor *(35)*. Maddux et al. described an endogenous cellular inhibitor of the insulin receptor tyrosine kinase, which appeared to be responsible for insulin resistance in several hyperandrogenic subjects *(36)*. This inhibitory factor was identified as a membrane glycoprotein known as PC-1, which confers insulin resistance on cells that are engineered to express it *(37)*. Further studies will be required to establish whether increased PC-1 expression in vivo, or that of other potential inhibitors of insulin action, are sufficient to account for substantial insulin resistance in HAIR-AN.

Patients with HAIR-AN syndrome are often treated, as are other patients with ovarian androgen excess (e.g., PCOS), with lifestyle modification and a combination of oral contraceptives, antiandrogens, and insulin sensitizers *(38–40)*. Nevertheless, estrogen–progestin preparations may aggravate the degree of metabolic dysfunction *(30)*. In some patients, their elevated insulin levels are sufficient to continue to stimulate excess theca cell androgen production, despite the relatively low gonadotropin levels achieved by oral contraceptive administration, and these women may benefit from treatment with a long-acting gonadotropin-releasing hormone (GnRH) analog *(38,41)* (Fig. 4). However, it may take 3–6 months before full suppression is achieved. It is also likely that HAIR-AN

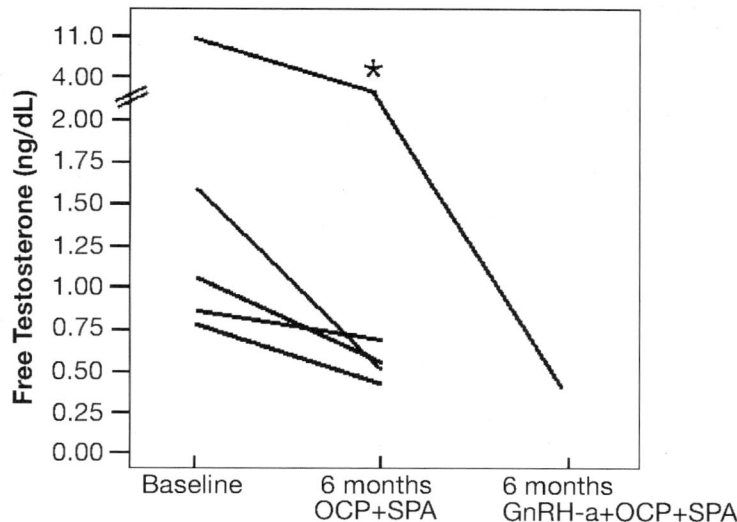

Fig. 4. Free testosterone levels before and 6 months after treatment with an oral contraceptive pill (OCP) and spironolactone (SPA) 200 mg per day. Patient marked with an asterisk received only an OCP and after inadequate suppression of free testosterone was then treated with a long-acting gonadotropin-releasing hormone (GnRH) analog (leuprolide), in combination with estrogen–progestin replacement and SPA. The free testosterone levels normalized within 6 months of this therapy. (Adapted from ref. *38.*)

patients are at greater risk for metabolic dysfunction and diabetes mellitus, although long-term studies are lacking.

The HAIR-AN syndrome appears to affect 2–3% of androgen excess patients *(42,43)*. However, the single greatest difficulty in understanding the HAIR-AN syndrome lies in its differentiation from PCOS, which is often also accompanied by insulin resistance. The distinction between HAIR-AN and PCOS is becoming even less clear as more patients with PCOS are being recognized as potentially having postreceptor abnormalities in insulin signaling. Consequently, until more detailed phenotyping and genotyping studies for the HAIR-AN syndrome become available, these patients will need to be defined by the presence of severe insulin resistance and hyperinsulinemia, determined by an arbitrary cutoff value, which currently is represented by a post-glucose-challenge insulin level of more than 300–500 µU/mL.

2.2. Other Syndromes of Severe Insulin Resistance That May Be Associated With Ovarian Dysfunction

Leprechaunism is a very rare autosomal recessive congenital syndrome of extreme insulin resistance characterized clinically by intrauterine and postnatal growth restriction with a large number of other dysmorphic features *(see* ref. *44).* Other important clinical findings include lipoatrophy, acanthosis nigricans, and glucose intolerance or overt diabetes in the presence of massive hyperinsulinemia and β-cell hypertrophy. Interestingly, affected female infants may also have hypertrichosis–hirsutism, with clitoromegaly and enlarged, cystic ovaries *(11,44).* In affected male patients, penile enlargement has also been observed. Insulin receptor mutations represent an apparently universal cause of this rare disorder.

The Rabson–Mendenhall syndrome is a rare recessive genetic disorder that is related to leprechaunism. This syndrome is characterized by extreme insulin resistance, acanthosis nigricans, and hyperplasia of the pineal gland and may include a variety of other characteristics, such as dysmorphic features, accelerated growth, phallic enlargement, and precocious pseudopuberty *(11)*. Clinical manifestations of hyperandrogenism or ovarian dysfunction have been reported in affected female subjects with this syndrome *(45)*. Like leprechaunism, the Rabson–Mendenhall syndrome appears to be exclusively associated with severe defects affecting the insulin receptor gene. These patients are at significant risk for developing metabolic complications *(6)*.

A heterogeneous group of lipodystrophic syndromes can also be associated with both insulin resistance and ovarian hyperandrogenism or dysfunction. These lipodystrophies can be classified according to whether there is complete (generalized) or partial absence of adipose tissue and whether the disorder is congenital (inherited) or acquired. The best known of these rare syndromes is congenital generalized lipodystrophy (also known as the Berardinelli–Seip syndrome or congenital total lipoatrophy), an autosomal-recessive condition *(11)*. In addition to lipoatrophy and severe insulin resistance with marked hyperlipidemia, affected female patients may reportedly develop clitoromegaly and the early development of secondary sexual characteristics *(11)*. Other women with severe insulin resistance in the context of partial lipodystrophy have been reported to develop clinical signs of ovarian hyperandrogenism *(11,46)*.

3. CONCLUSION

Insulin resistance syndromes include the type A (primarily affecting lean women and resulting from defects of the insulin receptor), type B (resulting from an autoimmune process affecting the insulin receptor), and type C (a variant of type A, the so-called HAIR-AN syndrome). More rare syndromes include leprechaunism, the Rabson–Mendenhall syndrome, and a heterogeneous group of lipodystrophic syndromes. The HAIR-AN syndrome generally is present in obese women who do not demonstrate defects of the insulin receptor, although they may exhibit postreceptor defects. Although type A and B insulin resistance syndrome are rare causes of ovarian hyperandrogenism, the HAIR-AN syndrome may affect up to 3% of women with androgen excess. These patients may be treated like other women with ovarian hyperandrogenism, although a few patients may require further suppression of their hypothalamic–pituitary–ovarian axis with a long-acting GnRH analog. Although it is likely that HAIR-AN patients are at greater risk for metabolic dysfunction and diabetes mellitus, long-term studies are lacking. Finally, the distinction between HAIR-AN and PCOS is becoming less clear, as more patients with PCOS are being recognized as potentially having postreceptor abnormalities in insulin signaling. Better classification awaits more detailed phenotyping and genotyping studies.

4. FUTURE AVENUES OF INVESTIGATION

Better characterization of the HAIR-AN syndrome is required to establish phenotypic and genotypic differences and to improve the distinction between this syndrome and PCOS, although it is possible that both syndromes form part of the same continuum. Likewise, long-term studies determining optimum therapy and prognosis for these patients is required. Finally, more detailed and specific molecular studies are required to better understand the pathophysiology of these disorders.

KEY POINTS

- The type A insulin resistance syndrome primarily affects lean women and results from defects of the insulin receptor.
- The type B insulin resistance syndrome is the consequence of an autoimmune process affecting the insulin receptor.
- The type C insulin resistance syndrome is considered a variant of the type A syndrome and is frequently termed the HAIR-AN syndrome.
- More rare syndromes include leprechaunism, Rabson–Mendenhall syndrome, and a heterogeneous group of lipodystrophic syndromes.
- HAIR-AN generally is present in obese women who do not demonstrate defects of the insulin receptor,

although they may exhibit postreceptor defects, and may affect up to 3% of women with androgen excess.
- HAIR-AN patients may be treated as are other women with ovarian hyperandrogenism, namely using lifestyle modification, oral contraceptives, insulin sensitizers, and antiandrogens, although a few patients may require further suppression of their hypothalamic–pituitary–ovarian axis with a long-acting GnRH analog.
- It is likely that HAIR-AN patients are at greater risk for metabolic dysfunction and diabetes mellitus, although long-term studies are lacking.
- The distinction between HAIR-AN and the PCOS is becoming less clear and awaits more detailed phenotyping and genotyping studies.

REFERENCES

1. Rogers DL. Acanthosis nigricans. Semin Dermatol 1991;10:160–163.
2. Flier JS. The metabolic importance of acanthosis nigricans. Arch Derm 1985;121:193–194.
3. Kahn CR., Flier JS, Bar RS, et al. The syndromes of insulin resistance and acanthosis nigricans: insulin receptor disorders in man. N Engl J Med 1976;294:739–745.
4. Moller DE, Cohen O, Yamaguchi Y, Azziz R, Eberle A, Flier JS. Prevalence of mutations in the insulin receptor gene in subjects with features of the Type A syndrome of insulin resistance. Diabetes 1994;43:247–255.
5. Young J, Morbois-Trabut L, Couzinet B, et al. Type A insulin resistance syndrome revealing a novel lamin A mutation. Diabetes 2005;54:1873–1878.
6. Musso C, Cochran E, Moran SA, et al. Clinical course of genetic diseases of the insulin receptor (type A and Rabson-Mendenhall syndromes): a 30-year prospective. Medicine 2004;83:209–222.
7. Flier JS, Young JB, Landsberg L. Familial insulin resistance with acanthosis nigricans, acral hypertrophy, and muscle cramps: a new syndrome. N Engl J Med 1980;390:970–973.
8. Barbieri, RL. Hyperandrogenism, insulin resistance, and acanthosis nigricans: 10 years of progress. J Reprod Med 1994;39:327–336.
9. Flier JS, Grigorescu F, Chaiken RL, et al. Pseudoacromegaly: a syndrome apparently caused by insulin resistance with a receptor kinase defect. Clin Res 1985;33:569A.
10. Holdaway IM, Frengley PA, Graham FM, Wong M. Insulin resistance with acanthosis nigricans and acral hypertrophy. NZ Med J 1984;97:286–288.
11. Moller DE, O'Rahilly S. Congenital syndromes of severe insulin resistance. In: Moller DM, ed. Insulin Resistance. New York: John Wiley and Sons, 1993.
12. Moller DE, Yokota A, White MF, Pazianos AG, Flier JS. A naturally occurring mutation of insulin receptor alanine 1134 impairs tyrosine kinase function and is associated with dominantly inherited insulin resistance. J Biol Chem 1990;265:14979–14985.
13. Flier JS, Kahn CR, Roth J. Receptors, antireceptor antibodies, and mechanisms of insulin resistance. N Engl J Med 1979;300:413–419.
14. Gorden P, Collier E, Roach P. Autoimmune mechanisms of insulin resistance and hypoglycemia. In: Moller DE, ed. Insulin Resistance. New York: John Wiley, 1993:123–141.
15. Arioglu E, Andewelt A, Diabo C, Bell M, Taylor SI, Gorden P. Clinical course of the syndrome of autoantibodies to the insulin receptor (type B insulin resistance): a 28-year perspective. Medicine 2002;81:87–100.
16. Eriksson JW, Bremell T, Eliasson B, Fowelin J, Fredriksson L, Yu ZW. Successful treatment with plasmapheresis, cyclophosphamide, and cyclosporin A in type B syndrome of insulin resistance. Case report. Diabetes Care 1998;21:1217–1220.
17. Gehi A, Webb A, Nolte M, Davis J Jr. Treatment of systemic lupus erythematosus-associated type B insulin resistance syndrome with cyclophosphamide and mycophenolate mofetil. Arthritis Rheum 2003;48:1067–1070.
18. Coll AP, Thomas S, Mufti GJ. Rituximab therapy for the type B syndrome of severe insulin resistance. N Engl J Med 2004;350:310–311.
19. Coll AP, Morganstein D, Jayne D, Soos MA, O'Rahilly S, Burke J. Successful treatment of Type B insulin resistance in a patient with otherwise quiescent systemic lupus erythematosus. Diabet Med 2005;22:814–815.
20. Taylor SI, Dons RF, Hernandez E, Roth J, Gorden P. Insulin resistance associated with androgen excess in women with autoantibodies to the insulin receptor. Ann Int Med 1982;97:851–855.
21. DeClue TJ, Shah SC, Marchese M, Malone JI. Insulin resistance and hyperinsulinemia induce hyperandrogenism in a young Type B insulin-resistant female. J Clin Endocrinol Metab 1991;72:1308–1311.
22. Flier JS, Eastman RC, Minaker KL, Matteson D, Rowe JW. Acanthosis nigricans in obese women with hyperandrogenism: characterization of an insulin-resistant state distinct from the type A and B syndromes. Diabetes 1985;34:101–107.
23. Brown J, Winkelmann RK. Acanthosis nigricans: a study of 90 cases. Medicine 1968;47:33–51.
24. Barbieri RL, Ryan KJ. Hyperandrogenism, insulin resistance, and acanthosis nigricans syndrome: a common endocrinopathy with distinct pathophysiologic features. Am J Obstet Gynecol 1983;147:90–101.

25. Legro R, Finegood D, Dunaif A. A fasting glucose to insulin ratio is a useful measure of insulin sensitivity in women with polycystic ovary syndrome. J Clin Endocrinol Metab 1998;83:2694–2698.

26. Robbins DC, Andersen L, Bowsher R, et al. Report of the American Diabetes Association's task force on standardization of the insulin assay. Diabetes 1996;45:242–256.

27. Dunaif A, Hoffman AR, Scully RE, et al. Clinical, biochemical, and ovarian morphologic features in women with acanthosis nigricans and masculinization. Obstet Gynecol 1985;66:545–552.

28. Nagamani M, Van Dinh T, Kelver ME. Hyperinsulinemia in hyperthecosis of the ovaries. Am J Obstet Gynecol 1986;154:384–389.

29. Richards GE, Cavallo A, Meyer WJ 3rd, et al. Obesity, acanthosis nigricans, insulin resistance, and hyperandrogenemia: pediatric perspective and natural history. J Pediatr 1985;107:893–897.

30. Vaidya RA, Shringi MS, Vaidya AB, Gogate JG. Development of diabetes mellitus and hypertension due to aggravation of insulin resistance after estrogen—progestogen in HAIR-AN syndrome. J Assoc Physicians India 1997;45:883–884.

31. Tsokos GC, Gorden P, Antonovych T, Wilson CB, Balow JE. Lupus nephritis and other autoimmune features in patients with diabetes mellitus due to autoantibody to insulin receptors. Ann Int Med 1985;102:176–181.

32. Globerman H, Karnieli E. Analysis of the insulin receptor gene tyrosine kinase domain in obese patients with hyperandrogenism, insulin resistance, and acanthosis nigricans (type C insulin resistance). Int J Obes Relat Metab Disord 1998;22:349–353.

33. Globerman H, Zauberman Y, Makarov T, et al. Analysis of the peroxisome proliferator activated receptor gamma (PPARgamma) gene in HAIRAN syndrome with obesity. Clin Endocrinol (Oxf) 2000;52:479–485.

34. Peters EJ, Stuart CA, Prince MJ. Acanthosis nigricans and obesity; acquired and intrinsic defects in insulin action. Metabolism 1986;35:807–813.

35. Harrison LC, Dean B, Peluso I, Clark S, Ward G. Insulin resistance, acanthosis nigricans, and polycystic ovaries associated with a circulating inhibitor of postbinding insulin action. J Clin Endocrinol Metab 1985;60:1047–1052.

36. Maddux BA, Sbraccia P, Reaven GM, Moller DE, Goldfine ID. Inhibitors of insulin receptor tyrosine kinase in fibroblasts from diverse patients with impaired insulin action: evidence for a novel mechanism of postreceptor insulin resistance. J Clin Endocrinol Metab 1993;77:73–79.

37. Maddux BA, Sbraccia P, Kumakura S, et al. Membrane glycoprotein PC-1 and insulin resistance in non–insulin-dependent diabetes mellitus. Nature 1995;373:448–451.

38. Azziz R. The hyperandrogenic-insulin-resistant acanthosis nigricans syndrome: therapeutic response. Fertil Steril 1994;61:570–572.

39. Zemtsov A, Wilson L. Successful treatment of hirsutism in HAIR-AN syndrome using flutamide, spironolactone, and birth control therapy. Arch Dermatol 1997;33:431–433.

40. Moore DC. Prolonged suppression of hirsutism with combination therapy in an adolescent with insulin resistance and acanthosis nigricans. J Adolesc Health Care 1987;8:445–448.

41. Corenblum B, Baylis BW. Medical therapy for the syndrome of familial virilization, insulin resistance, and acanthosis nigricans. Fertil Steril 1990;53:421–425.

42. Barbieri RL. Hyperandrogenic disorders. Clin Obstet Gynecol 1990;33:640–654.

43. Azziz R, Sanchez LA, Knochenhauer ES, et al. Androgen excess in women: experience with over 1000 consecutive patients. J Clin Endocrinol Metab 2004;89:453–462.

44. Elsas LJ, Endo F, Priest JH, Strumlauf E. Leprechaunism: an inherited defect in insulin-receptor interaction. In: Wapnir RA, ed. Congenital Metabolic Disease: Diagnosis and Treatment. New York: Marcel Dekker, 1985:301–334.

45. Barnes ND, Palumbo PJ, Hayles AB, Folgar H. Insulin resistance, skin changes, and virilization: a recessively inherited syndrome possibly due to pineal gland dysfunction. Diabetologia 1974;10:285–289.

46. Kobberling J. Genetic syndromes associated with lipoatrophic diabetes. In: Kobberling J, Neel JV, eds. The Genetics of Diabetes Mellitus. New York: Springer-Verlag, 1976:147–154.

12

Idiopathic Hirsutism

Enrico Carmina

SUMMARY

During the past two decades, the definition of idiopathic hirsutism (IH) has changed many times along with our improved understanding of the different androgen excess disorders. Currently, the diagnosis of IH requires evidence of hirsutism, normal androgens, normal ovulatory cycles, and normal ovarian morphology. Using these criteria, the prevalence of IH is generally less than 10% of hirsute women. Differentiation from other causes of hirsutism is important because in patients with IH there is no need for further endocrine and metabolic evaluation.

Key Words: Hirsutism; androgen excess; hyperandrogenism; polycystic ovary syndrome; idiopathic hirsutism; 5α-reductase; androgens.

1. INTRODUCTION

In most hirsute women, excessive body hair is a sign of a complex disorder that may determine multiple endocrine and metabolic alterations. However, in a minority of patients, hirsutism is the only abnormality and no other alterations are found. *Idiopathic hirsutism* (IH) is the term generally used to describe this form of hirsutism *(1)*. During the past two decades, a better understanding of the phenotypic heterogeneity of androgen excess disorders has modified the criteria to define IH. In particular, changes in the criteria for polycystic ovary syndrome (PCOS) has strongly influenced the calculated prevalence, diagnosis, and treatment of IH.

2. BACKGROUND

2.1. The Changing Definition of IH

Initially, the diagnosis of IH was reserved for all conditions in which the pathogenesis of hirsutism was unknown, independent of the finding of normal or elevated androgen levels *(2,3)*. Most studies did not exclude patients with nonclassic adrenal hyperplasia or with severe insulin resistance. Subsequently, the term *idiopathic hirsutism* was reserved for those forms of hirsutism associated with regular menstrual cycles *(4)*. However, this definition also was inaccurate because eumenorrhea may be present in many patients with nonclassic adrenal enzymatic deficiency *(5)*, and regular vaginal bleeding does not exclude the presence of anovulation in the hirsute patient *(6)*. In our experience, 20% of patients with classic PCOS (National Institutes of Health [NIH] 1990 definition) present with apparently regular, but anovulatory cycles *(6)*.

Because patients with ovulatory dysfunction and hyperandrogenemia are now generally considered part of PCOS *(7)*, we and others suggested that the diagnosis of IH be restricted to those hirsute patients who have normal circulating androgens and normal ovulatory function *(1,8,9)*. While this

From: *Contemporary Endocrinology: Androgen Excess Disorders in Women:*
Polycystic Ovary Syndrome and Other Disorders, Second Edition
Edited by: R. Azziz et al. © Humana Press Inc., Totowa, NJ

definition has been largely used, it is clear that current commercial methods do not generally allow for accurate measurement of circulating androgens, and the distinction between patients with high and normal androgens is often difficult. Because of this limitation, some investigators and clinicians have continued to diagnose IH in all women with clinical or biological evidence of hyperandrogenism in which PCOS and other uncommon, well-defined, hyperandrogenic disorders are not found. However, we prefer to maintain the diagnostic separation between patients with normal and those with increased androgen levels and reserve the diagnosis of IH only for patients with hirsutism, normal ovulatory cycles, normal androgen levels, and probably, normal ovarian morphology (*see* next section). It is clear that this requires a careful and accurate evaluation of serum androgens.

2.2. Prevalence of IH

In the past, IH was considered the most common form of hirsutism, and in some studies up to 55% of hirsute patients were considered affected *(10)*. However, as we understand IH currently, the prevalence of this abnormality is much lower. Studying 588 consecutive Italian patients referred because of hirsutism, we found that only 6% had normal ovulatory cycles and normal circulating androgens and could be diagnosed as affected by IH *(6)*. In an initial study evaluating 132 consecutive hirsute women from Alabama and using the same criteria, Azziz et al. found that 17% had IH *(7)*. The same authors have recently published a much larger study including 883 patients referred because of possible hyperandrogenism, noting that only 4.7% of patients were diagnosed as having IH *(11)*. In a recent study of 168 hirsute women from Turkey, IH was observed in 16% of the patients *(12)*. Although it is probable that the ethnic composition of the population may influence the prevalence of IH, in most populations the prevalence of IH appears to be relatively low (<10%). We should note that these prevalences reflected the use of the NIH 1990 criteria for PCOS, which is relatively strict.

Newer criteria for the diagnosis of PCOS were proposed at an expert conference held in Rotterdam in 2003 *(13,14)*, including the ovarian finding of polycystic ovaries. As such, the use of the Rotterdam 2003 criteria has required a reassessment of the definition of IH. In fact, in the past, patients with hirsutism who had normal androgens and regular ovulatory cycles were considered affected by IH independent of their ovarian morphology. According to the 2003 Rotterdam criteria, patients presenting with hirsutism, normal androgens, and regular ovulatory cycles, but who have polycystic ovarian morphology on ultrasonography, could be considered affected by PCOS *(15)*. Consequently, it is possible that the diagnosis of IH today requires a normal ovarian morphology. Our current definition of IH (*see* Table 1) is as follows:

- Hirsutism: Criteria for definition of hirsutism may vary depending on studied populations. We consider women who have a modified Ferriman–Gallwey index of 6 or more as hirsute *(1)*.
- Normal ovulatory cycles: We generally assess ovulation by measuring serum progesterone on days 20–24 of the cycle. Levels of serum progesterone 3 ng/mL or more are consistent with ovulation in most laboratories, although in women with normal ovulatory function they generally are *no less than 7* ng/mL.
- Normal serum androgens: We generally measure total and free testosterone and dehydroepiandrosterone sulfate (DHEAS) (*see* Chapter 5 for more on androgen measures).
- Normal ovaries at the ultrasound: The criteria for the diagnosis of polycystic ovaries are relatively strict and have been recently reviewed *(16)*.

According to these criteria, in our population the prevalence of IH is 7.6% of patients referred because of clinical evidence of hyperandrogenism *(17)*. Interestingly, about 2% of patients previously diagnosed as having IH had polycystic ovaries and, according to current definition of PCOS, can now be considered to have ovulatory PCOS. Therefore, only a small quantity of patients previously diagnosable as having IH were excluded when their ovarian morphology was considered, and the prevalence of this disorder remains similar to that previously reported by Azziz et al. *(11)* and ourselves *(8)*. In Table 2 the prevalence of IH in different studies is compared.

Table 1
Criteria of Diagnostic of Idiopathic Hirsutism

- Hirsutism
- Normal androgen levels[a]
- Normal ovulatory cycles
- Normal ovarian morphology

[a]At a minimum a normal free androgen index and dehydroepiandrosterone sulfate level.

Table 2
Prevalence of Idiopathic Hirsutism in Selected Large Series

References	No. of total subjects studied	Prevalence of idiopathic hirsutism (%)	Ovarian sonography
Azziz et al., 2004 *(11)*	883	4.7	No
Unluhirzarci et al., 2004 *(12)*	168	16.0	No
Carmina et al., 2006 *(17)*	950	7.6	Yes

2.3. Insulin Resistance in Patients With IH

Recently, some studies have shown that insulin resistance, a well-known hallmark of PCOS *(18)*, may be present in women with IH *(19,20)*. However, at least in most patients, this association seems to be primarily linked to the presence of polycystic ovaries (increased ovarian size) *(20)*, and therefore these reports may be considered confirmatory of the finding that mild insulin resistance is also present in women with ovulatory PCOS *(15,21)*. However, it is possible that some patients with IH do exhibit very mild degrees of insulin resistance. In fact, the distinction between women with IH and those with idiopathic hyperandrogenism (i.e., increased androgen levels, hirsutism, but normal ovulatory function and ovarian morphology) is unclear, and we have demonstrated that patients with idiopathic hyperandrogenism also have mild degrees of insulin resistance *(15)*.

2.4. Pathogenesis of IH

The pathogenesis of IH remains unclear. Three principal hypotheses have been suggested to account for the development of this abnormality: (1) exaggerated activity of skin 5α-reductase (5α-RA), (2) androgen receptor modifications, and (3) altered skin androgen metabolism and/or altered regulation of hair growth. These are briefly reviewed here.

2.4.1. Exaggerated Activity of Skin 5α-RA

This hypothesis suggests that the activity of 5α-RA is increased in the affected skin and is based on the studies of Serafini and Lobo *(21)*. In fact, in this study skin 5α-RA activity was as high in IH patients as it was in frankly hyperandrogenic patients. It is also known that there at least two 5α-RA isoenzymes, type 1 and type 2, which have differing tissue distribution and regulation *(23)*. We have observed that in skin the most common isoenzyme is type 1 5α-RA *(24)*, which may in part explain the modest activity of drugs like finasteride on hirsutism, which principally block the type 2 isoenzyme.

Whereas in hyperandrogenic patients 5α-RA activity may be increased secondary to induction by the androgens themselves *(22)*, in IH it has been suggested that this enzyme is increased by a mechanism independent of androgens. 5α-RA activity is regulated not only by androgens, but also by a number of other hormones and mediators also present in skin, including insulin-like growth factor

(IGF)-1, insulin, and transforming growth factor (TGF)-β *(1)*. The possibility that an alteration in the local mechanism(s) regulating 5α-RA activity is the cause of IH is currently under investigation.

2.4.2. Androgen Receptor Modifications

Although no increase in androgen receptor content has been found in patients with IH, changes in androgen receptor activity have been suggested to be operant in hirsutism *(25)*, and genetic modifications affecting the activity of androgen receptors have been found in some patients with prostate cancer *(26)*. It is possible that the same polymorphisms of the androgen receptor may play a role in IH. In some studies an inverse correlation between the number of CAG repeat trinucleotides in exon 1 of the androgen receptor and the development of hirsutism has been reported *(27)*. Androgen receptors with a low number of CAG trinucleotides have a higher affinity for androgens and may therefore promote hirsutism in some patients whose circulating androgens are normal *(27)*.

2.4.3. Altered Skin Androgen Metabolism and/or Altered Regulation of Hair Growth

Peripheral androgen metabolism may be altered in some hirsute patients. Faredin and Toth reported that some patients with IH exhibited increased 17β-hydroxysteroid dehydrogenase activity in skin, which favored the formation of testosterone from androstenedione *(28)*. It has also been suggested that hormones or mediators present in the skin, such as insulin or TGF-β, may transform inactive precursors into active androgens *(29)*. This hypothesis is a variant of the first mechanism proposed above, but does not implicate changes in 5α-RA activity. Finally, altered regulation of hair growth independent of androgens but linked to the increased production of local hormones or mediators has been hypothesized *(1)*.

2.5. Serum Markers of IH

Several studies have shown that patients with IH demonstrate increased levels of several androgen metabolites—in particular 3α-androstanediol glucuronide (3α-Adiol-G), an androgen metabolite derived from the metabolism of dihydrotestosterone *(30)*. However, serum 3α-Adiol-G levels are increased not only in women with IH, but also in other hirsute hyperandrogenic patients, probably because of the increased activity of androgen precursors *(31)*. Therefore, the measurement of 3α-Adiol-G alone is not useful in the diagnosis of IH. In fact, in clinical practice no specific assay is used to diagnose IH; rather androgen and other hormone assays are used to exclude other causes of hirsutism.

2.6. Treatment of IH

The treatment of IH is generally similar to that of other forms of hirsutism (hormonal treatment of hirsutism is discussed further in Chapter 34.) As for other forms of hirsutism, nonpharmacological hair removal (e.g., by lasers or electrolysis) has an important role in treatment of the patient with IH. However, it is important to differentiate IH from other forms of hirsutism because patients with IH do not generally need screening for endometrial or metabolic alterations, as do patients with PCOS. In fact, although in some patients with IH mild insulin resistance may be found, an increased risk of glucose intolerance or cardiovascular disease in these patients has yet to be demonstrated *(15)*. In general, a diagnosis of IH indicates that the hirsute patient should be treated only because of excessive hair growth, and not because of other related morbidities. However, because some patients with IH may actually have subclinical PCOS, long-term follow-up is warranted.

3. CONCLUSIONS

Idiopathic hirsutism is a relatively uncommon disorder that should be diagnosed in patients in whom hirsutism is the only alteration present and no changes in androgen production or increased metabolic or cardiovascular risk may be found. In these patients nonpharmacological hair removal represents the first line of treatment, although hormonal suppression is also of significant benefit in many affected women.

4. FUTURE AVENUES OF INVESTIGATION

Many issues remain to be addressed, of which one of the most important may be the need for a better definition of the phenotype—i.e., establish the exact relationship between IH and idiopathic hyperandrogenism. In fact, most commercial androgen assays are not able to detect slightly increased androgen levels, and normal androgen levels are often observed even in patients with PCOS. Therefore, a distinction between these two disorders may be artificial and difficult to attain. In addition, future studies need to develop a better understanding of the molecular etiology and genetics of IH, which may guide the search for future and better treatments.

KEY POINTS

- IH should be diagnosed in hirsute patients who have normal serum androgens, normal ovulatory cycles, and normal ovarian morphology.
- The prevalence of IH is 8% in hirsute women, but may vary between 5 and 16%, depending on the population studied and the criteria used for diagnosis.
- The pathogenesis of IH is still unclear, but alterations of local mechanisms affecting androgen activation and/or hair growth are probably involved.
- It is important to differentiate IH from other forms of hirsutism, because in IH patients there is little need for further endocrine or metabolic evaluation.

REFERENCES

1. Azziz R, Carmina E, Sawaya ME. Idiopathic hirsutism. Endocr Rev 2000;21:347–362.
2. Kirschner MA, Zucker R, Jespersen D. Idiopathic hirsutism—an ovarian abnormality. N Engl J Med1976;294:637–641.
3. Pugeat M, Forest MG, Nisula BC, et al. Evidence of excessive androgen secretion by both the ovary and the adrenal in patients with idiopathic hirsutism. Obstet Gynecol 1982;59:46–51.
4. McKenna TJ, Miller RB, Liddle GW. Plasma pregnenolone and 17-OH-pregnenolone in patients with adrenal tumors, ACTH excess or idiopathic hirsutism. J Clin Endocrinol Metab 1977;44:231–236.
5. Carmina E. Pathogenesis and treatment of hirsutism in late-onset congenital adrenal hyperplasia. Reprod Med Rev 1995;4:179–187.
6. Carmina E, Lobo RA. Do hyperandrogenic women with normal menses have PCOS? Fertil Steril 1999;71:319–322.
7. Zawdaki JK, Dunaif A. Diagnostic criteria for polycystic ovary syndrome: towards a rationale approach. In: Dunaif A,Givens JR, Haseltine F, Merriam GR, eds. Polycystic Ovary Syndrome. Boston: Blackwell Scientific Publications; 1992:377–384.
8. Carmina E. Prevalence of idiopathic hirsutism. Eur J Endocrinol 1998;139:424–427.
9. Azziz R, Waggoner WT, Ochoa T, Knochenhauer ES, Boots LR. Idiopathic hirsutism: an uncommon cause of hirsutism in Alabama. Fertil Steril 1998;70:274–278.
10. Bernasconi D, Del Monte P, Meozzi M, et al. The impact of obesity on hormonal parameters in hirsute and nonhirsute women. Metab Clin Exp 1996;45:72–75.
11. Azziz R, Sanchez LA, Knochenhauer ES, et al. Androgen excess in women: experience with over 1000 consecutive patients. J Clin Endocrinol Metab 2004;89:453–462.
12. Unluhizarci K, Gokee C, Atmaca H, Bayram F, Kelestimur F. A detailed investigation of hirsutism in a Turkish population: idiopathic hyperandrogenemia as a perplexing issue. Exp Clin Endocrinol Diabetes 2004;112:504–509.
13. Rotterdam ESHRE-ASRM Sponsored PCOS Consensus Workshop Group. Revised 2003 consensus on diagnostic criteria and long-term health risks related to polycystic ovary syndrome. Hum Reprod 2004;19:41–47.
14. Rotterdam ESHRE/ASRM-Sponsored PCOS Consensus Workshop Group. Revised 2003 consensus on diagnostic criteria and long-term health risks related to polycystic ovary syndrome. Fertil Steril 2004;81:19–25.
15. Carmina E, Longo RA, Rini GB, Lobo RA: Phenotypic variation in hyperandrogenic women influences the finding of abnormal metabolic and cardiovascular risk parameters. J Clin Endocrinol Metab 2005;90:545–549.
16. Balen AH, Laven JS, Tan SL, Dewailly D. Ultrasound assessment of the polycystic ovary: international consensus definitions. Hum Reprod Update 2003;9:505–514.
17. Carmina E, Rosato F, Jannì A, Rizzo M, Longo RA. Relative prevalence of different androgen excess disorders in 950 women referred because clinical hyperandrogenism. J Clin Endocrinol Metab 2006;91:2–6.
18. Dunaif A, Segal KR, Futterweit W, Dobrjansky A. Profound peripheral insulin resistance, independent of obesity, in polycystic ovary syndrome. Diabetes 1989;38:1165–1174.
19. Unluhizarci K, Karababa Y, Bayram F, Kelestimur F. The investigation of insulin resistance in patients with idiopathic hirsutism. J Clin Endocrinol Metab 2004;89:2741–2744

20. Frajndlich R, Spritzer PM. Association between ovarian volume and serum insulin levels in ovulatory patients with idiopathic hirsutism. Fertil Steril 2005;83:1561–1564.
21. Serafini P, Lobo RA. Increased 5-alfa reductase in idiopathic hirsutism. Fertil Steril 1985;43:74–78.
22. Mauvais-Jarvis P. Regulation of androgen receptor and 5-reductase in the skin of normal and hirsute women. Clin Endocrinol Metab 1986;15:307–317.
23. Thigpen AE, Silver RI, Guilleyard JM, et al. Tissue distribution and ontogeny of steroid 5-reductase isoenzyme expression. J Clin Invest 1993;92:903–910.
24. Carmina E. Role of 5-reductase isoenzymes in the pathogenesis of acne and hirsutism. In: Dastidar KG, Dastidar SG, Chowdurry NMR, eds. Proceedings of International Conference on Advances in Reproductive Medicine—ICARM 97, Calcutta, India, 1997, pp. 154–162.
25. Sawaya MA, Salita AR. Androgen receptor polymorphisms (CAG repeat lengths) in androgenetic alopecia, hirsutism and acne. J Cutan Med Surg 1998;3:9–15.
26. Hardy DO, Scher HL, Rogenreider T, et al. Androgen receptor CAG repeat lengths in prostate cancer: correlation with the age of onset. J Clin Endocrinol Metab 1996;81:4400–4405.
27. Legro RS, Shahbahrami B, Lobo RA, Kovacs BW. Size polymorphisms of the androgen receptor among female Hispanics and correlation with androgen characteristics. Obstet Gynecol 1994;83:701–706.
28. Faredin I, Toth I. Metabolism and concentration of androgenic steroids in the abdominal skin of women with idiopathic hirsutism. Acta Med. Hung 1984;41:19–34.
29. Danilenko DM, Ring BD, Pierce GF. Growth factors and cytokines in hair follicle development and cycling: recent insights from animal models and the potentials for clinical therapy. Mol Med Today 1996;2:460–467.
30. Horton R, Hawks D, Lobo RA. 3α,17β-Androstanediol glucuronide in plasma. J Clin Invest 1982;69:1203–1206.
31. Carmina E, Gentzschein E, Stanczyk FZ, et al. Substrate dependency of C19 conjugates in hirsute hyperandrogenic women and the influence of adrenal androgen. Hum Reprod 1995;10:299–303.

IV

The Polycystic Ovary Syndrome

13

Definition and Epidemiology
of the Polycystic Ovary Syndrome

Ricardo Azziz

SUMMARY

Polycystic ovary syndrome (PCOS) is a heterogeneous disorder of functional androgen excess, although its definition remains fluid and controversial. PCOS is characterized by clinical and/or biochemical hyperandrogenism and is frequently accompanied by ovulatory dysfunction and polycystic ovaries. PCOS is a diagnosis of exclusion, with other androgen excess and related disorders to be excluded. Two principal definitions are in use today: one arising from an expert conference sponsored by the National Institutes of Health (NIH) in 1990 (NIH 1990 criteria) and the other from another expert conference sponsored by the European Society for Human Reproduction and Embryology and the American Society for Reproductive Medicine in 2003 in Rotterdam (Rotterdam 2003 criteria). The prevalence of PCOS among unselected reproductive-aged women is at least 6.5–8.0% using the NIH 1990 criteria. Future studies must establish the prevalence of the disorder in different populations and using the various criteria proposed to define PCOS. Finally, studies on the long-term development of reproductive and metabolic abnormalities in women with the two new phenotypes of PCOS defined by Rotterdam 2003 are needed to determine whether these have risks similar to classic PCOS.

Key Words: Epidemiology; polycystic ovary syndrome; phenotyping; diagnostic criteria.

1. INTRODUCTION

The polycystic ovary syndrome (PCOS) was first described by Stein and Leventhal in 1935 *(1)* and has since been recognized as one of the most common endocrine/metabolic disorders of women *(2)*. PCOS is a heterogeneous disorder, whose principal features include androgen excess, ovulatory dysfunction, and/or polycystic ovaries. Although it is widely recognized that PCOS is a diagnosis of exclusion, the definition of PCOS remains controversial. Here, we review the definition and epidemiology of PCOS as it currently stands.

2. BACKGROUND

2.1. Defining PCOS

Although the disorder is relatively heterogeneous at present, we recognize that a relatively consistent feature of PCOS is androgen excess, or hyperandrogenism *(3)*. Hyperandrogenism is detectable either by laboratory analysis, generally measuring circulating androgen levels, or by clinical exam, primarily in the form of hirsutism. Two other features often considered fundamental to the diagnosis include ovulatory dysfunction, often but not always associated with overt menstrual dysfunction, and polycystic ovarian morphology, generally detectable by ultrasonography.

From: *Contemporary Endocrinology: Androgen Excess Disorders in Women:*
Polycystic Ovary Syndrome and Other Disorders, Second Edition
Edited by: R. Azziz et al. © Humana Press Inc., Totowa, NJ

First we should note that PCOS is a diagnosis of exclusion, such that androgen excess disorders with clearly defined etiologies are to be excluded. One disorder that is difficult to distinguish clinically from PCOS is 21-hydroxylase-deficient nonclassic adrenal hyperplasia (NCAH) *(4)*, generally detected by a basal and/or ACTH-stimulated 17-hydroxyprogesterone level *(5)*. Other disorders that can be suspected clinically include those resulting in Cushing's syndrome with hyperandrogenic features, such as adrenocortical carcinomas, Cushing's disease, and ovarian and other adrenal androgen-secreting neoplasms. The history may reveal use or abuse of anabolic drugs. Although still controversial, many investigators also consider patients with the hyperandrogenism, severe insulin resistance, and acanthosis nigricans syndrome as distinct from PCOS (*see* Chapter 11), because these women will have extreme degrees of hyperinsulinism and insulin resistance far greater than the vast majority of PCOS patients and may have other unique features, including lipodystrophy.

If ovulatory dysfunction is identified, other disorders that may result in this abnormality, such as thyroid dysfunction and hyperprolactinemia, will also need to be excluded. Nonetheless, recent studies suggest that the prevalence of these endocrine abnormalities in patients with apparent PCOS is relatively low, on the order of 1–3% *(6–9)*. Likewise, investigators generally will also exclude those disorders that can result in polycystic-like ovarian morphology, such as hypothalamic amenorrhea *(10,11)*.

Women with PCOS demonstrate a rate of obesity higher than the general population (30–60%, depending on country of origin) *(9,12–14)*, insulin resistance and hyperinsulinism (present in 50–70%) *(15–18)*, and a luteinizing hormone (LH)–to–follicle-stimulating hormone (FSH) ratio of greater than 2 or 3 (in 30–50%) *(19–24)*. However, these features have not generally been included in the diagnostic criteria, as they are either highly prevalent in disorders other than PCOS (e.g., obesity and insulin resistance) or are not observable in the majority of patients with routine laboratory assessments (e.g., an elevated LH-to-FSH ratio, because LH levels are lower in obese individuals, which accounts for a large fraction of women with PCOS).

To date, two major criteria have been proposed, with other investigators proposing modifications of these. We will review the criteria arrived at a National Institutes of Health (NIH)/National Institute of Child Health and Human Development (NICHD) expert conference sponsored in April 1990 and another co-sponsored by the European Society for Human Reproduction and Embryology (ESHRE) and the American Society for Reproductive Medicine (ASRM) in May 2003, as well as a modification of these.

2.1.1. The 1990 NIH Criteria

The definition of PCOS most commonly used today arose from the proceedings of an expert conference sponsored by the NIH in April 1990 (Table 1). Participants were surveyed, and tabulation of the results indicated that most felt that the features of PCOS were (in order of importance) (1) hyperandrogenism and/or hyperandrogenemia; (2) chronic anovulation; and (3) exclusion of related disorders such as hyperprolactinemia, thyroid disorders, and congenital adrenal hyperplasia *(25)*. Conference participants felt that polycystic ovaries were suggestive, not diagnostic, of the syndrome. Notwithstanding this omission, we should note that the results of this survey had the lucidity of identifying PCOS as an androgen excess disorder of exclusion, with ovarian consequences.

We should note that these proceedings did not provide clear guidelines as to how to define each criterion. Clinical hyperandrogenism has generally been interpreted as hirsutism, because more than 70% of hirsute women have PCOS *(9)*. At a minimum, hyperandrogenemia generally is interpreted to be an elevated free testosterone (T) level, observable in approximately 70% *(26)* of affected women. Measuring total T, dehydroepiandrosterone sulfate, and androstenedione levels will increase the number of patients diagnosed as hyperandrogenemic *(26,27)*. As noted in Chapter 5, close attention must be paid to the quality of the androgen assays.

Chronic anovulation is generally defined as menstrual cycles (or more accurately vaginal bleeding episodes) at no less than 35-day intervals *(28,29)* or no more than 10 bleeds per year (i.e., 365 days/35

Table 1
Proposed Criteria for the Definition of PCOS

NIH, 1990 *(25)*:

    To include all of the following:

       1. Hyperandrogenism and/or hyperandrogenemia
       2. Chronic anovulation
       3. Exclusion of related disorders

ESHRE/ASRM (Rotterdam), 2003 *(36,37)*:

    To include at least two of the following, in addition to exclusion of related disorders:

       1. Oligo-anovulation
       2. Hyperandrogenism and/or hyperandrogenemia
       3. Polycystic ovaries

Modified NIH criteria[a] *(45)*:

    To include all of the following:

       1. Androgen excess (clinical and/or biochemical hyperandrogenism)
       2. Ovarian dysfunction (oligo-anovulation and/or polycystic ovarian morphology)
       3. Exclusion of other androgen excess or ovulatory disorders

PCOS, polycystic ovary syndrome; NIH, National Institutes of Health; ESHRE, European Society for Human Reproduction and Embryology; ASRM, American Society for Reproductive Medicine.

days = 10.4). Although other investigators define oligomenorrhea as cycles at no less than 43-day intervals or no more than 8 bleeds per year *(30)*, they also note that cycle length varies with age, such that women between the ages of 27 and 42 years (which includes most women with PCOS) usually have a maximum (within 3 standard deviations) cycle length of less than 35 days *(31)*. Although most women with PCOS will have overt oligomenorrhea, about 15% will present with a history of regular menses, and oligo-ovulation in these women is only discoverable by monitoring the luteal (cycle day 20–24) progesterone levels. The disorders that may be considered for exclusion already have been discussed.

Three principal phenotypes of PCOS are recognized using the NIH 1990 criteria, including women with (a) hirsutism, hyperandrogenemia, and oligo-ovulation; (b) hyperandrogenemia and oligo-ovulation; or (c) hirsutism and oligo-ovulation. The overall prevalence of these phenotypes, at least in one large study, was approximately 50%, 30%, and 20%, respectively *(26)*. There were no differences in mean age, mean body mass index (BMI), mean waist-to-hip ratio, racial distribution, severity of oligomenorrhea, or prevalence of family history for hyperandrogenism between the phenotypes. The exception was the fasting insulin levels, which were highest in patients with hirsutism, hyperandrogenemia, and oligo-ovulation and lowest in those women with oligo-ovulation and hirsutism only. Whether different mechanisms underlie the development of these phenotypes remains to be demonstrated.

Overall, the NIH criteria have proven extremely useful to begin to define and understand, among other features, the high prevalence of the disorder *(2,32,33)*, its high frequency of insulin resistance *(18,34)*, and the considerable risk of these women for developing type 2 diabetes mellitus (DM) *(16,35)*.

2.1.2. The 2003 ESHRE/ASRM (Rotterdam) Criteria

Another expert conference was organized in Rotterdam in May 2003 (Table 1), in part sponsored by the ESHRE and the ASRM. The proceedings of the conference noted that PCOS could be diagnosed, after the exclusion of related disorders, by at least two of the following three features: (a) oligo- or anovulation, (b) clinical and/or biochemical signs of hyperandrogenism, or (c) polycystic ovaries *(36,37)*. It should be noted that these recommendations did not replace the NIH 1990 criteria; rather they expanded the definition of PCOS. Additional phenotypes now to be considered as having PCOS included (1) women with polycystic ovaries with clinical and/or biochemical evidence of androgen excess, but no signs of ovulatory dysfunction, and (2) women with polycystic ovaries and ovulatory dysfunction, but no signs of androgen excess. Whether or not these additional phenotypes actually represent PCOS remains to be determined. However, it is useful to briefly review the current data supporting or refuting this argument.

One possible approach to determining whether these new phenotypes are part of the syndrome known as PCOS is to compare long-term morbidities. Ovulatory hyperandrogenic patients with polycystic-appearing ovaries appear to have features that approximate patients with PCOS defined by the NIH 1990 criteria, although of a milder nature. These include slight excess in circulating LH, insulin, and other markers for cardiovascular disease *(38–41)*. Essentially, these patients can be considered to have "mild PCOS" and, in the absence of long-term follow-up studies, may not have the same degree of reproductive or metabolic consequences as women with the full PCOS phenotype.

Alternatively, it is even less clear that women with ovulatory dysfunction and polycystic ovaries, but without any evidence of hyperandrogenism, as a group have a similar morbidity to patients with PCOS (e.g., insulin resistance). For example, in a study of 235 women aged 40–42 years, polycystic-appearing ovaries were observed in 20.8% *(42)*. Of women with polycystic ovaries, the differences in the prevalence of irregular cycles and infertility were not significantly different compared to subjects with normal ovaries (41 vs 27%, and 16 vs 15%, respectively). Only 14% of women with polycystic ovaries demonstrated hirsutism. These data would suggest that the prevalence of PCOS among women with polycystic ovaries might not be higher than 15–25%. In addition, the prevalence of polycystic-appearing ovaries does not appear to predict abnormalities in insulin sensitivity either in women with PCOS *(43)* or in their sisters *(44)*. Finally, it is uncertain how patients with hypothalamic amenorrhea and polycystic ovaries will be differentiated from non-obese women with PCOS, an important conundrum considering the significant differences in long-term morbidity (e.g., bone loss for hypothalamic amenorrhea and increased bone mass for PCOS, increased risk of diabetes in PCOS but not hypothalamic amenorrhea patients). Overall, it may be prudent to withhold expanding the definition of PCOS much beyond that originally suggested by the NIH 1990 expert conference until more complete epidemiological and long-term data are available.

2.1.3. Modified NIH Criteria

In order to accommodate current epidemiological data, a modification of the NIH criteria has been proposed (Table 1 *[45]*). This definition suggests that PCOS should have three features: (1) androgen excess (clinical and/or biochemical hyperandrogenism), (2) ovarian dysfunction (oligo-anovulation and/or polycystic ovarian morphology), and (3) exclusion of other androgen excess or ovulatory disorders *(45)*. This definition then recognizes one additional phenotype to those noted by the NIH 1990 criteria, namely those women with polycystic ovaries, hyperandrogenism, and apparently normal ovulation. This definition does not eliminate the possibility that future research may demonstrate that the subset of women with polycystic ovaries and ovulatory dysfunction and without overt androgen excess may actually have PCOS. However, expanding the definition of PCOS without good supporting data may have significant detrimental implications for research (e.g., increased heterogeneity of the study populations), clinical practice (e.g., requiring that all these patients undergo ultrasonography), and patient quality of life (e.g., requiring long-term monitoring for the development of associated metabolic morbidities and potentially adversely affecting health insurability).

2.2. Prevalence of PCOS

Clearly the prevalence of PCOS will depend to some degree on the criteria used to define this disorder. The prevalence of PCOS has been determined primarily using the NIH 1990 criteria and in populations of white or Caucasian women with a few exceptions. Studying 277 women seeking a preemployment physical in the southeastern United States, we initially reported a prevalence of PCOS

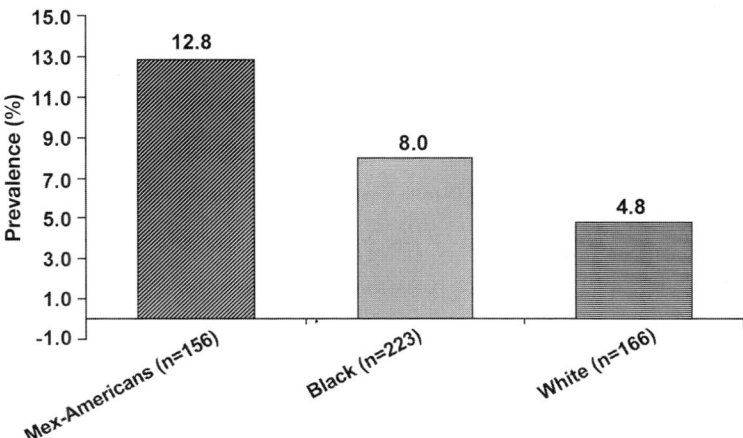

Fig. 1. Potential racial differences in the prevalence of PCOS in unselected women in the United States. Whether these data are confirmed in further studies and/or whether these differences are the product of differences in the population prevalence of insulin resistance remains to be confirmed. (Adapted from refs. *2* and *54*.)

of 4%, not significantly different between whites and blacks *(27)*. In a subsequent and more intensive study of 400 unselected consecutive women ages 18–45 years in the same setting (223 black, 166 white, 11 of other races), the prevalence of PCOS was observed to be 6.6%, still not significantly different between blacks and whites (8 and 4.8%, respectively) *(2)*. Nonetheless, the lack of a significant difference in the prevalence of PCOS between black and white women may solely reflect inadequate sample size.

Also using the 1990 NIH criteria, a study of 192 Greek women on the island of Lesbos, recruited through the promise of a free medical exam, reported a prevalence of PCOS of 6.8% *(32)*. Another study of 154 Caucasian blood donors in Madrid, Spain, found a similar prevalence (6.5%) *(33)*. Among 230 volunteers (97% white) recruited from two Oxford universities and two general practice surgeries who agreed to participate in "a study of women's health issues," the prevalence of PCOS using the NIH 1990 criteria was 8% *(46)*. These data indicate that the prevalence of clinically evident PCOS using the 1990 NIH criteria in unselected women of reproductive age ranges from 6.5 to 8.0%, affecting 1 in 13–15 unselected women. This translates to at least 5 million affected women in the United States and 105 million worldwide. Furthermore, as many women in the aforementioned studies were on hormonal contraceptives, it is probable that this figure underestimates the true prevalence.

A number of conditions are associated with increased prevalences of PCOS, including obesity *(47,48)*, insulin resistance *(48)*, type 1 or type 2 DM *(49–51)*, and oligo-ovulatory infertility *(40,52,53)*. The prevalence of PCOS also appears to be higher among Mexican-Americans than among white or black (African-American) women *(54)* (Fig. 1). Finally, the prevalence of PCOS seems higher among populations reporting premature adrenarche *(55)*, gestational diabetes *(56,57)*, and logically, in first-degree relatives of PCOS patients *(58,59)*.

The use of more expansive definitions for PCOS, such as the Rotterdam 2003 criteria, would logically raise the prevalence further. In their study, Michelmore and colleagues defined PCOS by the presence of polycystic ovaries on ultrasound plus one additional feature, including menstrual irregularity, acne, hirsutism, BMI greater than 25 kg/m^2, raised serum T, or raised LH (>10 IU/L) *(46)*. Applying these criteria, 26% of the 224 women undergoing a transabdominal sonography had evidence of PCOS, compared with only 8% if the NIH 1990 criteria were used. However, these features also occurred frequently in women without polycystic ovaries, and 112 of the 150 women (75%) with normal ovaries had the presence of one or more of these attributes. We can only conclude that the number of women diagnosable as having PCOS could possibly double or even triple if the Rotterdam 2003 criteria are used. Further studies comparing the prevalence of PCOS using the NIH 1990 and the Rotterdam 2003 criteria are needed.

3. CONCLUSIONS

PCOS is a heterogeneous disorder of functional androgen excess, detectable either by laboratory analysis or by clinical exam, with ovulatory dysfunction and polycystic ovarian morphology affecting a large proportion of patients. PCOS is a diagnosis of exclusion, with other androgen excess or related disorders to be ruled out. The first broadly used definition of PCOS arose from the proceedings of an expert conference sponsored by the NIH in 1990, which noted the features of PCOS to be (in order of importance) (1) hyperandrogenism and/or hyperandrogenemia; (2) chronic anovulation; and (3) exclusion of related disorders such as hyperprolactinemia, thyroid disorders, and congenital adrenal hyperplasia. Another expert conference held in Rotterdam in 2003 expanded the NIH 1990 criteria for PCOS, noting that the disorder could be diagnosed by having at least two of the following three features: (1) oligo- or anovulation, (2) clinical and/or biochemical signs of hyperandrogenism, or (3) polycystic ovaries, after the exclusion of related disorders. This definition created two new phenotypes for PCOS: (1) women with polycystic ovaries and ovulatory dysfunction, but no signs of androgen excess, and (2) women with polycystic ovaries with clinical and/or biochemical evidence of androgen excess, but no signs of ovulatory dysfunction. Whether or not these phenotypes actually represent PCOS remains to be determined.

To accommodate current epidemiological data, a modification of the NIH criteria has been proposed, defining PCOS by three features: (1) androgen excess (clinical and/or biochemical hyperandrogenism), (2) ovarian dysfunction (oligo-anovulation and/or polycystic ovarian morphology), and (3) exclusion of other androgen excess or ovulatory disorders. Clearly the prevalence of PCOS will depend to some degree on the criteria used to define this disorder, although using the NIH 1990 criteria, most studies have observed 6.5–8.0% prevalence in unselected reproductive-aged women. The prevalence of PCOS is increased in the presence of obesity, insulin resistance, type 1 or type 2 DM, oligo-ovulatory infertility, premature adrenarche, prior gestational diabetes, and first-degree relatives of PCOS. Further investigation is required to better establish the definition of PCOS.

4. FUTURE AVENUES OF INVESTIGATION

Studies of the prevalence of PCOS in large populations of unselected women from different ethnic, racial, and geographic groups is required and may yield important information that will assist in the elucidation of the respective roles of genetics and environment in the development of the disorder. The prevalence of the disorder using the various criteria proposed to define PCOS must also be established. Finally, studies on the long-term development of reproductive and metabolic abnormalities in women with the two new phenotypes of PCOS defined by Rotterdam 2003 are needed to determine whether these have risks similar to classic PCOS.

KEY POINTS

- PCOS is a heterogeneous disorder of functional androgen excess, ovulatory dysfunction, and polycystic ovarian morphology.
- PCOS is a diagnosis of exclusion, with other androgen excess or related disorders to be ruled out.
- Two major definitions are in use: one established by an expert conference sponsored by the NIH in 1990, and the other from an expert conference held in Rotterdam in 2003 and sponsored by the ESHRE and ASRM.
- The NIH 1990 definition states that the features of PCOS are (in order of importance) (1) hyperandrogenism and/or hyperandrogenemia; (2) chronic anovulation; and (3) exclusion of related disorders such as hyperprolactinemia, thyroid disorders, and congenital adrenal hyperplasia.

- Rotterdam 2003 defines PCOS as having two of the following three features: (1) oligo- or anovulation, (2) clinical and/or biochemical signs of hyperandrogenism, and (3) polycystic ovaries, after the exclusion of related disorders.
- A modification of the NIH criteria has been proposed, defining PCOS by three features: (1) androgen excess (clinical and/or biochemical hyperandrogenism), (2) ovarian dysfunction (oligo-anovulation and/or polycystic ovarian morphology), and (3) exclusion of other androgen excess or ovulatory disorders.
- While the prevalence of PCOS will vary according to the criteria used, using the NIH 1990 criteria, prevalences of 6.5–8.0% have been observed.
- The prevalence of PCOS is increased in the presence of obesity, insulin resistance, type 1 or type 2 DM, oligo-ovulatory infertility, premature adrenarche, prior gestational diabetes, and first-degree relatives of PCOS.

ACKNOWLEDGMENTS

Supported in part by NIH grants R01-HD29364 and K24-HD01346-01 (to RA).

REFERENCES

1. Stein IF, Leventhal NL. Amenorrhea associated with bilateral polycystic ovaries. Am J Obstet Gynecol 1935;29:181–191.
2. Azziz R, Woods KS, Reyna R, et al. The prevalence and features of the polycystic ovary syndrome in an unselected population. J Clin Endocrinol Metab 2004;89:2745–2749.
3. Azziz R. Androgen excess is the key element in the polycystic ovary syndrome. Fertil Steril 2003;80:252–254.
4. Moran C, Azziz R. 21-Hydroxylase deficient non-classic adrenal hyperplasia: the great pretender. Sem Reprod Med 2003;21:295–300.
5. Azziz R, Hincapie LA, Knochenhauer ES, et al. Screening for 21-hydroxylase deficient non-classic adrenal hyperplasia among hyperandrogenic women: a prospective study. Fertil Steril 1999;72:915–925.
6. Ferriman D, Purdie AW. The aetiology of oligomenorrhoea and/or hirsuties: a study of 467 patients. Postgrad Med J 1983;59:17–20.
7. O'Driscoll JB, Mamtora H, Higginson J, et al. A prospective study of the prevalence of clear-cut endocrine disorders and polycystic ovaries in 350 patients presenting with hirsutism or androgenic alopecia. Clin Endocrinol 1994;41:231–236.
8. Moran C, Tapia M del C, Hernandez E, et al. Etiological review of hirsutism in 250 patients. Arch Med Res 1994;25:311–314.
9. Azziz R, Sanchez LA, Knochenhauer ES, et al. Androgen excess in women: Experience with over 1000 consecutive patients. J Clin Endocrinol Metab 2004;89:453–462.
10. Futterweit W, Yeh HC, Mechanick JI. Ultrasonographic study of ovaries of 19 women with weight loss-related hypothalamic oligo-amenorrhea. Biomed Pharmacother 1988;42:279–283.
11. Ardaens Y, Robert Y, Lemaitre L, et al. Polycystic ovarian disease: contribution of vaginal endosonography and reassessment of ultrasonic diagnosis. Fertil Steril 1991;55:1062–1068.
12. Legro RS, Kunselman AR, Dunaif A. Prevalence and predictors of dyslipidemia in women with polycystic ovary syndrome. Am J Med 2001;111:607–613.
13. Carmina E, Legro RS, Stamets K, et al. Difference in body weight between American and Italian women with polycystic ovary syndrome: influence of the diet. Hum Reprod 2003;18:2289–2293.
14. Hahn S, Tan S, Elsenbruch S, et al. Clinical and biochemical characterization of women with polycystic ovary syndrome in North Rhine-Westphalia. Horm Metab Res. 2005;37:438–444.
15. Dunaif A, Segal KR, Futterweit W, et al. Profound peripheral insulin resistance, independent of obesity, in polycystic ovary syndrome. Diabetes 1989;38:1165–1174.
16. Legro RS, Kunselman AR, Dodson WC, et al. Prevalence and predictors of risk for type 2 diabetes mellitus and impaired glucose tolerance in polycystic ovary syndrome: a prospective, controlled study in 254 affected women. J Clin Endocrinol Metab 1999;84:165–169.
17. Carmina E, Lobo RA. Use of fasting blood to assess the prevalence of insulin resistance in women with polycystic ovary syndrome. Fertil Steril 2004;82:661–665.
18. DeUgarte CM, Bartolucci AA, Azziz R. Prevalence of insulin resistance in the polycystic ovary syndrome using the homeostasis model assessment. Fertil Steril 2005;83:1454–1460.
19. Rebar R, Judd HL, Yen SSC, et al. Characterization of the inappropriate gonadotropin secretion in polycystic ovary syndrome. J Clin Invest 1976;57:1320–1329.
20. Conway GS, Honour JW, Jacobs HS. Heterogeneity of the polycystic ovary syndrome: clinical, endocrine and ultrasound features in 556 patients. Clin Endocrinol (Oxf) 1989;30:459–470.
21. Anttila L, Ding Y-Q, Ruutiainen K, et al. Clinical features and circulating gonadotropin, insulin, and androgen interactions in women with polycystic ovarian disease. Fertil Steril 1991;55:1057–1061.

22. van Santbrink EJ, Hop WC, Fauser BC. Classification of normogonadotropic infertility: polycystic ovaries diagnosed by ultrasound versus endocrine characteristics of polycystic ovary syndrome. Fertil Steril 1997;67:452–458.

23. Arroyo A, Laughlin GA, Morales AJ, et al. Inappropriate gonadotropin secretion in polycystic ovary syndrome: influence of adiposity. J Clin Endocrinol Metab 1997;82:3728–3733.

24. Taylor AE, McCourt B, Martin KA, et al. Determinants of abnormal gonadotropin secretion in clinically defined women with polycystic ovary syndrome. J Clin Endocrinol Metab 1997;82:2248–2256.

25. Zawadzki JK, Dunaif A. Diagnostic criteria for polycystic ovary syndrome: towards a rational approach. In: Dunaif A, Givens JR, Haseltine FP, Merriam GR, eds. Polycystic Ovary Syndrome. Boston: Blackwell Scientific Publications, 1992:377–384.

26. Chang W, Knochenhauer ES, Bartolucci AA, Azziz R. Phenotypic spectrum of the polycystic ovary syndrome (PCOS): clinical and biochemical characterization of the major clinical subgroups. Fertil Steril 2005;83:1717–1723.

27. Knochenhauer ES, Key TJ, Kahsar-Miller M, et al. Prevalence of the polycystic ovary syndrome in unselected black and white women of the southeastern United States: a prospective study. J Clin Endocrinol Metab 1998;83:3078–3082.

28. Treloar AE, Boynton RE, Behn BG, et al. Variation of the human menstrual cycle through reproductive life. Int J Fertil 1967;12:77–126.

29. Chiazze L Jr, Brayer FT, Macisco JJ Jr, et al. The length and variability of the human menstrual cycle. JAMA 1968;203:377–380.

30. Harlow SD, Zeger SL. An application of longitudinal methods to the analysis of menstrual diary data. J Clin Epidemiol 1991;44:1015–1025.

31. Harlow SD, Lin X, Ho MJ. Analysis of menstrual diary data across the reproductive life span applicability of the bipartite model approach and the importance of within-woman variance. J Clin Epidemiol 2000;53:722–733.

32. Diamanti-Kandarakis E, Kouli CR, Bergiele AT, et al. A survey of the polycystic ovary syndrome in the Greek island of Lesbos: hormonal and metabolic profile. J Clin Endocrinol Metab 1999;84:4006–4011.

33. Asuncion M, Calvo RM, San Millan JL, et al. A prospective study of the prevalence of the polycystic ovary syndrome in unselected Caucasian women from Spain. J Clin Endocrinol Metab 2000;85:2434–2438.

34. Legro RS, Finegood D, Dunaif A. A fasting glucose to insulin ratio is a useful measure of insulin sensitivity in women with polycystic ovary syndrome. J Clin Endocrinol Metab 1998;83:2694–2698.

35. Ehrmann DA, Kasza K, Azziz R, Legro RS, Ghazzi MN; PCOS/Troglitazone Study Group. Effects of race and family history of type 2 diabetes on metabolic status of women with polycystic ovary syndrome. J Clin Endocrinol Metab 2005;90:66–71.

36. The Rotterdam ESHRE/ASRM-Sponsored PCOS Consensus Workshop Group. Revised 2003 consensus on diagnostic criteria and long-term health risks related to polycystic ovary syndrome. Fertil Steril 2004;81:19–25.

37. The Rotterdam ESHRE/ASRM-Sponsored PCOS Consensus Workshop Group. Revised 2003 consensus on diagnostic criteria and long-term health risks related to polycystic ovary syndrome (PCOS). Hum Reprod 2004;19:41–47.

38. Carmina E, Lobo RA. Polycystic ovaries in hirsute women with normal menses. Am J Med 2001;111:602–606.

39. Eden JA, Place J, Carter GD, et al. Is the polycystic ovary a cause of infertility in the ovulatory woman? Clin Endocrinol (Oxf) 1989;30:77–82.

40. Kousta E, White DM, Cela E, et al. The prevalence of polycystic ovaries in women with infertility. Hum Reprod 1999;14:2720–2723.

41. Carmina E, Chu MC, Longo RA, et al. Phenotypic variation in hyperandrogenic women influences the findings of abnormal metabolic and cardiovascular risk parameters. J Clin Endocrinol Metab 2005;90:2545–2549.

42. Cresswell JL, Barker DJ, Osmond C, et al. Fetal growth, length of gestation, and polycystic ovaries in adult life. Lancet 1997;350:1131–1135.

43. Legro RS, Chiu P, Kunselman AR, et al. Polycystic ovaries are common in women with hyperandrogenic chronic anovulation but do not predict metabolic or reproductive phenotype. J Clin Endocrinol Metab 2005;90:2571–2579.

44. Raskauskiene D, Jones PW, Govind A, et al. Do polycystic ovaries on ultrasound scan indicate decreased insulin sensitivity in sisters of women with polycystic ovary syndrome? J Clin Endocrinol Metab 2005;90:2063–2067.

45. Azziz R. Diagnostic criteria for polycystic ovary syndrome: a reappraisal. Fertil Steril 2005;83:1343–1346.

46. Michelmore KF, Balen AH, Dunger DB, Vessey MP. Polycystic ovaries and associated clinical and biochemical features in young women. Clin Endocrinol (Oxf) 1999;51:779–786.

47. Hartz AJ, Barboriak PN, Wong A, et al. The association of obesity with infertility and related menstrual abnormalities in women. Int J Obes 1979;3:57–73.

48. Heinonen S, Korhonen S, Hippelainen M, et al. Relationship of the metabolic syndrome and obesity to polycystic ovary syndrome: a controlled, population-based study. Am J Obstet Gynecol 2001;184:289–296.

49. Escobar-Morreale HF, Roldan B, Barrio R, et al. High prevalence of the polycystic ovary syndrome and hirsutism in women with type 1 diabetes mellitus. J Clin Endocrinol Metab 2000;85:4182–4187.

50. Conn JJ, Jacobs HS, Conway GS. The prevalence of polycystic ovaries in women with type 2 diabetes mellitus. Clin Endocrinol (Oxf) 2000;52:81–86.

51. Peppard HR, Marfori J, Iuorno M, et al. Prevalence of polycystic ovary syndrome among premenopausal women with type 2 diabetes. Diabetes Care 2001;24:1050–1052.
52. Hull MG. Ovulation failure and induction. Clin Obstet Gynaecol 1981;8:753–785.
53. Allen SE, Potter HD, Azziz R. Prevalence of hyperandrogenemia among nonhirsute oligo-ovulatory women. Fertil Steril 1997;67:569–572.
54. Goodarzi MO, Quinones MJ, Azziz R, Rotter JI, Hsueh WA, Yang H. Polycystic ovary syndrome in Mexican-Americans: prevalence and association with the severity of insulin resistance. Fertil Steril 2005;84:766–769.
55. Ibanez L, Dimartino-Nardi J, Potau N, et al. Premature adrenarche—normal variant or forerunner of adult disease? Endocr Rev 2000;21:671–696.
56. Holte J, Gennarelli G, Wide L, et al. High prevalence of polycystic ovaries and associated clinical, endocrine, and metabolic features in women with previous gestational diabetes mellitus. J Clin Endocrinol Metab 1998;83:1143–1150.
57. Anttila L, Karjala K, Penttila RA, et al. Polycystic ovaries in women with gestational diabetes. Obstet Gynecol 1998;92:13–16.
58. Legro RS, Driscoll D, Strauss JF 3rd, et al. Evidence for a genetic basis for hyperandrogenemia in polycystic ovary syndrome. Proc Natl Acad Sci USA 1998;95:14956–14960.
59. Kahsar-Miller M, Nixon C, Boots LR, Go RCP, et al. Prevalence of the polycystic ovary syndrome (PCOS) among first degree relatives of patients with PCOS. Fertil Steril 2001;75:53–58.

14

Clinical Features of the Polycystic Ovary Syndrome

Walter Futterweit, Evanthia Diamanti-Kandarakis, and Ricardo Azziz

SUMMARY

The features of polycystic ovary syndrome (PCOS) may be elicitable on obtaining the history or on physical exam. Biochemical or radiographic evaluation may also reveal additional features (*see* Chapters 16 and 17). The medical history will provide important information regarding the presenting complaint, the onset and progression of hyperandrogenic signs and symptoms, menstrual dysfunction and irregularity, and weight gain. Usually, the development of hyperandrogenism in PCOS is slowly progressive, with stable symptomatology by the mid to late 30s. As expected, the vast majority of patients with PCOS will present with a history of oligoamenorrhea, although a few may present with a history of eumenorrhea or polymenorrhea. Some may relate a history of peripubertal weight gain, and early development of obesity can be associated with the development of PCOS. In addition, many patients have symptoms suggestive of obstructive sleep apnea. The family history is frequently informative, indicating that relatives demonstrated hyperandrogenic symptoms reminiscent of PCOS or suffered from metabolic disorders most notably type 2 diabetes mellitus. Features of PCOS that may be elicited upon the physical exam include dermatological evidence of hyperandrogenism (hirsutism, acne, or alopecia), evidence of android body fat distribution, and evidence of metabolic dysfunction, including acanthosis and acrochordons. Clinical features of virilization or Cushing's syndrome are not seen in patients with PCOS, generally representing the presence of androgen-secreting neoplasms or Cushing's disease. In general, the findings elicited by the medical history and physical exam provide the strongest suggestion that the patient suffers from PCOS, and biochemical and radiographic evaluation only provide confirmatory evidence or exclude related disorders.

Key Words: Acne; alopecia; hirsutism; oligomenorrhea; amenorrhea; polycystic ovary syndrome; dysfunction; acanthosis nigricans; obesity; obstructive sleep apnea.

1. INTRODUCTION

Polycystic ovary syndrome (PCOS) may present with a variety of clinical manifestations (Table 1), which may differ according to ethnic, racial, and environmental factors, and the co-existing presence of, among other factors, obesity and insulin resistance. None of the clinical features observed, taken singly, are pathognomonic. However, the most commonly associated clinical abnormalities observed in PCOS are hyperandrogenism, oligomenorrhea, and polycystic ovaries, although the specific phenotypes considered to represent the disorder may vary according to the definition used (*see* Chapter 13). In addition, a number of other clinical traits may be evident in women with PCOS. We present here an overview of the various clinical features observable in the disorder and, when possible, provide an estimate of the frequency with which each occurs and the impact that race, obesity, and other factors may have on their prevalence or severity.

From: *Contemporary Endocrinology: Androgen Excess Disorders in Women:*
Polycystic Ovary Syndrome and Other Disorders, Second Edition
Edited by: R. Azziz et al. © Humana Press Inc., Totowa, NJ

Table 1
Clinical Manifestations of Polycystic Ovary Syndrome

Menstrual abnormalities[a]
Excess facial and body terminal hair growth, and hirsutism
Acne
Alopecia
Obesity and central (visceral) fat distribution
Obstructive sleep apnea
Acanthosis nigricans and acrochordons
Polycystic ovaries

[a]Including oligo-amenorrhea, polymenorrhea, and dysfunctional uterine bleeding.

2. BACKGROUND

The features of PCOS may be elicited when obtaining the medical history or upon the physical examination. The ovarian morphological and the biochemical and metabolic features of the disorder are discussed in Chapters 16 and 17. The evaluation of patients with androgen excess, including PCOS, is discussed in Chapter 33.

2.1. Features Elicitable by Medical History

In this section, we will outline important information that the medical history will elicit concerning the features of PCOS, including the presenting complaint, the onset and progression of hyperandrogenic signs and symptoms, menstrual dysfunction and irregularity, weight gain, symptoms of obstructive sleep apnea, and the family history.

2.1.1. Presenting Complaints

Patients with PCOS may present complaining of irregular or unpredictable menstrual cycles, unwanted hair growth, acne or scalp hair loss, or unexplained weight gain or overweight (*see* Section 2.1.4.). Another frequent presenting complaint of PCOS may be infertility, possibly associated with recurrent first trimester miscarriages. Approximately 30–50% of PCOS patients will complain of infertility at the time they are seen for their first visit *(1,2)*.

2.1.2. Onset and Progression of Hyperandrogenic Signs and Symptoms

The timing of the development of symptoms is important in assessing the etiology of hyperandrogenism. As such it is helpful to have the patient's mother present for at least initial part of the evaluation. A history of perimenarcheal changes in skin quality, such as the development of seborrhea or acne, the darkening and coarsening of hairs, or the appearance of new unwanted hairs, is common in patients with PCOS. A history of premature adrenarche or early pubarche may also be elicited *(3)* as well as a history of low birthweight *(4,5)*. Nevertheless, the prevalence of these abnormalities of birth and pubarche among women with PCOS remains to be determined.

The development of hyperandrogenism in PCOS is usually associated with a slowly progressive clinical history and stable symptomatology by the mid to late 30s. As patients enter the fourth decade of life, their symptoms may improve, possibly associated with a decline in circulating androgen levels *(6)*. Alternatively, the rapid development of symptoms, including hirsutism, oligo-amenorrhea, severe acne and alopecia, increased muscularity, and clitoromegaly, is indicative of a virilizing syndrome, most commonly caused by an androgen-producing neoplasm *(7)*. Finally, the use and abuse of anabolic or androgenic drugs and the use of medications that may promote hair growth, albeit primarily vellus in nature (e.g., phenytoin), should be considered.

2.1.3. Menstrual Dysfunction and Irregularity

PCOS accounts for 75% of patients with anovulatory infertility *(8,9)*, 30–40% of secondary amenorrhea, and 85–90% of women with oligomenorrhea *(10)*. In turn, ovulatory dysfunction affects a high proportion of patients with PCOS (80–100%, depending on criteria), and in a large series of patients diagnosed by varying criteria 75–85% of PCOS women demonstrated clinically evident menstrual irregularity *(2,11,12)*. In agreement, in a prospective study of PCOS patients diagnosed among 400 unselected women from the general population, approximately 60% of the patients had clinically evident menstrual dysfunction *(13)*. Menstrual dysfunction is usually characterized by infrequent or absent menstrual bleeding, usually not preceded by premenstrual symptoms that may include mood changes, bloating, or breast fullness or tenderness, suggesting anovulation *(14,15)*. Oligomenorrhea is generally defined as menstrual cycles (or more accurately vaginal bleeding episodes) at no less than 35-day intervals or no less than10 bleeds per year *(16,17)*. More subtle changes in menstrual cyclicity may occur in PCOS, including intermittent anovulation and associated dysfunctional uterine bleeding *(18–20)*. As PCOS patients enter the fourth decade of life, their menstrual function often becomes more regular *(21)*.

Not all PCOS patients demonstrate a clinically apparent abnormality in their cyclic vaginal bleeding pattern. Although many patients and clinicians refer to these bleeding episodes as "menses" or "periods," strictly speaking the term *menstrual period* properly refers to the cyclic vaginal bleeding that results from the estrogen and progestogen decline (withdrawal) that occurs at the end of the luteal phase of an ovulatory cycle in women who do not conceive. Up to 40% of women with oligo-ovulation secondary to PCOS can actually present with a history suggestive of "normal menstruation" (i.e., eumenorrhea) *(22,23)*. Consequently, the periodic vaginal bleeding that occurs in women suspected of PCOS cannot be taken as proof of ovulation until proven.

An adolescent history of erratic or infrequent menstrual cycles often predicts the development of PCOS. Van Hooff and colleagues followed a cohort of 766 adolescents beginning at age 15 through age 18 *(24)*. They noted that 2% of adolescents who initially had regular menstrual cycles developed oligomenorrhea, and 12% of those with irregular menstrual cycles did so; notably, 51% of the adolescents who were oligomenorrheic at 15 years old remained oligomenorrheic at follow-up. An increase in body mass index (BMI) or luteinizing hormone, androstenedione, or testosterone (T) levels and the presence of polycystic ovaries were associated with persistence of oligomenorrhea, although in a multivariate analysis only the BMI consistently contributed to predicting the persistence of oligomenorrhea.

2.1.4. Weight Gain

Peripubertal weight gain and the early development of obesity can be associated with the development of PCOS. The perimenarcheal onset of symptoms in PCOS is common, but this may occur later with development of a significant increased weight gain, and the latter is often an inciting factor in the development of menstrual dysfunction and cutaneous signs of hyperandrogenism. As noted by van Hooff and colleagues *(24)*, BMI is a predictor of the persistence of oligomenorrhea into adolescence and presumably the development of PCOS. However, in this study both obese and normal-weight oligomenorrheic adolescents were at high risk of remaining oligomenorrheic. Studying 230 young women ages 18–25 years, Michelmore and colleagues noted that the BMI was slightly, but significantly, higher among girls with compared to those without, PCOS *(25)*. Overall, peripubertal or adolescent obesity is a modest predictor of the development of PCOS.

Although any weight gain is of concern, the rate of gain that should elicit concern is less clear. In a study of 5117 individuals followed for 20 years beginning in 1971, white women ages 25–35 years at baseline gained 7.7 kg over the study period, or 0.39 kg/year (i.e., 0.85 lb/year), whereas black women of the same age gained 10.9 kg, or 0.54 kg/year (1.2 lb/year) *(26)*. In populations at high risk for obesity and insulin resistance (e.g., Pima Indians) this weight gain was higher (1.4 kg/year or 3 lb/year), a gain that was primarily attributable to baseline total energy intake *(27)*. In contrast, most women with PCOS will relate a history of either a sudden gain of 14–36 kg (30–80 lb) or more over a period of 6–12 months or, more commonly, a persistent gain of 4–7 kg/year (10–15 lb/year) for the previous 5–10 years. For example, Norman and colleagues found that obese, infertile women presenting to a weight-loss program aimed at improving fertility had experienced an average weight gain 10 times that of the normal population *(28)*. Weight gain in affected patients may be associated with the development of pilosebaceous symptoms and menstrual irregularity.

Weight gain that is primarily centripetal, especially if associated with extremity wasting, purple striae, easy bruisability, moon facies, and rubor, suggests the presence of Cushing's syndrome, and these patients should be appropriately screened using 24-hour urinary-free cortisol levels or a cortisol level following an overnight dexamethasone suppression test. Other information that should be sought includes the patient's awareness of her body fat distribution, as women with PCOS have a greater prevalence of abdominal obesity *(29)*. Weight gain may also be associated with carbohydrate craving and evidence of postprandial reactive hypoglycemia, particularly in mid-afternoons. For example, Holte and colleagues found that although insulin resistance in obese women with PCOS was reduced by weight loss to similar levels as BMI-matched controls, these patients continued to demonstrate an increased early insulin response to glucose, which could stimulate appetite and persistent weight gain *(30)*. A change in eating patterns or habits or smoking cessation, more sedentary lifestyle, and depression and self-image distortion should also be sought. Snoring and daytime somnolence may be indicative of obstructive sleep apnea *(see* Section 2.1.5.). Finally, the use of medications that may be associated with weight gain or obesity, including glucocorticoids (systemic, topical, or inhaled), phenothiazines, tricyclic antidepressants, lithium, and valproate, should be elicited.

2.1.5. Symptoms of Obstructive Sleep Apnea

The prevalence of sleep disorders, particularly sleep apnea, is high in women with PCOS *(31,32)*. Its principal symptoms are a significant degree of snoring and daytime sleepiness *(32)*. Sleep apnea is caused by repeated collapse of the pharynx airway during sleep, resulting in poor oxygen supply to the person involved. Its consequences include daytime sleepiness, reduced mental performance, and a subsequent effect on quality of life. Its incidence is 2–4% in normal men and women, and it has dangerous consequences, including the development of hypertension, heart attacks, and strokes *(33)*. It usually is associated with obesity, and its prevalence is 3- to10-fold higher in men than in women; in fact, excess male hormones may be associated with the development of sleep apnea *(34,35)*.

The incidence of laboratory-proven sleep apnea was 4- to 30-fold higher in women with PCOS compared to normal controls *(31,32)*. Although obesity, particularly android obesity, predisposes PCOS women to sleep apnea, this disturbance in PCOS does not appear to correlate strictly with body weight, and insulin resistance may be a major causative factor in its development *(32)*. It is important to identify women with sleep apnea because of the plethora of associated metabolic and cardiovascular morbidities. Treatment of the sleep apnea syndrome with continuous positive airway pressure (CPAP) is associated with reduced insulin resistance in women *(33)*. Unfortunately, 90% of women with sleep apnea remain undiagnosed, and a high degree of clinical suspicion is necessary to identify those women that may benefit from a formal sleep laboratory study (i.e., those who snore or awake frequently at night, those who have daytime somnolence and fatigue).

2.1.6. Family History

The family history is an important tool in predicting PCOS in women, as the risk of PCOS when a first-degree relative is affected exceeds 30% *(36,37)*. A history of menstrual irregularities, excess or unwanted hair growth, or polycystic ovaries, including the possibility that the mothers may have undergone an ovarian wedge resection in their youth, are suggestive of the diagnosis of PCOS. An increased incidence of premature balding (> 30 years of age) has also been noted in some male siblings of women with PCOS *(38)*.

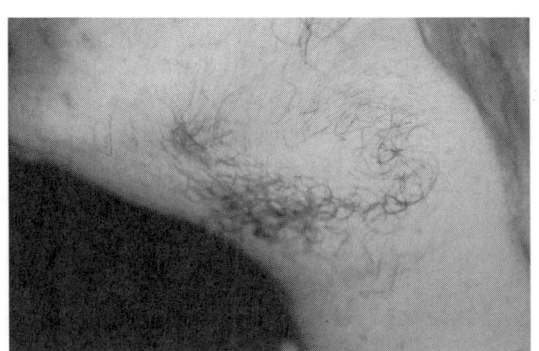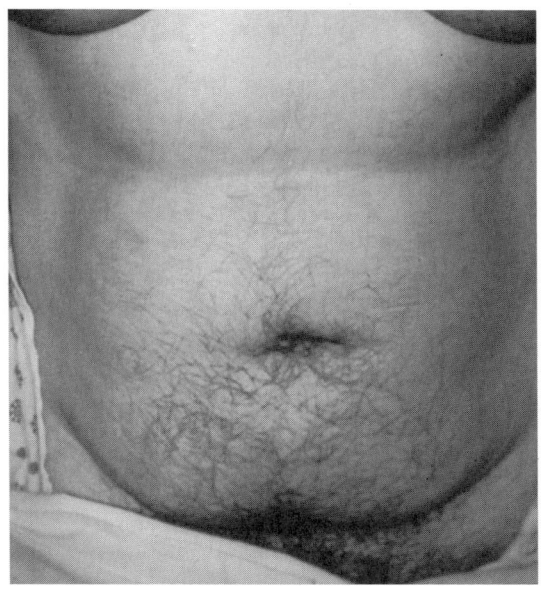

Fig. 1. Hirsutism of the chin (left) and abdomen (right), also called male escutcheon.

A family history of type 2 diabetes mellitus (DM) is more common in PCOS patients than in controls *(39,40)* and predicts glucose intolerance and DM in PCOS *(41)*. A family history of dyslipidemia, hypertension, and cardiovascular disease also occurs more frequently among PCOS patients *(39,42)*. Finally, a family history of obesity may predict the development of obesity in the patient.

2.2. Features Elicitable by the Clinical Exam

Features of PCOS that may be elicited upon the physical exam include dermatological evidence of hyperandrogenism (hirsutism, acne, or alopecia), evidence of android body fat distribution, and evidence of metabolic dysfunction, including acanthosis and acrochordons. Clinical features of virilization or Cushing's syndrome are not seen in patients with PCOS, generally representing the presence of androgen-secreting neoplasms or Cushing's disease, and are discussed further in Chapters 6 and 7.

2.2.1. Hirsutism

Hirsutism is defined as the presence of terminal hairs in a male-like pattern in women (Fig. 1). Excessive hair growth is a major symptom in PCOS and is present in approximately 75% of women with PCOS, at least of white or black race *(2,13,43,44)*. The sideburns of the face and chin areas are frequently involved. Other areas include anterior chest, midline abdominal area, and a triangular male-pattern pubic hair distribution in the lower abdomen. Increased hair growth may also be noted in the presacral and perineal areas and extremities. Frequently the onset of hirsutism, unlike acne, occurs several years after the onset of puberty *(45)*.

There is substantial phenotypic variation in the degree of hirsutism. The variability in the presence and degree of hirsutism in patients with PCOS is potentially the result of differences in the susceptibility of the pilosebaceous unit (PSU) to circulating androgens *(46,47)*. For example, coarse pigmented terminal hairs are infrequent in east Asian and Scandinavian women. Thus, genetic factors must be important determinants of the degree and distribution of hirsutism *(48)*. Although a correlation may be present between androgen levels, especially free T, and hirsutism *(49)*, we were not able to demonstrate a difference in androgen levels between PCOS women with and without hirsutism

(50). These data support the concept that a wide variation exists in the population with regard to end-organ PSU sensitivity to the effects of elevated free T in hyperandrogenic women *(46,49,51)*. Nonetheless, the measurable circulating androgen levels appear to play a limited role in determining the degree of hirsutism *(50)*.

Clinical assessment of the extent of hirsutism is generally based on the visual scoring method described by Ferriman and Gallwey in 1961 *(52)*. A score from 0 (*none*) to 4 (*maximum*) was assigned to each of 11 body areas, based on the amount of terminal hair growth in each area. These investigators noted that only nine body areas were useful for detecting abnormal hair growth (i.e., upper lip, chin, chest, upper abdomen, lower abdomen or male escutcheon, upper back, lower back, upper arms, and thighs, but not the lower legs and arms). Because many hirsute women demonstrate hair growth primarily in the sideburn area, neck, or perineum, areas not scored by Ferriman and Gallwey, later classification systems have included evaluation of these body areas *(53)*.

A modified Ferriman–Gallwey score (mF-G; comprising the nine body areas denoted above) of 8 or more was observed in about 5% of reproductive-aged Caucasian women, who were then defined as being hirsute *(54)* (*see* Chapter 33, Fig. 1). However, we should note that a recent study of more than 600 unselected women indicates that an mF-G of 3 or more appears to signal the population of women whose hair growth falls out of the norm—similar in black and white women *(55)*. These cutoff data are consistent with our report, noting that approximately 50% of women complaining of minimal unwanted hair growth (mF-G of 1–5) had PCOS *(44)*.

2.2.2. Acne

Acne is common among adolescent females, with a prevalence of 30–50% *(56)*. Severe postadolescent acne, however, is uncommon among Caucasian women between the ages of 18 and 21, with a frequency of less than 1% *(49)*. Between 20 and 40% of patients with treatment-resistant acne and without alopecia or hirsutism are reported to have androgen excess, principally PCOS *(57–59)*. Although acne may be the sole manifestation of androgen excess, the levels of plasma-free T in PCOS are similar in patients with or without acne *(60)*.

Hormonal evaluation is indicated if acne occurs before the age of 9 years, persists into the 20s or 30s, is resistant to conventional therapy, and recurs following one or several treatments with isotretinoin *(61)*. Acne rosacea may be present and may at times be confused with acne. Rosacea is less androgen dependent than acne vulgaris and is characterized as areas of redness, soft tissue thickening, and telangiectasia over the perinasal areas, tip of nose, and chin *(53)*. It is frequently associated with a prior history of blushing, easily associated with intermittent flushes. Consumption of heated drinks and alcohol worsens it.

Many factors are involved in the development of acne *(62)*. Acne is a manifestation of not only impaction and swelling of the PSU secondary to androgen effects, but also inflammatory changes in the follicular duct leading to increased sebum and keratin, which are released into the dermis. Inflammatory acne is characterized by erythematous papules, pustules, and nodular swellings that often lead to scarring *(63)*. Diagnosis is made by the presence of open and closed comedones (blackheads and whiteheads), which are the primary lesions of acne. They present alone or more frequently in combination with pustules and erythematous papules in the face and upper trunk. Progression of acne is associated with lateral extension from localized areas over the lower third of the face with development of large inflammatory lesions typical of cystic acne. The grading of acne is usually based on the number, type, and distribution of acneic comedones *(64)* (Table 2).

2.2.3. Androgenic Alopecia

Scalp hair loss in hyperandrogenic women is a distressing complaint with significant psychological morbidity. It may reflect the response of the PSU to endogenous androgens and may be associated with concomitant acne and hirsutism. As noted earlier, the sensitivity of the PSU to androgens varies widely, and there is poor correlation between clinical features and circulating androgen levels *(66,67)*.

Table 2
Grading of Acne

Grade (severity)	Description[a]
Mild	Noninflammatory lesions (comedones) are most frequently found in association with pustules (<10).
Moderate	Main lesions are erythematous papules (10–40) and pustules (10–40), predominantly on the face.
Moderate–severe	Numerous papules (40–100) and pustules (40–100), frequently with comedones (40–100), affecting the face and often the upper chest and back. Occasional larger nodular inflamed lesions (up to 5).
Severe	Nodulocystic acne with many painful nodules and pustules, most common on the chest and face. Many smaller papules, pustules, and comedones are also present.

[a]Numbers in parentheses refer to total number of lesions seen. (Adapted from ref. *64*.)

The presence of dihydrotestosterone, formed from the 5α-reduction of T in the dermal papilla, is associated with a higher 5α-reductase activity in the hairs plucked from a scalp presenting with androgenic alopecia *(68)*. Associated etiologies of alopecia in any woman may be genetic, environmental, and nutritional (poor protein intake and iron, vitamin B_{12}, or zinc deficiency). The presence of alopecia in PCOS appears to be common, even among young women. Androgen-related alopecia in women with PCOS often tends to be seen in the anterior mid-vertex area, starting as a "triangular" thinning patch with postero-lateral extension to the crown (Fig. 2). The anterior hairline generally remains intact in women with PCOS, and significant bitemporal scalp hair loss is unusual.

Studies defining the incidence of alopecia in PCOS or androgenic disorders have been sparse. In a report of 109 consecutive premenopausal women whose presenting complaint was alopecia, the incidence of hyperandrogenism in 42 patients was 38.5% *(69)*. This was compared to 24 control subjects with no PSU manifestations and no alopecia and normal pelvic ultrasonography. A diagnosis of PCOS was made in 40 women, 9 of who did not have hyperandrogenism at the time of presentation despite oligo-amenorrhea and bilaterally enlarged ovaries on ultrasonography. This represents a prevalence of 36.6% of PCOS in these women presenting with a chief complaint of alopecia. Scalp hair pattern findings were similar to those described by Ludwig *(70)*, which denoted preservation of the anterior hairline with thinning of the crown. Another study of 89 women with androgenic alopecia revealed a 67% prevalence of polycystic ovaries compared with 27% of control subjects *(71)*. However, no difference was found in the prevalance of menstrual irregularity between study and control groups (24 vs 15%). Correlation between the presence of alopecia and the presence or degree of hyperandrogenemia was also poor, probably secondary to the varying androgen sensitivity of the PSU. In epidemiological studies of women with PCOS, the prevalence of androgen alopecia varies widely. In one study of 257 hyperandrogenic patients undergoing treatment, most of whom had PCOS, only 12 (4.7%) presented with a complaint of hair loss *(2)*. Alternatively, in the clinical experience of another of the authors, encompassing almost 2000 women with PCOS over a period of 35 years, the presence of alopecia was close to 50%, representing both anterior midvertex and diffuse hair loss (W. Futterweit, August 2005, personal communication). These data attest to the significant differences in symptomatology that can be observed in this very heterogeneous syndrome, depending on referral patterns and clinical interest.

One must conclude that a finding of alopecia in any premenopausal woman warrants investigation for endocrine evidence of androgen excess. Because a loss of at least 25% of scalp hair is necessary before a woman becomes aware of thinning scalp hair *(70)*, it is also clear that clinicians should heed patients who complain of hair loss, even though obvious loss or balding may not be apparent.

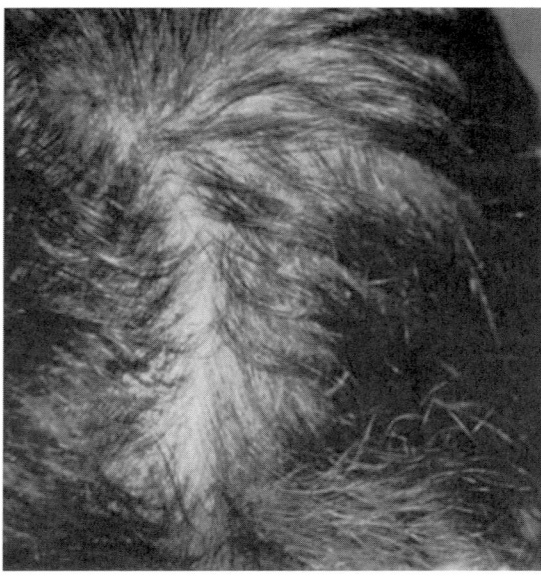

Fig. 2. Androgenic alopecia demonstrating the triangular hair thinning observed in the anterior midvertex area with postero-lateral extension to the crown.

2.2.4. Obesity

Obesity is observed in up to 60% of patients with PCOS seen in the United States *(2)* but is less prevalent in other populations *(72,73)*. Furthermore, other investigators have documented that obesity is less prevalent in adolescents with signs of PCOS compared with adults with PCOS in the same population *(74,75)*. Consequently, it should not be expected that adolescents with PCOS would have the same degree of obesity as older affected patients. Visceral adiposity is frequently noted, with an increase of the waist-to-hip ratio greater than 88 cm (35 in.), and is associated with a number of metabolic aberrations in PCOS, including higher degrees of hyperandrogenemia, insulin resistance, glucose intolerance, and dyslipidemia *(76)*.

2.2.5. Acanthosis Nigricans and Acrochordons

Although acanthosis nigricans (AN) can be a paraneoplastic cutaneous sign of malignancies, it is more commonly seen as a marker of hyperinsulinemia and the metabolic syndrome *(77,78)*. It is often noted in the creases of the nape of the neck, axilla, groin, and other intertrigenous zones, as well as exposed areas such as the elbows and knuckles (Fig. 3). AN appears as a hyperpigmented, possibly hyperkeratotic area of the skin. It may occur in many insulin-resistant states, including obesity and PCOS. Acrochordons (skin tags) are soft, pedunculated, flesh-colored to tan papules, usually ranging from 1 to 5 mm in diameter, and commonly occurring in areas that are exposed to a high degree of friction, such as the sides of the neck and axillae (Fig. 4). They are generally noted in women over the age of 40 years, and their presence prior to that time is distinctly abnormal.

In one small study, AN was present on clinical examination in 11 of 13 obese PCOS, 3 of 6 lean PCOS, 4 of 14 obese normal, and 0 of 4 lean normal women; histologically, AN was present in 13 of 13 obese PCOS, 5 of 6 lean PCOS, 13 of 14 obese normal, and 1 of 4 lean normal women *(77)*. The severity of histological AN was most highly correlated with insulin-mediated glucose disposal rather than fasting or glucose-stimulated insulin levels. Overall, AN appears to be detectable in a large proportion of women with PCOS, although large-scale studies are lacking. In general, its presence and that of acrochordons reflects impairments in insulin sensitivity.

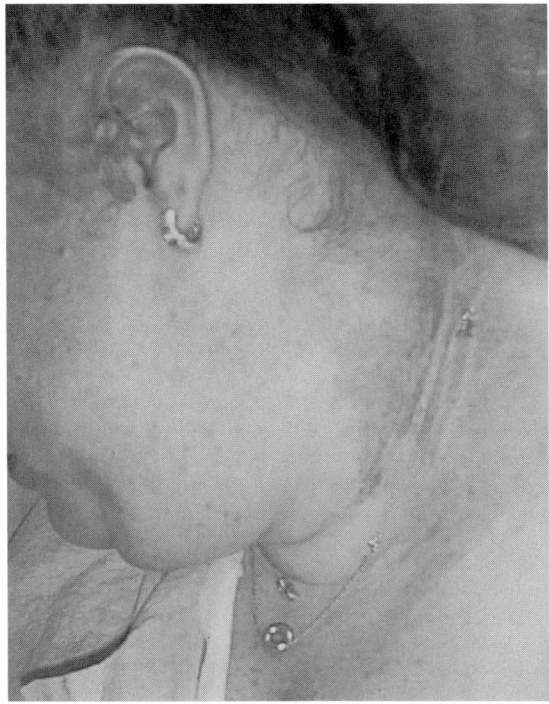

Fig. 3. Acanthosis nigricans of the nape of the neck.

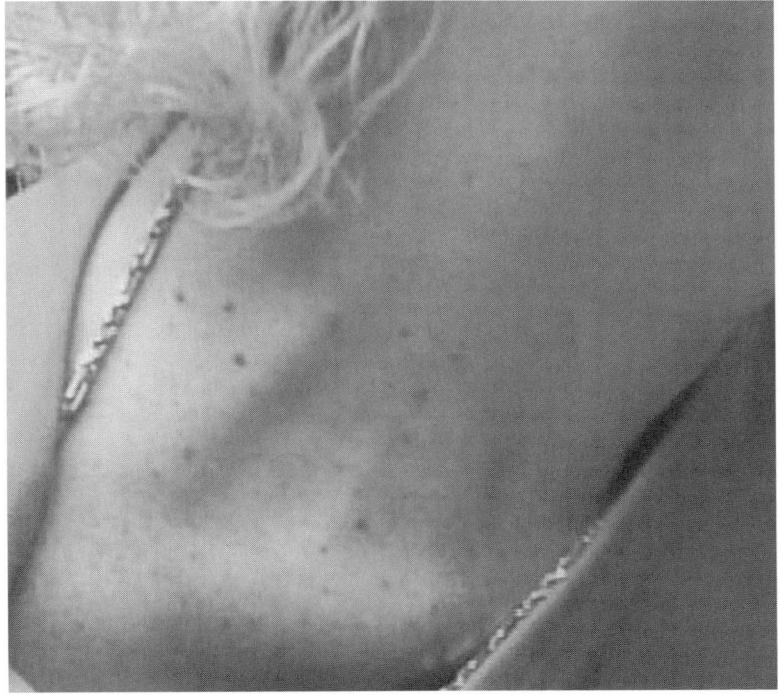

Fig. 4. Acrochordons of the neck.

3. CONCLUSIONS

Generally, the findings elicited by the medical history and physical exam provide the strongest suggestion that a patient suffers from PCOS. Clinical assessment of the hyperandrogenic woman requires a comprehensive evaluation of all signs and symptoms. Signs may be subtle but may yield clues as to the clinical presence of the insulin resistance syndrome and the frequent development of type 2 DM and possible cardiovascular disease. The role of the physician is to recognize the various presentations of this syndrome and not only to initiate treatment to minimize the reproductive and cosmetic features of PCOS, but to initiate appropriate treatment to minimize the associated risk factors leading to diabetes and cardiovascular events.

4. FUTURE AVENUES OF INVESTIGATION

Future investigation should focus on determining the clinical features of PCOS in the different ethnic/racial groups, including the prevalence and predictability of less common signs of PCOS, such as acne and androgenic alopecia. A greater degree of understanding of the predictive value of factors elicited during the medical history, particularly the predictability of symptoms suggestive of sleep apnea and the family history, remain to be established. The developmental features of PCOS, including sudden weight gain, need to be better established in a prospective fashion.

KEY POINTS

- The findings elicited by the medical history and physical exam provide the strongest suggestion that a patient suffers from PCOS; biochemical and radiographic testing only provide confirmatory evidence or exclude related disorders.
- The medical history will provide information regarding:
  o The presenting complaint, which may include irregular or unpredictable menstrual cycles, infertility, unwanted hair growth, acne or scalp hair loss, or unexplained weight gain or overweight.
  o The onset and progression of hyperandrogenic signs and symptoms, such that the development of hyperandrogenism in PCOS is usually slowly progressive, becoming stable by the mid- to late 30s.
  o Alternatively, the rapid development of symptoms, including hirsutism, oligo-amenorrhea, severe acne and alopecia, increased muscularity, and clitoromegaly, is indicative of a virilizing syndrome, most commonly caused by an androgen-producing neoplasms.
  o Patients or their mothers relating a history of premature adrenarche.
  o Menstrual dysfunction and irregularity, although up to 40% of anovulatory PCOS women may present with a history of "regular" cycles.
  o Excess weight gain, usually peripubertal, and sudden in nature.
  o Symptoms suggestive of obstructive sleep apnea, such as excessive snoring and daytime sleepiness.
  o Anabolic or androgenic drug use or abuse.
  o A family history of hyperandrogenic symptoms (e.g., unwanted hair growth) suggestive of PCOS, or of metabolic disorders such as type 2 DM.
- Features of PCOS that may be elicited upon the physical exam include:
- Dermatological evidence of hyperandrogenism, such as hirsutism, acne, or alopecia.
  o Android body fat distribution.
  o Evidence suggestive of metabolic dysfunction, including acanthosis and acrochordons.
- Clinical features of virilization or Cushing's syndrome are not seen in patients with PCOS and generally represent the presence of androgen-secreting neoplasms or Cushing's disease.

ACKNOWLEDGMENTS

Supported in part by an endowment from the Helping Hand of Los Angeles and by grants RO1-HD29364 and K24-D01346 from the National Institutes of Health.

REFERENCES

1. Balen AH, Conway GS, Kaltsas G, et al. Polycystic ovary syndrome: the spectrum of the disorder in 1741 patients. Hum Reprod 1995;10:2107–2111.
2. Azziz R, Sanchez LA, Knochenhauer ES, et al. Androgen excess in women: experience with over 1000 consecutive patients. J Clin Endocrinol Metab 2004;89:453–462.
3. Ibanez L, Dimartino-Nardi J, Potau N, Saenger P. Premature adrenarche—normal variant or forerunner of adult disease? Endocr Rev 2000;21:671–966.
4. Ibanez L, Potau N, Ferrer A, Rodriguez-Hierro F, Marcos MV, de Zegher F. Reduced ovulation rate in adolescent girls born small for gestational age. J Clin Endocrinol Metab 2002;87:3391–3393.
5. Ibanez L, Valls C, Potau N, Marcos MV, deZegher F. Polycystic ovary syndrome after precocious pubarche: ontogeny of the low-birthweight effect. Clin Endocrinol (Oxf) 2001;55:667–672.
6. Winters SJ, Talbott E, Guzick DS, Zborowski J, McHugh KP. Serum testosterone levels decrease in middle age in women with the polycystic ovary syndrome. Fertil Steril 2000;73:724–729.
7. Derksen J, Nagesser SK, Meinders AE, Haak HR, van de Velde CJH. Identification of virilizing adrenal tumors in hirsute women. N Engl J Med 1994;331:968–973.
8. Adams J, Polson DW, Franks S. Prevalence of polycystic ovaries in women with anovulation and idiopathic hirsutism. Br Med J 1986;293:355–359.
9. Hull MGR. Epidemiology of infertility and polycystic ovarian disease: endocrinological and demographic studies. Gynecol Endocrinol 1987;1:235–245.
10. Franks S, White DM. Prevalence of and etiological factors in polycystic ovarian syndrome. Ann NY Acad Sci 1993;687:112–114.
11. Goldzieher JW, Axelrod LR. Clinical and biochemical features of polycystic ovarian disease. Fertil Steril 1963;14:631–653.
12. Conway GS, Honour JW, Jacobs HS. Heterogeneity of the polycystic ovary syndrome: clinical, endocrine and ultrasound features in 556 patients. Clin Endocrinol (Oxf) 1989;30:459–470.
13. Azziz R, Woods KS, Reyna R, Key TJ, Knochenhauer ES, Yildiz BO. The prevalence and features of the polycystic ovary syndrome in an unselected population. J Clin Endocrinol Metab 2004;89:2745–2749.
14. Chang RJ, Katz SE. Diagnosis of polycystic ovary syndrome. Endocrinol Metab Clin North Am 1999;28:397–408.
15. Chang RJ. A practical approach to the diagnosis of polycystic ovary syndrome. Am J Obstet Gynecol 2004;191:713–717.
16. Treloar AE, Boynton RE, Behn BG, Brown BW. Variation of the human menstrual cycle through reproductive life. Int J Fertil 1967;12:77–126.
17. Chiazze L Jr, Brayer FT, Macisco JJ Jr, Parker MP, Duffy BJ. The length and variability of the human menstrual cycle. JAMA 1968;203:377–380.
18. Goldzieher JW. Polycystic ovarian disease. Fertil Steril 1981;35:371–394.
19. Young RL, Goldzieher JW. Clinical manifestations of polycystic ovarian disease. Endocrinol Metab Clin North Am 1988;17:621–633.
20. Deaton R. Hyperandrogenism and uterine bleeding. Infertil Reprod Med Clin North Am 1991;2:561–583.
21. Elting MW, Korsen TJ, Rekers-Mombarg LT, Schoemaker J. Women with polycystic ovary syndrome gain regular menstrual cycles when ageing. Hum Reprod 2000;15:24–28.
22. Azziz R, Waggoner WT, Ochoa T, Knochenhauer ES, Boots LR. Idiopathic hirsutism: an uncommon cause of hirsutism in Alabama. Fertil Steril 1998;70:274–278.
23. Carmina E. Prevalence of idiopathic hirsutism. Eur J Endocrinol 1998;139:421–423.
24. van Hooff MH, Voorhorst FJ, Kaptein MB, Hirasing RA, Koppenaal C, Schoemaker J. Predictive value of menstrual cycle pattern, body mass index, hormone levels and polycystic ovaries at age 15 years for oligo-amenorrhoea at age 18 years. Hum Reprod 2004;19:383–392.
25. Michelmore KF, Balen AH, Dunger DB, Vessey MP. Polycystic ovaries and associated clinical and biochemical features in young women. Clin Endocrinol (Oxf) 1999;51:779–786.
26. Sheehan TJ, DuBrava S, DeChello LM, Fang Z. Rates of weight change for black and white Americans over a twenty year period. Int J Obes Relat Metab Disord 2003;27:498–504.
27. Tataranni PA, Harper IT, Snitker S, et al. Body weight gain in free-living Pima Indians: effect of energy intake vs expenditure. Int J Obes Relat Metab Disord 2003;27:1578–1583.
28. Clark AM, Ledger W, Galletly C, et al. Weight loss results in significant improvement in pregnancy and ovulation rates in anovulatory obese women. Hum Reprod 1995;10:2705–2712.
29. Pasquali R, Casimirri F, Venturoli S, et al.. Body fat distribution has weight-independent effects on clinical, hormonal, and metabolic features of women with polycystic ovary syndrome. Metabolism 1994;43:706–713.
30. Holte J, Bergh T, Berne C, Wide L, Lithell H. Restored insulin sensitivity but persistently increased early insulin secretion after weight loss in obese women with PCOS. J Clin Endocrinol Metab 1995;80:2586–2593.

31. Fogel RB, Malhotra A, Pillar G, Pittman SD, Dunaif A, White DP. Increased prevalence of obstructive sleep apnea syndrome in women with polycystic ovary syndrome. J Clin Endocrinol Metab 2001;86:1175–1180.

32. Vgontzas AN, Legro RS, Bixler EO, et al. Polycystic ovary syndrome is associated with obstructive sleep apnea and daytime sleepiness: role of insulin resistance. J Clin Endocrinol Metab 2001;86:517–520.

33. American Academy of Sleep Medicine. Sleep-related breathing disorders in adults: recommendations for syndrome definition and measurement techniques in clinical research. The Report of an American Academy of Sleep Medicine Taskforce. Sleep 1999;22:667–689.

34. Zhou XS, Rowley JA, Demirovic F, Diamond MP, Badr MS. Effect of testosterone on the apneic threshold in women during NREM sleep. J Appl Physiol 2003;94:101–107.

35. Liu PY, Yee B, Wishart SM, et al. The short-term effects of high-dose testosterone on sleep, breathing, and function in older men. J Clin Endocrinol Metab. 2003;88:3605–3613.

36. Legro RS, Driscoll D, Strauss JF 3rd, Fox J, Dunaif A. Evidence for a genetic basis for hyperandrogenemia in polycystic ovary syndrome. Proc Natl Acad Sci USA 1998;95:14956–14960.

37. Kahsar-Miller M, Nixon C, Boots LR, Go RCP, Azziz R. Prevalence of the polycystic ovary syndrome (PCOS) among first degree relatives of patients with PCOS. Fertil Steril 2001;75:53–58.

38. Ferriman D, Purdie AW. The inheritance of polycystic ovarian disease and a possible relationship to premature balding. Clin Endocrinol (Oxf) 1979;11:291–300.

39. Atiomo WU, El Mahdi E, Hardiman P. Family association in women with polycystic ovary syndrome. Fertil Steril 2003;80:143–145.

40. Hahn S, Tan S, Elsenbruch S, et al. Clinical and biochemical characterization of women with polycystic ovary syndrome in North Rhine-Westphalia. Horm Metab Res 2005;37:438–444.

41. Ehrmann DA, Kasza K, Azziz R, Legro RS, Ghazzi MN; PCOS/Troglitazone Study Group. Effects of race and family history of type 2 diabetes on metabolic status of women with polycystic ovary syndrome. J Clin Endocrinol Metab 2005;90:66–71.

42. Yilmaz M, Bukan N, Ersoy R, et al. Glucose intolerance, insulin resistance and cardiovascular risk factors in first degree relatives of women with polycystic ovary syndrome. Hum Reprod 2005;20:2414–2420.

43. Diamanti-Kandarakis E, Kouli CR, Bergiele AT, et al. A survey of the polycystic ovary syndrome in the Greek island of Lesbos: hormonal and metabolic profile. J Clin Endocrinol Metab 1999;84:4006–4011.

44. Souter I, Sanchez A, Perez M, Bartolucci AA, Azziz R. The prevalence of androgen excess among patients with minimal unwanted hair growth. Am J Obstet Gynecol 2004;191:1914–1920.

45. Ruutiainen K, Erkkola R, Gronroos MA, Irjala K. Influence of body mass index and age on the grade of hair growth in hirsute women of reproductive ages. Fertil Steril 1998;50:260 –265.

46. Rosenfeld RL. Pilosebaceous physiology in relation to hirsutism and acne. Clin Endocrinol Metab 1986;15:341–362.

47. Ehrmann DA, Rosenfield RL. An endocrinologic approach to the patient with hirsutism. J Clin Endocrinol Metab 1990;71:1–4.

48. Carmina E, Koyama T, Chang L, Stanczyk FZ, Lobo RA. Does ethnicity influence the prevalence of adrenal hyperandrogenism in insulin resistance in the polycystic ovary syndrome? Am J Obstet Gynecol 1992;167:1807–1812.

49. Reingold SB, Rosenfield RL. The relationship of mild hirsutism or acne in women to androgens. Arch Dermatol 1987;123:209–212.

50. Chang W, Knochenhauer ES, Bartolucci AA, Azziz R. Phenotypic spectrum of the polycystic ovary syndrome (PCOS): clinical and biochemical characterization of the major clinical subgroups. Fertil Steril 2005;83:1717–1723.

51. Rosenfield RL, Deplewski D. Role of androgens in the developmental biology of the pilosebaceous unit. Am J Med 1995;98:80S–88S.

52. Ferriman D, Gallwey JD. Clinical assessment of body hair growth in women. J Clin Endocrinol Metab 1961;21:1440–1447.

53. Redmond GP. Clinical evaluation of the woman with an androgenic disorder. In: Redmond GP, ed. Androgenic Disorders. New York: Raven Press, 1995:1–20.

54. Hatch R, Rosenfield RL, Kim MH, Tredway D. Hirsutism: implications, etiology and management. Am J Obstet Gynecol 1981;140:815–830.

55. de Ugarte CM, Bartolucci AA, Woods KS, Azziz R. Degree of facial and body terminal hair growth in unselected black and white women: towards a populational definition of hirsutism J Clin Endocrinol Metab 2006. [Epub ahead of print].

56. Krowchuk DP, Lucky AW. Managing adolescent acne. Adolesc Med 2001;12:355–374.

57. Vexiau P, Husson C, Chivot M, et al. Androgen excess in women with acne alone compared with women with acne and/ or hirsutism. J Invest Dermatol 1990;94:279–283.

58. Slayden SM, Moran C, Sams WM Jr, Boots LR, Azziz R. Hyperandrogenemia is a frequent cause of acne, regardless of age of presentation. Fertil Steril 2001;75:889–892.

59. Borgia F, Cannavo S, Guarneri F, Cannavo SP, Vaccaro M, Guarneri B. Correlation between endocrinological parameters and acne severity in adult women. Acta Derm Venereol 2004;84:201–204.

60. Lucky AW, McGuire J. Rosenfield RL, Lucky PA, Rich BH. Plasma androgens in acne vulgaris. J Invest Dermatol 1983;81:70–74.
61. James WD. Acne. N Engl J Med 2005;352:1463–1472.
62. Lucky AW. Hormonal correlates of acne and hirsutism. Am J Med 1995;98:89S–94S.
63. Rosenfield RL, Lucky AW. Acne, hirsutism, and alopecia in adolescent girls. Endocrinol Metab Clin North Am 1993;22:507–532.
64. Pochi PE, Shalita AR, Strauss JS, et al. Report of the Consensus Conference on Acne Classification. Washington, DC, March 24 and 25, 1990. J Am Acad Dermatol 1991;24:495–500.
65. Cunliffe WJ, Meynadier J, Alirezai M, et al. Is combined oral and topical therapy better than oral therapy alone in patients with moderate to moderately severe acne vulgaris? J Am Acad Dermatol 2003;49:S218–S226.
66. Rosenfield RL. Polycystic ovary syndrome and insulin resistant hyperinsulinemia. J Am Acad Dermatol 2001;45:S95–104.
67. Deplewski D, Rosenfield RL. Role of hormones in pilosebaceous unit development. Endocr Rev 2000;21:363–392.
68. Schweikert H, Wilson JD. Regulation of human hair growth by steroid hormones. I. testosterone metabolism in isolated hairs. J Clin Endocrinol Metab 1974;38:811–819.
69. Futterweit W, Dunaif A, Yeh H-C, Kingsley P. The prevalence of hyperandrogenism in 109 consecutive female patients with diffuse alopecia. J Am Acad Dermatol 1988;19:831–836.
70. Ludwig E. Classification of the types of androgenetic alopecia (common baldness) occurring in the female sex. Br J Dermatol 1977;97:247–254.
71. Cela E, Robertson C, Rush K, et al. Prevalence of polycystic ovaries in women with androgenic alopecia. Eur J Endocrinol 2003;149:439–442.
72. Carmina E, Koyama T, Chang L, Stanczyk FZ, Lobo RA. Does ethnicity influence the prevalence of adrenal hyperandrogenism and insulin resistance in polycystic ovary syndrome? Am J Obstet Gynecol 1992;167:1807–1812.
73. Carmina E, Legro RS, Stamets K, Lowell J, Lobo RA. Difference in body weight between American and Italian women with polycystic ovary syndrome: influence of the diet. Hum Reprod 2003;18:2289–2293.
74. Gülekli B, Turhan NÖ, Senöz S, Kukner S, Oral H, Gokmen O. Endocrinological, ultrasonographic and clinical findings in adolescent and adult polycystic ovary patients: a comparative study. Gynecol Endocrinol 1993;7:273–277.
75. Dramusic V, Goh VHH, Rajan U, Wong YC, Ratnam SS. Clinical, endocrinologic, and ultrasound features of polycystic ovary syndrome in Singaporean adolescents. J Pediatr Adolesc 1997;10:125–132.
76. Ehrmann DA. Polycystic ovary syndrome. N Engl J Med 2005;352:1223–1236.
77. Dunaif A, Green G, Phelps RG, Lebwohl M, Futterweit W, Lewy L. Acanthosis nigricans, insulin action, and hyperandrogenism: clinical, histological, and biochemical findings. J Clin Endocrinol Metab 1991;73:590–595.
78. Hermanns-Le T, Scheen A, Pierard GE. Acanthosis nigricans associated with insulin resistance: pathophysiology and management. Am J Clin Dermatol 2004;5:199–203.

15

Biochemical Features of the Polycystic Ovary Syndrome

Adam H. Balen

SUMMARY

The polycystic ovary syndrome encompasses hyperandrogenism as the central biochemical disturbance, which has internal effects on ovarian function and metabolism and external manifestations on the skin. Ovarian dysfunction is associated with erratic menses and anovulation, and the ovaries have a characteristic morphological appearance. Hyperinsulinemia amplifies hyperandrogenism and is associated with both increasing body weight and worsening ovarian function. Within the ovary there are disturbances of steroid and nonsteroidal hormone production, which influence ovarian activity and feedback to the pituitary, where hypersecretion of luteinizing hormone is the major abnormality—itself affected by hypothalamic gonadotropin-releasing hormone pulsatility.

Key Words: Polycystic ovary syndrome; hyperandrogenism; testosterone; luteinizing hormone; insulin; inhibin.

1. INTRODUCTION

The polycystic ovary syndrome (PCOS) is a heterogeneous collection of signs and symptoms that gathered together form a spectrum of a disorder with a mild presentation in some and in others a severe disturbance of reproductive, endocrine, and metabolic function. There are many extra-ovarian aspects to the pathophysiology of PCOS, yet ovarian dysfunction is central. The definition of the syndrome has been much debated. At a recent joint European Society for Human Reproduction and Embryology/American Society for Reproductive Medecine consensus meeting, a refined definition of the PCOS was agreed: namely, the presence of at least two of the following three criteria: (1) oligo- and/or anovulation, (2) hyperandrogenism (clinical and/or biochemical), (3) polycystic ovaries, with the exclusion of other etiologies (1). The morphology of the polycystic ovary has been redefined as an ovary with 12 or more follicles measuring 2–9 mm in diameter and/or increased ovarian volume (>10 cm^3) (2). Polycystic ovaries are commonly detected by ultrasound or other forms of pelvic imaging, with estimates of the prevalence in the general population being in the order of 20–33% (3,4). Alternatively, not all women with polycystic ovaries demonstrate the clinical and biochemical features that define PCOS (2). There is considerable heterogeneity of symptoms and signs among women with PCOS, and for an individual these may change over time (4). A gain in weight is associated with a worsening of symptoms, whereas weight loss may ameliorate the endocrine and metabolic profile and symptomatology.

There appears to be national and racial differences in expression of PCOS. For example, the reported prevalence of PCOS among south Asian immigrants in Britain is 52% (6), and they had a comparable degree of insulin resistance to controls with established type 2 diabetes mellitus. Type 2 diabetes and insulin resistance have a high prevalence among indigenous populations in south Asia. We have also found that south Asians with anovulatory PCOS have greater degree of insulin resistance and more

From: *Contemporary Endocrinology: Androgen Excess Disorders in Women:*
Polycystic Ovary Syndrome and Other Disorders, Second Edition
Edited by: R. Azziz et al. © Humana Press Inc., Totowa, NJ

severe symptoms of the syndrome than anovulatory Caucasians with PCOS *(7)*. Furthermore, we have found that women from south Asia living in the United Kingdom appear to express symptoms at an earlier age than their Caucasian British counterparts and have greater risk factors for metabolic disease *(8)*. Thus, the biochemical features of PCOS may vary both within and between populations.

This chapter will discuss the biochemical features of the syndrome, namely elevated serum concentrations of testosterone, androstenedione, luteinizing hormone (LH), and insulin, which may vary between individuals and change with time. Adrenocortical abnormalities and adrenal-androgen excess will be discussed further in Chapter 20.

2. BACKGROUND

2. 1. Ovarian Biochemistry and Hyperandrogenism

Clinical phenotyping of PCOS involves determining the presence of clinical and/or biochemical androgen excess (hyperandrogenism), while excluding related disorders. The primary clinical signs of androgen excess are the presence of hirsutism, acne and, sometimes, alopecia (*see* Chapter 14). Whereas some studies have found that most patients with PCOS have evidence of hyperandrogenemia *(9)*, others have not *(5,10–12)*. In a study of more than 1700 women with PCOS, we found that one-third had an elevated serum total testosterone concentration and that the 95th percentile for total testosterone was 4.8 nmol/L *(5)*. We therefore use this value in practice as the cutoff for screening for other causes of androgen excess. If the value is greater than 4.8 nmol/L, it is only then necessary to assess the androgen profile in greater detail in order to exclude other causes such as androgen-secreting tumors of the ovary or adrenals (in which case the clinical history of hyperandrogenism is usually of more acute onset), nonclassic congenital adrenal hyperplasia (NC-CAH), or Cushing's syndrome. In populations with a high prevalence of NC-CAH, screening with a basal 17-hydroxyprogesterone and/or an adrenocorticotropic hormone (ACTH) stimulation test is advised.

The measurement of free testosterone or the free testosterone (free androgen) index are thought to be sensitive methods of assessing for hyperandrogenemia *(13,14)*. Methods for the assessment of free testosterone include equilibrium dialysis *(15,16)*, calculation of free testosterone from the measurement of sex hormone-binding globulin (SHBG) and total testosterone, or ammonium sulfate precipitation *(17)*. There are a number of different methods for the measurement of testosterone, but little in the way of standardization or normative data (*see* Chapter 5). Testosterone is bound to both SHBG and albumin. A measurement of SHBG can be used to calculate the free androgen index. The production of SHBG by the liver is suppressed by insulin, and women with hyperinsulinemia are likely to have a greater degree of hyperandrogenism.

Women with the classic syndrome have the highest levels of androgens, although even women with polycystic ovaries and mild or no symptoms have mean serum concentrations of testosterone higher than in those with normal ovaries *(4,18)*. The bulk of evidence points to the ovary being the source of excess androgens, which appears to result from an abnormal regulation (dysregulation) of steroidogenesis *(19)*. The ovary and adrenal cortex share the bulk of the steroid biosynthesis pathways by making equal contributions to the circulating concentrations of androstenedione and testosterone in a normal premenopausal woman. Both glands secrete androstenedione in significantly greater quantities than testosterone, while 50% of circulating testosterone is derived from the peripheral metabolism of androstenedione *(20)*. Androgen is produced in the ovary by the theca interna layer of the ovarian follicle, whereas the zona fasciculata of the adrenal cortex synthesizes adrenal androgens. The enzymes utilized in the formation of androstenedione from the initial substrate, cholesterol, are similar in both glands, under the endocrine control of LH in the ovary and ACTH in the adrenal glands *(21,22)*.

The initial step in the biosynthesis of all steroid hormones is the conversion of cholesterol to pregnenolone by a two-stage process involving the cholesterol side-chain cleavage enzyme and the acute steroidogenic regulatory protein. Along the Δ^5-steroid pathway, pregnenolone is first 17-hydroxylated to 17-hydroxypregnelonone, which is then converted to dehydroepiandrosterone (DHEA). This two-step process involves the activities of 17-hydroxylase and 17,20-lyase, respectively, both functions of cytochrome P450c17α (CYP17). DHEA serves as the principal precursor for androgens via its conversion to androstenedione under the action of 3β-hydroxysteroid dehydrogenase (3β-HSD). Androstenedione can then be catalyzed to testosterone by 17β-hydroxysteroid dehydrogenase (17β-HSD) or aromatized to estrone; in turn, testosterone can be subsequently 5α-reduced to dihydrotestosterone (DHT) or aromatized to estradiol. Pregnenolone can also first be catalyzed by 3β-HSD to progesterone, which is then 17-hydroxylated to 17-hydroxyprogesterone (the Δ^4-steroid pathway). In the adrenal gland, both progesterone and 17-hydroxyprogesterone proceed to the production of mineralocorticoids and glucocorticoids, respectively, first undergoing 21-hydroxylation and then 11-hydroxylation. Alternatively, little 17-hydroxyprogesterone is converted to androstenedione, because in humans Δ^4-17,20-lyase activity is minute. *See* Chapter 2 for additional details on the mechanisms of steroidogenesis.

Androgen secretion in normal women undergoes about twofold episodic, diurnal, and cyclic variation. The rate-limiting step in steroidogenesis is the formation of pregnenolone from cholesterol, by the action of P450scc, whose action is regulated by trophic hormones such as LH and ACTH. In turn, the rate-limiting step to androgen formation is P450c17α expression. The steroidogenic response to the trophic hormones is modulated by an array of small peptides, which include insulin and insulin-like growth factors (IGFs).

A certain amount of intraovarian androgen is essential for normal follicular growth and for the synthesis of estradiol. Nonetheless, when the synthesis of androgens is not coordinated with the needs of a developing follicle and is excessive, poor follicle maturation and increased folliclular atresia results. In the normal ovary LH acts on theca-interstitial-stromal cells, whereas follicle-stimulating hormone (FSH) acts on granulosa cells. According to the "two-gonadotropin, two-cell theory" of estrogen biosynthesis, the thecal compartment secretes androgens in response to LH, and the androstenedione thus formed is converted in the granulosa cell to estrogens by the action of aromatase, which in turn is under the influence of FSH. When a dominant follicle emerges, the estrogen content dominates over androgens and is not driven by long loop negative feedback effects. The intraovarian modulation of androgen synthesis by LH plays a critical regulatory role. As LH stimulation increases, a homologous desensitization sets in. Overstimulation with LH in a time- and dose-dependent manner causes downregulation of LH receptors, reduces cholesterol side-chain cleavage activity (P450scc), 17,20-lyase activity, and finally 17-hydroxylase activity. Thus, the ratio of 17-hydroxyprogesterone to androgen increases *(23)*.

Autocrine, paracrine, and hormonal factors modulate the coordination of thecal and granulosa cell function in terms of androgen synthesis. Androgens and estrogens are negative modulators of LH effects, whereas IGFs play a positive modulator role. Insulin also augments LH-stimulated androgen production, either via its own receptors or via IGF-1 receptors. Inhibin promotes androgen synthesis, whereas androgens stimulate inhibin production. Activin opposes the effects of inhibin. Furthermore, prostaglandins and angiotensin also play a promoter role, while corticotropin-releasing hormone, transforming growth factor-β, epidermal growth factor, tumor necrosis factor, and cytokines play an inhibitory role in androgen biosynthesis.

Granulosa cell development, and thereby the increase of aromatase activity, also determines androgen production. A healthy follicle that is 8 mm or more in diameter converts androstenedione to estradiol efficiently. Conversely, atretic and/or cystic follicles have a high androstenedione-to-estradiol ratio. The action of FSH on granulosa cells determines the growth of healthy follicles that are greater than 2–5 mm in diameter, partly mediated by the IGF system and insulin in physiological concentrations, all of which stimulate the production of estradiol. IGF-binding proteins inhibit FSH bioactivity and are markedly expressed in atretic follicles. Transforming growth factor and epidermal growth factor inhibit aromatase, whereas activin promotes granulosa cell estrogen production while inhibiting thecal androgen secretion.

Nearly half of the circulating testosterone in normal adult women is derived from the peripheral conversion of androstenedione, and the remainder is derived from the ovary and adrenal cortex. The important tissues in which this conversion takes place are the lung, liver, adipose tissue, and skin. Adipose tissue also forms estrone from androstenedione, which explains the mild estrogen excess of obesity. Plasma dihydrotestosterone is produced virtually entirely by 5α-reductase activity in the periphery, with plasma androstenedione being its major precursor *(24)*.

2.2. Ovarian Dysfunction in PCOS

The presence of enlarged polycystic ovaries suggests that the ovary is the primary site of the endocrine abnormality, particularly hyperandrogenism of PCOS. It has been suggested that derangement of P450c17α activity plays a central role in excess ovarian androgen production *(19)*. This was subsequently confirmed by other workers who assessed the response of the pituitary and ovary to a single dose of the gonadotropin-releasing hormone agonist (GnRHa) nafarelin in hyperandrogenemic women with PCOS in whom adrenal androgen production had been suppressed by administering dexamethasone *(23)*. Short-term stimulation with GnRHa yielded a significant elevation of 17-hydroxyprogesterone and, to a lesser extent, of androstenedione, testosterone, estrone, and estradiol *(23)*. This study was then extended to anovulatory and ovulating hyperandrogenemic women, and a small but significant increase in androstenedione levels was shown in both groups in response to GnRHa, and a similar response was found with 17-hydroxyprogesterone, which was significantly higher in the anovulatory women *(25)*. There was also no significant rise in these two hormones in response to ACTH injection, which excluded a significant role of adrenal androgen production *(25)*. These data indicate that hyperandrogenemia, in both ovulatory and anovulatory women with PCOS, is predominantly of ovarian origin. This also confirmed that the primary cause of excess androgen production by the polycystic ovary was not a result of hypersecretion of LH alone, and it was reasonable to conclude that the intrinsic defect was caused by ovarian theca-interstitial cell dysfunction or other stimulatory influences such as insulin or IGF-1, etc.

Both in vivo and in vitro data confirm that the theca cells of PCOS patients have a generalized overactive steroidogenesis. PCOS patients have a tendency to an excess of estradiol at all stages of follicular maturation. This is partly a result of the availability of excess androgen substrate for aromatase activity, as well as an excessive response of follicle development and estradiol secretion to FSH *(26)*. Granulosa cells from polycystic ovaries in vitro have also been reported to lose FSH responsiveness and produce low amounts of progesterone *(27)*.

The distinct ovarian morphology is pathognemonic for the syndrome, its major marker being hyperandrogenemia arising from the theca cells. Follicular development is disturbed, with antral follicles arrested at a diameter of 2–9 mm. It is thought that the abnormal endocrine environment adversely affects follicular maturation, although it is uncertain whether there is in addition an intrinsic abnormality within the follicle of polycystic ovaries. The whole process of follicle development from primordial to preovulatory takes about 6 months, with only the final 2 weeks being gonadotropin dependent. Preantral follicle development is dependent on local growth factors, which determine the growth and survival of those follicles that escape death by atresia. A recent study of follicle densities from normal and polycystic ovaries found that normal ovaries contained a mean of 11.4 small preantral follicles/m^3 (range 4–34), ovulatory polycystic ovaries had a mean density of 27.4 follicles (range 9–81), whereas anovulatory polycystic ovaries had a mean density of 73.0 follicles (range 31–94).

This significant difference was also demonstrated for primary follicles *(28)*. Anovulatory polycystic ovaries had the highest overall density of follicles, although there was no significant difference between those from anovulatory and ovulatory polycystic ovaries or between ovulatory polycystic ovaries and normal ovaries. Primordial follicle density was similar in all three groups, although those from polycystic ovaries were less likely to be healthy. Thus, there appears to be a significantly higher density of small preantral follicles, particularly in anovulatory polycystic ovaries. This is thought to be a result of a higher rate of recruitment from the resting follicle pool in polycystic ovaries, rather than a reduced rate of atresia (which, if anything, may be slightly increased). The observation that women with PCOS do not have an early menopause suggests that there may be a higher starting follicle pool, although this is yet to be proven.

LH excess is considered the cause of ovarian hyperandrogenism of PCOS in view of the stimulatory effect of LH on theca cells. Nevertheless, some women with PCOS have normal LH levels while being hyperandrogenic, whereas others who had downregulation of LH secretion with long-term GnRHa continued to display hyperresponsiveness of 17-hydroxyprogesterone to human chorionic gonadotropin (hCG) injection (i.e., challenge with LH). These findings argue against the sole role of LH in the androgen excess of PCOS. They favor the theory that theca cells of PCOS women hyperrespond to gonadotropins and produce excess androgens following an escape of their normal downregulation to gonadotropins, thereby linking this dysregulation to excess of insulin and IGF-1. Prelevic and colleagues supported this theory by demonstrating that suppression of insulin secretion by a somatostatin analog lowers serum LH and androgens in PCOS women *(29)*. Indeed, insulin acts as a "co-gonadotropin" and also amplifies the effects of testosterone by suppressing SHBG.

Inhibin is an FSH-inducible factor, which is capable of interfering with the downregulation of steroidogenesis. Plasma inhibin and androstenedione concentrations correlate, and women with PCOS have elevated serum inhibin-B *(30)*. This helps to explain the relatively low serum concentrations of FSH compared with LH in anovulatory women with PCOS. Because inhibin stimulates androgen production and androgens in turn stimulate inhibin secretion, there is a potential for the development of a vicious cycle within the ovary that would inhibit follicle development. Alternatively, a defect in the IGF system could cause an alteration of the set point for the response of the granulosa cell to FSH. It has been suggested that LH acts on granulosa cells concomitantly with insulin, leading to premature luteinization, maturational arrest, and excess androgen production *(27)*.

In summary, as a consequence of dysregulation of androgen synthesis within the ovary, women with PCOS have ovarian hyperresponsiveness to gonadotropins, with that of thecal cells to LH explaining the excess androgens, and that of granulosa cells to FSH leading to increased estrogens.

2.3. The Hypothalamic–Pituitary–Ovarian Axis

Serum LH concentrations can be significantly elevated in PCOS women as compared with controls *(31,32)*, as the result of increased amplitude and frequency of LH pulses *(33)*. Elevated LH concentrations above the 95th percentile of normal can be observed in approximately 40–60% of PCOS women *(5,12,34)*. An elevated serum LH concentration has been associated with a reduced chance of conception and an increased risk of miscarriage *(35)*. LH levels are influenced by the temporal relation to ovulation, which transiently normalizes LH because of the suppressive effect of progesterone, and by body weight, being higher in lean women with PCOS. Although an elevation in serum LH concentration occurs only in PCOS (in the absence of the midcycle preovulatory LH surge or the menopause transition), a measured elevation of LH is not required to make the diagnosis *(1)*. Consequently, no longer is an elevated LH-to-FSH ratio required or useful for the diagnosis of PCOS *(1)*.

The pituitary gonadotroph is central to reproductive function—its production and secretion of FSH and LH is directly stimulated by hypothalamic GnRH and is also influenced by integrated feedback mechanisms. The sensitivity of the pituitary to GnRH varies during the menstrual cycle in synchrony with changes in circulating estradiol (E2) concentrations *(36)*. In the early follicular phase, when E2 levels are low, pituitary sensitivity and gonadotropin content are at a minimum; as E2 levels rise, consequent upon follicular development, both sensitivity and content increase—particularly the latter, as E2 has a stimulating effect on pituitary gonadotropin synthesis and storage and promotes the self-priming effect of GnRH on the pituitary *(36)*. At the time of the midcycle surge, sensitivity to GnRH is maximal, with the resultant release of large amounts of gonadotropins. Estradiol also potentiates GnRH responsiveness, increasing the number of GnRH receptors by directly stimulating the protein synthesis required for receptor formation. FSH provides the initial stimulus for follicular development and also promotes granulosa cell conversion of androgens to estrogens by stimulating the aromatase enzymes. LH, classically known for its role in the luteal phase by promoting progesterone secretion, also has a vital role in the follicular phase, inducing thecal androgen production (the substrate for E2 synthesis) and initiating oocyte maturation at midcycle.

Some of the factors that influence GnRH activity include β-endorphin and opiate peptides, angiotensin II, serotonin, neuropeptide Y, neurotensin, somatostatin, corticotropin-releasing factor, dopamine, melatonin, noradrenaline, oxytocin, and substance P. The interrelationship of these factors is unclear. Endogenous opioid tone is important in the regulation of LH and prolactin secretion. Opioids, such as β-endorphin, inhibit GnRH release from the human mediobasal hypothalamus. It has been postulated that withdrawal of endogenous opioid tone in the presence of sufficient quantities of E2 may contribute to the initiation of the LH surge. When opioid tone decreases, a chain of neurosecretory events is initiated, which in the rat activates neuropeptide-Y neurons, which in turn, either alone or together with adrenergic transmitters, stimulate secretion of GnRH. The effects of opioids appear to be dependent on the steroid hormone environment, in particular estrogen, whose effect is augmented by progesterone *(37)*. Thus, the administration of an opioid antagonist, such as naloxone, during the early follicular phase has little effect on gonadotropin levels, whereas greater effects are observed midcycle, and the greatest effects are seen in the luteal phase *(37)*.

Tonic hypersecretion of LH in women with the PCOS has been suggested as being caused by, at least in part, a combination of diminished opioid and dopaminergic tone *(38,39)*. There is also evidence that adrenergic activity is altered in women who hypersecrete LH *(40,41)*. Women with PCOS were found to be very sensitive to exogenous dopamine, and it was proposed that these women had a deficiency in endogenous dopaminergic inhibition of GnRH secretion *(38)*. In normal women both dopamine receptor antagonists, such as metoclopramide, and opiate receptor antagonists, such as naloxone, elicit a rise in serum LH concentrations *(39)*. Conversely, administration of synthetic β-endorphin elicits a fall in serum LH concentration. In women with PCOS, the administration of metoclopramide, naloxone, and β-endorphin did not alter LH secretory activity. It was therefore proposed that an underlying hypothalamic defect might lead to hypersecretion of LH through a reduction in endogenous dopaminergic and opioid control of GnRH secretion *(39)*.

Naloxone infusion experiments performed in women with the PCOS and weight-matched controls showed that LH responses were similar *(42)*. Pretreatment with L-dopa-carbidopa for 1 week resulted in an absence of naloxone-stimulated LH increase in normal women but an exaggerated response of the rise in LH after naloxone in women with PCOS. It was suggested that central opioid tone is not decreased in PCOS and that dopaminergic tone and/or the interaction between the dopamine and opioid system might be altered in the PCOS *(42)*. Further studies failed to demonstrate major alterations in brain dopaminergic activity in women with the PCOS *(43)*. Berga and Yen *(44)* administered both progestogens and opioid antagonists to women with PCOS and found that there was an apparent link between an impairment of opioid and progesterone secretion in the genesis of hypersecretion of LH. Yoshino et al. studied women with the PCOS and found abnormalities not only in dopamine metabolites but also in adrenergic metabolites *(41)*. It was also demonstrated that 4 weeks of treatment with naltrexone reduced pituitary sensitivity to GnRH in PCOS patients *(45)*, and so the question of the precise role of endogenous opioids in the control of LH secretion is unresolved *(46)*.

The interactions of factors at the level of the hypothalamus are therefore complex, and the factors that predominate in influencing LH secretion are unknown. The central disturbances in the PCOS (at the level of hypothalamus and pituitary) appear to be secondary to one or more peripheral factors, which may be ovarian in origin *(47)*. An area of some further controversy is whether there is an increase in GnRH pulse frequency in women who hypersecrete LH. This is important because if steroids are the main ovarian product to influence LH secretion, they are able to cross the blood–

brain barrier (BBB) and so might be expected to also affect GnRH pulsatility. If, however, the primary defect is through perturbed secretion of an ovarian peptide, it would not be predicted to cross the BBB to affect GnRH pulse frequency. While there is no disputing the increase in pulse amplitude, some studies (33,48.49), but not all (50,51), have also described an increase in pulse frequency of LH (33,48,49). Murdoch and co-workers (52), who extensively investigated the variability of LH measurements in women with the PCOS, found good reproducibility with repeated studies over a 1-year period. They also assessed LH pulsatility by time-series analysis, which takes into consideration the complicated patterns of LH secretion that occur as superimposed pulses of differing frequency. No difference in pulse frequency was detected between 9 patients with PCOS and 12 normal women. Some studies have also demonstrated an alteration of the circadian rhythm of LH secretion, with a persistence of high-amplitude LH pulses during the night (51).

Some women with hypogonadotropic hypogonadism (HH) also have polycystic ovaries detected by pelvic ultrasound, and when these women are treated with pulsatile GnRH to induce ovulation they have significantly higher serum LH concentrations than women with HH and normal ovaries (53). Furthermore, the elevation in LH concentration was observed before serum E2 concentrations rose. Thus, hypersecretion of LH occurred in these women when the hypothalamus was replaced by an artificial GnRH pulse generator (i.e., the GnRH pump), with a fixed GnRH pulse interval of 90 minutes (equivalent to the pulse interval in the early follicular phase). These results suggest that the cause of hypersecretion of LH involves a perturbation of ovarian–pituitary feedback rather than a primary disturbance of hypothalamic pulse regulation. These findings are also consistent with the notion that there may be a nonsteroidal factor(s) that disturbs ovarian–pituitary feedback control of LH secretion.

The data collected in women with PCOS undergoing laparoscopic ovarian diathermy are also consistent with the hypothesis that it is altered ovarian–pituitary feedback that causes hypersecretion of LH. In these patients, LH pulse amplitude decreased with the procedure, although no change in the (normal) pulse frequency was observed (54). Rossmanith et al. (55) found an attenuation of GnRH-stimulated LH secretion after laparoscopic ovarian diathermy, a result consistent with abnormalities in the production of an ovarian factor(s) that regulates LH secretion rather than with the theory that the disorder starts at the level of either the hypothalamus or pituitary.

There is an alternative body of evidence, however, that is consistent with a persistently rapid frequency of pulsatile GnRH and hence LH secretion (56,57) in some women with PCOS, with absence of the normal luteal phase slowing of LH pulsatility—a function of progesterone suppression of LH. It has been proposed that during puberty the hypothalamus becomes less sensitive to ovarian steroidal feedback in girls with PCOS, which may be a particular effect of hyperandrogenemia (56). Body weight also has an effect, with pituitary responsiveness to GnRH being inversely related to the body mass index (BMI) (58).

2.3.1. Glycosylation of LH

LH exists in multiple forms in both the pituitary gland and the peripheral circulation, primarily because of the considerable variations in the oligosaccharide side chains, which may result in numerous LH "glycoforms" (59). The α and β subunits of LH have two N-linked glycosylation sites, and oligosaccharides form about 30% of each molecule. Individual glycoforms cannot be isolated, and even the recently available recombinant gonadotropins are likely to have a variety of glycoforms. The glycosylation isoforms that are more basic have a shorter half-life and a lower in vivo activity than the acidic isoforms. Conversely, the basic isoforms have a higher biopotency, with higher receptor binding and steroidogenic and intracellular cyclic adenosine monophosphate-stimulating ability in vitro. These differences are not recognized by either immunoassays (because antibodies do not recognize oligosaccharides) or in vitro bioassays, which are not dependent on in vivo clearance mechanisms. Lectins bind to oligosaccharides and might have a future role to play in two-site lectin/antibody assays.

LH is probably modified in the circulation, for example, by proteolytic cleavage of part of the β-chain ("nicking") *(60)*; hCG differs from LH by having a single additional C-terminal peptide. There is evidence that hCG itself is secreted by the normal pituitary and may circulate at concentrations of about 1% that of LH *(59)*. Exposure of the pituitary to different serum concentrations of steroids may affect glycosylation and hence bioactivity of LH, which in turn may influence ovarian steroid production. Although animal studies have shown that gonadal steroids may act directly at the pituitary to control the biopotency of stored and secreted gonadotropins, it is uncertain whether LH bioactivity is affected by the chronic alterations in steroid hormone secretion that are seen in women with PCOS. Some studies have shown that bioactive serum concentrations of LH are elevated in these patients *(31,61)*. Lobo et al. *(61)* suggested that the level of bioactive LH may be a more useful marker than the immunoactive LH concentration or the LH:FSH ratio, although the degree of bioactivity did not correlate with dopaminergic activity or serum E2 concentrations *(61)*. Fauser et al. *(31)* observed that bioactive LH levels correlated better with symptoms of PCOS (oligo/amenorrhoea) than did serum LH concentrations as measured by an immunoradiometric assay (IRMA). It has been suggested that women with PCOS may secrete LH isoforms with a high biological activity *(62)*. Further evidence for this was obtained by Ding et al. *(63)* who performed isoelectric focusing on serum from women with normal and elevated concentrations of LH. They found that those with high serum LH concentrations had the majority of LH isoforms distributed with an alkaline isoelectric point, and this correlated with a high biological activity.

Circulating serum concentrations of E2, estrone, androstenedione, testosterone, or dehydroepiandrosterone sulfate (DHEAS) were not found to correlate with either the bioactive or immunoreactive LH levels *(64)*. Experiments in nonhuman primates have demonstrated that the administration of different doses of GnRH results in pituitary secretion of LH isoforms of varying bioactivity *(65)*. The pituitaries of women with PCOS might be sensitized to GnRH such that higher activity LH isoforms are secreted than in women with normal ovaries. It might be that a nonsteroidal messenger from the ovary affects this phenomenon.

A genetic variant of LH (vLH) has been described with two missense point mutations in the LHβ gene (Trp[8]Arg and Ile[15]Thr), initially reported from Finland but now recognized as a common polymorphism with worldwide distribution and a mean population carrier frequency of 18.5% (Finland, 28.9%; United Kingdom, 16.8%, United States, 14.1%; Netherlands, 11.2%) *(66)*. The biological activity of vLH is greater than wild-type LH in vitro, whereas its half-life in the circulation is shorter and the overall effect on in vivo bioactivity is unclear. The vLH carrier frequency appears similar in obese and non-obese controls but lower in obese subjects with PCOS. The authors suggest that obese women with vLH might be protected from developing symptomatic PCOS, whereas those with wild-type LH may be more liable to develop the syndrome. Interestingly, of the four countries studied, this relationship was true for all but those from the United Kingdom, who exhibited a higher frequency of vLH in obese PCOS women than obese controls.

Insulin stimulates pituitary gonadotropin secretion, at least in vitro *(67)*. It has therefore been proposed that the hyperinsulinemia that is seen in many women with the PCOS might have a causal relationship with hypersecretion of LH. Antilla et al. *(62)* studied obese and non-obese women with PCOS and found that the non-obese women had elevated serum concentrations of bioactive LH and the obese women tended to have normal serum concentrations of bioactive LH. These differences could not be attributed to serum androgen concentrations. It was noted that bioactive, but not immunoactive, LH levels related to the serum insulin concentration, and Antilla proposed that the degree of hyperinsulinemia in obese, insulin-resistant subjects had a direct affect on the glycosylation of LH.

Women with PCOS often hypersecrete LH, and this may result in increased theca-cell androgen secretion. A study of 556 women with polycystic ovaries demonstrated that the patients with the highest serum LH concentrations did not have the highest serum testosterone concentrations *(68)*. Indeed, it has been shown that lean women with PCOS and normal fasting serum insulin concentrations have higher serum LH concentrations than lean or obese women with polycystic ovaries and elevated fasting insulin levels *(69)*. Furthermore, it was found that lean and obese women with polycystic ovaries and elevated fasting insulin levels had higher serum testosterone concentrations than women with normal insulin levels, suggesting a stronger relationship between androgen secretion and circulating insulin levels than with LH.

2.3.2. Nonsteroidal Feedback on LH Secretion

Since the isolation and characterization of inhibin it has become apparent that not only are there several members of the inhibin family of glycoprotein hormones, but there are also other nonsteroidal gonadal signals that influence gonadotropin secretion and help to fine-tune reproductive function. It has been established that ovarian inhibin exerts a negative feedback on pituitary gonadotropin production, preferentially effecting FSH. More recently a feedback pathway that influences pituitary LH secretion has been proposed, following in vivo and in vitro evidence that has suggested the presence of a putative inhibitory peptide, which has been named gonadotropin surge inhibiting or attenuating factor (GnSIF/GnSAF) *(70)*. The proposed existence of GnSAF is an attractive, although unproven hypothesis for the differential regulation of LH secretion. GnSAF is presumed to be an inhibitory factor, different from inhibin, that is produced by ovarian follicles and which suppresses GnRH-stimulated LH, and possibly also FSH, secretion. The hypothesis is that GnSAF antagonizes the priming of the pituitary gland by GnRH and raises its threshold to the action of GnRH, preventing a premature LH surge. The surge may then occur when elevated serum estradiol concentrations increase the sensitivity of the pituitary to GnRH, so overriding the inhibitory factor.

Abnormalities of inhibin secretion have long been implicated in the pathogenesis of PCOS, with the notion that hypersecretion of inhibin B by the ovary suppresses pituitary secretion of FSH to cause the relative imbalance in gonadotropin concentrations observed in these patients *(71)*. A series of experiments have demonstrated significantly elevated serum levels of inhibin B in women with PCOS, which has been postulated as being responsible both for the reversed FSH-to-LH ratio and for the increased sensitivity of the polycystic ovary to exogenous FSH *(72)*.

2.4. Hyperinsulinemia

The association between insulin resistance, compensatory hyperinsulinemia, and hyperandrogenism has provided insight into the pathogenesis of the PCOS. The cellular and molecular mechanisms of insulin resistance in PCOS have been extensively investigated (*see* Chapter 26), and it is evident that the major defect is a decrease in insulin sensitivity secondary to a postbinding abnormality in insulin receptor-mediated signal transduction, with a less substantial, but significant, decrease in insulin responsiveness *(73)*. It appears that decreased insulin sensitivity in PCOS is potentially an intrinsic defect in genetically susceptible women because it is independent of obesity, metabolic abnormalities, body fat topography, and sex hormone levels. There may be genetic abnormalities in the regulation of insulin receptor phosphorylation, resulting in increased insulin-independent serine phosphorylation and decreased insulin-dependent tyrosine phosphorylation *(73)*.

Although the insulin resistance may occur irrespective of BMI, the common association of PCOS and obesity has a synergistic deleterious impact on glucose homeostasis and can worsen both hyperandrogenism and anovulation. Insulin acts through multiple sites to increase endogenous androgen levels. Increased peripheral insulin resistance results in a higher serum insulin concentration. Excess insulin binds to the IGF-1 receptors that enhance the theca cells' androgen production in response to LH stimulation *(74)*. Hyperinsulinemia also decreases the synthesis of SHBG by the liver. Therefore, there is an increase in serum free testosterone concentration and consequent peripheral androgen action. In addition, hyperinsulinemia inhibits the hepatic secretion of insulin-like growth factor-binding protein (IGFBP)-1, leading to increased bioavailability of IGF-1 and -2 *(75)*, the important regulators of ovarian follicular maturation and steroidogenesis. Together with more IGF-2 secretion from the theca cells, IGF-1 and -2 further augment ovarian androgen production by acting on IGF-1 receptors *(76)*.

Insulin may also increase endogenous androgen concentrations by increased cytochrome P450c17α enzyme activity, which is important for ovarian and adrenal steroid hormone biosynthesis. Insulin-induced overactivity of P450c17α and exaggerated serum 17-hydroxyprogesterone response to stimulation by GnRHa have also been demonstrated *(19,77)*. Intraovarian androgen excess is responsible for anovulation by acting directly on the ovary, promoting the process of follicular atresia *(78)*. This latter process is characterized by apoptosis of granulosa cells. As a consequence there is an increasingly larger stromal compartment, which retains LH responsiveness and continues to secrete androgens.

Insulin resistance is defined as a reduced glucose response to a given amount of insulin and may occur secondary to resistance at the insulin receptor, decreased hepatic clearance of insulin, and/or increased pancreatic sensitivity. Both obese and non-obese women with PCOS are more insulin-resistant and hyperinsulinemic than age- and weight-matched women with normal ovaries *(79)*. Thus, there appear to be factors in women with PCOS that promote insulin resistance and are independent of obesity *(80)*. Pancreatic β-cell dysfunction has been described in women with PCOS, whereby there is increased basal secretion of insulin yet an inadequate post-prandial response. This defect remains even after weight loss *(81)*, despite an improvement in glucose tolerance.

Insulin acts through its receptor to initiate a cascade of postreceptor events within the target cell. Phosphorylation causes insulin receptor substrates (IRS1–4) to promote glucose uptake via the trans-membrane glucose transporter (GLUT4) and also intracellular protein synthesis. Tyrosine phosphorylation increases the tyrosine kinase activity of the insulin receptor, whereas serine phosphorylation inhibits it, and it appears that at least 50% of women with PCOS have excessive serine phosphorylation and inhibition of normal signaling *(80)*. This affects only glucose homeostasis and not the other pleotropic actions of insulin, so that cell growth and protein synthesis may continue. Serine phosphorylation also increases activity of P450c17 in both the ovary and adrenal, thus promoting androgen synthesis, and this may be a mechanism for both insulin resistance and hyperandrogenism in some women with PCOS.

3. CONCLUSIONS

PCOS is a heterogeneous condition with ovarian dysfunction leading to the main signs and symptoms and being influenced by external factors, in particular the gonadotropins, insulin, and other growth factors, which are dependent on both genetic and environmental influences. At the heart of the pathophysiology for many is insulin resistance and hyperinsulinemia, and even if this is not the initiating cause in some, it is certainly an amplifier of hyperandrogenism in those that gain weight. Gonadotropin biosynthesis and secretion are influenced by hypothalamic, paracrine, and endocrine factors, and there is considerable overlap among the three. The influence of nonsteroidal factors on pituitary and hypothalamic function is still being elucidated. Further work is required to unravel the pathophysiology of PCOS and its evolution through adolescence, and more research is required to examine both the pathophysiology of hypersecretion of LH and its effects at the level of the oocyte.

4. FUTURE AVENUES OF INVESTIGATION

A number of areas remain to better explored and understood, including (1) the effects on biochemistry of the polycystic ovary on oocyte competence and subsequent embryo development, (2) the interrelationship between fat metabolism and hyperandrogenism, (3) the critical times during life that affect the setting of homeostatic mechanisms in the hypothalamus, pituitary, and ovary, and (4) the evolution of normal and polycystic ovarian function longitudinally throughout life.

KEY POINTS

- The biochemical features of PCOS include elevated serum concentrations of testosterone, androstenedione, LH, and insulin, which may vary between individuals—both within and between populations—and may change with time.

- In understanding the biochemistry of PCOS one has to consider both the nature of the dysfunction within the ovary and the external influences that modify ovarian behavior.
- In vivo and in vitro data confirm that the theca cells of PCOS patients have a generalized overactive steroidogenesis.
- LH and insulin stimulate theca-cell androgen production, and insulin also amplifies the effects of test-osterone by suppressing SHBG.
- Serum LH concentrations are significantly elevated in PCOS as a result of increased amplitude and frequency of LH pulses.
- Hyperinsulinemia is a key component to the pathogenesis of PCOS but is not ubiquitous.

ACKNOWLEDGMENT

I am grateful to Dr. Chandrika Wijeyeratne, Professor Roy Homburg, Professor Rick Legro, and Dr. Gerry Conway for assistance with some aspects of the text.

REFERENCES

1. Fauser B, Tarlatzis B, Chang J, et al. The Rotterdam ESHRE/ASRM-sponsored PCOS consensus workshop group. Revised 2003 consensus on diagnostic criteria and long-term health risks related to polycystic ovary syndrome (PCOS). Hum Reprod 2004;19:41–47.
2. Balen AH, Laven JSE, Tan SL, Dewailly D. Ultrasound assessment of the polycystic ovary: international consensus definitions. Hum Reprod Update 2003;9:505–514.
3. Polson DW, Adams J, Wadsworth J, Franks S. Polycystic ovaries—a common finding in normal women. Lancet 1988;1:870–872.
4. Michelmore KF, Balen AH, Dunger DB, Vessey MP. Polycystic ovaries and associated clinical and biochemical features in young women. Clin Endocrinol Oxf 1999;51:779–786.
5. Balen AH, Conway GS, Kaltsas G, et al. Polycystic ovary syndrome: the spectrum of the disorder in 1741 patients. Human Reprod 1995;10:2705–2712.
6. Rodin DA, Bano G, Bland JM, Taylor K, Nussey SS. Polycystic ovaries and associated metabolic abnormalities in Indian subcontinent Asian women. Clin Endocrinol 1998;49(1):91–99.
7. Wijeyaratne CN, Balen AH, Barth J, Belchetz PE. Clinical manifestations and insulin resistance (IR) in polycystic ovary syndrome (PCOS) among South Asians and Caucasians: Is there a difference? Clin Endocrinol 2002;57:343–350.
8. Wijeyaratne CN, Nirantharakunar K, Balen AH, Barth JA, Sheriff R, Belchetz PE. Plasma homocysteine in polycystic ovary syndrome: does it correlate with insulin resistance and ethnicity? Clin Endocrinol 2004;60:560–567.
9. Legro RS, Driscoll D, Strauss JF 3rd, Fox J, Dunaif A. Evidence for a genetic basis for hyperandrogenemia in polycystic ovary syndrome. Proc Natl Acad Sci USA 1998;95:14956–14960.
10. Knochenhauer ES, Key TJ, Kahsar-Miller M, Waggoner W, Boots LR, Azziz R. Prevalence of the polycystic ovary syndrome in unselected black and white women in the Southeastern United States: a prospective study. J Clin Endocrinol Metab 1988;83:3078–3082.
11. Pugeat M, Nicolas MH, Craves JC, et al. Androgens in polycystic ovarian syndrome. Ann NY Acad Sci 1993;687:124–135.
12. Laven JS, Imani B, Eijkemans MJ, Fauser BC. New approaches to PCOS and other forms of anovulation. Obstet Gynecol Surv 2002;57:755–767.
13. Cibula D, Hill M, Starka L. The best correlation of the new index of hyperandrogenism with the grade of increased hair. Eur J Endocrinol 2000;143:405–408.
14. Imani B, Eijkemans MJ, de Jong FH, et al. Free androgen index and leptin are the most prominent endocrine predictors of ovarian response during clomiphene citrate induction of ovulation in normogonadotropic oligoamenorrheic infertility. J Clin Endocrinol Metab 2000;85:676–682.
15. Rosner W. Errors in the measurement of plasma free testosterone. J Clin Endocrinol Metab 1997;82:2014–2015.
16. Vermeulen A, Verdonck L, Kaufman JM. A critical evaluation of simple methods for the estimation of free testosterone in serum. J Clin Endocrinol Metab 1999;84:3666–3672.
17. Tremblay RR, Dube JY. Plasma concentration of free and non-TeBG bound testosterone in women on oral contraceptives. Contraception 1974;10:599–605.
18. Franks S, White D, Gilling-Smith C, Carey A, Waterworth D, Williamson R. Hypersecretion of androgens by polycystic ovaries: the role of genetic factors in the regulation of cytochrome P450c17α. Balliere's Clin Endocrinol Metab 1996;10(2):193–203.
19. Rosenfield RL, Barnes RB, Cara JF, Lucky AW. Dysregulation of cytochrome P450c17α as the cause of polycystic ovarian syndrome. Fertil Steril 1990;53(5):785–791.
20. Ehrmann DA, Barnes RB, Rosenfield RL. Polycystic ovary syndrome as a form of functional ovarian hypernadrogenism due to dysregulation of androgen secretion. Endocr Rev 1995;16:322–353.

21. Kirschner MA, Bardin CW Androgen production and metabolism in normal and virilized women. Metabolism 1972;21:667–688.
22. Rivarola M, Singleton R, Migoen C. Splanchnic extraction and interconversion of testosterone and androstenedione in man. J Clin Invest 1967;46:2095–2099.
23. White DW, Leigh A, Wilson C, et al. Gonadotrophin and gonadal steriod response to a single dose of a long-acting agonist of gonadotrophin-releasing hormone in ovulatory and anovulatory women with polycystic ovary syndrome. Clin Endocrinol 1995;42:475–481.
24. Barnes, RB, Rosenfield RL, Burstein S, et al. Pituitary-ovarian response to nafarelin testing in the polycystic ovary syndrome. N Engl J Med 1989;320:559–563
25. Franks S. Polycystic ovary syndrome. N Engl J Med 1995;333:853–861.
26. Erickson GF, Magoffin DA, Garza VG, et al. Granulosa cells of polycystic ovaries: are they normal or abnormal? Hum Reprod 1992;7:293.
27. Mason HD, Willis DS, Beard RW, et al. Estradiol production by granulosa cells of normal and polycystic ovaries (PCO): relationship to menstrual cycle history and to concentrations of gonadotrophins and sex steroids in follicular fluid. J Clin Endocrinol Metab 1994;79:1355.
28. Webber LJ, Stubbs S, Stark J, Trew GH, Franks S. Formation andf early development of follicles in the polycystic ovary. Lancet 2003;362:1017–1021.
29. Prelevic G, Wurzburger M, Balint-Peri C, et al. Inhibitory effect of sandostatin on secretion of luteinizing hormone and ovarian steroids in polycystic ovary syndrome. Lancet 1990;336:900–906.
30. Anderson R, Groome N, Baird D. Inhibin A and inhibin B in women with polycystic ovarian syndrome during treatment with FSH to induce mono-ovulation. Clin Endocrinol 1998;48:577–582
31. Fauser BCJM, Pache TD, Lamberts WJ, Hop WCJ, de Jong FH, Dahl KD. Serum bioactive and immunoreactive luteinising hormone and follicle stimulating hormone levels in women with cycle abnormalities, with or without polycystic ovary disease. J Clin Endocrinol Metab 1991;73:811–817.
32. Taylor AE, McCourt B, Martin K, et al. Determinants of abnormal gonadotropin secretion in clinically defined women with PCOS. J Clin Endocrinol Metab 1997;82:2248–2256.
33. Waldstreicher J, Santoro NF, Hall HJE, Filicori M, Crowley WF. Hyperfunction of the hypothalamic-pituitary axis in women with polycystic ovarian disease: Indirect evidence of partial gonadotroph desensitization. J Clin Endocrinol Metab 1998;66:65–172.
34. Van Santbrink EJ, Hop WC, Fauser BC. Classification of normogonadotropin infertility: Polycystic ovaries diagnosed by ultrasound versus endocrine characteristics of PCOS. Fertil Steril 1997;67:452–458.
35. Balen AH, Tan SL, Jacobs HS. Hypersecretion of luteinising hormone—a significant cause of subfertility and miscarriage. Br J Obstet Gynaecol 1993;100:1082–1089.
36. Wang CF, Lasley BL, Lein A, Yen SSC. The functional changes of the pituitary gonadotrophs during the menstrual cycle. J Clin Endocrinol Metab 1976;42:718–728.
37. Rosmanith WG, Wirth U, Sterzik K, Yen SS. The effects of prolonged opioidergic blockade on LH pulsatile secretion during the menstrual cycle. J Endocrinol Invest 1989;12:245–252.
38. Quigley M, Rakoff J,Yen SSC. Increased luteinising hormone sensitivity to dopamine inhibition in the polycystic ovary syndrome. J Clin Endocrinol Metab 1981;52:231.
39. Cumming DC, Reid RL, Quigley ME, Rebar RW, Yen SS. Evidence for decreased endogenous dopamine and opioid inhibitory influences on LH secretion in polycystic ovary syndrome. Clin Endocrinol (Oxf) 1984;20:643–648.
40. Shoupe D, Lobo RA. Evidence for altered catecholamine metabolism in polycystic ovary syndrome. Am J Obstet Gynecol 1984;150:566–570.
41. Yoshino K, Takahashi K, Shirai T, Nishigaki A, Araki Y, Kitao M. Changes in plasma catecholamines and pulsatile patterns of gonadotropins in subjects with a normal ovulatory cycle and with polycystic ovary syndrome. Int J Fertil 1990;35:34–39.
42. Barnes R, Lobo R. Central opioid activity in the polycystic ovary syndrome with and without dopaminergic modulation. J Clin Endocrinol Metab 1985;61:779.
43. Barnes RB, Mileikowsky GN, Cha KY, Spencer CA, Lobo RA. Effects of dopamine and metoclopramide in polycystic ovary syndrome. J Clin Endocrinol Metab 1986;63:506–509.
44. Berga SL, Yen SSC. Opioidergic regulation of LH pulsatility in women with polycystic ovary syndrome. Clin Endocrinol 1989;30:177–184.
45. Lanzone A, Apa R, Fulghesu AM, Cutillo G, Caruso A, Mancuso S. Long-term naltrexone treatment normalizes the pituitary response to gonadotrophin-releasing hormone in polycystic ovarian syndrome. Fertil Steril 1993;59:734–737.
46. Lobo RA, Granger LR, Paul WL, Goebelsmann U, Mishell DR. Psychological stress and increases in urinary norepinephrine metabolites, platelet serotonin and adrenal androgens in women with polycystic ovary syndrome. Am J Obstet Gynecol 1983;145:496–503.
47. Schoemaker J. Neuroendocrine control in polycystic ovary-like syndrome. Gynecol Endocrinol 1991;5:277–288.
48. Rebar R, Judd HL, Yen SCC, Rakoff J, Vandenberg G, Naftolin F. Characterization of the inappropriate gonadotropin secretion in polycystic ovary syndrome. J Clin Invest 1976;57:1320–1329.

49. Burger CW, Korsen T, Van Kessel H, Van Dop PA, Caron FJM, Schoemaker J. Pulsatile luteinizing hormone patterns in the follicular phase of the menstrual cycle, polycystic ovarian disease (PCOD) and non PCOD secondary amenor-rhoea. J Clin Endocrinol Metab 1985;61:1126–1132.

50. Baird DT, Corker CS, Davison DW, Hunter WM, Michie EA, Van Look PFA. Pituitary ovarian relationships in poly-cystic ovary syndrome. J Clin Endocrinol Metab 1977;45:798–809.

51. Venturoli S, Porcu E, Fabbri R, et al. Episodic pulsatile secretion of FSH, LH, prolactin, oestradiol, oestrone and LH circadian variations in polycystic ovary syndrome. Clin Endocrinol 1988;28:93–107.

52. Murdoch AP, Diggle PJ, White MC, Kendall-Taylor P, Dunlop W. LH in polycystic ovary syndrome: reproducibility and pulsatile secretion. J Endocrinol 1989;121:185–191.

53. Schachter M, Balen AH, Patel A, Jacobs HS. Hypogonadotrophic patients with ultrasonographically diagnosed poly-cystic ovaries have aberrant gonadotropin secretion when treated with pulsatile gonudatrophin-releasing hormone—a new insight into the pathophysiology of polycystic ovary syndrome. Gynecol Endocrinol 1996;10:327–335.

54. Abdel Gadir A, Mowafi RS, Alnaser HMI, Alnaser HMI, Alrashid AH, Shaw RW. Ovarian electrocautery versus hMG and pure FSH therapy in the treatment of patients with polycystic ovarian disease. Clin Endocrinol 1990;33:585–592.

55. Rossmanith WG, Keckstein J, Spatzier K, Lauritzen C. The impact of ovarian laser surgery on the gonadotrophin secretion in women with PCOD. Clin Endocrinol 1991;34:223–230.

56. McCartney CR, Prendergast K, Chhabra S, Chopra C, Marshall JC. Neuroendocrine connection in PCOS. In: Filicori M, ed. Updates in Infertility. Medimond International Proceedings, Bologna, 2004,.pp. 427–440.

57. Marshall JC, Eagleson CA. Neuroendocrine aspects of PCOS. Endocrinol Metab Clin North Am 1999;28:295–324.

58. Hall JE. Gonadotropin secretion in PCOS. In: Filicori M, ed. Updates in Infertility Treatment 2004. Medimond Interna-tional Proceedings, Bologna, 2004, pp. 440–457.

59. Jeffcoate SL. Analytical and clinical significance of peptide hormone heterogeneity with particular reference to growth hormone and luteinising hormone in serum. Clin Endocrinol 1993;38:113–121.

60. Iles RK, Lee CL, Howes I, Davies S, Edward R, Chard T. Immunoreactive ß-core-like material in normal postmeno-pausal urine: human chorionic gonadotrophin or LH origin? Evidence for the existence of LH core. J Endocrinol 1992;133:459–466.

61. Lobo RA, Kletzky OA, Campeau J, diZerega G. Elevated bioactive luteinising hormone in women with polycystic ovary syndrome. Fertil Steril 1983;39:674–679.

62. Antilla L, Ding Y-Q, Ruutiainen K, Erkkola R, Irjala K, Huhtaniemi I. Clinical features and circulating gonadotropin, insulin and androgen interactions in women with polycystic ovarian disease. Fertil Steril 1991;55:1057–1061.

63. Ding Y-Q, Huhtaniemi I. Preponderance of basic isoforms of serum LH is associated with the high bio/immuno ratio of LH in healthy women and in women with polycystic ovarian disease. Human Reprod 1991;6:346–350.

64. Fauser BCJM, Pache TD, Hop WCJ, de Jong FH, Dahl KD. The significance of a single serum LH measurement in women with cycle disturbances: discrepancies between immunoreactive and bioactive hormone estimates. Clin Endocrinol 1992;37:445–452.

65. Matteri RL, Djiershke DJ, Bridson WE, Rhutasel NS, Robinson JA. Regulation of the biopotency of primate LH by GnRH in-vitro and in-vivo. Biol Reprod 1990;43:1045–1049.

66. Tapanainen JS, Koivunen Riitta, Fauser BCJM, et al. A new contributing factor to polycystic ovary syndrome: the genetic variant of luteinizing hormone. J Clin Endocrinol Metab 1999;84:1711–1715.

67. Adashi EY, Hsueh AJW, Yen SSC. Insulin enhancement of luteinising hormone and follicle stimulating hormone release by cultured pituitary cells. Endocrinology 1981;108:1441–1449.

68. Conway GS, Honour JW, Jacobs HS. Heterogeneity of the polycystic ovary syndrome: clinical, endocrine and ultra-sound features in 556 patients. Clin Endocrinol 1989;30:459–470.

69. Conway GS, Clark PMS, Wong D. Hyperinsulinaemia in the polycystic ovary syndrome confirmed with a specific immunoradiometric assay for insulin. Clin Endocrinol 1993;38:219–222.

70. Sopelak VM, Hodgen GD. Blockade of the estrogen-induced LH surge in monkeys: a nonsteroidal, antigenic factor in porcine follicular fluid. Fertil Steril 1984;41:108–113.

71. Lockwood GM, Muttukrishna S, Groome NP, Matthews DR, Ledger WL. Mid-follicular phase pulses of inhibin B are absent in polycystic ovary syndrome and are initiated by successful laparoscopic ovarian diathermy: a possible mecha-nism for the emergence of the dominant follicle. J Clin Endocrinol Metab 1998;83:1730–1735.

72. Lockwood GM. The role of inhibin in PCOS. Human Fertil 2000;3:86–92.

73. Dunaif A. Insulin resistance and the polycystic ovary syndrome: mechanisms and implication for pathogenesis. Endocrinol Rev 1997;18:774–800.

74. Bergh C, Carlsson B, Olsson JH, Selleskog U, Hillensjo T. Regulstion of androgen production in cultured human thecal cells by insulin-like growth factor I and insulin. Fertil Steril 1993;59:323–331.

75. De Leo V, la Marca A, Orvieto R, Morgante G. Effect of metformin on insulin-like growth factor (IGF) I and IGF-binding protein I in polycystic ovary syndrome. J Clin Endocrinol Metab 2000;85:1598–1600.

76. Voutilainen R, Franks S, Mason HD, Martikainen H. Expression of insulin-like growth factor (IGF), IGF-binding protein, and IGF receptor messenger ribonucleic acids in normal and polycystic ovaries. J Clinl Endocrinol Metab 1996;81:1003

77. la Marca A, Egbe TO, Morgante G, Paglia T, Ciani A, De Leo V. Metformin treatment reduces ovarian cytochrome P450c17α response to human chorionic gonadotrophin in women with insulin resistance-related polycystic ovary syndrome. Hum Reprod 2000;15:21–23.

78. Hsueh ADW, Billig H, Tsafiri A. Ovarian follicle atresia: A hormonally controlled apoptotic process. Endocrinol Rev 1994;15:707–724.

79. Dunaif A, Segal KR, Shelley DR, Green G, Dobrjansky A, Licholai T. Evidence for distinctive and intrinsic defects in insulin action in polycystic ovary syndrome. Diabetes 1992;41:1257–1266.

80. Tsilchorozidou T, Overton C, Conway GS. The pathophysiology of polycystic ovary syndrome. Clin Endocrinol 2004;60:1–17.

81. Holte J, Bergh T, Berne C, Wide L, Lithell H. Restored insulin sensitivity but persistently increased early insulin secretion after weight loss in obese women with polycystic ovary syndrome. J Clin Endocrinol Metab 1995;80:2586–2593.

16

Ovarian Histology, Morphology, and Ultrasonography in the Polycystic Ovary Syndrome

Sophie Jonard, Yann Robert, Yves Ardaens, and Didier Dewailly

SUMMARY

The need for calibrated imaging of polycystic ovaries is now stronger than ever since the consensus conference held in Rotterdam in May 2003 to establish the diagnosis of the polycystic ovary syndrome. However, imaging polycystic ovaries is not an easy procedure, and it requires a thorough technical and medical background. Two-dimensional ultrasonography remains the standard for imaging polycystic ovaries, and the current consensus definition of polycystic ovaries determined at the joint Rotterdam meeting rests on this technique. The definition includes either 12 or more follicles measuring 2–9 mm in diameter and/or increased ovarian volume (>10 cm^3) in at least one ovary. Other imaging techniques, such as Doppler and three-dimensional ultrasonography, can help in the diagnosis, but should be considered second-line techniques.

Key Words: Polycystic ovary syndrome; two-dimensional ultrasonography; ovarian volume; follicle number.

1. INTRODUCTION

The need for calibrated imaging of polycystic ovaries is now more urgent than ever since the consensus conference held in Rotterdam in 2003 to define the polycystic ovary syndrome (PCOS). Indeed, the subjective criteria that were proposed 20 years ago and were in use until recently by the vast majority of authors are now being replaced by a stringent definition using objective criteria (1,2).

Imaging for polycystic ovaries is not an easy procedure. It requires a thorough technical and medical background. The goal of this chapter is to provide the reader with an understanding of the main issues ensuring a well-controlled imaging for the diagnosis of PCOS. Histological and morphological data will be first addressed. Thus, two-dimensional (2D) ultrasonography (U/S) will be extensively detailed because it remains the standard for imaging polycystic ovaries. Other techniques, such as Doppler and three-dimensional (3D) U/S, will be more briefly described.

2. BACKGROUND

2.1. Ovarian Histology and Morphology in PCOS

Since the very comprehensive review of Hughesdon (3) in 1982, few additional data have been reported concerning the gross morphological features of polycystic ovaries (Table 1). This is mainly a result of the unavailability of anatomical pieces. However, the follicular excess present in the disorder has recently been re-examined by studying ovarian cortical biopsies (4). These studies suggest that the mechanism of PCOS involves a primary follicle dysfunction (4,5). One important component of this dysfunction is the observation by Hughesdon (3) that PCOS ovaries contain two- to threefold

From: *Contemporary Endocrinology: Androgen Excess Disorders in Women:
Polycystic Ovary Syndrome and Other Disorders, Second Edition*
Edited by: R. Azziz et al. © Humana Press Inc., Totowa, NJ

Table 1
Histomorphological Abnormalities of the Ovaries in PCOS

Histological features of PCOS	Detection by U/S
Total ovarian hypertrophy	Yes
Thickened capsule (> 100Δ)	No
Increased number of subcapsular follicle cysts	Yes
Scarcity of corporea lutea or albicantia	Yes
Hyperplasia and fibrosis of ovarian stroma	Yes
Decreased thickness of granulosa layer	No
Atretic pattern of granulosa layer	No
Increased thickness of theca interna	No
Premature luteinization of theca cells	No

PCOS, polycystic ovary syndrome; U/S, ultrasonography. (Adapted from refs. *3* and *41*).

the normal number of follicles from the time they start to grow up to the size of 2–5 mm (antral follicles).

Recently, Webber et al. *(4)* revisited these data by examining ovarian cortical biopsies from normal women and women with PCOS. They found an even greater (sixfold) increase in the number of primary growing follicles in polycystic ovaries from anovulatory women in comparison to normal ovaries. Maciel et al. *(6)* also confirmed that PCOS ovaries are associated with a dramatic increase in the number of growing follicles, particularly at the primary follicle stage. In both studies the absolute number of primordial follicles was not different from the normal ovaries. Another component of follicle dysfunction in polycystic ovaries is that follicles stop growing and developing when they reach 4–7 mm in diameter. Therefore, the follicular problem of PCOS seems to be twofold (for review, *see* ref. *7*). First, early follicular growth is excessive, which is presumably the consequence of intra-ovarian hyperandrogenism. Second, the selection of one follicle from the increased pool and its further maturation to a dominant follicle does not occur (follicular arrest) because of an impaired action of follicle-stimulating hormone (FSH) and/or premature luteinizing hormone (LH) action.

2.2. Two-Dimensional Ultrasonography

Prior to transvaginal U/S, a transabdominal U/S examination should always be performed. Of course, a full bladder is required for visualization of the ovaries, although it should be noted that an overfilled bladder could compress the ovaries, yielding a falsely increased length. The main advantage of this route is that it offers a panoramic view of the pelvic cavity. Therefore, it allows the exclusion of associated uterine or ovarian abnormalities with an abdominal development. Indeed, lesions with cranial growth could be missed if transvaginal U/S is used exclusively.

Using the transvaginal route, high-frequency probes (>6 MHz), with better spatial resolution but less examination depth, can be used because the ovaries are close to the vagina and because the presence of fatty tissue is usually less. With this technique, not only the size and the shape of the ovaries are visualized, but also their internal structure, namely the follicles and stroma. It is now possible to obtain pictures that have a definition close to anatomical cuts. However, the evaluation of the ovarian size via the transvaginal approach is still difficult. To be the most accurate, it requires choosing meticulously the picture where the ovary appears the longest and the widest. This picture must then be frozen and the distance measured. Two means can be used for calculating the ovarian area: either fitting the outline of the ovary to a predrawn ellipse or outlining the ovary by hand. In both circumstances the area is calculated automatically by the computer. The latter technique of tracing the outline of the ovary by hand is preferred in cases of nonellipsoid ovaries, as is sometimes observed.

The ovarian volume is the parameter of greatest importance. Generally, the ovarian volume can be estimated from the length, width, and thickness of the ovary, using the classical formula for a prolate ellipsoid: $L \times W \times T \times 0.523$ *(8–10)*. However, the ovaries have to be studied in three orthogonal planes, a condition that is not always obtained. Three-dimensional U/S is an attractive alternative for the accurate assessment of ovarian volume, although this technique is not yet commonly available (*see* Subsection 2.5.1.).

In order to count the total number of "cysts" and to evaluate their size and position, each ovary should be scanned in longitudinal and/or transversal cross sections from the inner to outer margins. It is still unclear whether the counting of follicles should be performed on the whole ovary or on a selected longitudinal ovarian cut. Although both options are valid, it must be stressed that most physicians use the first, which is the only one to have been submitted to statistical estimation of its sensitivity and specificity (*see* Subsection 2.3.2.).

2.3. The Consensual Definition of Polycystic Ovaries

According to the literature review dealing with all available imaging systems and the discussion at the joint American Society for Reproductive Medicine (ASRM)/European Society for Human Reproduction and Embryology (ESHRE) consensus meeting on PCOS held in Rotterdam, May 1– 2003, the current consensus definition of polycystic ovaries is (1) either 12 or more follicles measuring 2–9 mm in diameter and/or (2) increased ovarian volume (>10 cm^3). The ovarian volume and the follicle number were selected as the diagnostic parameters because both have the advantage of being physical entities that can be measured in real-time conditions, and both are still considered key and consistent features of polycystic ovaries.

2.3.1. Increased Ovarian Volume

Many studies have reported an increased mean ovarian volume in series of patients with PCOS. However, the upper normal limit of the ovarian volume suffers from some variability in the literature (8–15 cm^3) (Table 2). Such variability may be explained by the following:

1. The small number of controls in some studies.
2. Differences in inclusion or exclusion criteria for control women.
3. Operator-dependent technical reasons; it is difficult indeed to obtain strictly longitudinal ovarian cuts, which is an absolute condition for accurate measures of the ovarian axis (length, width, thickness).

The consensual volume threshold to discriminate a normal ovary from a polycystic ovary was proposed to be 10 cm^3 *(1)*. This volume was felt to be the best compromise considering published studies *(11,12)*. Indeed, no study published to date has used an appropriate statistical appraisal of sensitivity and specificity of the volume threshold. This prompted us to recently revisit this issue through a prospective study including 154 women with PCOS and 57 controls. The receiver operating characteristic (ROC) curves indicated that a threshold at 10 cm^3 yielded a good specificity (98.2%), albeit poor sensitivity (39%). Setting the threshold at 7 cm^3 offered the best compromise between specificity (94.7%) and sensitivity (68.8%) *(13)*. Thus, in our opinion the threshold of 10 cm^3 for ovarian volume should be lowered to 7–8 cm^3 in order to increase the sensitivity of the polycystic ovary definition.

2.3.2. Increased Follicle Number

The polyfollicular pattern (i.e., excessive number of small hypoechoic areas < 10 mm in diameter) is strongly suggestive of PCO, because it is reminiscent of the syndrome label (i.e., polycystic). It is now broadly accepted that most of these cysts are in fact healthy oocyte-containing follicles and are not atretic.

Initially, only a descriptive definition of this feature was used to define polycystic ovaries *(9)*: "presence of 10 or more cysts measuring 2–8 mm in diameter arranged peripherally around a dense core of stroma or scattered through an increased amount of stroma." This definition has been used for many years, although no study was undertaken to estimate its sensitivity and specificity. Furthermore, the threshold for the follicle number applied to a single ovarian slice and not to the whole ovary. Currently, the consensus definition for polycystic ovaries is a whole ovary that contains 12 or more follicles 2–9 mm in diameter, omitting the follicle distribution. Again, the expert panel for the Rotterdam consensus considered this threshold as being the best compromise between the available

Table 2
Results of Selected Studies of Use of Ovarian Volume Determined by Ultrasound as a Discriminator for PCOS

Reference	Ultrasound examination route	No. of patients studied	No. of studied controls	Volume threshold indicative of PCOS (cm^3)	% Patients with clinical PCOS meeting volume criteria (%)	Controls meeting volume criterion (%)
Adams et al., 1985 (9)	TA	76	17	>15	33	0
Yeh et al., 1987 (11)	TA	108	25	>10	70	0
Pache et al., 1992 (22)	TV	52	29	>8	~70	0
Van Santbrink et al., 1997 (12)	TV	330	48	>10.7	41	0
Atiomo et al., 2000 (42)	TV	32	40	>9	~70	~45
Fulghesu et al., 2001 (23)	TV	53	30	>13.2	21	5
Jonard et al., 2006 (13)	TV	154	57	>7	70	5

PCOS, polycystic ovary syndrome;TA, transabdominal; TV, transvaginal.

186

Table 3
Receiver Operating Characteristic (ROC) Curve Data for Assessment of Polycystic Ovaries

Follicle number per ovary (mm)	Area under the ROC curve)	Threshold	Sensitivity (%)	Specificity (%)
2–5	0.924	10	65	97
		12	57	99
		15	42	100
6–9	0.502	3	42	69
		4	32.5	80
		5	24	89
2–9	0.937	10	86	90
		12	75	99
		15	58	100

Adapted from ref. *14*.

studies, including the one in which we compared 214 patients with PCOS to 112 women with normal ovaries *(14)*. By ROC analysis, a follicle number per ovary (FNPO) of 12 or more follicles of 2–9 mm diameter yielded the best compromise between sensitivity (75%) and specificity (99%) for the diagnosis of polycystic ovaries (Table 3). It is not possible to compare these data to the recent modification of their criteria proposed by Adams et al. *(9)*, that is, eight or more cysts 2–8 mm in diameter in a single plane with a peripheral distribution and the impression of increased stroma in the absence of such statistical approach in this last study *(14a)*.

The Rotterdam consensus meeting did not address the difficult issue concerning the presence of multifollicular ovaries (MFO) observed in clinical conditions other than PCOS. Again, these ovaries might be more correctly termed as multifollicular rather than multicystic. There is no consensual definition for the MFO, although they have been described as ovaries in which there are multiple (≥ 6) follicles, usually 4–10 mm in diameter, with normal stromal echogenicity *(9)*. No histological data concerning MFO are available. The MFO are characteristically seen during puberty and in women recovering from hypothalamic amenorrhea—both situations being associated with follicular growth without consistent recruitment of a dominant follicle *(15,16)*. Although the clinical pictures are theoretically different, there may be some overlap—hence the confusion between polycystic ovaries and MFO by inexperienced ultrasonographers. This stresses the need for considering carefully the other clinical and/or biological components of the consensual definition for PCOS. We recently re-examined the ovarian follicular pattern in a group of women with hypothalamic amenorrhea. About one-third had an FNPO higher than 12 (unpublished personal data). Because they were oligo- or anovulatory, they could be considered as having PCOS if one applied too inflexibly the Rotterdam definition! Some of these patients may truly have PCOS, whose clinical and biological expression has been modified by the chronically suppressed LH levels observed in secondary hypothalamic dysfunction *(17)*. In others, however, such an overlap in the FNPO emphasizes the need for a wise and careful utilization of the Rotterdam criteria as well as for considering other ultrasound criteria for polycystic ovaries in difficult situations.

2.4. Other Criteria and Definitions

2.4.1. External Morphological Signs of Polycystic Ovaries

In the 1970s, the weak resolution of U/S abdominal probes only allowed the detection exclusively of external morphological ovarian features, which were used as the first criteria for defining polycystic ovaries:

1. An ovarian length greater than 4 cm was a simple criterion. However, this unidimensional approach may lead to many false-positive results, particularly when a full bladder compresses the ovary (with the transabdominal route), or false-negative results when the ovaries are spherical with a relatively short length.

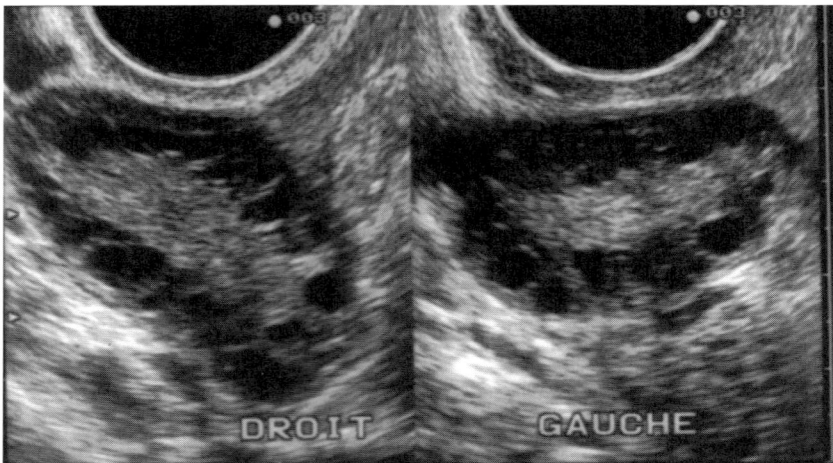

Fig. 1. Polycystic ovary (B mode). The right and left ovarian outlined areas are increased (9.7 and 8.9 cm^2, respectively). The follicle number, with a diameter between 2 and 9 mm, is more than 12. The small follicles display a typical peripheral pattern around the hyperechoic stroma.

2. The uterine width-to-ovarian length (U:O) ratio was reported to be decreased (<1) in polycystic ovaries, because the ovaries are normally longer, although the uterus is unchanged.
3. Polycystic ovaries often display a spherical shape in contrast to normal ovaries, which are ellipsoid. This morphological change can be evaluated by the sphericity index (ovarian width/ovarian length), which is generally greater than 0.7 in polycystic ovaries.

However, we should note that these parameters are less used today because of their low sensitivity and specificity *(18)*.

2.4.2. Ovarian Area

The ovarian area is less often used than ovarian volume and was not retained in the consensus definition, but in our recent study the diagnostic value of the ovarian area (assessed by ROC curves) was slightly better than that of the ovarian volume (sensitivity 77.6%; specificity 94.7% for a threshold at 5 cm^2/ovary) *(13)*. We also observed that the measured ovarian area (obtained after outlining the ovary by hand) was more informative than the calculated ovarian area (by using the formula for an ellipse: $L \times W \times W \times /4$). Indeed, ovaries are not strictly ellipsoid. We previously reported that the sum of the area of both ovaries was less than 11 cm^2 in a large group of normal women *(19,20)*, but a threshold of 5 cm^2 per ovary seems to offer the best compromise between sensitivity and specificity. Consequently, if the ovarian area (for a single ovary) is >5 cm^2, the diagnosis of polycystic ovaries is suggested *(13)*.

2.4.3. Ovarian Stroma

Stromal hypertrophy is frequently present in PCOS and is characterized by an increased central ovarian component, which is rather hyperechoic (Fig. 1). In others *(21)* and our *(18,19)* opinion, stromal hypertrophy and hyperechogenicity are helpful in distinguishing between polycystic ovaries and MFO, because these features are specific for the former. However, the estimation of hyperechogenicity is considered highly subjective, mainly because it depends on the settings of the U/S machine. Likewise, in the absence of a precise quantification, stromal hypertrophy is a relatively subjective sign.

For standardizing the assessment of stromal hypertrophy, we designed a computerized quantification of ovarian stroma, allowing for selective calculation of the stromal area by subtraction of the cyst area from the total ovarian area on a longitudinal ovarian cut *(19,20).* By this means we were able to set the upper normal limit of the stromal area (i.e., 95th percentile of a control group of 48 normal women) at 380 mm^2 per ovary. However, providing a precise outlining of the ovarian shape on a strictly longitudinal cut of the ovaries, the diagnostic value of the total ovarian area equaled that of stromal area, as both were highly correlated.

Fulghesu et al. *(22)* proposed the ratio of the ovarian stroma-to-total area as a reliable criterion for the diagnosis of PCOS. The ovarian stromal area was evaluated by outlining the peripheral profile of the stroma with the caliper, identified by a central area slightly hyperechoic with respect to the other ovarian area. However, this evaluation is not easy to reproduce in routine practice.

Stromal echogenicity has been described by others *(23)* in a semi-quantitative manner, where a score was assigned for normal (1), moderately increased (2), or frankly increased (3) stroma. In this study the total follicle number of both ovaries combined correlated significantly with stromal echogenicity. Stromal echogenicity has also been quantified by Al-Took et al. *(24)* as the sum of the product of each intensity level (ranging from 0 to 63 on the scanner) and the number of pixels for that intensity level divided by the total number of pixels in the measured area. Buckett et al. *(25)* used this same formula, but found no difference of the stromal echogenicity between women with PCOS and those with normal ovaries. Their conclusion was that the subjective impression of increased stromal echogenicity is a result of both increased stromal volume and reduced echogenicity of the multiple surrounding follicles.

In summary, because ovarian volume or area correlate well with ovarian function and it is easier to reliably measure these than ovarian stroma in routine practice, neither qualitative nor quantitative assessment of the ovarian stroma is required to define polycystic ovaries.

2.4.4. Follicle Distribution

In polycystic ovaries the follicular distribution is predominantly peripheral, typically with an echoless peripheral array, as initially described by Adams et al. *(9)* (Fig. 1). Younger patients more often display this peripheral distribution, whereas a more generalized pattern, with small cysts in the central part of the ovary, is noticed in older women *(26).* At the Rotterdam meeting, this subjective criterion was judged too inconstant and subjective to be retained for the consensus definition of polycystic ovaries *(1).*

2.5. Other U/S Techniques for Imaging Polycystic Ovaries

2.5.1. 3D Ultrasound

To avoid the difficulties and pitfalls in outlining or measuring the ovarian shape, the use of 3D U/S has been proposed *(27–29).* Using a dedicated volumetric probe and a manual survey of the ovary, the scanned ovarian volume is displayed on the screen in three adjustable orthogonal planes. This allows the three ovarian dimensions, and subsequently the ovarian volume, to be more accurately calculated. In a study by Kyei-Mensah et al. *(30),* three groups of patients were defined: (1) those with normal ovaries, (2) those with asymptomatic polycystic ovaries, and (3) those with PCOS. The ovarian and stromal volumes were similar in groups 2 and 3, and both had greater volumes than group 1. Stromal volume was positively correlated with serum androstenedione levels in group 3 only. The mean total volume of the follicles was similar in all groups, indicating that increased stromal volume is the principal cause of ovarian enlargement in polycystic ovaries. Nardo et al. *(31)* observed a tight correlation between 2D and 3D ultrasound measurements of ovarian volume and polycystic ovary morphology. However, in this prospective study total ovarian volume, ovarian stromal volume, follicular volume, and follicle numbers did not correlate with testosterone concentration.

Because 3D ultrasound requires expensive equipment, intensive training, and significant storage and data analysis, its superiority over 2D ultrasound for imaging polycystic ovaries in clinical practice is not evident.

2.5.2. Doppler Ultrasound

Color (or power) Doppler U/S allows for detection of the vascularity network within the ovarian stroma. Power Doppler is more sensitive to slow blood flow and consequently demonstrates more vascular signals within the ovaries, although it does not discriminate between arteries and veins. Moreover, the sensitivity of the machines currently marketed differs from one to another. The pulsed Doppler focuses on the hilum or internal ovarian arteries and offers a more objective approach. Because of the slow blood flow in the ovary, the pulse repetition frequency should be set at minimum (~400 Hz) with the lowest frequency filter (~50 Hz).

The study of ovarian vascularity by these techniques is still highly subjective. Blood flow is more frequently visualized in PCOS (88%) than in normal patients (50%) in early follicular phase and seems to be greater in the former *(32)*. No significant difference was found between obese and lean women with polycystic ovaries, but the stroma was less vascularized in patients displaying a general cystic pattern than in those with peripheral cysts. In the latter group, the pulsatility index (PI) values were significantly lower and inversely correlated with the FSH-to-LH ratio *(33)*. In another study *(34)*, the resistive index (RI) and PI were significantly lower in PCOS (RI = 0.55 ± 0.01 and PI = 0.89 ± 0.04) than in normal patients (RI = 0.78 ± 0.06 and PI = 1.87 ± 0.38), and the peak systolic velocity was greater in PCOS (11.9 ± 3.2) than in normal women (9.6 ± 2.1). No correlation was found between the number of follicles and the ovarian volume, although there was a positive correlation between LH levels and increased peak systolic velocity. In a study by Zaidi et al. *(35)*, no significant differences in PI values were found between normal and PCOS patients, while the ovarian flow, as reflected by the peak systolic velocity, was increased in the former. It is possible that Doppler-detected blood flow may have some value in predicting the risk for ovarian hyperstimulation during gonadotropin therapy *(36)*. Increased stromal blood flow has also been suggested as a more relevant predictor of ovarian response to ovulation induction compared to other parameters such as ovarian or stromal volume *(25,37)*.

In summary, with ovarian Doppler U/S the increased stromal component in polycystic ovaries appears to be accompanied by an increased peak systolic velocity and a decreased PI. However, in all studies the values in patients with polycystic ovaries overlapped widely with those of normal women, and there are few data to date to support the diagnostic usefulness of Doppler U/S in polycystic ovaries or PCOS.

3. CONCLUSIONS

The U/S study of polycystic ovaries has now advanced beyond the era of artistic haziness. It must be viewed as a diagnostic tool that requires the same quality controls as other biological measures, such as the plasma LH, insulin, or androgen assays. This supposes that the results of U/S for polycystic ovaries are expressed as quantitative, not descriptive, variables. Finally, the clinician can use U/S for polycystic ovaries only if the ultrasonographer is sufficiently trained, yielding reproducible results. By its sensitivity (providing that sufficient specificity is guaranteed), 2D U/S has widened the clinical spectrum of PCOS, and this has led to a reduction in the numbers of cases diagnosed as having "idiopathic hirsutism" and "idiopathic anovulation."

Ovarian ultrasonography is now more accurate and describes with significant precision ovarian size, shape, and internal structure, with definitions that approximate those of anatomical cuts. Thus, this tool can serve as an informative technique for pathophysiological studies. For example, we observed two different categories of follicle size (2–5 and 6–9 mm) using a 7 MHz transvaginal ultrasound scan *(14)*. The mean FNPO in the 6- to 9-mm range was similar in polycystic ovaries compared with normal ovaries; alternatively, the FNPO in the 2- to 5-mm range was significantly higher in polycystic ovaries. Therefore, these data suggested that it is the accumulation of 2- to 5-mm follicles, but not those 6–9 mm in diameter, that give rise to the typical multifollicular ovaries seen at U/S in women with polycystic ovaries. We also demonstrated that this follicle excess was in close relationship with the androgen serum level, in agreement with morphological *(38)* and experimental studies. In another study *(39)*, we observed that the FNPO in the 6- to 9-mm-diameter range was very closely related to the serum levels of anti-Müllerian hormone, an excess of which might be involved in the follicular arrest in PCOS *(7)*. Indeed, the discrepancy between the FNPO in the 2- to 5- and the 6- to 9-mm-diameter ranges might reflect the mechanism of follicular arrest in polycystic ovaries. In fact, that the FNPO of 6-9 mm follicles was negatively associated with overweight and/or hyperinsulinism *(14)* is consistent with the well-known detrimental effect of hyperinsulinemia on the anovulation of PCOS.

4. FUTURE AVENUES OF INVESTIGATION

The predictive value of polycystic ovaries for PCOS and the development of the related morbidities remains to be better determined. Likewise, methods for distinguishing polycystic ovaries in PCOS from polycystic ovaries in other ovulatory disorders, such as hypothalamic amenorrhea, remain to be established. Finally, it should be remembered that endovaginal U/S is an improving technique, becoming more accurate over time. Therefore, the thresholds of the currently used criteria for polycystic ovaries are liable to change, and new diagnostic criteria will probably become evident in the near future.

KEY POINTS

- 2D U/S is the standard for imaging polycystic ovaries.
- The transabdominal route should always be the first step in a pelvic sonographic examination, followed by the transvaginal route (excepted in virginal women or those refusing the exam), which offers a better spatial resolution.
- The current consensus definition of polycystic ovaries by U/S determined at the joint ASRM/ESHRE Rotterdam 2003 consensus meeting on PCOS is as follows: (1) 12 or more follicles measuring 2–9 mm in diameter and/or (2) increased ovarian volume (>10 cm^3).
- Other U/S techniques, such as Doppler and 3D U/S, may be helpful in the diagnosis of polycystic ovaries, but are only used as second-line techniques.

REFERENCES

1. Balen AH, Laven JSE, Tan SL, Dewailly D. Ultrasound assessment of the polycystic ovary: international consensus definitions. Hum Reprod Update 2003;9:505–514.
2. The Rotterdam ESHRE/ASRM-sponsored PCOS consensus workshop group. Revised 2003 consensus on diagnostic criteria and long-term health risks related to polycystic ovary syndrome (PCOS). Hum Reprod 2004;19:41–47.
3. Hughesdon PE. Morphology and morphogenesis of the Stein-Leventhal ovary and of so-called "hyperthecosis." Obstet Gynecol Survey 1982;37:59–77.
4. Webber LJ, Stubbs S, Stark J, et al. Formation and early development of follicles in the polycystic ovaries. Lancet 2003;362:1017–1021.
5. Jacobs HS. Polycystic ovaries and polycystic ovary syndrome. Gynecol Endocrinol 1987;1:113–131.
6. Maciel G, Chang RJ, Erickson GF. Evidence for stockpiling of classic primary preantral follicles in PCOS ovaries. Endocrine Society Congress, New Orleans, June 16–19, 2004, Poster 1-531.
7. Jonard S, Dewailly D. The follicular excess in polycystic ovaries, due to intra-ovarian hyperandrogenism, may be the main culprit for the follicula arrest. Hum Reprod Update 2004;10:107–117.
8. Sample WF, Lippe BM, Gyepes MT. Grey-scale ultrasonography of the normal female pelvis. Radiology 1977;125:477–483.
9. Adams Jm, Polson DW, Abulwadi N, et al.. Multifollicular ovaries: clinical and endocrine features and response to pulsatile gonadotropin-releasing hormone. Lancet 1985;2:1375–1378.
10. Orsini LF, Venturoli S, Lorusso R. Ultrasonic findings in polycystic ovarian disease. Fertil Steril 1985;43:709–714.
11. Yeh HC, Futterweit W, Thornton JC. Polycystic ovarian disease: US features in 104 patients. Radiology 1987;163:111–116.
12. Van Santbrink EJP, Hop WC, Fauser BCJM. Classification of normogonadotropic infertility: polycystic ovaries diagnosed by ultrasound versus endocrine characteristics of polycystic ovary syndrome. Fertil Steril 1997;67:452–458.
13. Jonard S, Robert Y, Dewailly D. Revisiting the ovarian volume as a diagnostic criterion for polycystic ovaries. Hum Reprod 2005;20:2893–2898.

14. Jonard S, Robert Y, Cortet-Rudelli C, Pigny P, Decanter C, Dewailly D. Ultrasound examination of polycystic ovaries: is it worth counting the follicles? Hum Reprod 2003;18:598–603.

14a. Adams JM, Taylor AE, Crowley WF Jr, Hall JE. Polycystic ovarian morphology with regular ovulatory cycles: insights into the pathophysiology of polycystic ovarian syndrome. J Clin Endocrinol Metab 2004;89:4343–4350.

15. Venturoli S, Porcu E, Fabbri R, Paradisi R, Orsini LF, Flamigni C. Ovaries and menstrual cycles in adolescence. Gynecol Obstet Invest 1983;17:219–223.

16. Stanhope R, Adams J, Jacobs HS, Brook CG. Ovarian ultrasound assessment in normal children, idiopathic precocious puberty, and during low dose pulsatile gonadotrophin releasing hormone treatment of hypogonadotrophic hypogonadism. Arch Dis Child 1985;60:116–119.

17. Reyss AC, Merlen E, Demerle C, Dewailly D. Revelation of a polymicrocystic ovary syndrome after one month's treatment by pulsatile GnRH in a patient presenting with functional hypothalamic amenorrhea. Gynecol Obstet Fertil. 2003;31:1039–1042.

18. Ardaens Y, Robert Y, Lemaitre L, Fossati P, Dewailly D. Polycystic ovarian disease: contribution of vaginal endosonography and reassessment of ultrasonic diagnosis. Fertil Steril 1991;55:1062–1068.

19. Dewailly D, Robert Y, Helin I, et al. Ovarian stromal hypertrophy in hyperandrogenic women. Clinical Endocrinology 1994;41:557–562.

20. Robert Y, Dubrulle F, Gaillandre G, et al. Ultrasound assessment of ovarian stroma hypertrophy in hyperandrogenism and ovulation disorders: visual analysis versus computerized quantification. Fertil Steril 1995;64:307–312.

21. Pache TD, Wladimiroff JW, Hop WCJ, Fauser BCJM. How to descriminate between normal and polycystic ovaries: transvaginal US study. Radiology 1992;183:421–423.

22. Fulghesu AM, Ciampelli M, Belosi C, Apa R, et al. A new ultrasound criterion for the diagnosis of polycystic ovary syndrome: the ovarian stroma/ total area ratio. Fertil Steril 2001;76:326–331.

23. Pache TD, Hop WC, Wladimiroff JW, Schipper J, Fauser BCJM. Transvaginal sonography and abnormal ovarian appearance in menstrual cycle disturbances. Ultrasound Med Biol 1991;17:589–593.

24. Al-Took S, Watkin K, Tulandi T, Tan SL. Ovarian stromal echogenicity in women with clomiphene citrate-sensitive and clomiphene citrate-resistant polycystic ovary syndrome. Fertil Steril 1999;71:952–954.

25. Buckett WM, Bouzayen R, Watkin KL, Tulandi T, Tan SL. Ovarian stromal echogenicity in women with normal and polycystic ovaries. Hum Reprod 1999;14:618–621.

26. Battaglia C, Artini PG, Salvatori M, Giulini S, Petraglia F, Maxia N, Volpe A. Ultrasonographic pattern of polycystic ovaries: color Doppler and hormonal correlations. Ultrasound Gynaecol Obstet 1998;11:332–336.

27. Wu M-H, Tang H-H, Hsu C-C, Wang S-T, Huang K-E. The role of three-dimensional ultrasonographic imaging in ovarian measurment. Fertil Steril 1998;69:1152–1155.

28. Kyei-Mensah A, Maconochie N, Zaidi J, Pittrof R, Campbell S, Tan SL. Transvaginal three-dimensional ultrasound: accuracy of ovarian follicular volume measurements. Fertil Steril 1996;65:371–376.

29. Kyei-Mensah A, Zaidi J, Campbell S. Ultrasound diagnosis of polycystic ovary syndrome. Bailliere's Clin Endocrinol Metab 1996;10:249–262.

30. Kyei-Mensah A, Tan SL, Zaidi J, Jacobs HS. Relationship of ovarian stromal volume to serum androgen concentrations in patients with polycystic ovary syndrome. Hum Reprod 1998;13:1437–1441.

31. Nardo LG, Buckett WM, Khullar V. Determination of the best-fitting ultrasound formulaic method for ovarian volume measurement in women with polycystic ovary syndrome. Fertil Steril 2003;79:632–633.

32. Battaglia C, Artini PG, Genazzani AD, Sgherzi MR, Salvatori M, Giulini S, Volpe A. Color Doppler analysis in lean and obese women with polycystic ovaries. Ultrasound in Gynaecology and Obstetrics 1996;7:342–346.

33. Battaglia C, Genazzani AD, Salvatori M, Giulini S, et al. Doppler, ultrasonographic and endocrinological environment with regard to the number of small subcapsular follicles in polycystic ovary syndrome. Gynecol Endocrinol 1999;13:123–129.

34. Aleem FA, Predanic MP. Transvaginal color Doppler determination of the ovarian and uterine blood flow characteristics in polycystic ovary disease. Fertil Steril 1996;65:510–516.

35. Zaidi J, Campbell S, Pittrof R, et al. Ovarian stromal blood flow in women with polycystic ovaries: a possible new marker for diagnosis? Hum Reprod 1995,10:1992–1996.

36. Agrawal R, Conway G, Sladkevicius P, et al. Serum vascular endothelial growth factor and Doppler blood flow velocities in in vitro fertilization: relevance to ovarian hyperstimulation syndrome and polycystic ovaries. Fertil Steril 1998;70:651–658.

37. Engmann L, Sladkevicius P, Agrawal LR, Bekir JS, Campbell S, Tan SL. Value of ovarian stromal blood flow velocity measurement after pituitary suppression in the prediction of ovarian responsiveness and outcome of in vitro fertilization treatment. Fertil Steril 1999;71:22–29.

38. Takayama K, Fukaya T, Sasano H, Funayama Y, et al. Immunohistochemical study of steroidogenesis and cell proliferation in polycystic ovarian syndrome. Hum Reprod 1996;11:1387–1392.

39. Pigny P, Merlen E, Robert Y, Cortet-Rudelli C, et al. Elevated serum level of anti-mullerian hormone in patients with polycystic ovary syndrome: relationship to the ovarian follicle excess and to the follicular arrest. J Clin Endocrinol Metab 2003;88:5957–5962.
40. Goldzieher JW, Green JA. The polycystic ovary. I. Clinical and histologic features. J Clin Endocrin Metabol 1962;22:325–338.
41. Atiomo WU, Pearson S, Shaw S, Prentice A, Dubbins P. Ultrasound criteria in the diagnosis of polycystic ovary syndrome (PCOS). Ultrasound Med Biol 2000;26:977–980.

17

Hypothalamic–Pituitary–Gonadotropic Dysfunction in the Polycystic Ovary Syndrome

R. Jeffrey Chang and Rinku V. Mehta

SUMMARY

Almost all women with the polycystic ovary syndrome exhibit increased luteinizing-hormone (LH) secretion and decreased levels of serum follicle-stimulating hormone. By inference, hypothalamic gonadotropin-releasing hormone (GnRH) activity is increased, although other factors appear to affect the relationship between GnRH and LH interaction. The precise mechanism(s) responsible for inappropriate gonadotropin release are not known, but they appear to involve the effects of abnormal ovarian steroidogenesis, including androgen excess. Other factors may include abnormal insulin secretion and possibly extrahypothalamic neurotransmitters.

Key Words: Gonadotropin-releasing hormone; GnRH; luteinizing hormone; LH; follicle-stimulating hormone; FSH; androgen; estrogen; insulin.

1. INTRODUCTION

Gonadotropic abnormalities in the polycystic ovary syndrome (PCOS), mainly elevated luteinizing hormone (LH) levels, were reported as early as 1958 using urinary bioassays. Subsequently, radioimmunoassays were developed that allowed the detection and measurement of circulating gonadotropins, suggesting that a significant fraction of women with PCOS had an elevated (>2–3) LH–to–follicle-stimulating hormone (FSH) ratio. Following, we review our current understanding of the prevalence, mechanism, and role of gonadotropic abnormalities in PCOS.

2. BACKGROUND

Women with PCOS are distinctive with respect to reproductive neuroendocrine function characterized by elevated 24-hour mean serum LH concentrations, greater LH responses to gonadotropin-releasing hormone (GnRH), and increased LH pulse frequency and amplitude compared to those found in normal women (1). The mechanism(s) responsible for this increased release of LH are not well understood. A particular characteristic of LH secretion in PCOS is increased pulse frequency, the periodicity of which is approximately 1 hour. This rapid rate of LH release does not appear to be altered by experimental manipulation under physiological conditions (2,3). Previous in vivo studies in animals have demonstrated that individual LH pulses are correlated to isolated bursts of endogenous GnRH release (4). Correspondingly, these observations imply that in PCOS the episodic pattern of pituitary LH secretion reflects a parallel increase in hypothalamic GnRH activity.

In PCOS the relationship between GnRH pulse frequency and gonadotrope responsiveness may be a primary mechanism underlying inappropriate gonadotropin secretion. First, previous studies in rodents have demonstrated a preference for the expression of the gene for the β-subunit of LH (LHβ) in response

From: *Contemporary Endocrinology: Androgen Excess Disorders in Women:*
Polycystic Ovary Syndrome and Other Disorders, Second Edition
Edited by: R. Azziz et al. © Humana Press Inc., Totowa, NJ

to rapid rates of GnRH delivery, whereas β-subunit of FSH (FSHβ) expression is suppressed *(5)*. Second, not only does GnRH drive LH release, it has been shown in normal women to self-prime the pituitary and thereby contributes to the increased LH sensitivity to subsequent GnRH stimulation *(6)*. Collectively, these findings suggest that the profound abnormality of gonadotropin secretion in PCOS may be a primary consequence of increased hypothalamic GnRH activity.

2.1. GnRH in PCOS

In humans it has been shown that administration of intravenous GnRH at varying frequencies may dictate the rate of LH release as well as alter basal concentrations *(7,8)*. In hypogonadotropic hypogonadal individuals, an increase in the rate of GnRH administration from every 90 minutes to every 60 minutes was not associated with corresponding increases in serum LH or changes in pulse amplitude, whereas increasing GnRH frequency to 30-minute intervals resulted in elevated LH levels and reduction of pulse amplitude *(7)*. When the rate of GnRH administration was increased to every 15 minutes, both the concentration and amplitude of LH release were markedly diminished as a result of receptor desensitization. Thus, although consistent with the primacy of increased hypothalamic GnRH, an LH pulse frequency of 1 hour may represent a physiological limit beyond which more frequent pulses in women do not occur.

In normal ovulatory women during the late follicular phase and at midcycle, as well as in post-menopausal women, it has been documented that LH pulse frequency approximates 60 minutes *(9)*. In addition, previous studies have shown that the GnRH release from the medial basal hypothalamus of the fetus and adult has a periodicity of about 1 hour *(10)*. These findings suggest that in PCOS women, the pulse frequency and, to some degree, magnitude of gonadotropin secretion are established by hypothalamic GnRH activity. However, beyond the essential role of GnRH, LH responsiveness to GnRH and, accordingly, maximal increases in LH pulse amplitude are probably influenced by other factors.

2.2. Role of Estrogen and Progesterone

It has been suggested that the positive feedback effects of chronic estrogen secretion associated with this disorder may bring about an increase of LH either by a direct effect on gonadotrope sensitivity to GnRH or indirectly by facilitating GnRH pulse frequency. In vitro, estrogen has been shown to increase the fraction of individual gonadotropes responding to GnRH that is consistent with amplification of LH responses to GnRH in normal women receiving estradiol benzoate *(11)*. In PCOS, baseline levels of estrone and estradiol have been correlated with LH responses to GnRH *(1)*. However, prolonged administration of estrone to PCOS patients failed to raise circulating levels of LH beyond baseline values or increase GnRH-stimulated LH responses *(12)*. In animals it has been demonstrated that estrogen enhances GnRH pulse frequency, and in women with PCOS, serum GnRH levels are increased. That estrogen may exert an effect at the hypothalamus in PCOS is supported by the strong positive correlation of mean serum estradiol levels to GnRH pulse frequency *(13)*. Despite these findings, the capacity of exogenous estrogen, administered at physiological levels, to alter LH pulse frequency has not been demonstrated.

In contrast to estrogen, progesterone, either alone or in combination with estrogen (oral contraceptive), was shown to suppress circulating LH levels and LH pulse frequency in both PCOS and normal women *(14)*. The observation that suppression of LH release was more pronounced in normal women than women with PCOS suggested to the investigators that increased LH pulse frequency might reflect a fundamental property of the hypothalamic pulse generator in PCOS. Moreover, in a study of women with PCOS that employed the administration of progesterone and estradiol at physiological concentrations equivalent to those found in the midluteal phase of normal women, LH pulse frequency was not diminished compared to the reduced rate found in a group of normal women studied similarly *(15)*.

2.3. Role of Androgens

As the outstanding abnormality of ovarian steroid production in PCOS, hyperandrogenemia has been implicated as a potential cause of increased LH secretion. In vitro, it has been shown that androgen administration increased hypothalamic GnRH pulse generator activity in rodents. Examination of LH secretion in hyperandrogenic patients with congenital adrenal hyperplasia revealed that mean LH levels and LH responses to GnRH were increased, which tended to normalize with the onset of treatment and corresponding lowered androgen levels *(16)*. By comparison, other studies have not been able to detect an increase in LH following the administration of androgen. Short-term infusion of androgen to both normal women and women with PCOS failed to alter basal LH secretion *(17)*. In addition, high-dose androgen infusion in normal women appeared to result in an acute reduction of serum LH levels *(18)*.

Notwithstanding these past findings, recent studies have indicated that excess androgen production may have a significant influence on LH pulse frequency in women with this disorder. In PCOS women pretreated with an androgen-blocking agent, administration of estrogen and progesterone at physiological concentrations was associated with a reduction of LH pulse frequency comparable to that observed in normal women treated similarly *(19)*. These findings suggested that in PCOS, high circulating levels of androgen prevent the negative feedback effects of estrogen and progesterone on LH pulse release, as noted in earlier studies. Moreover, the physiological relevance in women relates to a rapid rate of spontaneous hypothalamic GnRH activity, which is regulated by the feedback effects of ovarian steroids.

2.4. Role of Insulin

Insulin has also been implicated as a potential regulator of LH secretion in PCOS. The notion that insulin may contribute to inappropriate gonadotropin secretion in PCOS has been largely based on in vitro studies, which demonstrated that rat pituitary cells preincubated with insulin exhibited increased LH responsiveness following administration of GnRH in a dose-dependent manner compared to untreated cells *(20)*. Interestingly, an effect of insulin was not observed when these studies were performed in the presence of serum-supplemented media. Efforts to determine an effect of insulin in PCOS women have not documented consistent alterations in LH secretion or LH release following GnRH stimulation. Reduction of hyperinsulinemia by administration of insulin-lowering drugs to PCOS patients or dietary restriction resulted in decreased mean serum levels of androgens and LH in some cases, whereas in others an accompanying decrease in LH was not found. Interestingly, the failure to document a decline in LH levels despite improved insulin sensitivity was independent of whether ovulatory activity had resumed in these studies.

Recently we have explored the role of insulin on gonadotropin secretion in PCOS and normal women. Using the hyperinsulinemic-euglycemic clamp technique, our results demonstrated that episodic gonadotropin secretion and LH responses to multidose GnRH stimulation were not altered by insulin infusion over an interval of 12 hours in both groups *(21)*. In particular, endogenous serum LH levels were unchanged prior to and immediately following initiation of the insulin clamp (Table 1). In an extension of these studies, episodic LH secretion and LH responses to GnRH stimulation were not influenced by an insulin-lowering drug, pioglitazone, administered for 3 months (Fig. 1) *(22)*. These findings confirm and clarify previous studies, which have been unable to document consistent alterations in LH secretion or LH release following GnRH stimulation following insulin administration in normal women and women with PCOS. In normal women undergoing long-term insulin administration by a 16-hour hyperinsulinemic-euglycemic clamp, mean serum LH levels, measured every hour, remained unchanged during the entire course of infusion *(23)*.

During 6-hour insulin infusions randomized to either of 2 consecutive days in women with and without PCOS, consistent alterations in mean serum LH, LH pulse frequency or pulse amplitude, and LH release following GnRH could not be documented *(24)*. Collectively, these results may explain

Table 1
Effect of Insulin Infusion on 12-Hour Composite Mean LH, Pulse Frequency, and LH Pulse Amplitude in Women With and Without PCOS

	Normal (*n* = 9)	PCOS (*n* = 11)
12-hour composite LH (mIU/mL)		
No insulin	3.5 ± 0.4	6.7 ± 0.1[a]
Insulin infusion	3.2 ± 0.3	4.9 ± 0.9
LH pulse frequency (no./12 hours)		
No insulin	8.8 ± 0.8	10.2 ± 0.4[b]
Insulin infusion	8.1 ± 0.8	10.3 ± 0.5[b]
LH pulse amplitude (mIU/mL)		
No insulin	1.6 ± 0.2	1.8 ± 0.3
Insulin infusion	1.6 ± 0.2	1.4 ± 0.2

Mean ± SE depicted.
Normal vs PCOS; [a]$p < 0.01$; [b]$p < 0.006$.
LH, luteinizing hormone; PCOS, polycystic ovary syndrome. (From ref. *21*.)

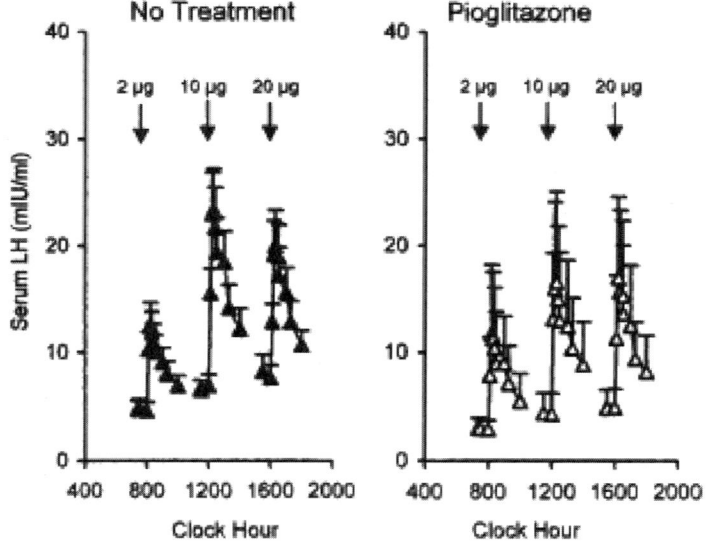

Fig. 1. Time-course of mean (±SE) serum luteinizing hormone (LH) concentrations after intravenous administration of three successive doses of gonadotropin-releasing hormone (GnRH) given at 4-hour intervals during the hyperinsulinemic-euglycemic clamp (80 mU/m^2) in women with polycystic ovary syndrome (PCOS) before and after treatment with pioglitazone. No significant differences were detected in the percent increment of LH from baseline as a result of treatment. (From ref. *22*.)

why changes in serum LH were not observed in women with PCOS treated with insulin-lowering drugs despite significant reductions in circulating androgen levels. Alternatively, interpretation of these clinical trials is potentially confounded by several factors. First, most of the patients studied were obese, and it has been shown that obesity is inversely correlated to LH secretion in PCOS. Second, hyperinsulinemia is positively correlated with the body mass index (BMI) in women with this syndrome. Third, the occurrence of ovulation in PCOS is associated with a lowering of LH levels into the normal range. Although additional studies are necessary, the evidence to date has not clearly demonstrated a functional interaction between insulin and LH in this disorder.

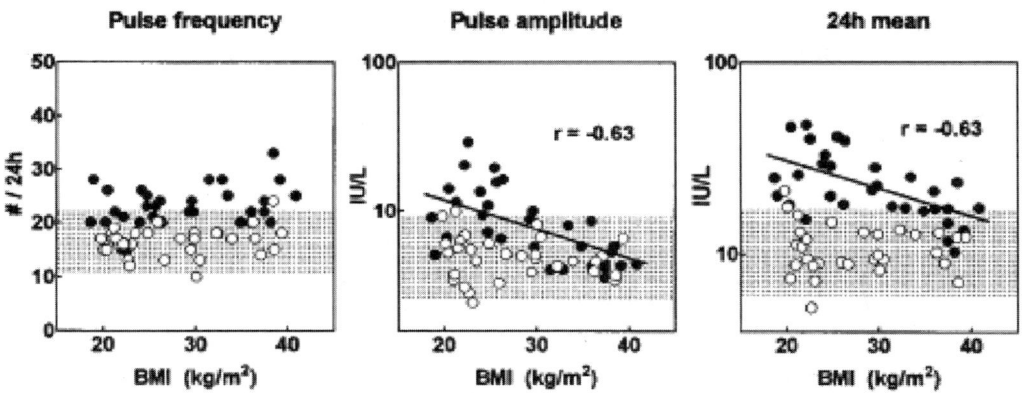

Fig. 2. Regression of body mass index (BMI) for polycystic ovary syndrome (PCOS) (●) and normal control (NC) (○) women against luteinizing hormone (LH) pulse frequency (PCOS and NC: *p* = NS); LH pulse amplitude: r = -0.63; *p* < 0.001; NC: *p* = NS), and 24-hr mean LH levels (PCOS: r = - 0.63; *p* < 0.001; NC: *p* = NS). Values for LH pulse amplitude and 24-hour mean are $\log_{10}$ transformed. Shaded area represents 95% confidence interval for NC (frequency: 11–22 pulses/24 hours; amplitude: 2.6–9.2 IU/L; 24-hour mean: 6.1–8.2 IU/L). (From ref. 2.)

2.5. Role of Obesity

The inability of insulin to induce alterations in LH secretion in women with PCOS is relevant to studies that have shown that both 24-hour insulin levels and BMI are inversely correlated to serum LH and LH pulse amplitude (Fig. 2) *(2,3)*. In an effort to address the relationship between body mass and LH secretion, obese women with PCOS were subjected to extreme caloric restriction over 7 days, during which a small but significant weight loss occurred. Daily LH secretion was increased, as evidenced by significantly greater LH levels and increased LH pulse amplitude, although changes in LH pulse frequency were not observed *(25)*. Similar findings were reported in obese women with PCOS who were placed on very low-calorie intake for 6 weeks and lost 8% of their original weight. Nocturnal LH secretion increased after weight loss compared with pretreatment values *(26)*. These changes were accompanied by improved insulin sensitivity and glucose utilization that correlated with increased LH secretion. Whether the weight loss influenced gonadotropin secretion directly or indirectly through alterations of metabolic function induced by caloric restraint such as reduced insulin secretion or decreased availability of glucose for oxidation in the brain could not be determined. Nevertheless, these results suggest that obesity may contribute to diminished LH pulse amplitude in women with PCOS compared with that observed in women without the syndrome.

2.6. Antiepileptic Drugs, Gonadotropic Abnormalities, and PCOS

Increased LH pulse frequency, as well as other features of PCOS, has been described in women with epileptic disorders or women treated with antiepileptic drugs *(27,28)*. These observations have led to the intriguing consideration that epilepsy or treatment with antiepileptic drugs (sodium valproate in particular) and PCOS are causally related (*see also* Chapter 10). The link between epileptic and postseizure ictal states may involve stimulation of excitatory neurotransmitters, the receptors of which exist in hypothalamic nuclei that influence GnRH release. Thus, the consequences of epileptic activity may result in increased GnRH activity and simulate the pattern of increased LH secretion in PCOS. In addition to altered LH secretion, polycystic ovaries and hyperandrogenism have been reported in untreated and treated women with epileptic seizures, which strengthen the possible association between PCOS and epilepsy *(28)*.

A mechanistic role for antiepileptic medication, including sodium valproate, in the development of excess androgen production or follicle cyst formation in treated subjects has not been elucidated. Unfortunately, the vast majority of reports linking epilepsy or antiepileptic medication to PCOS have been beset by poor experimental design and insufficient rigor to determine whether causality exists.

2.7. FSH Secretion in PCOS

In contrast to LH, FSH secretion in PCOS is decreased, as indicated by significantly lower serum concentrations compared with those found in normal women during the early follicular phase of the menstrual cycle *(1)*. In addition, FSH responses to GnRH stimulation are reduced, as shown in some, but not all studies. The precise underlying basis for the decreased FSH secretion in PCOS has not been determined, although the negative-feedback effect of chronic unopposed estrogen secretion in these women has been implicated as a mechanism. Support for this concept has been demonstrated by a study in which women with were treated with estradiol benzoate for 2 weeks *(12)*. Daily measurement of serum gonadotropin levels revealed a progressive decline of circulating FSH, whereas serum LH concentrations remained unaltered, resulting in a decrease in the LH:FSH ratio. The reduction in serum FSH may also reflect the activity of hypothalamic GnRH. As mentioned earlier, increased frequency of pulsatile GnRH predisposes to a preference for LHβ gene expression at the expense of the FSHβ gene *(5)*.

3. SUMMARY/CONCLUSIONS

The characteristic neuroendocrine alterations in PCOS are increased LH secretion and decreased circulating FSH levels, which likely reflect increased activity of the hypothalamic GnRH pulse generator. Increased GnRH release favors the gene expression of LHβ and inhibits the expression of FSHβ. Heightened GnRH release and inappropriate gonadotropin secretion may also reflect the influence of abnormal ovarian steroid production, including chronic estrogen secretion in the absence of progesterone and excess androgen production. In contrast to studies in animals, clinical studies in women with PCOS to date have not demonstrated a role for insulin on gonadotropin secretion in this disorder. A role for excitatory hypothalamic neurotransmitters in LH secretion has not been established, although the occurrence of PCOS symptomatology in women with epileptic disorders or women treated with antiepileptic drugs is provocative.

4. FUTURE AVENUES OF INVESTIGATION

Studies are needed to further establish which factors may be responsible for the mechanism of inappropriate gonadotropin section in PCOS, including steroid hormones, obesity, insulin, and possibly neurotransmitters. Moreover, the role of enhanced GnRH–LH interaction in the genesis of this disorder has not been clarified. That PCOS appears to be a heritable condition speaks to a genetic basis for the disruption of reproductive physiology, including the possibility of abnormal activity of the GnRH pulse generator.

KEY POINTS

- PCOS is a heterogeneous disorder in which the characteristic neuroendocrine abnormality is hypersecretion of LH and decreased circulating FSH.
- The precise etiology of the increased LH secretion is yet to be elucidated, but ovarian steroids and insulin appear to be involved.
- Recent studies have suggested that PCOS is a heritable condition; therefore, the abnormal functioning of the GnRH pulse generator may be genetically programmed.

REFERENCES

1. Rebar R, Judd HL, Yen SS, Rakoff J, Vandenberg G, Naftolin F. Characterization of the inappropriate gonadotropin secretion in polycystic ovary syndrome. J Clin Invest 1976;57:1320–1329.

2. Arroyo A, Laughlin GA, Morales AJ, Yen SS. Inappropriate gonadotropin secretion in polycycstic ovary syndrome: influence of adiposity. J Clin Endocrinol Metab 1997;82:3728–3733.
3. Taylor AE, McCourt B, Martin KA, et al. Determinants of abnormal gonadotropin secretion in clinically defined women with polycystic ovary syndrome. J Clin Endocrinol Metab 1997;82:2248–2256.
4. Clarke IJ, Cummins JT. GnRH pulse frequency determines LH pulse amplitude by altering the amount of releasable LH in the pituitary glands of ewes. J Reprod Fertil 1985;73:425–431.
5. Kaiser UB, Sabbagh E, Katzenellenbogen RA, Conn PM, Chin WW. A mechanism for the differential regulation of gonadotropin subunit gene expression by gonadotropin-releasing hormone. Proc Natl Acad Sci USA 1995;92:12280–12284.
6. Hoff JD, Lasley BL, Yen SS. The functional relationship between priming and releasing actions of luteinizing hormone-releasing hormone. J Clin Endocrinol Metab 1979;49:8–11.
7. Hall JE, Taylor AE, Hayes FJ, Crowley WF Jr. Insights into hypothalamic-pituitary dysfunction in polycystic ovary syndrome. J Endocrinol Invest 1998;21:602–611.
8. Spratt DI, Finkelstein JS, Butler JP, Badger TM, Crowley WF Jr. Effects of increasing the frequency of low deses of gonadotropin-releasing hormone (GnRH) on gonadotropin secretion in GnRH-deficient men. J Clin Endocrinol Metab 1987;64:1179–1186.
9. Rossmanith WG, Liu CH, Laughlin GA, Mortola JF, Suh BY, Yen SS. Relative changes in LH pulsatility during the menstrual cycle: using data from hypogonadal women as a reference point. Clin Endocrinol 1990;32:647–660.
10. Rasmussen DD, Gambacciani M, Swartz W, Tueros VS, Yen SS. Pulsatile gonadotropin-releasing hormone release from the human mediobasal hypothalamus in vitro: opiate receptor-mediated suppression. Neuroendocrinology 1989;49:150–156.
11. Smith PF, Frawley LS, Neill JD. Detection of LH release from individual pituitary cells by the reverse hemolytic plaque assay: estrogen increases the fraction of gonadotropes responding to GnRH. Endocrinology 1984;115:2484–2486.
12. Chang RJ, Mandel FP, Lu JK, Judd HL. Enhanced disparity of gonadotropin secretion by estrone in women with polycystic ovarian disease. J Clin Endocrinol Metab 1982;54:490–494.
13. Waldstreicher J, Santoro NF, Hall JE, Filicori M, Crowley WF Jr. Hyperfunction of the hypothalamic-pituitary axis in women with polycystic ovarian disease: indirect evidence for partial gonadotroph desensitization. J Clin Endocrinol Metab 1988;66:165–172.
14. Daniels TL, Berga SL. Resistance of gonadotropin releasing hormone drive to sex steroid-induced suppression in hyperandrogenic anovulation. J Clin Endocrinol Metab 1997;82:4179–4183.
15. Pastor CL, Griffin-Korf ML, Aloi JA, Evans WS, Marshall JC. Polycystic ovary syndrome: evidence for reduced sensitivity of the gonadotropin-releasing hormone pulse generator to inhibition by estradiol and progesterone. J Clin Endocrinol Metab 1998;83:582–590.
16. Barnes RB, Rosenfield RL, Ehrmann DA, et al. Ovarian hyperandrogynism as a result of congenital adrenal virilizing disorders: evidence for perinatal masculinization of neuroendocrine function in women. J Clin Endocrinol Metab 1994;79:1328–1333.
17. Dunaif A. Do androgens directly regulate gonadotropin secretion in the polycystic ovary syndrome? J Clin Endocrinol Metab 1986;63:215–221.
18. Serafini P, Silva PD, Paulson RJ, Elkind-Hirsch K, Hernandez M, Lobo RA. Acute modulation of the hypothalamic-pituitary axis by intravenous testosterone in normal women. Am J Obstet Gynecol 1986;155:1288–1292.
19. Eagleson CA, Gingrich MB, Pastor CL, et al. Polycystic ovarian syndrome: evidence that flutamide restores sensitivity of the gonadotropin-releasing hormone pulse generator to inhibition by estradiol and progesterone. J Clin Endocrinol Metab 2000;85:4047–4052.
20. Adashi EY, Hsueh AJ, Yen SS. Insulin enhancement of luteinizing hormone and follicle-stimulating hormone release by cultured pituitary cells. Endocrinology 1981;108:1441–1449.
21. Patel K, Coffler MS, Dahan MH, et al. Increased luteinizing hormone secretion in women with polycystic ovary syndrome is unaltered by prolonged insulin infusion. J Clin Endocrinol Metab 2003;88:5456–5461.
22. Mehta RV, Patel KS, Coffler MS, et al. Luteinizing hormone secretion is not influenced by insulin infusion in women with polycystic ovary syndrome despite improved insulin sensitivity during pioglitazone treatment. J Clin Endocrinol Metab 2005;90:2136–2141.
23. Nestler JE, Clore JN, Strauss JF III, Blackard WG. The effects of hyperinsulinemia on serum testosterone, progesterone, dehydroepiandrosterone sulfate, and cortisol levels in normal women and in a woman with hyperandrogenism, insulin resistance, and acanthosis nigricans. J Clin Endocrinol Metab 1987;64:180–184.
24. Dunaif A, Graf M. Insulin administration alters gonadal steroid metabolism independent of changes in gonadotropin secretion in insulin-resistant women with the polycystic ovary syndrome. J Clin Invest 1989;83:23–29.
25. Van Dam EW, Roelfsema F, Veldhuis JD, et al. Increase in daily LH secretion in response to short-term calorie restriction in obese women with PCOS. Am J Physiol Endocrinol Metab 2002;282:E865–E872.
26. Butzow TL, Lehtovirta M, Siegberg R, et al. The decrease in luteinizing hormone secretion in response to weight reduction is inversely related to the severity of insulin resistance in overweight women. J Clin Endocrinol Metab 2000;85:3271–3275.

27. Bauer J, Jarre A, Klingmuller D, Elger CE. Polycystic ovary syndrome in patients with focal epilepsy: a study in 93 women. Epilepsy Res 2000;41:163–167.
28. Isojarvi JI, Laatikainen TJ, Pakarinen AJ, Juntunen KT, Myllyla VV. Polycystic ovaries and hyperandrogenism in women taking valproate for epilepsy. N Engl J Med 1993;329:1383–1388.

18

Ovarian Steroidogenic Abnormalities in the Polycystic Ovary Syndrome

Denis A. Magoffin

SUMMARY

Excess androgen biosynthesis is a diagnostic feature of polycystic ovary syndrome. The excess circulating androstenedione and testosterone is produced primarily by the ovary. The ovarian theca cells, the site of *de novo* androgen biosynthesis, are increased in number in polycystic ovaries, and they have increased steroidogenic capacity. The increase in steroidogenic capacity is caused by overexpression of steroidogenic enzymes because of increased transcription and mRNA stability. Primary factors in the hyperstimulation of thecal androgen production appear to be increased luteinizing hormone (LH) concentrations in some women and elevated insulin concentrations secondary to insulin resistance. Additional contributions may be made by other intraovarian factors that can augment the stimulatory effects of LH on thecal androgen biosynthesis.

The granulosa cells in arrested follicles in polycystic ovaries fail to increase the expression of aromatase, causing markedly decreased estrogen secretion. They also express higher concentrations of 5α-reductase enzymes, leading to the production of 5α-androstane-3,17-dione, a competitive inhibitor of aromatase activity. The granulosa cells prematurely express the cholesterol side-chain cleavage enzyme and LH receptors, hence they are overresponsive to LH and produce increased amounts of progesterone compared to granulosa cells from follicles at a similar developmental stage in regularly cycling control women. Thus, polycystic ovaries produce increased concentrations of androgens and progesterone, decreased concentrations of estrogens, and abnormally high concentrations of 5α-reduced androgens, compared to normal ovaries.

Key Words: Ovary; theca cell; granulosa cell; steroidogenesis; androgens; estrogens; progesterone; polycystic ovary syndrome.

1. INTRODUCTION

The polycystic ovary syndrome (PCOS) is a heterogeneous disorder in which abnormalities of steroidogenesis are inherent to the etiology. Excess circulating androgen concentrations are a cardinal feature of PCOS that is central to the diagnosis *(1)*. Evidence supporting a central role for hyperandrogenism as a cause of PCOS includes the development of polycystic ovaries in female-to-male transsexuals treated with high concentrations of testosterone and in girls with excess adrenal androgen secretion because of congenital adrenal hyperplasia, and the observation that PCOS can be reversed in some women by reducing circulating androgen concentrations using insulin-sensitizing therapy. Therefore, understanding ovarian steroidogenic abnormalities in PCOS is central to understanding the etiology of the disorder.

2. BACKGROUND

Ovarian steroidogenesis in PCOS affects androgen and estrogen—and possibly progesterone—production.

From: *Contemporary Endocrinology: Androgen Excess Disorders in Women:*
Polycystic Ovary Syndrome and Other Disorders, Second Edition
Edited by: R. Azziz et al. © Humana Press Inc., Totowa, NJ

2.1. Abnormal Androgen Production

2.1.1. Contribution of the Ovary to Abnormal Androgen Production in PCOS

The serum concentrations of androstenedione and testosterone are elevated in women with PCOS. In up to 50% of hyperandrogenic women, dehydroepiandrosterone sulfate is also increased *(2)*. Although it is clear that the ovary and the adrenal gland are the principal sources of circulating androgens in women, it is important to understand the relative contributions of each gland. Direct measurement of androgens in selectively catheterized ovarian and adrenal veins and the use of ovarian and adrenal stimulation and suppression have not unambiguously defined the relative contributions of each gland to elevated circulating androgens. However, selective suppression of ovarian androgen production with long-acting gonadotropin-releasing hormone agonists that do not affect adrenal androgen production indicate that the ovary is the principal source of the excessive androstenedione and testosterone in hyperandrogenic women *(3)*. In hyperandrogenic women in whom congenital adrenal hyperplasia involving 3β-hydroxysteroid dehydrogenase (3β-HSD), 21-hydroxylase, or 11-hydroxylase deficiency has been ruled out, up to 50% have both ovarian and adrenal hyperandrogenism, and the remainder have exclusively ovarian hyperandrogenism. These data lead to the conclusions that the ovary is the principal source of excessive androstenedione and testosterone in hyperandrogenic women and that the ovary plays a major, although not exclusive, role in female hyperandrogenism.

2.1.2. Excessive Androgen Production by Theca Cells From Polycystic Ovaries

In the ovarian follicle steroidogenesis is accomplished through the cooperation of the theca and granulosa cells (Fig. 1). The theca cells are the exclusive source of androstenedione in women. The 17β-hydroxysteroid dehydrogenase enzyme is predominantly expressed in the granulosa cells; hence much of the testosterone is produced in the granulosa cells. Regardless of the site of metabolism of androstenedione to testosterone, the *de novo* production of androgens in the ovaries is exclusive to the theca cells.

Clinically, women with PCOS respond to human chorionic gonadotropin (hCG) challenge with increased testosterone production compared with control women. The increased responsiveness of the polycystic ovary to gonadotropin stimulation implies an increased steroidogenic capacity of the theca interna. In vitro studies confirm that theca cells from polycystic ovaries produce more androgen both in the unstimulated and in the gonadotropin-stimulated state *(4)*. Thus, it is clear that the polycystic ovary has enhanced capacity to secrete androgens. Two principal factors influence the total amount of androgen secreted by the ovary: the total number of theca cells in the ovary and the steroidogenic capacity of each of the theca cells.

2.1.3. Excessive Theca-Cell Proliferation in Polycystic Ovaries

In the polycystic, unlike the normal, ovary there is an accumulation of small antral follicles 3–7 mm in diameter forming the classic "string-of-pearls" morphology. Thus, the ovaries in women with PCOS contain more antral follicles than normal. Many of the follicles in polycystic ovaries demonstrate a hypertrophied theca interna containing many more layers of differentiated steroidogenic cells than the three to five layers in the normal theca interna (Fig. 2). Because of these factors, the polycystic ovary contains more steroidogenic cells in the theca interna than normal, which is consistent with the excessive androgen secretion in PCOS. In contrast, the granulosa cell layers are underdeveloped, and the polycystic ovary contains fewer potential estrogen-producing cells than would be expected for the stage of follicle development.

Little is known about the factors regulating human theca-cell proliferation. Most of the data originate from studies in animals. Chronic exposure of theca cells to luteinizing hormone (LH) (or hCG) leads to a marked increase in the number of cells in the theca interna. There is also evidence that intraovarian growth factors can support theca-cell proliferation. Among these are activin, epidermal growth factor or transforming growth factor (TGF)-α, insulin and insulin-like growth factor (IGF)-1,

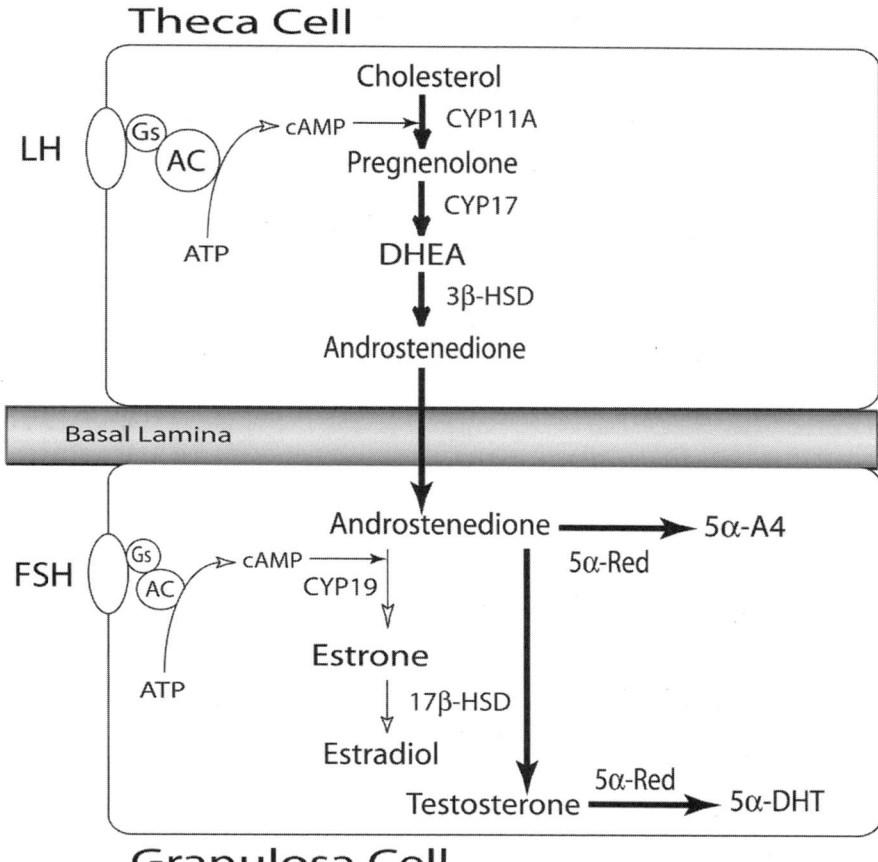

Fig. 1. Steroidogenic capabilities of theca and granulosa cells. Under the influence of luteinizing hormone (LH) theca cells can produce androstenedione de novo. The theca cells express little 17β-hydroxysteroid dehydrogenase and produce very little testosterone. The androstenedione is metabolized in the granulosa cells to estradiol in normal ovaries, but in polycystic ovaries the androstenedione is metabolized to testosterone, 5α-dihydrotestosterone, and 5α-androstane-3,17-dione. Gs, Stimulatory guanosine triphosphate-binding protein; AC, adenyl cyclase; cAMP, cyclic adenosine monophosphate; CYP11A, cholesterol side-chain cleavage cytochrome P450; CYP17, 17α-hydroxylase/C_{17-20} lyase cytochrome P450; 3β-HSD, 3β-hydroxysteroid dehydrogenase; DHEA, dehydroepiandrosterone; FSH, follicle-stimulating hormone; CYP19, aromatase cytochrome P450; 17β-HSD, 17β-hydroxysteroid dehydrogenase; 5α-Red, 5α-reductase; 5α-A4, 5α-androstane-3,17-dione; 5α-DHT, 5α-dihydrotestosterone.

stem cell factor or kit ligand, and tumor necrosis factor (TNF)-α *(5)*. There is evidence that insulin and IGF-1 can stimulate human theca-cell proliferation in vitro, but the extent to which these findings apply to normal follicle development and/or PCOS remains uncertain.

2.1.4. Theca-Cell Overexpression of Steroidogenic Enzyme mRNAs in Polycystic Ovaries

In addition to the increase in the number of theca cells in polycystic ovaries, overexpression of steroidogenic enzymes also contributes to ovarian hyperandrogenism. The mRNAs for LH receptor, steroidogenesis acute regulatory protein (StAR), cholesterol side-chain cleavage enzyme P450scc (*CYP11A*), and the 17α-hydroxlase/C_{17-20} lyase enzyme P450scc (*CYP17*) are overexpressed compared to theca cells from regularly cycling women *(6)*. The relative overexpression of theca-cell

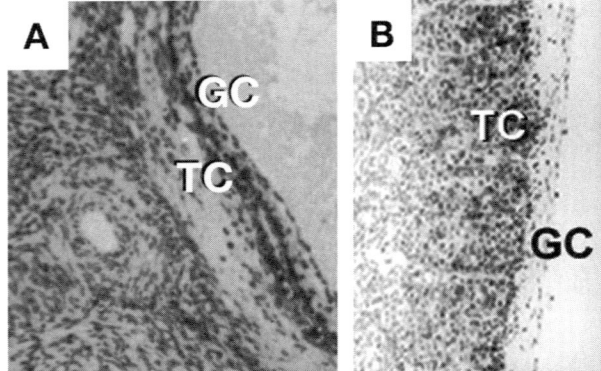

Fig. 2. Ovarian follicle cells in normal and polycystic ovaries. (**A**) Section of follicle wall from a normal ovary containing approximately three layers of healthy granulosa cells (GC) and three to four layers of theca cells (TC). (**B**) Section of a follicle wall from a polycystic ovary. The GC are fewer in number and loosely organized. The theca interna contains many more layers of highly differentiated theca cells.

mRNAs persists, and there continues to be increased 17α-hydroxylase activity when theca cells are cultured in vitro *(7)*, indicating that the signals regulating theca-cell steroidogenesis are increased in PCOS. The increased rate of CYP17 gene transcription in theca cells from polycystic ovaries demonstrates that both thecal differentiation and steroidogenesis are hyperstimulated in PCOS. The heightened state of thecal differentiation appears to involve not only increased transcription of steroidogenic enzyme genes, but also increased mRNA stability, indicating multifactorial abnormalities in theca-cell regulation.

2.1.5. Abnormal Androgen Metabolism in Polycystic Ovaries

The normal ovary metabolizes the vast majority of the androstenedione produced by the theca cells to estrone and estradiol in the granulosa cells. The arrested small antral follicles in polycystic ovaries do not express the high concentrations of aromatase that are present in larger dominant follicles *(8)*, hence there is an accumulation of androstenedione. Small antral follicles from normal and polycystic ovaries express both isoforms of 5α-reductase *(9)*. At approximately the time when aromatase begins to be expressed in the granulosa cells, 5α-reductase expression declines to undetectable levels. In PCOS, follicle development is arrested at this stage. Hence, there is increased expression of 5α-reductase activity and the concentration of 5α-androstane-3, 17-dione is elevated in both follicular fluid and serum in PCOS *(10)*. Although the concentration of 5α-androstane-3,17-dione in normal developing follicles has little effect, the concentrations in the follicular fluid of polycystic ovaries is high enough to suppress aromatase activity by a competitive mechanism *(10)*, potentially contributing to the failure to develop dominant follicles.

2.1.6. Abnormalities in the Regulation of Androgen Production

Within the ovary there are many regulatory molecules that have the potential to contribute to hyperstimulation of theca-cell differentiation and androgen production in PCOS.

2.1.6.1. GONADOTROPINS

The literature is replete with in vivo and in vitro studies demonstrating that LH is the principal regulator of theca-cell differentiation and function. The actions of LH are mediated through activation of the cyclic adenosine monophosphate (cAMP)/protein kinase A (PKA) signaling pathway involving activation of either type I or type II PKA. LH stimulates the expression of *CYP11A*, 3β-HSD, and *CYP17* mRNAs and translation of the mRNAs into functional proteins. Although LH can stimulate inositol phosphate formation in granulosa and luteal cells, there is no evidence that inositol phosphate signaling is involved in mediating the actions of LH in theca cells. Direct activation of PKC with ligands such as angiotensin II sensitizes rat theca cells to LH, and depletion of PKC with long-term treatment with phorbol ester inhibits LH stimulation of thecal steroidogenesis, suggesting that protein kinase C (PKC) may play a role in modulating the sensitivity of theca cells to LH. The role of PKC in ovarian hyperandrogenism has yet to be explored. LH stimulates marked increases in progesterone and androgen production in differentiated theca cells from all mammalian species, including rats, mice, sheep, swine, cows, horses, nonhuman primates, and women. Thus, LH is a key regulator that must be considered when attempting to understand the genesis of ovarian hyperandrogenism.

2.1.6.2. INTRAOVARIAN GROWTH FACTORS

LH is clearly not the only regulator that can modulate the steroidogenic capacity of theca cells. There are a host of stimulatory and inhibitory factors within the ovary with the potential to play a role in PCOS. Of these, only a few have been investigated with respect to their role in PCOS.

2.1.6.2.1. The Transforming Growth Factor Family

The TGF-β family of proteins includes TGF-β, inhibin, activin, bone morphogenetic protein (BMP)-4, and growth differentiation factor (GDF)-9. The role of these molecules, and related proteins such as follistatin, have been studied with respect to PCOS. Both TGF-β1 and TGF-β2 are secreted in the human ovary, with TGF-β1 being present in both theca and granulosa cells and TGF-β2 being localized exclusively in theca cells. All of the inhibin subunits (α, β_A, and β_B) are expressed in human follicles, indicating that various inhibin isoforms are produced, but there is no evidence that follistatin is produced in the ovary. GDF-9 is produced exclusively by the oocyte.

TGF-β inhibits StAR protein expression in human theca tumor cells in vitro and blocks LH-dependent androgen production in rat theca cells through a direct noncompetitive inhibition of CYP17 activity *(11)*. Inhibin stimulates, whereas activin inhibits thecal CYP17 mRNA expression and androgen production *(12)*. GDF-9 has been reported to promote the proliferation of theca cells and inhibit androgen production *(13)*. BMP-4 inhibits androgen production and CYP17 mRNA expression in human theca tumor cells *(14)*. Thus, the evidence is consistent with a potential role for members of the TGF-β family in the genesis of steroidogenic abnormalities in PCOS.

There appears to be consensus that circulating inhibin B concentrations are not abnormal in women with PCOS in aggregate, but lean women with PCOS appear to have increased circulating inhibin B concentrations in contrast to obese women with PCOS *(15)*. This observation does not appear to be relevant to PCOS because there is a negative effect of obesity on inhibin B concentrations independent of PCOS *per se (16)*. On the other hand, the pulsatility of inhibin B secretion in regularly cycling women is not present in PCOS. It is unclear whether there are differences in circulating activin A concentrations in PCOS. Activin concentrations were found to be similar between blood and follicular fluid. Neither inhibin B nor activin A were altered in follicles from PCOS, but there was a decrease in inhibin A in PCOS follicles compared to similar sized follicles in regularly cycling women *(17)*. Follistatin in the circulation is increased in PCOS whether the women were lean or obese *(18)*, but the alterations in circulating concentrations do not occur in the follicular fluid, raising doubts as to the role of follistatin in steroidogenic abnormalities in PCOS. Taken together, they do not support a major role for inhibin and activin in causing ovarian hyperandrogenism.

It has been reported that GDF-9 is decreased in polycystic ovaries *(19)*, but the significance of this finding is uncertain. Considering that GDF-9 appears to be essential for the formation of the theca interna, a decrease in GDF-9 would be inconsistent with the theca hypertrophy observed in PCOS. On the other hand, a decrease in GDF-9 would be in agreement with the increased androgen production. Thus, the possibility exists that stage-specific alterations in GDF-9 secretion may contribute to altered steroidogenesis. Further research needs to be performed before a role for GDF-9 can be established in PCOS.

2.1.6.2.2. Transforming Growth Factor-α

TGF-α is an inhibitor of thecal androgen biosynthesis that has been implicated in PCOS. TGF-α has been localized to both theca and granulosa cells in developing human follicles, but there are no differences between TGF-α concentrations in follicular fluid and polycystic ovaries, and the production of TGF-α in vitro is comparable between tissues from polycystic ovaries and control ovaries *(20)*. The concentrations of TGF-α in follicular fluid vary considerably, but they are always less than 1 ng/mL. The available evidence does not support a role for TGF-α in PCOS.

2.1.6.2.3. Tumor Necrosis Factor-α

The ovary, and the theca interna in particular, contains a population of resident macrophages that play an important role in a variety of ovarian functions. TNF-α is a cytokine with potent inhibitory effects on LH-dependent androgen production by theca cells. TNF-α is not present in theca cells, and it is uncertain whether granulosa cells synthesize TNF-α, but TNF-α is produced by oocytes and resident ovarian macrophages. Although the concentration of TNF-α in serum increases with increasing obesity in both control and women with PCOS *(21)*, women with PCOS have increased serum TNF-α concentrations compared to weight-matched controls. In contrast, there are no differences in follicular fluid TNF-α concentrations between women with PCOS and regularly cycling controls *(22)*. TNF-α secretion by human ovarian tissues in vitro was undetectable *(22)*. Thus, the evidence does not support a role for intraovarian effects of TNF-α in the genesis of ovarian hyperandrogenism.

2.1.6.3. ADIPOCYTOKINES

2.1.6.3.1. Leptin

Obesity is commonly associated with PCOS and ovarian hyperandrogenism. The amount of body fat in humans is strongly correlated with circulating leptin concentrations, raising the question of whether leptin might play a role in abnormal steroidogenesis. The follicular fluid concentration of leptin is equivalent to the plasma concentration, but there is less leptin binding in the follicular fluid than in plasma. The circulating concentration of leptin in women with PCOS is equivalent to the concentration in control women when adjusted for obesity *(23)*, demonstrating that there are no abnormalities of leptin secretion associated with PCOS. There do not appear to be direct effects of leptin alone on ovarian steroidogenesis, but leptin inhibits the stimulatory effects of IGF-1 on thecal androstenedione production and granulosa estrogen production with no effects on progesterone synthesis *(24)*. This combination of effects, taken together with the lack of alteration of leptin concentrations in PCOS, indicates that leptin does not play a significant role in the steroidogenic abnormalities in PCOS.

2.1.6.3.2. Resistin

Resistin is a recently described protein secreted by cells in the adipose tissue. In mice, treatment with recombinant resistin caused impaired insulin action and glucose intolerance, whereas administration of antiresistin antibodies improved insulin action, thus linking resistin with insulin resistance. Because of the association between insulin resistance and PCOS, resistin is a candidate to play a role with respect to abnormal steroidogenesis in PCOS.

The human homolog of resistin exhibits 53% homology to murine resistin and resides on chromosome 19p13.3 in a region not previously linked to obesity, insulin resistance, or diabetes. Several investigations have shown that resistin concentrations are elevated in the circulation of women with PCOS, and a recent study demonstrated a direct stimulatory action of resistin on thecal 17α-hydroxylase activity in vitro *(25)*. Although a role for resistin in PCOS has yet to be established, these data indicate that elevated resistin may play a role in ovarian hyperandrogenism in PCOS.

2.1.6.4. INSULIN

Insulin resistance associated with PCOS *(1)* causes a mild compensatory increase in insulin concentrations in order to control blood glucose concentrations within normal limits. There is a large body of evidence supporting the concept that the increase in insulin concentrations is related to the elevated androgen production characteristic of PCOS. The elevated insulin in vivo is thought to contribute to the elevated basal androgen secretion by cultured human theca cells from polycystic ovaries, and it is clear that the stimulatory effect of insulin on thecal steroidogenesis is augmented in PCOS *(26)*. These data suggest that theca cells may be hyperresponsive to insulin in PCOS.

Insulin hyperresponsiveness of the theca cells in otherwise insulin-resistant women seems to present a paradox. The concept that increased insulin could activate the IGF-1 receptor and thus overstimulate androgen production is not supported by the evidence. Although the theca cells contain IGF-1 receptors, the affinity for insulin is significantly lower than for the insulin receptor. Thus, hyperinsulinemia greater than that observed in most women with PCOS would be required to significantly activate the type 1 receptor. The evidence shows that the effects of insulin on thecal androgen biosynthesis are mediated through the insulin receptor and that blocking the IGF-1 receptor with neutralizing antibody does not prevent the insulin stimulation of androgen production *(27)*. Interestingly, theca cells from polycystic ovaries appear to have increased expression of insulin receptor substrates 1 and 2, suggesting that there may be an abnormal amplification of insulin signaling in PCOS *(28)*.

There is growing consensus that insulin per se has an important contribution to excess androgen production in PCOS. Inhibition of insulin secretion with diazoxide causes a decline in circulating androgen concentrations, indicating that insulin has a direct effect on ovarian androgen production in vivo. Further evidence is provided by clinical studies with the insulin-sensitizing agents such as metformin and the thiazolidinediones. Metformin reduces fasting plasma glucose concentrations and increases oral glucose tolerance in women with PCOS, decreases plasma insulin and improves insulin sensitivity in type 2 diabetic subjects, and improves insulin-stimulated glucose transport in skeletal muscle from subjects with type 2 diabetes. Recent studies demonstrate that metformin not only improves peripheral insulin sensitivity in women with PCOS, but also decreases androgen concentrations in both lean and obese women with PCOS and can restore menses in approximately one-third of previously amenorrheic women *(29)*.

Thiazolidinediones are a class of insulin-sensitizing agents that act by a different mechanism from metformin, potentially involving peroxisome proliferator-activated receptor-γ, which is involved in stimulating transcription of factors related to glucose disposal in skeletal muscle. Thiazolidinediones decrease hyperglycemia in subjects with type 2 diabetes and increase glucose uptake in skeletal muscle. In PCOS there are dramatic reductions in androgen excess and insulin resistance and improvement in metabolic and endocrine dysfunctions *(29)*.

On the other hand, reduction of androgen concentrations yields only a modest improvement in insulin resistance. Neither long-term treatment with gonadotropin-releasing hormone analogs nor use of anti-androgens was able to normalize insulin sensitivity in women with PCOS. Thus, reduction of androgen secretion or blocking androgen action has little effect on insulin sensitivity, implying that excess androgen secretion is secondary to increased insulin concentrations in insulin-resistant women with PCOS.

3. ABNORMAL ESTROGEN PRODUCTION

In normally cycling women the predominant site of estrogen production is the granulosa cells of the ovarian follicle. In ovulatory cycles in PCOS, the dominant follicle secretes estradiol in sufficient quantities to trigger the midcycle surge of LH, but in the anovulatory cycles typical of PCOS, follicle development is arrested at the small antral stage when selection of a dominant follicle would normally occur and the aromatase enzyme would be expressed *(30)*. Consequently, the granulosa cells fail to express aromatase, and estrogen production from the ovary is minimal.

PCOS, however, is not a hypoestrogenic condition. The high concentrations of androstenedione secreted by the theca cells are metabolized to estrone in the peripheral tissues, in particular the adipose stromal cells, leading to a chronic elevation of estrogen in the circulation. The chronic elevation of estrogen in PCOS contributes to the frequently observed increase in LH-to-follicle-stimulating hormone (FSH) ratio and can promote the development of reproductive tract cancer.

4. ABNORMAL PROGESTERONE PRODUCTION

Progesterone production by the ovary in PCOS has not been thoroughly studied. Progesterone can be produced by theca and granulosa cells in developing follicles and by the corpus luteum in ovulatory cycles. It is likely that progesterone production from the theca cells is increased in PCOS because of the increase in steroidogenic capacity of the theca cells.

There is evidence that the granulosa cells in arrested follicles in polycystic ovaries prematurely express genes associated with luteinization, namely cytochrome P450scc and the LH receptor *(8)*. The premature luteinization causes the granulosa cells to respond to LH and produce excessive amounts of progesterone compared to cells from control ovaries at a similar developmental stage *(31)*. It is unclear whether abnormal progesterone production plays an important role in the pathophysiology of PCOS.

5. FUTURE AVENUES OF INVESTIGATION

Our current understanding of steroidogenic abnormalities in PCOS is the result of recent studies that have characterized the phenotype of the polycystic ovary. However, little is known regarding the relative contributions of various regulatory molecules to the steroidogenic abnormalities, and the molecular mechanisms that are dysregulated in PCOS have not been worked out. Active areas of investigation include intracellular signaling mechanisms by which transcription and translation of steroidogenically relevant genes are controlled. Additional work is needed that defines the mechanisms and regulators, both intraovarian and extragonadal, that cause hyperandrogenism. An area that has not received much attention is the mechanism of how elevated androgen concentrations lead to follicle arrest. Finally, research into the mechanisms of granulosa cell dysfunction would be a fruitful area of investigation. It would be helpful to understand if there are unifying defects in theca and granulosa cells that lead to the abnormal phenotypes of these cells in PCOS.

KEY POINTS

- The ovarian theca cells in PCOS appear to be hyperstimulated by abnormal concentrations of LH and insulin.
- The increased number and steroidogenic capacity of the theca cells in PCOS leads to excessive production of androstenedione, which is metabolized to testosterone and 5α-androstane-3,17-dione rather than estrogen by the granulosa cells.
- The granulosa cells in PCOS fail to express adequate amounts of aromatase, but instead prematurely express LH receptors and produce progesterone.

REFERENCES

1. The Rotterdam ESHRE/ASRM-sponsored PCOS consensus workshop group. Revised 2003 consensus on diagnostic criteria and long-term health risks related to Polycystic Ovary Syndrome. Fertil Steril 2004;81:19–25.
2. Kumar A, Woods KS, Bartolucci AA, Azziz R. Prevalence of adrenal androgen excess in patients with the polycystic ovary syndrome (PCOS). Clin Endocrinol (Oxf) 2005;62:644–649.
3. Chang RJ, Laufer LR, Meldrum DR, et al. Steroid secretion in polycystic ovarian disease after ovarian suppression by a long-acting gonadotropin-releasing hormone agonist. J Clin Endocrinol Metab 1983;56:897–903.
4. Gilling-Smith C, Willis DS, Beard RW, Franks S. Hypersecretion of androstenedione by isolated thecal cells from polycystic ovaries. J Clin Endocrinol Metab 1994;79:1158–1165.
5. Magoffin DA. The ovarian androgen-producing cells. A 2001 perspective. Rev Endocr Metab Disord 2002;3:47–53.
6. Jakimiuk AJ, Weitsman SR, Navab A, Magoffin DA. Luteinizing hormone receptor, steroidogenesis acute regulatory protein and steroidogenic enzyme messenger ribonucleic acids are overexpressed in theca and granulosa cells from polycystic ovaries. J Clin Endocrinol Metab 2001;86:1318–1323.
7. Nelson VL, Legro RS, Strauss III JF, McAllister JM. Augmented androgen production is a stable steroidogenic phenotype of propagated theca cells from polycystic ovaries. Mol Endocrinol 1999;13:946–957.
8. Jakimiuk AJ, Weitsman SR, Brzechffa PR, Magoffin DA. Aromatase messenger ribonucleic acid expression in individual follicles from polycystic ovaries. Mol Human Reprod 1998;4:1–8.

9. Jakimiuk AJ, Weitsman SR, Magoffin DA. 5α-Reductase activity in women with polycystic ovary syndrome. J Clin Endocrinol Metab 1999;84:2414–2418.

10. Agarwal SK, Judd HL, Magoffin DA. A mechanism for suppression of estrogen production in women with polycystic ovary syndrome. J Clin Endocrinol Metab 1996;81:3686–3691.

11. Fournet N, Weitsman SR, Zachow RJ, Magoffin DA. Transforming growth factor-β inhibits ovarian 17α-hydroxylase activity by a direct non-competitive mechanism. Endocrinology 1996;137:166–174.

12. Hillier SG, Yong EL, Illingworth PJ, Baird DT, Schwall RH, Mason AJ. Effect of recombinant activin on androgen synthesis in cultured human thecal cells. J Clin Endocrinol Metab 1991;72:1206–1211.

13. Yamamoto N, Christenson LK, McAllister JM, Strauss JF. Growth differentiation factor-9 inhibits 3'5'-adenosine monophosphate-stimulated steroidogenesis in human granulosa and theca cells. J Clin Endocrinol Metab 2002;87:2849–2856.

14. Dooley CA, Attia GR, Rainey WE, Carr BR. Bone morphogenetic protein inhibits ovarian androgen production. J Clin Endocrinol Metab 2000;85:3331–3337.

15. Pigny P, Cortet-Rudelli C, Decanter C, et al. Serum levels of inhibins are differentially altered in patients with polycystic ovary syndrome: effects of being overwieght and relevance to hyperandrogenism. Fertil Steril 2000;73:972–977.

16. Cortet-Rudelli C, Pigny P, Decanter C, et al. Obesity and serum luteinizing hormone level have an independent and opposite effect on the serum inhibin B level in patients with polycystic ovary syndrome. Fertil Steril 2002;77:281–287.

17. Magoffin DA, Jakimiuk AJ. Inhibin A, inhibin B and activin concentrations in follicular fluid from women with polycystic ovary syndrome. Hum Reprod 1998;13:2693–2698.

18. Eldar-Geva T, Spitz IM, Groome NP, Margalioth EJ, Homburg R. Follistatin and activin A serum concentrations in obese and non-obese patients with polycystic ovary syndrome. Hum Reprod 2001;16:2552–2556.

19. Teixeira Filho FL, Baracat EC, Lee TH, et al. Aberrant expression of growth differentiation factor-9 in oocytes of women with polycystic ovary syndrome. J Clin Endocrinol Metab 2002;87:1337–1344.

20. Mason HD, Carr L, Leake R, Franks S. Production of transforming growth factor-α by normal and polycystic ovaries. J Clin Endocrinol Metab 1995;80:2053–2056.

21. Gonzalez F, Thusu K, Abdel-Rahman E, Prabhala A, Tomani T, Dandona P. Elevated serum levels of tumornecrosis factor alpha in normal-weight women with polycystic ovary syndrome. Metabolism 1999;48:437–441.

22. Jasper M, Norman RJ. Immunoreactive interleukin-I beta and tumour necrosis factor-alpha in thecal, stromal and granulosa cell cultures from normal and polycystic ovaries. Hum Reprod 1995;10:1352–1354.

23. Magoffin DA, Duggal PS, Norman RJ. The role of leptin in polycystic ovary syndrome. In: Castracane VD, Henson MC, eds. Leptin and Reproduction. Dordrecht: Kluwer Academic/Plenum Publishers, 2003:333–345.

24. Agarwal SK, Vogel K, Weitsman SR, Magoffin DA. Leptin antagonizes the insulin-like growth factor-I augmentation of steroidogenesis in granulosa and theca cells of the human ovary. J Clin Endocrinol Metab 1999;84:1072–1076.

25. Munir I, Yen H-W, Baruth T, et al. Resistin stimulation of 17α-hydroxylase activity in ovarian theca cells in vitro: relevance to polycystic ovary syndrome. J Clin Endocrinol Metab 2005;90:4852–4857.

26. Barbieri RL, Makris A, Randall RW, Daniels G, Kistner RW, Ryan KJ. Insulin stimulates androgen accumulation in incubations of ovarian stroma obtained from women with hyperandrogenism. J Clin Endocrinol Metab 1986;62:904–910.

27. Willis D, Franks S. Insulin action in human granulosa cells from normal and polycystic ovaries is mediated by the insulin receptor and not the type-I insulin-like growth factor receptor. J Clin Endocrinol Metab 1995;80:3788–3790.

28. Yen HW, Jakimiuk AJ, Munir I, Magoffin DA Selective alterations in insulin receptor substrates-1, -2 and -4 in theca but not granulosa cells from polycystic ovaries. Mol Hum Reprod 2004;10:473–479.

29. Lord JM, Flight IH, Norman RJ. Insulin-sensitising drugs (metformin, troglitazone, rosiglitazone, pioglitazone, D-chiro-inositol) for polycystic ovary syndrome. Cochrane Database Syst Rev 2003;3:CD003053.

30. Jakimiuk AJ, Weitsman SR, Brzechffa PR, Magoffin DA. Aromatase messenger ribonucleic acid expression in individual follicles from polycystic ovaries. Mol Hum Reprod 1998;4:1–8.

31. Willis DS, Watson H, Mason HD, Galea R, Brincat M, Franks S. Premature response to luteinizing hormone of granulosa cells from anovulatory women with polycystic ovary syndrome: relevance to mechanism of anovulation. J Clin Endocrinol Metab 1998;83:3984–3991.

19

Hypothalamic–Pituitary–Adrenal Dysfunction in the Polycystic Ovary Syndrome

Bulent O. Yildiz, Enrico Carmina, and Ricardo Azziz

SUMMARY

Between 20 and 30% of patients with polycystic ovary syndrome (PCOS) demonstrate adrenal androgen (AA) excess, detectable primarily by elevated dehydroepiandrosterone sulfate (DHEAS) levels. Generalized adrenocortical hyperresponsivity to adrenocorticotropic hormone (ACTH) stimulation is also observed and may be the principal mechanism determining AA excess in PCOS. The causes of this abnormality are unclear, but increased peripheral metabolism of cortisol, altered factors regulating glucose-mediated glucose disposal, and perhaps ovarian sex steroids may in different ways contribute to the AA excess in PCOS. Additionally, DHEAS levels and the response of AAs to ACTH are relatively constant over time and may be a genetically determined trait.

Key Words: Androgen; adrenal; hyperandrogenism; cortisol.

1. INTRODUCTION

The polycystic ovary syndrome (PCOS) is the most common endocrine disorder of reproductive-aged women, with an estimated prevalence of 6–7% (1). Although the ovaries are the main source of androgen excess in PCOS, excess adrenal androgen (AA) levels (e.g., dehydroepiandrosterone [DHEA] and adrenal-secreted androsteredione [A4]) and adrenocortical dysfunction have also been observed in many patients with PCOS (2–6). Proof of the role of AA excess in the development of PCOS is circumstantial at best. For example, peripubertal AA excess is linked to the development of PCOS-like symptomatology in patients with 21-hydroxylase (21-OH)-deficient classic (CAH) (7) and nonclassic adrenal hyperplasia (NCAH) (8–11), including the development of polycystic-appearing ovaries on ultrasound, elevated luteinizing hormone levels, and ovarian hyperandrogenism. Other investigators have noted that patients with PCOS have a greater incidence of peripubertal stress, resulting in exaggerated AA secretion during this vulnerable period (12). Finally, patients with premature adrenarche are at higher risk for the development of PCOS (13–16). In this chapter we review the prevalence of AA excess and hypothalamic–pituitary–adrenal (HPA) axis dysfunction in PCOS and the potential underlying mechanisms.

2. BACKGROUND

2.1. Epidemiology of Adrenal Androgen Excess in PCOS

Clinically, the measurement of circulating levels of the AA metabolite DHEA sulfate (DHEAS) has been traditionally used as a marker for AA excess (17–19) because this steriod is (1) 97–99% of adrenocortical origin (20–22), (2) the second most abundant steroid after cortisol (F), (3) relatively stable throughout the day and the menstrual cycle (23–25), because of its relatively long half-life (26–30), and (4) easily measured.

From: *Contemporary Endocrinology: Androgen Excess Disorders in Women:*
Polycystic Ovary Syndrome and Other Disorders, Second Edition
Edited by: R. Azziz et al. © Humana Press Inc., Totowa, NJ

Excess AA levels, particularly elevations in the levels of the dehydroepiandrosterone (DHEA) metabolite DHEAS and 11-hydroxyandrostenedione (11OHA4), were initially reported in 40–60% of patients with PCOS *(3–6)*. However, in most of these early studies criteria for the selection of PCOS patients were different from those currently used. Moreover, it is clear that several factors, including age and race, should be considered when estimating the prevalence of AA excess in PCOS, because AAs begin to decline after the age of 30 years in both normal women and women with PCOS *(31,32)*. In a retrospective study of 145 hyperandrogenic patients, we found that hyperandrogenic patients with high DHEAS levels were younger, in addition to being thinner and more hirsute, than hyperandrogenic women with lower DHEAS levels *(33)*. The impact of race on the prevalence of AA excess in PCOS is unclear. In one report the prevalence of AA excess among PCOS patients was found to be similar among Italian, US, Hispanic-American, and Japanese women *(34)*. However, only small groups of patients were compared.

To revaluate the prevalence of AA excess in PCOS taking into account race and age-related changes in AAs, we undertook a study of 213 (27 black and 186 white) women with PCOS and 182 (88 black and 94 white) age-matched healthy eumenorrheic nonhirsute women (controls) *(31)*. The diagnosis of PCOS was based on hyperandrogenism and chronic anovulation, consistent with the National Institutes of Health 1990 criteria. DHEAS levels were significantly lower in black than white controls, whereas fasting insulin and body mass index (BMI) were higher in black controls, and DHEAS levels decreased similarly with age in control and PCOS women of either race (Fig. 1). Body mass and fasting insulin had little impact on circulating DHEAS levels in healthy women. Among PCOS patients, these parameters were negatively associated with circulating DHEAS levels among white, but not black patients. For each race and age group, the upper 95% normative values for log DHEAS was calculated, and the number of PCOS subjects with log DHEAS values above this level were assessed. The prevalence of supranormal DHEAS levels was 33 and 20% among black and white women with PCOS, respectively—not a significant difference.

These data indicate that AA excess, defined by the circulating level of DHEAS, is somewhat less common in PCOS than previously reported, affecting between 20 and 30% of affected women when using age- and race-adjusted normative values. However, it also appears that women with absolute DHEAS excess simply represent the upper edge of the normal DHEAS distribution in the general population, not a separate population. For example, cluster analysis failed to reveal any specific subpopulations of DHEAS levels among our patients with PCOS *(31)*. Finally, there may be significant differences in mean DHEAS levels between white and black control women. Whether these differences are also present among women of other racial or ethnic groups (e.g., Hispanics and Asians) remains to be deter- mined. However, because AA secretion appears to have a strong genetic component (*see* Subsection 2.7.), it is not surprising that race and ethnicity play a role in determining the prevalence of AA excess in PCOS.

Finally, we should note that DHEAS, does not uniformly reflect AA secretion in response to adrenocorticotropic hormone (ACTH) in normal or hyperandrogenic patients. For example, only 50% of patients with 21-OH-deficient NCAH have a supranormal DHEAS levels *(35)*. Witness also the profound suppression in DHEAS levels that occurs in NCAH patients treated with glucocorticoids despite the still elevated production of low-dose A4 *(36)*. Likewise, note should be taken of the increase in DHEAS in response to exogenous testosterone administration to oophorectomized women, despite the absence of any change in the AA response to acute ACTH stimulation *(37)*. Consequently, it is apparent that a number of factors may alter DHEAS levels without modifying adrenocortical AA production, most likely through regulation of DHEA sulfotransferase (DHEA-ST) activity. Hence, the investigation of those mechanisms underlying the AA excess of PCOS requires evaluation not only of DHEAS levels, but also of adrenocortical biosynthesis (usually measurable by the response to acute ACTH stimulation).

2.2. Hypothalamic–Pituitary Regulation and Adrenal Androgen Excess in PCOS

There appears to be no difference in basal morning plasma levels of ACTH *(38,39)* or the circadian or diurnal variation of this hormone *(39)* between PCOS or hyperandrogenic women and healthy

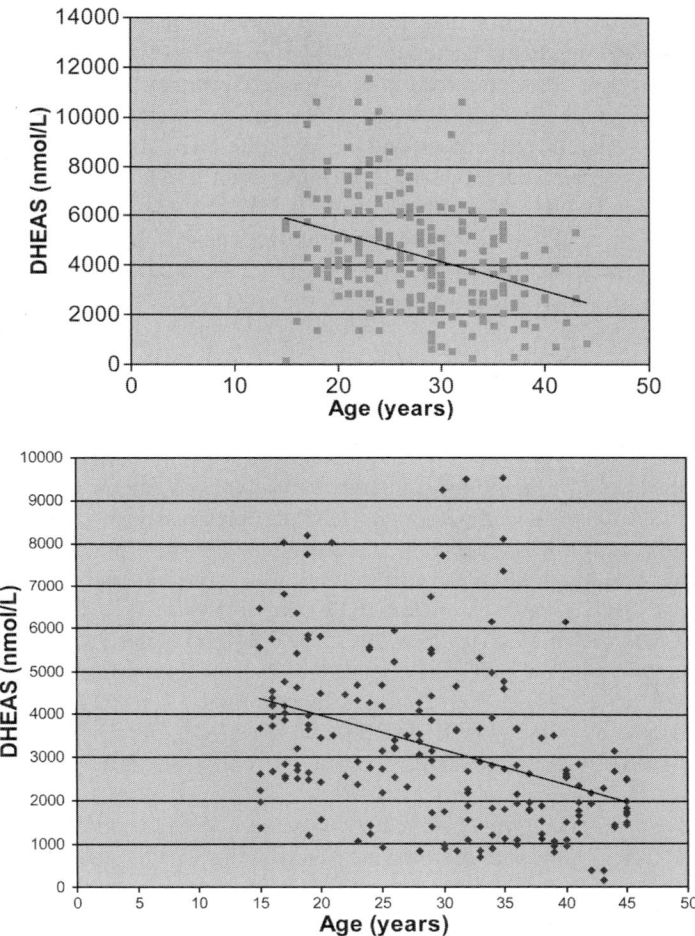

Fig. 1. Scatter grams and linear regression trend line for dehydroepiandrosterone sulfate (DHEAS) levels according to age in polycystic ovary syndrome (PCOS) (top) (r = –0.34, *p* < 0.0001) and healthy controls (bottom) (r = –0.38, *p* < 0.0001). Note that DHEAS levels in PCOS and controls decrease with age at similar rates. (From ref. *31*.)

controls. The response of ACTH to endogenous or exogenous corticotrophin-releasing hormone is also normal in PCOS *(40,41)*, with and without DHEAS excess *(41)*. These data suggest that pituitary responsivity is not altered in PCOS, regardless of the presence of AA excess.

Because hypothalamic–pituitary alterations appear to be mostly excluded, AA excess in PCOS may be a result of adrenocortical steroidogenic abnormalities, alterations in the sensitivity or responsivity of the adrenal to ACTH stimulation, or abnormalities in the metabolism of adrenal products, including DHEA and cortisol.

2.3. Alteration in Adrenocortical Biosynthesis in PCOS

Estimation of the relative adrenocortical enzymatic activities in vivo can be performed using acute adrenal stimulation by intravenous or intramuscular administration of ACTH(1-24). Acute administration of commercially available doses of 0.25 mg provides maximum adrenal stimulation, regardless of body weight *(42)*. Several studies have shown that PCOS women demonstrate a generalized hypersecretion of adrenocortical products, basally and in response to ACTH, including pregnenolone (PREG), 17-hydroxypregnenolone (17-OH-PREG), DHEA, A4, 11-deoxycortisol (S), and F *(40,41)*.

The observed exaggerated secretion in response to ACTH may be because of a change in the responsiveness or sensitivity of the adrenal cortex to ACTH stimulation. We observed that excess DHEAS levels in PCOS patients were associated with an exaggerated secretory response of the adrenal to ACTH stimulation for DHEA and A4 *(41)*. However, no differences in the sensitivity of AAs to ACTH stimulation was observed between PCOS women with and without DHEAS excess and controls.

Importantly, the exaggerated adrenocortical response to ACTH does not appear to be the result of genetic defects affecting 21-OH (P450c21, encoded by *CYP21*), 11β-hydroxylase (11-OH P450c11, encoded by *CYP11B1*), or 3β-hydroxysteroid dehydrogenase (3β-HSD; encoded by *HSD3B2*) *(42,43)*. Overall, the prevalence of NCAH resulting from 21-OH deficiency in the populations studied is 1–2% *(44)*, with NCAH as a result of 3β-HSD *(45)* or 11-OH *(46)* occurring very rarely, on the order of less than 1%.

Other investigators have observed defects in 3β-HSD activity among women with PCOS *(47–50)*, although these do not appear to be related to inherited defects of the *HSD3B2* gene *(45)*. Pang and colleagues suggested that the exaggerated 3β-HSD activity observed was secondary to a variant of insulin-resistant PCOS *(51)*. However, these investigators were not able to observe a difference in insulin resistance (measured by the frequently sampled intravenous glucose tolerance test [FSIVGTT]) between women with PCOS with and without 3β-HSD deficiency, suggesting that a mechanism other than a defect in insulin action was responsible for the perceived steroidogenic abnormality.

In fact, the single steroidogenic difference that we were able to observe between PCOS women and healthy controls was a greater estimated $\Delta^5$17-α-hydroxylase (17-OH) activity, primarily observed in PCOS patients with high DHEAS levels *(52)*. We should note that $\Delta^5$17-OH in vivo activity is estimated from the ratio of the 17-OH-PREG to PREG (i.e., poststimulated levels), such that women with PCOS and DHEAS excess demonstrated a supranormal 17-HPREG$_{60}$:PREG$_{60}$ ratio. Consequently, if only 17α-hydroxylated (17-OH-PREG) and not 17-deoxy (e.g., PREG) products are measured, the higher 17-HPREG levels observed could suggest 3β-HSD deficiency, as observed by some investigators *(51)*.

The exaggerated $\Delta^5$17-OH activity observed in women with PCOS and DHEAS excess may reflect an abnormality in P450c17 function, the enzyme determining 17-OH and 17,20-lyase activity. The *CYP17* gene encodes this enzyme. However, we were unable to demonstrate a difference in the prevalence of a common polymorphism of *CYP17*, hypothesized to increase transcription of the gene, between women with PCOS with and without DHEAS excess *(53)*. We should note, however, that this does not necessarily exclude the presence of other mutations or variants that may affect the function of this gene. What factors may result in the exaggerated $\Delta^5$17-OH activity remain to be elucidated, although potential candidates include various extra-adrenal factors (*see* Subsections 2.5. and 2.6.).

2.4. Alterations in Cortisol Metabolism in PCOS

Increased peripheral metabolism of F has been observed in some patients with PCOS. An increase in F metabolism would potentially result in decreased negative feedback of ACTH, such that hypothalamic–pituitary–adrenal (HPA) axis activity would increase to maintain normal F levels at the expense of AA excess *(54,55)*. The increased metabolism of F may be a result of enhanced inactivation of this steroid by 5α-reductase (5α-RA) *(55,56)* or impaired reactivation of F from cortisone by 11β-hydroxysteroid dehydrogenase type 1 (11β-HSD1) *(54,55)*.

Studying 11 PCOS patients, Stewart and colleagues observed that total urinary cortisol metabolites were higher in patients than controls, suggesting increased breakdown of this steroid *(55)*. They also observed a higher ratio of 5α- to 5β-cortisol metabolites in the urine of these women, consistent with enhanced activity of 5α-RA. Tsilchorozidou and colleagues also noted enhanced 5α-RA and reduced 11β-HSD1 activities, as assessed by the urinary profile, in 18 lean PCOS women compared with 19 lean controls *(56)*. In this study, insulin seemed to enhance 5α-reduction of steroids in PCOS, but was not associated with the elevated F production rate or with enhanced 11β-HSD1 activity observed.

Taken together, these data suggest that increased F metabolism may be present in PCOS and appears independent of the presence of obesity, but may be enhanced by hyperinsulinemia. However, the role of these metabolic abnormalities in the AA excess of PCOS remains to be demonstrated.

2.5. The Role of Extra-Adrenal Sex Steroids in the HPA Dysfunction of PCOS

Adrenocortical dysfunction in PCOS patients may represent an acquired defect secondary to abnormal ovarian secretion. In general, most investigators have observed a 20–25% decrease in mean DHEAS levels following long-acting GnRH analog (GnRHa) suppression in women with PCOS and elevated levels of this metabolite, although the elevated AA levels rarely normalize *(57,58)*. Consequently, it is possible that ovarian factors, possibly including androgens and estrogens, may increase AA secretion in PCOS.

Ditkoff et al. observed that transdermal estradiol (E2) administration restored the exaggerated responses of 11OHA4 and DHEA to ACTH stimulation following GnRHa suppression in women with PCOS *(59)*. Although estrogen may be operant in maintaining the exaggerated adrenocortical function observed in PCOS, we were unable to observe a similar effect on the adrenocortical steroidogenesis of postmenopausal women receiving transdermal E2 *(60)*.

Investigating the effect of extra-adrenal testosterone, we administered exogenous parenteral testosterone for 3 weeks to seven healthy oophorectomized women and did not observe significant changes in adrenocortical steroidogenesis, measured by the response to acute ACTH stimulation *(37)*. Alternatively, we were able to observe an increase in the DHEAS-to-DHEA ratio, suggesting increased DHEA-ST activity. Alternatively, studying 31 male-to-female and 22 female-to-male transsexual patients before and during sex steroid treatment, Polderman and colleagues reported that basal AA levels decreased by 27–48% in males treated with ethinylestradiol and cyproterone acetate and increased by 23–70% in females treated with testosterone *(61)*. Alternatively, studying four female-to-male transsexuals before and after 12 months of testosterone enanthate treatment, Futterweit did not observe any difference in the 17-OH-PREG, 17-hydroxyprogesterone (17-HP), DHEA, A4, S, and F levels, basally and after acute ACTH stimulation *(62)*.

In conclusion, available data suggest that extra-adrenal sex steroids may alter AA secretion in response to ACTH stimulation, and DHEA-ST activity and circulating DHEAS levels. However, there is significant conflict in the published data, and better designed and longer-term studies are needed to definitively answer this question.

2.6. Glucose/Insulin Axis and HPA Dysfunction in PCOS

Approximately 50–70% of PCOS patients have insulin resistance and hyperinsulinemia *(63,64)*. Insulin resistance in young women generally results in the development of compensatory hyperinsulinemia, which stimulates androgen secretion by the ovary *(65)* and perhaps the adrenal *(66)*. Studying minced human adrenal tissues, we observed that insulin (0.2–5.0 nmol/L) decreased DHEA and increased DHEAS, suggesting that at a minimum insulin stimulates DHEA-ST activity *(57)*. Additional evidence that hyperinsulinism plays a role in regulating adrenocortical biosynthesis arises from studies examining the effect of insulin sensitizers. We studied 305 women with PCOS randomized to receive placebo or the insulin sensitizer troglitazone in doses of 150, 300, or 600 mg/day for 20 weeks *(67)*. In these women, circulating DHEAS levels decreased 18–26% with troglitazone 600 mg/day, but not placebo, regardless of basal DHEAS level (Fig. 2). Nevertheless, troglitazone alters androgen biosynthesis directly *(68)* and ovarian androgen secretion indirectly *(69)*, which may have contributed to the observed results.

In contrast, studying 213 consecutive untreated women with PCOS and 165 age- and race-matched healthy eumenorrheic nonhirsute controls, Kumar et al. were unable to detect a significant association between circulating DHEAS and fasting insulin levels *(31)*. Moreover, PCOS patients with increased DHEAS tend to be leaner than patients with normal DHEAS levels *(33)*, a finding that seems to exclude a major role for circulating insulin in AA excess in PCOS.

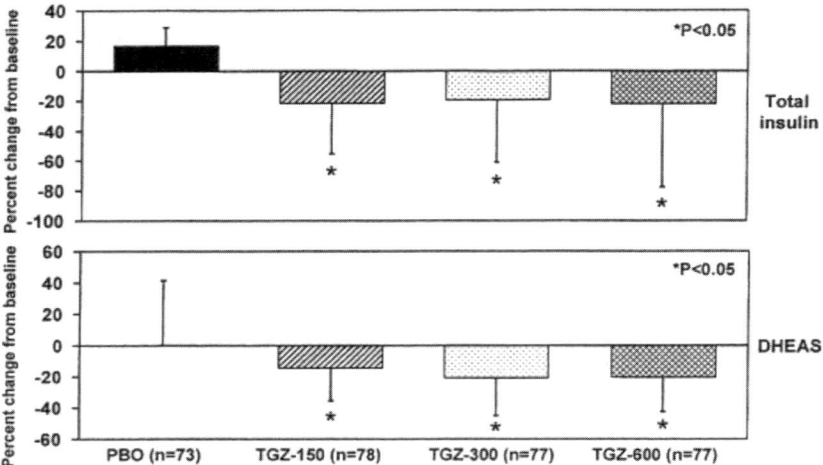

Fig. 2. Change in basal fasting total insulin and dehydroepiandrosterone sulfate (DHEAS) levels with placebo (PBO) or troglitazone 150 mg/day (TGZ-150), 300 mg/day (TGZ-300), or 600 mg/day (TGZ-600). Mean ± SD noted. (From ref. *67*.)

In order to determine whether more subtle abnormalities of insulin action are related to the adrenocortical dysfunction of PCOS, Falcone and colleagues studied 19 women with PCOS and nine age- and weight-matched controls using a tolbutamide-modified FSIVGTT analyzed by the minimal model *(70)*. A significant decline in DHEA levels was observed in control subjects and in PCOS women with normal insulin sensitivity 3 hours after glucose administration. Alternatively, no change in DHEA was observed in insulin-resistant PCOS subjects. DHEAS levels were not measured in this study. The investigators hypothesized that failure of glucose-stimulated endogenous insulin secretion to significantly depress DHEA levels in insulin-resistant women with PCOS may account in part for their androgen excess.

Farah-Eways and colleagues studied nine reproductive-aged patients with PCOS and nine age-, race-, and body mass indexed-matched controls with an insulin-modified FSIVGTT and an acute 60-minute ACTH(1-24)-stimulation test *(71)*. The fasting insulin and fasting glucose levels were not correlated to any of the adrenal parameters studied, with the exception of a positive association between the basal 17-OHP and the fasting glucose level ($r = 0.85$, $p < 0.004$). The insulin sensitivity index (S_I) and the acute insulin response to glucose (AIR_G) also had a limited correlation with adrenocortical parameters in both groups. Alternatively, glucose effectiveness (S_G), a measure of the ability of glucose to control its own production/uptake (i.e., glucose- or non-insulin-mediated glucose disposal), was strongly and positively associated in PCOS patients, but not controls, with the basal levels of F, DHEA, and DHEAS, the ACTH-stimulated peak levels of F, DHEA, and 17-OH-PREG, and the net increment following ACTH administration for F, DHEA, and 17-OH-PREG. These results suggest that adrenocortical biosynthesis, basally and in response to ACTH, may be more closely associated with glucose-mediated glucose disposal, or the mechanisms determining it, than with the degree of hyperinsulinemia or the sensitivity of insulin-mediated glucose disposal.

Overall, available data suggest that there may be a complex interaction between the glucose–insulin axis and HPA dysfunction in PCOS. Insulin itself may have a modest stimulatory effect on DHEA-ST, which may increase the correct levels of DHEAS at the expense of DHEA. It is also possible that AA excess may be the result of a decreased ability of insulin to suppress DHEA production. Finally, our data suggest that adrenocortical dysfunction in PCOS may be closely linked to those mechanisms underlying glucose-mediated glucose disposal (i.e., glucose effectiveness).

2.7. *Heritability of Adrenal Androgen Excess and HPA Dysfunction in PCOS*

Inheritance plays a significant role in determining the circulating AA levels in normal individuals *(72)*. In addition, circulating AA levels and their response to ACTH are highly individualized, compared to their secretion of glucocorticoids in normal women *(73)*. For example, both basal and ACTH-stimulated levels of DHEA demonstrated a high level of between-subject variability (60–70%) compared to the between-subject variability for F of 15–40% *(73)*. Essentially, the population could be divided into significant numbers of "high and low AA secretors." Furthermore, in PCOS the within-subject responses of DHEA, A4, or F to ACTH stimulation remain relatively unchanged over time *(74)*. Taken together, these data suggest that ACTH-stimulated AA secretion is highly individualized but relatively constant within individuals, consistent with the behavior of an inherited factor. Whether this represents the inheritance of factors regulating adrenocortical biosynthesis or of factors determining AA metabolism or clearance remains unclear.

The heritability of AA secretion was also demonstrated by Legro and colleagues, who studied 119 brothers of 87 unrelated women with PCOS and 68 weight- and ethnicity-comparable unrelated control men *(75)*. Brothers of women with PCOS had significantly higher DHEAS levels compared to controls, with a significant positive linear relationship in DHEAS levels between PCOS probands and their brothers. These data suggested a familial clustering of elevated DHEAS levels in the families of PCOS women, suggesting that this may reflect an inherited abnormality in the disorder.

Overall, these data indicate that a wide variation in the ability of the adrenal to secrete DHEA, basally and in response to ACTH and compared to that of F, exists in the normal population. Consequently, it is possible that those women with a greater ability to secrete AAs may also be at greater risk for developing PCOS. Consequently, AA excess in PCOS may result from overrepresentation by those individuals who are "high AA secretors" among PCOS women. Overall, AA secretion, basally and in response to ACTH stimulation, is relatively constant over time, potentially representing an inherited trait and a risk factor for PCOS.

3. CONCLUSIONS

AA excess affects approximately 20–30% of PCOS patients, as reflected by the circulating DHEAS levels using age-matched normative values. Patients with PCOS demonstrate a generalized hypersecretion of adrenocortical products in response to ACTH stimulation, primarily because of hyperresponsivity of AAs to ACTH stimulation and to increased $\Delta^5$17-OH activity. The mechanisms underlying these abnormalities remain unclear. However, several factors, including altered F metabolism, increased ovarian steroids and hyperinsulinemia, or factors regulating glucose-mediated glucose disposal (glucose effectiveness) may play a role. Finally, we should note that circulating AA levels and their response to ACTH are highly individualized and relatively constant over time, such that the population has significant numbers of "high and low AA secretors." Consequently, AA excess and HPA axis dysfunction in PCOS may reflect inherited factors and hypothetically may reflect an overrepresentation by high AA secretors in this population, essentially indicating that individuals who are genetically determined to secrete high levels of AAs will be at greater risk for developing PCOS.

4. FUTURE AVENUES OF INVESTIGATION

Despite extensive work by many investigators, this field remains fertile for new and innovative research. Areas that should be explored include (1) ethnic and racial differences in AA secretion, (2) extra-adrenal regulation of AA biosynthesis and metabolism (e.g., DHEA-ST), including the long-term effect of sex steroids on HPA and AA secretion, (3) the role of insulin, insulin-mediated glucose disposal, and glucose-mediated glucose disposal (i.e., glucose effectiveness) in AA steroidogenesis in normal and PCOS individuals, (4) the role that altered F metabolism plays in the AA excess and HPA axis dysfunction observed in PCOS, (5) the molecular mechanism underlying the steroidogenic abnormalities observed in vivo (e.g., exaggerated $\Delta^5$17-OH activity observed), (6) the heritability of adrenocortical function in normal and PCOS individuals, and (7) the role of peripubertal AA excess in the development of PCOS.

KEY POINTS

- HPA dysfunction in the form of AA excess (generally indicated by elevated circulating DHEAS levels) is present in 20–30% of PCOS patients compared to age- and race-matched healthy women.
- PCOS women with AA excess have a generalized adrenocortical hypersecretion of adrenocortical products, basally and in response to ACTH stimulation, similar to the hypersecretion in response to human chorionic gonadotropin or GnRH analogs observed in the ovaries of these patients.
- AA excess in PCOS appears to be partially related to multiple extra-adrenal factors, including factors regulating glucose-mediated glucose disposal (glucose effectiveness), cortisol metabolism dysregulation, extra-adrenal sex steroids, or factors regulating insulin and glucose homeostasis.
- Circulating AA levels and their response to ACTH stimulation are highly individualized and relatively constant over time, suggesting that AA hypersecretion may be an inherited trait. Whether AA hypersecretion represents an inherited factor and whether it predisposes to PCOS remain to be determined.

REFERENCES

1. Azziz R, Woods KS, Reyna R, Key TJ, Knochenhauer ES, Yildiz BO. The prevalence and features of the polycystic ovary syndrome in an unselected population. J Clin Endocrinol Metab 2004;89:2745–2749.
2. Gallagher TF, Kappas A, Hellman L, Lipsett MB, Pearson OH, West CD. Adrenocortical hyperfunction in idiopathic hirsutism and the Stein-Leventhal syndrome. J Clin Invest 1958;37:794–799.
3. Wild RA, Umstot ES, Andersen RN, Ranney GB, Givens JR. Androgen parameters and their correlation with body weight in one hundred thirty-eight women thought to have hyperandrogenism. Am J Obstet Gynecol 1983;146:602–606.
4. Hoffman DI, Klove K, Lobo RA. The prevalence and significance of elevated dehydroepiandrosterone sulfate levels in anovulatory women. Fertil Steril 1984;42:76–81.
5. Steinberger E, Smith KD, Rodriguez-Rigau LJ. Testosterone, dehydroepiandrosterone, and dehydroepiandrosterone sulfate in hyperandrogenic women. J Clin Endocrinol Metab 1984;59:471–477.
6. Carmina E, Rosato F, Janni A. Increased DHEAs levels in PCO syndrome: evidence for the existence of two subgroups of patients. J Endocrinol Invest 1986;9:5–9.
7. Price FV, Legro RS, Watt-Morse M, Kaplan SS. Chediak–Higashi syndrome in pregnancy. Obstet Gynecol 1992;79:804–806.
8. Levin JH, Carmina E, Lobo RA. Is the inappropriate gonadotropin secretion of patients with polycystic ovary syndrome similar to that of patients with adult-onset congenital adrenal hyperplasia? Fertil Steril 1991;56:635–640.
9. Carmina E, Lobo RA. Ovarian suppression reduces clinical and endocrine expression of late-onset congenital adrenal hyperplasia due to 21-hydroxylase deficiency. Fertil Steril 1994;62:738–743.
10. Dewailly D, Vantyghem-Haudiquet MC, Sainsard C, et al. Clinical and biological phenotypes in late-onset 21-hydroxylase deficiency. J Clin Endocrinol Metab 1986;63:418–423.
11. Moran C, Azziz R, Carmina E, et al. 21-Hydroxylase-deficient nonclassic adrenal hyperplasia is a progressive disorder: a multicenter study. Am J Obstet Gynecol 2000;183:1468–1474.
12. Lobo RA, Goebelsmann U, Horton R. Evidence for the importance of peripheral tissue events in the development of hirsutism in polycystic ovary syndrome. J Clin Endocrinol Metab 1983;57:393–397.
13. Ibanez L, Potau N, Virdis R, et al. Postpubertal outcome in girls diagnosed of premature pubarche during childhood: increased frequency of functional ovarian hyperandrogenism. J Clin Endocrinol Metab 1993;76:1599–1603.
14. Miller D, Emans SJ, Kohane I. Follow-up study of adolescent girls with a history of premature pubarche. J Adolesc Health 1996;18:301–305.
15. Ibanez L, Potau N, Zampolli M, Street ME, Carrascosa A. Girls diagnosed with premature pubarche show an exaggerated ovarian androgen synthesis from the early stages of puberty: evidence from gonadotropin-releasing hormone agonist testing. Fertil Steril 1997;67:849–855.
16. Meas T, Chevenne D, Thibaud E, et al. Endocrine consequences of premature pubarche in post-pubertal Caucasian girls. Clin Endocrinol (Oxf) 2002;57:101–106.
17. Korth-Schutz S, Levine LS, New MI. Dehydroepiandrosterone sulfate (DS) levels, a rapid test for abnormal adrenal androgen secretion. J Clin Endocrinol Metab 1976;42:1005–1013.
18. Lobo RA, Paul WL, Goebelsmann U. Dehydroepiandrosterone sulfate as an indicator of adrenal androgen function. Obstet Gynecol 1981;57:69–73.
19. Feher T, Poteczin E, Bodrogi L. Relationship between serum dehydroepiandrosterone sulphate and urinary 17-ketosteroid values. Exp Clin Endocrinol 1985;85:209–216.

20. Abraham GE, Chakmakjian ZH. Serum steroid levels during the menstrual cycle in a bilaterally adrenalectomized woman. J Clin Endocrinol Metab 1973;37:581–587.
21. Abraham GE. Ovarian and adrenal contribution to peripheral androgens during the menstrual cycle. J Clin Endocrinol Metab 1974;39:340–346.
22. Vermeulen A. Androgen secretion by adrenals and gonads. In: Mahesh VB, Greenblatt RM, eds. Hirsutism and Virilism. Boston: John Wright/PSG Inc., 1983:17.
23. Nieschlag E, Loriaux DL, Ruder HJ, Zucker IR, Kirschner MA, Lipsett MB. The secretion of dehydroepiandrosterone and dehydroepiandrosterone sulphate in man. J Endocrinol 1973;57:123–134.
24. Mikulecky M, Kreze A, Putz Z, Moravcik M. Daily variation of serum cortisol, 17-hydroxyprogesterone and five androgens in healthy women. Braz J Med Biol Res 1995;28:485–490.
25. Rosenfeld RS, Rosenberg BJ, Fukushima DK, Hellman L. 24-Hour secretory pattern of dehydroisoandrosterone and dehydroisoandrosterone sulfate. J Clin Endocrinol Metab 1975;40:850–855.
26. Wang DY, Bulbrook RD, Sneddon A, Hamilton T. The metabolic clearance rates of dehydroepiandrosterone, testosterone and their sulphate esters in man, rat and rabbit. J Endocrinol 1967;38:307–318.
27. Gant NF, Hutchinson HT, Siiteri PK, MacDonald PC. Study of the metabolic clearance rate of dehydroisoandrosterone sulfate in pregnancy. Am J Obstet Gynecol 1971;111:555–563.
28. Baulieu EE, Corpechot C, Dray F, et al. An adrenal-secreted "androgen": dehydroisoandrosterone sulfate. Its metabolism and a tentative generalization on the metabolism of other steroid conjugates in man. Recent Prog Horm Res 1965;21:411–500.
29. Longcope C. Dehydroepiandrosterone metabolism. J Endocrinol 1996;150(suppl):S125–127.
30. Legrain S, Massien C, Lahlou N, et al. Dehydroepiandrosterone replacement administration: pharmacokinetic and pharmacodynamic studies in healthy elderly subjects. J Clin Endocrinol Metab 2000;85:3208–3217.
31. Kumar A, Woods KS, Bartolucci AA, Azziz R. Prevalence of adrenal androgen excess in patients with the polycystic ovary syndrome (PCOS). Clin Endocrinol (Oxf) 2005;62:644–649.
32. Azziz R, Koulianos G. Adrenal androgens and reproductive aging in females. Semin Reprod Endo 1991;9:249–260.
33. Moran C, Knochenhauer E, Boots LR, Azziz R. Adrenal androgen excess in hyperandrogenism: relation to age and body mass. Fertil Steril 1999;71:671–674.
34. Carmina E, Koyama T, Chang L, Stanczyk FZ, Lobo RA. Does ethnicity influence the prevalence of adrenal hyperandrogenism and insulin resistance in polycystic ovary syndrome? Am J Obstet Gynecol 1992;167:1807–1812.
35. Huerta R, Dewailly D, Decanter C, Knochenhauer ES, Boots LR, Azziz R. 11-Beta-hydroxyandrostenedione and delta5-androstenediol as markers of adrenal androgen production in patients with 21-hydroxylase-deficient nonclassic adrenal hyperplasia. Fertil Steril 1999;72:996–1000.
36. Sanchez LA, Moran C, Reyna R, Ochoa T, Boots LR, Azziz R. Adrenal progestogen and androgen production in 21-hydroxylase-deficient nonclassic adrenal hyperplasia is partially independent of adrenocorticotropic hormone stimulation. Fertil Steril 2002;77:750–753.
37. Azziz R, Gay FL, Potter SR, Bradley E Jr, Boots LR. The effects of prolonged hypertestosteronemia on adrenocortical biosynthesis in oophorectomized women. J Clin Endocrinol Metab 1991;72:1025–1030.
38. Carmina E, Levin JH, Malizia G, Lobo RA. Ovine corticotropin-releasing factor and dexamethasone responses in hyperandrogenic women. Fertil Steril 1990;54:245–250.
39. Horrocks PM, Kandeel FR, London DR, et al. Acth function in women with the polycystic ovarian syndrome. Clin Endocrinol (Oxf) 1983;19:143–150.
40. Carmina E, Lobo RA. Pituitary-adrenal responses to ovine corticotropin-releasing factor in polycystic ovary syndrome and in other hyperandrogenic patients. Gynecol Endocrinol 1990;4:225–232.
41. Azziz R, Black V, Hines GA, Fox LM, Boots LR. Adrenal androgen excess in the polycystic ovary syndrome: sensitivity and responsivity of the hypothalamic-pituitary-adrenal axis. J Clin Endocrinol Metab 1998;83:2317–2323.
42. Azziz R, Bradley E, Jr., Huth J, Boots LR, Parker CR Jr, Zacur HA. Acute adrenocorticotropin-(1-24) (ACTH) adrenal stimulation in eumenorrheic women: reproducibility and effect of ACTH dose, subject weight, and sampling time. J Clin Endocrinol Metab 1990;70:1273–1279.
43. Azziz R. Abnormalities of adrenocortical steroidogenesis in PCOS. In: Azziz R, Nestler JE, Dewailly D, eds. Androgen Excess Disorders in Women. Philadelphia: Lippincott-Raven, 1997:403–414.
44. Azziz R, Sanchez LA, Knochenhauer ES, et al. Androgen excess in women: experience with over 1000 consecutive patients. J Clin Endocrinol Metab 2004;89:453–462.
45. Lutfallah C, Wang W, Mason JI, et al. Newly proposed hormonal criteria via genotypic proof for type II 3beta-hydroxysteroid dehydrogenase deficiency. J Clin Endocrinol Metab 2002;87:2611–2622.
46. Joehrer K, Geley S, Strasser-Wozak EM, et al. CYP11B1 mutations causing non-classic adrenal hyperplasia due to 11 beta-hydroxylase deficiency. Hum Mol Genet 1997;6:1829–1834.
47. Lobo RA, Goebelsmann U. Evidence for reduced 3 beta-ol-hydroxysteroid dehydrogenase activity in some hirsute women thought to have polycystic ovary syndrome. J Clin Endocrinol Metab 1981;53:394–400.
48. Pang SY, Lerner AJ, Stoner E, et al. Late-onset adrenal steroid 3 beta-hydroxysteroid dehydrogenase deficiency. I. A cause of hirsutism in pubertal and postpubertal women. J Clin Endocrinol Metab 1985;60:428–439.

49. Siegel SF, Finegold DN, Lanes R, Lee PA. ACTH stimulation tests and plasma dehydroepiandrosterone sulfate levels in women with hirsutism. N Engl J Med 1990;323:849–854.

50. Azziz R, Bradley EL, Jr., Potter HD, Boots LR. 3 Beta-hydroxysteroid dehydrogenase deficiency in hyperandrogenism. Am J Obstet Gynecol 1993;168:889–895.

51. Carbunaru G, Prasad P, Scoccia B, et al. The hormonal phenotype of Nonclassic 3 beta-hydroxysteroid dehydrogenase (HSD3B) deficiency in hyperandrogenic females is associated with insulin-resistant polycystic ovary syndrome and is not a variant of inherited HSD3B2 deficiency. J Clin Endocrinol Metab 2004;89:783–794.

52. Moran C, Reyna R, Boots LS, Azziz R. Adrenocortical hyperresponsiveness to corticotropin in polycystic ovary syndrome patients with adrenal androgen excess. Fertil Steril 2004;81:126–131.

53. Kahsar-Miller M, Boots LR, Bartolucci A, Azziz R. Role of a CYP17 polymorphism in the regulation of circulating dehydroepiandrosterone sulfate levels in women with polycystic ovary syndrome. Fertil Steril 2004;82:973–975.

54. Walker BR, Rodin A, Taylor NF, Clayton RN. Endogenous inhibitors of 11beta-hydroxysteroid dehydrogenase type 1 do not explain abnormal cortisol metabolism in polycystic ovary syndrome. Clin Endocrinol (Oxf) 2000;52:77–80.

55. Stewart PM, Shackleton CH, Beastall GH, Edwards CR. 5 Alpha-reductase activity in polycystic ovary syndrome. Lancet 1990;335:431–433.

56. Tsilchorozidou T, Honour JW, Conway GS. Altered cortisol metabolism in polycystic ovary syndrome: insulin enhances 5alpha-reduction but not the elevated adrenal steroid production rates. J Clin Endocrinol Metab 2003;88:5907–5913.

57. Hines GA, Smith ER, Azziz R. Influence of insulin and testosterone on adrenocortical steroidogenesis in vitro: preliminary studies. Fertil Steril 2001;76:730–735.

58. Carmina E, Gonzalez F, Chang L, Lobo RA. Reassessment of adrenal androgen secretion in women with polycystic ovary syndrome. Obstet Gynecol 1995;85:971–976.

59. Ditkoff EC, Fruzzetti F, Chang L, Stancyzk FZ, Lobo RA. The impact of estrogen on adrenal androgen sensitivity and secretion in polycystic ovary syndrome. J Clin Endocrinol Metab 1995;80:603–607.

60. Slayden SM, Crabbe L, Bae S, Potter HD, Azziz R, Parker CR, Jr. The effect of 17 beta-estradiol on adrenocortical sensitivity, responsiveness, and steroidogenesis in postmenopausal women. J Clin Endocrinol Metab 1998;83:519–524.

61. Polderman KH, Gooren LJ, van der Veen EA. Effects of gonadal androgens and oestrogens on adrenal androgen levels. Clin Endocrinol (Oxf) 1995;43:415–421.

62. Futterweit W, Green G, Tarlin N, Dunaif A. Chronic high-dosage androgen administration to ovulatory women does not alter adrenocortical steroidogenesis. Fertil Steril 1992;58:124–128.

63. Marin CM, Bartolucci A, Azziz R. Prevalence of insulin resistance in polycystic ovary syndrome (PCOS) patients using the homeostatic measurement assessment (HOMA-IR). Fertil Steril 2003;80:274–275.

64. Carmina E, Lobo RA. Use of fasting blood to assess the prevalence of insulin resistance in women with polycystic ovary syndrome. Fertil Steril 2004;82:661–665.

65. Barbieri RL, Makris A, Randall RW, Daniels G, Kistner RW, Ryan KJ. Insulin stimulates androgen accumulation in incubations of ovarian stroma obtained from women with hyperandrogenism. J Clin Endocrinol Metab 1986;62:904–910.

66. Moghetti P, Castello R, Negri C, et al. Insulin infusion amplifies 17 alpha-hydroxycorticosteroid intermediates response to adrenocorticotropin in hyperandrogenic women: apparent relative impairment of 17,20-lyase activity. J Clin Endocrinol Metab 1996;81:881–886.

67. Azziz R, Ehrmann DA, Legro RS, Fereshetian AG, O'Keefe M, Ghazzi MN. Troglitazone decreases adrenal androgen levels in women with polycystic ovary syndrome. Fertil Steril 2003;79:932–937.

68. Arlt W, Auchus RJ, Miller WL. Thiazolidinediones but not metformin directly inhibit the steroidogenic enzymes P450c17 and 3beta-hydroxysteroid dehydrogenase. J Biol Chem 2001;276:16767–16771.

69. Azziz R, Ehrmann D, Legro RS, et al. Troglitazone improves ovulation and hirsutism in the polycystic ovary syndrome: a multicenter, double blind, placebo-controlled trial. J Clin Endocrinol Metab 2001;86:1626–1632.

70. Falcone T, Finegood DT, Fantus IG, Morris D. Androgen response to endogenous insulin secretion during the frequently sampled intravenous glucose tolerance test in normal and hyperandrogenic women. J Clin Endocrinol Metab 1990;71:1653–1657.

71. Farah-Eways L, Reyna R, Knochenhauer ES, Bartolucci AA, Azziz R. Glucose action and adrenocortical biosynthesis in women with polycystic ovary syndrome. Fertil Steril 2004;81:120–125.

72. Rotter JI, Wong FL, Lifrak ET, Parker LN. A genetic component to the variation of dehydroepiandrosterone sulfate. Metabolism 1985;34:731–736.

73. Azziz R, Fox LM, Zacur HA, Parker CR, Jr., Boots LR. Adrenocortical secretion of dehydroepiandrosterone in healthy women: highly variable response to adrenocorticotropin. J Clin Endocrinol Metab 2001;86:2513–2517.

74. Yildiz BO, Woods KS, Stanczyk F, Bartolucci A, Azziz R. Stability of adrenocortical steroidogenesis over time in healthy women and women with polycystic ovary syndrome. J Clin Endocrinol Metab 2004;89:5558–5562.

75. Legro RS, Kunselman AR, Demers L, Wang SC, Bentley-Lewis R, Dunaif A. Elevated dehydroepiandrosterone sulfate levels as the reproductive phenotype in the brothers of women with polycystic ovary syndrome. J Clin Endocrinol Metab 2002;87:2134–2138.

20

The Genetic Basis of the Polycystic Ovary Syndrome

Mark O. Goodarzi

SUMMARY

Polycystic ovary syndrome (PCOS) is a common, complex genetic disorder. Its inherited basis was established by studies demonstrating an increased prevalence of PCOS, hyperandrogenemia, insulin resistance, and disordered insulin secretion in relatives of women with PCOS. To date, efforts to elucidate the genetic basis of PCOS have focused on candidate genes chosen from logical pathways, including steroid synthesis and action, insulin sensitivity and secretion, obesity and fuel regulation, gonadotropin production and action, and, most recently, cardiovascular risk modifiers. Although several positive results have been reported, no gene or genes are universally accepted as important in PCOS pathogenesis, largely because of lack of replication of positive results. This has resulted, in part, from various factors, most importantly inadequate coverage of genes by the analysis of only one or two variants and of small study cohorts in many studies. In the future, optimal application of the candidate gene approach using haplotype-based analyses, intermediate phenotypes, and internal replication of positive results will enhance gene discovery in PCOS, as will the application of pharmacogenetics and whole-genome analysis.

Key Words: Polycystic ovary syndrome; candidate gene; genetic association; genetic linkage; single nucleotide polymorphism; haplotype; pharmacogenetics.

1. INTRODUCTION

Polycystic ovary syndrome (PCOS) is considered a common, complex genetic disorder, as are conditions such as schizophrenia, asthma, and type 2 diabetes (1). Such common diseases, including PCOS, appear to have a complex, multifactorial etiology, in which a variety of predisposing genes, not just one gene, interact with environmental factors to produce disease. Studies in families have demonstrated the heritable nature of PCOS itself as well as the component phenotypes of PCOS. Subsequently, a large number of population studies have attempted to discover genes that influence PCOS using the candidate gene approach.

2. BACKGROUND

2.1. Heritability of PCOS and Hyperandrogenemia

Family studies demonstrate that PCOS is significantly more prevalent among family members than in the general population (2). Among first-degree female relatives, of 93 patients with PCOS, 35% of premenopausal mothers and 40% of sisters on no hormonal therapy were also affected with the disorder (3). These affection rates are significantly higher than the 6–7% observed in the general population (4). In another study, 115 sisters of 80 women with PCOS were evaluated. Of these, 22% met the criteria for PCOS (5). An additional 24% of sisters had hyperandrogenemia with normal

From: *Contemporary Endocrinology: Androgen Excess Disorders in Women:*
Polycystic Ovary Syndrome and Other Disorders, Second Edition
Edited by: R. Azziz et al. © Humana Press Inc., Totowa, NJ

menses. Total and free testosterone levels were similar between the sisters with hyperandrogenemia only and the sisters and probands with PCOS. A bimodal distribution of testosterone levels in the sisters of women with PCOS was observed, suggesting a major genetic component to hyperandrogenemia *(5)*. Brothers of women with PCOS also display abnormal androgens: a study of such brothers found them to have elevated levels of dehydroepiandrosterone sulfate (DHEAS) *(6)*.

Physiological evidence that hyperandrogenemia in PCOS is genetically controlled is supported by the observation that ovarian theca cells removed from women with PCOS and propagated in culture display persistently elevated testosterone secretion compared to cells from unaffected women *(7)*. Because removal from the body and multiplication in culture removes the influence of an abnormal hormonal milieu from these cells, it is likely that persistent differences from normal represent intrinsic (genetic) defects.

2.2. Heritability of Insulin-Related Traits in PCOS

Not only is PCOS itself a heritable condition, but within PCOS insulin resistance and insulin secretion also appear to be under significant genetic control. Among sisters of women with PCOS, those who had PCOS or hyperandrogenemia with regular menses had lower insulin sensitivity than unaffected sisters, assessed by fasting insulin and glucose measurements *(8)*. Likewise, in families of Australian patients with PCOS, hyperinsulinemia was found to occur in 69% of all family members, suggesting that this trait was inherited *(9)*. One hundred and two relatives of 52 Turkish women with PCOS underwent assessment of insulin resistance in the fasting and post-glucose-challenge states; compared to population controls, mothers, sisters, and brothers of PCOS subjects had greater insulin resistance *(10)*. Brothers of Indian women with PCOS had insulin resistance and endothelial dysfunction *(11)*. In studies of families of women with PCOS, insulin secretion levels, quantified directly by the frequently sampled intravenous glucose tolerance test, displayed significant heritability, suggesting a genetic component to β-cell dysfunction in PCOS *(12)*. The most abnormal insulin secretion was observed in women with PCOS and a history of type 2 diabetes in a first-degree relative *(13)*.

Abnormal responses to insulin in cells that have been removed from women with PCOS suggest the presence of intrinsic cellular defects because these cells have been removed from the hormonal milieu present in these women. Lower insulin receptor substrate (IRS)-1-associated phosphatidylinositol 3-kinase (PI3K) activity and higher IRS-2 content were observed in the myocytes of PCOS patients compared to controls, in the face of similar amounts of IRS-1 and the p85 subunit of PI3K *(14)*. In PCOS adipocytes, the maximum glucose uptake stimulated by insulin was found by some investigators *(15)* to be lower compared to controls. Although others *(16)* found the maximum response to be normal, these latter investigators found that the sensitivity of the response to insulin was deficient in PCOS. The amount of glucose transporter-4 in adipocytes was lower in PCOS (on either a membrane protein or cell surface basis) than in controls *(15)*. Such defects are likely caused by abnormal expression or function of genes encoding products of the insulin-signaling pathway.

2.3. The Principle of Population Genetics

Given that investigators often start with no knowledge of genetic variants that lead to disease, they must take advantage of chromosomal markers. Markers, such as microsatellites and single-nucleotide polymorphisms (SNPs), are polymorphic variants interspersed throughout the genome. Microsatellites are tandem repeats of short nucleotide sequences, occurring with variable numbers of the repeated unit; SNPs are changes at a single genomic base pair, comprising two possible alleles. These markers are used as tags to track disease-causing variants or mutations. The underlying principle is that markers that are close to disease-causing variants tend to be inherited on the same chromosomes. Linkage refers to the situation wherein markers in a region of the genome are inherited in families in a nonrandom fashion in relation to a particular phenotype. Association refers to the situation wherein a particular allele of a marker is found with greater frequency in those with a particular phenotype.

2.4. Overview of the Candidate Gene Approach

One method of identifying disease genes is the candidate gene approach, in which common polymorphic genetic markers within a gene of interest, selected based on its hypothesized role in the disease, are evaluated to determine whether the polymorphisms are associated with the phenotype in populations or in families. This approach works because about 90% of variation is a result of common polymorphisms, most of which arose from single historical mutation events *(17)*. Each variant is associated with nearby variants that were present on the ancestral chromosome (or haplotype, the collection of variants existing together) on which the mutation occurred. These associations (referred to as linkage disequilibrium) allow the use of known markers to track down unknown disease mutations. In recent years, the availability of large numbers of DNA polymorphisms, both microsatellite repeats and SNPs, the development of the technology to perform relatively inexpensive, high-throughout genotyping, and improved statistical analysis approaches have made it feasible to comprehensively investigate the role of multiple candidate genes in disease susceptibility.

The candidate gene approach has resulted in the identification of a significant number of genes that contribute to susceptibility to various diseases. In type 2 diabetes mellitus (DM), for example, positive associations have been reported with the genes for the Kir6.2 subunit of the adenosine triphosphate (ATP)-sensitive potassium (K_{ATP}) channels of pancreatic βcells (*KCNJ11*), glycogen synthetase, IRS-1, peroxisome proliferator-activated receptor-γ (PPAR-γ 2), and PPAR-γ coactivator 1, among others *(18)*. Although none of these genes has been singly demonstrated to account for a large portion of diabetes susceptibility, replication in several studies suggests that these genes are indeed playing important roles in this disease.

2.5. Candidate Genes in PCOS: Progress and Challenges

To date, efforts to identify genes that influence PCOS susceptibility have largely utilized the candidate gene approach. Candidate genes studied in PCOS have generally targeted loci regulating five areas: (1) steroid biosynthesis and action (Table 1), (2) gonadotropic action (Table 2), (3) insulin secretion and action (Table 3), (4) weight and energy regulation (Table 3), and (5) cardiovascular factors (Table 4). The field has been comprehensively reviewed *(19)*. Several provocative genetic associations with PCOS have been reported that are slowly starting to illuminate the underlying causes of PCOS.

Despite repeated attempts to identify the putative gene or genes responsible for this disorder, the PCOS gene(s) remain elusive. As is evident in Table 4, despite many positive results, no gene or genes has clearly emerged as most important in PCOS, and many positive results were not confirmed in subsequent studies. Studies of the genetic etiology of PCOS have been hampered by various limitations, including (1) only one or two variants genotyped in each gene; (2) incomplete characterization of the phenotype in family members; (3) inability to assign a PCOS phenotype to prepubertal girls, postmenopausal women, and men; (4) possible inclusion of patients with nonclassic adrenal hyperplasia; (5) lack of appropriate controls; (6) unclear ethnic/racial composition; (7) varying criteria used to diagnose PCOS in different studies (in part because of the lack of universally accepted diagnostic criteria); and (8) small numbers of subjects in most studies. Regarding the latter, many of the studies in Tables 1–4 report results on fewer than 100 women with PCOS. Therefore, it is likely that many underpowered studies resulted in false-negative reports and that several small studies produced false-positive results.

The power issue is particularly relevant to common disease genetics. Validated genetic determinants of type 2 DM, such as the Pro12Ala variant of the *PPARG* gene and the Glu23Lys variant of the *KCNJ11* gene, only modestly alter risk for type 2 DM, on the order of 10–20% *(18)*. If genes with similar magnitude of effect influence PCOS risk, then many of the studies to date were seriously underpowered and inadequate to detect genetic variants predisposing to PCOS.

Table 1
Published Candidate Gene Studies in PCOS: Concerning Steroid Metabolism and Action

Genes	Comments
CYP17 (17α-hydroxylase/17,20-lyase)	A promoter polymorphism was associted with PCOS and hyperandrogenism in PCOS; however, this was not confirmed in several other studies. A microsatellite near *CYP17* was not associated with PCOS. No coding variants in *CYP17* were found in PCOS.
CYP11A (cholesterol side-chain cleavage enzyme)	A pentanucleotide VNTR was associated with PCOS in independent studies; however, other studies did not confirm this association, including one with a very large sample size.
AR (androgen receptor)	Decreased length of a polymorphic CAG repeat in exon 1 was found to be associated with hirsutism in Hispanic women and anovulation/infertility in women with PCOS. However, this polymorphism was not linked or associated with PCOS in other studies.
CYP21 (21-hydroxylase)	Heterozygosity for *CYP21* mutations was found more frequently in hyperandrogenic adolescents than controls. *CYP21* genotype was not found to be predictive of ovarian or adrenal hyperandrogenism in PCOS.
CYP19 (aromatase)	A linkage study and mutation screening were negative; however, an association study utilizing haplotypes showed association with precocious pubarche, PCOS, symptom score and testosterone level.
SHBG	Longer alleles of a promoter VNTR were associated with PCOS and lower SHBG levels.
H6PD (hexose-6-phosphate dehydrogenase)	Associated with PCOS and increased cortisol and 17-hydroxyprogesterone levels in one study.

Negative studies for linkage or association with PCOS or androgen levels have also been published regarding *HSD3B2*, *HSD17B3*, *UGT2B15*, SF-1, *StAR*, *DAX-1*, and *HSD11B1*.

VNTR, variable number tandem repeat; PCOS, polycystic ovary syndrome; SHBG, sex hormone-binding globulin.

Additional challenges face PCOS genetics as well as the genetics of other common, complex genetic disorders. Heterogeneity—in terms of multiple susceptibility genes, multiple alleles at a given gene, and possibly multiple subsets of PCOS with differing underlying pathophysiologies—confounds gene discovery. Gene-by-gene interactions that are likely to be important are difficult to model in genetic analyses of limited sample size. Environmental influences and gene-by-environment interactions are even more difficult to factor into genetic analyses. Local environmental and genetic differences between populations may explain why many results were found in one population but not another. For example, women with PCOS in the United States tend to have a greater body mass index than women with PCOS in Europe *(20)*. Because adiposity likely influences the expression and impact of various genes, this difference in body mass may have important implications for comparison of genetic results from the different regions.

Finally, inherent to the candidate gene approach are assumptions regarding the underlying pathophysiology of PCOS. This is a particular problem in PCOS, as the underlying causes are still fundamentally unknown. Candidate genes evaluated so far were selected from pathways affecting components of PCOS. Genes coding for transcription factors or signaling pathway components that may globally affect the organs involved in PCOS (pituitary, ovaries, adrenals, pancreatic β cells, insulin-responsive tissues) are unlikely to be selected as candidate genes by this approach.

Table 2
Published Candidate Gene Studies in PCOS: Gonadotropin Action and Regulation

Genes	Comments
Follistatin	Linkage of follistatin to PCOS has been demonstrated. Subsequent studies of follistatin in PCOS failed to discover any associated functional variants.
LH β-subunit	Coding variants were associated with elevated testosterone in healthy women and protection from developing PCOS in obese women. Three other studies showed no association of these variants with PCOS.
FSH β-subunit	A polymorphism in exon 3 was associated with PCOS; however, this variant does not alter the amino acid sequence.
Dopamine D3 receptor	Association with PCOS was found in a Hispanic population, but not in non-Hispanic white women.
FSH receptor	Several negative association studies have been reported. A positive study demonstrated association of coding variants with PCOS and responses to fertility treatment.

Studies have found no evidence for linkage and association of PCOS with several activin receptor subtypes, inhibins, the LH receptor, and the GnRH receptor.

PCOS, polycystic ovary syndrome; LH, luteinizing hormone; FSH, follicle-stimulating hormone, GnRH, gonadotropin-releasing hormone.

3. CONCLUSIONS

The inherited nature of PCOS, as well as hyperandrogenism and abnormal insulin action and metabolism in PCOS, has been firmly established by family studies. This led to over 100 publications focusing on logically chosen candidate genes. Unfortunately, most of these studies were in small cohorts. Additional issues such as only one or two variants genotyped for each candidate gene, phenotypic misclassification, genetic and phenotypic heterogeneity, and ignorance of the underlying causes of PCOS have confounded many genetic studies of PCOS. As a result, despite a large number of positive reports, no particular gene is universally recognized as importantly contributing to PCOS risk.

4. FUTURE AVENUES OF INVESTIGATION

Various new approaches may assist in the elucidation of involved genes, including haplotype-based analyses, inclusion of intermediate phenotypes, internal replication, whole-genome scan, and pharmacogenetics.

4.1. Advantages of the Haplotype-Based Approach to Genetic Association Studies

Almost every candidate gene study in PCOS has assessed the effect of one or two variants in each gene. This provides only partial information on whether a gene is associated with PCOS. There is increasing evidence that genetic variation is best described by groups of associated polymorphisms (inherited together on the same chromosome), referred to as haplotypes. Haplotypes reflect global gene structure, encompassing chromosomal blocks that have remained unbroken by recombination during the population history of the gene. These haplotypes encompass both coding and regulatory elements that individually or in combination could be responsible for the differences in phenotype and therefore can serve as important surrogates in the early stages of gene finding. Identification of a haplotype associated with increased or decreased disease risk should facilitate identification of the actual functional variant that affects disease risk, because this variant should lie on chromosomes identified by that haplotype.

Haplotypes carry more information than the individual SNPs that comprise them. Haplotypes capture the majority of common variation in a gene; consequently, the use of haplotypes is more likely to

Table 3
Published Candidate Gene Studies in PCOS: Insulin Secretion and Action, Obesity, and Fuel Metabolism

Genes	Comments
Insulin	A promoter VNTR was linked and associated with PCOS and/or insulin sensitivity in women with polycystic ovaries in some studies but not others. Analysis of multiple data sets showed no association of the VNTR with PCOS or testosterone levels.
Insulin receptor	An SNP in the tyrosine kinase domain was associated with PCOS, particularly in lean women. Other studies of the insulin receptor in women with PCOS have identified only common, silent polymorphisms.
Microsatellite D19S884 (located 1 cM from the insulin receptor gene)	Two independent studies have reported linkage and association of the microsatellite marker D19S884 with PCOS. A study of Italian and Spanish women with PCOS failed to confirm this association.
IRS1 and IRS2 (insulin receptor substrates-1 and -2)	Variants of both IRS1 and IRS2 were found to influence fasting insulin and postload glucose levels, respectively, in women with PCOS. IRS1 was associated with PCOS and with adolescent hyperandrogenism and obese insulin-resistant PCOS. IRS1 genotype influenced response to metformin in PCOS. IRS1 was not found to be linked or associated with PCOS or hyperandrogenemia in other studies.
CAPN10 (calpain-10)	One study found the 112/121 haplotype combination associated with a 2-fold increased risk of PCOS, but a larger study showed no such association. Two different studies from Spain produced conflicting results regarding association of SNP-44 with PCOS.
PPARG (PPAR-γ)	In Caucasian but not African American women with PCOS, the Pro12Ala variant was found to be associated with less insulin resistance. A marker near PPARG was not linked or associated with PCOS. Pro12Ala influenced body mass index in hyperandrogenic adolescents. Pro12Ala was associated with insulin sensitivity and lower hirsutism in German PCOS patients. Pro12Ala associated with obesity and increased insulin sensitivity in Turkish PCOS cases and controls. Pro12Ala associated with PCOS in a Finnish study. Negative association studies of Pro12Ala were also reported.
Resistin	A promoter variant was not associated with PCOS; however, it was associated with body mass index in PCOS.
IGF2 (insulin-like growth factor-2	An SNP in IGF2 was associated with PCOS in a small study.
PPP1R3	A single variant was associated with insulin response to glucose challenge and hyperandrogenemia in PCOS.
Plasma cell membrane glycoprotein-1 (PC-1)	A coding SNP was associated with PCOS.

Negative studies have also been published concerning IGF-1, IGF-1 receptor, IGFBP1, IGFBP3, Leydig insulin-like protein 3, SORBS1, IGF-2 receptor, and PTP1B. Negative linkage and association studies of genes relating to obesity and fuel metabolism in PCOS have been published concerning leptin receptor, glucocorticoid receptor, glycogen synthase, melanocortin 4 receptor, propiomelanocortin, and uncoupling proteins 2 and 3. PCOS, polycystic ovary syndrome; VNTR, variable number tandem repeat; SNP, single nucleotide polymorphism.

Table 4
Published Candidate Gene Studies in PCOS: Cardiovascular Disease

Genes	Comments
Paraoxonase	An SNP in paraoxonase was associated with PCOS in a small study.
PAI-1 (plasminogen activator inhibitor-1)	A promoter variant was associated with PCOS and increased PAI-1 levels. Negative association of PAI-1 with PCOS was also reported.
IL-6 (interleukin-6)	Two promoter SNPs were associated with PCOS. In another study, one of these SNPs was not associated with PCOS but was associated with BMI, testosterone, and response to glucose challenge.
IL-6 receptor complex	A polymorphism in the gp130 subunit was associated with hyperandrogenism.
Adiponectin	Positive associations with PCOS and fasting insulin in PCOS and negative association have been reported. Not associated with PCOS but associated with body mass index and hyperinsulinemia within PCOS.
EPHX (microsomal epoxide hydrolase)	A haplotype made up of only two coding SNPs was associated with PCOS.
Aldosterone synthatase	A promoter SNP was associated with PCOS and increased renin-angiotensin system activity.
Tumor necrosis factor receptor-2	A coding SNP was associated with PCOS in a small study.
Matrix metalloproteinase-1	Associated with PCOS in a small study.
Factor V	Factor V Leiden mutation found associated and not associated with PCOS in small studies from the same center.

Negative association studied have been reported for the β-adrenergic receptor, TNF-α (tumor necrosis factor-α), methylene tetrahydrofolate reductase, prothrombin, and apolipoprotein E genes.
PCOS, polycystic ovary syndrome; SNP, single nucleotide polymorphism.

identify gene variations than is the use of random SNPs *(21)*. Gabriel et al. *(22)* sequenced 13 megabases across the genome in subjects from Africa, Europe, and Asia. They showed that the human genome is organized in haplotype blocks (most of which are longer than 10 kilobases), with three to five commonly occurring (>5%) haplotypes per block. Only six to eight variants were sufficient to define the most common haplotypes in each block. Thus, a manageable number of appropriately chosen SNPs (termed haplotype-tagging SNPs [htSNPs]) can be genotyped to identify the most common haplotypes in a population, providing critical tools for association studies. The goal of the International HapMap Project is to delineate haplotype-tagging SNPs in all human genes *(23)*, which will greatly facilitate future haplotype-based association studies.

As discussed above, prior candidate gene studies in PCOS genotyped only one or a few SNPs per gene, indicating only incomplete coverage of each candidate gene. This is particularly true for larger genes that may contain multiple haplotype blocks. Application of the haplotype approach to PCOS genetics, particularly for genes wherein functional variants are unknown, should reduce the number of false-negative studies and may allow more positive findings to be replicated. To date, only studies of the calpain-10 gene in PCOS utilized haplotypes, based on the htSNPs from the original report associating calpain-10 with type 2 diabetes *(24)*. A few other PCOS studies constructed haplotypes from only two variants, unlikely to fully characterize haplotype blocks. Only recently did another gene, *CYP19* (aromatase), undergo haplotype analysis in PCOS. Previously, a study utilizing microsatellites in and around *CYP19* did not show linkage with PCOS *(25)*, and mutation screening in the exons and promoter failed to discover any variants *(26)*. In contrast, the association study utilizing haplotypes showed association with precocious pubarche (an antecedent of PCOS) as well

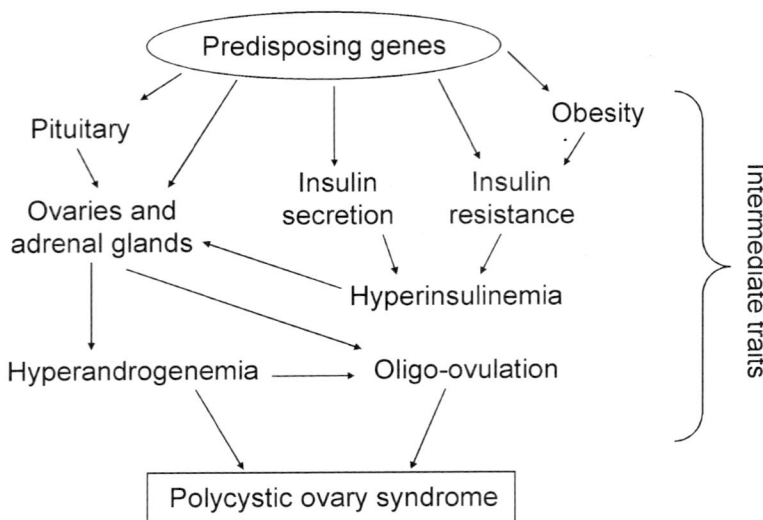

Fig. 1. Component (intermediate) traits in polycystic ovary syndrome (PCOS). In this conception of the pathophysiology of PCOS, it is apparent that intermediate traits are closer in the pathway to causative genes. In addition, these are quantitative variables that avoid the arbitrary nature of the qualitative variable of presence or absence of PCOS. Thus, genetic analyses utilizing intermediate traits may be powerful tools in gene discovery in PCOS.

as PCOS *(27)*. This demonstrates the potentially greater ability to discover genetic effects when using haplotypes to capture global variation across a gene.

4.2. Intermediate Phenotypes as the Focus of Genetic Study

Given that there are no universally accepted diagnostic criteria for PCOS, it is prudent for genetic analyses to focus on component (intermediate) phenotypes such as androgen levels, insulin resistance, and their related traits. Such intermediate phenotypes are likely to be closer to the basic abnormality, that is, action of predisposing genes, than the end-point disease (*see* Fig. 1). Another advantage of using intermediate phenotypes is that they can be analyzed as continuous variables. Thus, there is no arbitrary cut point to be imposed, and everyone with the quantitative trait measures (including healthy relatives and controls and males for insulin-related traits) can be used in the analyses. The latter increases the statistical power in identifying susceptibility genes in association studies. An advantage to measuring intermediate phenotypes in healthy subjects, especially at-risk relatives, is that it avoids secondary changes in phenotype caused by the disease itself or its therapy, which can obscure the detection of genetic influences.

Androgen-related intermediate phenotypes in PCOS include free or bioavailable testosterone, dehydroepiandrosterone (DHEA) and DHEAS, androstenedione and other androgens, sex hormone-binding globulin, and the modified Ferriman–Gallwey hirsutism score. Insulin-related phenotypes include fasting and postload glucose and insulin measurements, insulin sensitivity, and insulin secretion indices. Obesity is quantified by waist-to-hip ratio, body mass index, and body fat distribution measured by dual x-ray absorptiometry or computed tomography scanning. Of note, the most accurate phenotyping (but also the most expensive and labor-intensive) is performed with detailed physiological studies such as the euglycemic hyperinsulinemic clamp or frequently sampled intravenous glucose tolerance test for insulin-related traits (as opposed to fasting or oral glucose tolerance test measurements). There is a clear trend in the more recent candidate gene studies that association with PCOS is evaluated along with association with component phenotypes within PCOS (and sometimes also within controls).

4.3. Internal Replication as Validation of Positive Results

Genetic epidemiology is often criticized because positive reports of linkage or association are not subsequently confirmed by others. Besides false-positive and false-negative studies, such lack of replication may be a result of the study of different ethnic groups: a certain genetic variant may interact with other variants and local environmental influences such that it alters phenotype only in a particular group. Thus, replication studies should first be carried out in the same ethnic group, with the goal of validating the initial result. Subsequent replication attempts in other populations would serve to determine whether the particular genetic variant affects disease susceptibility universally. However, even when ostensibly the same ethnic group is studied, subtle differences in the history of the population may lead to ethnic differences such that two nominally similar cohorts are sufficiently different in genetic background to limit replication.

Additionally, a valid confirmation test should be carried out with the same phenotypes in a cohort from the same population, ascertained by identical criteria. This has been done only rarely in the field of common metabolic disorders. One example is the FUSION (Finland–United States Investigation of NIDDM Genetics) study, wherein an initial genome scan for type 2 diabetes was carried out in almost 500 affected sib-pair Finnish families (FUSION 1) *(28)* with the identification of 10 suggestive regions linked to type 2 diabetes. In FUSION 2, an independent set of 242 families ascertained in the same way *(29)*, four regions (on chromosomes 6, 11, 14, X) replicated loci found in FUSION 1, providing compelling evidence that these regions harbor genes for type 2 diabetes. Genetic studies in PCOS have generally not been replicated in this way, largely because of sample size limitations. With time and increasing subject recruitment, high-volume centers and/or collaborative consortia will be able to reassess their own positive results, providing critical validation of these results and prioritizing them for study by other groups.

4.4. The Whole-Genome Scan Approach to Gene Discovery

Linkage scans of the entire genome are often carried out using a panel of microsatellites (~400) that cover the whole genome. The key benefit of whole-genome approaches is that no prior knowledge or assumption regarding the underlying pathogenetic mechanism of disease is required (in contrast to the candidate gene approach). A challenge in conducting whole-genome linkage analyses is the need to recruit large numbers of families. In addition, linkage analyses are statistically less sensitive than association analyses. Thus, if genetic variants have small to modest effects on disease risk, they may not be detected by linkage analyses. Furthermore, positive linkage signals identify large chromosomal regions that often contain hundreds of genes. Fine-mapping with additional markers or selection of positional candidate genes is often performed to pursue positive linkage signals.

To date, no genome-wide linkage analysis has been published for PCOS. The main challenge has been recruiting enough families to conduct a linkage analysis with sufficient power. Investigators are actively pursuing this goal; genome-wide scans in PCOS are anticipated in the near future.

With further advances in high-throughput genotyping and statistical analysis, another option in the future will be whole-genome association studies, combining an assumption-free systematic search of the whole genome with the increased power of linkage disequilibrium (association) over allele sharing (linkage) for gene discovery.

4.5. Pharmacogenetics as a Probe to Understanding Disease Pathophysiology

Pharmacogenetics seeks to determine whether genetic variation influences response to drug therapy. Variants in drug-metabolizing enzymes may alter therapeutic response. Alternatively, variants in genes coding for components of key biological pathways may affect drug response. Knowledge of such variants allows prediction of therapeutic response and adverse effects. Pharmacogenetics may also be used as a probe to understanding disease pathophysiology. Drug therapy is a controlled environmental stimulus designed to provoke a genetically determined response. Without this physiological challenge, these genetic variants may go undetected. In PCOS, the first pharmacogenetic study published observed that the Gly972Arg variant in the insulin receptor substrate-1 (*IRS1*) gene influenced response to metformin in terms of insulin and androgen levels *(30)*. This implicates IRS-1 as an important factor in the pathogenesis of PCOS, particularly as a participant in insulin resistance in PCOS. Future application of pharmacogenetics to PCOS will shed further light on its genetic determinants and may in the future lead to clinical genotyping to assist in decisions as to whom to treat and what agent(s) to use.

KEY POINTS

- PCOS clearly runs in families.
- Within PCOS, hyperandrogenemia and abnormalities in insulin action and secretion are also heritable.
- The candidate gene approach has been extensively applied to PCOS, with many intriguing positive results; however, to date, no genes have emerged as predominant.
- In the future, application of haplotypes, intermediate phenotypes, internal replication, genome-wide scans, and pharmacogenetics will significantly assist gene-discovery efforts in PCOS.

ACKNOWLEDGMENTS

Supported in part by an endowment from the Helping Hand of Los Angeles and by grants RO1-HD29364 and K24-D01346 from the National Institutes of Health.

REFERENCES

1. King RA, Rotter JI, Motulsky AG. The Genetic Basis of Common Diseases, 2nd ed. New York: Oxford University Press, 2002.
2. Legro RS, Strauss JF. Molecular progress in infertility: polycystic ovary syndrome. Fertil Steril 2002;78:569–576.
3. Kahsar-Miller MD, Nixon C, Boots LR, Go RC, Azziz R. Prevalence of polycystic ovary syndrome (PCOS) in first-degree relatives of patients with PCOS. Fertil Steril 2001;75:53–58.
4. Azziz R, Woods KS, Reyna R, Key TJ, Knochenhauer ES, Yildiz BO. The prevalence and features of the polycystic ovary syndrome in an unselected population. J Clin Endocrinol Metab 2004;89:2745–2749.
5. Legro RS, Driscoll D, Strauss JF 3rd, Fox J, Dunaif A. Evidence for a genetic basis for hyperandrogenemia in polycystic ovary syndrome. Proc Natl Acad Sci USA 1998;95:14956–14960.
6. Legro RS, Kunselman AR, Demers L, Wang SC, Bentley-Lewis R, Dunaif A. Elevated dehydroepiandrosterone sulfate levels as the reproductive phenotype in the brothers of women with polycystic ovary syndrome. J Clin Endocrinol Metab 2002;87:2134–2138.
7. Nelson VL, Legro RS, Strauss JF, McAllister JM. Augmented androgen production is a stable steroidogenic phenotype of propagated theca cells from polycystic ovaries. Mol Endocrinol 1999;13:946–957.
8. Legro RS, Bentley-Lewis R, Driscoll D, Wang SC, Dunaif A. Insulin resistance in the sisters of women with polycystic ovary syndrome: association with hyperandrogenemia rather than menstrual irregularity. J Clin Endocrinol Metab 2002;87:2128–2133.
9. Norman RJ, Masters S, Hague W. Hyperinsulinemia is common in family members of women with polycystic ovary syndrome. Fertil Steril 1996;66:942–947.
10. Yildiz BO, Yarali H, Oguz H, Bayraktar M. Glucose intolerance, insulin resistance, and hyperandrogenemia in first degree relatives of women with polycystic ovary syndrome. J Clin Endocrinol Metab 2003;88:2031–2036.
11. Kaushal R, Parchure N, Bano G, Kaski JC, Nussey SS. Insulin resistance and endothelial dysfunction in the brothers of Indian subcontinent Asian women with polycystic ovaries. Clin Endocrinol (Oxf) 2004;60:322–328.
12. Colilla S, Cox NJ, Ehrmann DA. Heritability of insulin secretion and insulin action in women with polycystic ovary syndrome and their first degree relatives. J Clin Endocrinol Metab 2001;86:2027–2031.
13. Ehrmann DA, Sturis J, Byrne MM, Karrison T, Rosenfield RL, Polonsky KS. Insulin secretory defects in polycystic ovary syndrome. Relationship to insulin sensitivity and family history of non-insulin-dependent diabetes mellitus. J Clin Invest 1995;96:520–527.
14. Dunaif A, Wu X, Lee A, Diamanti-Kandarakis E. Defects in insulin receptor signaling in vivo in the polycystic ovary syndrome (PCOS). Am J Physiol Endocrinol Metab 2001;281:E392–E399.
15. Rosenbaum D, Haber RS, Dunaif A. Insulin resistance in polycystic ovary syndrome: decreased expression of GLUT-4 glucose transporters in adipocytes. Am J Physiol 1993;264:E197–E202.
16. Ciaraldi TP, el-Roeiy A, Madar Z, Reichart D, Olefsky JM, Yen SS. Cellular mechanisms of insulin resistance in polycystic ovarian syndrome. J Clin Endocrinol Metab 1992;75:577–583.

17. Kruglyak L, Nickerson DA. Variation is the spice of life. Nat Genet 2001;27:234–236.

18. Parikh H, Groop L. Candidate genes for type 2 diabetes. Rev Endocr Metab Disord 2004;5:151–176.

19. Escobar-Morreale HF, Luque-Ramirez M, San Millan JL. The molecular-genetic basis of functional hyperandrogenism and the polycystic ovary syndrome. Endocr Rev 2005;26:251–282.

20. Carmina E, Legro RS, Stamets K, Lowell J, Lobo RA. Difference in body weight between American and Italian women with polycystic ovary syndrome: influence of the diet. Hum Reprod 2003;18:2289–2293.

21. Judson R, Salisbury B, Schneider J, Windemuth A, Stephens JC. How many SNPs does a genome-wide haplotype map require? Pharmacogenomics 2002;3:379–391.

22. Gabriel SB, Schaffner SF, Nguyen H, et al. The structure of haplotype blocks in the human genome. Science 2002;296:2225–2229.

23. The International HapMap Consortium. The International HapMap Project. Nature 2003;426:789–796.

24. Horikawa Y, Oda N, Cox NJ, et al. Genetic variation in the gene encoding calpain-10 is associated with type 2 diabetes mellitus. Nat Genet 2000;26:163–175.

25. Gharani N, Waterworth DM, Batty S, et al. Association of the steroid synthesis gene CYP11a with polycystic ovary syndrome and hyperandrogenism. Hum Mol Genet 1997;6:397–402.

26. Soderlund D, Canto P, Carranza-Lira S, Mendez JP. No evidence of mutations in the P450 aromatase gene in patients with polycystic ovary syndrome. Hum Reprod 2005;20:965–969.

27. Petry CJ, Ong KK, Michelmore KF, et al. Association of aromatase (CYP 19) gene variation with features of hyperandrogenism in two populations of young women. Hum Reprod 2005;20:1837–1843.

28. Ghosh S, Watanabe RM, Valle TT, et al. The Finland-United States investigation of non-insulin-dependent diabetes mellitus genetics (FUSION) study. I. An autosomal genome scan for genes that predispose to type 2 diabetes. Am J Hum Genet 2000;67:1174–1185.

29. Silander K, Scott LJ, Valle TT, et al. A large set of Finnish affected sibling pair families with type 2 diabetes suggests susceptibility loci on chromosomes 6, 11, and 14. Diabetes 2004;53:821–829.

30. Ertunc D, Tok EC, Aktas A, Erdal EM, Dilek S. The importance of IRS-1 Gly972Arg polymorphism in evaluating the response to metformin treatment in polycystic ovary syndrome. Hum Reprod 2005;20:1207–1212.

21

Pediatric Hyperandrogenism and the Polycystic Ovary Syndrome in Adolescence

Kathleen M. Hoeger

SUMMARY

Signs of androgen excess in young women, such as hirstutism and acne, are not uncommon. Although causes of hyperandrogenism in young women include congenital adrenal hyperplasia, Cushing's syndrome, and androgen-producing tumors, the most common cause of androgen excess in this age group is polycystic ovary syndrome (PCOS). The same metabolic and endocrine abnormalities seen in adult women with PCOS are also manifest in adolescence. Evidence suggests that PCOS has a perimenarchal onset but may also be linked to endocrine dysfunction prior to adolescence. As in adults, insulin resistance may play a key role in adolescent PCOS. Further investigation of the impact of treatment options in adolescents with PCOS, particularly in the presence of obesity, is needed.

Key Words: Adolescence; polycystic ovary syndrome; obesity; insulin sensitizers; premature adrenarche.

1. INTRODUCTION

Polycystic ovary syndrome (PCOS) is one of the most common endocrine disorders in women of reproductive age, affecting 6–8% of women. Although the causes of hyperandrogenism in young women include congenital adrenal hyperplasia, Cushing's syndrome, and androgen-producing tumor, these make up a small minority of diagnoses. Most young women with hyperandrogenic symptoms will have PCOS. Most women diagnosed with PCOS have evidence of menstrual irregularities and hyperandrogenism that are evident around the time of menarche *(1)*. The etiology of the syndrome is unclear, although there is strong evidence for a genetic basis for the disease, with 50% of sisters of women with PCOS demonstrating hyperandrogenism or PCOS *(2)*. Although symptoms are manifest in adolescence, the syndrome may have its genesis in earlier development. Evidence suggests that low birthweight can be associated with precocious pubarche, hyperinsulinemia, ovarian hyperandrogenism, and anovulation, consistent with the diagnosis of PCOS in adolescence *(3)*.

The metabolic disturbances seen in adult women with PCOS, including insulin resistance, are also seen in young women. Adolescents with PCOS are insulin resistant compared to control adolescents matched for age, body fat distribution, and obesity *(4)*. A high prevalence of glucose intolerance is seen in obese adolescents with PCOS, similar to that demonstrated in adult women with the disorder *(5)*. Adolescents presenting with hyperandrogenic symptoms should therefore undergo both endocrine and metabolic evaluation for appropriate diagnosis and treatment early in the course of disease.

From: *Contemporary Endocrinology: Androgen Excess Disorders in Women:*
Polycystic Ovary Syndrome and Other Disorders, Second Edition
Edited by: R. Azziz et al. © Humana Press Inc., Totowa, NJ

2. BACKGROUND

There are a number of etiologies of hyperandrogenism in pediatric and adolescent populations, as detailed here.

2.1. Premature Adrenarche

Premature adrenarche (also known as premature pubarche) is defined as the development of pubic hair prior to age 8 years. Additionally, tall stature, axillary hair, apocrine odor, or acne may also be present. Premature adrenarche is distinguished from precocious puberty by the lack of accelerated bone maturation, virilization, and breast development. Premature adrenarche can present as the earliest sign of abnormal adrenal androgen production or androgen-secreting tumors, but is most commonly attributed to benign, early maturation of the zona reticularis of the adrenal gland. The precise cause of premature adrenarche is unknown, but retrospective data suggest that many girls initially evaluated for precocious pubarche show signs and symptoms of PCOS in midadolescence (6). In that series, 45% of the 35 girls initially evaluated for premature adrenarche had evidence of oligomenorrhea and evidence of ovarian hyperandrogenism when reexamined after menarche. This and other investigations suggest that premature adrenarche may be a precursor to the development of PCOS, although not all affected girls will develop PCOS. Long-term, prospective, controlled studies of girls with premature adrenarche have not been reported, and therefore the natural history of benign premature adrenarche as well as the link to, and risk factors for, the development of PCOS in adolescence remain to be clarified.

The development of premature adrenarche has been linked to higher insulin levels throughout puberty. Girls with premature adrenarche also have low circulating levels of sex hormone-binding globulin (SHBG) and insulin-like growth factor-binding proteins, which are associated with hyperinsulinemia (7). Many girls with premature adrenarche have acanthosis nigricans, which is also linked to insulin resistance. Studies of prepubertal girls with premature adrenarche and acanthosis nigricans demonstrate reduced insulin sensitivity (8). The majority of the study subjects, however, had an elevated body mass index (BMI), and changes in BMI may be an important regulator of adrenarche (9). In summary, girls with benign premature adrenarche have evidence of hyperinsulinemia, and in predisposed girls this may lead to the development of PCOS. Careful follow-up of these girls for signs of PCOS, with particular attention to prevention of obesity, is warranted.

2.2. Other Etiologies of Hyperandrogenism

Hyperandrogenism resulting from PCOS in pediatric and adolescent populations should be distinguished from other causes of androgen excess. In general, most are rare causes of androgen excess in younger populations. Clinical presentation and a careful history will distinguish most of these diagnoses. These are detailed below.

2.2.1. Nonclassic Congenital Adrenal Hyperplasia

Late-onset or nonclassic 21-hydroxylase-deficient congenital adrenal hyperplasia (NCAH), is recognized as a cause of androgen excess in adolescents, and the presenting features of NCAH are often similar to PCOS. In contrast to patients with classic congenital adrenal hyperplasia, most NCAH patients do not demonstrate cortisol deficiency, although there is excess accumulation of 21-hydroxylase precursors when adrenocorticotropic hormone (ACTH) stimulates the adrenal gland. These precursors are converted to androgens, resulting in a hyperandrogenic picture. Although the exact incidence of NCAH in a hyperandrogenic adolescent population varies by ethnic group and geographic region, it is estimated that 1–6% of patients presenting with hyperandrogenism may have NCAH. It may present as premature adrenarche or mimic PCOS in an adolescent population. Virilization is uncommon, and regular menstrual cycles are seen in approximately 50% of patients (10).

Evaluation of pediatric or adolescent patients presenting with hyperandrogenism should include a basal unstimulated 17-hydroxyprogesterone level. Levels should be obtained in the morning in the follicular phase of menstruating girls. Data suggest that a value of less than 2 ng/mL under these circumstances has a low false-negative rate. Patients having values greater than this under basal circumstances should undergo an ACTH-stimulation test, obtaining a 17-hydroxyprogesterone level 30–60 minutes after the administration of ACTH (1–24) *(11)*.

2.2.2. Rare Causes of Hyperandrogenism

Abrupt onset of hirsutism or virilization in a pediatric or adolescent patient should raise the suspicion of rare causes of hyperandrogenism such as androgen-secreting tumors of the adrenal gland or ovary. Such tumors are exceedingly rare, with a mean age of diagnosis in one series of 23.4 years *(12)*. Generally, a very high level of dehydroepiandrosterone sulfate (DHEAS) (>700 µg/dL) or testosterone (>200 ng/dL) may prompt an investigation, although not all virilizing tumors will be detected using serum hormone concentration cutoffs *(13)*. Rapid progression of androgenic symptoms, particularly in young patients, should also trigger an evaluation.

Cushing's syndrome is a rare cause of hyperandrogenism in young patients. In an 11-year period in Denmark, 168 cases of Cushing's syndrome were diagnosed. Only 9 occurred in patients younger than 20 years, and they were evenly split between males and females *(14)*. Clinical features often include those associated with androgen excess, but additionally, characteristic centripetal weight gain with abdominal striae, hypertension, as well as growth abnormalities is noted. Osteopenia may be seen. The need for further evaluation for Cushing's syndrome in young patients with hyperandrogenism is guided by the clinical presentation.

2.3. Polycystic Ovary Syndrome

2.3.1. Prevalence and Pathophysiology

There is growing recognition that onset of clinical PCOS occurs in the perimenarchal period. Studies in adult women suggest the prevalence in an unselected cohort to be about 6–8% *(15)*. The disorder is difficult to characterize in adolescence because of the higher prevalence of menstrual irregularity observed in the initial years following menarche. In a population sample of adolescents with a mean age of 15.3 years, 18.7% had oligomenorrhea, secondary amenorrhea, or irregular menstrual cycles. Within this group, 5.5% reported either amenorrhea or menstrual cycles of more than 42 days. Less than 2% of the amenorrheic or oligomenorrheic adolescents demonstrated a hormonal pattern consistent with hypogonadism. The majority of adolescent girls with oligo- or amenorrhea in this population had endocrine findings consistent with PCOS *(16)*.

Although the prevalence of PCOS in the population is high, the pathophysiology of PCOS remains controversial. The mechanism behind ovarian androgen excess is not entirely clear. It may be driven by abnormal neuroendocrine signaling, resulting in hypersecretion of LH or by peripheral metabolic disturbances of insulin resistance. Additionally, there may be intrinsic abnormalities in ovarian steroidogenesis that are exaggerated by both insulin and gonadotropin abnormalities.

Abnormal gonadotropin secretion is seen in a majority of women with PCOS, with elevated LH secretion as a characteristic finding. Adolescent women with hyperandrogenism have evidence of similar neuroendocrine abnormalities, with acceleration of LH pulse frequency seen in the perimenarchal period *(17)*. Mean 24-hour LH concentrations are increased, and an increased LH-to-follicle-stimulating hormone (FSH) ratio is noted. This LH hypersecretion is present in both thin and obese women with PCOS compared with weight-matched controls, although it appears abrogated in obese women *(18)*. Nonetheless, it is not clear whether the neuroendocrine disturbances are primary in the development of PCOS or secondary to the metabolic or hyperandrogenic disturbances seen.

It is now well accepted that a majority of women with PCOS, irrespective of body weight, are significantly insulin resistant. Insulin resistance increases during normal puberty, and the expression

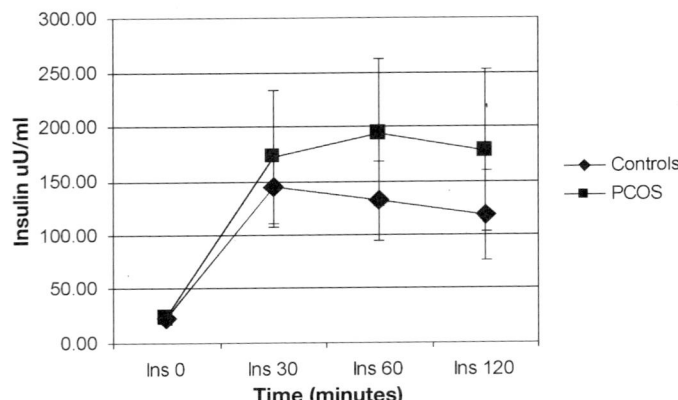

Fig. 1. Insulin concentrations for 2-hour oral glucose tolerance test from obese adolescents with polycystic ovary syndrome (PCOS) ($n = 39$) and age and body mass index (BMI)-matched control adolescents ($n = 28$). $p = 0.001$ for area under the curve insulin, PCOS vs controls. *$p < 0.05$ between groups for time point. (From Sukalich et al., unpublished data.)

of PCOS during adolescence may be facilitated by an exaggerated insulin response. Early differences in insulin resistance measures are noted in adolescents with PCOS compared to weight- and age-matched controls. In one study, 12 PCOS adolescents and 10 weight-matched control adolescents were studied with euglycemic-hyperinsulinemic clamps. Both groups were obese, and they had similar fasting glucose levels. Compared to the controls, adolescents with PCOS had significantly greater fasting insulin levels, and insulin sensitivity was significantly reduced *(4)*. Figure 1 presents data from our group on insulin area under the curve from an oral glucose tolerance test in a larger group of 39 obese adolescents with PCOS and 28 age- and weight-matched controls. The glycemic profiles of the two groups differed only slightly at the 2-hour time point (data not shown), but adolescents with PCOS demonstrated significantly higher insulin levels as measured by area under the curve.

The trigger for PCOS in the perimenarchal period is unclear. Puberty is normally associated with an increase in insulin secretion, and many features of normal puberty, such as increased LH pulsatility, increased adrenal and ovarian androgen production, and changes in SHBG, are exaggerated in PCOS adolescents. Additionally, there is an association with low birthweight and premature adrenarche with hyperinsulinism, both of which predispose to development of PCOS at puberty. This is true even in girls who are not obese *(19)*. Importantly, apparent restoration of insulin sensitivity in these adolescents reduces ovarian hyperandrogenism and restores regular menstrual cycles *(20)*. The role of hyperinsulinemia and insulin resistance in the development of PCOS, however, remains controversial.

2.3.2. Endocrine Changes and Clinical Features

Adolescents with PCOS demonstrate endocrine disturbances similar to those of their adult counterparts. Typical clinical features associated with the presentation of PCOS in adolescence include hirsutism and acne as manifestations of hyperandrogenism. These features cause significant distress in adolescence, and there is evidence that it is the adolescent's perception of the severity of her PCOS features, rather than the clinician's clinical score, that has the most impact on her quality of life *(21)*. The distribution and severity of hirsutism is generally quantified by use of the modified Ferriman–Gallwey score *(22)*. This scoring system was developed in a homogeneous ethnic population, and ethnic and racial variation in the expression of terminal hair is known to occur. Adjustments for these ethnic and racial differences should be taken into consideration in the clinical setting. Acne is another common clinical expression of hyperandrogenism seen in PCOS in adolescence. It may be the presenting feature in many adolescents, just as significant hirsutism may be a later finding.

Elevated total and/or free testosterone is a characteristic finding in adult women with PCOS, but normative data for androgen concentrations in adolescent females are sparse. When controlled for age and body weight, adolescents with PCOS have increased total and free testosterone as well as the adrenal androgen metabolite, DHEAS and reduced SHBG compared to adolescents with normal menstrual cycles and no clinical evidence of hyperandrogenism *(23)*. The range of concentration for these hormones in adolescents with PCOS overlaps with the normal range reported in adult women. Reliance on absolute endocrine markers for diagnosis of PCOS in adolescence is therefore not recommended.

Adolescents with PCOS exhibit increases in LH pulsatility compared with those shown in adult women with PCOS *(17)*. Generally, LH pulse frequency is increased and the amplitude of pulses is exaggerated. This is attenuated somewhat by increased BMI *(18)*, but adolescents with PCOS demonstrate hypersecretion of LH throughout adolescence compared to control adolescents. The increased LH secretion may compound the abnormal ovarian steroidogenesis by inducing stimulation of ovarian theca cells to overproduce androgens.

Menstrual disturbance is a classic hallmark of PCOS, although the normal maturational changes of the hypothalamic–pituitary axis during puberty that result in initially irregular cycles may make the identification of disturbance associated with PCOS difficult. Menarche can be delayed in adolescents with PCOS despite normal age of thelarche. In a prospective study of pubertal onset of menstrual disturbance from the Netherlands, approximately 10% of adolescent girls demonstrated irregular menstrual cycles *(16)*. However, in adolescent girls who were at least 6 months out from menarche, only 5.5% demonstrated menstrual cycles at greater than 42-day intervals. This study excluded girls on oral contraceptives and therefore may underestimate the prevalence of menstrual disturbance in this population. Nonetheless, when studied at least 6 months postmenarche, most adolescent girls will demonstrate relatively regular menstrual cycles, and disturbances beyond this point should warrant evaluation.

The 2003 jointly sponsored (European Society of Human Reproduction and Endocrinology [ESHRE]/American Society for Reproductive Medicine [ASRM]) consensus conference on the diagnostic criteria and sequelae of PCOS considered the appearance of the ovary as one of the key diagnostic features. The definition of polycystic ovaries included the "presence of 12 or more follicles in each ovary measuring 2–9 mm in diameter, and/or increased ovarian volume (>10 mL)" *(11)*. However, it is recognized that polycystic ovaries can be found in up to one-third of normally menstruating adolescents, and it is not yet clear that this finding predicts metabolic or reproductive consequences *(24)*. This finding alone in adolescent women should not be the sole criterion for a diagnosis of PCOS.

2.3.3. Metabolic Disturbances

Adult women with PCOS are at increased risk for type 2 diabetes mellitus (DM) and impaired glucose tolerance. Metabolic disturbance is already noted in adolescents with PCOS. Hyperinsulinemia and insulin resistance exists in young adolescents with PCOS compared with control adolescents *(4)*. In a study of 27 obese (mean BMI 38.4 kg/m^2) adolescents with PCOS (mean age 16.7 years) 33% demonstrated abnormal glucose tolerance on a 2-hour oral glucose challenge (i.e., a 2-hour glucose of no less than 140 mg/dL) *(5)*. Unpublished data from our recent examination of 35 obese (mean BMI 34.6 kg/m^2) adolescents with PCOS (mean age 15.6 years) demonstrated a 31.4% incidence of abnormal glucose tolerance defined by an abnormal 2-hour glucose of no less than 140 mg/dL or a fasting glucose of more than 100 mg/dL, with one individual demonstrating previously undiagnosed diabetes (Fig. 2). A fasting glucose cutoff of 100 mg/dL would have identified fewer individuals, and it is suggested, similar to the recommendation for adult women with PCOS, that a 2-hour oral glucose challenge may be a better assessment of glucose tolerance than fasting glucose alone in obese adolescents.

Dyslipidemia is a feature of adults with PCOS. Abnormalities in lipids have also been observed in adolescents with PCOS. Girls with a history of premature adrenarche and hyperandrogenism have

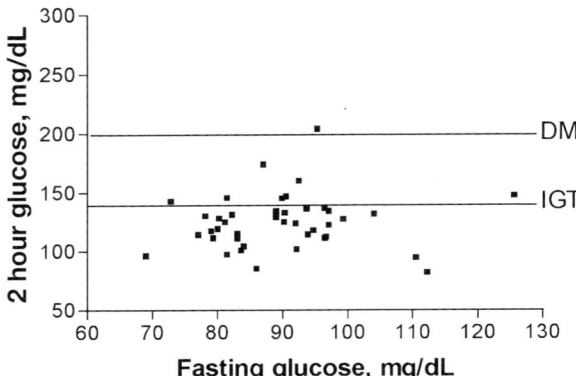

Fig. 2. Fasting glucose vs 2-hour glucose during an oral glucose tolerance test in 35 obese adolescents with polycystic ovary syndrome (PCOS). Cutoff points on 2-hour glucose were 140 mg/dL for impaired glucose tolerance (IGT) and 200 mg/dL for diabetes mellitus (DM). (From Hoeger et al., unpublished data.)

demonstrated abnormal lipid profiles compared to control subjects matched for age *(25)*. However, not all investigators have found abnormalities in serum lipids in adolescents with PCOS when compared with BMI-matched controls *(4)*. It may be that the dyslipidemia in adult women may be a later development in PCOS.

Although a clear consensus does not exist based on available data, cardiovascular risk may be increased in women with PCOS. Cardiovascular risk parameters have been evaluated in small series of adolescents with PCOS. Lipids may not be altered in adolescents with PCOS, but other abnormal vascular changes are seen early in the course of the disease. PCOS adolescents lack the normal nocturnal dipping in diastolic blood pressure, and the lack of nocturnal dipping has been associated with an increased risk of cardiovascular disease in adult life *(26)*. Additionally, endothelial dysfunction has been noted in young women (mean age 22.2 years) with PCOS compared with control women, suggesting vascular changes consistent with increased cardiovascular risk beginning at an early age in PCOS *(27)*.

In summary, metabolic abnormalities, particularly disorders of glucose metabolism, are evident early on in adolescents with PCOS. Cardiovascular disease risk factors may also be increased in adolescents with PCOS, but this is not a consistent finding and may be influenced by BMI and body fat distribution. Further evaluation of larger cohorts of adolescents who are followed over time is needed.

2.3.4. Impact of Obesity

Obesity is highly prevalent in PCOS. Estimates in some studies of women with PCOS are as high as 70%. Adipose tissue is a highly active endocrine tissue and is an important site of the production and metabolism of androgens and estrogen. As a result, in addition to the marked insulin resistance that occurs with increased body weight, there is evidence of endocrine changes related to obesity in PCOS. The prevalence and role of obesity in the clinical manifestations of PCOS in adolescence is not well studied. In one small series, the rate approached that seen in adult populations, but large series have not been reported *(28)*. The onset of adrenarche is closely tied to a change in BMI in adolescence, and extremes of weight gain prior to puberty may be associated with the development of premature adrenarche *(9)*. Premature adrenarche, in turn, may be linked to subsequent development of PCOS *(19)*.

The impact of obesity in adolescents with PCOS has been studied in small series. In one series comparing 11 nonobese adolescents with PCOS (mean BMI 22.5) with 22 obese adolescents with PCOS (mean BMI 35.9 kg/M2), the obese population was significantly more insulin resistant. Three

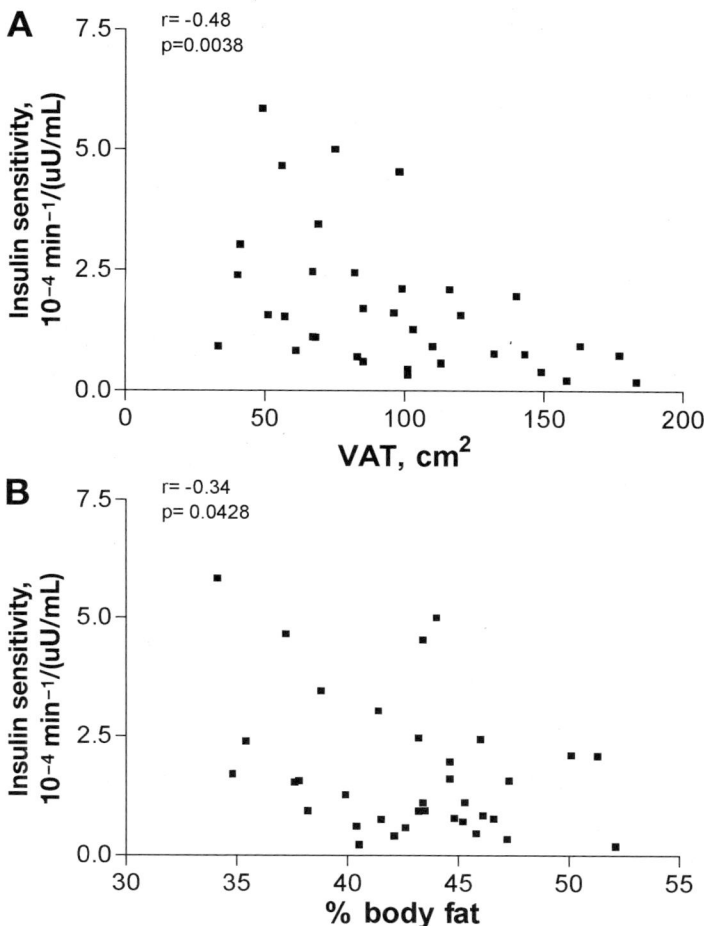

Fig. 3. Body composition vs insulin sensitivity in 35 obese adolescents with polycaystic ovary syndrome (PCOS). Insulin sensitivity determined by frequently sampled iv glucose tolerance test (MINMOD). Visceral adipose tissue (VAT) determined by abdominal computed tomography (CT) scan at L4-5. Percent body fat determined by dual x-ray absorptiometry. (From Hoeger et al., unpublished data.)

of the obese adolescents, but none of the non-obese, had evidence of impaired glucose tolerance. Low-density lipoprotein cholesterol was higher and high-density lipoprotein cholesterol was lower in the obese group. No difference in total testosterone was seen, but SHBG was significantly lower in the obese group *(23)*. These differences between obese and non-obese adolescent women with PCOS mimic what is seen in the adult population. Long-term follow-up of obese adolescents with PCOS as they mature through adulthood is not available, but concern regarding the significant impact of obesity at an early age in PCOS is warranted.

In addition to general measures of obesity, the measurement of the distribution of adipose tissue may be important in predicting metabolic consequences. Visceral adipose tissue has been associated with significant metabolic disturbance in adults. Total body adiposity is correlated with insulin sensitivity in obese adolescents with PCOS. However, abdominal adiposity, particularly visceral fat, shows a stronger correlation to insulin resistance in obese adolescents with PCOS (Fig. 3). Further investigation is needed to determine the link between visceral adiposity and the endocrine and metabolic disturbances seen in PCOS in adolescence.

Table 1
Initial Evaluation of Hyperandrogenism

Physical exam

  Height and weight
  Waist circumference
  Hirsutism score (modified Ferriman–Gallwey) *(22)*
  Presence and severity of acne
  Presence of acanthosis nigricans

Laboratory evaluation

Total testosterone	
SHBG (free androgen index) or free testosterone	Calculated free androgen index = total
DHEAS	testosterone/SHBG (nmol/L)
Basal 17-hydroxyprogesterone	
Fasting glucose or 2-hour OGTT	OGTT if obese
TSH	
Prolactin	} If oligo-amenorrhea present
FSH	

Imaging

Pelvic ultrasound	Not required for diagnosis
	Note: PCOS-appearing ovaries may be found
	in normal adolescents.

SHBG, sex hormone-binding globulin; DHEAS, dehydroepiandrosterone sulfate; OGTT, oral glucose tolerance test; TSH, thyroid-stimulating hormone; FSH, follicle-stimulating hormone; PCOS, polycystic ovary syndrome.

2.4. Evaluation

The majority of adolescents presenting for evaluation of hyperandrogenism will have PCOS, but there is no single definitive test for diagnosis. The 2003 ESHRE/ASRM consensus conference report detailed the diagnostic criteria for PCOS that are currently accepted, but it recognizes the difficulty of using some of these criteria in the evaluation of adolescents *(11)*. Importantly, normal range values for androgen concentrations have not been well studied by age or BMI. Additionally, menstrual irregularity and anatomically polycystic ovaries may be seen in normal adolescents in early puberty. Nevertheless, the evaluation of a young women presenting with hyperandrogenism is similar to that of the adult patient.

Table 1 details the work-up for hyperandrogenism in the pediatric/adolescent population. Clinical suspicion of Cushing's syndrome should prompt additional evaluation of the adrenal axis, such as a 24-hour urinary-free cortisol and/or overnight dexamethasone suppression test. An ACTH stimulation test would be indicated for an abnormal basal 17-hydroxyprogesterone concentration. Additional pelvic and adrenal imaging would be indicated if suspicion of androgen-secreting tumor was present. As in adult patients diagnosed with PCOS, a 2-hour oral glucose tolerance test is indicated for obese adolescents and is preferred over a fasting glucose alone.

2.5. Treatment Options

PCOS in adolescence presents numerous management challenges for the clinician. Adolescence is a time of maturation of self-concept, and the development of physical features of hyperandrogenism, such as acne and hirsutism, particularly in association with obesity, can result in significant negative impact on self-esteem. Adolescents with PCOS experience lower health-related quality of life, and medical intervention in this period may be helpful to emotional development *(21)*. Overall studies of treatments for PCOS in adolescence are limited to small series, many of which are uncontrolled.

However, they include options similar to those used in adult patients. Long-term outcome data on these treatments, however, are lacking in adolescent populations.

Treatment options include hormonal manipulation with oral contraceptives, use of insulin-sensitizing agents (principally metformin), antiandrogens, and lifestyle modification. No drug therapy is Food and Drug Administration-approved for the treatment of PCOS. None of these agents have been studied in large numbers, and few have included obese subjects, making efficacy comparisons difficult. Table 2 summarizes published trials of treatment for adolescent women with PCOS or premature adrenarche in the premenarchal patient.

The therapeutic goals in adolescence are both short-term, such as reduction in hyperandrogenic symptoms and regulation of menses, and long-term, including the reduction in diabetes risk and prevention of cardiovascular disease. Most of the agents studied have focused on reduction in androgens, menstrual cycle regularity, and changes in putative serum cardiovascular risk markers, but the longest follow-up is 12 months. Four studies have looked at the use of oral contraceptives either alone *(29,30)* or in combination with metformin/antiandrogens *(31,32)*. All of these studies were conducted in normal-weight adolescents with PCOS for a duration of 3–9 months. Menstrual irregularity consistently improved, as did androgens and SHBG. Cholesterol increased with oral contraceptive treatment alone, but addition of metformin or flutamide plus metformin resulted in improved dyslipidemia and body composition.

Few studies of metformin alone in adolescents with PCOS have been randomized *(20,33,34)*. Treatment doses range from 1275 to 2550 mg/day for 3–10.5 months. Two of the studies focused on obese adolescents *(33,34)*. These studies demonstrated reductions in serum androgens and improvements in both menstrual cyclicity and glucose tolerance. One study included weight loss as a component, but this was not studied independent of metformin *(34)*. A single randomized trial of metformin for 12 weeks demonstrated improvement in androgens but no weight loss was seen with metformin *(35)*.

One study of metformin in combination with the antiandrogen flutamide included a no-treatment arm *(36)*. Improvements were noted in body composition and metabolic and endocrine features when using low-dose flutamide and metformin compared to no treatment in normal-weight adolescents.

Premenarchal girls with premature adrenarche have also been treated with metformin in low doses (425–850 mg/day) *(37,38)*. Significant improvements were noted in androgens as well as body composition. These girls were considered at risk for development of PCOS. These results suggest that improvements in insulin sensitization at this stage may prevent development of PCOS, but long-term follow-up is needed.

Overall, evidence suggests that use of hormonal manipulation; insulin sensitization, and/or antiandrogens improves the endocrine and metabolic state of PCOS in adolescents. No large-scale, head-to-head trials are available, and most studies have not included obese adolescents. Additional evaluation, including studies with lifestyle modification, is needed to examine efficacy in this population.

3. CONCLUSIONS

Hyperandrogenism in pediatric and adolescent populations is not uncommon. The majority of patients presenting with hyperandrogenism will eventually be diagnosed with PCOS. The diagnostic criteria for PCOS in adolescence are challenging because of the high incidence of menstrual irregularity in early adolescence and the lack of normative clinical and endocrine parameters in ethnic heterogeneous populations. The role for insulin resistance as a primary pathophysiological finding is emerging. The consequences of insulin resistance, including impact on diabetes risk and future cardiovascular health, are already evident in adolescents with PCOS, particularly if they are obese. Treatment options, focusing on the management of the presenting signs of PCOS as well as attention to the long-term sequelae, are imperative at this critical time point in adolescent development.

Table 2
Studies of Treatment for Hyperandrogenism in Pediatric and Adolescent Populations

Study (ref.)	Number of subjects	Mean BMI	Mean age	Treatment	Randomized (Y/N)	Duration	Findings
Ibanez et al., 2000 (19)	10	21.9	16.8	Metformin 1275 mg/d	N	6 months	↓ insulin, FG score, serum androgens; improvement in menstrual cyclicity
Glueck et al., 2001 (34)	11	33.8	16.2	Metformin 1500–2550 mg/d	N	Mean 10.5 months	Resumption of regular menses in most; ↓ cholesterol
Arslanian et al., 2002 (33)	15 (all with IGT)	38.1	14.0	Metformin 850 mg bid	N	3 months	Improvement in glucose tolerance; ↓ androgens
Mastorakis et al., 2002 (29)	28	25	17.5	Oral contraceptive with desogestrel or cyproterone acetate	Y	9 months	Both formulations comparable with regards to ↓ androgens, ↓ FG score, ↑ SHBG and ↑ cholesterol
Ibanez et al., 2003 (35)	21	21.2	15.0	Flutamide 62.5 mg/d + metformin 1275mg/d or no treatment	Y	3 months	Improved insulin, testosterone, SHBG; ↓ abdominal fat mass; ↓ lipids with treatment
Ibanez et al., 2004 (30)	32	21.9	14.6	Oral contraceptive or flutamide + metformin	Y	9 months	↑ SHBG on oral contraceptive; ↓ Total cholesterol, IL-6, fat mass and ↑ adiponectin on flutamide + metformin
Ibanez et al., 2004 (31)	40	22.3	16.7	Oral contraceptive with drosperinone + metformin ± flutamide	Y	3 months	↓ LDL, IL-6 and fat mass with addition of flutamide
Ibanez et al., 2005 (32)	31	22.2	16.0	Flutamide + oral contraceptive ± metformin	Y	3 months	Addition of metformin reduced IL-6, improved body composition; both groups had ↓ androgens
Bridger et al., 2006 (35)	22	32.2	16.1	Metformin or placebo	Y	3 months	Metformin ↓ testosterone 38%; no change in BMI; improved menses
Precocious pubarche							
Ibanez et al., 2004 (36)	33	18.5	8	Metformin 425 mg/d or no treatment	Y	6 months	↓ androstenedione, DHEAS, lipids IL-6 and fat mass; ↑ SHBG and adiponectin with metformin
Ibanez et al., 2004 (37)	24	21.0	12.4	Metformin 850 mg/d or no treatment	Y	12 months	↑ HDL, ↓ testosterone, LDL and fat mass with metformin

BMI, body mass index; IGT, impaired glucose tolerance, FG, Ferriman-Gallwey hirsutism score; SHBG, sex hormone-binding globulin; IL, interleukin; LDL, low-density lipoprotein, DHEAS, dehydroepiandrosterone sulfate; HDL, high-density lipoprotein.

4. FUTURE AVENUES OF INVESTIGATION

The emergence of PCOS as a perimenarchal disease raises a number of questions for future investigation. What are the factors that lead to development of PCOS in an at-risk population? Given the uncertainty of diagnosis in adolescence, normative endocrine data that are age and weight matched, with attention to racial and ethnic variation, are needed. It is clear that the different metabolic and endocrine components of PCOS seen in adult women are already present in adolescents. What interventions could prevent progression of endocrine and metabolic disease, and at what age should these interventions be considered? Long-term large-scale clinical trials of effective treatments for PCOS in adolescents need to be done. The impact on quality of life and future reproductive function needs to be studied. Finally, what is the impact of the growing trends in adolescent obesity on the incidence, consequences, and treatment of PCOS?

KEY POINTS

- Hyperandrogenic symptoms are not uncommon in adolescent populations.
- PCOS is the most common diagnosis of hyperandrogenism in adolescence.
- Insulin resistance is highly prevalent in adolescents with PCOS and is associated with increased risk of abnormal glucose tolerance and diabetes, particularly if obesity is present.
- Treatments for PCOS in adolescents are primarily based on clinical presentation and include oral contraceptives, insulin-sensitizing agents, and antiandrogens.
- Future large-scale clinical trials are needed to define optimum long-term management.

ACKNOWLEDGMENTS

This work was supported in part by NIH grant 5 R03 HD41989-02, General Clinical Research Center (GCRC) grant, 5 M01 RR00044 from the National Center for Research Resources, NIH, Women's Reproductive Health Research (WRHR) grant K12 HD01332-03 to the University of Rochester and 1 K23 HD043881-01A1.

REFERENCES

1. Franks S. Adult polycystic ovary syndrome begins in childhood. Best Pract Res Clin Endocrinol Metab 2002;16(2):263–272.
2. Legro RS, Driscoll D, Strauss JF, 3rd, Fox J, Dunaif A. Evidence for a genetic basis for hyperandrogenemia in polycystic ovary syndrome. Proc Natl Acad Sci USA 1998;95(25):14956–14960.
3. Ibanez L, de Zegher F, Potau N. Anovulation after precocious pubarche: early markers and time course in adolescence. J Clin Endocrinol Metab 1999;84(8):2691–2695.
4. Lewy VD, Danadian K, Witchel SF, Arslanian S. Early metabolic abnormalities in adolescent girls with polycystic ovarian syndrome. J Pediatr 2001;138(1):38–44.
5. Palmert MR, Gordon CM, Kartashov AI, Legro RS, Emans SJ, Dunaif A. Screening for abnormal glucose tolerance in adolescents with polycystic ovary syndrome. J Clin Endocrinol Metab 2002;87(3):1017–1023.
6. Ibanez L, Potau N, Virdis R, et al. Postpubertal outcome in girls diagnosed of premature pubarche during childhood: increased frequency of functional ovarian hyperandrogenism. J Clin Endocrinol Metab 1993;76(6):1599–1603.
7. Vuguin P, Linder B, Rosenfeld RG, Saenger P, DiMartino-Nardi J. The roles of insulin sensitivity, insulin-like growth factor I (IGF-I), and IGF-binding protein-1 and -3 in the hyperandrogenism of African-American and Caribbean Hispanic girls with premature adrenarche. J Clin Endocrinol Metab 1999;84(6):2037–2042.
8. Oppenheimer E, Linder B, DiMartino-Nardi J. Decreased insulin sensitivity in prepubertal girls with premature adrenarche and acanthosis nigricans. J Clin Endocrinol Metab 1995;80(2):614–618.
9. Remer T, Manz F. Role of nutritional status in the regulation of adrenarche. J Clin Endocrinol Metab 1999;84(11):3936–3944.
10. Azziz R, Dewailly D, Owerbach D. Clinical review 56: Nonclassic adrenal hyperplasia: current concepts. J Clin Endocrinol Metab 1994;78(4):810–815.
11. Revised 2003 consensus on diagnostic criteria and long-term health risks related to polycystic ovary syndrome. Fertil Steril 2004;81(1):19–25.
12. Cordera F, Grant C, van Heerden J, Thompson G, Young W. Androgen-secreting adrenal tumors. Surgery 2003;134(6):874–880.
13. Sciarra F, Tosti-Croce C, Toscano V. Androgen-secreting adrenal tumors. Minerva Endocrinol 1995;20(1):63–68.

14. Lindholm J, Juul S, Jorgensen JO, et al. Incidence and late prognosis of cushing's syndrome: a population-based study. J Clin Endocrinol Metab 2001;86(1):117–123.
15. Azziz R, Woods KS, Reyna R, Key TJ, Knochenhauer ES, Yildiz BO. The prevalence and features of the polycystic ovary syndrome in an unselected population. J Clin Endocrinol Metab 2004;89(6):2745–2749.
16. van Hooff MH, Voorhorst FJ, Kaptein MB, Hirasing RA, Koppenaal C, Schoemaker J. Endocrine features of polycystic ovary syndrome in a random population sample of 14–16 year old adolescents. Hum Reprod 1999;14(9):2223–2229.
17. Apter D, Butzow T, Laughlin GA, Yen SS. Accelerated 24-hour luteinizing hormone pulsatile activity in adolescent girls with ovarian hyperandrogenism: relevance to the developmental phase of polycystic ovarian syndrome. J Clin Endocrinol Metab 1994;79(1):119–125.
18. Morales AJ, Laughlin GA, Butzow T, Maheshwari H, Baumann G, Yen SS. Insulin, somatotropic, and luteinizing hormone axes in lean and obese women with polycystic ovary syndrome: common and distinct features. J Clin Endocrinol Metab 1996;81(8):2854–2864.
19. Ibanez L, Dimartino-Nardi J, Potau N, Saenger P. Premature adrenarche—normal variant or forerunner of adult disease? Endocr Rev 2000;21(6):671–696.
20. Ibanez L, Valls C, Potau N, Marcos MV, de Zegher F. Sensitization to insulin in adolescent girls to normalize hirsutism, hyperandrogenism, oligomenorrhea, dyslipidemia, and hyperinsulinism after precocious pubarche. J Clin Endocrinol Metab 2000;85(10):3526–3530.
21. Trent ME, Rich M, Austin SB, Gordon CM. Quality of life in adolescent girls with polycystic ovary syndrome. Arch Pediatr Adolesc Med 2002;156(6):556–560.
22. Ferriman D, Gallwey JD. Clinical assessment of body hair growth in women. J Clin Endocrinol Metab 1961;21:1440–1447.
23. Silfen ME, Denburg MR, Manibo AM, et al. Early endocrine, metabolic, and sonographic characteristics of polycystic ovary syndrome (PCOS): comparison between nonobese and obese adolescents. J Clin Endocrinol Metab 2003;88(10):4682–4688.
24. Legro RS, Chiu P, Kunselman AR, Bentley CM, Dodson WC, Dunaif A. Polycystic ovaries are common in women with hyperandrogenic chronic anovulation but do not predict metabolic or reproductive phenotype. J Clin Endocrinol Metab 2005;90(5):2571–2579.
25. Ibanez L, Potau N, Chacon P, Pascual C, Carrascosa A. Hyperinsulinaemia, dyslipaemia and cardiovascular risk in girls with a history of premature pubarche. Diabetologia 1998;41(9):1057–1063.
26. Arslanian SA, Lewy VD, Danadian K. Glucose intolerance in obese adolescents with polycystic ovary syndrome: roles of insulin resistance and beta-cell dysfunction and risk of cardiovascular disease. J Clin Endocrinol Metab 2001;86(1):66–71.
27. Orio F Jr, Palomba S, Cascella T, et al. Early impairment of endothelial structure and function in young normal-weight women with polycystic ovary syndrome. J Clin Endocrinol Metab 2004;89(9):4588–4593.
28. Rosenfield RL, Ghai K, Ehrmann DA, Barnes RB. Diagnosis of the polycystic ovary syndrome in adolescence: comparison of adolescent and adult hyperandrogenism. J Pediatr Endocrinol Metab 2000;13 Suppl 5:1285–1289.
29. Mastorakos G, Koliopoulos C, Creatsas G. Androgen and lipid profiles in adolescents with polycystic ovary syndrome who were treated with two forms of combined oral contraceptives. Fertil Steril 2002;77(5):919–927.
30. Ibanez L, de Zegher F. Ethinylestradiol-drospirenone, flutamide-metformin, or both for adolescents and women with hyperinsulinemic hyperandrogenism: opposite effects on adipocytokines and body adiposity. J Clin Endocrinol Metab 2004;89(4):1592–1597.
31. Ibanez L, Valls C, Cabre S, De Zegher F. Flutamide-metformin plus ethinylestradiol-drospirenone for lipolysis and antiatherogenesis in young women with ovarian hyperandrogenism: the key role of early, low-dose flutamide. J Clin Endocrinol Metab 2004;89(9):4716–4720.
32. Ibanez L, de Zegher F. Flutamide-metformin plus ethinylestradiol-drospirenone for lipolysis and antiatherogenesis in young women with ovarian hyperandrogenism: the key role of metformin at the start and after more than one year of therapy. J Clin Endocrinol Metab 2005;90(1):39–43.
33. Arslanian SA, Lewy V, Danadian K, Saad R. Metformin therapy in obese adolescents with polycystic ovary syndrome and impaired glucose tolerance: amelioration of exaggerated adrenal response to adrenocorticotropin with reduction of insulinemia/insulin resistance. J Clin Endocrinol Metab 2002;87(4):1555–1559.
34. Glueck CJ, Wang P, Fontaine R, Tracy T, Sieve-Smith L. Metformin to restore normal menses in oligo-amenorrheic teenage girls with polycystic ovary syndrome (PCOS). J Adolesc Health 2001;29(3):160–169.
35. Bridger T, MacDonald S, Baltzer F, Rodd C. Randomized placebo-controlled trial of metformin for adolescents with polycystic ovary syndrome. Arch Pediatr Adolesc Med 2006;160:241–246.
36. Ibanez L, De Zegher F. Flutamide-metformin therapy to reduce fat mass in hyperinsulinemic ovarian hyperandrogenism: effects in adolescents and in women on third-generation oral contraception. J Clin Endocrinol Metab 2003;88(10):4720–4724.
37. Ibanez L, Valls C, Marcos MV, Ong K, Dunger DB, De Zegher F. Insulin sensitization for girls with precocious pubarche and with risk for polycystic ovary syndrome: effects of prepubertal initiation and postpubertal discontinuation of metformin treatment. J Clin Endocrinol Metab 2004;89(9):4331–4337.
38. Ibanez L, Ferrer A, Ong K, Amin R, Dunger D, de Zegher F. Insulin sensitization early after menarche prevents progression from precocious pubarche to polycystic ovary syndrome. J Pediatr 2004;144(1):23–29.

22

Environmental Factors in the Polycystic Ovary Syndrome

Marie-France Hivert and Jean-Patrice Baillargeon

SUMMARY

Polycystic ovary syndrome (PCOS) is a prevalent disorder that is still not well understood. Genetic predisposition is important, but neonatal, prepubertal, pubertal, and postpubertal factors participate in its clinical expression. In this chapter we will review these factors and the evidence that supports causal association with the development of PCOS. Unfortunately, little evidence exists for many of these factors. Obesity and its determinant, birthweight, progression of weight after birth, and precocious pubarche have been studied more extensively. Therefore, these factors will be reviewed in greater detail.

Key Words: Polycystic ovary syndrome; environment; obesity; birthweight; weight gain; puberty.

1. INTRODUCTION

The polycystic ovary syndrome (PCOS) has an important genetic background *(1)*. Family studies have shown that mothers of women with PCOS have a 35% chance of being affected by the syndrome, and sisters have a 40–50% chance of developing it *(2)*, which is definitely higher than the prevalence in the general female population of 6–10% *(3)*. Brothers of women with PCOS are also more likely to be affected by insulin resistance *(4)*. Despite this evident genetic background, PCOS is probably not caused by a single gene defect and, as is the case for many other metabolic conditions, is highly influenced by environmental factors. Twin studies have shown that the presence of polycystic ovaries on ultrasound is very unlikely to be inherited via a single autosomal genetic defect, and authors have hypothesized that environmental factors, perhaps both intrauterine and extrauterine, are involved in the pathogenesis of PCOS *(5)*.

In this chapter we will discuss the influence of obesity on PCOS and the factors that predispose to obesity. We will look at the relationships among birthweight, maturation, puberty, and PCOS. Insulin resistance is extensively discussed in Chapters 24 and 26 of this book, but the early predictors of insulin resistance will be highlighted here. Finally, we will evaluate how dietary intake, eating disorders, and stress are related to PCOS.

2. BACKGROUND

2.1. Obesity and PCOS

Overweight has been linked to PCOS since its very first description by Stein and Leventhal, who noticed the association of obesity, hirsutism, anovulation, and infertility in women. In fact, approximately 50% of women affected by PCOS are overweight or obese *(6)*, and usually the obesity is characterized by a central distribution. Even in lean women matched for body mass index (BMI

From: *Contemporary Endocrinology: Androgen Excess Disorders in Women:*
Polycystic Ovary Syndrome and Other Disorders, Second Edition
Edited by: R. Azziz et al. © Humana Press Inc., Totowa, NJ

Table 1
Clinical and Biochemical Features of Obese vs Non-Obese Patients With PCOS

Features	Obese PCOS	Lean PCOS
Hirsutism	++	+
Total testosterone:androstenedione ratio	↑	↑
Sex hormone-binding globulin	↓	Normal
Free testosterone	↑↑	↑
Androsterone glucuronide	↑↑	↑
Oligomenorrhea/anovulation	++	+
Infertility	++	+
Acanthosis nigricans	++	+
Insulin resistance	++	+
IGT/diabetes mellitus	++	+
Hyperlipidemia	+	–
Hypertension	+ (?)	–
Cardiovascular risk	++ (?)	+ (?)
Endometrial carcinoma risk	++	+

PCOS, polycystic ovary syndrome; IGT, impaired glucose tolerance. (Adapted from ref. *8*.)

[weight in kg divided by height in m^2]), women with PCOS have a higher percentage of body fat and a larger waist-to-hip ratio *(7)* than their matched controls.

Clinical manifestations of PCOS are different in obese women with PCOS compared with lean women with PCOS *(8)* (Table 1). Obese women with PCOS report more menstrual irregularities and more oligo-/amenorrhea than their lean counterparts. Obesity is also associated with an increased prevalence of infertility in PCOS and in the general population as well. The risk of miscarriage is also increased in obese women, whether or not they have PCOS. Moreover, obesity increases the risk of complications during pregnancy, such as gestational diabetes and pre-eclampsia *(9)*. Finally, obese women with PCOS tend to have higher hirsutism and acne scores than lean PCOS women.

Biochemical features of PCOS also differ between obese and non-obese women affected by the syndrome. Insulin resistance is associated with PCOS, but obesity seems to amplify the degree of insulin resistance and hyperinsulinemia *(10)* (Fig. 1). Obese women with PCOS also have a higher risk of developing glucose intolerance or diabetes than lean women with PCOS *(11)*. Dyslipidemia is more frequent in PCOS and is often characterized by high triglycerides, high cholesterol, and low high-density lipoprotein levels *(3)*. Obese women with PCOS usually have higher levels of triglycerides than normal-weight women with PCOS *(8)*.

Growth hormone (GH) and insulin-like growth factors (IGF)-1 and -2 are involved in ovarian function and possibly in the pathogenesis of PCOS. GH pulse amplitude and GH mean levels over 24 hours are reduced in obese PCOS compared with lean women with PCOS *(6)*. IGF-1 and IGF-2 levels in women with PCOS do not seem to differ from those in women without PCOS, but the bioavailability of these hormones seems to be influenced by body weight in women with PCOS. Lean women with PCOS appear to have a higher IGF bioavailability (by IGF binding protein [IGFBP]-1 suppression and greater GH stimulation of IGF), whereas obese women with PCOS seem to have lower IGF bioavailability (because of higher insulin levels and reduced GH stimulation) *(6)*.

Sex hormone levels are also influenced by the degree of overweight. women with PCOS have lower sex hormone-binding globulin (SHBG) levels, with more pronounced SHBG reduction in obese women with PCOS, especially if they present with abdominal obesity *(6)*. Lower SHBG levels increase the bioavailability of sex hormones, and therefore increase hyperandrogenemia in obese PCOS women. Both total and free testosterone levels are increased in obese PCOS compared to normal-weight women with PCOS because of the combination of increased androgen production and lower SHBG *(6)*.

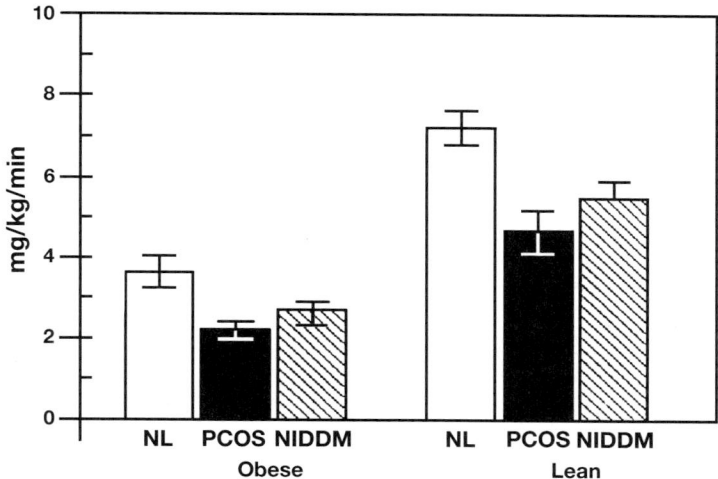

Fig. 1. Insulin sensitivity in lean and obese women with polycystic ovary syndrome (PCOS) or subjects with type 2 diabetes mellitus (NIDDM). Women with PCOS are less insulin sensitive than normal women, and obese women with PCOS are the most insulin resistant of all groups. (Adapted from ref. *10*).

Estrogens are elevated in obese women in the general population. In obese women with PCOS, the abundance of adipose tissue induces high peripheral aromatization of androgens into estrogens. This effect, combined with low SHBG levels, increases the free estradiol fraction, which probably accounts for the increased risk of endometrial cancer observed in women with PCOS. Moreover, for reasons not completely understood, obese women with PCOS have lower luteinizing hormone (LH) levels than lean women with PCOS, and very obese women with PCOS often have LH levels comparable with those of women without PCOS *(6)*.

2.1.2. Factors Predisposing to Obesity

Knowing that obesity is highly prevalent in PCOS and influences the clinical and biochemical manifestations of the syndrome, it would be important to summarize environmental factors predisposing to it. Obesity has increased dramatically in the past few decades. Human beings have been evolving for thousands of years in a world where the individuals who were capable of surviving despite long winters, wars, famines, and other starvation situations had a genetic advantage. In the past 50 years or so, our environment has changed to a world of high food availability—often densely caloric—and low energy expenditure. The consequent obesity epidemic has led the scientific community to try to identify the factors predisposing to obesity *(12)*.

Parental weight is positively related to child's body weight, especially if both parents are obese *(12)*. However, being born from a diabetic mother increases the risk of obesity in teenage years, even after adjustment for the mother's BMI *(13)*. Indeed, a higher birthweight has led to a higher BMI in childhood and in adulthood in many longitudinal studies *(14)*. Finally, many studies have also shown that BMI at any age usually tracks into the subsequent years *(15–17)*.

Normally, BMI increases rapidly after birth, then decreases over a few years to reach a minimum (between 3 and 7 years), and thereafter increases slowly until adulthood (but seems to continue to increase in adulthood according to more recent findings in adult cohorts). Age at adiposity rebound is defined as the age where the minimum BMI is reached, before the second rise of BMI (Fig. 2). An early adiposity rebound has been associated with a higher BMI later in life *(18)*. Others have argued that this effect was simply related to BMI tracking, because a higher BMI in young age is associated with early adiposity rebound and by consequence to a higher BMI in the following years *(19)*.

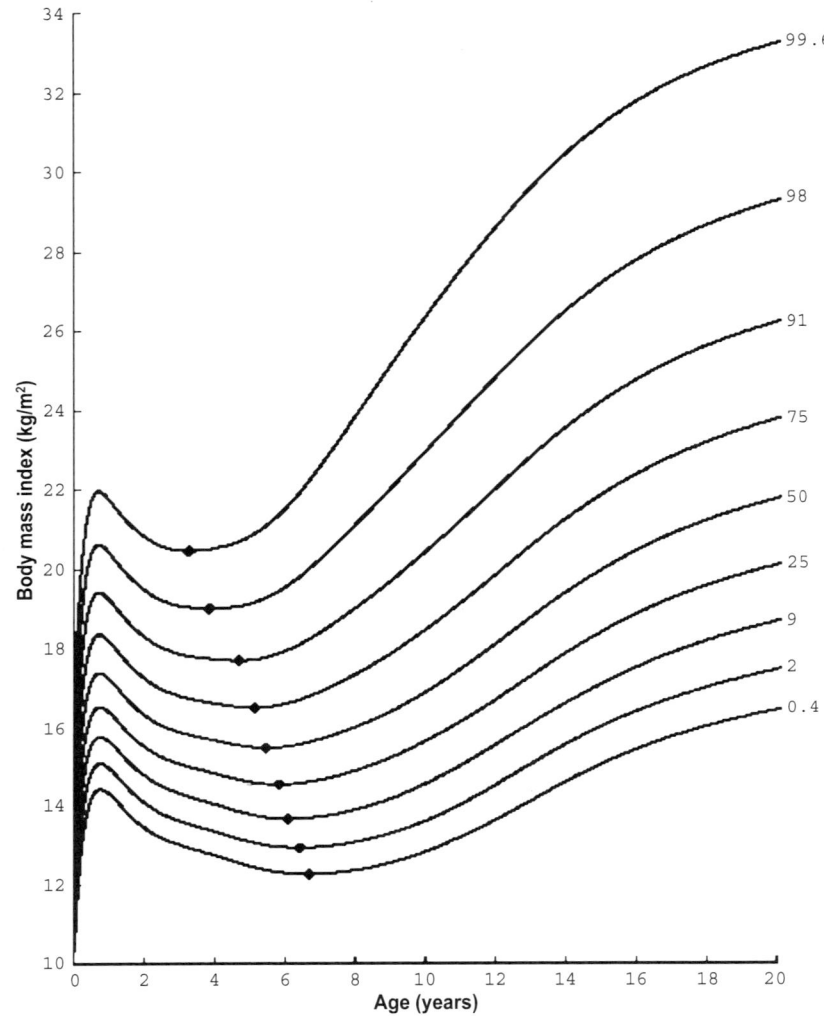

Fig. 2. Body mass index (BMI) growth charts for British girls, 1990. The age at adiposity rebound is illustrated by a dot on each BMI centile line. (Adapted from ref. *19*.)

Another factor predisposing to obesity might be low socioeconomic status *(20)*. Similarly, individuals with higher stress levels and lower level of life satisfaction have shown greater weight gain in longitudinal studies, especially in women *(20)*. Former smokers usually have higher BMI than nonsmokers or current smokers *(21)*.

Instinctively, dietary intake and physical activity should be the two main environmental determinants of overweight and obesity. Unfortunately, both are very difficult to measure with precision, and studies have not always been able to prove the contribution of one or the other. In cross-sectional studies, frequent consumption of vegetables is usually associated with lower risk of obesity *(21)*. Conversely, high dietary fat intake has been associated with greater weight gain in some longitudinal studies *(15)*. However, a systematic review of the literature has underlined the inconsistency of these studies, the difficulty of estimating food intake from questionnaires, and the problem of underreporting in food diaries *(12)*.

Physical activity level offers the same challenges concerning estimation by questionnaires, but shows a little more consistency. Many longitudinal studies have shown that higher level of physical activity is associated with less weight gain *(16,17)*. Conversely, inactivity was associated with weight gain in other cohort studies. Finally, sedentary lifestyle, often measured as time spent watching television, has been related to higher risk of developing obesity and diabetes in both men *(22)* and women *(23)*.

Overall, overweight and obesity are prevalent in women with PCOS and might play an important role in its pathophysiology. Furthermore, when women with PCOS are obese, they usually present with more severe clinical and biochemical features of the syndrome. Accordingly, prevention of obesity is critical in order to prevent PCOS and/or its consequences, and factors that predispose to obesity may start very early in life and are numerous throughout the life cycle.

2.2. Birthweight and PCOS

As mentioned previously, birthweight may be related to body weight later in life. Some authors have tried to determine if a direct link exists between birthweight and PCOS. In a population of women aged 18–25 years , Michelmore et al. *(7)* showed that an increased birthweight was predictive of polycystic ovaries on ultrasound, but not related to symptoms of PCOS. On the other hand, in the same cohort, low birthweight was related to insulin resistance (which is consistent with many other cohort studies). Using birth records from 1952 to 1953, Cresswell et al. *(24)* traced 235 women in their 40s. Polycystic ovaries were present on ultrasound in 21% of them, who were found to have a slightly but nonsignificant higher birthweight. A longer gestation, however, was significantly related to polycystic ovaries on ultrasound, especially in lean women. Furthermore, women born with heavier weights and from mothers with greater BMIs were more likely to have polycystic ovaries on ultrasound and were more likely to have a greater BMI themselves.

Sadrzadeh et al. *(25)* assessed the population of an infertility clinic and compared women with PCOS with women with diminished ovarian reserve capacities or tubal obstruction (as the reference group). They gathered information retrospectively using questionnaires. In this study, birthweight was not related to PCOS, but higher actual BMI (as we already know) and delayed menarche were. In a prospective cohort followed up from birth and re-assessed at 31 years old, Laitinen et al. *(26)* did not find a relationship between PCOS symptoms (hirsutism or oligomenorrhea) and birthweight, prematurity, or growth retardation. However, they did find that obesity, especially of central distribution, was related to PCOS symptoms.

Alternatively, in a recent report from a UK birth cohort, Ong et al. *(27)* showed that a low birthweight was associated with greater weight gain in subsequent years and a higher androgen level at 8 years in both boys and girls. Similarly, Ibanez et al. *(28)* found that being small for gestational age increased the risk of anovulation at 15 years (Fig. 3).

Overall, these studies are not convincing for a direct link between birthweight and PCOS. However, they underscore again the importance of a higher actual body weight in the clinical manifestations of PCOS, which might be related to birthweight. They also highlight the relationship between low birthweight and important characteristics of PCOS, particularly the development of insulin resistance.

2.3. Insulin Resistance

2.3.1. Insulin Resistance and PCOS

There is growing evidence that insulin resistance plays a key role in the pathophysiology of PCOS *(3)*. Hyperinsulinemia increases androgen levels by several mechanisms *(29)*, including by increasing androgen production and greater bioavailability by lowering SHBG. It has also been shown that women with PCOS are characterized by both increased peripheral insulin resistance and β-cell dysfunction. For the same level of glycemia, women with PCOS have higher basal plasma insulin levels but lower insulin secretory response to meals compared with weight-matched control women *(30)*. Hyperinsulinemic euglycemic glucose clamp studies have demonstrated a significant decrease in

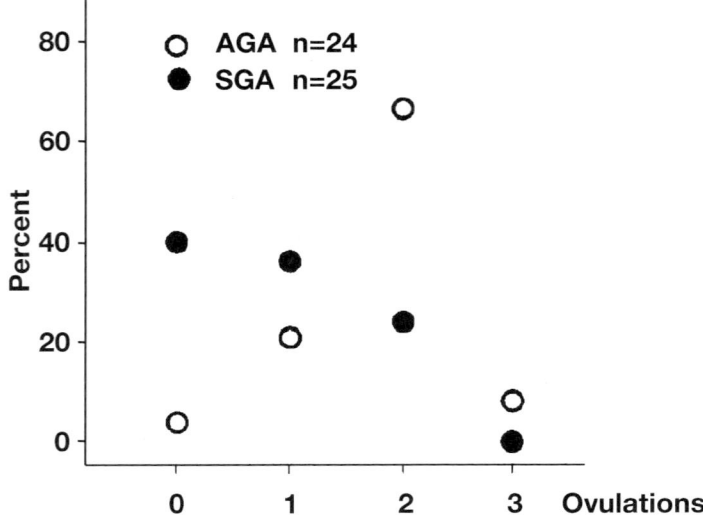

Fig. 3. Fractions of appropriate for gestational age (AGA) and small for gestational age (SGA) subpopulations distributed by number of ovulations detected over 3 months of study at age 15 years. Girls born SGA have significantly fewer ovulations over 3 months at age 15 years compared with girls born AGA. (Adapted from ref. *28*.)

insulin-mediated glucose disposal in women with PCOS *(10)*. Hyperinsulinemia results from increased basal insulin secretion and reduced hepatic clearance of insulin in both lean and obese PCOS women, but obesity also brings on increased hepatic glucose production *(30)*.

We will not review here the cause nor the implication of insulin resistance in PCOS because it will be extensively discussed in Chapters 24 and 26. But we will review environmental and prenatal factors that might be implicated in the development of the insulin resistance typical of PCOS.

2.3.2. Early Predictors of Insulin Resistance

Many studies have confirmed that low birthweight is associated with central obesity and insulin resistance after adjustment for BMI. Hofman et al. *(31)* showed that prematurity or being small for gestational age was related to diminished insulin sensitivity measured at 4–10 years old. In the Nurses Health Study, low birthweight was associated with higher risk of type 2 diabetes, even after adjustment for BMI and all the other confounders *(32)*.

Adiposity rebound has also been linked to development of insulin resistance. In Helsinki, a birth cohort followed up from 1934 showed that an early adiposity rebound (before 5 years vs after 7 years old) was related to an increased risk of type 2 diabetes *(33)*. In India, a 30-year follow-up of a birth cohort also showed that early adiposity rebound was associated with the development of impaired glucose tolerance and diabetes mellitus *(34)*. In this cohort, birthweight was not statistically predictive of the oral glucose tolerance test result, but a low weight in infancy (1 or 2 years old) was associated with abnormal glucose tolerance later in life, especially if followed by a rapid weight gain and higher weight at 12 years of age. However, as noted earlier, the adiposity rebound might also be a manifestation of higher BMI tracking into childhood.

In general, BMI tracks into childhood, and longitudinal studies have shown that a high BMI, even in childhood, was predictive of higher BMI and insulin resistance in adolescence and adulthood. The Minneapolis Children's Blood Pressure Study *(35)* followed individuals from 7 to 23 years of age and showed that BMI in childhood was predictive of early adulthood BMI, fasting insulin levels, lipid profile, and systolic blood pressure. The Bogalusa Heart Study *(36)* followed individuals (8–17 years old at baseline) over 11 years to identify the factors related to the clustering of features of the metabolic syndrome into adulthood. The strongest predictor was the initial BMI, which remained statistically significant even after adjustment for initial fasting insulin levels.

Overall, insulin resistance is strongly related to weight changes. A low birthweight is probably predictive of higher risk of impaired glucose tolerance and diabetes mellitus, especially if followed by a rapid weight gain, early adiposity rebound, and higher BMI into childhood and adolescence. However, weight gain in adulthood, overweight, and central obesity are still independently and strongly associated with insulin resistance and usually have a greater influence on the actual risk of impaired glucose metabolism. Insulin resistance alone does not cause PCOS in every woman. This might be explained by genetic predisposition(s) or perhaps by *in utero* imprinting or the timing of development and maturation of those women.

2.4. Precocious Pubarche and PCOS

Puberty is a critical period for the future appearance of PCOS because it involves sexual maturation and the initiation of ovulation and ovarian steroidogenesis. Moreover, it has been shown that the timing of menarche is influenced by body weight and adiposity and that insulin resistance typically increases during puberty.

Precocious pubarche (PP; also called precocious adrenarche) has been associated with higher risk of developing PCOS symptoms, but the link with PP, which is defined by appearance of pubic hair before 8 years of age in girls, has been more extensively investigated, especially by Ibanez and colleagues. PP is usually followed by normal puberty, menarche, and growth; but appears to predispose to functional ovarian hyperandrogenism *(37)*. Early pubarche has been associated with increased risk of polycystic ovaries on ultrasound, hyperandrogenism, hyperinsulinism, and altered lipid profile *(38)* during and after pubertal development. Girls with PP also have a higher risk of developing anovulation a few years after menarche *(39)*. Finally, Ibanez et al. *(40)* have found that girls 6–18 years old who had presented with PP displayed higher waist circumference, waist-to-hip ratio, total fat mass, and percentage fat mass than girls matched for age, pubertal stage, and BMI. These girls also presented with higher abdominal fat, which was positively correlated to the level of insulinemia and androgenemia.

Birthweight also influences pubertal evolution. Girls with PP had a lower birthweight than matched control girls, and those with hyperandrogenism had an even lower birthweight. If they presented with hyperinsulinism on top of that, their mean birthweight was lower again *(41)* (Fig. 4). Prospective follow-up of girls with PP until after menarche showed that the subgroup with lower birthweight had decreased insulin sensitivity, abnormal ovarian function, and higher triglycerides and low-density lipoprotein levels compared to girls with PP and normal birthweight *(42)*.

In summary, these studies have shown that girls having a low birthweight followed by early pubarche are at increased risk of developing PCOS. Those who also presented with hyperinsulinemia were at highest risk to develop PCOS after puberty, and prepurbertal intervention with an insulin-sensitizing drug in these individuals prevented the progression towards PCOS *(43)*. Therefore, birthweight and/or PP might be significant modifiers of the relationship between insulin resistance and PCOS.

2.5. Dietary Intake and Other Lifestyle Factors

There are few published studies of the influence of diet on PCOS. Wild et al. *(44)* investigated the difference in lipid content of the diet between women with PCOS and a control group of eumenorrheic women of similar age, as well as other lifestyle habits. Women with PCOS were heavier than controls, had a diet characterized by more saturated fat and less fiber, and were more sedentary. They did not find any difference in alcohol consumption and smoking habits, which is consistent with other observations. On the other hand, Wright et al. *(45)* did not find any difference in the composition of the diet or physical activity level of women with PCOS compared with age- and race-matched control women. Intriguingly, lean women with PCOS had a lower total caloric intake than lean control

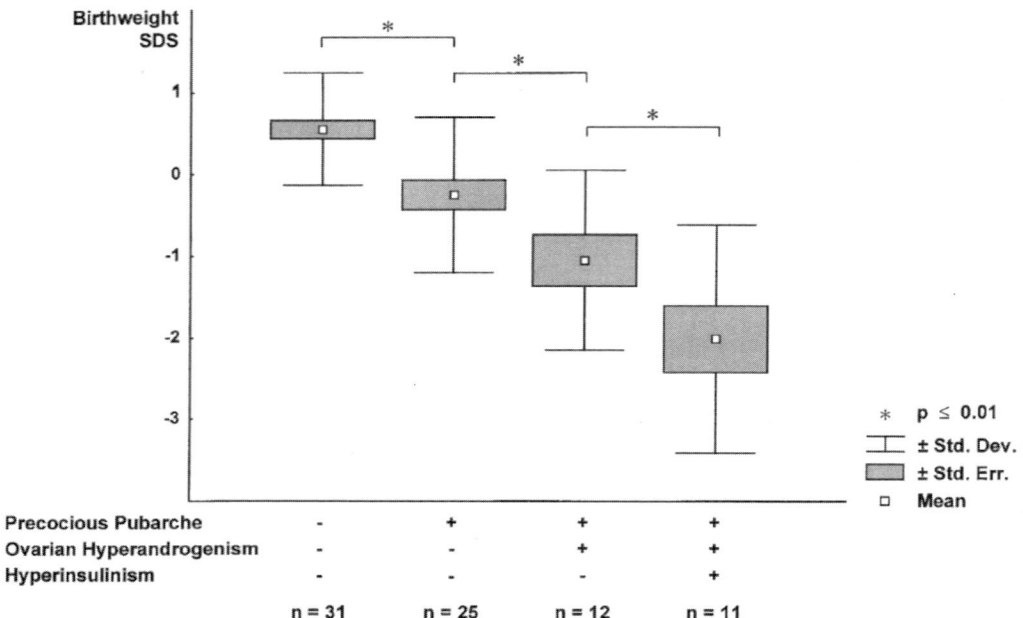

Fig. 4. Precocious pubarche, hyperinsulinism, and ovarian hyperandrogenism in girls—relation to reduced fetal growth. (Adapted from ref. *41*.)

women. The authors suspected that women with PCOS had to restrict themselves to maintain normal body weight.

Carmina et al. *(46)* looked at the differences between women with PCOS from the United States and Italy. American women with PCOS had a higher BMI and worse insulin resistance and lipid profile than the Italian women with PCOS. The only difference in diet found was the higher consumption of saturated fat in the American women. Therefore, diet composition might influence the manifestations of PCOS rather than causing PCOS.

2.5.1. Eating Disorders and PCOS

Some authors have noticed a link between binge eating or bulimia nervosa (BN) and PCOS. In a twin cohort *(47)* the presence of polycystic ovaries on ultrasound was associated with a higher score on the bulimia investigation test–Edinburg (BITE) questionnaire. Using the same questionnaire, McCluskey et al. *(48)* compared women with PCOS with women suffering from organic endocrinopathy. They found that women with PCOS had higher scores on the BITE questionnaire and that about 6% of them had scores suggestive of BN compared with only 1% in the control group. McCluskey also found that among patients suffering from BN, 56% had menstrual irregularities and 76% had polycystic ovaries on ultrasound *(49)*. Moreover, when patients with BN who had polycystic ovaries on ultrasound were successfully treated, they normalized their ovarian morphology *(50)*.

In contrast to those findings, Michelmore et al. used the Eating Disorders Examination questionnaire in a group of young women (18–25 years old) who volunteered for a study on women's health and did not find any association between binge eating or overeating and polycystic ovaries on ultrasound or PCOS (defined as polycystic ovaries on ultrasound plus one feature of the syndrome) *(51)*.

In summary, binge eating might be related to the development of polycystic ovaries or even PCOS, but more data are needed to confirm such an association. Also, the mechanisms by which periods of binge eating could induce PCOS are not known. It is possible that such eating behavior causes a physical or psychological stress that induces insulin resistance independent of body weight, which in turn contributes to PCOS.

2.6. Stress and PCOS

Psychological stress has been associated with PCOS in a few studies. Lobo et al. *(52)* showed that women with PCOS had higher scores on the Life Events Inventory questionnaire and had higher numbers of major life events. They also demonstrated that women with PCOS had higher urinary levels of norepinephrine metabolites, which is a reflection of norepinephrine turnover and, according to the authors, a surrogate of psychological stress. Trent et al. *(53)* investigated the quality of life of adolescent girls with PCOS using the Child Health Questionnaire–Child Self-Report Form. They found that teenagers with PCOS scored lower on general health perception, physical functioning, general behavior, and limitations in family activities compared to healthy age-matched adolescents.

Based on these limited number of studies, it is reasonable to conclude that women with PCOS usually have a higher level of stress. However, because all the evidence is cross-sectional, it is impossible to assume any causal relationship, that is, is it the stress that increases the risk of PCOS or simply that women with PCOS suffer more stress because of their condition. In fact, McCook et al. *(54)* used the Health-Related Quality of Life Questionnaire adapted for women with PCOS to evaluate the impact of overweight, hyperandrogenism, menstrual problems, and infertility on the quality of life of women with PCOS. They found that weight was the major concern for those women and that this level of distress was positively correlated to BMI. The other specific categories of distress (infertility, body hair) were also correlated with each respective clinical manifestation.

3. CONCLUSIONS

Twin studies have shown that genetic factors alone cannot adequately explain the development of PCOS in women. Many environmental factors probably modulate the clinical expression of genetic predisposition. Obesity is the most important of these factors and contributes directly to insulin resistance and hyperinsulinemia, two key factors in the pathogenesis of PCOS. We have reviewed the most important environmental and neonatal predictors of obesity and insulin resistance, which include low birthweight. Low birthweight seems also to modify the relation between insulin resistance and PCOS in that girls with premature pubarche and hyperinsulinism are at higher risk of developing PCOS if they were born with a lower weight. Finally, diet composition, eating disorders, and psychological stress have been associated with PCOS, but available studies are conflicting and not conclusive for causality.

4. FUTURE AVENUES OF INVESTIGATION

Many environmental factors have been associated with the development of PCOS. Some of them might be causal and others might significantly modify the relationship between causal factors and the clinical development of PCOS. However, most of the studies reviewed in this chapter were cross-sectional and only defined associations. In order to critically assess the causal or modifying effect of environmental factors in PCOS, more long-term and large prospective studies are needed. These studies should focus on the relationship between postpurbertal development of PCOS and birthweight, prepubertal progression of weight, characteristics of puberty, stress or stress management, eating behaviors, and evolution of insulin resistance. Finally, long-term intervention studies might help elucidate the possible causal effect of diet composition in the phenotypic expression of PCOS.

KEY POINTS

- Obesity is probably the most important environmental factor in the development of PCOS.
- Significant predictors of postpubertal obesity and insulin resistance are birthweight, age at adiposity rebound, as well as weight gain and obesity during childhood.
- Caloric intake and levels of physical activity are probably important predictors, but they are difficult to assess in large studies.
- Low birthweight and/or PP seem to significantly modify the relationship between insulin resistance and the progression to PCOS.

- Diet composition, eating disorders, and psychological stress have also been associated with PCOS, but causal relationships cannot be ascertained.

REFERENCES

1. Franks S, Gharani N, Waterworth D, et al. Genetics of polycystic ovary syndrome. Mol Cell Endocrinol 1998;145:123–128.
2. Kahsar-Miller MD, Nixon C, Boots LR, Go RC, Azziz R. Prevalence of polycystic ovary syndrome (PCOS) in first-degree relatives of patients with PCOS. Fertil Steril 2001;75:53–58.
3. Baillargeon JP, Iuorno MJ, Nestler JE. Insulin sensitizers for polycystic ovary syndrome. Clin Obstet Gynecol 2003;46:325–340.
4. Baillargeon JP, Carpentier A. Determination of insulin sensitivity in brothers of women with polycystic ovary syndrome—preliminary result. Program and Abstracts of the 87th Annual Scientific Meeting of The Endocrine Society, 2005.
5. Jahanfar S, Eden JA, Warren P, Seppala M, Nguyen TV. A twin study of polycystic ovary syndrome. Fertil Steril 1995;63:478–486.
6. Gambineri A, Pelusi C, Vicennati V, Pagotto U, Pasquali R. Obesity and the polycystic ovary syndrome. Int J Obes Relat Metab Disord 2002;26:883–896.
7. Michelmore K, Ong K, Mason S, et al. Clinical features in women with polycystic ovaries: relationships to insulin sensitivity, insulin gene VNTR and birthweight. Clin Endocrinol (Oxf) 2001;55:439–446.
8. Salehi M, Bravo-Vera R, Sheikh A, Gouller A, Poretsky L. Pathogenesis of polycystic ovary syndrome: What is the role of obesity? Metabolism 2004;53:358–376.
9. Linne Y. Effects of obesity on women's reproduction and complications during pregnancy. Obes Rev 2004;5:137–143.
10. Dunaif A, Segal KR, Futterweit W, Dobrjansky A. Profound peripheral insulin resistance, independent of obesity, in polycystic ovary syndrome. Diabetes 1989;38:1165–1174.
11. Legro RS, Kunselman AR, Dodson WC, Dunaif A. Prevalence and predictors of risk for type 2 diabetes mellitus and impaired glucose tolerance in polycystic ovary syndrome: a prospective, controlled study in 254 affected women. JCEM 1999;84:165–169.
12. Parsons TJ, Power C, Logan S, Summerbell CD. Childhood predictors of adult obesity: a systematic review. Int J Obes Relat Metab Disord 1999;23 Suppl 8:S1–107.
13. Rosenbaum M, Leibel RL. The physiology of body weight regulation: relevance to the etiology of obesity in children. Pediatrics 1998;101:525–539.
14. Whitaker RC, Wright JA, Pepe MS, Seidel KD, Dietz WH. Predicting obesity in young adulthood from childhood and parental obesity. N Engl J Med 1997;337:869–873.
15. Sherwood NE, Jeffery RW, French SA, Hannan PJ, Murray DM. Predictors of weight gain in the Pound of Prevention study. Int J Obes Relat Metab Disord 2000;24:395–403.
16. Schmitz KH, Jacobs DR Jr, Leon AS, Schreiner PJ, Sternfeld B. Physical activity and body weight: associations over ten years in the CARDIA study. Coronary Artery Risk Development in Young Adults. Int J Obes Relat Metab Disord 2000;24:1475–1487.
17. Wenche DB, Holmen J, Kruger O, Midthjell K. Leisure time physical activity and change in body mass index: an 11-year follow-up study of 9357 normal weight health women 20–49 years old. J Womens Health (Larchmt) 2004;13:55–62.
18. Guo SS, Huang C, Maynard LM, et al. Body mass index during childhood, adolescence and young adulthood in relation to adult overweight and adiposity: the Fels Longitudinal Study. Int J Obes Relat Metab Disord 2000;24:1628–1635.
19. Cole TJ. Children grow and horses race: is the adiposity rebound a critical period for later obesity? BMC Pediatr 2004;4:6.
20. Korkeila M, Kaprio J, Rissanen A, Koshenvuo M, Sorensen TI. Predictors of major weight gain in adult Finns: stress, life satisfaction and personality traits. Int J Obes Relat Metab Disord 1998;22:949–957.
21. Lahti-Koski M, Pietinen P, Heliovaara M, Vartiainen E. Associations of body mass index and obesity with physical activity, food choices, alcohol intake, and smoking in the 1982-1997 FINRISK Studies. Am J Clin Nutr 2002;75:809–817.
22. Hu FB. Sedentary lifestyle and risk of obesity and type 2 diabetes. Lipids 2003;38:103–108.
23. Hu FB, Li TY, Colditz GA, Willett WC, Manson JE. Television watching and other sedentary behaviors in relation to risk of obesity and type 2 diabetes mellitus in women. JAMA 2003;289:1785–1791.
24. Cresswell JL, Barker DJ, Osmond C, Egger P, Phillips DI, Fraser RB. Fetal growth, length of gestation, and polycystic ovaries in adult life. Lancet 1997;350:1131–1135.
25. Sadrzadeh S, Klip WA, Broekmans FJ, et al. Birthweight and age at menarche in patients with polycystic ovary syndrome or diminished ovarian reserve, in a retrospective cohort. Hum Reprod 2003;18:2225–2230.
26. Laitinen J, Taponen S, Martikainen H, et al. Body size from birth to adulthood as a predictor of self-reported polycystic ovary syndrome symptoms. Int J Obes Rel Metab Disord 2003;27:710–715.
27. Ong KK, Potau N, Petry CJ, et al. Opposing influences of prenatal and postnatal weight gain on adrenarche in normal boys and girls. JCEM 2004;89:2647–2651.

28. Ibanez L, Potau N, Ferrer A, Rodriguez-Hierro F, Marcos MV, de Zegher F. Reduced ovulation rate in adolescent girls born small for gestational age. JCEM 2002;87:3391–3393.

29. Nestler JE. Role of hyperinsulinemia in the pathogenesis of the polycystic ovary syndrome, and its clinical implications. Semin Reprod Endocrinol 1997;15:111–122.

30. Schroder AK, Tauchert S, Ortmann O, Diedrich K, Weiss JM. Insulin resistance in patients with polycystic ovary syndrome. Ann Med 2004;36:426–439.

31. Hofman PL, Regan F, Jackson WE, et al. Premature birth and later insulin resistance. N Engl J Med 2004;351:2179–2186.

32. Rich-Edwards JW, Colditz GA, Stampfer MJ, et al. Birthweight and the risk for type 2 diabetes mellitus in adult women. Ann Intern Med 1999;130:278–284.

33. Eriksson JG, Forsen T, Tuomilehto J, Osmond C, Barker DJ. Early adiposity rebound in childhood and risk of Type 2 diabetes in adult life. Diabetologia 2003;46:190–194.

34. Bhargava SK, Sachdev HS, Fall CH, et al. Relation of serial changes in childhood body-mass index to impaired glucose tolerance in young adulthood. N Engl J Med 2004;350:865–875.

35. Sinaiko AR, Donahue RP, Jacobs DR Jr, Prineas RJ. Relation of weight and rate of increase in weight during childhood and adolescence to body size, blood pressure, fasting insulin, and lipids in young adults. The Minneapolis Children's Blood Pressure Study. Circulation 1999;99:1471–1476.

36. Srinivasan SR, Myers L, Berenson GS. Predictability of childhood adiposity and insulin for developing insulin resistance syndrome (syndrome X) in young adulthood: the Bogalusa Heart Study. Diabetes 2002;51:204–209.

37. Ibanez L, Potau N, Virdis R, et al. Postpubertal outcome in girls diagnosed of premature pubarche during childhood: increased frequency of functional ovarian hyperandrogenism. JCEM 1993;76:1599–1603.

38. Ibanez L, de Zegher F, Potau N. Premature pubarche, ovarian hyperandrogenism, hyperinsulinism and the polycystic ovary syndrome: from a complex constellation to a simple sequence of prenatal onset. J Endocrinol Invest 1998;21:558–566.

39. Ibanez L, de Zegher F, Potau N. Anovulation after precocious pubarche: early markers and time course in adolescence. JCEM 1999;84:2691–2695.

40. Ibanez L, Ong K, de Zegher F, Marcos MV, del Rio L, Dunger DB. Fat distribution in non-obese girls with and without precocious pubarche: central adiposity related to insulinaemia and androgenaemia from prepuberty to postmenarche. Clin Endocrinol (Oxf) 2003;58:372–379.

41. Ibanez L, Potau N, Francois I, de Zegher F. Precocious pubarche, hyperinsulinism, and ovarian hyperandrogenism in girls: relation to reduced fetal growth. JCEM 1998;83:3558–3562.

42. Ibanez L, Valls C, Potau N, Marcos MV, de Zegher F. Polycystic ovary syndrome after precocious pubarche: ontogeny of the low-birthweight effect. Clin Endocrinol (Oxf) 2001;55:667–672.

43. Ibanez L, Ferrer A, Ong K, Amin R, Dunger D, de Zegher F. Insulin sensitization early after menarche prevents progression from precocious pubarche to polycystic ovary syndrome. J Pediatr 2004;144:23–29.

44. Wild RA, Painter PC, Coulson PB, Carruth KB, Ranney GB. Lipoprotein lipid concentrations and cardiovascular risk in women with polycystic ovary syndrome. JCEM 1985;61:946–951.

45. Wright CE, Zborowski JV, Talbott EO, McHugh-Pemu K, Youk A. Dietary intake, physical activity, and obesity in women with polycystic ovary syndrome. Int J Obes Relat Metab Disord 2004;28:1026–1032.

46. Carmina E, Legro RS, Stamets K, Lowell J, Lobo RA. Difference in body weight between American and Italian women with polycystic ovary syndrome: influence of the diet. Hum Reprod 2003;18:2289–2293.

47. Jahanfar S, Eden JA, Nguyent TV. Bulimia nervosa and polycystic ovary syndrome. Gynecol Endocrinol 1995;9:113–117.

48. McCluskey S, Evans C, Lacey JH, Pearce JM, Jacobs H. Polycystic ovary syndrome and bulimia. Fertil Steril 1991;55:287–291.

49. McCluskey SE, Lacey JH, Pearce JM. Binge-eating and polycystic ovaries. Lancet 1992;340:723.

50. Morgan JF, McCluskey SE, Brunton JN, Hubert LJ. Polycystic ovarian morphology and bulimia nervosa: a 9-year follow-up study. Fertil Steril 2002;77:928–931.

51. Michelmore KF, Balen AH, Dunger DB. Polycystic ovaries and eating disorders: Are they related? Hum Reprod 2001;16:765–769.

52. Lobo RA, Granger LR, Paul WL, Goebelsmann U, Mishell DR Jr. Psychological stress and increases in urinary norepinephrine metabolites, platelet serotonin, and adrenal androgens in women with polycystic ovary syndrome. Am J Obstet Gynecol 1983;145:496–503.

53. Trent ME, Rich M, Austin SB, Gordon CM. Quality of life in adolescent girls with polycystic ovary syndrome. Arch Pediatr Adolesc Med 2002;156:556–560.

54. McCook JG, Reame NE, Thatcher SS. Health-related quality of life issues in women with polycystic ovary syndrome. J Obstet Gynecol Neonatal Nurs 2005;34:12–20.

23

Animal Models and Fetal Programming of the Polycystic Ovary Syndrome

David H. Abbott, Daniel A. Dumesic, Jon E. Levine, Andrea Dunaif, and Vasantha Padmanabhan

SUMMARY

At least 28 animal models provide insight into the etiological and pathophysiological basis of polycystic ovary syndrome (PCOS). About 50% of them, however, either do not show sufficient traits meriting designation of a PCOS phenotype or exhibit alternate features mimicking other disorders, such as hyperprolactinemia. In contrast, animal models of fetal programming through androgen excess show remarkable resilience and reliability in replicating PCOS, including metabolic defects in males, and therefore strongly implicate a fetal etiology in the developmental origins of PCOS. This chapter reviews the relevance of animal models for PCOS and their potential value for providing insight into the etiology and pathophysiology of this disorder.

Key Words: Polycystic ovary syndrome; rhesus monkeys; androgens; prenatal; fetal programming; animal model; anovulation; hyperandrogenism.

1. INTRODUCTION

The etiology and pathophysiology of polycystic ovary syndrome (PCOS) in women are poorly understood. PCOS is multifaceted and includes reproductive, metabolic, and general health disorders (Table 1). The syndrome is strongly familial in origin, with 67–93% of daughters born to women with PCOS developing the PCOS syndrome as adults (1–3). Clinical or biochemical manifestation of androgen excess is the most reliably transmitted PCOS trait (4). PCOS is the most common endocrinopathy of women in their reproductive years (5,6), with a prevalence of 6–7% (7), and one of its most troubling general health disorders is early-onset type 2 diabetes (8,9).

Nevertheless, PCOS has a heterogeneous and unpredictable clinical presentation (5,6), and most putative gene candidates studied to date have been unable to adequately explain its phenotype (10), suggesting that PCOS has multiple (albeit undiscovered) genetic origins modified by environmental factors and perhaps fetal programming (11). Therefore, the development and application of animal models for PCOS provide timely insight into the origins and pathophysiological mechanisms that are difficult to resolve from human studies. This chapter reviews the many animal models proposed as relevant to PCOS, with particular emphasis on models that reliably and reproducibly emulate the PCOS syndrome through androgen excess fetal programming.

From: *Contemporary Endocrinology: Androgen Excess Disorders in Women:*
Polycystic Ovary Syndrome and Other Disorders, Second Edition
Edited by: R. Azziz et al. © Humana Press Inc., Totowa, NJ

Table 1
Common Signs and Symptoms of PCOS

A. Consensus diagnostic criteria[a]

Two out of three of the following:

1. Clinical or biochemical hyperandrogenism, as determined by elevated circulating levels of total or unbound testosterone or hirsutism
2. Intermittent or absent menstrual cycles
3. Polycystic ovaries (as visualized by ultrasound)

The following conditions must also be excluded: classical and nonclassical congenital adrenal hyperplasia, Cushing's syndrome, thyroid dysfunciton, hyperprolactinemia, androgen-secreting tumors, and drug-induced androgen excess.

B. PCOS signs and symptoms outside those required for diagnosis (some, all, or none of these may be present in an individual)

Reproductive and endocrine
Luteinizing hormone (LH) hypersecretion
Reduced steroid-negative feedback on LH release
Increased recruitment and persistence of ovarian follicles
Ovarian hyperresponsiveness to gonadotropic therapy for in vitro fertiliztion (IVF)
High rates of miscarriage
Endometrial hyperplasia and cancer
Adrenal hyperandrogenism
Gestational diabetes

Metabolic
Insulin resistance and compensatory hyperinsulinemia
Imparied glucose tolerance
Type 2 diabetes
Obesity (including abdominal adiposity)
Pancreatic impairments in insulin responses to glucose
Hyperlipidemia

General health disorders
Cardiovascular disease
Sleep apnea
Acne
Chronic inflammation
Intra-uterine growth retardation

[a]These criteria *(13)* are extented from the previous 1990 NIH Consensus diagnosis *(12)* which specified the first two criteria, alone, as a basis for PCOS diagnosis, following exclusion of conditions that mimic PCOS (listed above). Revisions of these criteria by Azziz *(14)* suggest requiring criterion 1 together with either criteria 2 or 3, whereas revisions by Chang *(15)* relegate criterion 3 to nondiagnostic status. PCOS, polycystic ovary syndrome. (Modified from ref. *11*.)

2. BACKGROUND

2.1. Overview of PCOS

Because PCOS is a diagnosis of exclusion, animal models need to replicate traits consistent with PCOS without having features that might mimic other clinical diseases. Such a goal is difficult and compounded by recent changes in the manner in which PCOS is clinically diagnosed. For example, the 1990 National Institutes of Health (NIH) consensus diagnosis for PCOS specifies hyperandrogenism accompanied by oligo- or amenorrhea, excluding conditions that mimic PCOS, such as classic and nonclassical congenital adrenal hyperplasia, thyroid dysfunction, hyperprolactinemia, androgen-producing tumors, and drug-induced androgen excess *(12)*.

The Revised 2003 Rotterdam consensus diagnosis adds polycystic ovaries, as diagnosed by tranvaginal ultrasound (TVUS), to the above 1990 NIH consensus diagnosis for PCOS, specifying that two of the three criteria are required to diagnose PCOS *(13)*. A recent reappraisal of the 1990 NIH and Revised 2003 Rotterdam consensus diagnoses *(14)* combines criteria from both consenses and specifies PCOS as androgen excess accompanied by oligo- or amenorrhea or by TVUS-confirmed polycystic ovaries while still excluding clinical conditions mimicking PCOS. An additional suggestion *(15)* is to continue using the 1990 NIH consensus for the diagnosis of PCOS, while relegating sonographic imaging of polycystic ovaries to confirmatory, rather than diagnostic, status. Equally complex, individual women with PCOS can exhibit different combinations of diagnostic criteria with varying degrees of severity, can show abnormalities other that those used for diagnosing PCOS *(5,6)*, and can experience onset of symptoms at puberty *(16,17)*, which can resolve during middle age *(18)*.

Thus, for an animal model to truly approximate the complexities of the PCOS phenotype, it must not only exhibit ovarian hyperandrogenism, oligo- or amenorrhea, and/or an increased number of medium-sized ovarian follicles in the absence of features mimicking other clinical diseases, but also show an extraordinary array of relevant traits, permitting a heterogeneous phenotype.

2.2. Animal Models for PCOS

A variety of mammalian species have been employed as animal models of PCOS, ranging from rodents to nonhuman primates. Each species has differences in reproductive function compared to humans, and such reproductive differences need to be considered when translating experimental findings into clinical applications *(19)*. For example, rats and mice undergo spontaneous ovulation approximately every 4–5 days (not every 26–34 days), and they complete follicle luteinization and form corpora lutea only if mating occurs, with the luteotropic support of prolactin (not luteinizing hormone [LH]).

2.2.1. Animal Models for PCOS That Produce Large Ovarian Follicular Cysts

Much has been made of the experimental induction of ovarian cysts in animal models for PCOS. Unlike PCOS ovarian morphology, follicular cysts in animal models are frequently large (preovulatory sized or larger) and are thus not a diagnostic trait for the PCOS syndrome. These animal models, while providing insight into the development of cystic ovarian morphology, may not be relevant in resolving the etiology and pathophysiology of PCOS. Models inducing large ovarian follicular cysts are found in a variety of experimental treatment paradigms (summarized in Table 1), with and without additional induction of PCOS and non-PCOS diagnostic traits. In this context it is important to note that rodents are generally multiovular and thus normally exhibit multifollicular ovarian morphology.

2.2.2. Animal Models for PCOS That Fail to Demonstrate PCOS Diagnostic Traits

When animal models lack elevated androgen levels or exhibit regular ovulatory cycles, they fail to demonstrate the basic tenets of PCOS. Those that fall into this category (Table 1) include treatment of adult females with constant exposure to light (probably the oldest animal model proposed for PCOS), acute estrogen, and valproic acid, as well as treatment of newborn females with testosterone. Estradiol valerate treatment of female rats, while inhibiting ovulatory cycles, also induces traits that are uncharacteristic of PCOS: growth hormone excess and hypothalamic degeneration. Such animal models hold little promise for understanding etiological and pathophysiological mechanisms of PCOS.

2.2.3. Animal Models for PCOS That Exhibit Diagnostic Traits But Show Other Traits Excluding a PCOS Diagnosis

Four animal models, while exhibiting PCOS diagnostic traits, show other traits that are inconsistent with PCOS in women, for example, thyroid dysfunction in hypothyroid rats treated with human chorionic gonadotropin (hCG) (Table 2). The remaining three models, transgenic overexpression of

Table 2
Characteristics of Animal Models of PCOS Relevant to PCOS Diagnostic Criteria and Criteria for Diagnostic Exclusion from PCOS

Model	Species	Criteria for PCOS diagnosis			Criteria that exclude a PCOS diagnosis			Model characteristics not found in PCOS	Ref.
		Androgen excess	Intermittent or absent ovulatory cycles	Multiple, medium-sized ovarian antral follicles	Adrenal [17]OHP excess	Thyroid dysfunction	Hyperprolactinemia		
Models with insufficient traits for a PCOS diagnosis and no known exclusion traits									
Constant light	Rat	–	+	–	N/A	?	+	Many large, cystic ovarian follicles, loss of endogenous endocrine rhythms	35
Estradiol valerate	Rat	?	+	–	N/A	?	+	Elevated growth hormone, large ovarian follicular cysts, hypothalamic degeneration	68
Chronic estrogen	Rat	?	+	+	N/A	?	?		22
Acute estrogen	Guinea pig	–	–	–	?	?	?	Large ovarian follicular cysts	69
Valproic acid treatment (anticonvulsant and fatty acid analog)	Rat	–	–	+	N/A	?	?		70
Valproic acid treatment	Rhesus monkey	–	–	–	–	?	?		71
PMSG treatment (pregnant mare serum gonadotropin)	Rat	?	+	–	N/A	?	?	Large ovarian follicular cysts	72
Neonatal testosterone treatment	Newborn rat	–	+	–	N/A	?	?	Large ovarian follicular cysts, small ovaries	73
Models with PCOS diagnostic traits, but also diagnostic traits that exclude a PCOS diagnosis									
Hypothyroid treatment with hCG	Rat	+	?	+	N/A	+	+	Hypothyroidism	67
Transgenic overexpression of LH	Mouse	+	+	–	N/A	?	+?	Ovarian cancer, renal pathology, large, blood-filled ovarian follicular cysts	20
DHEA treatment	Immature rat	+[b]	+	–	N/A	?	+	Large ovarian follicular cysts	21,22
	Adult rat	+[b]	+	–	N/A	?	+	Large ovarian follicular cysts	
RU486 treatment (antiprogestagenic and anti-glucocorticoid)	Rat	+	+	–	N/A	?	+	Large ovarian follicular cysts with precocious luteinization	23

262

Models with PCOS diagnostic traits and no known exclusion traits

Model	Species					Other traits	Refs.
Nymphomania	Cow	+	-	?	?	Excessive sexual behavior	22
Immunization against testosterone	Rat	+[a]	-	N/A	-	Large ovarian follicular cysts	24
Letrozole treatment (aromatase inhibitor)	Rat	+	+	N/A	?		30
Aromatase knockout (ArKO)	Mouse	+	-	N/A	?	Ovaries and uterus immature	31
Dexamethasone treatment (synthetic glucocorticoid)	Pig	+	+	?	?	Glucocorticoid excess	29
Chronic testosterone treatment	Immature rat	+[b]	+	N/A	?	Large ovarian follicular cysts	25
Chronic testosterone treatment	Rhesus monkey	+[b]	-	?	?		33
Chronic androstenedione treatment	Rhesus monkey	+[b]	(+) Seasonal anovulation only	-	?	?	34
Chronic estrone treatment	Rhesus monkey	+	-	?	?		35
Chronic hCG treatment	Rat	+	?	N/A	?	Large ovarian follicular cysts	26
Chronic insulin and hCG treatment	Rat	+	-	N/A	?	Large ovarian follicular cysts	27
Chronic IGF-1 and hCG treatment	Rat	+	-	N/A	?	Large ovarian follicular cysts	28
Fetal testosterone treatment	Mouse	+	?	?	?	Virilized genitalia	48
Fetal testosterone treatment	Rat	+	?	?	?	Virilized genitalia	49
Fetal testosterone treatment	Sheep	+[c]	+	?	?	Virilized genitalia only in females exposed during early gestation	42
Fetal testosterone treatment	Rhesus monkey	+	+	-	-	Virilized genitalia only in females exposed during early gestation	11,40

[a]Low serum levels of free unbound testosterone.

[b]Exogenous androgen excess.

[c]Functional ovarian hyperandrogenism (V. Padmanabhan, unpublished results).

N/A, not applicable—rodent adrenal glands do not synthesize corticosterone from ^{17}OHP.

PCOS, polycystic ovary syndrome; LH, luteinizing hormone; hCG, human chorionic gonadotropin; DHEA, dehydroepiandrosterone; IGF, insulin-like growth factor.

LH *(20)*, and treatment with either dehydroepiandrosterone *(21,22)* or RU486 (an antiprogestagen and antiglucocorticoid) *(23)*, all induce hyperprolactinemia, a well-known mimic of PCOS in women *(12,13)*. All but the hypothyroid rats also exhibit large ovarian follicular cysts (Table 2). These animal models might thus provide useful information in determining mechanisms underlying the pathophysiology of conditions that resemble PCOS, but are not PCOS itself.

2.2.4. Animal Models for PCOS That Exhibit Diagnostic Traits and Lack Traits Excluding a PCOS Diagnosis

The remaining 16 animal models all demonstrate traits consistent with those for PCOS diagnosis, and none exhibit additional traits for exclusion (Table 2). The nymphomaniac cow is the only naturally occurring animal model for PCOS *(22)*, but its unpredictable occurrence and unknown mechanism have yet to make it useful. Although manipulation of adult female rats by immunization against testosterone *(24)*, chronic treatment with testosterone *(25)*, hCG *(26)*, insulin and hCG *(27)*, and insulin-like growth factor-1 and hCG *(28)* all induce hyperandrogenic females that have intermittent or absent ovulatory cycles, the treatments induce large ovarian follicular cysts, unlike the smaller-sized cysts found in women with PCOS. Treatment of adult female pigs with dexamethasone, a synthetic glucocorticoid, induces PCOS diagnostic criteria provided that adrenergic innervation to the ovaries remains intact *(29)*, yet glucocorticoid excess is not a common symptom accompanying PCOS. Use of an aromatase inhibitor, letrozole, on adult female rats produces a phenotype remarkably similar to that of PCOS, including LH hypersecretion *(30)*. Not surprisingly, however, serum levels of estradiol are greatly diminished, a steroidogenic abnormality not found in PCOS. Although complete aromatase knockout female mice are hyperandrogenic and anovulatory, they do not represent a PCOS phenotype because their ovaries and uteri fail to mature *(31)*. In this context it is interesting to note that fetal female monkeys exposed to a highly specific aromatase inhibitor have greatly diminished ovarian follicular development that is prevented by simultaneous treatment with estradiol *(32)*.

Chronic exposure of adult female monkeys to testosterone *(33)*, androstenedione *(34)*, or estrone *(35)* induce females with hyperandrogenism and intermittent or absent menstrual cycles (Table 1). None of these adult models, however, exhibit traits commonly associated with PCOS beyond the diagnosis, such as PCOS-like ovarian morphology, LH hypersecretion, or metabolic dysfunction *(36)*.

Such chronic manipulation of the adult steroid hormone environment, including that induced by aromatase inhibition, has generated four animal models that more closely approximate the symptomology of PCOS than the models previously discussed. Neither these nor the previous models, however, match the replication of PCOS phenotype generated by animal models of fetal androgen excess that reprogram differentiation and development of multiple organ systems. Acute exposure of normal adult female monkeys to testosterone, accelerates the early stages of ovarian follicular development *(37)* and may mimic accelerated the early follicular development found in women with PCOS associated with diminished intraovarian expression of anti-Müllerian hormone *(38)*.

2.3. Animal Models of Androgen Excess Fetal Programming of PCOS

Compared with other animal models for PCOS, models of fetal programming have one clear advantage: simultaneous exposure of multiple organ systems to a specific developmental insult during differentiation and maturation (Fig. 1). Altered structure and function is commonly permanent, such as fetal androgen excess virilization of the female urogenital tract, resulting in expression of fetal programming in adulthood *(11)*. The best known example of fetal programming is described by Barker and colleagues *(39)*, in which human fetal undernutrition and low birthweight are associated with adult cardiovascular disease, hypertension, insulin resistance, and type 2 diabetes, some key hallmarks of PCOS outside of those required for its diagnosis.

In the previous edition of this book, Abbott and colleagues *(40)* provided the first description of a fetal programming model for PCOS: the prenatally androgenized female rhesus monkey. As illustrated in Fig. 1 and in Tables 2 and 3, early gestation exposure of the female monkey fetus to fetal

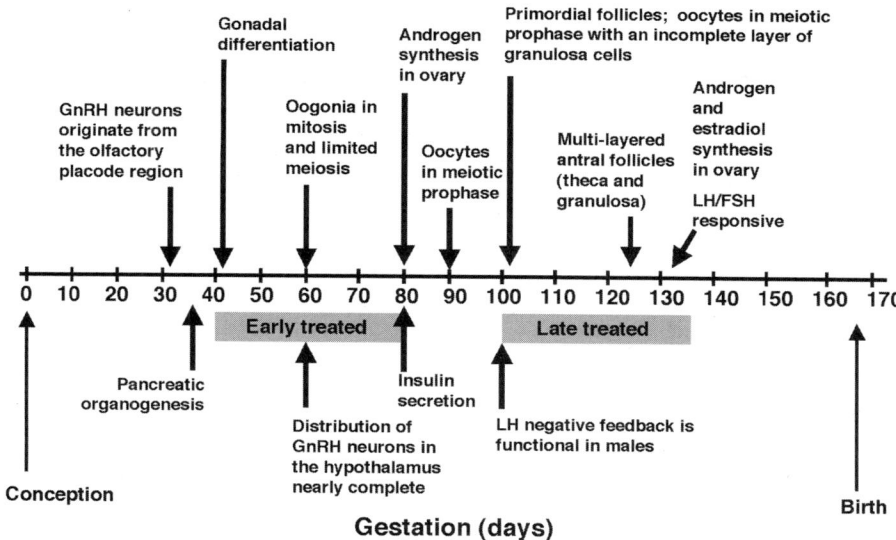

Fig. 1. Gestational progression of aspects of differentiation and maturation of hypothalamic–pituitary–ovarian function and pancreas and β-cell function in rhesus monkeys. The timing of exposure of females to androgen excess (early or late in gestation) is indicated in relation to fetal developmental progress. GnRH, gonadotropin-releasing hormone; LH, luteinizing hormone; FSH, follicle-stimulating hormone. (Modified from ref. *11*.)

male levels of testosterone results in adult females with PCOS traits. Early gestation in the rhesus monkey provides a stage in development when multiple organ systems regulating reproductive and metabolic function are undergoing differentiation. A single insult during such a sensitive stage of development can permanently alter disparate organ systems, producing a phenotypic mimic of PCOS.

2.3.1. Fetal Androgen Programming of Reproductive Defects

Prenatally androgenized female monkeys exhibit heterogeneity in their presentation of PCOS traits, an inherent complexity in the human syndrome. Approximately 70% of prenatally androgenized female monkeys exposed to androgen excess during early gestation have serum testosterone levels in excess of the mean value in normal adult female monkeys of the same age, weight, and body mass index; approximately 40% are anovulatory (~10 times the normal rate, while the remainder have mostly intermittent menstrual cycles); and approximately 40% have increased numbers of medium-sized ovarian antral follicles (approximately twice the normal incidence) *(41)*. Differing degrees of virilization of both internal and external genitalia, however, are found in all female monkeys exposed to androgen excess during early gestation. Because virilized genitalia is not a feature of PCOS in women, exposure of female monkeys to androgen excess during late gestation, when the urogenital tract is no longer responsive to androgen reprogramming, produces a closer PCOS phenotype that retains heterogeneity of trait expression, but without genital virilization (Fig. 1) *(11,41)*. Such results from early and late gestationally exposed prenatally androgenized female monkeys, confirmed in prenatally androgenized ewes *(42)*, suggest that exposure of human female fetuses to androgen excess during the latter part of gestation (second to third trimesters) from genetic and/or environmental factors might induce organ system reprogramming, causing PCOS. Thus, the key element in the etiology and initial pathophysiology of PCOS may be appropriately timed fetal hyperandrogenism derived from hyperandrogenic fetal ovaries *(43)*, fetal adrenal cortex *(44)*, or from hyperandrogenemia of PCOS mothers *(45)* reflected in the fetal circulation *(46)*.

Table 3
**Common Signs and Symptoms Associated With PCOS and Shown
by Animal Models of Androgen Excess Fetal Programming of PCOS**

	Mouse[a]	Rat[b]	Sheep[c]	Monkey (early/late gestation)[d]
Reproductive and endocrine				
Ovarian hyperandrogenism	+	+	+	+
Intermittent or absent ovulatory cycles	+	+	+	+
Multiple medium-sized ovarian follicles	???	???	+	+
LH hypersecretion	+	+	+	+
Reduced steroid negative feedback on LH	???	???	+	+
Ovarian endocrine hyper-responsiveness to gonadotropic hyperstimulation for IVF	???	???	???	(???) Poor embryo development
High rates of miscarriage				
Endometrial hyperplasia and cancer	???	???	???	(+) Hyperplasia
Adrenal hyperandrogenism	N/A	N/A	N/A	+
Gestational diabetes	???	???	???	???
Metabolic				
Insulin resistance and compensatory hyperinsulinemia	???	+	+	+
Impaired glucose tolerance	???	+	???	+
Type 2 diabetes	???	???	???	+
Obesity (including abdominal adiposity)	???	???	???	+
Pancreatic impairments in insulin responses to glucose	???	+	???	+
Hyperlipidemia	???	???	???	+
General health disorders				
Cardiovascular disease	???	???	+	???
Sleep apnea	???	???	???	???
Chronic inflammation	???	???	???	???
Low birthweight	???	+	+	–
Heterogeneity of PCOS trait expression	???	???	+	+

N/A, not applicable because adrenal glands from nonprimate species do not normally synthesize androgens.
[a]From ref. *48*.
[b]From ref. *49* and J. E. Levine, unpublished results.
[c]From refs. *51*, *56*, and *62*, and V. Padmanabhan, unpublished results.
[d]From refs. *11*, *41*, *50*, and *66* and D. H. Abbott, unpublished results.
LH, luteinizing hormone; IVF, in vitro fertilization.

Recent findings from prenatally androgenized female sheep *(47)*, mice *(48)*, and rats *(49)* confirm and extend those obtained from prenatally androgenized female monkeys (Tables 2 and 3). All express PCOS diagnostic traits, as well as traits commonly associated with the syndrome (e.g., LH hypersecretion). Anovulation in prenatally androgenized female mice *(48)* and rats *(49)* is induced by fetal exposure to the nonaromatizable androgen dihydrotestosterone (DHT), suggesting an androgen receptor-mediated neuroendocrine defect. In prenatally androgenized monkeys, although prenatal DHT exposure can induce similar behavioral outcomes to those achieved by testosterone *(11)*, prenatal DHT-exposed animals are not available for PCOS studies. Certainly in prenatally androgenized female mice, treatment of such adults with the antiandrogen flutamide restores ovulatory cycles and implicates adult excess androgen, acting via the androgen receptor, in the neuroendocrine mechanism of anovulation *(48)*.

2.3.2. Fetal Androgen Programming of Oocyte Quality

Beyond ovulatory dysfunction, diminished oocyte quality in both prenatally androgenized female monkeys and women with PCOS provides an additional barrier to fertility *(50)*. Following controlled ovarian hyperstimulation for in vitro fertilization (IVF), retrieved oocytes from either early or late gestation exposed, prenatally androgenized female monkeys exhibit reduced competence as defined by the ability of the resulting diploid zygotes to reach the blastocyst stage. In women with PCOS, diminished quality of retrieved oocytes contributes to implantation failure and pregnancy loss. Both prenatally androgenized female monkeys and PCOS women exhibit abnormal intrafollicular steroidogenic responses to controlled ovarian hyperstimulation, which in the former is associated with an inability to normally suppress circulating insulin levels between the first day of the recombinant human follicle-stimulating hormone (rhFSH) treatment and the day of oocyte retrieval *(50)*. Women with PCOS, however, are hyperresponsive to rhFSH and exhibit intrafollicular hyperandrogenism at oocyte retrieval following recombinant human chorionic gonadotropin (rhCG) injection. Prenatally androgenized female monkeys, however, are hyporesponsive to rhFSH relative to normal females, and early gestation-exposed female monkeys exhibit diminished intrafollicular androgen and estrogen levels along with an exaggerated shift in intrafollicular steroidogenesis from androgen and estrogen to progesterone at oocyte retrieval following rhCG injection. The early gestation-exposed female monkey response to controlled ovarian hyperstimulation, therefore, is reminiscent of that shown by normoandrogenic ovulatory women with reduced ovarian responsiveness to rhFSH *(50)*. Collectively, in addition to ovulatory defects, the timing of fetal androgen excess exposure impairs oocyte quality through ovarian and/or metabolic dysfunction, with such oocyte defects possibly having transgenerational consequences for female offspring of prenatally androgenized monkeys and for daughters of women with PCOS.

2.3.3. Fetal Androgen Programming of Metabolic Defects

Insulin resistance and diminished pancreatic insulin response to glucose are integral defects in the development of type 2 diabetes in women with PCOS and are exacerbated by obesity *(5,6)*. Prenatally androgenized female monkeys *(11)*, sheep *(51)*, and rats (J. E. Levine, unpublished results) all exhibit such insulin dysfunction, while early gestation-exposed androgenized female monkeys also exhibit abdominal obesity, hyperlipidemia, and an increased incidence of type 2 diabetes (Table 3). Insulin sensitivity in early, but not late, gestationally exposed, prenatally androgenized female monkeys is reduced to that found in normal male monkeys and in normal females during the luteal phase of the menstrual cycle, and a similar degree of insulin resistance is found in prenatally androgenized sheep *(51)*. Such parallels in metabolic dysfunction between androgen excess fetal programming models of PCOS and women with PCOS provide strong evidence for a fetal origin of metabolic defects in both cases, possibly through fetal programming of preferential accumulation of abdominal fat (Fig. 2).

2.3.4. Fetal Androgen Programming of General Health Disorders

Poor intrauterine growth and low birthweight are associated with the development of precocious puberty and PCOS in northern Spanish women *(52)* and with PCOS pregnancies in Chilean women *(53)*, but not in larger studies of Finnish *(54)* and Dutch *(55)* women. Prenatally androgenized female sheep *(56)* and rats *(57)* exhibit clear evidence of intrauterine growth restriction and low birthweight, whereas prenatally androgenized female monkeys do not *(41)*. Prenatally androgenized sheep and rats may thus provide more suitable animal models for women with PCOS who have placental insufficiency. Perhaps not surprisingly, in this context, prenatally androgenized sheep have enlarged left ventricles of the heart, kidneys, and adrenal glands suggestive of developing cardiovascular disease *(42)*, and prenatally androgenized female rats have increased mortality *(58)*.

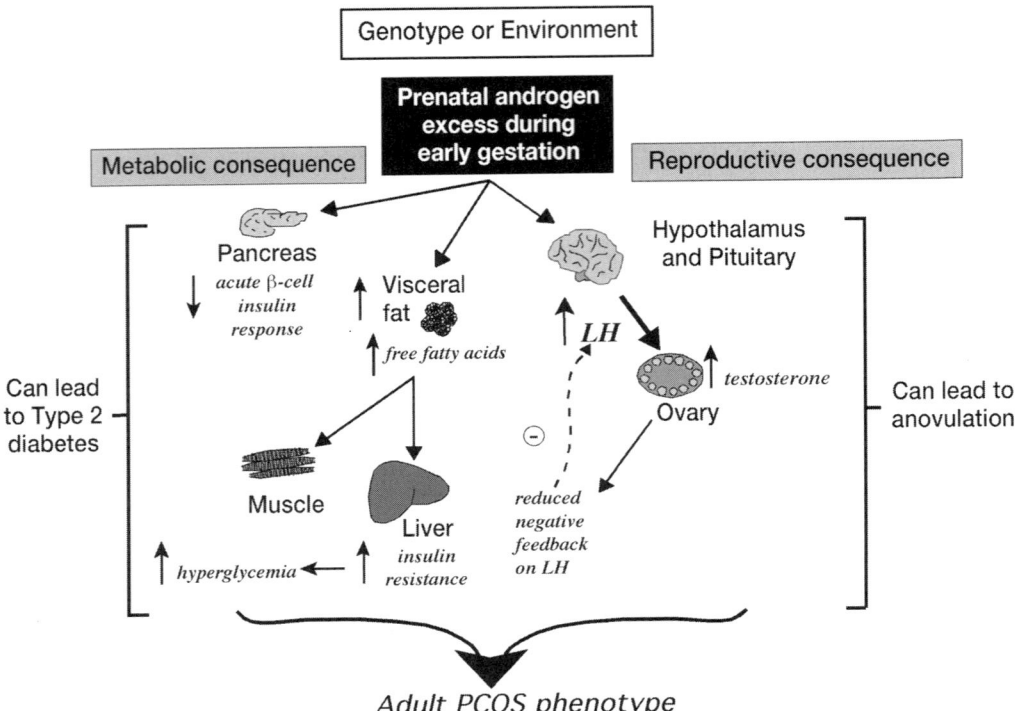

Fig. 2. Diagrammatic representation of our hypothesis for early gestation, fetal androgen excess programming of adult polycystic ovary syndrome traits. Genetic or environmental mechanisms induce fetal hyperandrogenism (*see* text) that result in permanent changes in both reproductive and metabolic function. Reproductive consequences include (1) altered hypothalamic–pituitary function leading to luteinizing hormone (LH) hypersecretion, (2) ovarian hyperandrogenism that may or may not be the result of LH hypersecretion, (3) reduced steroid hormone negative-feedback regulation of LH, which may be a component of the initial permanent alteration in hypothalamic–pituitary function, and (4) increased anovulation. Metabolic consequences include (1) increased abdominal adiposity, which may be responsible for increased circulating total free fatty acid levels, (2) impaired pancreatic insulin secretory response to glucose, (3) impaired insulin action and compensatory hyperinsulinemia, (4) hyperglycemia, and (5) increased incidence of type 2 diabetes. Insulin resistance and compensatory hyperinsulinemia may be functionally implicated in the anovulatory mechanism. (Modified from ref. *41*.)

2.3.5. Comparison of Androgen Excess Fetal Programming Models for PCOS

Table 3 illustrates the comparability of androgen excess fetal programming models for PCOS. Although most PCOS traits to date have been identified in prenatally androgenized female monkeys (*11*), increasing numbers of reports are describing PCOS traits in prenatally androgenized females in nonprimate species (*42,49,48*). The only clear difference in symptomology between primate and nonprimate models involves low birthweight in the latter and not the former, indicating an otherwise remarkable degree of concurrence among models. Because low birthweight is found in some (*52,53*), but not all (*54,55*) populations of women with PCOS, this heterogeneity in animal models may prove extremely useful in identifying mechanisms underlying the different fetal phenotypes found in PCOS. Rodent models, nevertheless, may not be ideal for ovarian phenotype determination as they normally develop multiple large follicles prior to ovulation, unlike sheep or monkeys. On the other hand, transgenic rodent models hold much promise for determining fetal programming consequences of specific gene changes, such as those involving neuroendocrine regulation of ovarian function *(48)*. Sheep models are particularly useful for understanding abnormal ovarian follicular recruitment and persistence through sequential ultrasonography, repeated blood sampling, and ovarian manipulation. Nonhuman primate models, with more than 90% genetic similarity to humans and the closest symptomology to PCOS, provide the most straightforward translation of experimental findings into improved clinical application.

The only major difference in trait defects between prenatally androgenized female rhesus monkeys and women with PCOS involves their responses to ovarian hyperstimulation for IVF. Women with PCOS are hyperandrogenic and hyperresponsive to rhFSH, whereas prenatally androgenized female monkeys are not (Table 3). It is tempting to speculate that this differential response to gonadotropin stimulation may reflect different regulatory mechanisms governing hyperandrogenism in ovarian theca cells of women with PCOS vs prenatally androgenized female monkeys, perhaps reflecting an intrinsic, genetically determined hyperandrogenism in the former *(59)*, but not the latter.

2.4. Animal Models for a Male Phenotype of PCOS

Close male relatives of women with PCOS present with metabolic dysfunction similar to that of their female kin *(60–62)*. Consistent with a fetal origins hypothesis for PCOS, male monkeys exposed to androgen excess during gestation develop insulin resistance and diminished insulin response to glucose as adults *(63)*. Because fetal monkey androgen treatments induce circulating testosterone levels in male and female fetuses only within the normal range for fetal males *(64)*, fetal programming of male (and female?) metabolic dysfunction may be induced by mechanisms beyond androgen-mediated action in the fetus, perhaps involving an estrogenic metabolite of testosterone, or the action of androgen on the placenta or the mother. In support of potential estrogenic involvement in PCOS fetal programming, exposure of fetal ewes to bisphenol-A, a plasticizer and estrogen mimic, leads to intrauterine growth restriction and LH surge defects similar to those produced by fetal exposure to testosterone (V. Padmanabhan et al., unpublished).

3. CONCLUSIONS

A variety of animal models have been proposed for PCOS, but only models of androgen excess fetal programming have reliably produced the spectrum and heterogeneity of traits that closely reflect PCOS in women (Tables 2 and 3). Animal models of fetal programming may thus provide the elusive etiology and initial pathophysiology for PCOS in women and for metabolic dysfunction in their close male kin. Such fetal models suggest that permanent alteration of gene expression in multiple organ systems may provide a closer approximation to PCOS phenotypes than differing genotypes alone.

4. FUTURE AVENUES OF INVESTIGATION

Until the etiology of PCOS is determined and the cellular and molecular mechanisms of its pathophysiology are understood, animal models for the syndrome will be essential. Animal models will continue to lead the way in identifying a probable fetal origin for PCOS until sampling from human fetuses becomes a low-risk routine procedure and pregnancies at risk for PCOS are identified prospectively. Such PCOS animal models will also continue to be essential in ascertaining the otherwise ethically unattainable goal of normalized oocyte and embryo quality in women with PCOS. Fetal programming models thus hold promise for determining the gestational timing of crucial changes in specific organ and system functions that ultimately result in specific PCOS defects and for determining transcriptional, translational, and posttranscriptional levels of dysfunction that can be targeted for therapeutic amelioration during pre- and postconception.

KEY POINTS

- Animal models provide insight into the etiology and pathophysiology of PCOS unattainable in human studies.
- Many animal models either fail to exhibit PCOS diagnostic traits or exhibit traits that resemble other clinical disorders, including thyroid dysfunction and hyperprolactinemia (Table 2).
- Androgen excess fetal programming of PCOS in female mice, rats, sheep, and monkeys (Table 3) provides compelling evidence for fetal origins of the syndrome in humans.
- Androgen excess fetal programming of a male PCOS metabolic phenotype suggests fetal origins for male PCOS beyond the direct effects of androgens on the fetus.
- The intrauterine environment conducive to PCOS may be induced by a variety of genetic or environmental factors, or a combination of both, resulting in fetal androgen excess.

ACKNOWLEDGMENTS

We thank the many staff members of our respective laboratories and institutions for their multiple contributions to the work reported here and A.D.M. Abbott for his contribution towards compilation of Table 2. This work was supported by NIH grants P50 HD044405, U01 HD044650, R01 RR013635, R21 RR014093, T32 AG000268, P51 RR000167, R01 HD041098, and P01 HD044232 and was partly conducted at a facility constructed with support from Research Facilities Improvement Program grant numbers RR15459 and RR020141.

REFERENCES

1. Hague WM, Adams J, Reeders ST, Peto TEA, Jacobs HS. Familial polycystic ovaries: a genetic disease? Clin Endocrinol 1988;29:593–605.
2. Govind A, Obhrai MS, Clayton RN. Polycystic ovaries are inherited as an autosomal dominant trait: analysis of 29 polycystic ovary syndrome and 10 control families. J Clin Endocrinol Metab 1999;84:38–43.
3. Legro RS. The genetics of polycystic ovary syndrome. Am J Med 1995;16;98(1A):9S–16S.
4. Legro RS, Spielman R, Urbanek M, Driscoll D, Strauss JF 3rd, Dunaif A. Phenotype and genotype in polycystic ovary syndrome. Recent Prog Horm Res 1998;53:217–256.
5. Ehrmann DA, Barnes RB, Rosenfield RL. Polycystic ovary syndrome as a form of functional ovarian hyperandrogenism due to dysregulaton of androgen secretion. Endocr Rev 1995;16:322–353.
6. Dunaif A. Insulin resistance and the polycystic ovary syndrome: mechanism and implications for pathogenesis. Endocr Rev 1997;18:774–800.
7. Azziz R, Woods KS, Reyna R, Key TJ, Knochenhauer ES, Yildiz BO. The prevalence and features of the polycystic ovary syndrome in an unselected population. J Clin Endocrinol Metab 2004;89:2745–2749.
8. Arslanian SA, Lewy VD, Danadian K. Glucose intolerance in obese adolescents with polycystic ovary syndrome: roles of insulin resistance and beta-cell dysfunction and risk of cardiovascular disease. J Clin Endocrinol Metab 2001;86:66–71.
9. Peppard HR, Marfori J, Iuorno MJ, Nestler JE. Prevalence of polycystic ovary syndrome among premenopausal women with type 2 diabetes. Diabetes Care 2001;24:1050–1052.
10. Legro RS, Strauss JF. Molecular progress in infertility: polycystic ovary syndrome. Fertil Steril 2003;78:569–576.
11. Abbott DH, Barnett DK, Bruns CM, Dumesic DA. Androgen excess fetal programming of female reproduction: a developmental aetiology for polycystic ovary syndrome? Hum Reprod Update 2005;11:357–374.
12. Zawadzki JA, Dunaif A. Diagnostic criteria for polycystic ovary syndrome: towards a rational approach. In: Dunaif A, Givens JR, Haseltine FP, Merriam GR, eds. Polycystic Ovary Syndrome. Boston: Blackwell Scientific, 1992:377–384.
13. The Rotterdam ESHRE/ASRM-Sponsored PCOS consensus workshop group. Revised 2003 consensus on diagnostic criteria and long-term health risks related to polycystic ovary syndrome (PCOS). Hum Reprod 2004;19:41–47.
14. Azziz R. Diagnostic criteria for polycystic ovary syndrome: a reappraisal. Fertil Steril 2005;83:1343–1346.
15. Chang RJ. A practical approach to the diagnosis of polycystic ovary syndrome. Am J Obstet Gynecol 2004;191:713–717.
16. Arslanian SA, Witchell S. Premature pubarche, insulin resistance, and adolescent polycystic ovary syndrome. In: Chang RJ, Heindel JJ, Dunaif, A, eds. Polycystic Ovary Syndrome. New York: Marcel Dekker, Inc., 2002:37–53.
17. Rosenfield RL, Ghai K, Ehrmann DA, Barnes RB. Diagnosis of the polycystic ovary syndrome in adolescence: comparison of adolescent and adult hyperandrogenism. J Pediatr Endocrinol Metab 2000;13(Suppl 5):1285–1289.
18. Taylor AE, McCourt B, Martin KA, et al. Determinants of abnormal gonadotropin secretion in clinically defined women with polycystic ovary syndrome. J Clin Endocrinol Metab 1997;82:2248–2256.
19. Abbott DH, Foong SC, Barnett DK, Dumesic DA. Nonhuman primates contribute unique understanding to anovulatory infertility in women. ILAR J 2004;45:116–131.

20. Risma KA, Hirshfield AN, Nilson JH. Elevated luteinizing hormone in prepubertal transgenic mice causes hyperandrogenemia, precocious puberty, and substantial ovarian pathology. Endocrinology 1997;138:3540–3547.
21. Ward RC, Costoff A, Mahesh VB. The induction of polycystic ovaries in mature cycling rats by the administration of dehydroepiandrosterone (DHA). Biol Reprod 1978;18:614–623.
22. Mahesh VB, Mills TM, Bagnell CA, Conway BA. Animal models for study of polycystic ovaries and ovarian atresia. In: Mahesh VB, Dhindsa DS, Anderson E, Katra SP, eds. Regulation of Ovarian and Testicular Function. New York: Plenum Press, 198, pp. 237–257.
23. Ruiz A, Aguilar R, Tébar M, Gaytán F, Sánchez-Criado E. RU486-treated rats show endocrine and morphological responses to therapies analogous to responses of women with polycystic ovary syndrome treated with similar therapies. Biol Reprod 1996;55:1284–1291.
24. Hillier SG, Groom GV, Boyns AR, Cameron EH. Development of polycystic ovaries in rats actively immunised against T-3-BSA. Nature 1974;250:433–434.
25. Beloosesky R, Gold R, Almog B, et al. Induction of polycystic ovary by testosterone in immature female rats: modulation of apoptosis and attenuation of glucose/insulin ratio. Int J Mol Med 2004;14:207–215.
26. Bogovich K. Induction of ovarian follicular cysts in the pregnant rat by human chorionic gonadotropin. Biol Reprod 1991;45:34–42.
27. Poretsky L, Clemons J, Bogovich K. Hyperinsulinemia and human chorionic gonadotropin synergistically promote the growth of ovarian follicular cysts in rats. Metabolism 1992;41:903–910.
28. Damario MA, Bogovich K, Liu HC, Rosenwaks Z, Poretsky L. Synergistic effects of insulin-like growth factor-I and human chorionic gonadotropin in the rat ovary. Metabolism 2000;49:314–320.
29. Dzienis A, Majewski M, Wojtkiewicz J, Piskula M, Jana B. Adrenergic innervation and steroidogenic activity of cystic porcine ovaries. Rocz Akad Med Bialymst 2004;49(Suppl 1):114–116.
30. Kafali H, Iriadam M, Ozardah I, Demir N. Letrozole-induced polycystic ovaries in the rat: a new model for cystic ovarian disease. Arch Med Res 2004;35:103–108.
31. Fisher CR, Graves KH, Parlow AF, Simpson ER. Characterization of mice deficient in aromatase (ArKO) because of targeted disruption of the *cyp19* gene. Proc Natl Acad Sci USA 1998;95:6965–6970.
32. Zachos NC, Billiar RB, Albrecht ED, Pepe GJ. Developmental regulation of baboon fetal ovarian maturation by estrogen. Biol Reprod 2002;67:1148–1156.
33. Faiman C, Reyes FI, Dent DW, Fuller GB, Hobson WC, Thliveris JA. Effects of long-term testosterone exposure on ovarian function and morphology in the rhesus monkey. Anat Rec 1988;222:245–251.
34. Billiar RB, Richardson D, Anderson E, Mahajan D, Little B. The effect of chronic and acyclic elevation of circulating androstenedione or estrone concentrations on ovarian function in the rhesus monkey. Endocrinology 1985;16:2209–2220.
35. Mahajan DK. Polycystic ovarian disease: animal models. Endocrinol Metab Clin North Am 1988;17:705–732.
36. Billiar RB, Richardson D, Schwartz R, Posner B, Little B. Effect of chronically elevated androgen or estrogen on the glucose tolerance test and insulin response in female rhesus monkeys. Am J Obstet Gynecol 1987;157:1297–1302.
37. Vendola K, Zhou J, Wang J, Famuyiwa OA, Bievre M, Bondy CA. Androgens promote oocyte insulin-like growth factor I expression and initiation of follicle development in the primate ovary. Biol Reprod 1999;61:353–357.
38. Stubbs SA, Hardy K, Da Silva-Buttkus P, Stark J, Webber LJ, Flanagan AM, Themmen AP, Visser JA, Groome NP, Franks S. Anti-Mullerian hormone (AMH) protein expression is reduced during the initial stages of follicle development in human polycystic ovaries. J Clin Endocrinol Metab 2005;90:5536–5543.
39. Barker DJP. Mothers, Babies and Health in Later Life. Edinburgh: Churchill Livingstone, 1994.
40. Abbott DH, Dumesic DA, Eisner, Kemnitz JW, Goy RW. The prenatally androgenized female rhesus monkey as a model for polycystic ovarian syndrome. In: Azziz R, Nestler JE, Dewailly D, eds. Androgen Excess Disorders in Women. Philadelphia: Lippincott-Raven Press, 1997:369–382.
41. Abbott DH, Bruns CM, Barnett DK, Dumesic DA. Fetal programming of polycystic ovary syndrome. In: Kovacs WG, Norman RL, eds. Polycystic Ovary Syndrome, 2nd ed. Cambridge: Cambridge University Press, 2006.
42. Padmanabhan V, Manikkam M, Recabarren S, Foster D. Prenatal testosterone programs reproductive and metabolic dysfunction in the female. Mol Cell Endocrinol 2005;246:165–174.
43. Barbieri RL, Saltzman DH, Torday JS, Randall RW, Frigoletto FD, Ryan KJ. Elevated concentrations of the beta-subunit of human chorionic gonadotropin and testosterone in the amniotic fluid of gestations of diabetic mothers. Am J Obstet Gynecol 1986;154:1039–1043.
44. Barnes RB, Rosenfield RL, Ehrmann DA, et al. Ovarian hyperandrogynism as a result of congenital adrenal virilizing disorders: evidence for perinatal masculinization of neuroendocrine function in women. J Clin Endocrinol Metab 1994;79:1328–1333.
45. Sir-Petermann T, Maliqueo M, Angel B, Lara HE, Perez-Bravo F, Recabarren SE. Maternal serum androgens in pregnant women with polycystic ovarian syndrome: possible implications in prenatal androgenization. Hum Reprod 2002;17:2573–2579.
46. Gitau R, Adams D, Fish NM, Glover V. Fetal plasma testosterone correlates positively with cortisol. Arch Dis Child Fetal Neonatal Ed 2005;90:F166–F169.

47. West C, Foster DL, Evans NP, Robinson J, Padmanabhan V. Intrafollicular activin availability is altered in prenatal-androgenized lambs. Mol Cell Endocrinol 2001;185:51–59.
48. Sullivan SD, Moenter SM. Prenatal androgens alter GABAergic drive to gonadotropin-releasing hormone neurons: implications for a common fertility disorder. Proc Natl Acad Sci USA 2004;101:7129–7134.
49. Foecking EM, Szabo M, Schwartz NB, Levine JF. Neuroendocrine consequences of prenatal androgen exposure in the female rat: absence of luteinizing hormone surges, suppression of progesterone receptor gene expression, and acceleration of the gonadotropin-releasing hormone pulse generator. Biol Reprod 2005;72:1475–1483.
50. Dumesic DA, Schramm RD, Abbott DH. Early origins of polycystic ovary syndrome (PCOS). Reprod Fertil Dev 2005;17:349–360.
51. Recabarren SE, Padmanabhan V, Codner E, et al. Postnatal developmental consequences of altered insulin sensitivity in female sheep treated prenatally with testosterone. Am J Physiol Endocrinol Metab 2005;E807–E806.
52. Ibanez L, Potau N, Zampolli M, et al. Hyperinsulinemia in postpubertal girls with a history of premature pubarche and functional ovarian hyperandrogenism. J Clin Endocrinol Metab 1996;81:1237–1243.
53. Sir-Petermann T, Hitchsfeld C, Maliqueo M, et al. Birth weight in offspring of mothers with polycystic ovarian syndrome. Hum Reprod 2005;20:2122–2126.
54. Laitinen J, Taponen S, Martikainen H, et al. Body size from birth to adulthood as a predictor of self-reported polycystic ovary syndrome symptoms. Int J Obes Relat Metab Disord 2003;27:710–715.
55. Sadrzadeh S, Klip WA, Broekmans FJ, et al.; OMEGA Project Group. Birth weight and age at menarche in patients with polycystic ovary syndrome or diminished ovarian reserve, in a retrospective cohort. Hum Reprod 2003:18:2225–2230.
56. Manikkam M, Crespi EJ, Doop DD, et al. Fetal programming: prenatal testosterone excess leads to fetal growth retardation and postnatal catch-up growth in sheep. Endocrinology 2004;145:790–798.
57. Slob AK, den Hamer R, Woutersen PJ, van der Werff ten Bosch JJ. Prenatal testosterone propionate and postnatal ovarian activity in the rat. Acta Endocrinol (Copenh) 1983;103:420–427.
58. Wolf CJ, LeBlanc GA, Gray LE Jr. Interactive effects of vinclozolin and testosterone propionate on pregnancy and sexual differentiation of the male and female SD rat. Toxicol Sci 2004;78:135–143.
59. Wood JR, Ho CK, Nelson-Degrave VL, McAllister JM, Strauss JF III. The molecular signature of polycystic ovary syndrome (PCOS) theca cells defined by gene expression profiling. J Reprod Immunol 2004;63:51–60.
60. Yildiz BO, Yarali H, Oguz H, Bayraktar M. Glucose intolerance, insulin resistance, and hyperandrogenemia in first degree relatives of women with polycystic ovary syndrome. J Clin Endocrinol Metab 2003;88:2031–2036.
61. Fox R. Prevalence of a positive family history of type 2 diabetes in women with polycystic ovarian disease. Gynecol Endocrinol 1999;13:390–393.
62. Sir-Petermann T, Angel B, Maliqueo M, Carvajal F, Santos JL, Perez-Bravo F. Prevalence of Type II diabetes mellitus and insulin resistance in parents of women with polycystic ovary syndrome. Diabetologia 2002;45:959–964.
63. Bruns CM, Baum ST, Colman RJ, et al. Insulin resistance and impaired insulin secretion in prenatally androgenized male rhesus monkeys. J Clin Endocrinol Metab 2004;89:6218–6223.
64. Resko JA, Buhl AE, Phoenix CH. Treatment of pregnant rhesus macaques with testosterone propionate: observations on its fate in the fetus. Biol Reprod 1987;37:1185–1191.
65. Sarma HN, Manikkam M, Herkimer C, Dell'Orco J, Foster DL, Padmanabhan V. Fetal programming: excess prenatal testosterone reduces postnatal LH, but not FSH responsiveness to estradiol negative feedback in the female. Endocrinology 2005;146:4281–4291.
66. Dumesic DA, Schramm RD, Peterson E, Paprocki AM, Zhou R, Abbott DH. Impaired developmental competence of oocytes in adult prenatally androgenized female rhesus monkeys undergoing gonadotropin stimulation for in vitro fertilization. J Clin Endocrinol Metab 2002;87:1111–1119.
67. Bagavandoss P, England B, Asirvatham A, Bruot BC. Transient induction of polycystic ovary-like syndrome in immature hypothyroid rats. Proc Soc Exp Biol Med 1998;219:77–84.
68. Brawer JR, Munoz M, Farookhi R. Development of the polycystic ovarian condition (PCO) in the estradiol-valerate treated rat. Biol Reprod 1986;35:647–655.
69. Quant LM, Hutz RJ. Induction by estradiol 17β of polycystic ovaries in the guinea pig. Biol Reprod 1993;48:1088–1094.
70. Lagace DC, Nachtigal MW. Valproic acid fails to induce polycystic ovary syndrome in female rats. Prog Neuropsychopharmacol Biol Psychiatry 2003;27:587–594.
71. Ferin M, Morrell M, Xiao E, et al. Endocrine and metabolic responses to long-term monotherapy with antiepileptic drug valproate in the normally cycling rhesus monkey. J Clin Endocrinol Metab 2003;88:2908–2915.
72. Furudate S, Nakano T. PMSG-induced persistent estrus in rats as a model for polycystic ovary disease: characteristics and restoration to the normal cycle. Exp Anim 1989;38:121–126.
73. Jones HM, Vernon MW, Rush ME. Systematic studies invalidate the neonatally androgenized rat as a model for polycystic ovary disease. Biol Reprod 1987;36:1253–1265.

24

Insulin Resistance and Hyperinsulinism in the Polycystic Ovary Syndrome

Paulina A. Essah and John E. Nestler

SUMMARY

Insulin resistance with compensatory hyperinsulinemia has been demonstrated to occur in 50–70% of women with polycystic ovary syndrome (PCOS), regardless of weight. Hyperinsulinemia stimulates excess ovarian androgen production, thereby contributing to the hyperandrogenism and chronic anovulation characteristic of PCOS. The exact cause of insulin resistance in PCOS is unknown, but it appears to be related to a postbinding defect in insulin receptor-mediated signal transduction. Because of insulin resistance, women with PCOS are at risk for several long-term metabolic complications, including type 2 diabetes and cardiovascular disease.

Key Words: Insulin resistance; hyperinsulinism; insulin receptor-mediated signal transduction.

1. INTRODUCTION

Although the pathogenesis of polycystic ovary syndrome (PCOS) remains unknown, the discovery that many women with PCOS have underlying insulin resistance and compensatory hyperinsulinemia has led to a much better understanding of the syndrome. This metabolic abnormality contributes to the hyperandrogenism that characterizes PCOS and leads to its clinical signs and symptoms. The recognition of insulin resistance in PCOS has also influenced our understanding of the metabolic complications associated with PCOS. This chapter will provide an overview of the current body of knowledge regarding the prevalence, causes, and consequences of insulin resistance and hyperinsulinism in PCOS.

2. BACKGROUND

2.1. Overview of Insulin Resistance in PCOS

The relationship between PCOS and insulin resistance was first described by Burghen et al. in 1980 (1), with reports of a significant positive correlation between levels of androgens (testosterone and androstenedione) and insulin in a small number of obese women with PCOS. Several studies subsequently supported these findings (2–5).

Insulin resistance is defined as decreased sensitivity of target organ tissues to the action of insulin. Another way to describe insulin resistance is a decreased glucose response to a given amount of insulin, also known as decreased insulin-mediated glucose uptake. Hyperinsulinism refers to a state of elevated insulin expression, either clinically or biochemically (hyperinsulinemia).

Despite the compensatory hyperinsulinemia that accompanies insulin resistance in nondiabetic individuals, insulin-mediated glucose uptake remains subnormal (6). Interestingly, insulin resistance in PCOS does not occur in all tissues, but rather appears to be tissue-specific (7). Skeletal muscle and

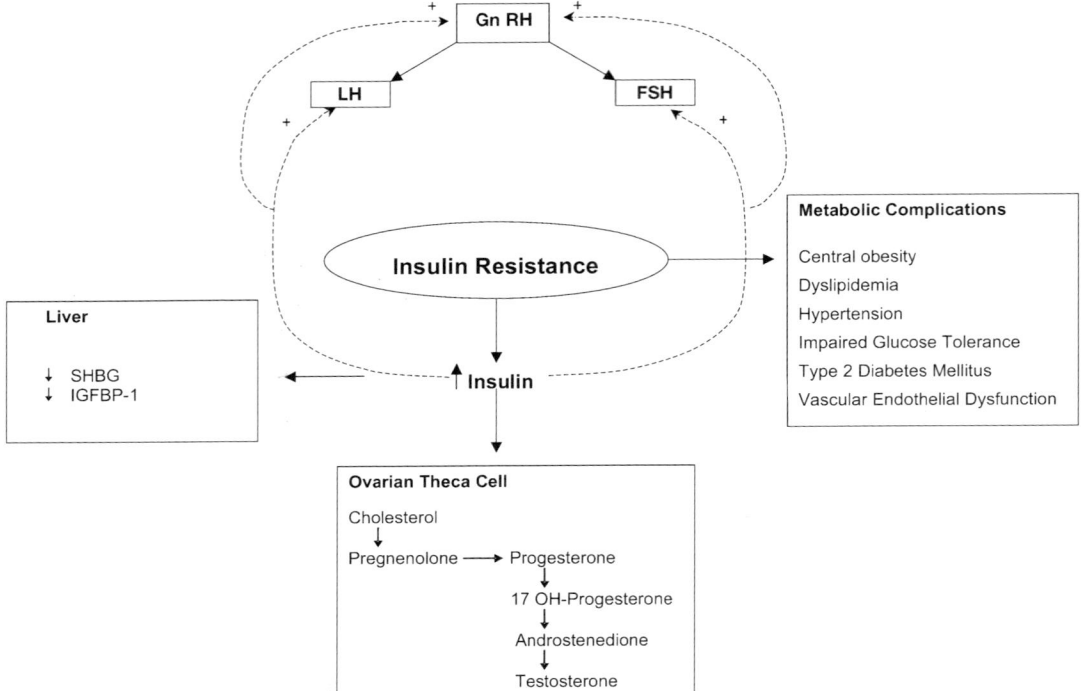

Fig. 1. Overview of the role of insulin resistance and hyperinsulinemia in the polycystic ovary syndrome (PCOS). FSH, follicle-stimulating hormone; GnRH, gonadotropin-releasing hormone; IGFBP-1, insulin-like factor-binding protein-1; LH, luteinizing hormone; SHBG, sex hormone-binding globulin.

adipose tissue become insulin resistant, resulting in decreased glucose uptake and increased lipolysis, respectively, whereas the ovary, adrenal, liver, and skin remain insulin sensitive *(7)*.

In PCOS, hyperinsulinemia occurs as a compensatory response to insulin resistance. This resulting hyperinsulinemia has a stimulatory effect on the ovaries and adrenal glands that leads to enhanced androgen production by these organs. More specifically, excess insulin enhances androgen production in ovarian theca cells in response to luteinizing hormone (LH) stimulation, resulting in follicular arrest and anovulation. In addition, hyperinsulinemia stimulates proliferation of the pilosebaceous unit and sebum production, resulting in hirsutism and acne. In contrast, hyperinsulinemia acts to suppress hepatic production of sex hormone-binding globulin, the primary binding protein for testosterone in the serum. Therefore, insulin resistance with compensatory hyperinsulinemia results in hyperandrogenemia (Fig. 1).

There is also evidence that women with PCOS may have pancreatic β-cell dysfunction, as occurs in type 2 diabetes *(6,8)*. It has been reported that women with PCOS secrete an inadequate amount of insulin for the degree of peripheral insulin resistance that they experience *(6)*.

2.2. Measurement of Insulin Resistance in PCOS

The gold standard method for measuring insulin sensitivity is the hyperinsulinemic-euglycemic clamp technique, first described by De Fronzo et al. *(9)*, which measures insulin-mediated glucose uptake primarily in skeletal muscle. With this method, insulin is administered intravenously in one arm at a steady rate, while a variable glucose infusion is administered in the other arm in order to "clamp" the serum glucose level at a normal fasting concentration. Blood samples are taken frequently to monitor serum glucose levels in order to sustain a steady fasting level. The continuous insulin infusion suppresses hepatic glucose production, so that the amount of glucose infused to maintain a steady fasting level must equal the amount of glucose taken up by the cells of the body (after a small correction for urinary glucose losses). The degree of insulin resistance is inversely proportional to the glucose uptake. Therefore, low glucose uptake reflects lower insulin sensitivity, which is equivalent to greater insulin resistance. A variation of the clamp technique is the hyperinsulinemic hyperglycemic clamp, which takes less time to perform and better measures pancreatic β-cell function, but is less physiological than the euglycemic technique *(7,10)*.

Unfortunately, the hyperinsulinemic-euglycemic clamp technique is expensive, time-consuming, and labor intensive. Therefore, several other measures of insulin sensitivity have been devised. These include the insulin tolerance test (ITT), frequently sampled intravenous glucose tolerance test (FSIVGTT), oral glucose tolerance test (OGTT) with determinations of both insulin and glucose, fasting glucose-to-insulin ratio (G:I ratio), homeostasis model assessment (HOMA), and quantitative sensitivity check index (QUICKI), all of which have been shown to correlate well with the hyperinsulinemic-euglycemic clamp technique *(7)*. The FSIVGTT and OGTT methods are considered to be minimal models, meaning that, unlike the clamp technique and the ITT, they require a minimum of glucose administration only without any insulin administration (although insulin is used in the modified FSIVGTT method). A detailed discussion of each of these techniques is beyond the scope of this chapter. Table 1 summarizes these measures of insulin sensitivity and their applicability in women with PCOS.

2.3. Prevalence of Insulin Resistance in PCOS

Approximately 50–70% of women with PCOS have some degree of insulin resistance *(7)*, ranging from 20 to 95% (Table 2). Both obese and lean women with PCOS have been demonstrated to have some degree of insulin resistance *(5)*. However, obese women with PCOS are more insulin resistant than either lean women with PCOS or obese, women without PCOS *(5,11)*.

Some, but not all studies have reported a significant influence of ethnicity on the prevalence of insulin resistance in PCOS. In a study of 37 Mexican-American women and 65 Caucasian American women, all with PCOS, Kauffman et al. *(12)* reported that, using HOMA to estimate insulin sensitivity, Mexican-American women were more insulin resistant than age- and weight-matched Caucasian women, with prevalences of insulin resistance of 73.1 vs 43.8%, respectively. In a controlled study of 36 women, Dunaif et al. *(13)* also found a higher prevalence of insulin resistance in Caribbean-Hispanic women with PCOS compared with Caucasian women.

Larger studies have not detected a significant influence of ethnicity on the presence of insulin resistance in PCOS. Carmina et al. *(14)* evaluated 75 women with PCOS—25 each from the United States, Italy, and Japan—and reported that insulin resistance, as measured by an ITT, was present similarly in all three groups of women, with a prevalence of 68–76% in the women. In a larger study of 267 women with PCOS and 50 ovulating controls, Carmina and Lobo *(15)* determined the prevalence of insulin resistance to be approximately 80% using the homeostasis model assessment and quantitative assessment check indices and 65% using G:I ratios.

Similarly, Azziz and colleagues *(16)* recently observed a 64% prevalence of insulin resistance in 271 consecutive patients with PCOS using the homeostasis model assessment of insulin resistance (HOMA-IR) calculation to estimate insulin sensitivity, with no significant association between HOMA-IR and race among the black and white women with PCOS studied. They also reported that patients with insulin resistance were more obese, had greater degrees of android body fat, had higher serum androgen levels, and displayed more hirsutism and acne compared with their non-insulin-resistant counterparts.

In terms of the prevalence of insulin resistance in women with PCOS, studies have revealed that the majority of, but not all, women with PCOS are insulin resistant.

Table 1
Methods of Measuring Insulin Sensitivity

Test	Correlation with clamp technique	iv access needed	Simple mathematical model	Insulin resistance cutoff in PCOS	Comments
Hyperinsulinemic-euglycemic clamp	Gold standard	Yes	None	None	Hyperglycemic version allows for best assessment of β-cell function
Insulin tolerance test (ITT)	Good	Yes	None	None	Numerous modifications
Insulin sensitivity test (IST)	Good	Yes	None	None	Numerous modifications
Frequently sampled intravenous glucose tolerence test (FSIVGTT)	Good	Yes	None	None	Numerous modifications
Oral glucose tolerance test (OGTT)	Good	No	G_{120}/I_{120}	≤1.0	Modifications using this method include are under the curve (AUC) insulin and AUC glucose
Fasting insulin	Good	No	I_0	≥20 white	Loss of accuracy with hyperglycemia
Fasting glucose:insulin ratio (G:I ratio)	Good	No	G_0/I_0	≥23 Mexican-American ≤4.5 white	Loss of accuracy with hyperglycemia; different values in different ethnic groups
Homeostatic model assessment (HOMA)	Good	No	$I_0 \times G_0 / 405$	≥7.2 white ≤4.0 Mexican-American	Use 22.5 instead of 40.5 when glucose expressed as mmol/L
Quantitative insulin sensitivity check index (QUICKI)	Good	No	$1/[10 g (I_0) + 10 g (G_0)]$	None	Applicable to hyperglycemic patients

Adapted with permission from ref. 7.

Table 2
Prevalence of Insulin Resistance in Women With Polycystic Ovary Syndrome

Study (ref.)	Population	Measure of insulin resistance	Prevalence (%)
Dunaif et al., 1989 *(5)*	Obese	Euglycemic glucose clamp	26
	Nonobese	Euglycemic glucose clamp	60
Carmina et al., 1992 *(14)*	U.S.	Insulin tolerance test	76
	Italy	Insulin tolerance test	72
	Japan	Insulin tolerance test	68
Meirow et al., 1995 *(48)*	Obese	Glucose/insulin ratio	64
	Nonobese	Glucose/insulin ratio	20
Legro et al., 1998 *(49)*	Obese, white	Modified FSIVGTT	53
Kauffman et al., 2002 *(12)*	U.S., all	HOMA-IR	54.50
	White	HOMA-IR	43.80
	Mexican-American	HOMA-IR	73.10
Carmina and Lobo, 2004 *(15)*	Italy, all	HOMA-IR	77
		QUICKI	79.20
		Glucose/insulin ratio	65.40
	Italy, obese	HOMA-IR	95.30
		QUICKI	95.30
		Glucose/insulin ratio	76.70
DeUgarte et al., 2005 *(16)*	Southeast U.S., all races	HOMA-IR	64

FSIVGTT, frequently sampled intravenous glucose tolerance test; HOMA-IR, homeostatic model assessment for insulin resistance; QUICKI, quantitative insulin sensitivity check index.

2.4. Pathophysiology of Insulin Resistance in PCOS

Over the past decade, accumulating evidence has indicated that insulin resistance plays a major role in the pathogenesis of PCOS. Moreover, it has been hypothesized that the cellular defect that causes insulin resistance in PCOS may simultaneously result in increased ovarian androgen production. There is also emerging evidence that a form of insulin resistance present in PCOS may be intrinsic (and possibly unique) to the disorder, and that it is acquired only by women who are genetically susceptible *(17)*. The insulin resistance of PCOS is independent of obesity, metabolic abnormalities, and sex hormone levels *(17)*.

2.4.1. The Role of the Insulin Receptor and Hyperinsulinemia

The cause of insulin resistance in PCOS appears to be a postbinding defect in insulin receptor-mediated signal transduction *(18)*. However, the exact nature of this defect remains largely unknown and is being explored. A subsequent chapter of this book elaborates on this concept in a summary of the molecular mechanisms of insulin resistance in PCOS.

The outcome of the postreceptor defect in insulin receptor-mediated signal transduction is peripheral tissue resistance to insulin, which in turn results in higher insulin production and secretion from the pancreas. Insulin may then act both directly and indirectly to increase endogenous androgen levels via several mechanisms:

1. In the typical case of PCOS, insulin binds directly to insulin receptors in the ovary to increase ovarian androgen production by theca cells in response to LH stimulation *(19)*.

2. If hyperinsulinemia is marked (as in rare syndromes of type A or B insulin resistance), extreme hyperinsulinemia may result in insulin "spillover" activation of ovarian insulin-like growth factor (IGF)-1 receptors and thereby stimulate androgen production *(20,21)*. In addition, atypical IGF receptor subtypes may be present in the ovary that have extremely high affinity for insulin binding *(22)*.

3. Hyperinsulinemia inhibits synthesis of sex hormone-binding globulin by the liver, which leads to elevated levels of serum-free testosterone *(17)*.

4. Hyperinsulinemia causes a decrease in secretion of IGF binding protein-1 in both the liver and the ovary, which in turn results in increased intraovarian bioavailability of IGF-1 and IGF-2 *(23,24)*, two important regulators of ovarian follicular maturation and steroidogenesis *(17,25)*.

Genetic defects have been implicated in the insulin signal transduction abnormality that leads to insulin resistance and hyperinsulinemia. Dunaif et al. reported genetic abnormalities in the regulation of insulin receptor phosphorylation, causing an increase in insulin-dependent serine phosphorylation and a decrease in insulin-dependent tyrosine phosphorylation *(29)*. These abnormalities in insulin receptor phosphorylation result in reduced insulin responsiveness. Several polymorphisms have been identified at the coding region of the insulin receptor gene in women with PCOS *(30)*. However, most of these polymorphisms have also been demonstrated in normal subjects and do not appear to result in significant dysfunction of the insulin receptor *(30)*. Extensive research is ongoing in locating major insulin receptor genetic defects and other insulin-related genetic defects that may cause the insulin resistance and hyperinsulinemia of PCOS.

2.4.2. The Role of Obesity

Obesity may also contribute to the insulin resistance of PCOS. However, it has been suggested that it is the distribution of fat, rather than the mere presence of obesity or increased body mass index, that is mainly significant *(31)*. Regardless of ethnicity, most overweight women with PCOS have central, or android (visceral), obesity, resulting in a waist-to-hip ratio greater than 0.85 *(32)*. Visceral adiposity is known to be metabolically active and is more highly associated with hyperinsulinemia than subcutaneous fat *(33)*. Of note, 70% of lean women with PCOS also have an android distribution of fat *(34)*.

The mechanism(s) by which increased visceral adipose tissue results in hyperinsulinemia are not well understood. Altered lipolysis in visceral fat cells appears to result in increased free fatty acids that drain via the portal vein to the liver and subsequently affect the secretion, metabolism, and peripheral actions of insulin *(35)*. Notably, defects in visceral adipose cell lipolysis have been demonstrated even in non-obese women with PCOS *(36)*. Insulin resistance in central obesity can also be associated with tumor necrosis factor (TNF)-α and leptin, both produced from adipose tissue *(35)*. Both TNF-α and leptin are involved in mediating serine phosphorylation of insulin receptor substrate-1, which interferes with the action of both insulin and IGF-1. TNF-α has also been reported to inhibit insulin signaling through peroxisome proliferators-activated receptor-γ.

2.5. Complications of Insulin Resistance in PCOS

Because of its association with insulin resistance, PCOS can lead to several metabolic complications. The metabolic syndrome has been reported to occur at an increased overall prevalence rate of 43–47% in women with PCOS *(37–39)* compared with the 24% prevalence rate in US women *(40)*. Studies have shown that atherosclerosis is more prevalent in women with PCOS *(5)*. Women with PCOS are also at risk for developing gestational diabetes and type 2 diabetes mellitus (DM) *(41,42)*. In women with PCOS, there is a reported prevalence of 30–40% of glucose intolerance *(43)* and prevalence of 5–10% of type 2 diabetes *(44)*. In addition, women with PCOS have a 5- to 10-fold increased rate of conversion from impaired glucose tolerance to type 2 DM *(44,45)*. A recent study revealed that women with PCOS and baseline normal glucose tolerance have a 16% conversion rate per year to type 2 DM *(46)*.

Other long-term risks resulting from insulin resistance include hypertension, atherogenic dyslipidemia, hypercoagulability, and vascular endothelial dysfunction, all risk factors for cardiovascular disease. Indeed, women with PCOS have been estimated to have a sevenfold increased risk for myocardial infarction, with increased cardiovascular risk independent of obesity *(47)*.

3. CONCLUSION

Insulin resistance with compensatory hyperinsulinemia occurs in the majority of women with PCOS, both obese and lean. This hormone abnormality contributes to the hyperandrogenism and chronic anovulation characteristic of PCOS. The pathophysiology of insulin resistance in PCOS is related in part to a postbinding defect in insulin receptor-mediated signal transduction, but the cause of this defect remains unknown. Nevertheless, because of insulin resistance, women with PCOS are at risk for several long-term metabolic complications, most notably type 2 DM and cardiovascular disease.

4. FUTURE AVENUES OF INVESTIGATION

Much is still unknown about the pathogenesis of insulin resistance and hyperinsulinism in PCOS. Future avenues of investigation include identification of genetic defects that might lead to the insulin resistance present in PCOS and further delineation of the cellular and molecular pathways that result in insulin resistance and hyperinsulinemia. In addition, investigation is needed to determine why some tissues, such as skeletal muscle, are insulin resistant whereas other tissues, particularly the ovaries, remain sensitive to the effects of insulin in PCOS.

KEY POINTS

- Approximately 50–70% of women with PCOS have some degree of insulin resistance that is independent of weight.
- The pathogenesis of insulin resistance in PCOS appears to involve a postbinding receptor defect in insulin receptor-mediated signal transduction, which in turn results in higher insulin secretion from the pancreas.
- Because of its association with insulin resistance, PCOS can lead to several metabolic complications, including type 2 DM, dyslipidemia, hypertension, and cardiovascular disease.

REFERENCES

1. Burghen GA, Givens JR, Kitabchi AE. Correlation of hyperandrogenism with hyperinsulinism in polycystic ovarian disease. J Clin Endocrinol Metab 1980;50:113–116.
2. Reaven GM. Banting lecture 1988. Role of insulin resistance in human disease. Diabetes 1988;37:1595–1607.
3. Nestler JE, Clore JN, Strauss JF, III, Blackard WG. Effects of hyperinsulinemia on serum testosterone, progesterone, dehydroepiandrosterone sulfate, and cortisol levels in normal women and in a woman with hyperandrogenism, insulin resistance and acanthosis nigricans. J Clin Endocrinol Metab 1987;64:180–184.
4. Nestler JE, Barlascini CO, Matt DW, et al. Suppression of serum insulin by diazoxide reduces serum testosterone levels in obese women with polycystic ovary syndrome. J Clin Endocrinol Metab 1989;68:1027–1032.
5. Dunaif A, Segal KR, Futterweit W, Dobrjansky A. Profound peripheral insulin resistance, independent of obesity, in polycystic ovary syndrome. Diabetes 1989;38:1165–1174.
6. Dunaif A, Finegood DT. Beta-cell dysfunction independent of obesity and glucose intolerance in the polycystic ovary syndrome. J Clin Endocrinol Metab 1996;81:942–947.
7. Legro RS, Castracane VD, Kauffman RP. Detecting insulin resistance in polycystic ovary syndrome: purposes and pitfalls. Obstet Gynecol Surv 2004;59(2):141–154.
8. Ehrmann DA, Sturis J, Byrne MM, Karrison T, Rosenfield RL, Polonsky KS. Insulin secretory defects in polycystic ovary syndrome. Relationship to insulin sensitivity and family history of non- insulin-dependent diabetes mellitus. J Clin Invest 1995;96:520–527.
9. DeFronzo RA, Tobin JD, Andres R. Glucose clamp technique: a method for quantifying insulin secretion and resistance. Am J Physiol 1979;237(3):E214–E223.
10. Wallace TM, Matthews DR. The assessment of insulin resistance in man. Diabet Med 2002;19(7):527–534.
11. Dunaif A, Graf M, Mandeli J, Laumas V, Dobrjansky A. Characterization of groups of hyperandrogenemic women with acanthosis nigricans, impaired glucose tolerance, and/or hyperinsulinemia. J Clin Endocrinol Metab 1987; 65:499–507.
12. Kauffman RP, Baker VM, Dimarino P, Gimpel T, Castracane VD. Polycystic ovarian syndrome and insulin resistance in white and Mexican American women: a comparison of two distinct populations. Am J Obstet Gynecol 2002;187(5):1362–1369.

13. Dunaif A, Sorbara L, Delson R, Green G. Ethnicity and polycystic ovary syndrome are associated with independent and additive decreases in insulin action in Caribbean-Hispanic women. Diabetes 1993;42:1462–1468.

14. Carmina E, Koyama T, Chang L, Stanczyk FZ, Lobo RA. Does ethnicity influence the prevalence of adrenal hyperandrogenism and insulin resistance in polycystic ovary syndrome? Am J Obstet Gynecol 1992;167:1807–1812.

15. Carmina E, Lobo RA. Use of fasting blood to assess the prevalence of insulin resistance in women with polycystic ovary syndrome. Fertil Steril 2004;82(3):661–665.

16. DeUgarte CM, Bartolucci AA, Azziz R. Prevalence of insulin resistance in the polycystic ovary syndrome using the homeostasis model assessment. Fertil Steril 2005;83(5):1454–1460.

17. Balen A. The pathophysiology of polycystic ovary syndrome: trying to understand PCOS and its endocrinology. Best Pract Res Clin Obstet Gynaecol 2004;18(5):685–706.

18. Dunaif A. Insulin resistance and the polycystic ovary syndrome: mechanism and implications for pathogenesis. Endocr Rev 1997;18(6):774–800.

19. Nestler JE, Jakubowicz DJ, de Vargas AF, Brik C, Quintero N, Medina F. Insulin stimulates testosterone biosynthesis by human thecal cells from women with polycystic ovary syndrome by activating its own receptor and using inositolglycan mediators as the signal transduction system. J Clin Endocrinol Metab 1998;83:2001–2005.

20. Bergh C, Carlsson B, Olsson JH, Selleskog U, Hillensjo T. Regulation of androgen production in cultured human thecal cells by insulin-like growth factor I and insulin. Fertil Steril 1993;59:323–331.

21. Poretsky L, Cataldo NA, Rosenwaks Z, Giudice LC. The insulin-related ovarian regulatory system in health and disease. Endocr Rev 1999;20(4):535–582.

22. Ehrmann DA, Barnes RB, Rosenfield RL. Polycystic ovary syndrome as a form of functional ovarian hyperandrogenism due to dysregulation of androgen secretion. Endocr Rev 1995;16:322–353.

23. LeRoith D, Werner H, Beitner-Johnson D, Roberts CT Jr. Molecular and cellular aspects of the insulin-like growth factor I receptor. Endocr Rev 1995;16(2):143–163.

24. De Leo, V, la Marca A, Orvieto R, Morgante G. Effect of metformin on insulin-like growth factor (IGF) I and IGF-binding protein I in polycystic ovary syndrome. J Clin Endocrinol Metab 2000;85(4):1598–1600.

25. Adashi EY. Intraovarian regulation: the proposed role of insulin-like growth factors. Ann NY Acad Sci 1993;687:10–12.

26. Zhang LH, Rodriguez H, Ohno S, Miller WL. Serine phosphorylation of human P450c17 increases 17,20-lyase activity: implications for adrenarche and the polycystic ovary syndrome. Proc Natl Acad Sci USA 1995;92:10619–10623.

27. Rosenfield RL, Barnes RB, Cara JF, Lucky AW. Dysregulation of cytochrome P450c 17 alpha as the cause of polycystic ovarian syndrome. Fertil Steril 1990;53(5):785–791.

28. la Marca A, Egbe TO, Morgante G, et al. Metformin treatment reduces ovarian cytochrome P-450c17alpha response to human chorionic gonadotrophin in women with insulin resistance-related polycystic ovary syndrome. Hum Reprod 2000;15(1):21–23.

29. Dunaif A, Xia J, Book CB, Schenker E, Tang Z. Excessive insulin receptor serine phosphorylation in cultured fibroblasts and in skeletal muscle. A potential mechanism for insulin resistance in the polycystic ovary syndrome. J Clin Invest 1995;96:801–810.

30. Diamanti-Kandarakis E, Piperi C. Genetics of polycystic ovary syndrome: searching for the way out of the labyrinth. Hum Reprod Update 2005;11(6):631–643.

31. Lord J, Wilkin T. Polycystic ovary syndrome and fat distribution: the central issue? Hum Fertil (Cambr) 2002;5(2):67–71.

32. Strowitzki T, Halser B, Demant T. Body fat distribution, insulin sensitivity, ovarian dysfunction and serum lipoproteins in patients with polycystic ovary syndrome. Gynecol Endocrinol 2002;16(1):45–51.

33. Wajchenberg BL. Subcutaneous and visceral adipose tissue: their relation to the metabolic syndrome. Endocr Rev 2000;21(6):697–738.

34. Kirchengast S, Huber J. Body composition characteristics and body fat distribution in lean women with polycystic ovary syndrome. Hum Reprod 2001;16(6):1255–1260.

35. Gambineri A, Pelusi C, Vicennati V, Pagotto U, Pasquali R. Obesity and the polycystic ovary syndrome. Int J Obes Relat Metab Disord 2002;26(7):883–896.

36. Ek I, Arner P, Ryden M, et al. A unique defect in the regulation of visceral fat cell lipolysis in the polycystic ovary syndrome as an early link to insulin resistance. Diabetes 2002;51(2):484–492.

37. Apridonidze T, Essah PA, Iuorno MJ, Nestler JE. Prevalence and characteristics of the metabolic syndrome in women with polycystic ovary syndrome. J Clin Endocrinol Metab 2005;90(4):1929–1935.

38. Dokras A, Bochner M, Hollinrake E, Markham S, Vanvoorhis B, Jagasia DH. Screening women with polycystic ovary syndrome for metabolic syndrome. Obstet Gynecol 2005;106(1):131–137.

39. Glueck CJ, Papanna R, Wang P, Goldenberg N, Sieve-Smith L. Incidence and treatment of metabolic syndrome in newly referred women with confirmed polycystic ovarian syndrome. Metabolism 2003;52(7):908–915.

40. Mokdad AH, Ford ES, Bowman BA, et al. Prevalence of obesity, diabetes, and obesity-related health risk factors, 2001. JAMA 2003;289(1):76–79.

41. De Leo, V, Musacchio MC, Morgante G, la Marca A, Petraglia F. Polycystic ovary syndrome and type 2 diabetes mellitus. Minerva Ginecol 2004;56(1):53–62.

42. Glueck CJ, Streicher P, Wang P. Treatment of polycystic ovary syndrome with insulin-lowering agents. Expert Opin Pharmacother 2002;3(8):1177–1189.
43. Legro RS, Kunselman AR, Dodson WC, Dunaif A. Prevalence and predictors of risk for type 2 diabetes mellitus and impaired glucose tolerance in polycystic ovary syndrome: a prospective, controlled study in 254 affected women. J Clin Endocrinol Metab 1999;84(1):165–169.
44. Ehrmann DA, Barnes RB, Rosenfield RL, Cavaghan MK, Imperial J. Prevalence of impaired glucose tolerance and diabetes in women with polycystic ovary syndrome. Diabetes Care 1999;22(1):141–146.
45. Norman RJ, Masters L, Milner CR, Wang JX, Davies MJ. Relative risk of conversion from normoglycaemia to impaired glucose tolerance or non-insulin dependent diabetes mellitus in polycystic ovarian syndrome. Hum Reprod 2001;16(9):1995–1998.
46. Legro RS, Gnatuk CL, Kunselman AR, Dunaif A. Changes in glucose tolerance over time in women with polycystic ovary syndrome: a controlled study. J Clin Endocrinol Metab 2005;90(6):3236–3242.
47. Lobo RA, Carmina E. The importance of diagnosing the polycystic ovary syndrome. Ann Intern Med 2000;132(12):989–993.
48. Meirow D, Yossepowitch O, Rosler A, et al. Insulin resistant and non-resistant polycystic ovary syndrome represent two clinical and endocrinological subgroups. Hum Reprod 1995;10(8):1951–1956.
49. Legro RS, Finegood D, Dunaif A. A fasting glucose to insulin ratio is a useful measure of insulin sensitivity in women with polycystic ovary syndrome. J Clin Endocrinol Metab 1998;83(8):2694–2698.

25

The Endocrine Impact of Obesity and Body Habitus in the Polycystic Ovary Syndrome

Renato Pasquali and Alessandra Gambineri

SUMMARY

Obesity, particularly the abdominal type, is associated with several abnormalities of sex steroid balance in women, including an increased androgen production rate and decreased serum levels of sex hormone-binding globulin. It is hypothesized that the increasing epidemic of obesity worldwide may also favor the high prevalence of obesity among women with polycystic ovary syndrome (PCOS). Obesity has a profound impact on the PCOS phenotype, being associated with more severe hyperandrogenism and insulin-resistant state and fertility disorders.

Key Words: Obesity; fat distribution; hyperandrogenism; insulin resistance; infertility.

1. INTRODUCTION

Polycystic ovary syndrome (PCOS), one of the most common causes of infertility resulting from anovulation, affects 4–7% of women *(1–3)*. Although it was considered that PCOS may have some genetic component and that the clinical features of this disorder may change throughout the life span, starting from adolescence to postmenopausal age, no effort has been made to define differences in phenotype and clinical presentation according to age. Indeed, in the past decade it has been widely recognized that several features of the metabolic syndrome, particularly insulin resistance and hyperinsulinemia, are inconsistently present in the majority of women with PCOS. This represents an important factor in the evaluation of PCOS throughout life and implies that PCOS by itself may not be a hyperandrogenic disorder exclusively restricted and relevant to young and fertile aged women, but may also have health implications later in life.

In young women with PCOS, hyperandrogenism, menstrual irregularities, and insulin resistance may occur together, emphasizing the pathophysiological role of excess androgen and insulin on PCOS *(1)*. Symptoms related to androgen excess, oligo- or amenorrhea, and infertility represent the major complaints of adult women with PCOS of reproductive age. In addition, obesity and the metabolic syndrome may affect more than half of these women. Later in life it becomes clear that the association between obesity (particularly the abdominal phenotype) and PCOS renders affected women more susceptible to developing type 2 diabetes, with some differences in the prevalence rates between countries, suggesting that environmental factors are important in determining individual susceptibility *(4)*.

2. BACKGROUND

2.1. Early Onset of Obesity Favors Development of PCOS

· Overweight and obesity represent a rapidly growing threat to the health of the populations of an increasing number of countries worldwide *(5)*. As expected, there is an increasing prevalence of overweight and obesity not only in the population at large, but also in adolescent and young women

From: *Contemporary Endocrinology: Androgen Excess Disorders in Women:*
Polycystic Ovary Syndrome and Other Disorders, Second Edition
Edited by: R. Azziz et al. © Humana Press Inc., Totowa, NJ

with PCOS *(4)*. Whether the high prevalence of obesity in women with PCOS may depend, at least in part, on the increasing epidemic of obesity *per se* worldwide is unknown, although this represents an attractive hypothesis.

The association between obesity and alterations of reproductive function in women was recognized long ago. In Stein and Leventhal's original description, obesity, together with hirsutism and infertility, represented one of the characteristics of the syndrome that eventually bore their names. Much later, Rogers and Mitchell *(6)* demonstrated that 43% of women affected by various menstrual disorders, infertility, and recurrent miscarriages were overweight or obese. Furthermore, Hartz and colleagues *(7)* demonstrated that the presence of anovulatory cycles, oligomenorrhea, and hirsutism, separately or in association with one another, was significantly higher in obese than in normal-weight women. In adolescent and young women, the age of onset of obesity and that of menstrual irregularities are significantly correlated *(8)*. There are also data indicating that the association with menstrual disorders may be more frequent in girls with onset of excess body weight during puberty than in those who were obese during infancy. These findings have been substantially confirmed in a large study performed in approximately 6000 women by Lake et al. *(9)*, who found that obesity in childhood and the early 20s increased the risk of menstrual problems. It is therefore likely that overweight and obesity do contribute to a significant proportion of menstrual disorders in young women.

2.2. Obesity as a Condition of Sex Hormone Imbalance in Women

It is well known that an increase in body weight and fat tissue is associated with several abnormalities of sex steroid balance, particularly in women of reproductive age. Such alterations involve both androgens and estrogens and, overall, their carrier protein, sex hormone-binding globulin (SHBG). Changes in SHBG concentrations lead to an alteration of androgen and estrogen delivery to target tissues. SHBG levels are regulated by a complex of factors, including estrogens, iodothyronines, and growth hormone as stimulating agents and androgens and insulin as inhibiting factors. The net balance of this regulation is probably responsible for the decrease in SHBG concentration observed in obesity.

Body fat distribution has also been demonstrated to substantially affect SHBG concentrations. In fact, female subjects with central obesity usually have lower serum SHBG concentrations in comparison to their age- and weight-matched counterparts with peripheral obesity *(4)*. This seems to be dependent on higher circulating insulin in abdominally obese women and on the inhibitory capacity of insulin on SHBG synthesis by the liver. Reduction of circulating SHBG results in an increase in the metabolic clearance rate of circulating SHBG-bound steroids, specifically testosterone, dihydrotestosterone, and androstenediol, the principal active metabolite of dihydrotestosterone *(10)*. However, this effect is compensated by consequent increases in their respective production rates. Obesity also affects the metabolism of the androgens not bound to SHBG. In fact, both production rates and metabolic clearance rates of dehydroepiandrostenedione and androstenedione are equally increased in obesity *(11)*.

The role of adipose tissue is crucial in controlling the balance of sex hormone availability in nonadipose target tissues. In fact, adipose tissue is able to store various lipid-soluble steroids, including androgens. Most sex hormones appear to be preferentially concentrated within the adipose tissue rather than in the blood. As a consequence, because the amount of fat in obesity is larger than the intravascular space and the steroid concentration in adipose tissue is much higher than in plasma, the steroid pool in obese individuals is greater than that found in normal-weight individuals.

The pattern of body fat distribution can regulate androgen production and metabolism to a significant extent. In fact, women with central obesity have higher testosterone production rates than those with peripheral obesity *(4)*. Accordingly, metabolic clearance rates of testosterone and dihydrotestosterone are significantly higher in women with central obesity than in those with peripheral obesity. The maintenance of normal circulating levels of these hormones in obesity predicts the presence of a sophisticated regulatory system that can adjust both the production rate and the metabolic clearance rate of these hormones according to body size. Resulting from the greater reduction of SHBG concentrations, the free testosterone fraction tends to be higher in women with central obesity than in those with peripheral obesity *(12)*. An inverse correlation exists between waist-to-hip ratio (or other indices of body fat distribution) and testosterone and SHBG concentrations, regardless of body mass index (BMI) values *(12)*. Therefore, a condition of "relative functional hyperandrogenism" appears to be associated with the central obesity phenotype in women *(13)*.

2.3. Impact of Obesity on PCOS: Pathophysiological Aspects

Mechanisms by which obesity influences the pathophysiology and clinical expression of PCOS are complex and not completely understood *(4)*. However, in women with PCOS, obesity is believed to play a distinct pathophysiological role in the development of hyperandrogenism. In an obese woman with PCOS, the presence of obesity in her mother during pregnancy appears to influence the susceptibility to develop hyperandrogenism and the PCOS phenotype of the daughter later in time, although pathophysiological mechanisms have not been defined *(13)*. On the other hand, it has been hypothesized that *in utero* androgen excess may be an important factor programming subsequent PCOS development during puberty *(14)*. This theory appears to be substantiated by several studies performed in nonhuman primates *(15)* and sheep *(16)*. By this mechanism, it has been suggested that a primary ovarian disorder may occur early in the woman's life, leading to the development of a hyperandrogenized ovary later in life.

Along with this line of thinking, it is also possible that the early onset of overweight or obesity, that is, during peripubertal age, may play a role in the development of hyperandrogenism by the intervention of multiple interrelated mechanisms, which primarily involve inappropriate signals from different hormones and/or alterations of specific hormone-regulatory pathways. These include insulin, the insulin growth factor system, the opioid system, estrogens, and several cytokines, particularly leptin.

It is well known that obesity, particularly the abdominal phenotype, is a condition of insulin resistance and compensatory hyperinsulinemia. Contrary to what occurs in the classic target tissues (i.e., muscle, liver, adipose tissue) of insulin action, which become resistant to insulin, the ovaries remain responsive to insulin throughout the interaction with its own receptor. In the last two decades a large number of in vitro studies have demonstrated that in the ovaries of women with PCOS, excess insulin is capable of stimulating steroidogenesis and excessive androgen production by the theca-cell system. In addition, by inhibiting SHBG synthesis by the liver, excess insulin may further increase the delivery of free androgens to target tissues. In vivo, numerous studies have subsequently demonstrated that both acute and chronic hyperinsulinemia can stimulate testosterone production and that suppression of insulin levels can conversely decrease blood androgen concentrations (reviewed in refs. *3* and *4*). The excess in local ovarian androgen production induced by excess circulating insulin may also cause premature follicular atresia and thus favor anovulation *(3)*. It can therefore be speculated that insulin resistance and hyperinsulinemia, which develop together with the obesity state, play a dominant role in favoring hyperandrogenism in women susceptible to the development of PCOS, particularly during pubertal age.

The influence of obesity on hyperandrogenism can also be mediated by other factors and mechanisms. As in simple obesity, a hyperestrogenic state is present also in obese women with PCOS. Excess estrogens may exert positive feedback regulation on gonadotropin release, triggering in turn a rise in ovarian androgen production, according to a still valid theory proposed many years ago by Yen *(17)*. An additional factor involved in the dysregulation of this complex circuit may be an increased tone of the opioid system, which has been demonstrated to be present in obesity as well as in women with PCOS *(4)*. Several studies have shown that β-endorphin is able to stimulate insulin secretion *(4)*. The possibility that increased opioid activity may favor the development of hyperinsulinemia and, in turn, of hyperandrogenemia is further supported by the finding that acute and chronic administration of opioid antagonists, such as naloxone and naltrexone, suppress both basal and glucose-stimulated insulin blood concentrations *(18)*. Whether alterations of the opioidergic system play a causative role in the infertility of some women is, however, still undefined. On the other hand, there are studies showing that opioid antagonists given to obese women with PCOS may improve menses *(18)*.

Finally, several peptides, particularly leptin, are currently emerging as potential candidates involved in the pathogenesis of hyperandrogenism and infertility in women with PCOS. Leptin is considered one of the major peripheral signals that affect food intake and energy balance, and obesity is a classic condition of circulating leptin excess *(19)*. The discrepancy between high leptin blood levels and its central effects represents the basis for the concept that most forms of obesity may represent a condition of leptin resistance. On the other hand, many tissues other than fat mass have been shown to express leptin and its receptors. Specifically, several lines of evidence indicate that leptin acts directly on the ovaries. In particular, functional leptin receptors have been detected on the surface of ovarian follicular cells, including granulosa, theca, and interstitial cells *(20)*, and recent in vivo data indicate that leptin may exert a direct inhibitory effect on ovarian function by inhibiting both granulosa and thecal cell steroidogenesis, probably by antagonizing stimulatory factors, such as insulin-like growth factor-1, transforming growth factor-β, insulin, and luteinizing hormonee *(21)*. Moreover, high leptin concentrations in the ovary may interfere with the development of a dominant follicle and oocyte maturation, as demonstrated by in vitro and in vivo studies *(22)*.

Finally, exogenous infusion of leptin has been shown to significantly decrease ovulation rate in the female rat *(22)*. Taking into account the involvement of leptin in the control of ovulation and reproduction, much interest has been focused on leptin levels in women with PCOS because they are generally obese. To date, higher circulating levels of leptin than those expected for the BMI or normal concentrations of leptin have been reported *(4)*.Whether high leptin levels in the peripheral circulation and/or in ovarian tissues may play a role in determining anovulation in obese women with PCOS is presently unknown. However, it cannot be excluded that this mechanism may somehow be involved in the development of infertile ovaries.

2.4. Impact of Obesity on PCOS Phenotype

2.4.1. Androgen Abnormalities

Various studies have evaluated the impact of obesity on the phenotype of women with PCOS. They have uniformly demonstrated that obese women with PCOS are characterized by significantly lower SHBG plasma levels and worsened hyperandrogenism in comparison with their normal-weight counterparts *(4)*. In addition, a negative correlation between body fat mass and circulating androgens has been reported *(4)*. It has also been repeatedly reported that a higher proportion of obese women with PCOS complain of hirsutism and menstrual disturbances than do normal-weight women *(4)*. Therefore, there is consistent evidence that the increase in body weight may worsen the hyperandrogenic state in women with PCOS (Tables 1 and 2).

2.4.2. Metabolic Abnormalities

Women with PCOS are characterized by a high prevalence of several metabolic abnormalities that are strongly influenced by the presence of obesity. Adequate confirmation of the role of obesity in determining hyperinsulinemia and insulin resistance in women with PCOS derives from studies comparing groups of normal-weight and obese women with PCOS. Both fasting and glucose-stimulated insulin concentrations were significantly higher in obese than in non-obese PCOS subgroups (reviewed in refs. *2* and *4*). Accordingly, studies examining insulin sensitivity by using different methods such as the euglycemic-hyperinsulinemic clamp technique, the frequent-sample intravenous glucose test (FSIVGT), and the intravenous insulin test have further demonstrated that obese women with PCOS have significantly lower insulin sensitivity than their non-obese counterparts with PCOS and, therefore, a more severe insulin-resistant state (reviewed in ref. *4*).

The percentage of women affected by PCOS and obesity who present with glucose intolerance is rather high, ranging from 20 to 49% *(2)*, which is substantially above the prevalence rates reported in

Table 1
The Obese PCOS Phenotype

- Hyperandrogenemia
- Oligo-amenorrhea (moderate to severe)
- Hirsutism
- Insulin resistance (moderate to severe)
- Hyperinsulinemia
- Abdominal/visceral body fat distribution (>50%)
- Features of the metabolic syndrome
- Early development of type 2 diabetes mellitus

PCOS, polycystic ovary syndrome.

Table 2
Clinical, Hormonal, and Metabolic Features of Obese
and Non-Obese Women With PCOS

Features	Difference	p-value[a]
BMI	Higher	<0.0001
WHR	Higher	<0.001
Age of menarche	—	NS
Body weight at menarche	Higher	<0.05
Hirsutism (F-G score)	Higher	<0.05
Acanthosis nigricans	More frequent	<0.05
Oligo-amenorrhea	—	NS
Androgens (FT, A, DHEAS)	Higher	0.05–0.1
SHBG	—	NS
Estrogens	—	NS
Insulin (fasting/stimulated)	Higher	<0.001
Triglycerides	Higher	<0.001
Total cholesterol	Higher	<0.05
HDL cholesterol	—	NS

[a]p-values refer to a comparison by ANOVA of three groups of women with PCOS and different BMI values.

PCOS, polycystic ovary syndrome; BMI, body mass index, WHR, waist:hip ratio; F-G, Ferrimen-Gallwey; FT, free testosterone; A, androgen, DHEAS, dehydroepiandrosterone sulfate; SHBG, sex hormone-binding globulin, HDL, high-density lipoprotein.

See ref. *36* for further details; data obtained from refs. *4* and *8*.

premenopausal women in population-based studies. In contrast, glucose intolerance in normal-weight women with PCOS is uncommon *(2,23)*. Collectively, this suggests that obesity *per se* plays an important role in altering the insulin–glucose system in PCOS (*see* Tables 1 and 2).

In addition, several recent studies have identified defects of insulin secretion in obese women with PCOS *(2)*. Using the FSIVGT, Dunaif and Finegood *(24)* reported that obese women with PCOS mount an inadequate insulin secretory response to compensate for the peripheral insulin resistance state, suggesting relative β-cell dysfunction. However, regardless of alterations in insulin secretion, in a 10-year follow-up study we found that both fasting and glucose-stimulated insulin and C-peptide levels tended to increase spontaneously and significantly in women with PCOS, suggesting a worsened insulin resistant state over time *(25)*. In the same study we also found that several women developed impaired glucose tolerance. Longitudinal data are therefore warranted to investigate which factors, namely progressive insulin resistance and/or subtle alterations of insulin secretion, can predict the well-documented susceptibility of obese women with PCOS to the development of type 2 diabetes.

Although PCOS *per se* may be associated with alterations of both lipid and lipoprotein metabolism, the coexistence of obesity usually leads to a more atherogenic lipoprotein pattern. A greater reduction of high-density lipoprotein (HDL), together with a greater increase in both triglycerides and total cholesterol levels have in fact been observed in obese with respect to normal-weight women with PCOS (reviewed in ref. *4*) (Table 2).

2.4.3. Menstrual Abnormalities and Infertility

PCOS is one of the most common causes of anovulation and endocrine infertility in women. Several studies have clearly demonstrated that menstrual abnormalities are more frequent in obese than in normal-weight women with PCOS *(4)*. Moreover, there is evidence that a reduced incidence of pregnancy and blunted responsiveness to pharmacological treatments to induce ovulation may be more common in obese PCOS *(26)*. In a prospective study carried out among 158 anovulatory women, the dose of clomiphene required to achieve ovulation was positively correlated with body weight *(27)*. Both insulin resistance and hyperinsulinemia, which parallel the increase in body fat, may be responsible for the alteration of both spontaneous and induced ovulation observed in obese women with PCOS.

Administration of insulin-sensitizing agents, such as metformin and troglitazone, was in fact associated with improved menstrual cyclicity in women with PCOS *(28)*. Moreover, a double-blind placebo-controlled collaborative study, performed in a large cohort of women with PCOS *(29)*, demonstrated that short-term metformin treatment increased both spontaneous and low-dose (50 mg daily for 5 days) clomiphene-induced ovulation rates. It has also been reported that, compared to normal-weight women, obese women with PCOS may have lower ovulatory responses to pulsatile gonadotropin-releasing hormone analog administration *(30)*. Accordingly, the pregnancy rate after a low-dose human menopausal gonadotropin or pure follicle-stimulating hormone administration may be significantly lower in obese than in normal-weight women with PCOS *(31)*. Finally, in recent studies performed in women with PCOS conceiving after in vitro fertilization or intracytoplasmatic sperm injection, it was observed that those with obesity had higher gonadotropin requirements during stimulation, fewer oocytes, a higher abortion rate, and a lower live birth rate than their non-obese counterparts *(32)*. In conclusion, a decreased efficiency of the different treatments for anovulation and infertility may be expected in obese women with PCOS. The presence of hyperinsulinemia is probably the major factor responsible for this undesirable condition.

2.5. Impact of Body Fat Distribution on PCOS Phenotype

It is well documented that women with PCOS have a high prevalence of abdominal distribution of body fat, even if they are normal in weight (Fig. 1) *(23,33)*. The impact of abdominal obesity on PCOS may be greater than expected because this phenotype is associated with more pronounced hyperandrogenism and insulin resistance than is the peripheral body fat phenotype. We have repeatedly demonstrated that the androgen profile and basal insulin levels, as well as the insulin response to a glucose load, are significantly higher in the subgroup of women with PCOS with abdominal body fat distribution than in the group with the peripheral type, regardless of BMI *(4)*. This has been confirmed in studies using dual-energy x-ray absorptiometry to define different obesity phenotypes *(34)*.

Moreover, Holte et al. *(35)* found a significant association between abdominal fat mass and insulin resistance determined by the euglycemic-hyperinsulinemic clamp technique. They also found a highly significant correlation between plasma free fatty acid (FFA) concentrations and insulin resistance, which supports the concept that an increase of FFA flux from the highly lypolitic abdominal fat to the liver and muscles may represent the most important link between abdominal obesity and the insulin resistance state *(35)*. Moreover, this subgroup of women with PCOS may have a more unfavorable lipid profile, namely higher triglyceride and very-low-density lipoprotein levels and lower

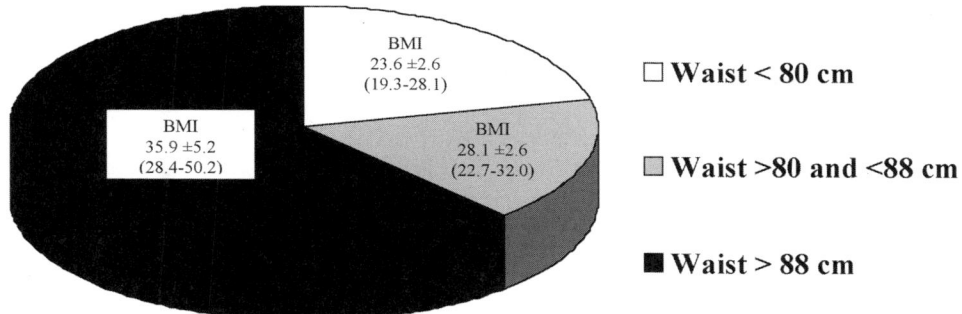

Fig. 1. Prevalence of different body fat distribution patterns: peripheral (waist < 80 cm), intermediate (waist 80–88 cm), and abdominal (>88 cm) in a group of consecutive women with polycystic ovary syndrome ranging from normal weight to obesity (*see* ref. *23*).

HDL cholesterol concentrations *(4)*. In addition, women with PCOS with the abdominal phenotype present with a higher prevalence of menstrual abnormalities and acanthosis nigricans (a cutaneous marker of insulin resistance) and a tendency to more severe hirsutism *(4)*.

As discussed earlier, abdominal obesity is associated with profound alterations of both production and metabolic clearance rates of major androgens and reduced SHBG blood levels. In abdominally obese women with PCOS, androgens could, in turn, play a role in regulating tissue metabolism. In fact, at the level of visceral depots, testosterone stimulates lipolysis and, therefore, increases FFA efflux *(4)*. In addition, at the level of the muscle, testosterone modifies the histological structure by increasing type II, less insulin-sensitive fibers. These androgen-dependent mechanisms may have a further important impact on the insulin resistance state.

In summary, in women with PCOS, abdominal obesity *per se* may play a key role in determining both altered androgen metabolism and insulin resistance. This may be an important consideration when phenotyping PCOS and in devising therapeutic strategies to reduce both hyperinsulinism and hyperandrogenism.

3. CONCLUSIONS

Obesity, particularly the abdominal phenotype, is associated with several abnormalities of sex steroid balance in women, including an increased androgen-production rate and a decrease in serum SHBG. Although epidemiological data are lacking, it is hypothesized that the increasing epidemic of obesity worldwide may also favor the high prevalence of PCOS in the general population. Because PCOS is frequently associated with obesity, it is suggested that this plays an important role in the pathophysiology of PCOS. Obesity has a profound impact on the phenotypic expression of PCOS, being associated with more severe hyperandrogenism, insulin resistance, and fertility disorders. Whether obesity represents a factor amplifying intrinsic hormonal and metabolic components of PCOS or, alternatively, whether it has a direct pathophysiological role is still a matter of debate. Whatever the truth, correcting obesity significantly improves not only metabolic derangements and hyperandrogenism, but also menstrual abnormalities and reduced fertility capacity.

4. FUTURE AVENUES OF INVESTIGATION

As reported in this chapter, it can be suggested that the increasing prevalence of obesity among adolescent and young women with PCOS may partly depend on the increasing epidemic of obesity worldwide. This indicates the need for long-term prospective epidemiological trials. It may also have great relevance in preventive medicine and offer the opportunity to expand our still limited knowledge of the genetic background of this disorder.

PCOS also represents a relatively simple clinical condition, but with complex pathophysiological aspects. Further efforts should be made to investigate the potential mechanisms by which obesity affects PCOS, particularly during adolescence and early reproductive age.

Because studies performed in experimental animals suggest that androgen exposure during intrauterine life may affect a female's susceptibility to develop PCOS later in life, both retrospective and prospective studies in large cohorts or well-selected groups of women should be undertaken. This topic appears to be a promising way to expand our knowledge of the natural history of PCOS and may open new strategies to prevent it.

Finally, because there are data supporting the fact that obesity appears to represent a prerequisite for the high level of susceptibility that women with PCOS have of developing type 2 diabetes mellitus and, possibly, cardiovascular diseases, more clinical studies investigating the impact of obesity on risk for diabetes and cardiovascular disease should be performed in adult women, particularly after the menopause. This is consonant with the common belief that this disorder affects women throughout the life span and that the clinical phenotype on PCOS after menopause is still largely unknown.

KEY POINTS

- Obesity is a common condition in women with PCOS, affecting more than half of them.
- Most PCOS subjects, whether overweight, obese, or normal weight, may present with an abdominal body fat distribution.
- Obesity may play an important pathophysiological role in favoring PCOS, particularly during adolescence
- Obese women with PCOS are characterized by more severe hyperandrogenism, insulin resistance and hyperinsulinemia, and menstrual abnormalities.
- Obesity appears to represent a prerequisite for the high level of susceptibility that women with PCOS have of developing type 2 diabetes mellitus as adults.

REFERENCES

1. Ehrmann DA. Polycystic ovary syndrome. N Engl J Med 2005;352:1223–1236.
2. Dunaif A. Insulin resistance and the polycystic ovary syndrome: mechanisms and implications for pathogenesis. Endocr Rev 1997;18:774–800.
3. Poretsky L, Cataldo NA, Rosenwaks Z, Giudice LC. The insulin-related ovarian regulatory system in health and disease. Endocr Rev 1999;20:535–582.
4. Gambineri A, Pelusi C, Vicennati V, Pagotto, U, Pasquali R. Obesity and the polycystic ovary syndrome. Int J Obes Rel Metab Dis 2002;26:883–896.
5. WHO. Preventing and managing the global epidemic. Report of a WHO consultation on obesity. Geneva: WHO/NUT/NCD/98.1, 1997.
6. Rogers J, Mitchell GW. The relation of obesity to menstrual disturbances. N Engl J Med 1952;247:53–56.
7. Hartz AJ, Barboriak PN, Wong A, Katayama KP, Rimm AA. The association of obesity with infertility and related menstrual abnormalities in women. Int J Obes 1979;3:57–77.
8. Pasquali R, Pelusi C, Genghini S, Cacciari M, Gambineri A. Obesity and reproductive disorders in women. Hum Reprod Update 2003;9:359–372.
9. Lake JK, Power C, Cole TJ. Women's reproductive health—the role of body mass index in early and adult life. Int J Obes.Relat Metab Disord 1997;21:432–438.
10. Samojlik E, Kirschner MA, Silber D, Scneider G, Ertel NH. Elevated production and metabolic clearance rates of androgens in morbidly obese women. J Clin Endocrinol Metab 1984;59:949–954.
11. Kirschner MA, Samojlik E, Drejka M, Szmal E, Schneider G, Ertel N. Androgen-estrogen metabolism in women with upper body versus lower body obesity. J Clin Endocrinol Metab 1990;70:473–479.
12. Evans DJ, Hoffmann RG, Kalkhoff RK, Kissebah AH. Relationship of androgenic activity to body fat topography, fat cell morphology and metabolic aberrations in premenopausal women. J Clin Endocrinol Metab 1990;57:304–310.
13. Cresswell JL, Barker DJ, Osmond C, Egger P, Phillips DI, Fraser RB. Fetal growth, length of gestation, and polycystic ovaries in adult life. Lancet 1997;350:1131–1135.
14. Abbott DH, Dumesic Da, Franks S. Developmental origin of polycystic ovary syndrome—a hypothesis. J Endocrinol 2002;174:1–5.

15. Eisner JR, Dumesic DA, Kemnitz JW, Abbott DH. Timing of prenatal androgen excess determines differential impairment in insulin secretion and action in adult female rhesus monkeys. J Clin Endocrinol Metab 2000;85:1206–1210.
16. Robinson JE, Forsdike RA, Taylor JA. In utero exposure of female lambs to testosterone reduces the sensitivity of the gonadotropin-releasing hormone neuronal network to inhibition by progesterone. Endocrinology 1999;140:5797–5805.
17. Yen SSC. The polycystic ovary syndrome. Clin Endocrinol (Oxf) 1980;12:177–208.
18. Pasquali R, Casimirri F. The impact of obesity on hyperandrogenism and polycystic ovary syndrome in premenopausal women. Clin Endocrinol (Oxf) 1993;39:1–16.
19. Considine RV, Sinha MK, Heiman ML, Kriauciunas A, Stephens TW, Nyce MR. Serum immunoreactive leptin concentrations in normal-weight and obese humans. N Engl J Med 1996;334:292–295.
20. Wiesner G, Vaz M, Collier G, et al. Leptin is released from the human brain. Influence of adiposity and gender. J Clin Endocrinol Metab 1999;84:2270–2274.
21. Agarwal SK, Vogel K, Weitsman SR, Magoffin DA. Leptin antagonizes the insulin-like growth factor–I augmentation of steroidogenesis in granulosa and theca cells of the human ovary. J Clin Endocrinol Metab 1999;84:1072–1076.
22. Duggal PS, Van Der Hoek KH, Milner CR, et al. The in vivo and in vitro effects of exogenous leptin on ovulation in rat. Endocrinology 2000;141:1971–1976.
23. Gambineri A, Pelusi C, Manicardi E, et al. Glucose intolerance in a large cohort of Mediterranean women with polycystic ovary syndrome. Phenotype and associated factors. Diabetes 2004;53:2353–2358.
24. Dunaif A, Finegood DT. β-Cell dysfunction independent of obesity and glucose intolerance in the polycystic ovary syndrome. J Clin Endocrinol Metab 1996;81:942–947.
25. Pasquali R, Gambineri A, Anconetani B, et al. The natural history of the metabolic syndrome in young women with the polycystic ovary syndrome and the effect of long-term oestrogen-progestagen treatment. Clin Endocrinol 1999;50:517–527.
26. Galtier-Dereure F, Pujol P, Dewailly D, Bringer J. Choice of stimulation in polycystic ovarian syndrome: the influence of obesity. Human Reprod 1997;12:88–96.
27. Lobo RA, Gysler M, March CM, Goebelman U, Mischell D Jr. Clinical and laboratory predictors of clomiphene response. Fertil Steril 1982;37:168–174.
28. De Leo V, la Marca A, Petraglia F. Insulin-lowering agents in the management of polycystic ovary syndrome. Endocr Rev 2003;24:633–637.
29. Nestler JE, Jakubowicz DJ, Evans WS, Pasquali R. Effects of metformin on spontaneous and clomiphene-induced ovulation in the polycystic ovary syndrome. N Engl J Med 1998;25:1876–1880.
30. Filicori M, Flamigni C, Dellai P. Treatment of anovulation with pulsatile gonadotropin-releasing hormone: prognostic factors and clinical results in 600 cycles. J Clin Endocrinol Metab 1994;79:1215–1220.
31. White DM, Polson DW, Kiddy D, et al. Induction of ovulation with low-dose gonadotropins in polycystic ovary syndrome: an analysis of 109 pregnancies in 225 women. J Clin Endocrinol Metab 1996;81:3821–3824.
32. Fedorcsák P, Dale PO, Storeng R, Tanbo T, Abyholm T. The impact of obesity and insulin resistance on the outcome of IVF or ICSI in women with polycystic ovarian syndrome. Hum Reprod 2001;16:1086–1091.
33. Kirchengast S, Huber J. Body composition characteristics and body fat distribution in lean women with polycystic ovary syndrome. Hum Reprod 2001;16:1255–1260.
34. Douchi T, Ijuin H, Nakamura S, Oki T, Yamamoto S, Nagata Y. Body fat distribution in women with polycystic ovary syndrome. Obstet Gynecol 1995;86:516–519.
35. Holte J, Bergh T, Berne C, Wide L, Lithell H. Restored insulin sensitivity but persistently increased early insulin secretion after weight loss in obese women with polycystic ovary syndrome. J Clin Endocrinol Metab 1995;80:2586–2593.

Molecular Mechanisms of Insulin Resistance in the Polycystic Ovary Syndrome

Theodore P. Ciaraldi

SUMMARY

Whole body resistance to the glycemic actions of insulin is a frequent occurrence in individuals with polycystic ovary syndrome (PCOS). Reductions in insulin action on glucose metabolism have been demonstrated in a number of tissues or cells from PCOS subjects. These involve changes in either, or both, the sensitivity and maximal responsiveness to insulin. Meanwhile, the mitogenic and ovarian steriodogenic actions of insulin are intact. Altered phosphorylation of key insulin-signaling intermediates are prime candidates for impaired insulin signal transduction that persists after correction of the in vivo metabolic and hormonal environment.

Key Words: Glucose transport; insulin receptor; insulin receptor substrates-1/2; serine phosphorylation; skeletal muscle; adipose tissue; polycystic ovary syndrome; insulin resistance.

1. INTRODUCTION

The high frequency of impaired glucose tolerance and insulin resistance for control of glucose metabolism in polycystic ovary syndrome (PCOS) has long been recognized, and further understanding of the nature of this aspect of PCOS depends on the answers to a number of questions:

1. Which tissues and responses display insulin resistance? More specifically, does insulin resistance extend to the ovary and steroidogenesis?
2. Does insulin resistance in PCOS differ from that in type 2 diabetes?
3. Which aspects of insulin resistance in PCOS are acquired from the hyperinsulinemic, hyperandrogenic, and hyperlipidemic environment characteristic of the syndrome, and which might be intrinsic or genetic in cause?
4. What are the specific molecular mechanisms of this insulin resistance; what events in insulin action might be altered?

Because an earlier chapter (*see* Chapter 24) dealt with the confluence of insulin resistance and hyperinsulinemia, this chapter will focus on studies in tissues from subjects with PCOS, which reveal both acquired and intrinsic aspects, and in cells cultured from similar subjects, where the environment can be controlled and intrinsic features revealed. Findings from these in vitro investigations will also be related, when possible, to behaviors in the intact individual. One point that will soon become apparent is that insulin action in PCOS is highly heterogeneous, and care must be taken in generalizing from results in specific cohorts of subjects.

2. BACKGROUND

2.1. Insulin Sensitivity and Responsiveness

In the consideration of insulin action at either the whole body, tissue, or cellular level, attention must be paid to two concepts: sensitivity and responsiveness. Insulin sensitivity refers to the shape of

From: *Contemporary Endocrinology: Androgen Excess Disorders in Women:*
Polycystic Ovary Syndrome and Other Disorders, Second Edition
Edited by: R. Azziz et al. © Humana Press Inc., Totowa, NJ

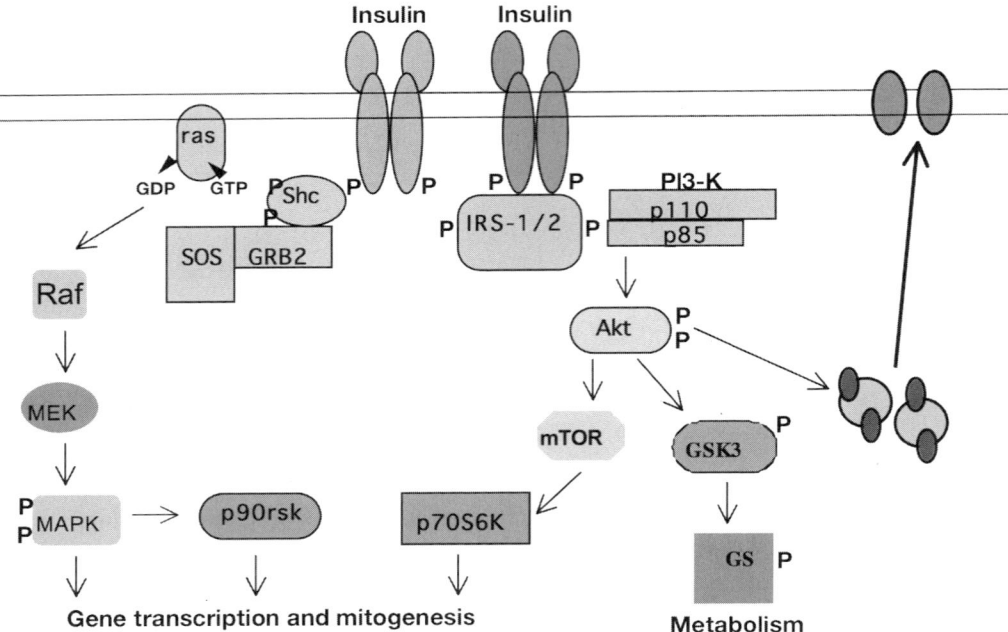

Fig. 1. Schematic for major pathways and events involved in insulin signaling for metabolic and mitogenic responses. IRS, insulin receptor substrate; PI3K, phosphotidylinositol 3-kinase; Akt, protein kinase B; GS, glycogen synthase; GSK3, glycogen synthase kinase 3; GLUT4, insulin-dependent glucose transporter; mTOR, mammalian target of rapomycin; p70S6K, ribosomal protein S6 kinase; SOS, son of sevenless; MAPK, mitogen-activated protein kinase.

the dose–response curve for a particular action and describes the effect that is attained at a certain hormone concentration, regardless of the final capacity of that system. Impairments in this parameter have been described as "receptor defects." Responsiveness describes the absolute change from baseline attained at a maximal hormone level; differences here are ascribed to "postreceptor" or "effector" defects. Such defects can be present independently or in concert.

Although insulin is a highly pleotrophic hormone, responses can be categorized in two general classes. Metabolic responses include control of glucose, fatty acid, and protein metabolism, often occuring after acute hormone exposure and involving changes in protein activity. Mitogenic responses include the control of proliferation, differentiation, cell survival, and modulation of gene expression.

2.2. The Basics of Insulin Signaling

Work over the last two decades has revealed much about the processes by which insulin signaling occurs to these ends (reviewed in ref. *1*). Figure 1 presents a simplified schematic illustration of the most studied pathways. A number of other pathways and intermediates have been identified, such as the APS/CAP/cbl pathway, or atypical protein kinase Cs, for the stimulation of glucose transport, but because there are no data regarding these alternative pathways in PCOS, they will not be considered here. There are several crucial things to understand about insulin signal transduction as it relates to insulin resistance. The initial event in the process is binding of insulin to its cell surface receptor. The receptor is a heterotetramer, consisting of two extracellular α-subunits, containing the hormone recognition sites, and two membrane-spanning β-subunits. Binding of the hormone causes a conformational change in the α-subunits that is transmitted to the β-subunits, resulting in activation of the tyrosine kinase activity intrinsic to the β subunit. The first substrate phosphorylated is the β-subunit itself (autophosphorylation), further activating the kinase activity. This autophosphorylation creates recognition sites for domains on other molecules, recruiting them to the receptor. Key among these are the insulin receptor substrate (IRS) proteins. There are four different IRS proteins with only partial overlap in function. IRS proteins can then be tyrosine phosphorylated by the receptor kinase, permitting them to dock with other proteins. Most important of these is the lipid kinase phosphatidylinositol 3-kinase (PI3-K), whose activity is increased when associated with IRS proteins. The lipid product of PI3-K, phosphatidylinositol-3,4,5-trisphosphate, is able to activate a further downstream lipid-dependent serine kinase, and this phosphorylation cascade continues on to endpoints such as activation of the rate-limiting enzyme for glycogen storage, glycogen synthase. Other kinases are involved in stimulating the translocation of a specific form of glucose transports protein from intracellular pools to the cell surface, where they can mediate increased glucose entry into the cell.

The IRS proteins are also involved in signaling to mitogenic responses, although other docking or scaffolding proteins function there as well. Phosphorylation of those docking proteins, IRS-1/2 and Shc, activates a separate phosphorylation cascade. These signaling pathways are not mutually exclusive—crosstalk also occurs. Much less is known about the behavior of the mitogernic pathway in PCOS.

An important general principle in insulin signal transduction is that these serine phosphorylations can be either activating or deactivating, depending on the target. Indeed, an important regulatory mechanism is serine phosphorylation of IRS-1 and IRS-2: both protein kinase C (PKC) ζ/λ and GSK3 can accomplish this feat *(2)*. Serine phosphorylation of IRS-1 reduces its ability to be tyrosine phosphorylated by the insulin receptor and to associate with PI3-K. This represents an important feedback mechanism for limiting insulin action. Serine phosphorylation of IRS-1 and IRS-2 is also elevated in a number of insulin-resistant conditions (e.g., diet-induced obesity, stress, hyperinsulinemia *[2]*), implicating it as an important control point.

2.3. Insulin Signaling in PCOS

In the vast majority of studies of PCOS, insulin action has been characterized from the homeostatic model assessment for insulin resistance (HOMA-IR), the 75-g oral glucose tolerance test (OGTT), or the modified frequently sampled intravenous glucose tolerance test. There is still a debate about the reliability of such measures *(3)* in different conditions, but one obvious shortcoming is that these approaches do not permit assessment of the role of different tissues in glucose homeostasis, as is possible with the glucose clamp procedure. In the most complete study of in vivo insulin action to date, Dunaif and colleagues performed multiple-dose hyperinsulinemic-euglycemic clamps in nonobese and obese subjects with PCOS, along with body mass index (BMI)-matched normal cycling controls *(4)*. With regard to insulin-mediated whole body glucose disposal (IMGD), individuals with PCOS displayed a reduced IMGD at the highest insulin infusion tested as well as rightward shifts in their dose–response curves. Thus, those with PCOS had impairments in both insulin responsiveness and sensitivity for IMGD.

A different set of results were obtained for insulin suppression of hepatic glucose production (HGP): maximal suppression was attained in all subject groups, whereas sensitivity was altered only in the obese PCOS group *(4)*, suggesting that the nature of insulin resistance may vary between tissues. It is interesting to note that subjects with type 2 diabetes, when compared with nondiabetic subjects matched for obesity, display insulin-responsiveness defects for both IMGD and HGP (reviewed in ref. *5*) but essentially normal insulin sensitivity. Thus, an added feature of PCOS is the impact on insulin sensitivity. Studies performed in isolated cells have provided additional insight into the nature of insulin action in PCOS. The literature will be discussed first from the perspective of how it relates to the issues of insulin sensitivity and responsiveness and potential differences between tissues. After that the impact of PCOS on individual steps in insulin signaling will be considered.

2.4. Insulin in Individual Tissues in PCOS

2.4.1. Studies in T Lymphocytes

One of the first such studies employed phytohemagglutinin-activated T lymphocytes *(6)*. While not a classic insulin target tissue, T lymphocytes do respond to insulin by increasing glucose utilization, including stimulation of pyruvate dehydrogenase (PDH). T lymphocytes were isolated from normal cycling controls, subjects with PCOS, and those with both PCOS and type 2 diabetes (PCOS-T2D): all subjects were obese, and samples were cultured for 3 days before acute exposure to varying concentrations of insulin. Stimulation of PDH in PCOS-T2D T lymphocytes was lower than control and PCOS groups at all insulin concentrations. In PCOS cells, stimulation of PDH was reduced at a submaximal hormone level but normal at the maximal insulin level *(6)*. Thus, T lymphocytes from PCOS subjects display reduced insulin sensitivity and normal responsiveness, whereas the combination of PCOS and diabetes introduced impaired responsiveness.

2.4.2. Studies in Adipocytes

Several reports in a more relevant insulin target, adipocytes isolated from the subcutaneous abdominal depot, followed soon after. One report studied obese PCOS and control subjects *(7)*, another used non-obese subjects *(8)*, and a third included both non-obese and obese women with PCOS *(4)*. Taking stimulation of glucose transport as the insulin response, all three laboratories found the dose–response curves for PCOS adipocytes to be right-shifted compared to the weight-matched controls. The reduction in insulin sensitivity, calculated from the half-maximally effective insulin concentration (ED_{50}), ranged from 50% *(4)* to eightfold (Fig. 2) *(7)*. The work of Dunaif et al. *(5)* was able to separate the PCOS-related reduction in sensitivity from that resulting from obesity. The impact on insulin responsiveness was more variable, with no difference from controls in one report *(7)* and a reduction in others *(4,8)*. Resistance was also present for the ability of insulin to suppress catecholamine-stimulated lipolysis, with reduced sensitivity seen by several groups *(8,9)*. Again there was heterogeneity regarding insulin responsiveness: normal in one report *(9)* and impaired in another *(8)*. Thus, impaired insulin sensitivity is a common feature of PCOS, both at the whole body level *(4)* and in isolated adipocytes *(4,7,8)*.

One possible source of variability in results on insulin responsiveness is that the PCOS subjects for these reports were classified on the basis of reproductive criteria and not segregated or stratified according to glucose tolerance. Although none of the PCOS subjects in the report of Ciaraldi et al. *(7)* were diabetic, the presence of impaired glucose tolerance was not ruled out, while a number of the obese PCOS subjects studied by Dunaif et al. *(4)* had abnormal responses to an OGTT, including several with type 2 diabetes. Although the findings in isolated adipocytes are generally reflective of insulin action on whole body glucose metabolism, it will be interesting to see if the behavior of adipocytes can distinguish between PCOS subjects with normal or abnormal glucose tolerance.

2.4.3. Studies in Fibroblasts

Because isolated adipocytes are studied within several hours after tissue collection, their behavior is greatly influenced by the in vivo metabolic environment: levels of substrates, metabolites, and hormones. Thus, the results from the above studies cannot distinguish between effects on insulin action intrinsic to PCOS, possibly genetic differences, and those acquired from the environment. Studies addressing this question employed skin fibroblasts cultured from PCOS and control subjects. Glucose metabolism in fibroblasts is responsive to insulin at the level of glycogen synthesis. Differences in this action are seen between fibroblasts from nondiabetic and type 2 diabetic subjects *(10)*, indicating that insulin resistance is retained in culture and supporting this cell type as a useful model. Two reports agreed that the mitogenic actions of insulin, measured as thymidine incorporation into DNA, were normal in PCOS fibroblasts *(10,11)*. Variability was again present for the metabolic action of insulin, measured as glucose incorporation into glycogen. One group showed reduced sensitivity and responsiveness to insulin *(11)*, and in the other the dose–response curves were

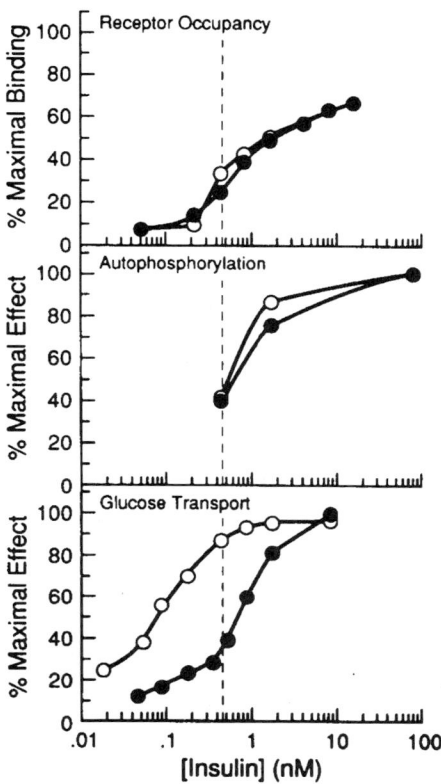

Fig. 2. Dose–response curves for insulin receptor occupancy (top), receptor autophosphorylation (middle), and glucose transport stimulation (bottom) in adipocytes and isolated receptors (middle) from age and body mass index (BMI)-matched normal cycling (O) and polycystic ovary syndrome (PCOS) (●) subjects. Results are normalized against the maximal activity for each function for each individual, presented as average values. Vertical dashed lines indicate a physiological insulin concentration (0.42 n*M*). (From ref. *8*.)

superimposable for control and PCOS fibroblasts, although differences were present in adipocytes from the same subjects *(10)*. Reasons for the differences in results are not readily apparent, but they are indicative of the heterogeneity seen both in vivo and in vitro.

2.4.4. Studies in Muscle

More relevant data about insulin action can be obtained using human skeletal muscle cells grown in culture. We and others have shown that satellite cells from human skeletal muscle tissue can be propagated in culture and differentiated so that they reflect the morphological and biochemical properties of mature skeletal muscle *(12)*. Perhaps most important, such cells from individuals with type 2 diabetes retain defects in glucose metabolism and insulin action similar to those observed in vivo *(12)*. Corbould et al. employed a modification of this system to investigate glucose metabolism in obese subjects with PCOS *(13)*. Surprisingly, they found that glucose uptake was elevated at all insulin levels in PCOS muscle cells, even as there were impairments in insulin signaling (*see* Subsection 2.5.). This contrasts with our initial work (currently available only in abstract form *[14]*), where we found defects in skeletal muscle cell glucose uptake that mirrored those seen for whole body glucose disposal. One possible reason for this discrepancy could be that the expression of several key markers of the myotube phenotype, creatine kinase activity, and a-myosin heavy-chain protein were threefold greater in cells from their subjects with PCOS than in controls *(13)*, suggesting a greater degree of differentiation. Because insulin responsiveness for glucose metabolism develops during myotube differentiation, this could complicate comparisons between populations at different stages of differentiation.

2.4.5. Studies in the Ovary

Studies on insulin action in the ovary show greater agreement, perhaps because of less methodological variation. Most of these studies have been performed in cultured granulosa cells. Wills and Franks showed that insulin stimulation of steroidogenesis, estradiol (E2), and progesterone (P) synthesis was mediated through the insulin receptor, not the insulin-like growth factor-1 receptor, and that the insulin receptor was also functioning in granulosa cells from PCOS ovaries *(15)*. Extending those observations, Franks and colleagues found that the insulin sensitivity and responsiveness of stimulation of E2 and P production was normal in PCOS granulosa cells *(16,17)*. In addition, the mitogenic actions of insulin (thymidine incorporation) are also normal in PCOS *(18)*. However, PCOS granulosa cells displayed insulin resistance for several aspects of glucose metabolism. Insulin stimulation of lactate production, a measure of glycolysis, was attenuated *(17,19)*, as was stimulation of glucose incorporation into glycogen *(18)*. Insulin responsiveness of glucose metabolism in PCOS granulose cells is so low that it is difficult to ascertain if sensitivity is also altered.

2.4.6. Summary

Even with the variability in data, contributed to by subject heterogeneity, as well as the differing methodologies and experimental systems, several conclusions can be drawn. One is that nonmetabolic actions of insulin, such as mitogenesis (fibroblasts and granulose cells) and steroidogenesis (granulosa cells), are normal in PCOS. The other is that insulin's effects on glucose (glucose transport, lactate production, glycogen synthesis) and lipid (antilipolysis) metabolism are impaired in multiple tissues, adipocytes, and granulosa cells, whereas the evidence is mixed in fibroblasts and skeletal muscle cells. This selective insulin resistance would mean that compensatory hyperinsulinemia induced to control circulating glucose levels would result in augmented stimulation of ovarian steroidogenesis. The next key questions are: What are the mechanisms for this selectivity? Where does insulin signaling diverge for metabolic and mitogenic/steroidogenic responses? For this reason, elucidating the mechanism(s) of insulin resistance in PCOS has the potential to greatly expand our understanding of insulin signal transduction.

2.5. Impact of PCOS on Individual Steps in Insulin Signaling

To understand the nature of selective changes in insulin signaling characteristic of PCOS, it will be useful to review step by step what is known about insulin signaling in tissues and cells from PCOS subjects. Starting at the initial event at the target cells, the recognition of insulin at the cell surface, the majority of investigators have found both the number and affinity of insulin receptors to be normal in PCOS. This result was observed in freshly isolated adipocytes *(4,7)*, cultured fibroblasts *(10,11)*, and granulosa cells *(18)*. Expression of the insulin receptor protein was also normal in skeletal muscle *(20)* and cultured muscle cells *(13)*. There are exceptions to this behavior, as reduced insulin binding has been reported in adipocytes *(8)* and T lymphocytes *(7)*. In general, however, the general consensus is that the hormone-binding function of the insulin receptor is intact in PCOS.

The second function of the insulin receptor is the tyrosine kinase activity of the β subunit, both for autophosphorylation and toward other substrates. Insulin-stimulated autophosphorylation in PCOS adipocytes displayed a normal sensitivity (Fig. 2), with a modestly reduced absolute maximal response *(7)*. Kinase activity of partially purified adipocyte insulin receptors against an artificial exogenous substrate was normal *(7)*, suggesting that this function was intact. That is in contrast to type 2 diabetes, where receptor kinase activity is greatly impaired (Table 1). Consistent with normal receptor function, a number of investigators have probed the insulin receptor gene in PCOS subjects and found either no missense or nonsense mutations *(21)*, or, if single-nucleotide polymorphisms were identified, they were not associated with insulin resistance *(22)*.

Table 1
**Comparison of Insulin Action and Signaling in Subcutaneous Adipocytes
From Obese Subjects With PCOS and Those With Type 2 Diabetes**

Activity	PCOS	Type 2 diabetes[a]
Glucose transport: maximal response	$\Rightarrow$ (7) $\Downarrow$ (4,8)	$\Downarrow$
Glucose transport: sensitivity	$\Downarrow$ (4,7,8)	$\Rightarrow$
Insulin receptor: binding	$\Rightarrow$ (4,7)	$\Downarrow$
Insulin receptor: autophosphorylation	$\Rightarrow$ (8)	$\Downarrow$
Insulin receptor: kinase activity	$\Rightarrow$ (8)	$\Downarrow$
IRS-1: tyrosine phosphorylation	$\Downarrow$	

Results presented as relative to activity in age and BMI-matched normal cycling individuals.
[a]From ref. *28.*
PCOS, polycystic ovary syndrome; IRS-1, insulin receptor substrate-1.

However, Dunaif and colleagues identified a subpopulation (6 of 12 subjects studied) of individuals with PCOS who displayed reduced insulin-stimulated autophosphorylation in part as a result of augmented basal phosphorylation *(23)*. Fibroblasts cultured from these individuals also showed reduced insulin stimulation of both autophosphorylation and receptor kinase activity against an exogenous substrate *(23)*; receptor phosphorylation was normal in the remainder of the study subjects. Interestingly, if fibroblast insulin receptors from those first individuals were purified before assay, insulin action on phosphorylation was normal, suggesting that a factor in fibroblasts extrinsic to the insulin receptor was responsible *(24)*. A defining feature of the receptors from these insulin-resistant subjects was that they had elevated serine phosphorylation in the basal state *(23)*; treatment of their fibroblasts with PKC inhibitors normalized insulin signaling *(24)*. Although the principle of increased serine phosphorylation of insulin signaling molecules *(2)* as a means of damping insulin signaling was noted earlier, the physiological impact of the increased receptor serine phosphorylation in these specific individuals is uncertain, as there were no correlations between in vitro measures of receptor phosphorylation and in vivo insulin action *(23)*: studies of greater numbers of subjects with this interesting behavior may be needed to clarify the causes and consequences of elevated receptor serine phosphorylation.

Additional data regarding insulin receptor phosphorylation in PCOS are limited. In contrast, receptor autophosphorylation was normal in cultured skeletal muscle cells *(13)*, as was insulin action. Thus, it appears that heterogeneity in PCOS extends to the molecular level, where it is possible to have altered insulin action on metabolic responses in the presence of either normal or impaired receptor kinase activity.

Analyzing the concentration dependence of the initial steps in insulin signaling, receptor binding, and autophosphorylation in comparison to a final response can highlight differences between PCOS and control subjects. Figure 2 shows that at a physiological insulin level (0.43 nM), 30–40% of insulin receptors are occupied and autophosphorylated in adipocytes from both control and PCOS subjects *(7)*. That same concentration resulted in approximately 90% of the maximal stimulation of glucose transport in control adipocytes, suggesting that an amplification of the insulin signal occurred distal to the receptor kinase. However, such amplification does not occur in PCOS adipocytes; the dose–response curves are superimposable for all three parameters. This comparison indicates that insulin resistance in PCOS involves impairment in the efficiency of insulin signal transduction. Regardless of the magnitude of the final response, to attain a set relative insulin response requires more occupied and activated insulin receptors in PCOS adipocytes. Hyperinsulinemia would be the expected response to compensate for this difference.

Except in the subset of individuals with elevated serine phosphorylation of their insulin receptors *(23)*, the presence of normal receptor function in PCOS places altered signaling further downstream. The existing data in PCOS extend only to the next two, related, steps; phosphorylation of IRS-1/2 and association with/activation of PI3-K. IRS-1-associated PI3-K activity was reduced in skeletal muscle from PCOS women when compared to age and BMI-matched control subjects *(20)*. The reduction was seen at both submaximal (100 mU/mL) and maximal (2000 mU/mL) circulating insulin levels. This occurred in the presence of normal tissue content of IRS-1 and the p85 regulatory subunit of PI 3-K. The PCOS subjects included in this study included those with both normal and impaired glucose tolerance.

What is unknown is the relationship between in vivo glucose tolerance and insulin action and differences in PI3-K activity. This information may be important because the same investigators reported that insulin-stimulated IRS-1-associated PI3-K activity (normalized to IRS-1 expression) was reduced in skeletal muscle cells cultured from PCOS subjects, even as insulin action on glucose metabolism was normal *(13)*. The reverse situation was observed in cultured fibroblasts; insulin-stimulated IRS-1-associated PI3-K activity was normal while there was insulin resistance for stimulation of glycogen synthesis *(11)*. Further studies will be needed to firmly establish the relationship between PI3-K activity and insulin action, if the relationship varies between tissues, and if any such relationship is acquired by the in vivo environment or intrinsic to PCOS and retained during culture.

Although the story regarding IRS-1/2-associated PI3-K activity in PCOS is incomplete, there is interesting information regarding IRS-1/2. Several investigators have reported that the incidence of the Gly972Arg polymorphism in IRS-1 is elevated two- to threefold in PCOS subjects compared to weight-matched controls *(25)*. Individuals homozygous for the Arg972 allele have higher insulin levels and HOMA-IR values, suggesting that altered IRS-1 function resulting from this amino acid substitution could contribute to insulin resistance. However, other investigators have found normal frequency of the Gly972Arg in PCOS *(26)*. Both results have been seen in populations of different ethnicities, so the differences may depend more on other variables, such as the specific definition of PCOS.

Several other lines of evidence suggest that the IRS proteins may play an important role in discriminating between metabolic and mitogenic responses and modulating insulin action in PCOS. One is that in peripheral tissues such as skeletal muscle and adipose tissue, the major portion of insulin signaling to metabolic responses is mediated through IRS-1, whereas IRS-2 plays a greater role in the ovary, especially with regard to reproductive function. Immunohistochemical staining showed a decreased expression of IRS-1 in granulosa cells from PCOS subjects compared to follicles from the same stage of development in ovulatory ovaries *(27)*. This would be consistent with a diminution of insulin action on metabolism and a preservation of mitogenic and steroidogenic responses. Cultured muscle cells from PCOS subjects display elevated phosphorylation of IRS-1 on Ser[312] *(13)*. Phosphorylation at this site is strongly associated with impaired IRS-1 function and insulin resistance *(2)*. However, it is difficult to reconcile the increased Ser[312]-IRS-1 phosphorylation and decreased insulin-stimulated IRS-1-associated PI3-K activity, both features of insulin-resistant states, with the normal insulin action on glucose metabolism seen in the same cells. Further studies will be needed to determine the relationships between signaling pathways and final responses.

One of the key questions in this field of study is if insulin resistance in PCOS differs from that in type 2 diabetes. The most complete information regarding such comparisons has been gathered in isolated adipocytes and is summarized in Table 1. In making such comparisons, it is crucial to match for other confounding variables such as age, obesity, and, of course, gender, also taking into account the considerable heterogeneity present in the syndrome. When this is done two conclusions become apparent. One is that insulin receptor function, binding and kinase activity, is essentially normal in PCOS whereas it is impaired in diabetes. Second, type 2 diabetes is primarily a state of impaired insulin responsiveness, whereasa major feature of PCOS is defects in insulin sensitivity. Further comparisons of signaling pathways between these two groups have the potential to reveal much about the roles of specific signal-transduction events.

3. CONCLUSIONS

Insulin resistance is a common, although not absolute, feature of PCOS, occurring in as much as 50% of subjects. The insulin resistance is present in both lean and obese individuals with PCOS and is exacerbated by obesity. At the level of the whole body, insulin resistance is displayed as a reduced sensitivity to the ability of insulin to stimulate glucose disposal, manifested primarily in skeletal muscle. Studies in freshly isolated adipocytes have shown reduced sensitivity for stimulation of glucose transport and suppression of lipolysis, with variable effects on insulin responsiveness. Resistance to the metabolic actions of insulin is seen in a number of other tissues, including skin fibroblasts and granulosa cells, with variable findings in skeletal muscle. Nonmetabolic actions of insulin, including stimulation of mitogenesis and E2 and P synthesis, appear to be normal. Insulin resistance in PCOS differs in some ways from that present in type 2 diabetes, being primarily a reduction in the efficiency of insulin signal transduction. Potential sites for defective insulin signaling implicated in PCOS include elevated serine phosphorylation of the insulin receptor and IRS-1/2, posttranslational modifications that impair the function of these proteins. Many populations of PCOS subjects also have an increased frequency of a polymorphism in IRS-1 that is associated with hyperinsulinemia and insulin resistance. Selective defects in insulin signaling to metabolic responses would lead to compensatory hyperinsulinemia, overstimulating steroid production in the normally insulin-sensitive ovary.

4. FUTURE DIRECTIONS

Although several key questions concerning insulin resistance in PCOS have been resolved to some extent—that resistance is present in multiple tissues and is selective for metabolic responses—other questions remain open. Among these are the role of different tissues in whole body insulin action, what aspects of resistance are acquired and which are intrinsic, and conclusive identification of the specific signaling events altered in PCOS. Future studies will need to investigate further downstream events as well as pathways other than the prototypical PI3-K pathway. Given the mounting evidence for improvement of the metabolic and reproductive features of PCOS by treatment with antidiabetic agents, it will be informative to determine the impacts of these treatments of insulin signaling. It will also be important to discriminate between PCOS subjects with normal insulin action and the approximately 50% who are insulin resistant: Is it possible to distinguish at the cellular and molecular level between resistant and sensitive states? In our laboratory we are employing isolated adipocytes and cultured muscle cells obtained from insulin-resistant and -sensitive subjects to address these open questions.

KEY POINTS

- Insulin resistance for metabolic responses (glucose transport, glycogen synthesis) is present in multiple tissues with PCOS.
- Insulin action for stimulation of mitogenesis and steroidogenesis is normal in multiple tissues, including the ovary.
- Insulin resistance for metabolic responses is, in part, an intrinsic property of PCOS tissues.
- Impaired insulin action involves posttranslational modification of signaling molecules, including altered phosphorylation of the insulin receptor and IRS-1/2.

REFERENCES

1. Virkamaki A, Ueki K, Kahn CR Protein-protein interaction in insulin signaling and the molecular mechanisms of insulin resistance. J Clin Invest 1999;103:931–943.
2. Gual P, Le Marchand-Brustel Y, Tanti J-F. Positive and negative regulation of insulin signaling through IRS-1 phosphorylation. Biochimie 2005;87:99–109.

3. Diamanti-Kandarakis E, Kouli C, Alexandraki K, Spina G. Failure of mathematical indices to accurately assess insulin resistance in lean, overweight, or obses women with polycystic ovary syndrome. J Clin Endocrinol Metab 2004;89:1273–1276.

4. Dunaif A, Segal KR, Shelly DR, Green G, Dobrjansky A, Licholai T. Evidence for distinctive and intrinsic defects in insulin action in polycystic ovary syndrome. Diabetes 1992;41:1257–1266.

5. Henry RR. Insulin resistance: from predisposing factor to therapeutic target in type 2 diabetes. Clin Ther 2003;25:B47–B63.

6. Buffington CK, Givens JR, Kitabchi AE. Sensitivity of pyruvate dehydrogenase to insulin in activated T lymphocytes. Lack of responsiveness to insulin in patients with polycystic ovarian disease and diabetes. Diabetes 1990;39:361–368.

7. Ciaraldi TP, El-Roeiy A, Madar Z, Reichart D, Olefsky JO, Yen SSC. Cellular mechanisms of insulin resistance in polycystic ovarian syndrome. J Clin Endocrinol Metab 1992;75:577–583.

8. Marsden PJ, Murdoch A, Taylor R. Severe impairment of insulin action in adipocytes from amenorrheic subjects with polcystic ovary syndrome. Metabolism 1994;43:1536–1542.

9. Ciaraldi TP, Morales AJ, Hickman MG, Odom-Ford R, Olefsky JM, Yen SSC. Cellular insulin resistance in adipocytes from obese polycystic ovary syndrome subjects involves adenosine modulation of insulin sensitivity. J Clin Endocrinol Metab 1997;82:1421–1425.

10. Ciaraldi TP, Morales AJ, Hickman MG, Odom-Ford R, Yen SSC, Olefsky JM. Lack of insulin resistance in fibroblasts from subjects with polycystic ovary syndrome. Metabolism 1998;47:940–946.

11. Book C-B DA. Selective insulin resistance in the polycystic ovary syndrome. J Clin Endocrinol Metab 1999;84:3110–3116.

12. Ciaraldi TP, Abrams L, Nikoulina S, Mudaliar S, Henry RR. Glucose transport in cultured human skeletal muscle cells. Regulation by insulin and glucose in nondiabetic and non-insulin-dependent diabetes mellitus subjects. J Clin Invest 1995;96:2820–2827.

13. Corbould A, Kim Y-B, Youngren JF, Pender C, Kahn BB, Lee A, Dunaif A. Insulin resistance in the skeletal muscle of women with PCOS involves intrinsic and acquired defects in insulin signaling. Am J Physiol Endocrinol Metab 2005;288:E1047–E1054.

14. Ciaraldi TP, Chu NV, Aroda V, et al. Insulin resistance in the polycystic ovary syndrome involves intrinsic defects in skeletal muscle glucose uptake and insulin action (abstr). Diabetes 2002;51:A298, 1211–P.

15. Willis D, Franks S. Insulin action in human granulosa cells from normal and polycystic ovaries is mediated by the insulin receptor and not the type-1 insulin-like growth factor receptor. J Clin Endocrinol Metab 1995;80:3788–3790.

16. Willis D, Mason H, Gilling-Smith C, Franks S. Modulation by insulin of follicle-stimulating hormone and luteinizing hormone action in human granulosa cells of normal and polycystic ovaries. J Clin Endocrinol Metab 1996;81:302–309.

17. Rice S, Christofordis N, Gadd C, et al. Impaired insulin-dependent glucose metabolism in granulosa-lutein cells from anovulatory women with polycystic ovaries. Hum Reprod 2004;20:373–381.

18. Wu XK, Zhou SY, Liu JX, et al. Selective ovary resistance to insulin signaling in women with polycystic ovary syndrome. Fert Steril 2003;80:954–965.

19. Lin Y, Fridstrom M, Hillensjo T. Insulin stimulation of lactate accumulation in isolated human granulosa-luteal cells: a comparison between normal and polycystic ovaries. Hum Reprod 1997;12:2469–2472.

20. Dunaif A, Wu XK, Lee A, Diamanti-Kandarakis E. defects in insulin receptor signaling in vivo in the polycystic ovary syndrome (PCOS). Am J Physiol Endocrinol Metab 2001;281:E392–E399.

21. Talbot JA, Bicknell EJ, Rajkhowa M, Krook A, O'Rahilly S, Clayton RN. Molecular scanning of the insulin receptor gene in women with polycystic ovarian syndrome. J Clin Endocrinol Metab 1996;81:1979–1983.

22. Conway GS, Avet C, Rumsby G. The tyrosine kinase domain of the insulin receptor gene is normal in women with hyperinsulinaemia and polycystic ovary syndrome. Hum Reprod 1994;9:1681–1683.

23. Dunaif A, Xia J, Book C-B, Schenker E, Tang Z. Excessive insulin receptor serine phosphorylation in cultured fibroblasts and in skeletal muscle. J Clin Invest 1995;96:801–810.

24. Li M, Youngren JF, Dunaif A, Goldfine ID, Maddux BA, Zhang BB, Evans JL. Decreased insulin receptor (IR) autophosphorylation in fibroblasts from patients with PCOS: effects of serine kinase inhibitors and IR activators. J Clin Endocrinol Metab 2002;87:4088–4093.

25. Sir-Peterman T, Perez-Bravo F, Angel B, Maliqueo M, Calvillan M, Palomino A. G972R polymorphism mof IRS-1 in women with polycystic ovary syndrome. Diabetologia 2001;44:1200–1201.

26. Villuendas G, Botella-Carretero JI, Roldan B, Sancho J, Escobar-Morreale HF, San Millan JL. Polymorphisisms in the insulin receptor substrate-1 (IRS-1) gene and the insulin receptor substrate-2 (IRS-2) gene influence glucose homeostasis and body mass index in women with polycystic ovary syndrome and non-hyperandrogenic controls. Hum Reprod 2005;20:3184–3191.

27. Wu X-K, Sallinen K, Anttila L, et al. Expression of insulin-receptor substrate-1 and -2 in ovaries from women with insulin resistance and from controls. Fertil Steril 2000;74:564–572.

28. Smith U, Axelsen M, Carvalho E, Eliasson B, Jansson PA, Wesslau C. Insulin signaling and action in fat cells: associations with insulin resistance and type 2 diabetes. Ann NY Acad Sci 1999;892:119–126.

27

Menstrual Dysfunction and Endometrial Neoplasia in the Polycystic Ovary Syndrome and Other Androgen Excess Disorders

Abdulkadir Wagley and Paul Hardiman

SUMMARY

About 80% of women with polycystic ovary syndrome (PCOS) suffer from menstrual dysfunction (oligomenorrhea, amenorrhea) at some point in their reproductive life. These conditions are associated with several factors, including obesity, degree of hyperandrogenism, and insulin insensitivity. Menstrual irregularity improves with advancing age. Women with menstrual irregularity are more likely to have an elevated luteinizing hormone (LH), altered LH-to-follicle-stimulating hormone (FSH) ratio, elevated androgens, and an altered uterine artery pulsatility index. Non-PCOS hyperandrogenemic conditions also affect menstrual function. Evidence for an association with endometrial neoplasia (endometrial hyperplasia and endometrial carcinoma) is at best incomplete. The altered hormonal environment of PCOS exerts some influence on endometrial gene expression with upregulation of human asialoglycoprotein receptor I, human secreted phosphoprotein I (osteopontin), and cytochrome P450 XVIIAI. Treatment for menstrual dysfunction must include lifestyle changes, not just in the obese woman. The low glycemic index diet has been shown to be effective in weight loss regimes, although the evidence for any benefit for insulin resistance and lipid metabolism is open to doubt. The oral contraceptive pill and progestagens are alternative means to regulating the menstrual cycle in appropriate patients.

Key Words: Polycystic ovary syndrome; menstrual dysfunction; endometrial hyperplasia; endometrial cancer; treatment.

1. INTRODUCTION

Irregular or absent menstruation were features of the disorder now known as the polycystic ovary syndrome (PCOS) when it was first described by Stein and Leventhal in 1935 (1). More recently, the 2003 Rotterdam Consensus criteria for PCOS diagnosis include oligo-ovulation or anovulation, although these terms are undefined and used interchangeably with oligomenorrhea and amenorrhea, respectively. Other criteria are biochemical or clinical hyperandrogenism and polycystic ovary morphology on ultrasound scan. Because only two of these three criteria have to be met, some women diagnosed with PCOS have regular menstrual cycles. There are still variations in clinical practice, however, and 70% of reproductive endocrinologists and 47% of gynecologists still consider menstrual irregularity a required criterion for diagnosing PCOS (2). Indeed, some authors question the 2003 consensus criteria, in that PCOS can be diagnosed in the absence of features considered, a least by them, central to the syndrome, such as menstrual irregularity, hyperandrogenism, or polycystic ovary morphology.

From: *Contemporary Endocrinology: Androgen Excess Disorders in Women:*
Polycystic Ovary Syndrome and Other Disorders, Second Edition
Edited by: R. Azziz et al. © Humana Press Inc., Totowa, NJ

Table 1
Menstrual Patterns in Women With PCOS

Study (ref.)	No. of women	% eumenorrhea	% oligomenorrhea	% amenorrhea	Criteria to diagnose PCOS
Balen et al. *(4)*	1741	34	47	19	Ultrasound PCO morphology
Franks et al. *(5)*	300	20	52	28	Ultrasound PCO morphology
Goldzieher et al. *(6)*	640	20	29	51	Surgical macroscopic appearance

PCOS, polycystic ovary syndrome.

Endometrial cancer was the first example of long-term morbidity associated with PCOS, and a study by Jackson and Dockerty *(3)* is often cited to confirm this association. As will be discussed here, evidence in support of this association is surprisingly inconclusive.

2. BACKGROUND

2.1. Menstrual Disturbances in PCOS

Women with PCOS may have normal, irregular, or absent menstrual bleeding (Table 1). The heterogeneous expression of symptoms is a reflection of the interplay among genetic variants impacting on androgen biosynthetic pathways, carbohydrate metabolism, and gonadotropin action (*CYP11A*; *CYP21*; *VTNR*; *calpain-10*) and environmental factors, notably nutrition and lifestyle factors (diet and exercise). Although studies of familial clustering of PCOS cases provide evidence for a genetic link, the variation in phenotypic expression within families and the discordance in PCOS expression in monozygotic twins illustrate the significant role of environmental factors.

The largest cohort of women with polycystic ovary morphology on ultrasound, investigated by Balen et al. *(4)*, showed that of 1871 subjects who had exhibited at least one further symptom of PCOS, 30% had a regular menstrual cycle, 50% oligomenorrhea, and 20% amenorrhea. Franks *(5)*, reporting on 300 women with an ultrasound diagnosis of PCO, attending a gynecological endocrine clinic, also noted that 50% had oligomenorrhea and 30% amenorrhea. Goldzieher *(6)* reviewed 1079 cases of polycystic ovaries identified at surgery by macroscopic appearance (a rather subjective method of diagnosis) and found that 85–90% of women had an abnormal menstrual cycle, with amenorrhea in up to 50% (*see* Table 1). Conversely, in a study by Gadir et al. *(7)* of 389 women presenting with menstrual cycle disturbances, 65% had polycystic ovaries on ultrasound scan. The variation in prevalence of menstrual cycle disturbance is likely to result from selection bias dependent on the clinics from which subjects were recruited, the criteria used to define menstrual irregularity, and the methods used to diagnose PCOS.

Numerous studies have reported the frequency of clinical symptoms and signs in women with or without polycystic ovary morphology. They show that menstrual cycle irregularities are more common in those with polycystic ovary morphology (*see* Table 2); however, there are again differences in findings. The large difference in the prevalence of menstrual disturbances in subjects with or without polycystic ovaries in the Polson study *(8)*, in which subjects were recruited by offering an ultrasound scan to hospital workers who had not previously attended a gynecology clinic, resulted not only from a high rate of menstrual irregularities in the polycystic ovary group, but also from an extremely low prevalence in women with normal ovaries. In contrast, Clayton *(9)*, recruiting from a single general practice list of all females of reproductive age, found no difference in the prevalence of menstrual dysfunction between those with and without polycystic ovaries. The explanation for this is unclear because the definition of irregular periods was similar to that used in other studies, but it may be attributed to the manner in which menstrual histories were recorded from those recruited.

Most women with menstrual irregularity and PCOS report that their menstrual symptoms have been present from menarche or soon thereafter. Irregular menstruation secondary to PCOS may,

Table 2
Frequency of Menstrual Disturbances in Women With and Without Polycystic Ovarian Morphology

Study (ref.)	No. normal ovaries	% normal ovaries	No. polycystic ovaries	% polycystic ovaries	Study population
Polson et al. *(8)*	116	1	33	76	Clinical and secretarial staff and volunteers
Clayton et al. *(9)*	165	27	43	29	General practice, 20–37 yr, and volunteers
Farquhar et al. *(73)*	144	20	39	46	Electoral Roll, 18–45 yr
Botsis et al. *(74)*	823	—	183	80	Attenders for routine Pap smear, 17–40 yr
Cresswell et al. *(75)*	186	27	49	41	Delivery register at a maternity hospital, 42 yr

however, be indistinguishable from that which precedes the onset of regular ovulatory function after the menarche in normal females. Consequently, doctors tend to reassure adolescents that their menstrual pattern disorder is transient and a result of incomplete maturation of the hypothalamic–pituitary–ovarian axis. Moreover, multiple ovarian follicles are seen on ultrasound scan in normal puberty, making ovarian morphology an unreliable marker in defining PCOS in adolescent females. Adams et al. *(10)* differentiated this "multicystic ovary" from the polycystic ovary by the absence of stromal hyperechogenicity and the presence of fewer cysts uniformly distributed throughout the ovary with a cyst diameter larger than 10 mm. However, the Rotterdam definition of polycystic ovary does not utilize information on the ovarian stroma or cyst distribution. Oligomenorrhoeic adolescents have higher mean serum luteinizing hormone (LH), androstenedione, testosterone, dehydroepi-androstenedione sulfate (DHEAS), and estradiol concentrations compared to girls with regular menstrual cycles *(11)*. Indeed, 57% of oligomenorrheic 15-year-olds had a serum LH and testosterone level above the 95th percentile of levels seen in girls with regular menstrual cycles *(11)*. Ventroli et al. *(12)* noted that adolescents with oligomenorrhea but normal plasma LH concentrations 2 years postmenarche had a higher chance of ovulating spontaneously within the following 40 months than those with oligomenorrhea and high LH levels. Consequently, it has been suggested that oligomenorrhea beyond 2 years postmenarche should be regarded as an early clinical sign of PCOS, rather than a transient stage in the normal physiological maturation of the hypothalamic–pituitary–ovarian axis, and that these girls require endocrine evaluation before reassurance or the prescribing of oral contraceptives. Although some women are genetically predisposed to developing PCOS, an additional event, such as an increase in insulin levels and insulin-like growth factor (IGF)-1 activity or prolonged stress or weight gain, is thought to trigger the development of the full syndrome.

PCOS may present as primary or secondary amenorrhea. Indeed, in female adolescents, PCOS is estimated to account for 3% of cases presenting with primary amenorrhea, absent breast development and low serum follicle-stimulating formone (FSH) *(13)*. PCOS is among the most common causes of secondary amenorrhea along with hypothalamic amenorrhea, prolactinomas, and ovarian failure. The possibility of excessive testosterone secretion in such cases is suggested by hirsutism, increased muscle mass, and/or other signs of virilization. Although PCOS is the most common disorder when amenorrhea is associated with hyperandrogenism, it is important to exclude other causes such as adrenal disease (e.g., adrenal hyperplasia, Cushing's syndrome, androgen-producing tumors) and ovarian androgen-producing tumors.

At the other end of reproductive life, the incidence of menstrual cycle irregularity in women with PCOS declines. This has been attributed to the drop in the ovarian follicle cohort and a decline in androgen levels. Studies of normally cycling women have shown that menstrual cycles become shorter as menopause approaches as a result of decreasing follicular phase length. With ovarian aging, a new balance between inhibin B and FSH is achieved. Although inhibin A and B are both lower in older women, a longitudinal study by Welt et al. *(14)* showed that the decline in inhibin B precedes that of inhibin A. In vitro studies have shown that inhibin stimulates androgen production, and hence with aging androgen synthesis declines. The normalization in menstrual cycle length in women with PCOS as they age is significant even after correction for possible confounders such as body mass index (BMI), weight loss, or hirsuitism.

Some women with premenstrual syndrome (PMS) also have clinical, endocrine, or ultrasound features of PCOS, and a causal relationship between the two conditions might be assumed. There are, however, very limited data to support such a relationship, although Iurassich et al. *(15)*, looking at PCOS women with acne, noted that 93% of his subjects suffered with PMS, diagnosed on the basis of a history of premenstrual breast or pelvic discomfort, change in mood, headache, or fever. It might seem reasonable, therefore, to offer women with PCOS and distressing PMS therapy for the former (perhaps metformin), but no clinical trials have investigated the efficacy of this approach. Conversely, it could be argued that the hormonal changes within the luteal phase of the menstrual cycle play a pivotal role in PMS, so that women with PCOS who suffer oligo- or anovulation should be protected from PMS. This is, however, likely to be an oversimplification because it fails to account for other hormonal and nonhormonal mechanisms involved in the pathogenesis of PMS.

2.2. Pathophysiology of Menstrual Dysfunction in PCOS

The pathophysiological mechanisms underlying menstrual dysfunction in PCOS are diagrammed in Fig. 1.

2.2.1. Body Mass Index

Women with PCOS frequently report that their oligomenorrhea is exacerbated by weight gain, and this is in accordance with current views of the pathogenesis of PCOS. Data from cross-sectional studies also show that menstrual irregularity in women with PCOS is more common in those who are overweight, and there is a progressive increase in body mass index (BMI) between women with PCOS who have normal menstrual cycles, oligomenorrhea, and amenorrhea. The influence of BMI on menstrual cyclicity in PCOS is likely to be important, since approximately 50% of women with PCOS are overweight or obese. Legro et al. *(16)* hypothesize that obesity, either of environmental origin or as an expression of certain genotypes, is responsible for modifying the phenotype of PCOS.

2.2.2. Ethnicity

Ethnicity could influence the prevalence or clinical presentation of PCOS either through genetic or environmental (diet, exercise) factors. In Britain, the prevalence of PCOS is 4–7% *(17)*. Rodin et al. *(18)* have shown that among south Asian immigrants in the united Kingdom, 52% exhibited PCOS. Of these women, 49% had menstrual irregularity. In contrast, Kousta et al. found there was no difference in the prevalence of PCOS in women of different ethnic origin attending a fertility clinic in the United Kingdom and that 26% of their subjects with PCOS had irregular cycles *(19)*. Indeed, Schmid et al., studying PCOS in local and immigrant Austrian women, found no difference in the prevalence of menstrual disorders between the two groups, althuogh the immigrant women perceived menstrual dysfunction as a "big problem" relative to the local women, based on the Cronin quality-of-life assessment *(20)*.

2.2.3. Diet

Diets high in poly- or monounsaturated fatty acids have been implicated in the expression of various metabolic derangements in PCOS because they may adversely affect glucose homeostasis. High glycemic-index foods produce a rapid rise in blood glucose and a compensatory increase in insulin. Diets that are high in such foods will eventually result in downregulation of the insulin receptor, worsening insulin resistance in PCOS. Thus, perturbations in insulin sensitivity may have an effect on menstrual function.

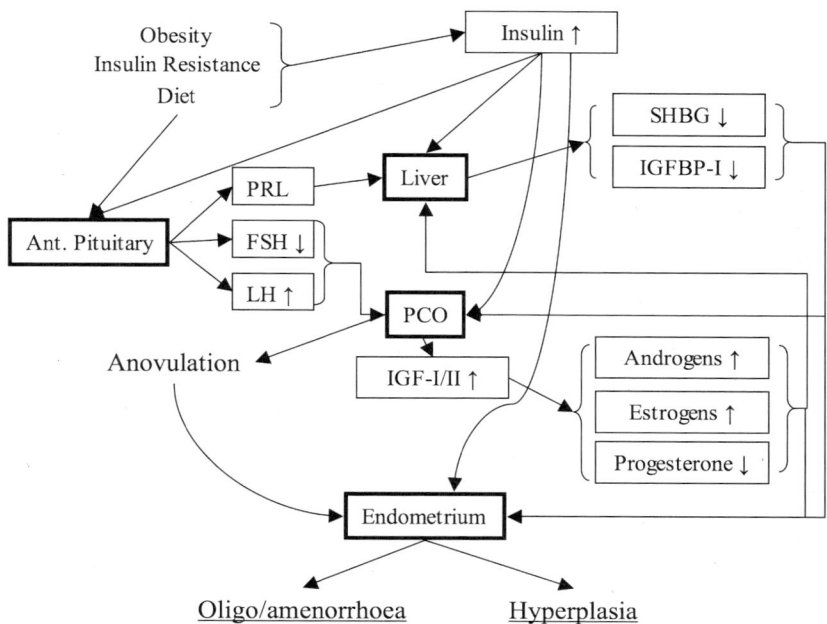

Fig. 1. Flow diagram of the pathophysiological mechanisms involved in menstrual dysfunction in women with PCOS.

2.2.4. Luteinizing Hormone

As already discussed, LH and testosterone levels increase progressively in the first 6 years after menarche, irrespective of menstrual cycle pattern. Van Hoof et al. showed, however, that teenage girls with oligomenorrhea have significantly higher serum LH and testosterone concentrations than do girls with regular menstruation. In contrast, Doi et al. *(21)* found that a normal LH does not predict menstrual status and that cycle regularity was unrelated to differences in LH from the normal range.

Serum LH levels are associated with menstrual disturbance in women with PCOS. For example, Botsis et al. *(22)* showed that LH-to-FSH ratios and testosterone levels were elevated in women with polycystic ovary morphology on ultrasound who had menstrual irregularities compared to those with regular cycles. Although an elevated serum LH concentration is inconsistently found in PCOS, increased LH bioavailability is an almost invariable feature of the syndrome *(23)*. The mechanism responsible for this alteration remains unclear. LH binding to receptors on the thecal cell surface activates cyclic adenosine monophosphate-mediated intracellular signaling, triggering the production of androstenedione and testosterone from cholesterol. The subsequent effect of androgens on the endometrium, causing menstrual irregularity, is discussed later.

Obese women with PCOS are more likely to suffer irregular periods, and women with PCOS who have irregular periods exhibit elevated serum LH concentrations. It is not the case, however, that obese women with PCOS have high serum LH levels. In fact, there is a negative correlation between LH and body weight in women with PCOS. Arroyo et al. *(24)* demonstrated that at a BMI of 20 kg/m^2, the serum LH was 2.5-fold higher in women with PCOS compared with BMI-matched euandrogenic women. They also noted a progressive decrease in LH pulse amplitude in women with PCOS as BMI increased to 40 kg/m^2, at which point the amplitude was comparable to that seen in the non-PCOS group. The BMI-related blunting of LH pulse amplitude in women with PCOS is accompanied by a parallel attenuation of the sensitivity of pituitary LH secretion to GnRH stimulation. The BMI-related reduction in LH release may reflect the effect of leptin at the hypothalamic or pituitary level, dampening LH secretion in the obese state. Leptin also directly increases ovarian androgen production by stimulating 17β-hydroxylase activity. Other contributory factors to menstrual irregularity in obesity include impaired estrogen metabolism, reduction in sex hormone-binding globulin (SHBG) levels, and hyperinsulinemia.

2.2.5. Sex Steroids

Menstrual irregularity in women with PCOS results, at least in part, from an increased serum estradiol-to-progesterone ratio, which arises as a consequence of impaired follicular maturation. Estrogen production rate positively correlates with body weight and proportion of fat content in women with PCOS. Altered estrogen metabolism in obese women as a result of decreased formation of inactive estradiol metabolites results in an increased balance of active to inactive estrogens, and hence an enhanced exposure of target tissues to estrogen.

In women with PCOS, endometrial stromal and epithelial cells show a 30% higher expression of estrogen receptor (ER) α and similar expression of ERβ and progesterone receptor (PR), compared with normal women *(25)*. ERα is the predominantly expressed ER isoform in the uterus and mediates estrogen sensitivity in the endometrium. ERβ has also been immunolocalized to the nuclei of multiple cell types within the endometrium, and the endometrial endothelial cell might represent a target for estrogen action mediated exclusively by ERβ. ERs belong to a class of ligand-dependent transcription factors that influence cell growth and differentiation via effects on gene expression. In women with PCOS, elevated ERα expression may be associated with implantation failure and loss of receptivity proteins.

The PR belongs to the same family of receptors as ER and is expressed in two isoforms, PR-A and PR-B, arising from a single gene. The action of progesterone is complex, with differential PR-A and PR-B expression reported during the menstrual cycle. Progesterone has the dual function of stimulating the expression of paracrine factors in the endometrial stroma while inhibiting ER and PR expression; it is therefore a physiological negative regulator of estrogen action in the endometrium. Downregulation of ERα releases from the inhibitory effects of estrogen the expression of several paracrine factors involved in priming the endometrium for implantation. In the absence of ovulation, for example, in PCOS, the corpus luteum fails to develop, and hence progesterone is not produced and the endometrium fails to undergo luteal phase differentiation but continues to be exposed to estrogen. The withdrawal bleed that marks the late luteal phase decline in progesterone levels fails to occur, and endometrial proliferation continues.

Apparao et al. *(26)* have demonstrated an overall increase in the androgen receptor (AR) content in the endometrium of women with hyperandrogenism and PCOS. In the normal menstrual cycle, ARs are upregulated by estrogen and downregulated by progesterone. Androgens also increase the expression of endometrial ARs, suggesting a synergistic effect with estrogens in regulating the receptor. The role of ARs in endometrial development and function during the menstrual cycle remains poorly understood. Some studies report relatively constant expression of ARs; others have noted increased AR expression during the proliferative phase. Iwai et al. *(27)* found that androgens inhibit the expression of ERs and PRs, but in PCOS it is believed that the stimulatory effects of estrogen on ER and PR expression outweigh androgenic inhibition. The overexpression of endometrial ARs in PCOS, as a result of estrogenic and androgenic stimulation, provides tangible evidence linking the abnormal hormonal milieu of PCOS with endometrial dysfunction *(28)*.

Overexpression of coactivators (factors that modulate the activity of activated steroid receptors) has been identified in the endometrium in the proliferative and secretory phases of the menstrual cycle in women with PCOS. Three coactivators have been identified within the p160 family (proteins of ~160 kDa encoded by a single gene): steroid receptor coactivator-1 (*SRC-1*); amplified in breast cancer-1 (*AIB1*); and transcriptional intermediary factor-2 (*TIF2*). Overexpression of *AIB1* and *TIF2* in women with PCOS has been shown to render the endometrium more sensitive to estrogen, contributing to their higher incidence of menstrual irregularities.

2.2.6. Insulin and IGF

Robinson et al. observed that non-obese women with PCOS and oligomenorrhea were more likely to be insulin resistant than those with regular cycles *(29)*. Norman et al. *(30)* similarly showed that in women with PCOS, the greater the abnormality in insulin resistance, assessed by fasting and stimulated insulin levels, the greater the menstrual irregularity, independent of BMI and androgens. Gonzalez et al. *(31)* reported the results of a retrospective study of women identified as having polycystic ovary morphology, aged 18–35 years, assessing them for insulin resistance (using homeostatic model assessment estimates for insulin resistance [HOMA-IR]), and comparing the results based on menstrual patterns. They noted that 78% of amenorrhoeic women with polycystic ovaries, compared with 41% of oligomenorrhoeic women, displayed insulin resistance, again suggesting that a higher degree of insulin resistance was associated with a greater prevalence of amenorrhea.

On the other hand, Chang et al. *(32)*, using the National Institutes of Health definition of PCOS, found that all PCOS phenotypes with oligomenorrhea had similar fasting glucose levels. HOMA-IR and percent β-cell function (HOMA-β-cell), while showing significant differences between those with biochemical and clinical hyperandrogenism vs those with hirsuitism only, revealed no differences in the degree of menstrual irregularity. Likewise, Elting et al. *(33)* found no difference in fasting insulin levels or insulin resistance (based on fasting glucose-to-insulin ratio) between irregularly and regularly menstruating women with PCOS, although serum androgens (androstenedione and testosterone) were lower in the regularly menstruating group.

Discrepant results in these studies may reflect small sample size (e.g., only 34 women in the study of Elting et al. *[33]*), varying definitions of PCOS (e.g., the study of Chang et al. *[32]* did not consider ultrasonographic ovarian appearance), failure to standardize for the effect of BMI on insulin resistance (e.g., the study of Chang et al. admitted obese women), varying techniques for assessment of insulin resistance, and a lack of effective control groups in some studies.

2.2.7. Vascular Endothelial Growth Factor

Vascular endothelial growth factor (VEGF) is known to be involved in endometrial angiogenesis *(34)*. It is currently unclear whether the increased serum level of VEGF in women with PCOS is involved in the pathogenesis of menstrual symptoms.

2.2.8. Plasminogen Activator Inhibitor

Women with PCOS have increased serum levels of plasminogen activator inhibitor (PAI)-1, an antagonist of plasminogen activation and resultant clot lysis, which has been correlated with insulin resistance. It is postulated that high PAI-1 levels contribute to abnormal follicle maturation and rupture and an increased risk of venous thromboembolism. No differences in intraovarian PAI-1 activity between normal and anovulatory polycystic ovaries have been demonstrated, however *(35)*.

2.2.9. Uterine Hemodynamics

Increased resistance to blood flow has been observed in the uterine arteries of women with PCOS and was not associated with the degree of insulin resistance. In trying to define factors that affect uterine perfusion in women with PCOS, estrogen has been considered to act as a moderator of uterine vascularity. Androgens play the lead role, however, exerting a direct vasoconstrictive effect, mediated by ARs in the artery wall, and an indirect effect through androgen-dependent collagen and elastin deposition in vascular smooth muscle cells. Studies have shown that women with PCOS who suffer with amenorrhea have a uterine artery pulsatility index (PI) higher than those with oligomenorrhea *(36)*; the significance of this is unclear. Interestingly, although BMI has been closely correlated with uterine arterial PI in normal women, the changes in PCOS occur independently of obesity.

2.3. Non-PCOS Hyperandrogenic Conditions Causing Menstrual Dysfunction

Androgen-producing granulosa cell tumors of the ovary can present at any age, although the median age at diagnosis is 50–54 years. Most cases tend to present early, as a result of the endocrine manifestations (menstrual dysfunction, abdominal bloating, virilization), with an estimated 50–60% exhibiting endometrial hyperplasia and 5–10% having concomitant uterine cancer. Malstrom et al., looking at presentations of granulosa cell tumors over a 34-year period in a German teaching hospital, noted that of premenopausal women, 6% complained of menstrual irregularity and 4% of secondary amenorrhea *(37)*. Unkila-Kallio et al. noted that 45% of women with a granulosa cell tumor had oligo-amenorrhea *(38)*. Congenital adrenal hyperplasia (salt-losing, mildly virilizing, and nonclassic) is also associated with menstrual irregularity. Rouach and Blumenfeld noted that of women suffering hirsuitism and menstrual disturbance, 1–35% may suffer from nonclassic adrenal hyperplasia *(39)*. After the diagnosis has been established, menstrual disorders are a sign of poor therapeutic control.

2.4. Epidemiology of Endometrial Cancer and PCOS

In 1957, Jackson and Dockerty *(3)* reported an association between endometrial cancer and PCOS in a study reporting on 43 women with Stein–Leventhal syndrome (SLS). Twenty-seven of the women in Jackson's study were symptomatic of SLS, with the diagnosis confirmed by ovarian tissue biopsy. The remaining 16 women were selected with symptoms of SLS from several thousand patients diagnosed histologically with endometrial cancer. From the former group, endometrial tissue was available in 15 of the 27 cases, of which 13 showed "thickening" and 2 were atrophic. They quoted a 37% (16 out of 43) prevalence of endometrial cancer in women with PCOS and concluded their "most important observation in women with SLS . . . concerns the complication of endometrial carcinoma." This was an influential study for its time and continued to be quoted for the next half century, despite its obvious flaws.

A number of studies have been performed to investigate the association between endometrial cancer and anovulation (these included some women with PCOS). Coulam et al. *(40)* calculated a relative risk of 3.1 for developing endometrial cancer after a diagnosis of chronic anovulation, with the risk greatest in a subgroup of obese women. Escobedo et al. *(41)* noted an odds ratio of 4.2 for endometrial cancer in women with ovarian factor infertility. Although PCOS is the most common cause of anovulation and ovarian factor infertility, these findings cannot be extrapolated to all women with PCOS. Wild et al. *(42)*, in a retrospective study, calculated an odds ratio of 5.3 for endometrial cancer in women with PCOS.

On the other hand, Ramzy and Nisker *(43)* compared ovaries from 15 women with endometrial carcinoma, 25 women with PCOS, and 21 controls. They found that the ovaries from women with endometrial cancer were morphologically more similar to those from the control group than to those from women with PCOS. Ho et al. *(44)* failed to show an association between endometrial cancer and PCOS, although they noted that polycystic ovary disease and subfertility were found significantly in the cases with endometrial cytological atypia. We concluded in a recent review *(45)* that the evidence for an increased risk of endometrial carcinoma in women with PCOS is at best incomplete.

2.5. Pathogenesis of Endometrial Neoplasia in PCOS

The association between endometrial cancer and PCOS is generally assumed to result from estrogenic stimulation of endometrial growth either unopposed by progestogens or opposed to a lesser extent than in normally cycling women. There is, however, no direct evidence that natural estrogens are carcinogenic in humans. The proliferative activity of estrogen and its action as a tumor promoter have been demonstrated, but there is no evidence that estrogen alone can immortalize cells. Instead, it is thought that estrogens act by genetic and epigenetic mechanisms on cancer cells, and a close relationship between estrogens, growth factors, and oncogenes is important in the development of human cancers *(46)*.

The etiology of endometrial tumorigenesis has not been clearly defined in women with PCOS. A recent study *(47)* found a threefold increased risk of endometrial cancer in obese compared with non-obese women, suggesting that anovulation may not be an independent risk factor for endometrial cancer.

Hypersecretion of LH has also been implicated in the development of endometrial cancer in women with PCOS. Receptors for LH and human chorionic gonadotrophin are overexpressed at both mRNA and protein levels in endometrial adenocarcinomas. Konishi et al. demonstrated overexpression of receptors for both these hormones in endometrial hyperplasia and endometrial carcinoma *(48)*. For 3 of the 15 women in their study, serum analysis revealed biochemical features of PCOS.

Cancer of the endometrium is thought to arise through progression via a series of precursor lesions, namely simple or complex hyperplasia without atypia, and atypical endometrial hyperplasia, or endometrial hyperplasia and endometrial intraepithelial neoplasia. The dualistic model of endometrial cancer proposed by Bockman *(49)*, supported by immunohistochemical and molecular genetic analyses, suggests that normal endometrium transforms into endometrial hyperplasia with or without atypia and then to endometrial cancer by two distinct mechanisms, one dependent on estrogen, the other not. This model applied to both premenopausal and postmenopausal women with endometrial cancer alike. PCOS is a cause of hyperestrogenism and is thus associated with a risk of developing type 1 endometrial carcinoma (i.e., estrogen dependent). It has, however, become progressively apparent that both groups overlap to some extent, making the dualistic model a guideline at best.

The structure of ERα receptor varies in normal, hyperplastic, and malignant endometrium. Sequence analysis indicates deletion of one exon, suggesting that ERα-splicing errors may occur in endometrial carcinogenesis. An exon 5 splice variant, detected in endometrial adenocarcinoma, has been shown to activate transcription of ERα-dependent genes in the absence of hormone. Its function in relation to the development and progression of endometrial neoplasms has yet to be examined.

When PR-A binds progesterone, it downregulates estrogen action by preventing transcription of ERα, whereas PR-B acts as an endometrial estrogen agonist. The antineoplastic effect of progesterone thus depends on tight regulation of PR-A and PR-B isoform balance. De Vivo et al. identified a functional polymorphism in the promoter region of the human progesterone receptor gene, which favors PR-B production, increasing the risk of endometrial neoplasia. This risk was found to be greater in obese carriers. The expression of these variants in a cohort of women with PCOS remains to be investigated. Li et al. have shown that the expression of PR in the stroma of hyperplastic endometrium from women with PCOS is unevenly distributed and significantly less than that seen in normal endometrium in non-PCOS women *(50)*.

As discussed earlier, the increase in p160 coactivators of steroid hormone receptors in endometrium from women with PCOS raises endometrial sensitivity to estrogen. This enhanced sensitivity results in an increase in ERα expression, stimulates endometrial proliferation, increasing the risk of estrogen-related simple endometrial hyperplasia. Progression to a more severe form of hyperplasia would occur against a background of further genetic and epigenetic insults.

In women with PCOS, the endometrium displays increased activity of Bcl-2 and Bax relative to endometrium from women without PCOS. In the normal menstrual cycle, during the late secretory phase, active endometrial cell death occurs via apoptosis. Bcl-2 is an apoptotic inhibitor that promotes cell survival, whereas Bax induces apoptosis. Endometrial apoptosis is regulated by estrogen and progesterone, which are known to modify the expression of apoptotic proteins, during the menstrual cycle. It has been noted that there is greater expression of Bcl-2 activity in relation to Bax in PCOS endometrium *(25)*, resulting in prolonged cell survival.

PCOS is associated with insulin resistance and hyperinsulinemia. It is therefore of interest that Nagamani et al. *(51)* found insulin-binding sites in the endometrial stroma of premenopausal women and women with endometrial cancer, and Bershtein et al. *(52)* found increased concentrations of plasma insulin in patients with endometrial cancer. Because insulin upregulates aromatase activity in endometrial glands and stroma, endogenous estrogen production is enhanced in women with high-circulating insulin. In contrast, the argument linking hyperinsulinemia with endometrial cancer is not supported by a study showing that the risk of endometrial cancer is greater in type 1 than type 2 diabetes *(53)*.

Recent work within our unit comparing gene expression in proliferative phase endometrium with or without simple hyperplasia derived from women with PCOS matched for age, parity, and biochemical profile (LH-to-FSH ratio, free testosterone, SHBG, free androgen index, androstenedione, prolactin, and insulin) showed differences in gene expression related to hyperplasia. By microarray analysis, 24 genes were found to be upregulated more than twofold in the hyperplastic group, of which 3 genes are known to have roles in endometrial carcinoma, namely human asialoglycoprotein receptor I, human-secreted phosphoprotein I (osteopontin) (OPN), and cytochrome P450 XVIIAI *(54)*. OPN is a glycophosphoprotein that is expressed and secreted by numerous human cancers, with roles in cell adhesion, chemotaxis, macrophage-directed interleukin-10 suppression, stress-dependent angiogenesis, apoptosis inhibition, and anchorage-independent tumor cell growth. It is a potent regulator of cell–matrix interactions and cellular signaling through binding with integrin and CD44 receptors. Its role in the endometrium is yet to be elucidated.

In summary, the evidence for increased risk of endometrial carcinoma in PCOS is incomplete and contradictory. Intuitively it seems likely that women with the syndrome who have irregular periods or amenorrhea are at greatest risk because of the stimulatory effects of unopposed estrogen activity. Although PCOS is the most common cause of anovulation and anovulation is associated with an increased risk of developing endometrial hyperplasia, it seems unlikely that all women with PCOS, as currently defined, are at increased risk of developing endometrial cancer, because the incidence of endometrial cancer is substantially less than that of anovulatory PCOS.

2.6. Non-PCOS Hyperandrogenic Conditions and Endometrial Cancer

Non-PCOS androgen excess disorders, including granulosa cell tumors of the ovary and adrenal hyperplasia and carcinoma, have been reported as causes of endometrial hyperplasia and uterine cancer. Malmstrom et al. *(37)* noted that one-third of 54 women with granulosa cell ovarian tumors had atypical endometrial cells, with endometrial cancer confirmed in 5 patients. A Finnish study describing 146 women diagnosed with ovarian granulosa cell tumor also noted a 30% rate of endometrial hyperplasia and an 8% rate of endometrial cancer *(38)*. However, because neither of these studies noted the androgen levels of these women, one cannot deduce that these effects are secondary to hyperandrogenemia as opposed to hyperestrogenism.

2.7. Treatment of Menstrual Disturbances in PCOS

The appropriate treatment for menstrual dysfunction in a woman with PCOS will be influenced by the need to treat other symptoms of the syndrome, such as subfertility, acne, hirsutism, or obesity. In general, the treatment used to induce ovulation for those complaining of subfertility will restore menstrual cyclicity. These treatments are discussed elsewhere; therefore, this discussion will be confined to women not desiring pregnancy.

2.7.1. Lifestyle Changes

In those women who are obese, weight reduction must be the first objective. This not only improves symptoms of the syndrome but also reduces long-term morbidity and mortality. Weight reduction decreases fasting insulin concentrations and improves hyperandrogenism in obese patients with PCOS; however, weight reduction is difficult to achieve in many patients. Diet and regular exercise are the most frequently recommended methods for accomplishing this. Cussons et al. *(2)* showed that, among gynecologists and endocrinologists in Australia and New Zealand, diet and exercise were recommended as first-line treatment in 90% of women with PCOS whose primary concern was oligomenorrhea.

Marsh and Brand-Millar *(55)* noted that research into the dietary management of PCOS is lacking and that most studies focus on energy restriction rather than dietary composition *per se*. A diet low in saturated fat and high in fiber from predominantly low-glycemic index (GI) carbohydrate foods is recommended. The GI is a classification index of carbohydrate foods based on their effects on postprandial blood glucose response. It is defined as the incremental area under the blood glucose curve produced by a standard amount (50 g) of carbohydrate in a food relative to the incremental area produced by the same amount of carbohydrate from a standard source (white bread or glucose). The GI of a food is affected by its physical structure and chemical composition. If a food has a GI of less than 55, it is considered to break down slowly, 55–70 is considered moderate, and above 70 is considered rapid. The glycemic load provides an estimate of quality and quantity of carbohydrate in the diet. Together with dieticians, patients can adjust their diet to encourage optimal weight loss using a strategy that will work gradually over time. The evidence for the impact low-GI diets have on metabolic parameters (insulin sensitivity, lipids) in PCOS is, however, inconclusive, and further research is required in this area.

Clark et al. *(56)* showed that weight reduction through dieting in oligomenorrheic, obese women with PCOS not only resulted in regular menstruation, but also improved self-esteem, ovulation, and pregnancy outcome, as well as endocrine parameters (serum SHBG, fasting insulin, and testosterone concentrations). Lifestyle modification, through diets and exercise, has been shown in the short term to improve menstrual cycle frequency in obese women with PCOS. Once weight is regained, however, the manifestations of PCOS return. Norman et al. *(57)* found that further lifestyle modifications are required for long-term weight loss in obese PCOS women. They recommend moderate exercise for at least 30 minutes a day, establishing an energy deficit of 500–1000 kcal per day, reduction in psychosocial stressors, cessation of smoking, moderation of alcohol and caffeine consumption, and the use of group interaction to provide support and assistance in implementing changes. There are, however, no long-term studies examining the effects of lifestyle interventions in PCOS women. The implementation of these changes in teenagers diagnosed with PCOS is vitally important for prevention of obesity, the long-term health consequences, and the emotional and financial strains of the syndrome.

2.7.2. Oral Contraceptive Pill

Ovarian androgen suppression can be achieved by administration of estrogens and progesterones, such as found in the combined oral contraceptive pill (COCP). For several decades this has been standard therapy for women with PCOS seeking to regularize their menses and as protection against the development of endometrial carcinoma. The COCP works by suppressing ovarian and adrenal androgen production and increasing SHBG levels. It is generally regarded as second-line treatment for menstrual dysfunction after lifestyle modifications in women for whom fertility is not an issue.

Concern has, however, been raised about the use of COCP in obese women in the light of evidence that it reduces insulin sensitivity and may worsen dyslipidemia associated with PCOS *(58)*. The effect on insulin sensitivity may occur only in obese PCOS women, as Elter et al. *(59)* showed that COCP use in non-obese PCOS women had no effect on glucose tolerance or insulin sensitivity. The use of COCP in teenagers with PCOS has the added advantage of increasing bone density and reducing follicular activity, the risk of ovarian and endometrial cancers, and the incidence of anemia, without increasing body weight or body fat content. Other effects of COCP on androgen metabolism include inhibiting 5α-reductase activity and androgen receptor binding, which may minimize virilization in women with PCOS.

The use of antiandrogens in women with PCOS has been the mainstay of treatment for acne and hirsuitism. Although these drugs are not routinely used for the sole purpose of regulating menstruation, they offer some benefit in this regard. Ethinyl estradiol and drosperinone (which has an antiandrogen effect) have been used in women with PCOS to treat acne and/or hirsuitism. A small observational study showed that all women on this drug had improvement in their menstrual cyclicity *(60)*.

2.7.3. Metformin

Metformin has been shown to improve ovarian function and glucose metabolism in women with PCOS. It also reduces plasma concentrations of LH, total and unbound testosterone, and DHEAS, and increases SHBG. In obese women with PCOS, the mechanism of metformin action is likely to involve an improvement in insulin sensitivity and hyperandrogenemia by a reduction in central obesity. In the non-obese patient, metformin improves the hepatic extraction of insulin.

Several studies have examined the effect of metformin on menstrual cyclicity in PCOS. Velazquez et al. *(61)* studied the use of metformin 500 mg three times a day for 8 weeks in 22 women with PCOS who suffered chronic oligomenorrhea. They found that 21 women reported regular menstruation at the close of the study, with 86% of these women having ovulatory progesterone levels. Glueck et al. *(62)* studied 43 women with PCOS and chronic oligomenorrhea on varying regimes of metformin and noted that the greatest drop in BMI was inversely related to the duration of therapy. Ninety-one percent of women reported regular menstruation during metformin therapy; treatment regimes had no differential effect.

The study of Kolodziejczyk et al. involved 39 women with PCOS taking a 12-week course of 500 mg metformin three times daily. They chose to assess menstrual regularity by average cycle length and noted that it decreased by 36%—from 56 to 36 days. They did not, however, state the number of women who had no improvement *(63)*. Moghetti et al., in the only randomized, double-blind, placebo-controlled study involving metformin and menstrual cyclicity, admitted 23 women with PCOS and chronic anovulation. Women were randomized to receive either metformin 500 mg, three times daily, or a placebo for 6 months. Those on metformin had improvement in the frequency of menstruation—most within 3 months *(64)*. A review of nine uncontrolled studies on metformin monotherapy, by Costello and Eden *(65)*, found that 62% of women with PCOS had regularization of their menses. Thus, by restoring menstrual cyclicity, metformin may reduce the likelihood of endometrial hyperplasia and carcinoma in women with PCOS.

There have been two studies showing no benefit from metformin use. The first, by Acbay et al. *(66)*, looked at 16 women with PCOS and chronic anovulation in a single-blinded study, in which participants received a placebo twice daily for 8 weeks, followed by 850 mg metformin twice daily over a further 10 weeks. No statistically significant improvements in insulin sensitivity, fasting insulin, or free testosterone were noted. Menstrual cyclicity was not an endpoint however. The other study, by Ehrmann et al. *(67)*, treated 20 morbidly obese women with PCOS with 850 mg metformin and increased the dose at weekly intervals to a maximum of 850 mg three times daily for 12 weeks. Calorific intake was monitored and adjusted so that each subject's weight remained constant over the course of the study. Again, insulin sensitivity was not improved, and menstrual cyclicity was not discussed as an endpoint. It has been reported that in women with a BMI greater than 30 kg/m^2, significant reduction in BMI is needed to improve insulin sensitivity.

Haas et al. *(68)* have calculated that metformin therapy alone can decrease BMI by 2.8%. Kiddy et al. *(69)* have shown that a 5% reduction in weight may lead to clinical improvements in women with PCOS, with regularization of their menses. From Ehrmann et al.'s study *(67)*, in which subjects were not allowed to lose weight while on metformin, no improvements in insulin sensitivity or serum hyperandrogenism were found. This suggests that some of the beneficial action of metformin arises secondary to weight loss.

Cussons et al. *(2)* noted that endocrinologists were three times more likely to prescribe metformin to women with PCOS who had oligomenorrhea as their main concern than were gynecologists. Overall, metformin rates highly as a treatment for oligomenorrhea in PCOS.

Like all drugs, however, metformin causes a variety of side effects, but although most are troublesome to the patient, few are dangerous or life-threatening. The most common side effects, affecting more than 1 in 10 women, are related to the gastrointestinal system, namely nausea, vomiting, diarrhea, abdominal pain, and loss of appetite. For the majority of patients, these side effects occur during initiation of treatment and resolve spontaneously. Manufacturers suggest that a slow increase in daily dose may avoid these side effects. One percent of patients notice a metallic taste, and those who are hypersensitive to the medication may develop mild erythema. Other side effects include decreased vitamin B_{12} absorption, although not sufficient to become a clinical problem. Overdosing with metformin can result in lactic acidosis.

2.7.4. Progestogens

There is evidence that use of cyclical progesterone prevents endometrial hyperplasia in women with PCOS. Although not often used as second-line treatment, progestagens have a role in patients for whom the COCP is contraindicated. In the United Kingdom, the Royal College of Obstetricians and Gynaecologists recommends that women experiencing oligomenorrhea for 3 months or longer should be offered progesterone to induce a withdrawal bleed, based on the work of Cheung et al. *(70)*. The College also suggests that in oligomenorrheic women with PCOS who are noted to have a thickened endometrium on ultrasound scan, an endometrial biopsy and hysteroscopy should be offered to rule out the possibility of endometrial hyperplasia *(71)*.

2.7.5. Future Strategies

Low levels of insulin growth factor binding protein-1 (IGFBP-1) in women with PCOS increase the bioavailability of IGF-1 and -2. Rajkumar et al. noted that IGFBP-1 inhibits endometrial estradiol action by inhibiting IGF receptor binding *(72)*. This could potentially be used to devise strategies to prevent or treat endometrial hyperplasia, for example, via progestogen treatment, which would stimulate IGFBP-1 production, IGFBP-1 agonists such as antiestrogens, or IGF-1 inhibitors.

3. CONCLUSION

Whereas menstrual irregularity is a common and defining feature of PCOS and other androgen excess conditions in women, the understanding of the pathophysiology of these conditions will help to create newer treatment modalities that address not only the patients' concerns for immediate care, but also the long-term risks associated with it, for which the physician is ultimately responsible. With PCOS now encompassing a wide variety of heterogeneous presentations, work needs to be done to identify which subsets are at greatest risk of endometrial neoplasia, and hence requiring prophylactic measures to prevent associated morbidity and mortality.

4. FUTURE AVENUES OF INVESTIGATION

The following areas are in need of further study:

1. The effect of androgens on a PCOS endometrial cell-line growth and development and on gene expression in these conditions
2. Microarray analysis of gene expression in endometrium from women with PCOS with differing clinical presentations and biochemical parameters
3. Creation of a central database of women with PCOS who develop endometrial hyperplasia or cancer, to permit advances in the pathogenesis and prevention of neoplasia in these women
4. Long-term studies examining the effects of lifestyle interventions in women with PCOS
5. Association between PMS and PCOS, as this will help the understanding of the much-studied, but little-understood subject of PMS

KEY POINTS

- Eighty percent of women with PCOS suffer with menstrual dysfunction.
- Adolescents with menstrual irregularity 2 years beyond the menarche should be investigated for PCOS before being prescribed medication.
- PCOS is a cause of primary amenorrhea and should be considered in the work-up of these patients.
- Granulosa cell tumors of the ovary and late-onset congenital adrenal hyperplasia are other androgen excess disorders causing menstrual dysfunction.
- Women with PCOS express increased steroid hormone receptor coactivators, raising endometrial sensitivity to estrogen, stimulating endometrial proliferation, and increasing the risk of simple endometrial hyperplasia.
- Lifestyle modifications must be first-line treatment in all women with PCOS, because it significantly improves the short-term effects of PCOS.
- Insulin-sensitizing drugs, such as metformin, are highly effective at regulating menstruation in women with PCOS.

REFERENCES

1. Stein IF, Leventhal ML. Amenorrhea associated with bilateral polycystic ovaries. Am J Obstet Gynecol 1935;29:181–191.
2. Cussons AJ, Stuckey BG, Walsh JP, Burke V, Norman RJ. Polycystic ovarian syndrome: marked differences between endocrinologists and gynaecologists in diagnosis and management. Clin Endocrinol (Oxf) 2005;62(3):289–295.
3. Dockerty MB, Jackson RL. The Stein-Leventhal syndrome: analysis of 43 cases with special reference to association with endometrial carcinoma. Am J Obstet Gynecol 1957;73(1):161–173.
4. Balen AH, Conway GS, Kaltsas G, et al. Polycystic ovary syndrome: the spectrum of the disorder in 1741 patients. Hum Reprod 1995;10(8):2107–2111.
5. Franks S. Polycystic ovary syndrome: a changing perspective. Clin Endocrinol (Oxf) 1989;31(1):87–120.
6. Goldzieher JW. Polycystic ovarian disease. Fertil Steril 1981;35(4):371–394.
7. Abdel GA, Khatim MS, Mowafi RS, Alnaser HM, Muharib NS, Shaw RW. Implications of ultrasonically diagnosed polycystic ovaries. I. Correlations with basal hormonal profiles. Hum Reprod 1992;7(4):453–457.
8. Polson DW, Adams J, Wadsworth J, Franks S. Polycystic ovaries—a common finding in normal women. Lancet 1988;1(8590):870–872.
9. Clayton RN, Ogden V, Hodgkinson J, et al. How common are polycystic ovaries in normal women and what is their significance for the fertility of the population? Clin Endocrinol (Oxf) 1992;37(2):127–134.
10. Adams J, Franks S, Polson DW, et al. Multifollicular ovaries: clinical and endocrine features and response to pulsatile gonadotropin releasing hormone. Lancet 1985;2(8469–70):1375–1379.
11. van Hooff MH, Voorhorst FJ, Kaptein MB, Hirasing RA, Koppenaal C, Schoemaker J. Endocrine features of polycystic ovary syndrome in a random population sample of 14-16 year old adolescents. Hum Reprod 1999;14(9):2223–2229.
12. Venturoli S, Porcu E, Fabbri R, et al. Longitudinal evaluation of the different gonadotropin pulsatile patterns in anovulatory cycles of young girls. J Clin Endocrinol Metab 1992;74(4):836–841.
13. Current evaluation of amenorrhea. Fertil Steril 2004;82(Suppl 1):S33–S39.
14. Welt CK, McNicholl DJ, Taylor AE, Hall JE. Female reproductive aging is marked by decreased secretion of dimeric inhibin. J Clin Endocrinol Metab 1999;84(1):105–111.
15. Iurassich S, Trotta C, Palagiano A, Pace L. [Correlations between acne and polycystic ovary. A study of 60 cases]. Minerva Ginecol 2001;53(2):107–111.
16. Legro RS, Bentley-Lewis R, Driscoll D, Wang SC, Dunaif A. Insulin resistance in the sisters of women with polycystic ovary syndrome: association with hyperandrogenemia rather than menstrual irregularity. J Clin Endocrinol Metab 2002;87(5):2128–2133.
17. Chang RJ. Polycystic ovary syndrome: diagnostic criteria. In: Chang RJ, Heinder JJ, Dunaif A, eds. Polycystic Ovary Syndrome. New York: Marcel Dekker, 2002:361–365.
18. Rodin DA, Bano G, Bland JM, Taylor K, Nussey SS. Polycystic ovaries and associated metabolic abnormalities in Indian subcontinent Asian women. Clin Endocrinol (Oxf) 1998;49(1):91–99.
19. Kousta E, White DM, Cela E, McCarthy MI, Franks S. The prevalence of19. polycystic ovaries in women with infertility. Hum Reprod 1999;14(11):2720–2723.
20. Schmid J, Kirchengast S, Vytiska-Binstorfer E, Huber J. Infertility caused by PCOS—health-related quality of life among Austrian and Moslem immigrant women in Austria. Hum Reprod 2004;19(10):2251–2257.
21. Doi SA, Al Zaid M, Towers PA, Scott CJ, Al Shoumer KA. Irregular cycles and steroid hormones in polycystic ovary syndrome. Hum Reprod 2005;20(9):2402–2408.
22. Botsis D, Kassanos D, Pyrgiotis E, Zourlas PA. Sonographic incidence of polycystic ovaries in a gynecological population. Ultrasound Obstet Gynecol 1995;6(3):182–185.
23. Gambineri A, Pelusi C, Vicennati V, Pagotto U, Pasquali R. Obesity and the polycystic ovary syndrome. Int J Obes Relat Metab Disord 2002;26(7):883–896.
24. Arroyo A, Laughlin GA, Morales AJ, Yen SS. Inappropriate gonadotropin 24. secretion in polycystic ovary syndrome: influence of adiposity. J Clin Endocrinol Metab 1997;82(11):3728–3733.
25. Maliqueo M, Clementi M, Gabler F, et al. Expression of steroid receptors and proteins related to apoptosis in endometria of women with polycystic ovary syndrome. Fertil Steril 2003;80 (Suppl 2):812–819.
26. Apparao KB, Lovely LP, Gui Y, Lininger RA, Lessey BA. Elevated endometrial androgen receptor expression in women with polycystic ovarian syndrome. Biol Reprod 2002;66(2):297–304.
27. Iwai M, Kanzaki H, Fujimoto M, et al. Regulation of sex steroid receptor gene expression by progesterone and testosterone in cultured human endometrial stromal cells. J Clin Endocrinol Metab 1995;80(2):450–454.
28. Gregory CW, Wilson EM, Apparao KB, et al. Steroid receptor coactivator expression throughout the menstrual cycle in normal and abnormal endometrium. J Clin Endocrinol Metab 2002;87(6):2960–2966.
29. Robinson S, Kiddy D, Gelding SV, et al. The relationship of insulin insensitivity to menstrual pattern in women with hyperandrogenism and polycystic ovaries. Clin Endocrinol (Oxf) 1993;39(3):351–355.
30. Norman RJ, Hague WM, Masters SC, Wang XJ. Subjects with polycystic ovaries without hyperandrogenemia exhibit similar disturbances in insulin and lipid profiles as those with polycystic ovary syndrome. Hum Reprod 1995;10(9):2258–2261.

31. Gonzalez CA, Nahum HP, Mendoza R, Ayala AR. [Correlation between menstruation disorders and insulin resistance]. Ginecol Obstet Mex 2003;71:312–317.

32. Chang WY, Knochenhauer ES, Bartolucci AA, Azziz R. Phenotypic spectrum of polycystic ovary syndrome: clinical and biochemical characterization of the three major clinical subgroups. Fertil Steril 2005;83(6):1717–1723.

33. Elting MW, Korsen TJ, Schoemaker J. Obesity, rather than menstrual cycle pattern or follicle cohort size, determines hyperinsulinemia, dyslipidaemia and hypertension in ageing women with polycystic ovary syndrome. Clin Endocrinol (Oxf) 2001;55(6):767–776.

34. Asahara T, Takahashi T, Masuda H, et al. VEGF contributes to postnatal neovascularization by mobilizing bone marrow-derived endothelial progenitor cells. EMBO J 1999;18(14):3964–3972.

35. Atiomo WU, Bates SA, Condon JE, Shaw S, West JH, Prentice AG. The plasminogen activator system in women with polycystic ovary syndrome. Fertil Steril 1998;69(2):236–241.

36. Chekir C, Nakatsuka M, Kamada Y, Noguchi S, Sasaki A, Hiramatsu Y. Impaired uterine perfusion associated with metabolic disorders in women with polycystic ovary syndrome. Acta Obstet Gynecol Scand 2005;84(2):189–195.

37. Malmstrom H, Hogberg T, Risberg B, Simonsen E. Granulosa cell tumors of the ovary: prognostic factors and outcome. Gynecol Oncol 1994;52(1):50–55.

38. Unkila-Kallio L, Tiitinen A, Wahlstrom T, Lehtovirta P, Leminen A. Reproductive features in women developing ovarian granulosa cell tumour at a fertile age. Hum Reprod 2000;15(3):589–593.

39. Rouach M, Blumenfeld Z. [Diagnosis of late onset congenital adrenal hyperplasia]. Harefuah 1993;124(5):249–53, 320.

40. Coulam CB, Annegers JF, Kranz JS. Chronic anovulation syndrome and associated neoplasia. Obstet Gynecol 1983;61(4):403–407.

41. Escobedo LG, Lee NC, Peterson HB, Wingo PA. Infertility-associated endometrial cancer risk may be limited to specific subgroups of infertile women. Obstet Gynecol 1991;77(1):124–128.

42. Wild S, Pierpoint T, Jacobs H, McKeigue P. Long-term consequences of polycystic ovary syndrome: results of a 31 year follow-up study. Hum Fertil (Cambr) 2000;3(2):101–105.

43. Ramzy I, Nisker JA. Histologic study of ovaries from young women with endometrial adenocarcinoma. Am J Clin Pathol 1979;71(3):253–256.

44. Ho SP, Tan KT, Pang MW, Ho TH. Endometrial hyperplasia and the risk of endometrial carcinoma. Singapore Med J 1997;38(1):11–15.

45. Hardiman P, Pillay OC, Atiomo W. Polycystic ovary syndrome and endometrial carcinoma. Lancet 2003;361(9371):1810–1812.

46. Lupulescu A. Estrogen use and cancer risk: a review. Exp Clin Endocrinol 1993;101(4):204–214.

47. Furberg AS, Thune I. Metabolic abnormalities (hypertension, hyperglycemia and overweight), lifestyle (high energy intake and physical inactivity) and endometrial cancer risk in a Norwegian cohort. Int J Cancer 2003;104(6):669–676.

48. Konishi I, Koshiyama M, Mandai M, et al. Increased expression of LH/hCG receptors in endometrial hyperplasia and carcinoma in anovulatory women. Gynecol Oncol 1997;65(2):273–280.

49. Bockman JV. Two pathogenetic types of endometrial carcinoma. Gynecol Oncol 1983;15:10–17.

50. Li L, Yang C, Qiaojie. [Pathological and immunohistochemical study on estrogen and progesterone receptors in endometrium of polycystic ovarian syndrome]. Zhonghua Fu Chan Ke Za Zhi 1998;33(2):89–91.

51. Nagamani M, Stuart CA, Dunhardt PA, Doherty MG. Specific binding sites for insulin and insulin-like growth factor I in human endometrial cancer. Am J Obstet Gynecol 1991;165(6 Pt 1):1865–1871.

52. Bershtein LM, Gamaiunova VB, Kvachevskaia I, Tsyrlina EV, Kovalenko IG. [The nature of hyperinsulinemia (insulin resistance) in endometrial carcinoma: of plasma levels of insulin and c-peptide]. Vopr Onkol 2000;46(2):191–195.

53. Weiderpass E, Persson I, Adami HO, Magnusson C, Lindgren A, Baron JA. Body size in different periods of life, diabetes mellitus, hypertension, and risk of postmenopausal endometrial cancer (Sweden). Cancer Causes Control 2000;11(2):185–192.

54. Pillay OC, Leonard A, Catalano R, Sharkey A, Hardiman P. Endometrial gene expression in women with polycystic ovary syndrome. Hum Reprod 2005;20 (Suppl 1):i96.

55. Marsh K, Brand-Miller J. The optimal diet for women with polycystic ovary syndrome? Br J Nutr 2005;94(2):154–165.

56. Clark AM, Ledger W, Galletly C, et al. Weight loss results in significant improvement in pregnancy and ovulation rates in anovulatory obese women. Hum Reprod 1995;10(10):2705–2712.

57. Norman RJ, Davies MJ, Lord J, Moran LJ. The role of lifestyle modification in polycystic ovary syndrome. Trends Endocrinol Metab 2002;13(6):251–257.

58. Prelevic GM, Wurzburger MI, Trpkovic D, Balint-Peric L. Effects of a low-dose estrogen-antiandrogen combination (Diane-35) on lipid and carbohydrate metabolism in patients with polycystic ovary syndrome. Gynecol Endocrinol 1990;4(3):157–168.

59. Elter K, Imir G, Durmusoglu F. Clinical, endocrine and metabolic effects of metformin added to ethinyl estradiol-cyproterone acetate in non-obese women with polycystic ovarian syndrome: a randomized controlled study. Hum Reprod 2002;17(7):1729–1737.

60. Palep-Singh M, Mook K, Barth J, Balen A. An observational study of Yasmin in the management of women with polycystic ovary syndrome. J Fam Plann Reprod Health Care 2004;30(3):163–165.
61. Velazquez E, Acosta A, Mendoza SG. Menstrual cyclicity after metformin therapy in polycystic ovary syndrome. Obstet Gynecol 1997;90(3):392–395.
62. Glueck CJ, Wang P, Fontaine R, Tracy T, Sieve-Smith L. Metformin-induced resumption of normal menses in 39 of 43 (91%) previously amenorrheic women with the polycystic ovary syndrome. Metabolism 1999;48(4):511–519.
63. Kolodziejczyk B, Duleba AJ, Spaczynski RZ, Pawelczyk L. Metformin therapy decreases hyperandrogenism and hyperinsulinemia in women with polycystic ovary syndrome. Fertil Steril 2000;73(6):1149–1154.
64. Moghetti P, Castello R, Negri C, et al. Metformin effects on clinical features, endocrine and metabolic profiles, and insulin sensitivity in polycystic ovary syndrome: a randomized, double-blind, placebo-controlled 6-month trial, followed by open, long-term clinical evaluation. J Clin Endocrinol Metab 2000;85(1):139–146.
65. Costello MF, Eden JA. A systematic review of the reproductive system effects of metformin in patients with polycystic ovary syndrome. Fertil Steril 2003;79(1):1–13.
66. Acbay O, Gundogdu S. Can metformin reduce insulin resistance in polycystic ovary syndrome? Fertil Steril 1996;65(5):946–949.
67. Ehrmann DA, Cavaghan MK, Imperial J, Sturis J, Rosenfield RL, Polonsky KS. Effects of metformin on insulin secretion, insulin action, and ovarian steroidogenesis in women with polycystic ovary syndrome. J Clin Endocrinol Metab 1997;82(2):524–530.
68. Haas DA, Carr BR, Attia GR. Effects of metformin on body mass index, menstrual cyclicity, and ovulation induction in women with polycystic ovary syndrome. Fertil Steril 2003;79(3):469–481.
69. Kiddy DS, Hamilton-Fairley D, Bush A, et al. Improvement in endocrine and ovarian function during dietary treatment of obese women with polycystic ovary syndrome. Clin Endocrinol (Oxf) 1992;36(1):105–111.
70. Cheung AP. Ultrasound and menstrual history in predicting endometrial hyperplasia in polycystic ovary syndrome. Obstet Gynecol 2001;98(2):325–331.
71. Long term consequences of polycystic ovary syndrome. Royal College of Obstetricians and Gynaecologists, 2003.
72. Rajkumar K, Dheen T, Krsek M, Murphy LJ. Impaired estrogen action in the uterus of insulin-like growth factor binding protein-1 transgenic mice. Endocrinology 1996;137(4):1258–1264.
73. Farquhar CM, Birdsall M, Manning P, Mitchell JM, France JT. The prevalence of polycystic ovaries on ultrasound scanning in a population of randomly selected women. Aust N Z J Obstet Gynaecol 1994;34(1):67–72.
74. Botsis D, Kassanos D, Pyrgiotis E, Zourlas PA. Sonographic incidence of polycystic ovaries in a gynecological population. Ultrasound Obstet Gynecol 1995;6(3):182–185.
75. Cresswell JL, Barker DJ, Osmond C, Egger P, Phillips DI, Fraser RB. Fetal growth, length of gestation, and polycystic ovaries in adult life. Lancet 1997;350(9085):1131–1135.

β-Cell Dysfunction, Glucose Intolerance, and Diabetes in the Polycystic Ovary Syndrome

David A. Ehrmann

SUMMARY

Among women with polycystic ovary syndrome (PCOS), the prevalence of impaired glucose tolerance is estimated to be between 30 and 40%, whereas that of type 2 diabetes mellitus has been placed at 5–10%. These prevalences are among the highest known among women of reproductive age. The predisposition to glucose intolerance in PCOS may be influenced by antenatal events, including the rate of *in utero* growth and development as well as exposure to excess androgen concentrations from the maternal circulation. Once defects in insulin secretion develop in the setting of the characteristic insulin resistance of PCOS, glucose intolerance becomes evident. Provocative testing of pancreatic β-cell function may provide insights into the future risk for glucose intolerance among women with PCOS.

Key Words: Polycystic ovary syndrome; glucose tolerance; β-cell; insulin resistance.

1. INTRODUCTION

It is now well established that women with polycystic ovary syndrome (PCOS) are predisposed to develop a number of metabolic abnormalities, including impaired glucose tolerance (IGT) and type 2 diabetes mellitus (DM) *(1,2)*. The prevalence of IGT has been estimated to be between 30 and 40%, whereas that of type 2 DM has been placed at 5–10% *(1,2)*. Given that PCOS is thought to affect between 5 and 8% (5.7–9.1 million) of reproductive-aged women in the United States *(3)*, it follows that at any given time an estimated 3 million women with PCOS will have IGT, whereas approximately 1 million women with PCOS will have type 2 DM. This chapter focuses on the origins of glucose intolerance in PCOS, with an emphasis on the role of the pancreatic β-cell in this process.

2. BACKGROUND

2.1. Antecedents to Glucose Intolerance in PCOS

Recent attention has focused on developmental origins of adult diseases, including both PCOS and type 2 DM. Evidence exists to support the hypothesis that low *(4,5)* birthweight and/or size for gestational age may lead to insulin resistance, obesity, and type 2 DM in later life. The mechanisms underlying these associations are unknown, but alterations in birthweight (reflecting in utero growth/ nutritional status) have also been implicated in the pathogenesis of PCOS per se and its associated insulin resistance and glucose intolerance in some *(6)*, but not all *(7,8)*, studies.

Another developmental factor that has been proposed to influence the phenotypic expression of PCOS is in utero androgen exposure (*see* Chapter 23). In nonhuman primates, fetal exposure to high levels of androgen during early in utero development is associated with defects in insulin secretion

From: *Contemporary Endocrinology: Androgen Excess Disorders in Women:
Polycystic Ovary Syndrome and Other Disorders, Second Edition*
Edited by: R. Azziz et al. © Humana Press Inc., Totowa, NJ

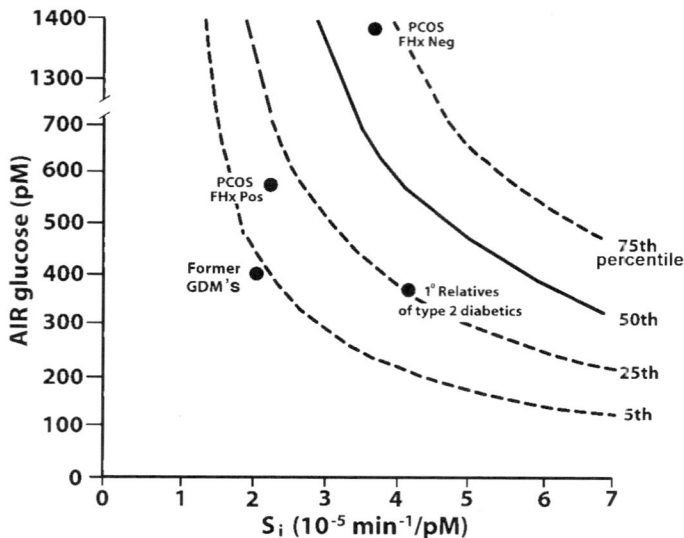

Fig. 1. Hyperbolic relationship between insulin secretion and insulin action in normoglycemic subjects studied with a frequently sampled intravenous glucose tolerance test reflecting the disposition index, based on data of Kahn et al. *(17)*. As insulin sensitivity (S_i) declines, insulin secretion (acute insulin response to glucose [AIR glucose]) must increase to maintain normal glucose tolerance. Shown are mean percentiles for PCOS subjects previously reported *(18)* as well as mean summary data derived from the literature for women with a prior history of gestational diabetes mellitus (Former GDMs) and subjects with a first-degree relative with type 2 diabetes mellitus. (From ref. *31*.)

and action in adult life *(9)*. Prenatally androgenized female rhesus monkeys exhibit glucoregulatory deficits similar to those seen in adult women with PCOS *(10)*. Of interest, the timing of the androgen exposure appears to differentially affect glucose regulation: early androgen exposure has been associated with impaired pancreatic β-cell function, whereas exposure later in gestation appears to primarily alter insulin sensitivity. The extent to which these hormonal factors relate to the pathogenesis of PCOS in the human is not known.

It is now well documented that glucose intolerance in PCOS can occur as early as during the second decade of life *(11–15)*. In one study, 27 adolescents with PCOS (mean age 16.7 ± 1.6 years; mean body mass index [BMI] 38.4 ± 8.8 kg/m^2) had an oral glucose tolerance test (OGTT): 8 (30%) were found to have IGT, 1 (4%) had undiagnosed DM, and the remaining 18 (66%) had normal glucose tolerance *(11)*. Studies by Arslanian et al. *(13)* have shown that metabolic precursors to type 2 DM (decreased first-phase insulin secretion, decreased glucose disposition index, and increased hepatic glucose production) are evident among obese adolescents with PCOS.

2.2. Relationship of Insulin Secretion to Insulin Action

Glucose intolerance typically develops when defects in insulin secretion are superimposed upon a background of insulin resistance *(16)*. Despite the fact that women with PCOS are characteristically insulin resistant, not all develop abnormalities in glucose tolerance. It has become evident that insulin secretory defects play an important role in the propensity to develop DM in PCOS.

Insulin secretion is most appropriately expressed in relation to the magnitude of ambient insulin resistance. The product of these measures can be quantified (the so-called "disposition index") and related as a percentile to the hyperbolic relationship for these measures (Fig. 1) established in normal subjects *(17)*. We *(18)*, as well as others *(19)*, have found that a subset of subjects with PCOS has β-cell secretory dysfunction. In absolute terms, women with PCOS had normal first-phase insulin secretion compared to controls. In contrast, when first-phase insulin secretion was analyzed in relation to the degree of insulin resistance, women with PCOS exhibited a significant impairment in β-cell function. This reduction was particularly marked in women with PCOS who had a first-degree relative with type 2 DM: the mean disposition index of women with PCOS and a family history of type 2 DM was in the 8th percentile, while that of those without such a family history was in the 33rd percentile ($p = 0.05$) (Fig. 1). We have additionally quantified β-cell function in PCOS by examining the insulin secretory response to a graded increase in plasma glucose and by the ability of the β-cell to adjust and respond to induced oscillations in the plasma glucose level *(18)*. Results from both provocative stimuli were consistent: when expressed in relation to the degree of insulin resistance; insulin secretion was impaired in PCOS subjects with a family history of type 2 DM when compared to controls.

Because these findings were consistent with studies showing a high degree of heritability of β-cell function (particularly when examined in relation to insulin sensitivity) among nondiabetic family members of individuals with type 2 DM *(20)*, we sought to examine this further. Using the frequently sampled intravenous glucose tolerance test, insulin secretion (the acute insulin response to glucose [AIR_g], insulin action (sensitivity to insulin [S_i], and their product (disposition index [DI] = $AIR_g \times S_i$) were quantified among women with PCOS ($n = 33$) and their nondiabetic first-degree relatives ($n = 48$) *(21)*. Heritability of these measures was calculated from familial correlations estimated within a genetic model. The sibling correlation for AIR_g was highly significant after adjustment for age and BMI, as was the disposition index, a measure of quantifying insulin secretion in relation to insulin sensitivity. This finding was supportive of a heritable component to β-cell dysfunction in families of women with PCOS.

Taken together, these results suggest that the risk imparted by insulin resistance to the development of type 2 DM in PCOS is enhanced by defects in insulin secretion. Furthermore, a history of type 2 DM in a first-degree relative appears to define a subset of PCOS subjects with the most profound defects in β-cell function.

Although nearly one-half of women with PCOS will ultimately develop glucose intolerance, most have been able to maintain glucose levels within the normal range at the time of initial clinical presentation *(1,2)*. This has been taken as evidence that their ability to adequately secrete insulin in compensation for the degree of insulin resistance is retained. However, most *(1,22)* but not all *(23)* studies have found that the rates of decline in glucose tolerance in PCOS are higher than expected compared to reference populations. This suggests that pancreatic β-cell dysfunction may supervene earlier in the evolution of glucose intolerance in women with PCOS compared to women without PCOS. The basis for this, however, remains unclear.

2.3. Provocative Assessment of Insulin Secretion

The development of transient DM in previously nondiabetic individuals treated with short-term glucocorticoids ("steroid diabetes") has been recognized for many years. The potential of this finding as a predictor for subsequent development of DM was first described by Fajans et al. *(24)*, who found that when normal glucose-tolerant individuals with a first-degree relative with DM were given small doses of cortisone acetate, 24% had IGT and 19% had DM on OGTT. In contrast, among those without a family history of DM, 3% developed IGT and 2% developed DM. In addition, an abnormal response to cortisone acetate was predictive of the subsequent development of DM over 7 years of follow-up: 35% of those with cortisone-induced glucose intolerance had developed DM compared with only 2% of those whose initial response was normal.

Henriksen et al. *(25)* have shown that nondiabetic first-degree relatives of type 2 diabetics with evidence of mild alteration of β-cell function at baseline are unable to enhance their β-cell response to dexamethasone-induced insulin resistance. Specifically, after treatment with dexamethasone (4 mg daily for 5 days), normoglycemic subjects with or without a first-degree relative with DM increased their first-phase insulin secretion to glucose on an intravenous glucose tolerance test. However, the disposition index ($AIR_g \times S_i$) was significantly lower in the relatives *(25)*.

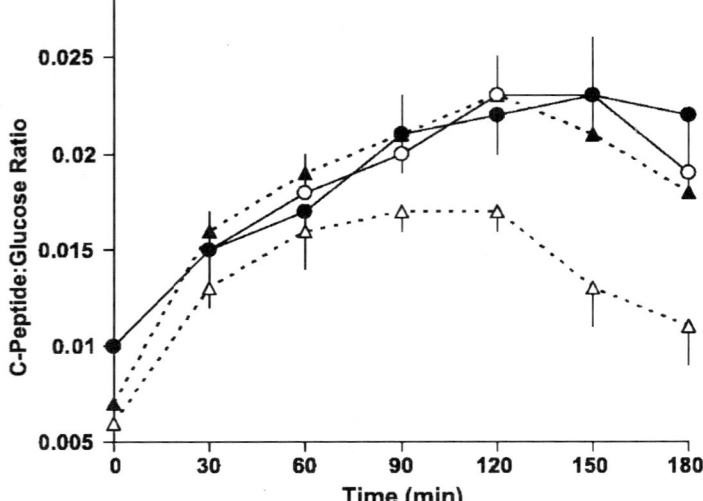

Fig. 2. C-peptide–to–glucose ratios obtained during the oral glucose tolerance tes:. When expressed in relation to the prevailing glucose concentration, those subjects with polycystic ovary syndrome (PCOS) were less
able than control subjects to mount a sufficient β-cell secretory response to an oral glucose load after dexamethasone administration. The increment in the ratio of C-peptide to glucose is significantly lower in PCOS than
in control subjects. PCOS subjects at baseline (○) and postdexamethasone (●); control subjects at baseline (△)
and postdexamethasone (▲). (From ref. *32.*)

The predominant mechanism responsible for glucocorticoid-induced glucose intolerance appears
to be related to the induction or exacerbation of insulin resistance *(26)*. Insulin-mediated peripheral
glucose disposal is markedly impaired at a post-insulin receptor level *(27)*. Both oxidative and
nonoxidative pathways of glucose disposal are reduced by glucocorticoids *(26)*, and muscle glycogen
synthase activity is reduced *(28,29)*. These alterations in insulin action resemble those observed in
type 2 DM. Although insulin secretion may be altered by glucocorticoids, this appears to occur only
at high doses *(30)*.

We postulated that women with PCOS with normal glucose tolerance would differ from control
women with normal glucose tolerance in their ability to secrete sufficient insulin and maintain normal glucose tolerance after administration of dexamethasone (2 mg orally over 12 hours). In the
baseline state (i.e., before the administration of dexamethasone), control women and women with
PCOS had normal fasting glucose concentrations (94 ± 2 vs 95 ± 2 mg/dL) as well as similar glucose
levels at 2 hours in response to a standard 75-g oral glucose load (120 ± 7 vs 124 ± 5 mg/dL).
However, when faced with a reduction in insulin sensitivity induced by the administration of dexamethasone, women with PCOS were significantly less able than control subjects to compensate with
adequate insulin secretion. This was evidenced by a relative attenuation in C-peptide levels relative
to plasma glucose during the OGTT (Fig. 2). These data suggest that short-term, low-dose glucocorticoid treatment augments insulin resistance sufficiently to reveal groups of patients in whom β-cell
compensation is inadequate. Thus, glucocorticoid administration may be a useful means by which to
determine whether the prevalence or magnitude of defects in insulin secretion differ between women
with PCOS and their controls and, likewise, to determine whether such defects are more profound or
present more often in a particular subset within a population of women with PCOS.

3. CONCLUSIONS

Women with PCOS have a substantial risk for the development of IGT and type 2 DM over their
life span. Several distinct genetic and environmental factors are emerging as important influences on

this predisposition to glucose intolerance in PCOS. Environmental factors include the rate of in utero growth and development as well as exposure to excess androgen concentrations from the maternal circulation. Although it has been well established that insulin resistance contributes to the reproductive and metabolic phenotypes of PCOS, the role of insulin secretory abnormalities in these phenotypes has received less attention. It is now known that the pancreatic β-cell does not respond normally to fluctuations in plasma glucose in a substantial subset of women with PCOS. Furthermore, these abnormalities appear to be heritable. Finally, although insulin-secretory defects may not be evident under basal conditions, with interventions such as the administration of low-dose glucocorticoids, latent defects can be demonstrated. Provocative testing of pancreatic β-cell function may provide insights into the future risk for glucose intolerance among women with PCOS.

4. FUTURE AVENUES OF INVESTIGATION

Future investigation into the pathogenesis of alterations in glucose tolerance in PCOS is likely to focus on genetic factors leading to specific defects in insulin secretion and insulin action. Through the identification of genetic markers, it may become possible to characterize those women with PCOS who are at high risk for the development of type 2 DM at a time when glucose tolerance is normal. In so doing, specific and targeted interventions may be used to minimize the risk for conversion from normal to abnormal glucose tolerance.

KEY POINTS

- IGT and type 2 DM occur with higher than expected frequency among women with PCOS.
- *In utero* growth retardation and exposure to excess levels of androgens may contribute to the subsequent development of glucose intolerance in PCOS.
- Defects in insulin action and insulin secretion contribute to glucose intolerance in PCOS and are evident as early as the second decade of life.
- Defects in insulin secretion appear to be heritable in families with PCOS and may identify those women with PCOS at highest risk for glucose intolerance.
- Provocative testing of β-cell function may identify a subset of women with PCOS at highest risk for future development of IGT and DM.

ACKNOWLEDGMENT

This work was supported by grants from the National Institutes of Health (M01-RR-00055, DK-41814, AG-11412, HL-075079, and P60-DK20595), a Clinical Research Award (to D.A.E.) from the American Diabetes Association, and a gift from the Blum-Kovler Foundation.

REFERENCES

1. Ehrmann D, Barnes R, Rosenfield R, Cavaghan M, Imperial J. Prevalence of impaired glucose tolerance and diabetes in women with polycystic ovary syndrome. Diabetes Care 1999;22:141–146.
2. Legro R, Kunselman A, Dodson W, Dunaif A. Prevalence and predictors of risk for type 2 diabetes mellitus and impaired glucose tolerance in polycystic ovary syndrome: a prospective, controlled study in 254 affected women. J Clin Endocrinol Metab 1999;84(1):165–169.
3. Knochenhauer E, Key T, Kahsar-Miller M, Waggoner W, Boots L, Azziz R. Prevalence of the polycystic ovary syndrome in unselected black and white women of the southeastern United States: a prospective study. J Clin Endocrinol Metab 1998;83(9):3078–3082.
4. Stocker CJ, Arch JR, Cawthorne MA. Fetal origins of insulin resistance and obesity. Proc Nutr Soc 2005;64(2):143–151.
5. Hofman PL, Regan F, Jackson WE, Jefferies C, Knight DB, Robinson EM, Cutfield WS. Premature birth and later insulin resistance. N Engl J Med 2004;351(21):2179–2186.
6. Cresswell JL, Barker DJ, Osmond C, Egger P, Phillips DI, Fraser RB. Fetal growth, length of gestation, and polycystic ovaries in adult life. Lancet 1997;350(9085):1131–1135.
7. Laitinen J, Taponen S, Martikainen H, et al. Body size from birth to adulthood as a predictor of self-reported polycystic ovary syndrome symptoms. Int J Obes Relat Metab Disord 2003;27(6):710–715.
8. Sadrzadeh S, Klip WA, Broekmans FJ, et al. Birthweight and age at menarche in patients with polycystic ovary syndrome or diminished ovarian reserve, in a retrospective cohort. Hum Reprod 2003;18(10):2225–2230.

9. Bruns CM, Baum ST, Colman RJ, et al. Insulin resistance and impaired insulin secretion in prenatally androgenized male rhesus monkeys. J Clin Endocrinol Metab 2004;89(12):6218–6223.

10. Eisner JR, Dumesic DA, Kemnitz JW, Abbott DH. Timing of prenatal androgen excess determines differential impairment in insulin secretion and action in adult female rhesus monkeys. J Clin Endocrinol Metab 2000;85(3):1206–1210.

11. Palmert MR, Gordon CM, Kartashov AI, Legro RS, Emans SJ, Dunaif A. Screening for abnormal glucose tolerance in adolescents with polycystic ovary syndrome. J Clin Endocrinol Metab 2002;87(3):1017–1023.

12. Lewy VD, Danadian K, Witchel SF, Arslanian S. Early metabolic abnormalities in adolescent girls with polycystic ovarian syndrome. J Pediatr 2001;138(1):38–44.

13. Arslanian SA, Lewy VD, Danadian K. Glucose intolerance in obese adolescents with polycystic ovary syndrome: roles of insulin resistance and beta-cell dysfunction and risk of cardiovascular disease. J Clin Endocrinol Metab 2001;86(1):66–71.

14. Silfen ME, Denburg MR, Manibo AM, et al. Early endocrine, metabolic, and sonographic characteristics of polycystic ovary syndrome (PCOS): comparison between nonobese and obese adolescents. J Clin Endocrinol Metab 2003;88(10):4682–4688.

15. Biro FM. Body morphology and its impact on adolescent and pediatric gynecology, with a special emphasis on polycystic ovary syndrome. Curr Opin Obstet Gynecol 2003;15(5):347–351.

16. Polonsky K, Sturis J, Bell G. Non-insulin-dependent diabetes mellitus—a genetically programmed failure of the beta cell to compensate for insulin resistance. N Engl J Med 1996;334:777–783.

17. Kahn S, Prigeon R, McCulloch D, et al. Quantification of the relationship between insulin sensitivity and B-cell function in human subjects Evidence for a hyperbolic function. Diabetes 1993;42:1663–1672.

18. Ehrmann DA, Sturis J, Byrne MM, Karrison T, Rosenfield RL, Polonsky KS. Insulin secretory defects in polycystic ovary syndrome. Relationship to insulin sensitivity and family history of non-insulin-dependent diabetes mellitus. J Clin Invest 1995;96(1):520–527.

19. Dunaif A, Finegood DT. Beta-cell dysfunction independent of obesity and glucose intolerance in the polycystic ovary syndrome. J Clin Endocrinol Metab 1996;81(3):942–947.

20. Elbein SC, Hasstedt SJ, Wegner K, Kahn SE. Heritability of pancreatic beta-cell function among nondiabetic members of Caucasian familial type 2 diabetic kindreds. J Clin Endocrinol Metab 1999;84(4):1398–1403.

21. Colilla S, Cox NJ, Ehrmann DA. Heritability of insulin secretion and insulin action in women with polycystic ovary syndrome and their first degree relatives. J Clin Endocrinol Metab 2001;86(5):2027–2031.

22. Norman RJ, Masters L, Milner CR, Wang JX, Davies MJ. Relative risk of conversion from normoglycaemia to impaired glucose tolerance or non-insulin dependent diabetes mellitus in polycystic ovarian syndrome. Hum Reprod 2001;16(9):1995–1998.

23. Legro RS, Gnatuk CL, Kunselman AR, Dunaif A. Changes in glucose tolerance over time in women with polycystic ovary syndrome: a controlled study. J Clin Endocrinol Metab 2005;90(6):3236–3242.

24. Fajans S, Conn J. An approach to the prediction of diabetes mellitus by modification of the glucose tolerance test with cortisone. Diabetes 1954;3:296–304.

25. Henriksen J, Alford F, Ward G, Beck-Nielsen H. Risk and mechanism of dexamethasone-induced deterioration of glucose tolerance in non-diabetic first-degree relatives of NIDDM patients. Diabetologia 1997;40(12):1439–1448.

26. Tappy L, Randin D, Vollenweider P, et al. Mechanisms of dexamethasone-induced insulin resistance in healthy humans. J Clin Endocrinol Metab 1994;79(4):1063–1069.

27. McMahon M, Gerich J, Rizza R. Effects of glucocorticoids on carbohydrate metabolism. Diabetes Metab Rev 1988;4(1):17–30.

28. Coderre L, Srivastava AK, Chiasson JL. Effect of hypercorticism on regulation of skeletal muscle glycogen metabolism by epinephrine. Am J Physiol 1992;262(4 Pt 1):E434–E439.

29. Coderre L, Srivastava AK, Chiasson JL. Effect of hypercorticism on regulation of skeletal muscle glycogen metabolism by insulin. Am J Physiol 1992;262(4 Pt 1):E427–E433.

30. Matsumoto K, Yamasaki H, Akazawa S, et al. High-dose but not low-dose dexamethasone impairs glucose tolerance by inducing compensatory failure of pancreatic beta-cells in normal men. J Clin Endocrinol Metab 1996;81(7):2621–2626.

31. Ehrmann D. Insulin resistance in PCOS: causes and consequences. In: Filicori M, ed. Updates in Infertility Treatment 2004. Bologna, Italy: Medimond International Proceedings; 2004:419–426.

32. Ehrmann DA, Breda E, Corcoran MC, et al. Impaired beta-cell compensation to dexamethasone-induced hyperglycemia in women with polycystic ovary syndrome. Am J Physiol Endocrinol Metab 2004;287(2):E241–E246.

29

Hypertension and Cardiovascular and Lipoprotein Abnormalities in the Polycystic Ovary Snydrome

Francesco Orio and Stefano Palomba

SUMMARY

Polycystic ovary syndrome (PCOS) should be considered not only a reproductive problem but a complex, endocrine, multifaceted disease with important health implications. Several lines of evidence suggest an increased cardiovascular risk and cardiovascular disease characterized by impairment of cardiac structure and function, endothelial dysfunction, lipid abnormalities, and chronic low-grade inflammation. All these features are probably linked to the insulin resistance often present in women with PCOS. Cardiovascular abnormalities represent important long-term sequelae of PCOS that warrant further critical investigation.

Key Words: Cardiovascular disease (CVD); cardiovascular risk (CVR); hypertension; lipoproteins; lipids; endothelial dysfunction; heart; inflammation.

1. INTRODUCTION

Polycystic ovary syndrome (PCOS) is a common endocrine-metabolic disorder associated with type 2 diabetes mellitus, hypertension, dyslipidemia, and insulin resistance (IR) *(1)*. The metabolic features of PCOS have led to widespread concern about concurrent increased cardiovascular risk (CVR). In fact, several studies have demonstrated various cardiovascular and lipid abnormalities in PCOS women. However, to date there is no definitive evidence of increased cardiovascular mortality in PCOS *(2)*. The main CVR factors associated with PCOS are obesity, IR, hypertension (HTN), and dyslipidemia (Fig. 1).

Visceral obesity, which is present in approximately 50% of women with PCOS *(1)*, is an established risk factor for IR/hyperinsulinemia, dyslipidemia, type 2 diabetes, HTN, coagulation abnormalities, and premature cardiovascular disease (CVD). It can aggravate all the adverse metabolic and cardiovascular features present in PCOS, but it does not represent the only and/or the primary etiology or pathogenetic factor for the increase of CVR in PCOS. Notably, although obesity and IR are associated with one another, IR is a determinant of overall CVR independent of obesity. In fact, the increased CVR is related to the degree of IR among women with PCOS*(3)*.

Women with PCOS, especially young women, generally do not manifest increased blood pressure values *(4)*. This is despite the fact that an increased prevalence of labile daytime blood pressure, which is a predisposing factor to sustained HTN later in life, has been reported in PCOS *(5)*. Furthermore, at the menopause women with PCOS develop a risk for HTN that is 2.5-fold higher than in age-matched controls, and this may be in part related to the obesity associated with PCOS *(6)*.

An abnormal lipid profile is a well-known and important CVR factor. Total cholesterol (TC), low-density lipoprotein (LDL), and triglycerides are commonly reported to be increased in PCOS, and, conversely, high-density lipoprotein (HDL) is reported to be reduced *(7,8)*. These CVR factors are

From: *Contemporary Endocrinology: Androgen Excess Disorders in Women:*
Polycystic Ovary Syndrome and Other Disorders, Second Edition
Edited by: R. Azziz et al. © Humana Press Inc., Totowa, NJ

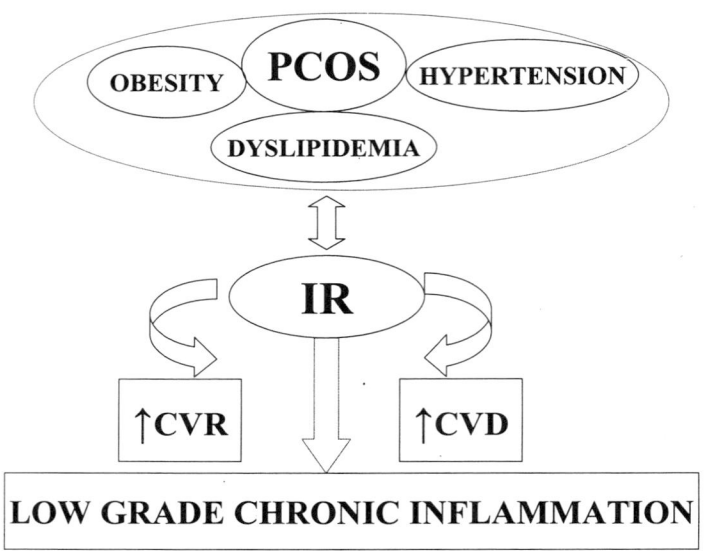

Fig. 1. Obesity, hypertension and dyslipidemia, together with insulin resistance (IR), participate in the development of an increased cardiovascular risk (CVR) and cardiovascular disease (CVD) in PCOS.

often evident at an early age, suggesting that women with PCOS represent a large group of women at increased risk for developing early-onset or premature CVD.

One of the early signs of CVD is endothelial injury and dysfunction. Precocious arterial anatomical and functional changes have been reported in young women with PCOS. As noted earlier, IR is likely a major risk factor for the development of CVD in PCOS and could play a key role in the development of endothelial damage, which itself represents an early sign of atherosclerosis.

The scientific interest in possible increased CVR in PCOS has mounted in the past few years because of the crucial consequence for the general health of women with this disorder. In particular, biochemical, morphological, and functional biomarkers of early CVD have been evaluated in women with PCOS in order to more precisely identify the cardiovascular morbidity of the syndrome. Recently, together with classical CVR factors such as TC, HDL cholesterol levels, and obesity, homocysteine, left ventricular hypertrophy (LVH), and low-grade chronic inflammation have also been shown to be independently associated with an increased CVR in PCOS *(9)*.

CVR factors and cardiovascular abnormalities are often evident at an early age, suggesting that the chronically abnormal hormonal and metabolic milieu found in women with PCOS, starting from adolescence, may predispose these women to premature atherosclerosis, hence making them candidates for early CVD.

One of several possible hypotheses to explain the long-term metabolic and CVR in PCOS is fetal programming, which is defined as the physiological "setting" that results in long-term consequences for function *(10)*. Furthermore, genetic factors intrinsic to or associated with PCOS could cause an increased CVR profile.

2. PCOS AND RISK FOR CARDIOVASCULAR DISEASE

At this moment there is no single and universally accepted definition for PCOS. This is probably the pivotal reason why published studies on PCOS cannot be easily reanalyzed in order to provide a conclusive assessment of the CVR or CVD in this group of patients. Nevertheless, several lines of evidence *(11–15)* indicate alterations in intermediate endpoints for CVR in women with PCOS and provide evidence for an association between CVR factors in PCOS and CVD (Table 1). Most studies

Table 1
Evidence for Association Between Polycystic Ovary Syndrome, Cardiovascular Risk (CVR) Factors and Cardiovascular Disease (CVD)

Recognized CVR factors	Atherosclerosis; coronary artery disease; myocardial infarction; atherogenic lipid profile ($\uparrow$TC, $\uparrow$LDL, $\uparrow$TG, or $\downarrow$HDL)
Emerging/Novel CVR factors	$\uparrow$CRP; $\uparrow$WBC (lymphocytes and monocytes)
Direct measurement of subclinical CVD	LVH and diastolic dysfunction; $\uparrow$IMT; endothelial dysfunction ($\downarrow$FMD, $\uparrow$ET-1); impaired fibrinolysis ($\uparrow$PAI-1)
Increased clinical CVD	No increased mortality from CVD documented to date in PCOS; however, insulin resistance is a frequent cause of increased CVR and CVD

TC, total cholesterol; LDL, low-density lipoprotein; TG, triglycerides; HDL, high-density lipoproteins; LVH, left ventricular hypertrophy; IMT, intima media thickness; CRP, C-reactive protein; WBC, white blood cell count; FMD, flow-mediated dilation; ET-1, endothelin-1; PAI-1, plasminogen activator inhibitor-1.

on CVD in PCOS used criteria adopted at the 1990 National Institute of Child Health and Human Development (NICHD) conference to diagnose PCOS *(16)*. Patients with PCOS diagnosed by the NICHD criteria, which include androgen excess as a *sine qua non*, should have a risk for CVD significantly higher in comparison with patients with PCOS who have their diagnosis made according to European Society for Human Reproduction and Embryology/American Society for Reproductive Medicine criteria *(17)*, which allow for normal androgen levels in women with oligo- or amenorrhea and anatomically polycystic ovaries. In fact, a crucial contributing factor to CVD in PCOS may be hyperandrogenemia.

Decreased sex hormone-binding globulin (SHBG), which is typical in PCOS and increases bioavailable testosterone, may be considered a surrogate marker of IR such that lower levels of SHBG are related to a greater degree of IR *(18)*. Therefore, IR may in part indirectly contribute to CVR in PCOS by amplifying androgen excess. In addition to this possible selection bias, large-scale clinical trials evaluating the morbidity and mortality for CVD in women with PCOS are lacking. No long-term data of well-characterized women with PCOS are present in the literature, and the link of PCOS to primary cardiovascular events, such as stroke or myocardial infarction (MI), remains to be demonstrated.

Epidemiological studies on isolated signs and stigmata of PCOS have produced mixed results. Some preliminary studies suggest a slight increase in cardiovascular events in women with PCOS. For example, an uncalculated but significantly increased risk of atherosclerosis *(19)* and a sevenfold increased risk of CVD *(20)* have been reported in subjects with PCOS. Notably, the risk of coronary artery disease and MI has been reported to be increased in patients with PCOS compared with regularly cycling women *(13)* even if mortality from circulatory disease does not seem to be increased. However, overall, the small size and limited number of these epidemiological studies have been inadequate to confirm an association between PCOS and CVD, especially because both are common conditions in women.

Moreover, as a consequence of IR, patients with PCOS often have an abnormal lipid profile and increased incidence of CVR factors *(21)*. As noted earlier, IR has been associated with elevated triglyceride levels, increased levels of LDL, and decreased levels of HDL *(7,8)*. Therefore, different confounding factors may co-exist in patients with PCOS, such as obesity, arterial HTN, impaired glucose tolerance (IGT), and/or type 2 diabetes mellitus, hyperinsulinemia, dyslipidemia, and coagulation disorders. Each of these factors could independently increase the risk for CVD in these subjects, making it difficult to quantify the increased CVR conferred by PCOS *per se* vs associated comorbidities. Interestingly, recent data have demonstrated a correlative and causative relationship between IR and inflammation *(22)*. In light of the role of IR in PCOS, and of the increased CVR of affected women, a relationship between inflammation and hormonal-metabolic features in women with PCOS has also been suggested *(23)*.

Women with PCOS appear to have significantly increased C-reactive protein (CRP) concentrations *(24)*, a marker of low-grade chronic inflammation and predictor of coronary heart disease and cardiovascular events, and indicated an independent inverse relationship between CRP and insulin insensitivity. The leukocyte count has been reported to be significantly higher in women with PCOS compared with healthy women, although frank leukocytosis was absent from both groups. Furthermore, in the leukocyte differential count, a significant increase in lymphocytes and monocytes was observed in women with PCOS with regard to controls *(23)*, as might be expected considering that they play a key role in the pathophysiological mechanism of atherosclerosis Although these findings are provocative, further studies are needed for confirmation.

The metabolic syndrome (MBS), as defined by The National Cholesterol Education Program Adult Treatment Panel, a condition of increased CVR thought to be linked to IR and visceral obesity, is more prevalent in PCOS than in the general female population. In a recent study by Apridonidze et al. *(25)*, the prevalence of the MBS in young women with PCOS was 43%. MBS contributes to an increased risk for CVD, and the presumed increased IR of women with PCOS with MBS should confer increased risk for glucose intolerance *(25)*. IR, as previously mentioned, is probably not the only major risk factor for the occurrence of CVD in PCOS, but may represent the most important and pathogenic feature of MBS.

It is well established that the presence of MBS correlates with CVD risk *(26)*; the finding of a markedly high prevalence of MBS even in young women with PCOS lends credence to previous reports of increased atherosclerosis and a sevenfold increased risk of CVD in women with PCOS *(20)*. Because MBS is associated with an increased risk for cardiovascular morbidity and mortality and the prevalence of MBS is significantly higher in PCOS women, it is reasonable to conclude that all women with PCOS, regardless of age and body mass index (BMI), should be screened for the individual components of MBS, other CVR factors, and CVD.

As noted above, in addition to classic CVR factors, other factors such as homocysteine, LVH, and oxidative stress seem to be independently associated with an increase in CVR *(9)*. Whereas data on homocysteine levels in women with PCOS are inconclusive and conflicting *(27,28)*, an early and IR-related impairment of endothelial structure and function *(12)*, as well as diastolic dysfunction and LVH *(11)*, have been demonstrated in young normal-weight women with PCOS. Both IR and hyperandrogenism have been widely accepted as risk factors for increased intima-media thickness, a potential marker of risk for CVD, which has also been shown to be present in women with PCOS *(12,19,29)*.

Impaired fibrinolysis, one of the more important cofactors in the development of fatal ischemic heart disease in women, has also been demonstrated in PCOS *(30)*. Specifically, a significant increase in serum PAI-1 activity *(31)* has been reported. This increase was positively related to IR, but independent of obesity *(31)*.

Finally, as documented from several studies, type 2 diabetes is present in 5–10% of women with PCOS, whereas IGT is found in 30–40% *(32,33)*. IGT is a major risk factor for type 2 diabetes mellitus and a strong independent risk factor for CVD. Hyperinsulinemia from IR or IGT, with or without diabetes, probably plays a pivotal role in the acceleration of macrovascular disease in women with PCOS*(21)*. Moreover, 50–60% of women with PCOS are obese *(34)*, and obesity could further worsen the metabolic pattern and even the CVR of this group of patients.

Does the putative increase in CVD in PCOS translate into increased mortality in PCOS women? One important study addressing this question reported no increased CVD mortality in PCOS *(2,35)*. The first report of the study *(2)* showed no increased mortality from cardiovascular-related causes, although there was an increased number of deaths following complications of diabetes in the PCOS group. The second *(35)* reported no increased long-term coronary heart disease mortality in the PCOS group, although there was evidence of increased stroke-related mortality even after adjustment for BMI.

It appears that PCOS may accelerate the development of an adverse CVR profile or even signs of subclinical atherosclerosis. Interestingly, IR is frequently cited as the cause of an increased risk for

CVD among women with PCOS, even if no measure of insulin sensitivity is included in current diagnostic criteria *(36)*. Finally, phenotypic variability has been reported to influence the findings of abnormal metabolic and CVR parameters in the general population of hyperandrogenic women *(35)*. For example, the prevalence of metabolic and cardiovascular abnormalities was highest in women who fulfilled the "classical" criteria for PCOS (which includes androgen excess and chronic oligo-amenorrhea), followed by normal ovulatory women with PCOS (i.e., with androgen excess and ana-tomically polycystic ovaries) and then women with idiopathic hyperandrogenism *(37)*. This is a crucial point because it demonstrates the difficulty of defining the true CVR and putative CVD of PCOS, since the lack of a validated definition results in PCOS currently consisting of a heterogeneous group of affected women. In the future, all evaluations of the CVR should not only specify the criteria used in establishing the diagnosis, but should also take into consideration the different phenotypes of this complex disorder.

3. CONCLUSIONS

Whereas consistent and well-characterized phenotypes of reproductive and metabolic abnormalities have been demonstrated in PCOS, suggesting an increased risk for CVD in the disorder, prospective studies and clinical trials from large-scale studies with long-term follow-up are lacking. Few studies of PCOS have selected mortality (cardiovascular or otherwise) as their endpoint, and at present these limited studies do not indicate an increased mortality for CVD in PCOS. Several parameters denoting increased CVR, such as lipoprotein abnormalities, HTN, LVH, endothelial dysfunction, and low-grade chronic inflammation, have been described in PCOS women, even at an early age, although in small sample studies. Therefore, larger prospective trials with long-term follow-up are needed to more accurately define the CVR and the putative subsequent CVD of PCOS.

4. FUTURE AVENUES OF INVESTIGATION

An early diagnosis of PCOS could provide an important opportunity to begin primary prevention of CVD. There are multiple and several surrogate variables in PCOS, including anthropometric and biochemical parameters, but the ability of these variables to predict primary clinical endpoints (such as cardiovascular mortality) remains to be determined, and represent a major problem when designing trials to assess CVR in PCOS women.

CVD is arguably the most important clinical endpoint in PCOS studies evaluating CVR, but few studies have addressed this endpoint. Future multicenter studies in women with PCOS should focus on CVD as the primary clinical endpoint.

The therapeutic effects of diet, physical exercise (i.e., lifestyle modifications), oral contraceptives, and insulin-sensitizing drugs (metformin, thiazolidinediones, etc.) on CVR factors should be carefully investigated in future studies of PCOS.

KEY POINTS

- PCOS appears to be associated with an increased CVR and/or a higher incidence of CVD.
- There is a relationship between CVD and CVR and PCOS, including increased prevalences of dyslipidemia, insulin resistance, increased intimal-media thickness, LVH, diastolic dysfunction, and markers of endothelial dysfunction, in women with PCOS.
- There are a few emerging novel factors to assess cardiovascular risk, including CRP and the white blood cell count.
- Prospective studies and clinical trials with large sample sizes from long-term follow-up are needed.

REFERENCES

1. Ehrmann D. The polycystic ovary syndrome. N Engl J Med 2005;352:1223–1236.
2. Pierpoint T, McKeigue PM, Isaacs AJ, Wild SH, Jacobs HS. Mortality of women with polycystic ovary syndrome at long-term follow-up. J Clin Epidemiol 1998;51:581–586.

3. Mather KJ, Kwan F, Corenblum B. Hyperinsulinemia in polycystic ovary syndrome correlates with increased cardiovascular risk independent of obesity. Fertil Steril 2000;73:150–156.

4. Zimmermann S, Phillips RA, Dunaif A, et al. Polycystic ovary syndrome: lack of hypertension despite profound insulin resistance. J Clin Endocrinol Metab 1992;75:508–513.

5. Holte J, Pennarelli G, Berne C, Bergh T, Lithell H. Elevated ambulatory day-time blood pressure in women with polycystic ovary syndrome: a sign of a pre-hypertensive state? Hum Reprod 1996;11:23–28.

6. Elting MW, Korsen TJM, Bezemer PD, Schoemaker J. Prevalence of diabetes mellitus, hypertension and cardiac complaints in a follow-up study of a Dutch PCOS population. Hum Reprod 2001;16:556–560.

7. Pirwany IR, Fleming R, Greer IA, Packard CJ, Sattar N. Lipids and lipoprotein subfractions in women with PCOS: relationship to metabolic and endocrine parameters. Clin Endocrinol 2001;54:447–453.

8. Wild RA, Painter PC, Coulson PB, Carruth KB, Ranney GB. Lipoprotein lipid concentrations and cardiovascular risk in women with polycystic ovary syndrome. J Clin Endocrinol Metab 1985;61:946–951.

9. Harjai KJ. Potential new cardiovascular risk factors: left ventricular hypertrophy, homocysteine, lipoprotein(a), triglycerides, oxidative stress, and fibrinogen. Ann Intern Med 1999;131:376–386.

10. Davies MJ, Norman RJ. Programming and reproductive functioning. Trends Endocrinol Metab 2002;13:386–392.

11. Orio F Jr, Palomba S, Spinelli L, et al. The cardiovascular risk of young women with polycystic ovary syndrome: an observational, analytical, prospective case-control study. J Clin Endocrinol Metab 2004;89:3696–3701.

12. Orio F Jr, Palomba S, Cascella T, et al. Early impairment of endothelial structure and function in in young normal-weight women with polycystic ovary yndrome. J Clin Endocrinol Metab 2004;89:4588–4593.

13. Cibula D, Cifkova R, Fanta M, Poledne R, Zivny J, Skibova J. Increased risk of non-insulin dependent diabetes mellitus, arterial hypertension and coronary artery disease in perimenopausal women with a history of the polycystic ovary syndrome. Hum Reprod 2000;15:785–789.

14. Yildiz BO, Haznedaroglu IC, Kirazli S, Bayraktar M. Global fibrinolytic capacity is decreased in polycystic ovary syndrome, suggesting a prothrombotic state. J Clin Endocrinol Metab 2002;87:3871–3875.

15. Taponen S, Martikainen H, Jarvelin MR, et al. Northern Finland Birth Cohort 1966 Study. Metabolic cardiovascular disease risk factors in women with self-reported symptoms of oligomenorrhea and/or hirsutism: Northern Finland Birth Cohort 1966 Study. J Clin Endocrinol Metab 2004;89:2114–2118.

16. Zawadzki JK, Dunaif A. Diagnostic criteria for polycystic ovary syndrome: towards a rational approach. Boston: Blackwell Scientific, 1992:377–384.

17. Rotterdam ESHRE/ASRM-Sponsored PCOS Consensus Workshop Group. Revised 2003 consensus on diagnostic criteria and long-term health risks related to polycystic ovary syndrome. Fertil Steril 2004;81:19–25.

18. Jayagopal V, Kilpatrick ES, Jennings PE, Hepburn DA, Atkin SL. The biological variation of testosterone and sex hormone-binding globulin (SHBG) in polycystic ovarian syndrome: implications for SHBG as a surrogate marker of insulin resistance. J Clin Endocrinol Metab 2003;88:1528–1533.

19. Talbott EO, Guzick DS, Sutton-Tyrrell K, et al. Evidence for association between polycystic ovary syndrome and premature carotid atherosclerosis in middle-aged women. Arterioscler Thromb Vasc Biol 2000;20:2414–2421.

20. Dahlgren E, Janson PO, Johansson S, Lapidus L, Oden A. Polycystic ovary syndrome and risk for myocardial infarction. Evaluated from a risk factor model based on a prospective population study of women. Acta Obstet Gynecol Scand 1992;71:599–604.

21. Amowitz LL, Sobel BE Cardiovascular consequences of polycystic ovary syndrome. Endocrinol Metab Clin North Am 1999;28:439–458.

22. Bloomgarden ZT. Inflammation and insulin resistance. Diabetes Care 2003;26:1922–1926.

23. Orio F Jr, Palomba S, Spinelli L, et al. The increase of leukocyte as a new marker of low grade chronic inflammation and early cardiovascular risk in the polycystic ovary syndrome. J Clin Endocrinol Metab 2005;90:2–5.

24. Kelly CC, Lyall H, Petrie JR, Gould GW, Connell JM, Sattar N. Low grade chronic inflammation in women with polycystic ovary syndrome. J Clin Endocrinol Metab 2001;86:2453–2455.

25. Apridonidze T, Essah PA, Iuorno MJ, Nestler JE. Prevalence and characteristics of the metabolic syndrome in women with polycystic ovary syndrome. J Clin Endocrinol Metab 2005;90:1929–1935.

26. Isomaa B, Almgren P, Tuomi T, et al. Cardiovascular morbidity and mortality associated with the metabolic syndrome. Diabetes Care 2001;24:683–689.

27. Orio F Jr, Palomba S, Di Biase S, et al. Homocysteine levels and C677T polymorphism of methylenetetrahydrofolate reductase in women with polycystic ovary syndrome. J Clin Endocrinol Metab 2003;88:673–679.

28. Yilmaz M, Biri A, Bukan N, et al. Levels of lipoprotein and homocysteine in non-obese and obese patients with polycystic ovary syndrome. Gynecol Endocrinol 2005;20:258–263.

29. Tarkun I, Arslan BC, Canturk Z, Turemen E, Sahin T, Duman C. Endothelial dysfunction in young women with polycystic ovary syndrome: relationship with insulin resistance and low-grade chronic inflammation. J Clin Endocrinol Metab 2004;89:5592–5596.

30. Meade TW, Cooper JA, Chakrabarti R, Miller GJ, Stirling Y, Howarth DJ. Fibrinolityc activity and clotting factor in ischaemic heart disease in women. Br Med J 1996;312:1581.

31. Orio F Jr, Palomba S, Cascella T, et al. Is plasminogen activator inhibitor-1 (PAI-1) a cardiovascular risk factor in young women with polycystic ovary syndrome? RBM Online 2004;9:505–510.
32. Legro RS, Kunselman AR, Dodson WC, Dunaif A. Prevalence and predictors of risk for type 2 diabetes mellitus and impaired glucose tolerance in polycystic ovary syndrome: a prospective, controlled study in 254 affected women. J Clin Endocrinol Metab 1999;84:165–169.
33. Ehrmann D, Barnes R, Rosenfield R, Cavaghan M, Imperial J. Prevalence of impaired glucose tolerance and diabetes in women with polycystic ovary syndrome. Diabetes Care 1999;22:141–146.
34. Dunaif A. Hyperandrogenic anovulation (PCOS) a unique disorder of insulin action associated with an increased risk of non-insulin dependent diabetes mellitus. Am J Med 1995;98 (suppl 1A):1A–33s.
35. Wild S, Pierpoint T, McKeigue P, Jacobs H. Cardiovascular disease in women with polycystic ovary syndrome at long-term follow-up: a retrospective cohort study. Clin Endocrinol (Oxf) 2000;52:595–600.
36. Legro RS, Castracane VD, Kauffman RP. Detecting insulin resistance in polycystic ovary syndrome: purposes and pitfalls. Obstet Gynecol Surv 2004;59:141–154.
37. Carmina E, Longo RA, Rini GB, Lobo RA. Phenotypic variation in hyperandrogenic women influences the findings of abnormal metabolic and cardiovascular risk parameters. J Clin Endocrinol Metab 2005;90:2545–2549. [Epub 2005 Feb 22.]

30

Reproductive and Obstetrical Consequences of the Polycystic Ovary Syndrome

Mary C. Johnson and Richard S. Legro

SUMMARY

Polycystic ovary syndrome (PCOS) is a disorder associated with infertility and subfecundity. There is evidence that both the oocyte and the endometrium have abnormalities that contribute to these disorders. These reproductive abnormalities include anovulation, increased early pregnancy wastage, and probable increased risk for a number of pregnancy complications including gestational diabetes and hypertensive disorders of pregnancy, such as pre-eclampsia. Ovulation induction holds the additional iatrogenic risks of multiple pregnancy and ovarian hyperstimulation syndrome. Gonadotropins and gonadotropin agonist therapies should be used with caution in this group of patients. There is emerging evidence that insulin-sensitizing therapy provides benefit for a wide variety of these reproductive abnormalities. In the long term, women with PCOS have multiple risk factors for the development of endometrial cancer, including centripetal obesity, hyperinsulinemia, diabetes, and chronic anovulation, although the association between PCOS and this cancer is less well established in the epidemiological literature. There is little evidence to date that insulin-sensitizing therapies protect against the development of endometrial hyperplasia and cancer.

Key Words: Hyperandrogenism; insulin resistance; pre-eclampsia; gestational diabetes; pregnancy.

1. INTRODUCTION

Polycystic ovary syndrome (PCOS) is an endocrine disorder of unexplained etiology found in approximately 5–7% of the female population (1). According to a 2003 expert consensus statement, the diagnosis of PCOS requires the patient to have at least two of the following three characteristics: (1) oligo- or anovulation, (2) clinical and/or biochemical signs of hyperandrogenism, and (3) polycystic ovaries. The diagnosis of PCOS also requires the exclusion of other possible etiologies, such as congenital adrenal hyperplasia, androgen-secreting tumors, and Cushing's syndrome (2). Women with PCOS typically present with menstrual dysfunction, hirsutism, and infertility.

This chapter focuses on the reproductive challenges of women with PCOS. Although there are PCOS-associated reproductive difficulties independent of obesity, obesity acts synergistically with PCOS to further confound attempts to successfully conceive and maintain a pregnancy to full term. It is estimated that approximately 50% of women with PCOS are obese, but more recent studies including multicenter trials and large family studies have shown that the vast majority are obese (up to 80%), and mean body mass index (BMI) approaches or exceeds 35 kg/M^2. The reproductive challenges associated with PCOS discussed in this chapter are infertility, multiple pregnancy, ovarian hyperstimulation syndrome (OHSS), gestational diabetes, miscarriage and pregnancy loss, pre-eclampsia and eclampsia, and endometrial cancer.

From: *Contemporary Endocrinology: Androgen Excess Disorders in Women:*
Polycystic Ovary Syndrome and Other Disorders, Second Edition
Edited by: R. Azziz et al. © Humana Press Inc., Totowa, NJ

2. BACKGROUND

2.1. Infertility and Anovulation

PCOS is the most common cause of anovulatory infertility and may account for more than 75% of cases of anovulatory infertility *(3)*. However, not all women with PCOS are completely amenorrheic. The majority of women with PCOS are oligomenorrheic and experience varying intervals of vaginal bleeding.

Abnormal ovulatory function associated with PCOS is distinguished by dysfunctional folliculogenesis and abnormal steroidogenesis. Follicular development within the ovary of a patient with PCOS occurs until an approximate follicular diameter of 5–8 mm, at which point follicular growth arrests. Proper evidence has yet to explain the cessation of development, but androgen excess in the intrafollicular milieu, exogenous follicle-stimulating hormone (FSH) deficiency, and the effects of hyperinsulinemia and insulin resistance in the follicle have all been identified as factors contributing to the follicular arrest and atresia that characterizes PCOS.

The premature discontinuation of follicle maturation occurs at a point when rising FSH levels should influence the developing follicles to produce increasing amounts of estrogen in the granulosa cells; exponential follicular growth should follow. It is postulated that an FSH deficiency prevents this next step and that the FSH deficiency may be secondary to increased inhibin production. Inhibin B, which is normally released in a pulsatile manner, selectively inhibits FSH secretion. In women with PCOS, inhibin B is secreted steadily, without pulsatility, and its abnormal secretory pattern has been postulated to induce the abnormal folliculogenesis *(4)*. The gonadotropic properties of insulin should also be taken into account. In vivo, hyperinsulinemia prematurely stimulates granulosa cells, arresting their growth and aromatase activity. Thus, progesterone production is increased, follicular development ceases, and follicular dominance fails to proceed.

It is interesting to note that of all the clinical and biochemical characteristics of PCOS, hyperinsulinemia is the most strongly associated with anovulation. Abnormal steroidogenesis seems to also be key in the pathogenesis of infertility in women with PCOS. Hyperandrogenemia remains one of the classic biochemical and clinical features of PCOS. Excess androgens have been found to inhibit endometrial epithelial cell growth and secretory activity in vitro *(5)*. By inhibiting growth, hyperandrogenemia may prevent fertility and also increase the probability of early termination of the pregnancy.

Especially with regard to infertility, obesity acts synergistically with PCOS to confound patients and their physicians; patients with PCOS who are overweight are less likely to conceive than those patients with PCOS of healthy weight. In fact, in one large in vitro fertilization series that examined factors associated with failure, obesity was found to significantly predict pregnancy failure and loss *(6)*. Obesity is further associated with an inadequate response to both ovulation-induction agents as well as insulin sensitizers *(7)*.

2.2. Iatrogenic Complications of Ovulation Induction in PCOS

Women with PCOS are at risk for iatrogenic complications of ovulation induction for anovulatory infertility. These include OHSS and multiple pregnancy. An inherent ovarian abnormality may contribute to these complications. One promise of ovulation induction with insulin-sensitizing therapies, as opposed to the more often utilized treatment with traditional ovulation-induction strategies such as clomiphene citrate and gonadotropins, is a lower rate of multiple follicular recruitment and lower risk of these iatrogenic complications (Fig. 1).

2.2.1. Ovarian Hyperstimulation Syndrome

The strongest association between OHSS and PCOS is with young, thin women with polycystic ovaries who undergo superovulation with urinary human menopausal gonadotropin (hMG) preparations and use a gonadotropin-releasing hormone (GnRH) agonist *(8)*. It must also be acknowledged

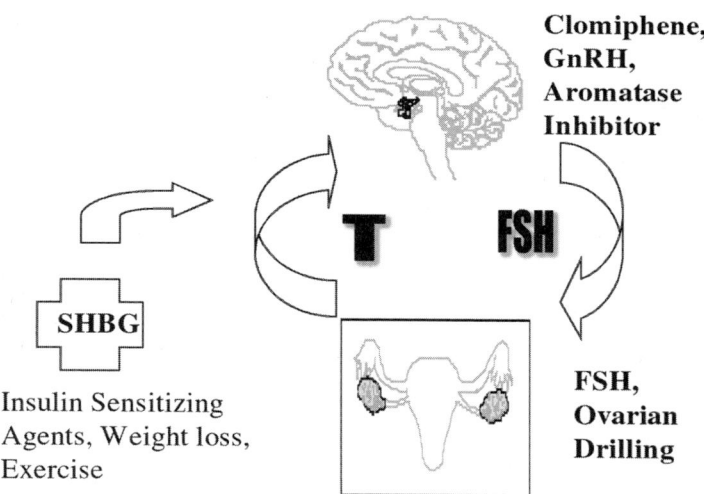

Fig. 1. There are multiple methods to induce ovulation in women with PCOS, which can alter testosterone levels (T), gonadotropins (FSH), and sex hormone-binding globulin (SHBG). GnRH, gonadotropin-releasing hormone.

that all women with PCOS are not at the same risk for OHSS because of the great confounder of obesity, that with progressive increases tends to ameliorate risk. Nevertheless, it is interesting to note that PCOS has been reported to be associated with OHSS developing in spontaneous cycles as well as with the use of clomiphene citrate, both of which are only rarely associated with OHSS in the larger population.

Severe OHSS results in massive enlargement of the ovaries, formation of multiple ovarian cysts, extravascular fluid accumulation, ascites, and intravascular volume depletion. In severe cases, hyper-coagulability and embolic events can develop, and systemic organ failure can occur *(9)*. To a degree, OHSS is a risk to any woman treated with ovulation induction. However, a consistent relationship has been found between the anatomical finding of polycystic ovaries and OHSS, designating poly-cystic ovaries as an additional risk factor *(10)*. The pathogenesis of OHSS is still unclear.

During ovulation induction with gonadotropins, women with PCOS and OHSS frequently are plagued with an inordinate number of midsize follicles, which are not thought to produce a compe-tent oocyte, but contribute significantly to the circulating estradiol pool and ultimately to the risk for OHSS. Women with polycystic ovaries in the baseline state appear to have an inordinate number of growing primary follicles—up to six times more than women with normal ovaries *(11)*. The increased density of small preantral follicles in polycystic ovaries could result from increased population of the fetal ovary by germ cells or from decreased rate of loss of oocytes during late gestation, childhood, and puberty. More recently, animal data have suggested that oocytes can develop from mesenchymal stem cells in the bone marrow *(12)*, which has challenged one of the central dogmas of ovarian physiology, which states that there are no new developments of oocytes postnatally and there is only atresia, albeit at varying rates, after birth. These data suggest that oocytes can develop from stem cells during the life span of the organism, and it awaits confirmation in human studies.

Patients with polycystic ovaries, which includes but is not limited to patients with PCOS, are a population that should receive special education about the risks of OHSS; primary prevention of OHSS could avoid uncomfortable and potentially life-threatening complications, such as renal fail-ure, hepatic failure, and thromboembolism. Elevated or rapidly rising estradiol levels have been iden-tified as predictive markers for the development of OHSS as well as the absolute number of follicles. Women with PCOS should be monitored very closely when undergoing ovulation induction with

Table 1
Maternal and Fetal Complications of Multiple Pregnancy

Maternal
- Need for fetal reduction
- Pregnancy complications:
- Anemia, PIH, gestational diabetes, IUGR
- Increased rate of cesarean delivery
- Increased risk of postpartum hemorrhage?
- Increased risk of postpartum stress/depression?
- Increased risk of miscarriage?

Fetal
- Prematurity
- Low birthweight
- Malformations
- Cerebral palsy from traumatic deliver
- Learning disabilities?
- Delayed development?
- Diabetes mellitus in the long term?

PIH, pregnancy-induced hypertension; IUGR, intrauterine growth restriction.

gonadotropins. A low-dose step-up ovulation-induction protocol, with the guiding principle of starting at a low dose (37.5–75 U/day) and increasing in small increments at 10- to 14-day intervals, appears effective at reducing the risk for OHSS *(13)*. A Cochrane Library meta-analysis has found that FSH-only gonadotropin preparations significantly lower the risk for OHSS compared with human menopausal gonadotropin preparations (odds ratio [OR] 0.20; 95% confidence interval [CI] 0.08–0.46), and the concomitant use of a GnRH agonist with gonadotropin therapy increases the risk for OHSS (OR 3.15; 95% CI 1.46–6.70) *(14)*.

2.2.2. Multiple Pregnancy (see Table 1)

Multiple pregnancy confers substantially increased fetal and maternal risk compared with a singleton pregnancy. It is assumed that women with PCOS are at increased risk for multiple pregnancy, but it is uncertain whether PCOS confers additional risk above and beyond that of ovulation induction. There is no evidence that the spontaneous twinning rate is higher in women with PCOS. In a large series of 1803 women who underwent ovarian drilling for anovulatory infertility, there were 1076 pregnancies and a 2.3% twinning rate *(15)*. Although this exceeds the expected spontaneous twinning rate of 0.8%, many of these resulted from subsequent ovulation induction *(15)*. In one large series of pregnancies resulting from ovulation induction with clomiphene citrate (PCOS probably was the largest underlying diagnosis), the twinning rate was 6.9% and the triplet rate 0.5% *(16)*.

In women with PCOS undergoing ovulation induction with gonadotropins, the multiple pregnancy rate was 20%, comparable with women undergoing the same treatment for unexplained infertility *(17)*. However, the same risk factors for OHSS, that is, multiple and excessive follicular development, probably also place women with PCOS undergoing gonadotropin stimulation at increased risk for multiple pregnancy. The use of insulin-sensitizing agents holds the promise of monofollicular ovulation and singleton pregnancy. A recent randomized trial did demonstrate no multiple pregnancies on metformin, but there were also no multiple pregnancies in the clomiphene arm *(18)*.

2.3. Other Pregnancy Complications

Anovulatory infertility is just one mechanism that contributes to subfecundity in women with PCOS. Other causes that lead to a decreased live birth rate and/or perinatal morbidity include higher rates of first-trimester pregnancy loss and an increased risk for later pregnancy complications, including pre-eclampsia and gestational diabetes.

2.3.1. Early Pregnancy Loss and Miscarriage

Women with PCOS appear to have increased rates of pregnancy loss, and this has been associated with increased luteinizing hormone (LH) secretion *(19)*, excess androgen production *(20)*, and most recently insulin resistance *(21)*. It is difficult to untangle the contributions of these reproductive and metabolic abnormalities in PCOS and their effect on early pregnancy loss. Because these women are subfecund and undergo infertility treatment and the accompanying increased surveillance, early non-viable pregnancies are more likely to be detected in this population. Depending on when the pregnancy is detected, pregnancy loss rates in humans have been estimated to range from 20 to 50% per conception *(22)*. Thus, we are not a particularly fecund species compared with some of our mammalian counterparts. There is likely a considerable detection bias in examining the early pregnancy outcomes in women with PCOS, and it is difficult to determine if pregnancy loss rates exceed those of other subfecund populations receiving similar surveillance. However, there remain concerns about the competency of oocytes that develop in the PCOS milieu, and clearly endometrial abnormalities appear common, such that implantation failure may further contribute to pregnancy loss.

Obesity, which affects up to 80–90% of women with PCOS in the United States, is also a major factor in pregnancy loss. The high risk of spontaneous abortion associated with PCOS is partially related to the increased prevalence of obesity in women with PCOS; the incidence of spontaneous abortion increases with increasing BMI *(6,23)*. Small randomized trials of pregnancy and pregnancy outcomes have shown lower pregnancy loss rates in pregnancies conceived with metformin *(18,24)* compared with conventional treatments, providing support for the idea of insulin resistance contributing to pregnancy loss in these women. Many researchers have begun empirically to use metformin throughout the first trimester (and longer) to prevent pregnancy loss and have reported a marked improvement in pregnancy rates *(21,25)*. These remain essentially large case series and await confirmation in prospective, appropriately powered, multicenter trials. A healthy skepticism is in order because the track record of interventions to prevent pregnancy loss has been marred with treatments causing fetal and maternal harm (e.g., diethylstilbestrol [DES], administration of intravenous γ-globulin, paternal leukocyte immunization).

2.3.2. Gestational Diabetes

Gestational diabetes affects about 4–5% of pregnancies in the United States and entails the risk of both maternal and fetal complications. Fetal risks include macrosomia, an increased perinatal morbidity rate, an increased likelihood of birth trauma, including shoulder dystocia and related brachial plexus injuries, and neonatal hypoglycemia and hypocalcemia. Maternal complications include birth trauma related to macrosomia, an increased risk for gestational hypertension, including pre-eclampsia, and complications of diabetes.

Gestational diabetes is usually detected at around 28 weeks gestation by routine glucose challenge testing. The etiology of gestational diabetes is thought to be primarily a result of an acquired insulin resistance that is induced by the hormonal changes of pregnancy. Women with PCOS are thought to be at a higher risk for developing gestational diabetes because they often exhibit insulin resistance even before pregnancy; up to 80% of obese and 30% of lean women with PCOS demonstrate insulin resistance prior to pregnancy *(26)*. There are a number of placental hormones, including human placental lactogen, and maternal hormones, such as glucocorticoids, whose ultimate physiological goal is to blunt maternal glucose uptake and shunt glucose to the fetus, because glucose remains the primary and preferred growth substrate of the developing fetus. In women who also have preexisting insulin resistance and some degree of β-cell dysfunction (thought to characterize the majority of women with PCOS), pregnancy and its accompanying acquired insulin resistance may exceed the compensatory ability of the pancreas. Gestational diabetes occurs when a woman's body cannot compensate for increased metabolic demands of pregnancy.

Although obesity is one of the primary predicting factors for gestational diabetes, the effects of PCOS have also been found to be significant, even after adjusting for the effect of increasing BMI *(27)*. PCOS carried a nearly twofold increased risk for developing gestational diabetes, with a rate of 20% compared to 9% in the control population *(27)*. In another report, 9 of 22 women with PCOS developed gestational diabetes *(28)*.

One critique of these and other studies involves the relatively small sample size. Some authors have recommended the use of metformin throughout pregnancy to reduce the risk of gestational diabetes, and in one small randomized trial there appeared to be a benefit, but this remains an extreme and unproven strategy for the prevention of gestational diabetes. The effects of developing gestational diabetes carry on well after the pregnancy, because women with a history of gestational diabetes mellitus have a 50% chance of developing type 2 diabetes later in life *(29)*. This necessitates periodically rescreening women with gestational diabetes for the development of type 2 diabetes.

2.3.3. Pre-Eclampsia and Eclampsia

Pre-eclampsia is characterized by maternal hypertension and proteinuria. Onset is typically after the 20th week of pregnancy, and pre-eclampsia complicates 3-4% of pregnancies. Eclampsia occurs when the pre-eclampsia condition results in seizures; eclampsia is the second leading cause of maternal death in the United States. Currently, the only mainstay therapy for severe pre-eclampsia is delivery of the baby.

A significant association has been found between first-trimester insulin resistance and risk of pre-eclampsia *(30)*. This is particularly notable for women with PCOS, who, as noted before, are typically insulin resistant before pregnancy. In addition, women with PCOS have been found to be at an increased risk of pre-eclampsia independent of BMI and ovulation induction *(31)*. Because elevated BMI, preexisiting insulin resistance and metabolic syndrome, the iatrogen contribution of multiple pregnancies, from fertility treatment are also associated with pre-eclampsia, of women with PCOS are at great risk for pre-eclampsia and should be monitored accordingly.

2.4. Endometrial Cancer

Endometrial cancer is the most common malignancy of the lower female genital tract but fortunately in its most common form, adenocarcinoma, has the best prognosis, with overall 5-year survival for stage 1 cancers well above 90%. Endometrial cancer is considered a late reproductive sequele of PCOS and is thought to be increased as a result of the history of chronic anovulation and unopposed estrogen *(32)*. However, it can equally be considered a late metabolic sequela of PCOS, as the classic risk factors for endometrial cancer are age, obesity (especially centripetal obesity), and diabetes *(33)*. In case series, women with PCOS have been overrepresented in developing endometrial cancer and often at an early age *(34)*.

However, when the hard evidence is closely observed, there is a paucity of studies, especially well-designed prospective studies, or even retrospective case-control studies documenting increased risk of endometrial cancer in women with PCOS *(34)*. This may be because endometrial cancer remains primarily a disease of postmenopausal women, with 90% of the cases occurring in the population past 50 years of age, and it is difficult to diagnose PCOS in this age range after natural ovarian failure and cessation of menses. A Scandinavian study that looked at a group of both premenopausal and postmenopausal women with endometrial carcinoma found hirsutism and obesity in both affected groups more often compared to controls *(35)*. In the younger group they additionally noted a recent history of anovulation and infertility, two of the most common presenting complaints of women with PCOS (in addition to hirsutism and obesity).

Adenocarcinoma of the endometrium is thought to be a progressive disease that originates from adenomatous hyperplasia, specifically atypical hyperplasia. Endometrial hyperplasia has also often been noted in association with anovulation and infertility *(32)*. In a large case–control study, increased endometrial cancer risk was noted in women with lower levels of sex hormone-binding globulin *(36)*

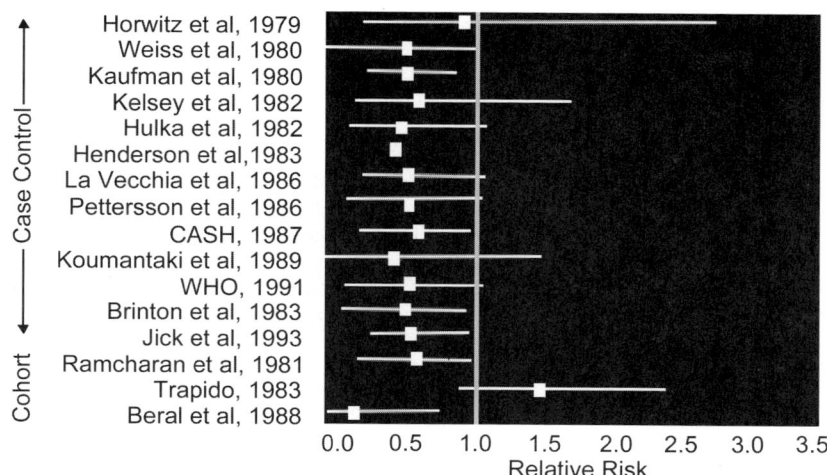

Fig. 2. Protective effect of oral contraceptives in preventing endometrial cancer in population-based epide-miological trials examining the association between ever use of the oral contraceptive pill and the development of endometrial cancer. (Adapted from ref. *39*.)

and elevated insulin levels *(37)*, both biochemical stigmata noted in women with PCOS. There are no systematic prospective studies of the prevalence of endometrial hyperplasia/neoplasia in a population with PCOS or conversion rates over time, although one single-center study noted a high prevalence of endometrial hyperplasia in women with PCOS (36%) seeking fertility *(38)*. No routine screening recommendations can be made at this time, and there have been no prospective preventive trials in at-risk women, although in epidemiological studies, use or any-time use of the oral contraceptive pill provides a strong preventive benefit against endometrial cancer (Fig. 2) as well as against the development of ovarian cancer *(39)*. *See* Chapter 27 for further discussion of the mechanisms underlying a potential increased rise of endometrial carcinoma in PCOS.

3. CONCLUSIONS

PCOS is a disorder associated with infertility and subfecundity. These reproductive abnormalities include anovulation, increased early pregnancy wastage, and increased probability for a number of pregnancy complications including gestational diabetes and hypertensive disorders of pregnancy, such as pre-eclampsia. There is evidence that both the oocyte and the endometrium have abnormalities that contribute to these disorders. In the long term, women with PCOS have multiple risk factors for the development of endometrial cancer, although the association between PCOS and this cancer is less well established in the epidemiological literature.

4. FUTURE AVENUES OF INVESTIGATION

A number of issues remain to be addressed in our understanding of the reproductive and obstetrical consequences of PCOS, including the need for the following:

1. Prospective trials examining pregnancy complications of weight-matched women with and without PCOS from conception to delivery
2. Randomized trials of metformin to prevent pregnancy loss in women with PCOS and a history of pregnancy loss
3. Improved understanding of the long-term cardiovascular sequelae of women with PCOS who develop gestational complications such as diabetes or pre-eclampsia
4. Prevention trials of endometrial hyperplasia and or cancer in obese women with PCOS.

KEY POINTS

- The strongest association between OHSS and PCOS is with young, thin women with polycystic ovaries who undergo superovulation with human menopausal gonadotropin preparations and concomitant use of a GnRH agonist.
- It is assumed that women with PCOS are at increased risk for multiple pregnancy, but it is uncertain whether PCOS confers additional risk above and beyond that of ovulation induction.
- Women with PCOS appear to have increased rates of pregnancy loss, and this has been associated with increased LH secretion, excess androgen production, and most recently with insulin resistance.
- Women iwth PCOS appear to be at increased risk for gestational diabetes and pre-eclampsia.
- Women with PCOS share a risk factor profile that overlaps with that for the development of endometrial cancer, although there is a paucity of well-designed studies indicating increased event rates in women with PCOS.

REFERENCES

1. Azziz R, Woods KS, Reyna R, Key TJ, Knochenhauer ES, Yildiz BO. The prevalence and features of the polycystic ovary syndrome in an unselected population. J Clin Endocrinol Metab 2004;89(6):2745–2749.
2. Revised 2003 consensus on diagnostic criteria and long-term health risks related to polycystic ovary syndrome. Fertil Steril 2004;81(1):19–25.
3. Laven JS, Imani B, Eijkemans MJ, Fauser BC. New approach to polycystic ovary syndrome and other forms of anovulatory infertility. Obstet Gynecol Surv 2002;57(11):755–767.
4. Lockwood GM, Muttukrishna S, Groome NP, Matthews DR, Ledger WL. Mid-follicular phase pulses of inhibin B are absent in polycystic ovarian syndrome and are initiated by successful laparoscopic ovarian diathermy: a possible mechanism regulating emergence of the dominant follicle. J Clin Endocrinol Metab 1998;83(5):1730–1735.
5. Tuckerman EM, Okon MA, Li T, Laird SM. Do androgens have a direct effect on endometrial function? An in vitro study. Fertil Steril 2000 Oct;74(4):771–779.
6. Wang JX, Davies MJ, Norman RJ. Obesity increases the risk of spontaneous abortion during infertility treatment. Obes Res 2002;10(6):551–554.
7. Azziz R, Ehrmann D, Legro RS, et al. Troglitazone improves ovulation and hirsutism in the polycystic ovary syndrome: a multicenter, double blind, placebo-controlled trial. J Clin Endocrinol Metab 2001;86(4):1626–1632.
8. Bayram N, van Wely M, van der Veen F. Pulsatile gonadotrophin releasing hormone for ovulation induction in subfertility associated with polycystic ovary syndrome. Cochrane Database Syst Rev 2004;(1):CD000412.
9. Kaiser UB. The pathogenesis of the ovarian hyperstimulation syndrome. N Engl J Med 2003;349(8):729–732.
10. Tummon I, Gavrilova-Jordan L, Allemand MC, Session D. Polycystic ovaries and ovarian hyperstimulation syndrome: a systematic review. Acta Obstet Gynecol Scand 2005;84(7):611–616.
11. Webber LJ, Stubbs S, Stark J, et al. Formation and early development of follicles in the polycystic ovary. Lancet 2003;362(9389):1017–1021.
12. Johnson J, Bagley J, Skaznik-Wikiel M, et al. Oocyte generation in adult mammalian ovaries by putative germ cells in bone marrow and peripheral blood. Cell 2005;122(2):303–315.
13. Homburg R, Levy T, Ben-Rafael Z. A comparative prospective study of conventional regimen with chronic low-dose administration of follicle-stimulating hormone for anovulation associated with polycystic ovary syndrome. Fertil Steril 1995;63:729–733.
14. Nugent D, Vandekerckhove P, Hughes E, Arnot M, Lilford R. Gonadotrophin therapy for ovulation induction in subfertility associated with polycystic ovary syndrome. Cochrane Database Syst Rev Issue 4, 2000.
15. Campo S. Ovulatory cycles, pregnancy outcome and complications after surgical treatment of polycystic ovary syndrome. Obstet Gynecol Surv 1998;53(5):297–308.
16. Asch RH, Greenblatt RB. Update on the safety and efficacy of clomiphene citrate as a therapeutic agent [review] [59 refs]J Reprod Med 1976;17(3):175–180.
17. Tadokoro N, Vollenhoven B, Clark S, et al. Cumulative pregnancy rates in couples with anovulatory infertility compared with unexplained infertility in an ovulation induction programme. Hum Reprod 1997;12(9):1939–1944.
18. Palomba S, Orio F Jr, Falbo A, et al. Prospective parallel randomized double-blind double-dummy controlled clinical trial comparing clomiphene citrate and metformin as the first-line treatment for ovulation induction in non-obese anovulatory women with polycystic ovary syndrome. J Clin Endocrinol Metab 2005;90(7):4068–4074.
19. Balen AH, Tan SL, Jacobs HS. Hypersecretion of luteinising hormone: a significant cause of infertility and miscarriage [see comments] [review]. Br J Obstet Gynaecol 1993;100:1082–1089.
20. Tulppala M, Stenman UH, Cacciatore B, Ylikorkala O. Polycystic ovaries and levels of gonadotrophins and androgens in recurrent miscarriage: prospective study in 50 women [see comments]. Br J Obstet Gynaecol 1993;100:348–352.

21. Jakubowicz DJ, Iuorno MJ, Jakubowicz S, Roberts KA, Nestler JE. Effects of metformin on early pregnancy loss in the polycystic ovary syndrome. J Clin Endocrinol Metab 2002;87(2):524–529.
22. Wilcox AJ, Baird DD, Weinberg CR. Time of implantation of the conceptus and loss of pregnancy. N Engl J Med 1999;340(23):1796–1799.
23. Hamilton-Fairley D, Kiddy D, Watson H, Paterson C, Franks S. Association of moderate obesity with a poor pregnancy outcome in women with polycystic ovary syndrome treated with low dose gonadotrophin. Br J Obstet Gynaecol 1992;99:128–131.
24. Palomba S, Orio F Jr, Nardo LG, et al. Metformin administration versus laparoscopic ovarian diathermy in clomiphene citrate-resistant women with polycystic ovary syndrome: a prospective parallel randomized double-blind placebo-controlled trial. J Clin Endocrinol Metab 2004;89(10):4801–4809.
25. Glueck CJ, Wang P, Goldenberg N, Sieve-Smith L. Pregnancy outcomes among women with polycystic ovary syndrome treated with metformin. Hum Reprod 2002;17(11):2858–2864.
26. Dunaif A, Segal KR, Futterweit W, Dobrjansky A. Profound peripheral insulin resistance, independent of obesity, in polycystic ovary syndrome. Diabetes 1989;38(9):1165–1174.
27. Mikola M, Hiilesmaa V, Halttunen M, Suhonen L, Tiitinen A. Obstetric outcome in women with polycystic ovarian syndrome. Hum Reprod 2001;16(2):226–229.
28. Radon PA, McMahon MJ, Meyer WR. Impaired glucose tolerance in pregnant women with polycystic ovary syndrome. Obstet Gynecol 1999;94(2):194–197.
29. Stowers JM, Sutherland HW, Kerridge DF. Long-range implications for the mother. The Aberdeen experience. Diabetes 1985;34 (Suppl 2):106–110.
30. Wolf M, Sandler L, Munoz K, Hsu K, Ecker JL, Thadhani R. First trimester insulin resistance and subsequent preeclampsia: a prospective study. J Clin Endocrinol Metab 2002;87(4):1563–1568.
31. de Vries MJ, Dekker GA, Schoemaker J. Higher risk of preeclampsia in the polycystic ovary syndrome. A case control study. Eur J Obstet Gynecol Reprod Biol 1998;76:91–95.
32. Coulam CB, Annegers JF, Kranz JS. Chronic anovulation syndrome and associated neoplasia. Obstet Gynecol 1983;61:403–407.
33. Brinton LA, Berman ML, Mortel R, et al. Reproductive, menstrual, and medical risk factors for endometrial cancer: results from a case-control study. Am J Obstet Gynecol 1992;167:1317–1325.
34. Hardiman P, Pillay OS, Atiomo W. Polycystic ovary syndrome and endometrial carcinoma. Lancet 2003;361(9371):1810–1812.
35. Dahlgren E, Friberg LG, Johansson S, Lindstrom B, Oden A, Samsioe G. Endometrial carcinoma; ovarian dysfunction—a risk factor in young women. Eur J Obstet Gynecol Reprod Biol 1991;41:143–150.
36. Potischman N, Hoover RN, Brinton LA, et al. Case-control study of endogenous steroid hormones and endometrial cancer. J Natl Cancer Inst 1996;88:1127–1135.
37. Troisi R, Potischman N, Hoover RN, Siiteri P, Brinton LA. Insulin and endometrial cancer. Am J Epidemiol 1997;146(6):476–482.
38. Cheung AP. Ultrasound and menstrual history in predicting endometrial hyperplasia in polycystic ovary syndrome. Obstet Gynecol 2001;98(2):325–331.
39. Grimes DA, Economy KE. Primary prevention of gynecologic cancers. Am J Obstet Gynecol 1995;172(1 Pt 1):227–235.

31

Psychosocial and Quality-of-Life Consequences of Androgen Excess and the Polycystic Ovary Syndrome

Onno E. Janssen, Susanne Hahn, and Sigrid Elsenbruch

SUMMARY

Quality of life is a multidimensional, subjective idea that defies exact definition. In androgen excess, and even more so in the polycystic ovary syndrome (PCOS), changes in outer appearance, menstrual disturbances, and infertility result in psychological distress, reduced quality of life, and a less satisfying sex life. Although obesity and hirsutism are major determinants of the physical component of quality of life, its psychological aspect appears to be inherent and specific for PCOS. Confirmation of the diagnosis and provision of detailed information to affected women, together with the availability of interdisciplinary treatment, including metformin or thiazolidinediones, have been shown to have positive psychological effects.

Key Words: Polycystic ovary syndrome; quality of life; psychological disturbances; sexual satisfaction; hirsutism; obesity; hyperandrogenism.

1. INTRODUCTION

To define quality of life is a many-sided task. Historically, the term quality of life appeared for the first time in Greek philosophy. Aristotle explained that happiness results from virtuous action of the soul and leads to a good life *(1)*. With their definition of health as "a state of physical, mental and social well being, and not only the absence of illness or fraility" members of the World Health Organization (WHO) added the term quality of life to the basic needs of humankind *(2)*. In 1978, the WHO specified explicitly that all human beings, in addition to physiological care, have a right to psychosocial care and adequate quality of life. The idea of quality of life did change with time *(3)*. Originally, this term was rather comprehensive and was used in a social context. Finally, however, it was adapted for application in the area of health care.

Social, cultural, and political factors have lent an increasing importance of the quality of life in health care. In 1990, Spilker described the assessment of quality of life by means of three interrelated parameters: global assessment of well-being, comprehensive domains (e.g., physical, psychological, or social domains), and the individual components of each domain *(4)*. These components classified the multidimensional character of quality of life. In 1993, the WHO defined quality of life as "the concept of an individual of its status in life in relation to the culture and the system of values in which it lives, and in relation to its goals, standards and needs" *(5)*. The definition comprises six large areas, namely physical health, emotional status, level of independence, social relations, enviromental characteristics, and spiritual needs.

From: *Contemporary Endocrinology: Androgen Excess Disorders in Women:*
Polycystic Ovary Syndrome and Other Disorders, Second Edition
Edited by: R. Azziz et al. © Humana Press Inc., Totowa, NJ

In 1990, Ferrans presented an overview of the literature on the quality of life and stressed five categories, into which the definitions of quality of life can be sorted *(6)*: ability to lead a normal life, happiness, contentedness, achievement of personal goals, ability to lead a socially useful life, and physical and/or mental abilities. Ross *(7)* introduced spiritual well-being to the concept of quality of life in nursing. In the City of Hope model *(8)* of quality of life, it is recognized that quality of life is subjective, based on the self-description of patients, multidimensional, and steadily changing with time and place. In the model of Ferrans, contentedness and quality of life are connected *(6)*. This model, created in 1990, has a solid base of terms, differentiates clearly between the domains, and offers a good example of the connection between theory and research. It classifies the relationships between the total construct of quality of life into four large domains, i.e., health, socioeconomics, emotions, and family. In 1990 Zhan developed a model that defined the dimensions of quality of life as contentedness with life, self concept, health, and socioeconomic factors *(9)*. According to this model, quality of life is also influenced by health and social situation, culture, and environment and is also age dependent.

A universally accepted definition of quality of life has not been achieved. The majority of researchers, however, agree that quality of life is a multidimensional, subjective idea. The variation with respect to the definition of this term is, last but not least, reflected in the various models of quality-of-life research. Any single tool to assess quality of life may not be sufficient to describe all aspects important for an individual person, and different tools are neither right nor wrong. Thus, quality of life may be compared among different patient groups or healthy controls by means of standardized questionnaires or within patient groups on follow-up with tools tailored to their special needs.

2. BACKGROUND

Polycystic ovary syndrome (PCOS) is a common endocrine disorder characterized by chronic anovulation and hyperandrogenism. The prevalence among women of reproductive age is at least 6% *(10)*. Its pathophysiology, most likely a combination of genetic disposition and environmental factors, is not completely understood *(11,12)*. PCOS is one of the leading causes of infertility and is also characterized by hirsutism, cystic acne, seborrhea, hair loss, and obesity *(13,14)*. A significant proportion of PCOS patients has been found to suffer from defective insulin secretion and insulin resistance *(15)*. Accordingly, PCOS patients may be expected to have a higher morbidity and mortality from the sequelae of the metabolic syndrome *(16,17)*.

2.1. Evaluation of Quality of Life in PCOS

Because PCOS often manifests at an age when finding a partner, sexual activity, and marriage are important, its cosmetic and psychosexual implications are thought to cause profound emotional distress in affected women *(18)*. The three most bothersome symptoms commonly reported by affected women are excess hair growth, irregular or absent menstruation, and infertility: as a result, PCOS patients express feeling "different" from other women and less "feminine" *(19)*. Almost all women associate negative emotions of frustration, anxiety, and, to a lesser extent, sadness with PCOS, although the self-reported knowledge level for the disorder is high *(20)*.

Furthermore, limitations in general health perceptions, physical functioning, general behavior, and family activities have been found in adolescents with PCOS *(21)*. Two case report studies have reported an association of an increased psychiatric morbidity in PCOS patients, implying an etiology of monoamine imbalances *(22)* or elevations of luteinizing hormone and androgens *(23)*, because monoamine oxidase inhibitor antidepressants and clomiphene citrate, respectively, ameliorated the psychiatric illness. Accumulating and published evidence on the long-term health risks associated with PCOS may also have a negative impact on psychosocial well-being, even without proof of higher morbidity or mortality from prospective studies *(24,25)*.

In 1998, a self-administered quality-of-life questionnaire (PCOSQ) for PCOS women was published by Cronin and coworkers *(26)*. The PCOSQ included 26 questions (items) that addressed five

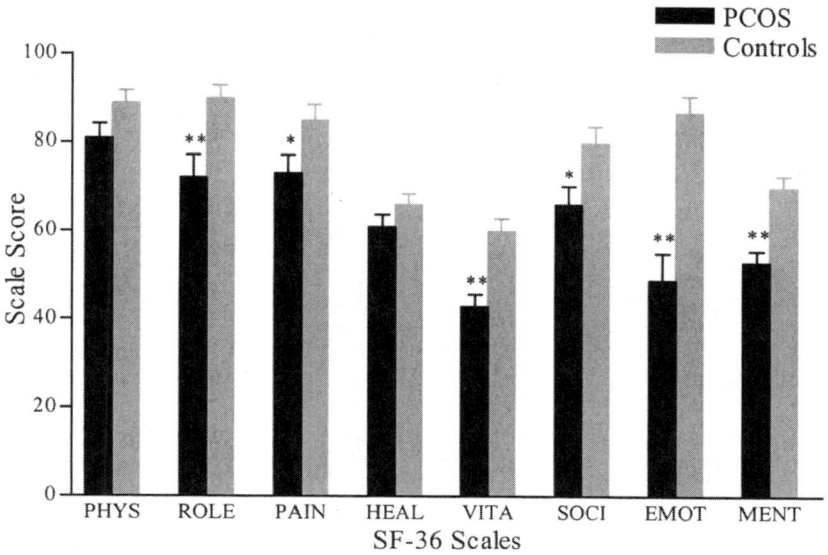

Fig. 1. Health-related quality of life measured with the German version of the SF-36 in women with PCOS and healthy controls. Patients reported significantly lower quality of life on the scales of physical role function (ROLE: **, $p < 0.001$), bodily pain (PAIN: *, $p < 0.05$), vitality (VITA: **, $p < 0.0001$), social function (SOCI: *, $p < 0.05$), emotional role function (EMOT: **, $p < 0.0001$), and mental health (MENT: **, $p < 0.0001$). No group differences were found for the scales of physical function (PHYS) and general health (HEAL). All data are presented as mean + SD. (From ref. *29*.)

areas of concern (domains), including emotions, body hair, body weight, fertility, and menstruation, rated on a 7-point scale in which lower scores denoted higher degrees of patient concern and a lower health-related quality of life (HRQL). The PCOSQ was validated and compared to the Short Form (SF)-36 in an outpatient clinic setting *(27)*. All five PCOSQ dimensions were found to be internally reliable and to correlate with similar scales of the SF-36. However, acne was identified as an important area of HRQL missing from the questionnaire. The PCOSQ was also used to evaluate HRQL in a multicenter, prospective, randomized, and placebo-controlled blinded troglitazone study of 393 patients presenting for the treatment of menstrual cycle irregularity and facial hair growth *(28)*. Cross-sectional correlations of the five domains of the PCOSQ were weak with all measures except for the hirsutism score and hair growth. Changes in the proportion of normal menstrual cycles correlated with change in the infertility domain and with the change in the menstruation domain. The PCOSQ proved to be as responsive as the hirsutism score and more responsive than the objective measures of hair growth, to salutary effects of treatment with troglitazone.

Elsenbruch et al. *(29)* investigated the impact of PCOS on HRQL, psychological well-being, and sexuality in 50 women with PCOS and 50 age-matched controls with three standardized validated questionnaires (SF-36) *(30,31)*, symptom checklist revised (SCL-90-R) *(32,33)*, and life satisfaction questionnaire (Fragebogen zur Lebenszufriedenheit, FLZ) *(34,35)* (Fig. 1). In addition, the impact of hirsutism, obesity, and infertility was assessed using 5-point rating scales, and sexual satisfaction was analyzed with visual analogue scales (VAS). Although PCOS patients and healthy controls did not differ with respect to sociodemographic variables, including education, family status, and employment, they showed significantly greater psychological disturbances, especially on the SCL-90-R dimensions: obsessive–compulsive, interpersonal sensitivity, depression, anxiety, aggression, and psychoticism. Women with PCOS further reported a significantly lower degree of life satisfaction in the FLZ scales: health, self, and sex compared with controls. HRQL measured with the SF-36 revealed significantly decreased scores for physical role function, bodily pain, vitality, social function, emotional role function, and mental health in PCOS women. For the physical function and general health scales no differences were found between the groups. Furthermore, although women with PCOS and control women were comparable with regard to partner status and frequency of sexual intercourse, they were significantly less satisfied with their sex life and found themselves less sexually attractive.

2.2. Modulators of Quality of Life in PCOS Patients

Several studies have proposed individual clinical and laboratory variables as potential contributors to psychological problems in women with PCOS: hyperandrogenism, hyperandrogenemia, menstrual cycle disturbances, infertility, obesity, and disturbed insulin regulation.

2.2.1. Hyperandrogenism

2.2.1.1. HIRSUTISM

For many PCOS patients, the most pronounced changes in physical appearance are related to excessive body hair. A psychometric evaluation by Sonino and coworkers *(36)* revealed more anxiety and psychotic symptoms on the Kellner's Symptom Rating Test and significantly higher social fears assessed with Marks' Social Situations Questionnaire in hirsute women compared to healthy nonhirsute control women, whereas no differences in depression, somatization, anger–hostility, and cognitive symptoms were found. Independent of PCOS, hirsutism also significantly impaired the quality of life of affected females in a study by Loo et al. *(37)*. In a German PCOS cohort, more severe hirsutism was associated with lower SF-36 scale scores for bodily pain, general health perception, and the physical sum scale, as well as with decreased sexual self-worth and sexual satisfaction *(38)*. On the other hand, there was no association between hirsutism and psychosocial distress, or with areas of quality of life representing emotional and social functioning *(38)*. These results are consistent with previous data documenting the lack of a relationship between the extent of hirsutism and the degree of psychological distress *(39–41)*. Together, these data support the idea that hirsutism does indeed negatively affect women with PCOS, but this appears to be limited to specific areas of quality of life, including the sexual domains, possibly linked with problems concerning female identity and sexual self-worth *(18,19,39,42)*.

2.2.1.2. ACNE

In patients with acne, substantially more pain and discomfort has been reported *(43)*. These data were recently confirmed (unpublished data), showing lower HRQL in the SF-36 pain subscale in PCOS women presenting with acne. Moreover, patients with acne have been reported to have a reduced quality of life *(44)*, especially a higher risk for depression and anxiety on the specific subscales of the Hospital Anxiety and Depression questionnaire, independent of the degree of severity *(45)*. A recent comparison of all psychosocial scales (SF-36, SCL, VAS) between patients with mild acne (*n* = 44) and those without acne (*n* = 76) revealed no significant differences in any scale *(38)*.

2.2.1.3. ALOPECIA

The negative psychological impact of hair loss in women has been well documented *(46)*. Women with androgenetic alopecia experience increased self-consciousness, feelings of unattractiveness, and emotional stress *(47,48)*. The lack of an association of quality of life with alopecia in available studies on PCOS is likely a result of the minor degree of alopecia and the low number of patients affected in the examined cohorts.

2.2.2. Hyperandrogenemia

In accordance with other study data *(28)*, Hahn et al. found no correlation of androgen levels with variables of quality of life in PCOS patients *(38)*. However, the occurrence of high testosterone levels was associated with higher distress scores in the SCL areas of somatization and obsessive–compulsive in a German cohort of PCOS patients, thus representing premenopausal women (unpublished data). In postmenopausal women, decreasing testosterone levels by cyproterone acetate administration significantly improved the SCL domains somatization, obsessive–compulsive, interpersonal sensitivity, and phobic anxiety *(49)*. Another study examined the relationship of mood dysfunction and androgens in PCOS subjects *(50)*, showing an association of free testosterone levels with negative mood dysfunction, especially depression. Several groups have focused on the impact of androgens on sexual behavior and satisfaction. In premenopausal women with low libido *(51)* or after oophorectomy *(52)*, transdermal testosterone therapy improved sexual function, including the amount of sexual fantasies as well as the frequency of sexual activity and masturbation. The addition of testosterone to estrogen replacement therapy in oophorectomized premenopausal women was shown to have a significantly better effect on the variables "enjoyment of sex," "satisfaction with frequency of sexual activity," and "interest in sex" *(53)*. Other studies produced divergent findings *(54)*, suggesting that the major impact of androgen levels in women is on sexual motivation and not on sexual activity *per se*. Thus, it appears that hyperandrogenemia has an impact on sexuality and quality of life in androgen-deficient women, but its impact on PCOS patients may be limited.

2.2.3. Menstrual Cycle Disturbances and Infertility

Although an early study on married women coping with infertility using the IPAT Anxiety Scale, the IPAT Depression Scale, the Tennessee Self-Concept Scale, and the Internal-External Scale found no emotional maladjustment *(55)*, recent studies on adolescent girls with PCOS using the Child Health Questionnaire–Child Self-Report Form (CHQ-CF-87) reported a 3.4-fold increased concern about their ability to become pregnant *(56)*. Girls with PCOS also had a 2.8-fold reduced likelihood to have sexual intercourse than healthy subjects, although the mean age at initiation of sexual intercourse among sexually active girls was not significantly different between the two groups, and the severity of PCOS and the worry about fertility were not associated with odds of being sexually active *(56)*. However, no consistent impact of infertility on quality of life has been reported *(57–59)*. The emotional consequences of infertility are modulated by several factors, including duration of infertility *(60–62)*, history of previous treatment failure *(62–64)*, age *(65)*, and cultural, ethnic, and societal background *(66)*. In PCOS patients, the high impact of menstrual problems on quality of life has been discussed *(67)*. These data are consistent with other findings comparing normally cycling women with oligo-/amenorrhoic women after metformin therapy, showing higher, and thus better SF-36 scales in the dimensions representing physical aspects of HRQL in women with normalized menstrual cycles *(82)*, whereas the type of menstrual cycle disturbances (amenorrhea or oligomenorrhea) had no impact on psychological well-being *(38)*.

2.2.4. Obesity

The potential importance of obesity in PCOS is supported by the finding that obesity can profoundly affect quality of life independent of the presence of other clinical symptoms in otherwise healthy subjects *(68)*. Interestingly, obesity is linked strongly to the physical dimension of quality of life, rather than with psychosocial status *(69)* and social adjustment *(70)*. A variety of studies demonstrated that body mass index (BMI) is a primary mediator in the relationship between PCOS and the reductions in HRQL *(67,71,72)*. Additionally, in obese patients the impact of weight reduction on HRQL has been well established *(70)*.

In a German PCOS cohort, higher BMI scores were associated with lower SF-36 scale scores, indicating decreased quality of life, for the scales of physical function, pain, general health perception, and physical sum *(29,38)*. Group comparisons of obese patients with PCOS (BMI ≥ 30) showed significantly decreased scores on the SF-36 scales of physical function, pain, general health perception, and physical sum compared with patients with PCOS and lower BMI. Higher BMI was further associated with decreased sexual satisfaction and self-worth. Interestingly, there were no consistent associations between BMI and either the social and emotional aspects of quality of life or psychological disturbances. This is of interest because the decrements in quality of life in women with PCOS were particularly pronounced in these areas. Furthermore, BMI had no impact on the VAS scales addressing sexual satisfaction: both obese and non-obese PCOS women reported being significantly less satisfied with their sex lives, found themselves significantly less sexually attractive, believed their partners to be significantly less satisfied with their sex lives, and reported that excessive body hair affected their sexuality *(29,38)*. BMI also had no significant effect on the frequency of sexual intercourse or sexual thoughts and fantasies *(29)*.

Taken together, these data are supportive of the notion that obesity is a major factor in decreased quality of life and impaired sexual self-worth in women with PCOS, which is consistent with the literature *(42,67,71,72)*. However, the effect appears to be limited to areas of quality of life concerning physical functions and general health. Obesity may not play a major role in emotional distress in PCOS, which is supported by two previous studies, showing that co-varying for BMI did not abolish significant differences between patients with PCOS and healthy controls *(29)* and that patients with PCOS matched for BMI with healthy controls had increased depression scores *(50)*. Epidemiological studies in the general population have also questioned a direct association between obesity and psychological distress, particularly with symptoms of depression, except for cases of severe obesity *(73–75)*.

2.2.5. Disturbed Insulin Regulation

As expected from the correlation of BMI with HRQL, indices of hyperinsulinemia and insulin resistance also correlate with variables of quality of life. However, two recent studies have shown that these correlations are abolished when BMI is used as a covariate *(28,38)*. Hence, current measures of hyperinsulinemia and insulin resistance may not be independent contributors to quality of life in PCOS.

2.3. Impact of Treatment on Quality of Life in PCOS

Based on data documenting the psychological and emotional consequences of changes in outer appearance, clinical interventions in women with PCOS that influence obesity, hirsutism, acne, menstrual disturbances, or infertility would be expected to improve overall HRQL. Traditional PCOS therapies based on oral contraceptives or clomiphene are judged unsatisfactory by many women *(20)*, who on average visit 4.5 physicians before the diagnosis of PCOS is made *(76)*. Independent of PCOS, any weight loss achieved by either dietary modification, physical activity, pharmacotherapy, surgery, or combinations thereof yields significant improvement of physical and social functioning in obese patients *(77–80)*. Laser treatment of hirsutism has been shown to improve quality of life in affected women *(37)*.

Confirmation of the diagnosis, provision of detailed information on PCOS, together with the availability of insulin sensitizing drugs such as metformin or glitazones *(13,81)* as a new therapeutic modality with well-established clinical benefits in specialized, interdisciplinary clinical settings would be expected to have positive psychological effects in women with PCOS. Guyatt et al. evaluated HRQL with the disease-specific PCOSQ in 393 patients with PCOS treated with troglitazone in a multicenter, randomized, placebo-controlled study *(28)*. Glitazone treatment was found to significantly improve HRQL in the domains of infertility, menstruation, and emotions (Fig. 2).

Recently, a prospective study investigating the effects of metformin treatment on HRQL, emotional well-being, and sexuality in PCOS has been published *(82)*. Compared to baseline, women with PCOS had significantly higher SF-36 scores reflecting improvement of all four psychosocial functions (vitality, social function, emotional role function, mental health) and the psychological sum scale after 6 months of metformin treatment. The physical dimensions of the SF-36 showed significant changes only in physical functioning. The increase in HRQL was associated with a significant reduction in body weight and an amelioration of menstrual irregularities. Analysis of the SCL results revealed significantly lower scores, indicating an improvement in all SCL scales after 6 months on metformin. In women with PCOS, hyperandrogenism was correlated with the SCL scales

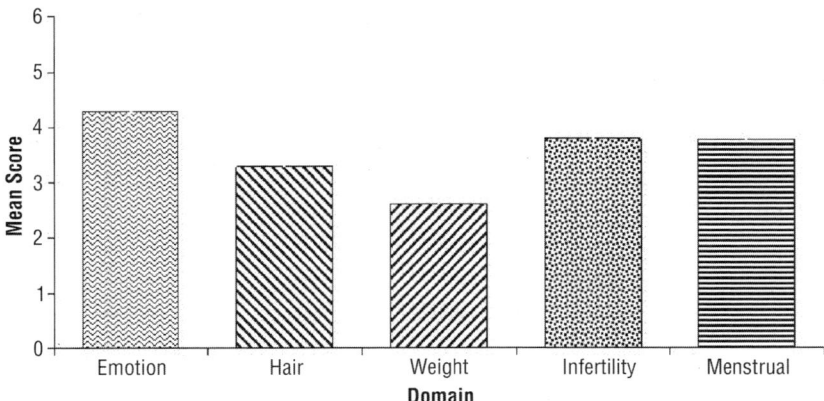

Fig. 2. Quality of life assessed using a self-administered questionnaire in 393 patients with polycystic ovary syndrome (the Polycystic Ovary Syndrome Questionnaire), determining degree of satisfaction with issues relating to emotions, hair growth, body weight, infertility, and menstrual disturbances. The PCOSQ includes 26 questions that assess patients' feelings by a 7-point scale in five areas (domains), including emotions, body hair, body weight, fertility, and menstruation. In the PCOSQ, lower scores denote higher degrees of patient concern and a lower quality of life. (Adapted from ref. *28*.)

of somatization and obsessive–compulsive. The decrease in androgens by treatment did not influence sexual fantasies. However, after 6 months women with PCOS were significantly more satisfied with their sex lives and had fewer difficulties in forming social contacts and an increased frequency of sexual intercourse *(82)*.

3. CONCLUSIONS

PCOS causes a major reduction in quality of life and psychological well-being and severely limits sexual satisfaction and self-worth in affected women. Changes in outer appearance, particularly obesity and excessive body hair—but not the presence of acne—were significantly associated with specific negative aspects of quality of life and sexual satisfaction. The role of biochemical, endocrine, and metabolic parameters, as well as menstrual irregularities and infertility, appear to be less important. Correlations between the degree of physical symptoms or endocrine disturbances and psychological distress or impairment of well-being are rather weak. It may not be possible to predict in any one individual the degree of psychological distress based on the presence or absence of individual symptoms or a specific constellation of symptoms alone. Therefore, clinicians should pay attention to the psychosocial dimension of PCOS on an individual basis, regardless of symptom severity or treatment response.

4. FUTURE AVENUES OF INVESTIGATION

Women with PCOS fail to conform to societal norms for outer appearance. Although the notion that patients with PCOS feel stigmatized appears reasonable, the concept of a loss of "feminine identity" in PCOS is as such difficult to assess using standardized measures. For this reason, psychosocial variables are not consistently integrated as outcome parameters in treatment studies. In addition to limitations in emotional well-being, quality of life, and life satisfaction, the diagnosis of PCOS clearly has a negative impact on sexual self-worth and sexual satisfaction, which deserves more attention in clinical practice and research. Although an effective medical treatment aimed at improving PCOS-related symptoms (especially hirsutism, obesity, menstrual irregularity, and infertility) will also reduce psychological distress and improve sexual self-worth, consideration of both the medical and psychological situation with the availability of additional psychological counseling and participation in a PCOS support group are likely to further improve life satisfaction and coping of affected women. Although the physical aspects of HRQL are clearly influenced by obesity and hirsutism, the determinants of psychological problems in androgen excess and women with PCOS remain elusive. Thus, more work is needed to better define the determinants of impaired quality of life in androgen excess and PCOS. Also, more attention needs to be given to the potential to improve quality of life of affected women by available treatment regimes.

KEY POINTS

- Apart from somatic impairments, PCOS is characterized by a substantial decrease in quality of life.
- Mood disturbances include symptoms of depression, decreased sexual satisfaction, and loss of feminine identity are prevalent.
- Obesity and hirsutism are the major modulators of quality of life in PCOS women.
- Treatment with insulin sensitizers and other therapies have been shown to improve quality of life in PCOS.

REFERENCES

1. McKeon R. Introduction to Aristotle. New York: The Modern Library, 1947.
2. WHO. Development and Constitution of the W.H.O. New York: United Nations, 1947.
3. Cooley ME. Quality-of-life in persons with non-small cell lung cancer: a concept analysis. Cancer Nurs 1998;21(3):151–161.
4. Spilker B. Quality-of-Life Assessment in Clinical Trials. New York: Raven Press; 1990.
5. WHO Division of Mental Health. WHO-QOL study protocol: The development of the World Health Organization quality-of-life assessment instrument. Geneva: United Nations, 1993.
6. Ferrans CE. Quality-of-life: conceptual issues. Semin Oncol Nurs 1990;6(4):248–254.
7. Ross LA. Spiritual aspects of nursing. J Adv Nurs 1994;19(3):439–447.
8. Ferrell B, Grant M, Padilla G, Vemuri S, Rhiner M. The experience of pain and perceptions of quality-of-life: validation of a conceptual model. Hosp J 1991;7(3):9–24.
9. Zhan L. Quality-of-life: conceptual and measurement issues. J Adv Nurs 1992;17(7):795–800.
10. Azziz R, Woods KS, Reyna R, Key TJ, Knochenhauer ES, Yildiz BO. The prevalence and features of the polycystic ovary syndrome in an unselected population. J Clin Endocrinol Metab 2004;89(6):2745–2749.
11. Dunaif A, Thomas A. Current concepts in the polycystic ovary syndrome. Annu Rev Med 2001;52:401–419.
12. Escobar-Morreale HF, Luque-Ramirez M, San Millan JL. The molecular-genetic basis of functional hyperandrogenism and the polycystic ovary syndrome. Endocr Rev 2005;26(2):251–282.
13. Ehrmann DA. Polycystic ovary syndrome. N Engl J Med 2005;352(12):1223–1236.
14. Hahn S, Tan S, Elsenbruch S, et al. Clinical and biochemical characterization of women with polycystic ovary syndrome in North Rhine-Westphalia. Horm Metab Res 2005;37(7):438–444.
15. Venkatesan AM, Dunaif A, Corbould A. Insulin resistance in polycystic ovary syndrome: progress and paradoxes. Recent Prog Horm Res 2001;56:295–308.
16. Talbott EO, Zborowski JV, Sutton-Tyrrell K, McHugh-Pemu KP, Guzick DS. Cardiovascular risk in women with polycystic ovary syndrome. Obstet Gynecol Clin North Am 2001;28(1):111–133, vii.
17. Christian RC, Dumesic DA, Behrenbeck T, Oberg AL, Sheedy PF, Fitzpatrick LA. Prevalence and predictors of coronary artery calcification in women with polycystic ovary syndrome. J Clin Endocrinol Metab 2003;88(6):2562–2568.
18. Eggers S, Kirchengast S. The polycystic ovary syndrome—a medical condition but also an important psychosocial problem. Coll Antropol 2001;25(2):673–685.
19. Kitzinger C, Willmott J. 'The thief of womanhood': women's experience of polycystic ovarian syndrome. Soc Sci Med 2002;54(3):349–361.
20. Sills ES, Perloe M, Tucker MJ, Kaplan CR, Genton MG, Schattman GL. Diagnostic and treatment characteristics of polycystic ovary syndrome: descriptive measurements of patient perception and awareness from 657 confidential self-reports. BMC Womens Health 2001;1(1):3.
21. Trent ME, Rich M, Austin SB, Gordon CM. Quality-of-life in adolescent girls with polycystic ovary syndrome. Arch Pediatr Adolesc Med 2002;156(6):556–560.
22. Bruce-Jones W, Zolese G, White P. Polycystic ovary syndrome and psychiatric morbidity. J Psychosom Obstet Gynaecol 1993;14(2):111–116.
23. Matsunaga H, Sarai M. Elevated serum LH and androgens in affective disorder related to the menstrual cycle: with reference to polycystic ovary syndrome. Jpn J Psychiatry Neurol 1993;47(4):825–842.
24. Wild S, Pierpoint T, McKeigue P, Jacobs H. Cardiovascular disease in women with polycystic ovary syndrome at long-term follow-up: a retrospective cohort study. Clin Endocrinol (Oxf) 2000;52(5):595–600.

25. Legro RS. Polycystic ovary syndrome and cardiovascular disease: a premature association? Endocr Rev 2003;24(3):302–312.

26. Cronin L, Guyatt G, Griffith L, et al. Development of a health-related quality-of-life questionnaire (PCOSQ) for women with polycystic ovary syndrome (PCOS). J Clin Endocrinol Metab 1998;83(6):1976–1987.

27. Jones GL, Benes K, Clark TL, et al. The Polycystic Ovary Syndrome Health-Related Quality-of-Life Questionnaire (PCOSQ): a validation. Hum Reprod 2004;19(2):371–377.

28. Guyatt G, Weaver B, Cronin L, Dooley JA, Azziz R. Health-related quality-of-life in women with polycystic ovary syndrome, a self-administered questionnaire, was validated. J Clin Epidemiol 2004;57(12):1279–1287.

29. Elsenbruch S, Hahn S, Kowalsky D, et al. Quality-of-life, psychosocial well-being, and sexual satisfaction in women with polycystic ovary syndrome. J Clin Endocrinol Metab 2003;88(12):5801–5807.

30. Ware JE, Jr., Sherbourne CD. The MOS 36-item short-form health survey (SF-36). I. Conceptual framework and item selection. Med Care 1992;30(6):473–483.

31. McHorney CA, Ware JE, Jr., Raczek AE. The MOS 36-Item Short-Form Health Survey (SF-36): II. Psychometric and clinical tests of validity in measuring physical and mental health constructs. Med Care 1993;31(3):247–263.

32. Schmitz N, Hartkamp N, Kiuse J, Franke GH, Reister G, Tress W. The Symptom Check-List-90-R (SCL-90-R): a German validation study. Qual Life Res 2000;9(2):185–193.

33. Derogatis LR, Rickels K, Rock AF. The SCL-90 and the MMPI: a step in the validation of a new self-report scale. Br J Psychiatry 1976;128:280–289.

34. Henrich G, Herschbach P. Questions on life satisfaction (FLZ)—a short questionnaire for assessing subjective quality-of-life. Eur J Psychol Assess 2000;16:150–159.

35. Goldbeck L, Schmitz TG. Comparison of three generic questionnaires measuring quality-of-life in adolescents and adults with cystic fibrosis: the 36-item short form health survey, the quality-of-life profile for chronic diseases, and the questions on life satisfaction. Qual Life Res 2001;10(1):23–36.

36. Sonino N, Fava GA, Mani E, Belluardo P, Boscaro M. Quality-of-life of hirsute women. Postgrad Med J 1993;69(809):186–189.

37. Loo WJ, Lanigan SW. Laser treatment improves quality-of-life of hirsute females. Clin Exp Dermatol 2002;27(6):439–441.

38. Hahn S, Janssen OE, Tan S, et al. Clinical and psychological correlates of quality-of-life in polycystic ovary syndrome. Eur J Endocrinol 2005;153(6):853–860.

39. Keegan A, Liao LM, Boyle M. 'Hirsutism': a psychological analysis. J Health Psychol 2003;8(3):327–345.

40. Barth JH, Catalan J, Cherry CA, Day A. Psychological morbidity in women referred for treatment of hirsutism. J Psychosom Res 1993;37(6):615–619.

41. Shulman LH, DeRogatis L, Spielvogel R, Miller JL, Rose LI. Serum androgens and depression in women with facial hirsutism. J Am Acad Dermatol 1992;27(2 Pt 1):178–181.

42. Coffey S, Mason H. The effect of polycystic ovary syndrome on health-related quality-of-life. Gynecol Endocrinol 2003;17(5):379–386.

43. Klassen AF, Newton JN, Mallon E. Measuring quality-of-life in people referred for specialist care of acne: comparing generic and disease-specific measures. J Am Acad Dermatol 2000;43(2 Pt 1):229–233.

44. Mallon E, Newton JN, Klassen A, Stewart-Brown SL, Ryan TJ, Finlay AY. The quality-of-life in acne: a comparison with general medical conditions using generic questionnaires. Br J Dermatol 1999;140(4):672–676.

45. Yazici K, Baz K, Yazici AE, et al. Disease-specific quality-of-life is associated with anxiety and depression in patients with acne. J Eur Acad Dermatol Venereol 2004;18(4):435–439.

46. Dolte KS, Girman CJ, Hartmaier S, Roberts J, Bergfeld W, Waldstreicher J. Development of a health-related quality-of-life questionnaire for women with androgenetic alopecia. Clin Exp Dermatol 2000;25(8):637–642.

47. Girman CJ, Hartmaier S, Roberts J, Bergfeld W, Waldstreicher J. Patient-perceived importance of negative effects of androgenetic alopecia in women. J Womens Health Gend Based Med 1999;8(8):1091–1095.

48. Van Der Donk J, Hunfeld JA, Passchier J, Knegt-Junk KJ, Nieboer C. Quality-of-life and maladjustment associated with hair loss in women with alopecia androgenetica. Soc Sci Med 1994;38(1):159–163.

49. Paoletti AM, Floris S, Mannias M, et al. Evidence that cyproterone acetate improves psychological symptoms and enhances the activity of the dopaminergic system in postmenopause. J Clin Endocrinol Metab 2001;86(2):608–612.

50. Weiner CL, Primeau M, Ehrmann DA. Androgens and mood dysfunction in women: comparison of women with polycystic ovarian syndrome to healthy controls. Psychosom Med 2004;66(3):356–362.

51. Goldstat R, Briganti E, Tran J, Wolfe R, Davis SR. Transdermal testosterone therapy improves well-being, mood, and sexual function in premenopausal women. Menopause 2003;10(5):390–338.

52. Shifren JL, Braunstein GD, Simon JA, et al. Transdermal testosterone treatment in women with impaired sexual function after oophorectomy. N Engl J Med 2000;343(10):682–688.

53. Floter A, Nathorst-Boos J, Carlstrom K, von Schoultz B. Addition of testosterone to estrogen replacement therapy in oophorectomized women: effects on sexuality and well-being. Climacteric 2002;5(4):357–365.

54. Sherwin BB, Gelfand MM, Brender W. Androgen enhances sexual motivation in females: a prospective, crossover study of sex steroid administration in the surgical menopause. Psychosom Med 1985;47(4):339–351.

55. Paulson JD, Haarmann BS, Salerno RL, Asmar P. An investigation of the relationship between emotional maladjustment and infertility. Fertil Steril 1988;49(2):258–262.
56. Trent ME, Rich M, Austin SB, Gordon CM. Fertility concerns and sexual behavior in adolescent girls with polycystic ovary syndrome. Implications for quality-of-life. J Pediatr Adolesc Gynecol 2003;16(1):33–37.
57. Monga M, Alexandrescu B, Katz SE, Stein M, Ganiats T. Impact of infertility on quality-of-life, marital adjustment, and sexual function. Urology 2004;63(1):126–130.
58. Wischmann T, Stammer H, Scherg H, Gerhard I, Verres R. Psychosocial characteristics of infertile couples: a study by the 'Heidelberg Fertility Consultation Service.' Hum Reprod 2001;16(8):1753–1761.
59. Ragni G, Mosconi P, Baldini MP, et al. Health-related quality-of-life and need for IVF in 1000 Italian infertile couples. Hum Reprod 2005;20(5):1286–1291.
60. Ardenti R, Campari C, Agazzi L, La Sala GB. Anxiety and perceptive functioning of infertile women during in-vitro fertilization: exploratory survey of an Italian sample. Hum Reprod 1999;14(12):3126–3132.
61. Chiba H, Mori E, Morioka Y, et al. Stress of female infertility: relations to length of treatment. Gynecol Obstet Invest 1997;43(3):171–177.
62. Guerra D, Llobera A, Veiga A, Barri PN. Psychiatric morbidity in couples attending a fertility service. Hum Reprod 1998;13(6):1733–1736.
63. Oddens BJ, den Tonkelaar I, Nieuwenhuyse H. Psychosocial experiences in women facing fertility problems—a comparative survey. Hum Reprod 1999;14(1):255–261.
64. Boivin J, Takefman JE, Tulandi T, Brender W. Reactions to infertility based on extent of treatment failure. Fertil Steril 1995;63(4):801–807.
65. Fekkes M, Buitendijk SE, Verrips GH, et al. Health-related quality-of-life in relation to gender and age in couples planning IVF treatment. Hum Reprod 2003;18(7):1536–1543.
66. Schmid J, Kirchengast S, Vytiska-Binstorfer E, Huber J. Infertility caused by PCOS—health-related quality-of-life among Austrian and Moslem immigrant women in Austria. Hum Reprod 2004;19(10):2251–2257.
67. McCook JG, Reame NE, Thatcher SS. Health-related quality-of-life issues in women with polycystic ovary syndrome. J Obstet Gynecol Neonatal Nurs 2005;34(1):12–20.
68. Stunkard AJ, Faith MS, Allison KC. Depression and obesity. Biol Psychiatry 2003;54(3):330–337.
69. Mannucci E, Ricca V, Barciulli E, et al. Quality-of-life and overweight: the obesity related well-being (Orwell 97) questionnaire. Addict Behav 1999;24(3):345–357.
70. Swallen KC, Reither EN, Haas SA, Meier AM. Overweight, obesity, and health-related quality-of-life among adolescents: the National Longitudinal Study of Adolescent Health. Pediatrics 2005;115(2):340–347.
71. Hashimoto DM, Schmid J, Martins FM, et al. The impact of the weight status on subjective symptomatology of the polycystic ovary syndrome: a cross-cultural comparison between Brazilian and Austrian women. Anthropol Anz 2003;61(3):297–310.
72. Trent M, Austin SB, Rich M, Gordon CM. Overweight status of adolescent girls with polycystic ovary syndrome: body mass index as mediator of quality-of-life. Ambul Pediatr 2005;5(2):107–111.
73. Faith MS, Matz PE, Jorge MA. Obesity-depression associations in the population. J Psychosom Res 2002;53(4):935–942.
74. Onyike CU, Crum RM, Lee HB, Lyketsos CG, Eaton WW. Is obesity associated with major depression? Results from the Third National Health and Nutrition Examination Survey. Am J Epidemiol 2003;158(12):1139–1147.
75. Derogatis LR, Rose LI, Shulman LH, Lazarus LA. Serum androgens and psychopathology in hirsute women. J Psychosom Obstet Gynaecol 1993;14(4):269–282.
76. http://scrambledthewebsite.com. 2001.
77. Dittmar M, Heintz A, Hardt J, Egle UT, Kahaly GJ. Metabolic and psychosocial effects of minimal invasive gastric banding for morbid obesity. Metabolism 2003;52(12):1551–1557.
78. Jalil RA, Manan WA, Bebakar WM, Halim R, Ooi GS, Othman R. Assessing changes in quality-of-life among obese participants in Kelantan, Malaysia. Asia Pac J Clin Nutr 2004;13(Suppl):S141.
79. Mathus-Vliegen EM, de Weerd S, de Wit LT. Health-related quality-of-life in patients with morbid obesity after gastric banding for surgically induced weight loss. Surgery 2004;135(5):489–497.
80. Ogden J, Clementi C, Aylwin S, Patel A. Exploring the impact of obesity surgery on patients' health status: a quantitative and qualitative study. Obes Surg 2005;15(2):266–272.
81. Lord JM, Flight IH, Norman RJ. Insulin-sensitising drugs (metformin, troglitazone, rosiglitazone, pioglitazone, D-chiro-inositol) for polycystic ovary syndrome. Cochrane Database Syst Rev 2003(3):CD003053.
82. Hahn S, Benson S, Elsenbruch S, et al. Metformin treatment of polycystic ovary syndrome improves health-related quality-of-life, emotional distress and sexuality. Hum Reprod 2006. [E-pub ahead of print].

32

Overview of Long-Term Morbidity and Economic Cost of the Polycystic Ovary Syndrome

Ricardo Azziz

SUMMARY

The long- and short-term morbidities associated with the polycystic ovary syndrome (PCOS) result in significant economic costs. Using figures obtained in the United States, it has been estimated that the total annual cost of evaluating and providing care to reproductive-aged women with PCOS in the United States is $4.37 billion. The initial diagnostic evaluation represented only 2.3% of costs; hormonally treating menstrual dysfunction/abnormal uterine bleeding represented 30.9%, infertility care 12.2%, type 2 diabetes mellitus treatment 40.4%, and treatment of hirsutism 14.2% of total costs. These data indicate that PCOS represents a significant economic burden to health care. Furthermore, because the cost of the diagnostic evaluation accounted for a relatively minor part of the total costs (~2%), more widespread and liberal screening for the disorder appears be a cost-effective strategy, leading to earlier diagnosis and intervention and possibly the amelioration and prevention of serious sequelae.

Key Words: Polycystic ovary; economic burden; health care costs; diabetes; hirsutism; menstrual irregularity; abnormal uterine bleeding; infertility; endometrial cancer.

1. INTRODUCTION

Clearly, the long- and short-term morbidities associated with the polycystic ovary syndrome (PCOS) result in significant economic costs. Calculations of health care-related economic burden are based on estimations of (1) prevalence of the disorder, (2) prevalence of the morbidities, and (3) cost of diagnosing and treating the disorder and its associated morbidities. In this chapter we briefly discuss the various morbidities and present an estimation of the associated health care-related economic burden.

2. BACKGROUND

2.1. Prevalence of PCOS

The prevalence of PCOS has been discussed in detail in Chapter 13. However, a brief overview here is worthwhile. The prevalence of PCOS has been determined in various populations, primarily white or Caucasian (1–5), and in two studies of African Americans in the same geographic area (1,5). These studies in different countries and ethnic populations suggest that the prevalence of PCOS among unselected reproductive-aged women of the general population ranges from 6.5% to 8.0%. We should also note that the use of more expansive criteria for PCOS, such as that presented by an expert conference in Rotterdam in 2003 (6,7), could result in a two- or even threefold increase in the reported prevalence of PCOS (2). For the purposes of calculating economic burden, we have used an overall prevalence of PCOS of 6.6% in reproductive-aged women, estimated using 1990 National

From: *Contemporary Endocrinology: Androgen Excess Disorders in Women:*
Polycystic Ovary Syndrome and Other Disorders, Second Edition
Edited by: R. Azziz et al. © Humana Press Inc., Totowa, NJ

Institutes of Health criteria. We have also restricted our calculations to the United States, because costs vary widely among different countries and health care-delivery systems. Considering that there are approximately 62 million women 15–44 years of age in the United States, we can conservatively estimate that there are at least 4 million affected patients in this country alone.

2.2. Prevalence of Principal Morbidities Associated with PCOS

A number of reproductive, metabolic, and dermatological morbidities are associated with PCOS, and have been reviewed in previous chapters. Here we will briefly summarize our current knowledge of these morbidities, including (1) menstrual dysfunction, abnormal uterine bleeding (AUB), endometrial carcinoma, infertility, and obstetrical complications, (2) type 2 diabetes mellitus (DM) and cardiovascular disease (CVD), and (3) hirsutism, acne, and alopecia.

2.2.1. Menstrual Dysfunction and Abnormal Uterine Bleeding

While ovulatory dysfunction affects a high proportion of patients with PCOS (80–100%, depending on criteria), we should note that not all patients demonstrate a clinically apparent abnormality in their cyclic vaginal bleeding pattern. Although many patients and clinicians refer to these bleeding episodes as "menses" or "periods," strictly speaking these terms properly refer to the cyclic vaginal bleeding that results from the estrogen and progestogen decline (withdrawal) at the end of the luteal phase of an ovulatory cycle in women who do not conceive. Up to 40% of women with oligo-ovulation secondary to PCOS can actually present with a history suggestive of "normal menstruation" (i.e., eumenorrhea) *(8,9)*. Consequently, the periodic vaginal bleeding that occurs in women suspected of PCOS cannot be taken as indicative of ovulation until proven.

In larger series of patients diagnosed with PCOS by varying criteria, 75–85% have clinically evident menstrual dysfunction or AUB *(8–12)*. In agreement, in a prospective study of PCOS patients diagnosed among 400 unselected women of the general population, approximately 60% of the patients had clinically evident menstrual dysfunction *(5)*. Overall, between 60 and 85% of patients with PCOS exhibit overt menstrual dysfunction, and we have used an average prevalence of 75% for calculating economic burden.

We should note that the prevalence of menstrual dysfunction in PCOS changes with age, decreasing as the patient approaches menopause. Studying 205 patients with PCOS in Finland, Elting and colleagues observed an inverse correlation between age and menstrual cycle length; the proportion of women with irregular menstrual cycles decreased from 60% in 30- to 35-year-old women to10% in patients who were 42–45 years of age *(13)*. Hence, although we may have underestimated the total costs of diagnosing and treating menstrual dysfunction and AUB in PCOS, it is possible that we may also have overestimated these costs by an unknown fraction in affected women 40 or older.

2.2.2. Endometrial Hyperplasia and Carcinoma

In addition to menstrual dysfunction and infertility, women with PCOS may be at increased risk for endometrial hyperplasia *(14)* and carcinoma *(15)*. However, the prevalence of these endometrial abnormalities in PCOS is still unclear because of the paucity of large-scale screening studies and the small number of patients actually identified with carcinoma (*see* Chapter 27).

It is assumed that the prevalence of endometrial carcinoma will drop to close to the background rate if patients are treated appropriately for menstrual and ovulatory dysfunction, such as with hormonal contraceptives or cyclic progestogens. Consequently, we have not considered the additional cost of treating endometrial abnormalities in PCOS in our calculation of economic burden, although we have included the costs of a screening transvaginal ultrasound and possibly an endometrial biopsy in estimating the costs of the initial evaluation of these patients.

2.2.3. Infertility

Depending on the diagnostic criteria, between 80 and 100% of patients with PCOS are considered to suffer from ovulatory dysfunction, affecting their fertility potential. However, we should note that not all patients with PCOS are necessarily infertile, and many women with PCOS will ovulate intermittently. In a study of more than 300 women with PCOS included in a prospective trial designed to study the effectiveness of troglitazone for the treatment of ovulatory dysfunction, among patients treated with a placebo, 32% ovulated over the 6 months of the study *(16)*. Whether or not patients with PCOS will suffer from infertility will depend on ovulatory rate, coital frequency, presence of concomitant infertility factors, overall expectations for fecundability and family size, and, ultimately, the desire for children. Overall, between 33 and 75% of PCOS patients report having infertility *(17–19)*, a variation largely explained by clinic referral or ascertainment bias. We have used an estimated prevalence of infertility of 50% for our calculations of economic burden of PCOS.

It is unclear what proportion of the overall infertility costs in the United States is accounted for by PCOS. In the United States in 1995 an estimated 6.2 million women between the ages of 15 and 44 were reported to have an impaired ability to have children, representing about 10% of the women in this age group *(20)*. Because we have estimated that 50% of women with PCOS will be infertile and PCOS represents about 7% of all women of reproductive age, we can then estimate that PCOS is present in one-third of all infertile women (i.e., [50% × 7%]/10%). Survey data of infertility support the high prevalence of PCOS as a cause of infertility. In western nations, including the United States and Australia, between 20 and 35% of all infertile women suffer from ovulatory disorders *(21–23)*. In turn, PCOS appears to be the most common cause of oligo-ovulatory infertility, accounting for approximately 70–90% of such patients *(24)*. Consequently, we can estimate from these data that 14–32% of infertile patients have PCOS, making this disorder the most important single cause of infertility in women.

2.2.4. Obstetrical Complications

Patients with PCOS may be at increased risk for obstetrical complications, including gestational diabetes mellitus (GDM) *(25–27)*, pre-eclampsia, pregnancy-induced hypertension (PIH) *(26–30)*, and, possibly, miscarriage *(31–33)*. However, we should also recognize that approximately 60% of patients with PCOS in the United States are obese *(10)*, and obesity itself is a risk factor for GDM and pre-eclampsia *(34)*. The increased risk of GDM and PIH in PCOS appears to be independent of body mass in some studies *(26,29)* but not others *(35)* and may be more closely related to the degree of insulin resistance (27). Obesity is also a risk factor for spontaneous abortion *(36,37)*, and none of the studies determining miscarriage rates in PCOS have used body mass-matched controls. Overall, the extent of the independent effect of PCOS on early pregnancy loss and obstetrical complications remains to be better determined, and we have not included the costs of diagnosing and caring for obstetrical complications in our calculation of economic burden (*see also* Chapter 30).

2.2.5. Type 2 Diabetes

Among US women with PCOS averaging 28–30 years of age, the prevalence of type 2 DM ranges from 4 to 10% *(38,39)*. Overall, women with PCOS are at a two- to sixfold higher risk of developing type 2 DM compared to age-matched average women, who have a prevalence of type 2 DM of 1.6–2.0% *(40,41)*. Consequently, for our calculations of economic burden we considered that PCOS patients have an overall prevalence of type 2 DM of 7.2% during their reproductive years, or about fourfold the prevalence in the general population of women of similar age (i.e., 1.8%).

We should note that the increased risk for type 2 DM observed in women with PCOS reflects, to a significant degree, the high prevalence of obesity observed in these patients. For example, 90% of diabetic PCOS patients diagnosed in the study by Legro and colleagues, who assessed 254 patients diagnosed either at the Pennsylvania State University College of Medicine or at Mount Sinai School of Medicine, had a body mass index (BMI) of 30 kg/M^2 or greater *(40)*. When the risk of type 2 DM in PCOS was adjusted for BMI, the increased prevalence of diabetes was no longer significantly different from controls, at least in one study assessing 319 women with PCOS and 1060 controls *(42)*.

2.2.6. Cardiovascular Disease

CVD, including coronary heart disease (CHD), stroke, and peripheral artery disease, is a leading cause of death, affecting 6% of the overall US population. Risk factors for CVD include hyperlipidemia, hypertension, diabetes, and obesity, all of which are present in PCOS, and several studies have found higher rates of CVD risk factors in women with PCOS when compared to a control group *(43,44)* (*see also* Chapter 29).

In 1999–2000, 28.7% of participants in the National Health and Nutrition Examination Survey (NHANES) reported having hypertension, with no significant difference between genders *(45)*. The prevalence of hypertension in PCOS ranges from 10 to 39%, depending on age, representing a 1.4- to 3.5-fold increase above controls *(42–44,46,47)*. Consequently, we can conservatively estimate that women with PCOS have a risk of hypertension that is about twofold that of the general population. However, we should note that the prevalence of hypertension in PCOS is significantly greater in obese patients *(46)*, and in one study, when the prevalence of hypertension was adjusted for BMI, the difference in risk between PCOS and age-matched controls was no longer significant *(42)*.

Alternatively, it is less clear whether the prevalence of dyslipidemia is grossly increased in PCOS. In a study of 398 women with PCOS screened for inclusion in a trial evaluating the effectiveness of troglitazone, Legro et al. *(48)* reported that the prevalence of abnormally high total cholesterol levels in this population was similar to, and the prevalence of low high-density lipoprotein (HDL) cholesterol levels was actually less than, that observed among women aged 20–39 in the NHANES IIII survey (i.e., 7.6% and 25.6%, respectively) *(49)*. In contrast, in another study including 195 women with PCOS and 62 controls, Legro and colleagues reported that the prevalence of borderline high total cholesterol (≥200 mg/dL) was higher among women with PCOS than controls (48% vs 22%, respectively), although the prevalence of abnormally low HDL cholesterol (<35 mg/dL) was similar in both groups (48 vs 45%, respectively) *(50)*. Overall, it remains unclear whether women with PCOS, at least of reproductive age, have a higher prevalence of dyslipidemia compared to matched controls or the general population.

Hypertension and dyslipidemia are both risk factors for CVD, either CHD or stroke. However, evidence suggesting that the incidence of premature CHD is higher in the population with PCOS is also conflicting. A higher prevalence of coronary artery and aortic calcification *(51)*, an increase in left ventricular mass index *(52)*, a decrease in diastolic filling *(52)*, a significantly lower flow-mediated dilation *(53)*, and a higher intima-media thickness (54) have been observed in PCOS. However, Legro *(55)* reviewed the current available evidence for an association between PCOS and CVD and noted that although existing data suggested that PCOS may adversely affect or accelerate the development of an adverse cardiovascular risk profile, and even of subclinical signs of atherosclerosis, it did not appear to lower the age of clinical presentation to a premenopausal age group.

A number of populational studies support the apparently modest role of PCOS in premature CHD *(42,46)*. It is possible that the lack of a significant association of PCOS with CVD- or CHD-related events or mortality in these studies may be a result of the short length of follow-up of the studies or the relatively young age of the patients with PCOS at the time of follow-up. Nevertheless, a study of 319 women with PCOS surveyed a mean of 31 years after diagnosis observed a significant increase in the incidence of cerebrovascular disease in PCOS compared with controls (3.1 vs 1.2%, respectively) *(42)*, although this association remains to be confirmed by other investigators.

Overall, although the long-term risk of hypertension and cerebrovascular disease appears to be increased in PCOS, these risks remain to be confirmed, and the extent to which the prevalence of dyslipidemia and CHD events are increased in the disorder is unclear. Most importantly, the incidence of CVD-associated events or mortality does not appear to be grossly increased in reproductive-aged PCOS patients compared to age-matched controls, although it is possible that these women may demonstrate an increased incidence of CVD as they age. Consequently, the present analysis does not include CVD and its associated morbidities in the calculation of economic burden.

2.2.7. Dermatological Abnormalities: Hirsutism, Acne, and Androgenic Alopecia

We should note that hirsutism is one of the more common signs of androgen excess, in particular in PCOS *(10)*, which has significant psychosocial and quality-of-life implications *(56,57)*. The prevalence and degree of hirsutism, however, is dependent on the ethnicity of the patients; apparently it is less prevalent in women with PCOS of East Asian extraction or Pacific Islanders *(58,59)* but more prevalent in women of Asian Indian origin (e.g., Bengali, Gujarati, or Dravidian Indian) *(60)*. Overall, approximately 70% of white and black women with PCOS will be hirsute, a figure we used to calculate economic burden.

Acne has been reported to affect 12–14% of white PCOS patients *(10,60)*, although the prevalence of this dermatological abnormality also varies with ethnicity. It is reportedly higher in Asian Indians *(60)* and lower in Pacific Islanders *(58)*. Androgenic alopecia is a recognized sign of PCOS *(61–63)*; however, in a study of 257 androgen excess patients undergoing treatment, only 12 (4.7%) complained of hair loss only *(10)*. Overall, acne and androgenic alopecia apparently have a low prevalence among patients with PCOS. Because studies quantifying and determining the prevalence of acne and androgenic alopecia in a significant number of unselected patients with PCOS are lacking, we did not include these disorders in our calculations of economic burden.

2.3. Estimating the Economic Burden of PCOS

We calculated the health care-related economic burden in PCOS based on the above prevalences of disease *(64)*. We restricted the calculation to the United States only, although we recognize that PCOS is an international disorder. However, we should note that our estimation is highly conservative because we did not include a number of costs for which we did not have accurate and present-day prevalence and monetary estimates (Table 1). We did include the costs of the initial evaluation, detailed in Table 2.

The overall costs associated with the treatment and, if appropriate, the diagnosis of the various morbidities evident in the premenopausal women with PCOS are outlined in Table 3. This conservative estimate of the health care-related economic burden of premenopausal women with PCOS exceeded $4 billion annually in the United States alone. Approximately 40% of the burden is a result of the increased prevalence of diabetes associated with PCOS; 30% arises from the treatment of the associated menstrual dysfunction/AUB, 14% from the treatment of hirsutism, and 12% the provision of infertility services. Notably, the costs of the diagnostic evaluation of all patients accounted for a relatively small portion of the calculated economic burden, about 2%. The calculated economic burden of patients with PCOS during their reproductive years is about threefold that of hepatitis C ($1 billion in 1998) *(65)* and about one-third that of morbid obesity ($11 billion in 2000) *(66)*.

According to factors not included in our model (*see* Table 1), it is highly likely that we are underestimating the economic burden of PCOS. Alternatively, a few factors in our analysis may have resulted in an overestimation of the economic burden. Most importantly, our calculations were based on the assumption that all patients with PCOS will seek and be willing to undergo the required evaluations and treatments. Second, the prevalence of menstrual dysfunction appears to decrease with in the later reproductive years *(13)*.

3. CONCLUSIONS

The costs of evaluating and caring for patients with PCOS during their reproductive years can be conservatively estimated to exceed $4 billion, although this estimate is likely an underestimation of the true cost. Importantly, the cost of the diagnostic evaluation accounts for a relatively minor part of

Table 1
Costs Not Included in the Present Calculation of Health Care-Related Economic Burden of PCOS

Not included were potential costs resulting from:
- Additional patients diagnosable as having PCOS with the use of more expansive criteria, compared to the NIH 1990 criteria used in this study (e.g., Rotterdam, 2003).
- Excess morbidity in the postmenopause (e.g., of type 2 DM, CVD, or neoplasm)
- Treatment of endometrial hyperplasia or carcinoma
- Diagnosis and treatment of early pregnancy loss
- Diagnosis and treatment of obstetrical complications (e.g., GDM, pre-eclampsia)
- Diagnosis and treatment of excess premenopausal morbidity resulting from CVD (including CVD, dyslipidemia, and hypertension)
- Treatment of acne and androgenic alopecia
- Repeat diagnostic procedures
- Use of hormonal medications other than OCPs and SPA that may be used for regulation of menstrual dysfunction and/or hirsutism (e.g., insulin sensitizers)
- Use of medications for pain relief and iron-replacement therapy (for treating menstrual pains and anemia, respectively)
- Use of laser hair removal
- Use of electrology beyond the first year of treatment
- Use of other cosmetic treatments (e.g., depilation, bleaching, shaving, cosmetics, or 13.9% eflornithine hydrochloride cream)
- Use of hygienic products (e.g., sanitary pads and tampons)
- Use of home pregnancy tests
- Diminished quality of life
- Indirect expenditures, such as that due to lost productivity
- Treatments occurring prior to the establishment of the diagnosis of PCOS

PCOS, polycystic ovary syndrome; DM, diabetes mellitus; CVD, cardiovascular disease; GDM,, gestational diabetes mellitus; OCPs, oral contraceptive pills; SPA, spironolactone.

Table 2
Cost of Initial Evaluation of Patients With PCOS in 2004 Dollars

Test	PCOS requiring test (%)	Cost/unit of service[a]	Total cost
Testosterone panel[b]	25	$ 71.65	$ 71.65
DHEAS	25	$ 31.07	$ 31.07
TSH	100	$ 23.47	$ 23.47
Prolactin	100	$ 27.08	$ 27.08
Basal 17-HP	100	$ 37.95	$113.85
ACTH test	6	$121.46	$121.46
TV-U/S	100	$126.28	$126.28
Endometrial biopsy	30	$123.87	$123.87
OGTT	100	$ 60.91	$ 60.91
Lipid profile	100	$ 37.90	$ 37.90
Total cost/initial evaluation		**$661.64**	**$737.54**

PCOS, polycystic ovary syndrome; DHEAS, dehydroepiandrosterone sulfate;TSH, thyroid-stimulating hormone; 17-HP, 17-hydroxyprogesterone; ACTH test, acute adrenocorticotropic hormone(1-24) stimulation test measuring 17-HP at baseline and 60 minutes; TV-U/S, a limited pelvic transvaginal ultrasonography; OGTT, 75-g oral glucose tolerance test measuring insulin and glucose at baseline and 2 hours.

[a]Based on the Centers for Medicare & Medicaid Services Clinical Lab and Physician Fee Schedules.

[b]Testosterone panel includes a total and free testosterone, generally including a sex hormone-binding globulin level. From ref. *64*.

Table 3
**Overall Health Care-Related Economic Burden of PCOS Patients
During Their Reproductive Years, 2004**

	Annual costs in millions of US dollars (% of total costs)
For the initial evaluation	99 (2.3)
For treatment of:	
Menstrual dysfunction/AUB	1,350 (30.9)
Infertility	533 (12.2)
Type 2 diabetes	1,766 (40.4)
Hirsutism	622 (14.2)
Total cost	**4,370 (100.0)**

PCOS, polycystic ovary syndrome; AUB, abnormal uterine bleeding.
From ref. *64*.

the costs, slightly more than 2%. This suggests that more widespread screening for the disorder is a potentially cost-effective strategy, leading to earlier diagnosis and intervention and possibly the amelioration and prevention of serious sequelae.

4. FUTURE AVENUES OF INVESTIGATION

Future investigations should concentrate on obtaining more accurate epidemiological data on PCOS as pertains to its prevalence and the prevalence of associated comorbidities. Particular emphasis should be placed on the study of different geographic and ethnic populations. Not only would this for more pertinent counseling and risk calculation, but population comparisons could also provide clues to the association of environmental and genetic factors to the development of the comorbidities. In the future, an updated calculation of economic burden should be made, especially including the excess morbidity of women with PCOS in the postmenopause.

KEY POINTS

- Of women with PCOS, 60–85% (average: 75%) have overt menstrual dysfunction.
- Of women with PCOS, 33–75% (average: 50%) report having infertility.
- Women with PCOS have a fourfold greater risk of type 2 DM than comparably aged women, with an estimated prevalence of 7.2% during the reproductive years.
- Although the long-term risk of hypertension and CVD appears to be increased in PCOS, these risks remain to be confirmed.
- The incidence of CVD-associated events or mortality does not appear to be grossly increased in reproductive-aged patients with PCOS compared to age-matched controls.
- Any increased risk of CVD among postmenopausal women with PCOS remains to be demonstrated.
- Approximately 70% of white and black women with PCOS will be hirsute, although the prevalence of this abnormality in other ethnic groups is less clear.
- The health care-related economic burden of premenopausal women with PCOS exceeded $4 billion annually in the United States alone.
- Approximately 40% of the economic burden is attributable to type 2 DM, 30% to menstrual dysfunction, 14% to hirsutism, and 12% to infertility.
- The costs of the diagnostic evaluation accounts for only approximately 2% of the total economic burden, suggesting that more liberal implementation of diagnostic screening is cost-effective.

ACKNOWLEDGMENTS

Supported in part by an endowment from the Helping Hand of Los Angeles and by grants RO1-HD29364 and K24-D01346 from the National Institutes of Health.

REFERENCES

1. Knochenhauer ES, Key TJ, Kahsar-Miller M, et al. Prevalence of the polycystic ovary syndrome in unselected black and white women of the southeastern United States: a prospective study. J Clin Endocrinol Metab 1998;83:3078–3082.
2. Michelmore KF, Balen AH, Dunger DB, Vessey MP. Polycystic ovaries and associated clinical and biochemical features in young women. Clin Endocrinol (Oxf) 1999;51:779–786.
3. Diamanti-Kandarakis E, Kouli CR, Bergiele AT, et al. A survey of the polycystic ovary syndrome in the Greek island of Lesbos: hormonal and metabolic profile. J Clin Endocrinol Metab 1999;84:4006–4011.
4. Asuncion M, Calvo RM, San Millan JL, et al. A prospective study of the prevalence of the polycystic ovary syndrome in unselected Caucasian women from Spain. J Clin Endocrinol Metab 2000;85:2434–2438.
5. Azziz R, Yildiz B, Woods KS, et al. The prevalence of polycystic ovary syndrome among unselected consecutive premenopausal women. J Clin Endocrinol Metab 2004;89:2745–2749.
6. Rotterdam ESHRE/ASRM-Sponsored PCOS Consensus Workshop Group. Revised 2003 consensus on diagnostic criteria and long-term health risks related to polycystic ovary syndrome. Fertil Steril 2004;81:19–25.
7. The Rotterdam ESHRE/ASRM-Sponsored PCOS Consensus Workshop Group. Revised 2003 consensus on diagnostic criteria and long-term health risks related to polycystic ovary syndrome (PCOS). Hum Reprod 2004;19:41–47.
8. Azziz R, Waggoner WT, Ochoa T, Knochenhauer ES, Boots LR. Idiopathic hirsutism: an uncommon cause of hirsutism in Alabama. Fertil Steril 1998;70:274–278.
9. Carmina E. Prevalence of idiopathic hirsutism. Eur J Endocrinol 1998;139:421–423.
10. Azziz R, Sanchez LA, Knochenhauer ES, et al. Androgen excess in women: experience with over 1000 consecutive patients. J Clin Endocrinol Metab 2004;89:453–462.
11. Goldzieher JW, Axelrod LR. Clinical and biochemical features of polycystic ovarian disease. Fertil Steril 1963;14:631–653.
12. Conway GS, Honour JW, Jacobs HS. Heterogeneity of the polycystic ovary syndrome: clinical, endocrine and ultrasound features in 556 patients. Clin Endocrinol (Oxf) 1989;30:459–470.
13. Elting MW, Korsen TJ, Rekers-Mombarg LT, Schoemaker J. Women with polycystic ovary syndrome gain regular menstrual cycles when ageing. Hum Reprod 2000;15:24–28.
14. Cheung AP. Ultrasound and menstrual history in predicting endometrial hyperplasia in polycystic ovary syndrome. Obstet Gynecol 2001;98:325–331.
15. Hardiman P, Pillay OC, Atiomo W. Polycystic ovary syndrome and endometrial carcinoma. Lancet 2003;361:1810–1812.
16. Azziz R, Ehrmann D, Legro RS, et al. Troglitazone improves ovulation and hirsutism in the polycystic ovary syndrome: a multicenter, double blind, placebo-controlled trial. J Clin Endocrinol Metab 2001;86:1626–1632.
17. Goldzieher JW. Polycystic ovarian disease. Fertil Steril 1981;35:371–394.
18. Balen AH, Conway GS, Kaltsas G, et al. Polycystic ovary syndrome: the spectrum of the disorder in 1741 patients. Hum Reprod 1995;10:2107–2111.
19. Azziz R, Woods KS, Reyna R, et al. The prevalence and features of the polycystic ovary syndrome in an unselected population. J Clin Endocrinol Metab 2004;89:2745–2749.
20. Chandra A, Stephen EH. Impaired fecundity in the United States: 1982-1995. Fam Plann Perspect 1998;30:34–42.
21. Hull MG, Glazener CM, Kelly NJ, et al. Population study of causes, treatment, and outcome of infertility. Br Med J (Clin Res Ed) 1985;291:1693–1697.
22. Haxton MJ, Black WP. The aetiology of infertility in 1162 investigated couples. Clin Exp Obstet Gynecol 1987;14:75–79.
23. Weiss TJ, Meffin EM, Jones RG, Jones WR. Trends in causes and treatment of infertility at Flinders Medical Centre, Adelaide, 1976–1989. Med J Aust 1992;156:308–311.
24. Hull MG. Epidemiology of infertility and polycystic ovarian disease: endocrinological and demographic studies. Gynecol Endocrinol 1987;1:235–245.
25. Radon PA, McMahon MJ, Meyer WR. Impaired glucose tolerance in pregnant women with polycystic ovary syndrome. Obstet Gynecol 1999;94:194–197.
26. Mikola M, Hiilesmaa V, Halttunen M, Suhonen L, Tiitinen A. Obstetric outcome in women with polycystic ovarian syndrome. Hum Reprod 2001;16:226–229.
27. Bjercke S, Dale PO, Tanbo T, et al. Impact of insulin resistance on pregnancy complications and outcome in women with polycystic ovary syndrome. Gynecol Obstet Invest 2002 54:94–98.
28. Diamant YZ, Rimon E, Evron S. High incidence of preeclamptic toxemia in patients with polycystic ovarian disease. Eur J Obstet Gynecol Reprod Biol 1982;14:199–204.
29. de Vries MJ, Dekker GA, Schoemaker J. Higher risk of preeclampsia in the polycystic ovary syndrome. A case control study. Eur J Obstet Gynecol Reprod Biol 1998;76:91–95.
30. Fridstrom M, Nisell H, Sjoblom P, Hillensjo T. Are women with polycystic ovary syndrome at an increased risk of pregnancy-induced hypertension and/or preeclampsia? Hypertens Pregnancy 1999;18:73–80.
31. Sagle M, Bishop K, Ridley N, et al. Recurrent early miscarriage and polycystic ovaries. BMJ 1988;297:1027–1028.
32. Homburg R, Armar NA, Eshel A, Adams J, Jacobs HS. Influence of serum luteinising hormone concentrations on ovulation, conception, and early pregnancy loss in polycystic ovary syndrome. BMJ 1988;297:1024–1026.

33. Jakubowicz DJ, Iuorno MJ, Jakubowicz S, Roberts KA, Nestler JE. Effects of metformin on early pregnancy loss in the polycystic ovary syndrome. J Clin Endocrinol Metab 2002;87:524–529.
34. O'Brien TE, Ray JG, Chan WS. Maternal body mass index and the risk of preeclampsia: a systematic overview. Epidemiology 2003;14:368–374.
35. Haakova L, Cibula D, Rezabek K, et al. Pregnancy outcome in women with PCOS and in controls matched by age and weight. Hum Reprod 2003;18:1438–1441.
36. Wang JX, Davies MJ, Norman RJ. Obesity increases the risk of spontaneous abortion during infertility treatment. Obes Res 2002;10:551–554.
37. Bellver J, Rossal LP, Bosch E, et al. Obesity and the risk of spontaneous abortion after oocyte donation. Fertil Steril 2003;79:1136–1140.
38. Harris MI, Hadden WC, Knowler WC, Bennett PH. Prevalence of diabetes and impaired glucose tolerance and plasma glucose levels in U.S. population aged 20–74 yr. Diabetes 1987;36:523–534.
39. Schiller JS, Coriaty Nelson Z, Hao C, Barnes P. Early release of selected estimates based on data from the January-March 2004 National Health Interview Survey. National Center for Health Statistics. http://www.cdc.gov/nchs/nhis.htm. June 2005.
40. Legro RS, Kunselman AR, Dodson WC, Dunaif A. Prevalence and predictors of risk for type 2 diabetes mellitus and impaired glucose tolerance in polycystic ovary syndrome: a prospective, controlled study in 254 affected women. J Clin Endocrinol Metab 1999;84:165–169.
41. Ehrmann DA, Barnes RB, Rosenfield RL, Cavaghan MK, Imperial J. Prevalence of impaired glucose tolerance and diabetes in women with polycystic ovary syndrome. Diabetes Care 1999;22:141–146.
42. Wild S, Pierpoint T, McKeigue P, Jacobs H. Cardiovascular disease in women with polycystic ovary syndrome at long-term follow–up: a retrospective cohort study. Clin Endocrinol (Oxf) 2000;52:595–600.
43. Talbott EO, Zborowski JV, Sutton-Tyrrell K, McHugh-Pemu KP, Guzick DS. Cardiovascular risk in women with polycystic ovary syndrome. Obstet Gynecol Clin North Am 2001;28:111–133, vii.
44. Dahlgren E, Janson PO, Johansson S, Lapidus L, Oden A. Polycystic ovary syndrome and risk for myocardial infarction. Evaluated from a risk factor model based on a prospective population study of women. Acta Obstet Gynecol Scand 1992;71:599–604.
45. Hajjar I, Kotchen TA. Trends in prevalence, awareness, treatment, and control of hypertension in the United States, 1988-2000. JAMA 2003;290:199–206.
46. Elting MW, Korsen TJ, Bezemer PD, Schoemaker J. Prevalence of diabetes mellitus, hypertension and cardiac complaints in a follow-up study of a Dutch PCOS population. Hum Reprod 2001;16:556–560.
47. Quinonez ZC, Silva RR, Torres Juarez JM. [Obesity, arterial hypertension, metabolic disorders, and polycystic ovary syndrome]. Ginecol Obstet Mex 2000;68:317–322.
48. Legro RS, Azziz R, Ehrmann D, et al. Minimal response of circulating lipids in women with polycystic ovary syndrome to improvement in insulin sensitivity with troglitazone. J Clin Endocrinol Metab 2003;88:5137–5144.
49. Brown CD, Higgins M, Donato KA, et al. Body mass index and the prevalence of hypertension and dyslipidemia. Obes Res 2000;8:605–619.
50. Legro RS, Kunselman AR, Dunaif A. Prevalence and predictors of dyslipidemia in women with polycystic ovary syndrome. Am J Med 2001;111:607–613.
51. Talbott EO, Zborowski JV, Rager JR, et al. Evidence for an association between metabolic cardiovascular syndrome and coronary and aortic calcification among women with polycystic ovary syndrome. J Clin Endocrinol Metab 2004;89:5454–5461.
52. Orio F Jr, Palomba S, Spinelli L, et al. The cardiovascular risk of young women with polycystic ovary syndrome: an observational, analytical, prospective case-control study. J Clin Endocrinol Metab 2004;89:3696–3701.
53. Tarkun I, Arslan BC, Canturk Z, et al. Endothelial dysfunction in young women with polycystic ovary syndrome: relationship with insulin resistance and low-grade chronic inflammation. J Clin Endocrinol Metab 2004 89:5592–5596.
54. Orio F Jr, Palomba S, Cascella T, et al. Early impairment of endothelial structure and function in young normal-weight women with polycystic ovary syndrome. J Clin Endocrinol Metab 2004;89:4588–4593.
55. Legro RS. Polycystic ovary syndrome and cardiovascular disease: a premature association? Endocr Rev 2003;24:302–312.
56. Guyatt G, Weaver B, Cronin L, Dooley JA, Azziz R. Health-related quality of life in women with polycystic ovary syndrome, a self-administered questionnaire, was validated. J Clin Epidemiol 2004;57:1279–1287.
57. Sonino N, Fava GA, Mani E, Belluardo P, Boscaro M. Quality of life of hirsute women. Postgrad Med J 1993;69:186–189.
58. Williamson K, Gunn AJ, Johnson N, Milsom SR. The impact of ethnicity on the presentation of polycystic ovarian syndrome. Aust N Z J Obstet Gynaecol 2001;41:202–206.
59. Carmina E, Koyama T, Chang L, Stanczyk FZ, Lobo RA. Does ethnicity influence the prevalence of adrenal hyperandrogenism and insulin resistance in polycystic ovary syndrome? Am J Obstet Gynecol 1992;167:1807–1812.
60. Wijeyaratne CN, Balen AH, Barth JH, Belchetz PE. Clinical manifestations and insulin resistance (IR) in polycystic ovary syndrome (PCOS) among South Asians and Caucasians: is there a difference? Clin Endocrinol (Oxf) 2002;57:343–350.

61. Futterweit W, Dunaif A, Yeh HC, Kingsley P. The prevalence of hyperandrogenism in 109 consecutive female patients with diffuse alopecia. J Am Acad Dermatol 1988;19:831–836.

62. O'Driscoll JB, Mamtora H, Higginson J, et al. A prospective study of the prevalence of clear-cut endocrine disorders and polycystic ovaries in 350 patients presenting with hirsutism or androgenic alopecia. Clin Endocrinol (Oxf) 1994;41:231–236.

63. Cela E, Robertson C, Rush K, et al. Prevalence of polycystic ovaries in women with androgenic alopecia. Eur J Endocrinol 2003;149:439–442.

64. Azziz R, Marin C, Hoq L, Badamgarav E, Song P. Economic burden of the polycystic ovary syndrome (PCOS) during the reproductive lifespan. J Clin Endocrinol Metab 2005;90:4650–4658.

65. Kim WR. The burden of hepatitis C in the United States. Hepatology 2002;36:S30–S34.

66. Arterburn DE, Maciejewski ML, Tsevat J. Impact of morbid obesity on medical expenditures in adults. Int J Obes Relat Metab Disord 2005;29:334–339.

V Evaluation and Treatment of Androgen Excess

33

Clinical and Hormonal Evaluation of Androgen Excess

Ricardo Azziz

SUMMARY

Androgen excess (hyperandrogenism) is one of the most common endocrine disorders of women. Clinical features of hyperandrogenism include hirsutism, ovulatory and menstrual dysfunction, and in some patients acne and androgenic alopecia. Biochemically, hyperandrogenism can be established by the presence of supranormal circulating levels of total or unbound testosterone (T), androstenedione, or the adrenal metabolite dehydroepiandrosterone sulfate (DHEAS). Specific disorders that result in androgen excess include the 21-hydroxylase-deficient nonclassic adrenal hyperplasia (NCAH), ovarian androgen-secreting neoplasms (ASNs), adrenal ASNs with or without Cushing's syndrome, adrenocorticotropin hormone (ACTH)-dependent Cushing's syndrome, the hyperandrogenism, insulin resistance, and acanthosis nigricans (HAIR-AN) syndrome, and the use or abuse of androgenic drugs. However, the bulk of androgen excess is secondary to the polycystic ovary syndrome (PCOS). Although not properly an androgen excess disorder, idiopathic hirsutism (IH) should be considered in the differential diagnosis of these women. Conceptually, the general approach to patients with androgen excess entails five steps: (1) screening for androgen excess; (2) confirmation of clinical and/or biochemical androgen excess; (3) exclusion of specific androgen excess or related disorders (e.g., 21-hydroxylase-deficient NCAH, ASNs, thyroid dysfunction, or hyperprolactinemia); (4) classification of those patients without a specific androgen excess disorder (i.e., those with "functional" hyperandrogenism) as having PCOS, IH, or occasionally unspecified functional androgen excess; and (5) identification of related morbidities, such as metabolic abnormalities in patients with PCOS or HAIR-AN syndrome, and endometrial abnormalities in patients with long-term ovulatory dysfunction.

Key Words: Polycystic ovary; adrenal hyperplasia; androgen-secreting neoplasms; hirsutism; hormonal testing; menstrual irregularity; clinical evaluation.

1. INTRODUCTION

Androgen excess (hyperandrogenism) is one of the most common endocrine disorders of women. Clinical features of hyperandrogenism include hirsutism, ovulatory and menstrual dysfunction, and in some patients acne and androgenic alopecia. Biochemically, hyperandrogenism can be established by the presence of supranormal circulating levels of total or unbound testosterone (T), androstenedione, or the adrenal androgen metabolite dehydroepiandrosterone sulfate (DHEAS). Specific disorders that result in androgen excess include the 21-hydroxylase-deficient nonclassic adrenal hyperplasia (NCAH), ovarian androgen-secreting neoplasms (ASNs), adrenal ASNs with or without Cushing's syndrome, adrenocorticotropic hormone (ACTH)-dependent Cushing's syndrome, the hyperandrogenic-insulin-resistant acanthosis nigricans (HAIR-AN) syndrome, and the use or abuse of androgenic drugs. However, the bulk of androgen excess is secondary to the polycystic ovary syndrome (PCOS). Although not properly an androgen excess disorder, idiopathic hirsutism (IH) should be considered in the differential diagnosis of these women.

From: *Contemporary Endocrinology: Androgen Excess Disorders in Women:*
Polycystic Ovary Syndrome and Other Disorders, Second Edition
Edited by: R. Azziz et al. © Humana Press Inc., Totowa, NJ

Table 1
Prevalence of Hyperandrogenism in Prospective Studies
of Reproductive-Aged Women in the General Population

Study (ref.)	Total no. (%)	Hirsute (incl. PCOS) (%)	Nonhirsute PCOS[a] (%)	Total hyperandrogenic (%)
Knochenhauer et al., 1998 *(2)*	369	7.6[b]	2.1	9.7
Diamanti-Kandarakis et al., 1999 *(3)*	192	29.2[b]	1.0	30.2
Michelmore et al., 1999 *(4)*	224	10.7[c]	—	—
Asuncion et al., 2000 *(5)*	154	5.2[c]	6.2	11.4
Azziz et al., 2004 *(6)*	400	6.8[b]	3.6	10.4

PCOS, polycystic ovary syndrome.
[a]Defined by the NIH 1990 criteria *(7)*.
[b]Hirsutism defined as modified Ferriman–Gallwey (mFG) score ≥ 6.
[c]Hirsutism defined as mFG score ≥ 8.

We should note that the cost of fully evaluating a patient with suspected androgen excess is minimal. For example, the cost of evaluating potentially hyperandrogenic women for PCOS was estimated to represent approximately 2% of the total economic burden of this disorder *(1)* (*see* Chapter 32). These disorders have been reviewed in detail in preceding chapters. Here we will discuss the prevalence and the clinical and hormonal evaluation of androgen excess.

2. BACKGROUND

2. 1. Prevalence of Androgen Excess

In prospective studies of unselected women, hyperandrogenism, including hirsutism and nonhirsute PCOS, was generally observed to affect 10–11% of study subjects, although in one study it was as high as 30% (Table 1). The high prevalence of hirsutism in this study, carried out on the Greek island of Lesbos *(3)*, can be in part explained by the recruitment method whereby women were invited to participate in a "free medical examination," potentially biasing the population toward those women who might be complaining of unwanted hair growth. Overall, about 1 in 10 women of reproductive age appear to demonstrate some form of androgen excess.

2.2. General Approach to the Patient With Suspected Androgen Excess

Conceptually, the general approach to patients with androgen excess entails five steps (Table 2). First, those individuals who may potentially suffer from androgen excess must be identified for further evaluation, primarily through the patient complaint or by eliciting suggestive historical or clinical features. Second, the presence of either clinical or biochemical androgen excess must be confirmed. Third, specific androgen excess or related disorders are excluded (e.g., 21-hydroxylase deficient NCAH, ASNs, thyroid dysfunction, or hyperprolactinemia). Fourth, patients without a specific androgen excess disorder (i.e., those with "functional" hyperandrogenism) are then classified (differentiated) as having PCOS, IH, or occasionally some other form of functional androgen excess. Finally, the presence of metabolic abnormalities in patients with PCOS or HAIR-AN syndrome and endometrial abnormalities in patients with long-term ovulatory dysfunction should be determined.

Clearly, these five steps in the evaluation process may occur simultaneously, as in the patient who is identified with hirsutism at her initial evaluation and whose history and physical exam excludes Cushing's syndrome and an ASN. Likewise, in the process of confirming the presence of androgen excess and excluding related disorders, the diagnosis and differentiation of PCOS and IH become evident.

Table 2
Conceptual Approach to the Patient With Androgen Excess: Five Steps

1. Screening (clinical suspicion)
   Patients who may have androgen excess are identified for further evaluation.[a]
2. Confirmation of androgen excess
   The presence of clinical and/or biochemical androgen excess is confirmed.
3. Exclusion of specific disorders
   Specific androgen excess and related disorders are excluded.[b]
4. Differentiation of functional disorders
   Patients in whom a specific hyperandrogenic disorder is not identified are classified (differentiated) into the various "functional" disorders of androgen excess.[c]
5. Identification of related morbidities
   The presence of metabolic abnormalities in patients with PCOS or HAIR-AN syndrome, an endome trial abnormality in patients with long-term ovulatory dysfunction, should be determined.

PCOS, polycystic ovary syndrome; HAIR-AN, hyperandrogenism, insulin resistance, and acanthosis nigricans.

[a]By the patient complaint, including unwanted excess facial or body hair growth, acne, hair loss, irregular, abnormal or absent periods, inability to conceive, or unexpected weight gain; or by the elucidation of historical or clinical features including hirsutism, seborrhea and/or acne, alopecia, oligo-ovulatory infertility, polycystic ovaries, or, rarely, masculinization or virilization.

[b]Including 21-hydroxylase-deficient nonclassic adrenal hyperplasia, the HAIR-AN syndrome, androgen-secreting neoplasms, Cushing's syndrome, thyroid dysfunction, and hyperprolactinemia, among others.

[c]Patients with functional androgen excess are differentiated into PCOS, idiopathic hirsutism, and, occasionally, nonspecific functional hyperandrogenism.

2.3. Screening and Evaluation of Androgen Excess

Patients may be suspected of having androgen excess when they present with varying complaints, including unwanted excess facial or body hair growth, seborrhea and/or acne, scalp hair loss, irregular, abnormal, or absent menstrual periods, inability to conceive, and unexpected weight gain. Likewise, clinical or historical features on examination that may also suggest androgen excess include evidence of hirsutism, acne, androgenic alopecia or male pattern balding, ovulatory or menstrual dysfunction, oligo-ovulatory infertility, polycystic ovaries, or, rarely, masculinization and virilization. A detailed medical history and physical exam are the cornerstone of patient screening and diagnosis.

2.3.1. Medical History and Physical Examination in Patients With Possible Androgen Excess

We outline here the basics of the medical history and physical exam in the patient suspected of suffering from androgen excess.

2.3.1.1. MEDICAL HISTORY

The timing and pace of pubertal development and its relation to complaints of unwanted hair growth, hair loss, acne, and/or obesity should be established. The onset and progression of these complaints should also be established. Drug or medication use and exposure or use of skin irritants should be elicited. A detailed menstrual history should be obtained, with an emphasis on determining whether evidence of ovulatory function (e.g., premenstrual molimina) is present. Change in skin pigmentation or texture, extremity or head size, and changes in facial contour should be noted. A detailed family history of endocrine, reproductive, or metabolic disorders should be obtained. A family history of similar hyperandrogenic signs and symptoms is a powerful clue to the inherited basis of the disorder, although a familial association can be noted for PCOS, HAIR-AN syndrome, NCAH, and IH patients. Clinicians should note that the etiology of hirsutism can often be suspected from the history alone. For example, rapid progression of androgenization, particularly appearing some time after puberty or in the menopause, is most suggestive of an ASN and not a functional cause of androgen excess.

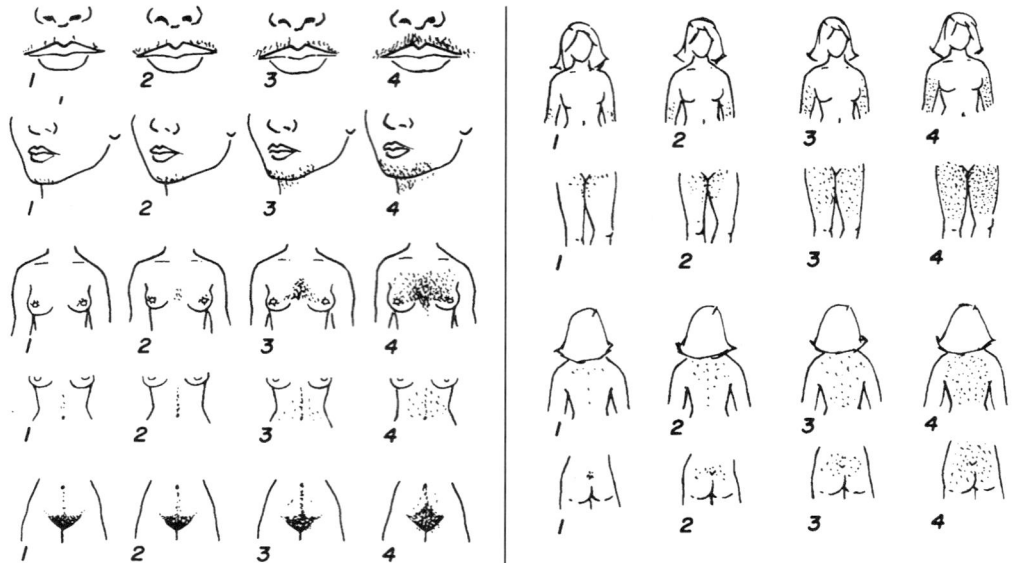

Fig. 1. Modified Ferriman–Gallwey scoring method. The total score is obtained by adding up the individual scores in each of nine body areas. If no excess terminal hairs are noted, the area is scored as zero. (Modified from ref. *8*.)

2.3.1.2. Physical Exam

During the physical exam it will be important to determine whether hirsutism or other hyperandrogenic features are actually present and whether there are signs or symptoms of related disorders. The type and pattern of excessive hair growth and/or acne should be noted and scored (Fig. 1). Hirsutism is defined as the presence of terminal hairs in a male-like pattern in women. Clinically, terminal hairs are generally pigmented, coarse (secondary to being medullated), and greater than 1 cm in length (if allowed to grow out). However, excessive terminal hair growth affecting only the lower legs or forearms does not constitute hirsutism, because a mixture of terminal and vellus hairs normally covers these areas. Alternatively, the excessive growth of vellus hairs, producing a "fuzzy" appearance, is termed vellus hypertrichosis and should not be considered hirsutism. Although a number of medical problems or medications can lead to vellus hypertrichosis, it is more commonly an ethnic variant (e.g., in individuals of Scandinavian or Mediterranean extraction). The presence of alopecia, acne, acanthosis nigricans, lipodystrophy, clitoromegaly, virilization and/or masculinization, pelvic and/or abdominal masses, abdominal and generalized obesity, cushingoid features, coarseness or bluntness of facial features, galactorrhea, thyroid enlargement, or signs of systemic illness should also be sought.

If the degree of androgenization is severe, resulting in clitoromegaly, severe male-pattern balding, and masculinization of the body (decrease in breast size, increase in muscle mass, loss of hip-to-waist discordance, etc.), an androgenic tumor or syndrome of severe insulin resistance (i.e., HAIR-AN syndrome) should be suspected. The presence of cushingoid features (i.e., centripetal obesity, muscle wasting of the extremities, moon facies, generalized facial rubor, or purple–red abdominal striae) in association with hirsutism is particularly worrisome, because it suggests the possibility of either an adrenocortical carcinoma or ACTH excess resulting from an ectopic or pituitary ACTH-producing tumor. The presence of acanthosis nigricans signals significant hyperinsulinism, observable in very insulin-resistant women with PCOS or in patients with the HAIR-AN syndrome.

Table 3
Pitfalls in the Biochemical Assessment of Androgen Excess

- Circulating androgen levels measure only one point in time; urinary measures are relatively nonspecific.
- Normal ranges may not be well established, particularly in adolescents and women in the later reproductive years.
- Wide variability in androgens levels in the normal population.
- The quality of most assays for total testosterone (T) and free T is poor and must be measured by relatively costly and time-consuming methods for improved accuracy.
- Multiple androgens are present; the greater the number of androgens measured, the higher the fraction of patients identified as androgen excess
- Androgens are suppressed rapidly by hormonal suppression and may remain suppressed even after discontinuation of hormonal treatment

2.3.2. Laboratory Analysis

The laboratory analysis has multiple objectives, including the confirmation of androgen excess (i.e., by the presence of hyperandrogenemia), which we briefly discuss here. We should note that laboratory testing will also be used for the exclusion of related or specific androgen excess disorders and the identification of associated morbidities. Androgen excess can be determined by measuring the circulating androgen levels, with supranormal levels (i.e., hyperandrogenemia) constituting evidence of androgen excess. However, a number of pitfalls should be considered when measuring androgens (Table 3). (The nuances and limitations of the currently used androgen assays are discussed in detail in Chapter 5.) Consequently, it is preferable to rely on the measurement of androgens primarily in those patients without clinical evidence of hyperandrogenism, i.e., without clinically apparent hirsutism. These may include adolescents and women of certain Asian ethnic origins.

Basal androgen levels have been suggested to be a good predictor of an ASN (9), with up to 80% of patients with an ovarian ASN demonstrating total T levels greater than 200 ng/dL. However, in patients with an ASN, total T (and DHEAS) levels often overlap those of hyperandrogenic women without a tumor (10,11) and consequently are of limited predictive value for these tumors. Overall, the clinical presentation is the single most important predictor of an ASN.

The choice of androgen measured clearly depends on the assays available. However, assuming that high-quality specialist assays are available, we should note that the more androgens, or androgen metabolites, measured, the higher the fraction of patients diagnosable as hyperandrogenic (12). In a study of 858 untreated consecutive women with androgen excess, although free T was elevated in 55.5%, alone or in combination with other androgens, 17% of women studied demonstrated only an elevated DHEAS level (Table 4). At a minimum, total and free T and DHEAS levels should be obtained.

DHEAS assays are generally robust, in part as a result of the very high levels of this metabolite in the circulation. This is not the case for the T assays, primarily because of the low concentration of these steroids in the circulation, their binding by sex hormone-binding globulin (SHBG) and albumin, and their relatively small and symmetrical molecular structure. Total T should be measured by a high-grade radioimmunoassay after extraction and chromatography or by liquid or gas chromatography and mass spectrometry. Free T should be determined by either equilibrium dialysis or ammonium sulfate precipitation or by calculation from SHBG measured by a high-quality immunoradiometric assay or by competitive binding. Direct assays for total and free T should not be used in women (13–23).

2.3.3. Confirmation of Ovulatory Function in Eumenorrheic Women

Approximately 75% women with androgen excess have overt oligo-menorrhea or menstrual dysfunction, signaling the presence of ovulatory abnormalities. Alternatively, a history of "regular

Table 4
Prevalence of Abnormally Elevated Androgen Measures in 858[a] Patients With Androgen Excess

Abnormal androgen(s)[b]	Number	Prevalence (%)
None	190	22.2
Total T only	20	2.4
Free T only	125	14.6
DHEAS only	146	17.0
Total T and free T	174	20.3
Total T and DHEAS	24	2.8
Free T and DHEAS	69	8.0
Total T, free T, and DHEAS	109	12.7
Total	**858**	**100.0**

T, testosterone; DHEAS, dehydroepiandrosterone sulfate.

[a]Note that 15 subjects did not have all three androgen results (total and free T and DHEAS) available, and thus were not included in this analysis.

[b]Abnormal androgen levels were defined as an androgen value above the 95th percentile of 98 healthy control women, i.e., a total T $\geq$ 2.94 nmol/L (88 ng/dL), free T $\geq$ 0.026 nmol/L (0.66 ng/dL), or DHEAS $\geq$ 6.64 μmol/L (2750 ng/mL), as previously reported (2).

From ref. 6.

menses" in women with other hyperandrogenic features is not always indicative of normal ovulatory function. Overall, about 40% of women with hyperandrogenic features (principally hirsutism) claiming to have regular menstrual cycles will be oligo-ovulatory on closer evaluation (24,25). Consequently, confirmation of ovulatory function should be performed in all eumenorrheic women suspected of having hyperandrogenism, most readily by measuring a progesterone (P4) level on days 22–24 of the cycle (day 1 being the first day of vaginal bleeding), preferably in two consecutive cycles. A P4 level of more than 3 or 4 ng/mL is generally indicative of ovulation.

2.3.4. Ovarian Sonography

The nuances of ovarian morphology have been reviewed in Chapter 16. It is important here to note that because polycystic ovarian morphology has become an integral part of the diagnosis of PCOS, patients suspected of having androgen excess should undergo transvaginal sonography when possible. Importantly, the definition of "polycystic-appearing" ovaries is relatively uniform and specific. It should also be understood that although virginal women and others who do not tolerate a transvaginal ultrasound probe can undergo transabdominal ultrasonography for ovarian examination, this latter method is considerably less sensitive than the transvaginal approach (26).

2.4. Predictive Value of the Clinical Features Observed in Patients With Possible Androgen Excess

Clinical experience has indicated that the majority of women with signs of virilization (i.e., masculinization of body muscular, severe or extreme male-pattern balding or hirsutism, clitoromegaly, etc.) will have androgen excess. Although rarely a sign of PCOS, virilization can be seen in patients with disorders of severe insulin resistance, androgen-secreting tumors, and androgenic substance abuse. Less clear is the predictive ability of less dramatic signs and symptoms. Although reviewed previously (see Chapter 14), it is still noteworthy to briefly summarize the predictive ability of the various historical and clinical features of androgen excess.

2.4.1. Hirsutism and Unwanted Hair Growth

Complaints of unwanted hair growth or clinical evidence of hirsutism are important predictors of an androgen excess disorder. In studies of large populations of hirsute women seeking care, 50–75% demonstrate PCOS *(12,25,27,28)*. It is also important to note that the sole complaint of "unwanted hair growth" in the absence of frank hirsutism on physical examination may also signal the presence of PCOS. In one study, approximately 50% of 288 women complaining of unwanted excess facial or body hair growth with minimal hair growth on examination (i.e., a modified Ferriman–Gallwey [mFG] score of 5 or less) demonstrated PCOS on further evaluation *(29)*.

2.4.2. Seborrhea and/or Acne

The prevalence of androgen excess among acneic-only patients (excluding patients with hirsutism) is less than among hirsute women. In small studies, between 20 and 40% of patients with treatment-resistant acne and without menstrual disturbance, alopecia, or hirsutism are reported to have androgen excess, principally PCOS *(30–33)*. Alternatively, data regarding the predictive ability of seborrhea for androgen excess is lacking. Large populational studies of acneic or hyperseborrheic patients, particularly those without other evidence of hyperandrogenism (e.g., hirsutism), are then still needed to better define this prevalence.

2.4.3. Alopecia

The prevalence of androgen excess among women with alopecia as the sole presenting hyperandrogenic symptom is unclear. However, in a study of 110 patients with alopecia and no other clinical signs of hyperandrogenism, Vexiau and colleagues observed that only 10% had PCOS *(30)*. Overall, it would appear that the proportion of women with alopecia only who have PCOS is considerably less than that of women with hirsutism, with or without alopecia.

2.4.4. Menstrual and Ovulatory Dysfunction

The prevalence of PCOS among women with clinically evident menstrual dysfunction can be estimated from four studies evaluating the prevalence of PCOS in the general population *(2,3,5,6)*. These studies noted that the overall prevalence of menstrual dysfunction approximated 20% of the women studied, very similar to the rate of 22.9% reported by 101,073 women participating in the Nurses Health Study II *(34)*. Of the women complaining of menstrual dysfunction, between one-fourth and one-third had androgen excess, notably PCOS (using the National Institutes of Health [NIH] 1990 criteria).

2.4.5. Polycystic Ovaries

Overall, polycystic ovaries are observed in 20–30% of the unselected female population *(4,35–38)*. The prevalence of polycystic ovaries in the general population decreases with age and was observed in only 7.8% of women older than 35 years *(37)*. The prevalence of androgen excess disorders among women with polycystic ovaries is less clear. Considering only the prevalence of PCOS, and assuming that 25% of unselected reproductive-aged women have polycystic ovaries, that the overall prevalence of PCOS in this population is 7%, and that about 75% of patients with PCOS will have polycystic ovaries on ultrasound, we can estimate that approximately 21% of women having polycystic ovaries will have PCOS. Although this represents about a threefold higher prevalence than in the general population (i.e., ~6.5%), it also indicates that about 80% of the pool of women with polycystic ovaries will not have PCOS. Consistent with this estimation, in their study of 224 college women Michelmore and colleagues reported that only 10.8% of women with polycystic ovaries detected by transabdominal sonography women had PCOS defined by the NIH 1990 criteria *(4)*.

2.4.6. Summary of Predictive Value of Clinical Markers

Overall, between 50 and 75% of women with evidence of hirsutism or the complaint of unwanted hair growth will have androgen excess, notably PCOS. Alternatively, only 20 and 40% of patients with acne as their sole presenting complaint and only about 10% of women complaining of hair loss will have androgen excess. Between one-fourth and one-third of women with oligo-/amenorrhea have androgen excess, and only about one-fifth of women with polycystic ovaries on ultrasonography will have androgen excess.

2.5. Exclusion of Specific Disorders

Approximately 6% of androgen excess patients suffer from a specific disorder, including classic and nonclassic 21-hydroxylase deficiencies, the HAIR-AN syndrome, or an ASN, among others *(12)*. In patients clinically suspected of having an ASN, a computed tomography or magnetic resonance imaging scan of the adrenals and transvaginal ovarian ultrasonography should be obtained to assess for adrenal or ovarian masses, respectively. Importantly, measurement of a basal 17-hydroxyprogesterone serum level should be obtained in the follicular phase of the menstrual cycle, preferably in the morning, to exclude 21-hydroxylase-deficient NCAH *(39)*. In patients suspected of having Cushing's syndrome, it will also include a 24-hour urinary free cortisol level or a cortisol level following an overnight dexamethasone (1.0 mg at 11 PM) test. If the HAIR-AN syndrome is suspected, a basal or preferably a glucose-stimulated insulin level should be obtained. Growth hormone levels should be obtained in patients suffering from acromegaly. In patients with a blind vaginal pouch and/or abnormality of the external genitalia, male pseudohermaphroditism or XY gonadal dysgenesis should be considered and a karyotype should be obtained.

As already mentioned, in eumenorrheic women with other hyperandrogenic features, a day 22–24 P4 level, preferably in two consecutive cycles, should be obtained. In patients demonstrating ovulatory dysfunction, thyroid-stimulating hormone and prolactin levels may also be obtained to exclude thyroid dysfunction and hyperprolactinemia, respectively. The prevalence of these two latter abnormalities among women with hyperandrogenic features is less than 2% *(12)*.

2.6. Differentiation of Functional Disorders

Once specific disorders are excluded, the remaining patients (the vast majority of patients with androgen excess) are considered to have androgen excess resulting from a functional abnormality, principally PCOS or IH. It is important to note that the classification of these patients will depend to a great extent on the definition used, particularly for PCOS (*see* Chapter 13). Hyperandrogenic women in whom specific disorders have been excluded and who have evidence of ovarian dysfunction (either oligo-anovulation or polycystic ovarian morphology) are generally considered to have PCOS. Only a minority of women suspected of androgen excess will have IH, if strictly defined (including the absence of ovulatory dysfunction, polycystic ovarian morphology, and hyperandrogenemia, in the face of evident hirsutism; *see* Chapter 12). Also, depending on diagnostic criteria, a variable proportion of androgen excess patients will be classified as having nonspecific hyperandrogenism (e.g., patients with hirsutism and hyperandrogenemia but normal ovarian morphology).

2.7. Identification of Associated Morbidities

Once the diagnosis (specific or functional) is established, related morbidities should be identified. For example, PCOS and the HAIR-AN syndrome are associated with important metabolic dysfunction, including the metabolic syndrome. As such, it is recommended that patients diagnosed with these disorders undergo measurement of fasting lipids and insulin and glucose levels following an oral glucose load (75 g).

The results of the oral glucose tolerance test (OGTT) will allow the detection of impaired glucose tolerance (IGT) or type 2 diabetes mellitus (DM). Although many patients can be diagnosed as having DM by a fasting glucose of greater than 126 mg/dL *(40)*, patients with PCOS frequently require an OGTT to diagnose more subtle forms of DM or IGT *(41)*. During the OGTT, IGT is diagnosed by a fasting glucose of less than 140 mg/dL and a 2-hour glucose level between 140 and 199 mg/dL *(42)*. Type 2 DM is diagnosed if the 2-hour glucose during the OGTT is 200 mg/dL or more. These diagnoses require that abnormal results be repeated for confirmation.

The OGTT results can also be used to estimate the degree of hyperinsulinism (and indirectly the degree of insulin resistance). Although a number of calculations to assess the degree of insulin secretion have been proposed *(43–45)*, clinically it is probably simplest and most informative to examine the peak poststimulation insulin levels. Roughly, patients with poststimulation insulin levels of less than 80–100 µIU/mL can be considered to have no significant hyperinsulinemia. Alternatively, women with poststimulation insulin levels of 100–150, 150–300, and greater than 300 µIU/mL generally can be considered to have mild, moderate, and severe hyperinsulinism, respectively.

Women with longstanding ovulatory dysfunction, particularly those not having received long-term progestogen therapy or hormonal contraception, will be at increased risk for endometrial hyperplasia and/or carcinoma (*see* Chapter 27), although the extent of this risk is unclear *(46)*. Current diagnostic standards suggest that an endometrial biopsy, particularly if the endometrial thickness on sonography is more than 4–8 mm, should be an integral part of the evaluation of women at moderate risk for endometrial cancer. This includes women who are younger than 40 years with additional risk factors such as PCOS *(47)*. In a small study, Cheung evaluated 36 patients with PCOS by endometrial biopsy and observed that 35.7% had endometrial hyperplasia and 25% of these had cytological atypia *(48)*.

3. CONCLUSIONS

Conceptually, investigators and clinicians alike need to recognize that there are five steps to the evaluation and diagnosis of patients with androgen excess: (1) screening of potentially affected women, (2) confirmation of androgen excess, (3) exclusion of specific disorders, (4) differentiation of functional disorders, and (4) diagnosis of related morbidities. A high degree of clinical suspicion, together with a systematic approach to the patient suspected of androgen excess, will elucidate the cause of the abnormality in the vast majority of women seen. Finally, greater agreement on the nature and definition of these disorders, in particular PCOS, HAIR-AN syndrome, and IH, will guide the development of more uniform and universally accepted diagnostic schemes.

4. FUTURE AVENUES OF INVESTIGATION

More extensive, detailed, and careful epidemiological and outcome studies are required to determine the most cost-effective and efficient approach to the patient with androgen excess. Furthermore, the impact of ethnicity and race on the effectiveness of any proposed diagnostic scheme must be established. Finally, research to improve the availability and quality of androgen assays is critically needed.

KEY POINTS

- Clinical features of hyperandrogenism include hirsutism, ovulatory and menstrual dysfunction, and in some patients acne and androgenic alopecia.
- Biochemically, hyperandrogenism can be established by the presence of supranormal circulating levels of total or unbound T, androstenedione, or the adrenal androgen metabolite DHEAS.
- Specific disorders that result in androgen excess include 21-hydroxylase-deficient NCAH, ovarian ASNs, adrenal ASNs with or without Cushing's syndrome, ACTH-dependent Cushing's syndrome, HAIR-AN syndrome, and the use or abuse of androgenic drugs; in turn, PCOS generally accounts for more than 80% of hyperandrogenic patients.
- Although not properly an androgen excess disorder, IH should be considered in the differential diagnosis of these women.
- Conceptually, the general approach to patients with androgen excess entails five steps:
  1. Screening for androgen excess based on patient complaint or medical or family history
  2. Confirmation of the presence of clinical and/or biochemical androgen excess

3. Exclusion of specific androgen excess or related disorders (e.g., NCAH. ASNs, and HAIR-AN syndrome)
4. Classification of patients without a specific androgen excess disorder into PCOS, IH, or nonspecific functional androgen excess
5. Identification of related morbidities (e.g., diabetes in PCOS)

ACKNOWLEDGMENTS

Supported in part by an endowment from the Helping Hand of Los Angeles and by grants RO1-HD29364 and K24-D01346 from the National Institutes of Health.

REFERENCES

1. Azziz R, Marin C, Hoq L, Badamgarav E, Song P. Economic burden of the polycystic ovary syndrome (PCOS) during the reproductive lifespan. J Clin Endocrinol Metab 2005;90:4650–4658.
2. Knochenhauer ES, Key TJ, Kahsar-Miller M, et al. Prevalence of the polycystic ovary syndrome in unselected black and white women of the southeastern United States: a prospective study. J Clin Endocrinol Metab 1998;83:3078–3082.
3. Diamanti-Kandarakis E, Kouli CR, Bergiele AT, et al. A survey of the polycystic ovary syndrome in the Greek island of Lesbos: hormonal and metabolic profile. J Clin Endocrinol Metab 1999;84:4006–4011.
4. Michelmore KF, Balen AH, Dunger DB, Vessey MP. Polycystic ovaries and associated clinical and biochemical features in young women. Clin Endocrinol (Oxf) 1999;51:779–786.
5. Asuncion M, Calvo RM, San Millan JL, et al. A prospective study of the prevalence of the polycystic ovary syndrome in unselected Caucasian women from Spain. J Clin Endocrinol Metab 2000;85:2434–2438.
6. Azziz R, Woods KS, Reyna R, et al. The prevalence and features of the polycystic ovary syndrome in an unselected population. J Clin Endocrinol Metab 2004;89:2745–2749.
7. Zawadzki JK, Dunaif A. Diagnostic criteria for polycystic ovary syndrome: towards a rational approach. In: Dunaif A, Givens JR, Haseltine FP, Merriam GR, eds. Polycystic Ovary Syndrome. Boston: Blackwell Scientific Publications, 1992:377–384.
8. Hatch K, Rosenfield RL, Kim MH, et al. Hirsutism: implications, eticology, and mnagmenet. Am J Obstet Gynecol 1981;140:815–830.
9. Meldrum DR, Abraham GE. Peripheral and ovarian venous concentrations of various steroid hormones in virilizing ovarian tumors. Obstet Gynecol 1979;53:36–43.
10. Derksen J, Nagesser SK, Meinders AE, Haak HR, van d, V. Identification of virilizing adrenal tumors in hirsute women. N Engl J Med 1994;331:968–973.
11. Waggoner W, Boots LR, Azziz R. Total testosterone and DHEAS levels as predictors of androgen-secreting neoplasms: a populational study. Gynecol Endocrinol 1999;13:394–400.
12. Azziz R, Sanchez LA, Knochenhauer ES, et al. Androgen excess in women: Experience with over 1000 consecutive patients. J Clin Endocrinol Metab 2004;89:453–462.
13. Fitzgerald RL, Herold DA. Serum total testosterone: immunoassay compared with negative chemical ionization gas chromatography-mass spectrometry. Clin Chem 1996;42:749–755.
14. Steinberger E, Ayala C, Hsi B, et al. Utilization of commercial laboratory results in management of hyperandrogenism in women. Endocr Pract 1998;4:1–10.
15. Boots LR, Potter S, Potter HD, Azziz R. Measurement of total serum testosterone levels using commercially available kits: high degree of between-kit variability. Fertil Steril 1998;69:286–292.
16. Vermeulen A, Verdonck L, Kaufman JM. A critical evaluation of simple methods for the estimation of free testosterone in serum. J Clin Endocrinol Metab 1999;84:3666–3672.
17. Escobar-Morreale HF, Asuncion M, Calvo RM, Sancho J, San Millan JL. Receiver operating characteristic analysis of the performance of basal serum hormone profiles for the diagnosis of polycystic ovary syndrome in epidemiological studies. Eur J Endocrinol 2001;145:619–624.
18. Taieb J, Mathian B, Millot F, et al. Testosterone measured by 10 immunoassays and by isotope-dilution gas chromatography-mass spectrometry in sera from 116 men, women, and children. Clin Chem 2003;49:1381–1395.
19. Wang C, Catlin DH, Demers LM, Starcevic B, Swerdloff RS. Measurement of total serum testosterone in adult men: comparison of current laboratory methods versus liquid chromatography-tandem mass spectrometry. J Clin Endocrinol Metab 2004;89:534–543.
20. Van Uytfanghe K, Stockl D, Kaufman JM, et al. Evaluation of a candidate reference measurement procedure for serum free testosterone based on ultrafiltration and isotope dilution-gas chromatography-mass spectrometry. Clin Chem 2004;50:2101–2110.
21. Christ-Crain M, Meier C, Huber P, Zimmerli L, Trummler M, Muller B. Comparison of different methods for the measurement of serum testosterone in the aging male. Swiss Med Wkly 2004;134:193–197.

22. Van Uytfanghe K, Stockl D, Kaufman JM, Fiers T, De Leenheer A, Thienpont LM. Validation of 5 routine assays for serum free testosterone with a candidate reference measurement procedure based on ultrafiltration and isotope dilution-gas chromatography-mass spectrometry. Clin Biochem 2005;38:253–261.

23. Miller KK, Rosner W, Lee H, et al. Measurement of free testosterone in normal women and women with androgen deficiency: comparison of methods. J Clin Endocrinol Metab 2004;89:525–533.

24. Azziz R, Waggoner WT, Ochoa T, Knochenhauer ES, Boots LR. Idiopathic hirsutism: an uncommon cause of hirsutism in Alabama. Fertil Steril 1998;70:274–278.

25. Carmina E. Prevalence of idiopathic hirsutism. Eur J Endocrinol 1998;139:421–423.

26. Ardaens Y, Robert Y, Lemaitre L, Fossati P, Dewailly D. Polycystic ovarian disease: contribution of vaginal endosonography and reassessment of ultrasonic diagnosis. Fertil Steril 1991;55:1062–1068.

27. Glintborg D, Henriksen JE, Andersen M, et al. Prevalence of endocrine diseases and abnormal glucose tolerance tests in 340 Caucasian premenopausal women with hirsutism as the referral diagnosis. Fertil Steril 2004;82:1570–1579.

28. Unluhizarci K, Gokce C, Atmaca H, Bayram F, Kelestimur F. A detailed investigation of hirsutism in a Turkish population: idiopathic hyperandrogenemia as a perplexing issue. Exp Clin Endocrinol Diabetes 2004;112:504–509.

29. Souter I, Sanchez A, Perez M, Bartolucci AA, Azziz R. The prevalence of androgen excess among patients with minimal unwanted hair growth. Am J Obstet Gynecol 2004;191:1914–1920.

30. Vexiau P, Husson C, Chivot M, et al. Androgen excess in women with acne alone compared with women with acne and/or hirsutism. J Invest Dermatol 1990;94:279–283.

31. Timpatanapong P, Rojanasakul A. Hormonal profiles and prevalence of polycystic ovary syndrome in women with acne. J Dermatol 1997;24:223–229.

32. Slayden SM, Moran C, Sams WM Jr, Boots LR, Azziz R. Hyperandrogenemia is a frequent cause of acne, regardless of age of presentation. Fertil Steril 2001;75:889–892.

33. Borgia F, Cannavo S, Guarneri F, Cannavo SP, Vaccaro M, Guarneri B. Correlation between endocrinological parameters and acne severity in adult women. Acta Derm Venereol 2004;84:201–204.

34. Solomon CG, Hu FB, Dunaif A, et al. Long or highly irregular menstrual cycles as a marker for risk of type 2 diabetes mellitus. JAMA 2001;286:2421–2426.

35. Polson DW, Wadsworth J, Adams J, Franks S. Polycystic ovaries: a common finding in normal women. Lancet 1988;1:870–872.

36. Clayton RN, Ogden V, Hodgkinson J, et al. How common are polycystic ovaries in normal women and what is their significance for the fertility of the population? Clin. Endocrinol (Oxf) 1992;37:127–134.

37. Koivunen R, Laatikainen T, Tomas C, Huhtaniemi I, Tapanainen J, Martikainen H. The prevalence of polycystic ovaries in healthy women. Acta Obstet Gynecol Scand 1999;78:137–141.

38. Lowe P, Kovacs G, Howlett D. Incidence of polycystic ovaries and polycystic ovary syndrome amongst women in Melbourne, Australia. Aust NZ J Obstet Gynecol 2005;45:17–19.

39. Azziz R, Hincapie LA, Knochenhauer ES, Dewailly D, Fox L, Boots LR. Screening for 21-hydroxylase deficient nonclassic adrenal hyperplasia among hyperandrogenic women: A prospective study. Fertil Steril 1999;72:915–925.

40. Expert Committee on the Diagnosis and Classification of Diabetes Mellitus. Report of the Expert Committee on the Diagnosis and Classification of Diabetes Mellitus. Diabetes Care 1997;20:1183–1197.

41. Ehrmann DA, Barnes RB, Rosenfield RL, Cavaghan MK, Imperial J. Prevalence of impaired glucose tolerance and diabetes in women with polycystic ovary syndrome. Diabetes Care 1999;22:141–146.

42. World Health Organization. Diabetes Mellitus: Report of a WHO Study Group. Geneva: WHO, 1985:727.

43. Hollenbeck CB, Chen N, Chen Y-DI, Reaven GM. Relationship between the plasma insulin response to oral glucose and insulin-stimulated glucose utilization in normal subjects. Diabetes 1984;33:460–463.

44. Matsuda M, DeFronzo RA. Insulin sensitivity indices obtained from oral glucose tolerance testing: comparison with the euglycemic insulin clamp. Diabetes Care 1999;22:1462–1470.

45. Yeni-Komshian H, Carantoni M, Abbasi F, Reaven GM. Relationship between several surrogate estimates of insulin resistance and quantification of insulin-mediated glucose disposal in 490 healthy nondiabetic volunteers. Diabetes Care 2000;23:171–175.

46. Hardiman P, Pillay OC, Atiomo W. Polycystic ovary syndrome and endometrial carcinoma. Lancet 2003;361:1810–1812.

47. Critchley HO, Warner P, Lee AJ, et al. Evaluation of abnormal uterine bleeding: comparison of three outpatient procedures within cohorts defined by age and menopausal status. Health Technol Assess 2004;8:iii–139.

48. Cheung AP. Ultrasound and menstrual history in predicting endometrial hyperplasia in polycystic ovary syndrome. Obstet Gynecol 2001;98:325–331.

34

Ovarian Suppression and Treatment of Hirsutism

Paolo Moghetti

SUMMARY

The treatment of hirsutism is based on a dual approach: pharmacological therapy of hyperandrogenism and removal of terminal hairs already present. Ovarian suppression of androgen secretion with oral contraceptives is widely used in these women, but its efficacy as sole agent appears limited. The most reliable medical therapy for hirsutism is the antiandrogen drugs. However, these drugs are only partially effective on terminalized hairs, and removal of these hairs is generally required. Electrolysis and laser photothermolysis appear to be the most effective cosmetic procedures, although the effects of these methods should not be considered permanent.

Key Words: Hirsutism; antiandrogen drugs; oral contraceptives; insulin sensitizers; cosmetic procedures.

1. INTRODUCTION

Several strategies may be considered in the treatment of hirsutism. They comprise cosmetic procedures and drugs, administered either topically or systemically. In many hirsute subjects, cosmetic procedures may work well. However, in most cases, especially in more hyperandrogenic subjects, all these procedures may fail to achieve satisfactory control of hirsutism. In these cases, drugs are an indispensable addition to correct the underlying causative factor.

The aim of pharmacological treatment is to lower circulating free androgen levels and/or block peripheral androgen action (Fig. 1). Overall, suppression of androgen secretion alone has limited effect on established hirsutism. The most reliable medical therapy in these women is antiandrogen drugs. Alternatively, these drugs are only partially effective on terminalized hairs. Removal of unwanted hairs is generally required in women with established hirsutism. Thus, treatment of hirsutism is often based on a dual approach, pharmacological therapy of hyperandrogenism, and removal of terminal hairs already present (1).

It is noteworthy that there are few licensed drugs for the treatment of hyperandrogenism. In fact, there is only the topical drug eflornithine—licensed for facial hirsutism—or estrogen–progestogen combinations. Oral contraceptives are widely used for this purpose, but their efficacy appears limited.

Spironolactone and cyproterone acetate are the antiandrogen drugs most commonly used in women with hirsutism. These compounds are not pure antiandrogens, as they interact with other steroid receptors and frequently show troublesome side effects. Other drugs, such as flutamide and finasteride, have been assessed in the treatment of this condition, but experience is still limited. It should be borne in mind that the use of any antiandrogens implies the need to avoid a pregnancy, given the potential risk of feminization of male fetuses. For this reason and for potential synergic effects of combined therapy, antiandrogens are often administered in combination with nonandrogenic oral contraceptives.

From: *Contemporary Endocrinology: Androgen Excess Disorders in Women:
Polycystic Ovary Syndrome and Other Disorders, Second Edition*
Edited by: R. Azziz et al. © Humana Press Inc., Totowa, NJ

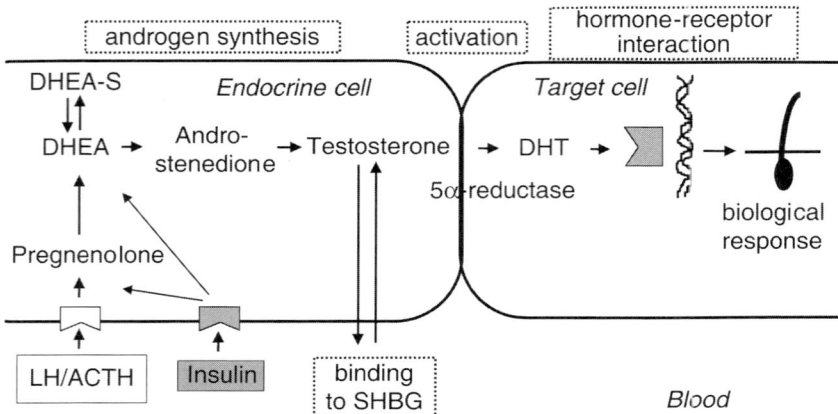

Fig 1. Sequence of steps between androgen synthesis in endocrine glands and androgen action in the peripheral tissues. Each of these steps is a potential target for the treatment of hyperandrogenic women. DHEAS, dehydroepiandrosterone sulfate; DHT, dihydrotestosterone; LH, luteinizing hormone; ACTH, adrenocorticotropic hormone; SHBG, sex hormone-binding globulin.

Hirsutism treatment is symptomatic. There is a latency of some months in the appearance of clinical effects of drugs, which can take more than 1 year of therapy before the full benefits are observed. In addition, in general, hair growth resumes after treatment is stopped, especially in more hyperandrogenic subjects. Therefore, therapy should be maintained indefinitely, or at least for long periods.

The risk of teratogenicity in case of pregnancy is the main reason that pharmaceutical companies are not willing to fund trials to evaluate the efficacy of antihirsutism treatments. In turn, this may explain why few large-scale, prolonged clinical trials have assessed the use of antiandrogen drugs in women with hirsutism. Indeed, the large majority of studies concerning hirsutism treatment demonstrate considerable shortcomings, including lack of control group, assessment of results only by subjective methods, or too short a duration of therapy relative to the physiology of hair growth *(2)*. These limitations make it difficult to establish relative efficacy of antiandrogen drugs.

Apart from contraceptive pills, reduction of ovarian androgen secretion may be efficaciously obtained by using gonadotropin-releasing hormone analogs *(3)*. However, these drugs also markedly reduce estrogen secretion, causing estrogen deficiency-related symptoms and, in particular, appreciable loss of bone mass. Concomitant estrogen replacement has been proposed to avoid this inconvenience. However, side effects and the cost of these drugs limit their use to highly selected cases.

Increased secretion of adrenal androgens may be observed in congenital adrenal hyperplasia, Cushing's syndrome, and the rare androgen-secreting adrenal tumors. In addition, adrenal hyperandrogenemia may be seen in many subjects with polycystic ovary syndrome (PCOS), as well as in several women with hirsutism not easily classified into a definite diagnosis. Glucocorticoids may be helpful in reducing adrenal secretion. However, adrenal androgen suppression has very limited effects on hirsutism, and titration of glucocorticoids is not easy. Thus, this treatment should be limited to women with congenital adrenal hyperplasia.

Ketokonazole is an antifungal drug that inhibits steroidogenesis. It may attenuate both ovarian and adrenal hyperandrogenism, and some studies reported significant improvement in hirsutism during treatment *(4)*. However, this drug also impairs cortisol secretion, and dose-dependent liver toxicity is common. Overall, this drug is indicated for hirsutism therapy only in subjects with Cushing's syndrome.

A limited number of centers have been using laparoscopic ovarian drilling—the modern evolution of wedge resection—for reproductive goals in patients with PCOS (*see* Chapter 37). This method induces transient suppression of ovarian androgens. However, evidence of efficacy on hair growth is very limited. As such, improvement in hair growth should not be considered a primary outcome when using this surgical procedure.

Another novel method to attenuate hyperandrogenism in patients with PCOS is through reduction in plasma insulin levels, because insulin resistance and the associated hyperinsulinemia commonly found in these women may contribute to their androgen excess *(5)*. Lifestyle changes and insulin-sensitizing drugs consistently lower androgen levels in PCOS patients. Limited evidence suggests that these procedures may, to some extent, improve hirsutism. Reduction in insulin levels may also be obtained through drugs that block insulin secretion, such as somatostatin. This treatment may also act through its effects on the growth hormone/insulin-like growth factor system. Preliminary data have demonstrated an improvement in the hirsutism score in patients with PCOS treated with long-acting derivatives of somatostatin *(6)*.

At present, several of these pharmacological options are of experimental interest. Only estrogen–progestogen combinations, eflornithine, antiandrogens, finasteride, and metformin are generally used in clinical practice. Therefore, only these drugs will be discussed here. Table 1 summarizes the main characteristics of peripheral-acting drugs.

2. BACKGROUND

2.1. Cosmetic Procedures

Cosmetic procedures may be very helpful in controlling mild-to-moderate degrees of unwanted hair and are virtually always required as an adjunct in women being treated medically to remove existing terminal hair. Reduction in the frequency with which women use cosmetic procedures is also reliable evidence of drug efficacy. In general, these procedures are time-consuming and may be expensive. Complications include pain, discomfort, dyspigmentation, and scarring.

Several different techniques may be used, according to the characteristics of the individual patient (Table 2). Shaving does not increase the rate of hair growth, as erroneously thought by many patients, but it leaves an unpleasant sharp stubble. Therefore, other procedures are generally preferred. Waxing and plucking may be effective, but there is the risk of folliculitis and in-grown hairs. Furthermore, skin irritation may sometimes induce a paradoxical increase in local hair growth. In addition, these complications may subsequently make more difficult the removal of hairs by electrolysis. For these reasons, several authors strongly discourage their use, in particular in women with clinically significant degrees of hirsutism.

Electrolysis and laser photothermolysis appear to be the most effective procedures, although with these methods hair removal should not be considered permanent. Moreover, multiple treatments are usually necessary because of the nature of the hair growth cycle. Electrolysis uses an electric current transmitted through a fine needle inserted into the hair follicle. This results in destruction of the follicle. There are different electrolysis techniques using direct (galvanic), high-frequency alternating current, or a combination *(7)*. Pain is common, and scarring may occur if the operator is unskilled or if the current used is too high.

Laser and light-assisted hair removal is based on the principle of selective photothermolysis. Selective absorption by hair chromophores of energy from lasers and broadband light sources may result in destruction of hair follicles while leaving the skin undamaged, although there are significant differences according to the specific characteristics of the patients and techniques used *(8)*. In general, laser therapy is more effective in women with darker hair and lighter skin. Efficacy is lower and complications are more common in women with darker skin. Transient erythema and edema are common after laser therapy, and blistering, crusting, or alterations in skin pigmentation may also occur.

Evidence regarding the efficacy of these procedures is mostly anecdotal, and there are a few controlled studies supporting their efficacy. Laser treatment was effective in reducing, over 6 months,

Table 1
Main Characteristics of Peripheral-Acting Drugs Used for Hirsutism Therapy

Drugs (schedule)	Pros	Cons
Antiandrogens		
Cyproterone acetate (2–100 mg/day, usually on cycle days 5–14)	Large experience Licensed for this use[a]	Combination with estrogens needed Frequent side effects, usually mild (e.g., metabolic abnormalities, mood changes, weight gain, edema, headache, and rare liver toxicity) Necessary to avoid pregnancies if not used in oral contraceptive form
Spironolactone (50–200 mg, continuously)	Large experience Low cost May improve menses	Necessary to avoid pregnancy Frequent side effects, usually mild (e.g., polymenorrhea, diuresis, mastodynia, abdominal discomfort, and rare hyperkalemia)
Flutamide (62.5–500 mg/day, continuously)	Pure antiandrogen No adverse effect on ovulation	Necessary to avoid pregnancy Usually mild side effects (dry skin) Rare but occasionally severe liver toxicity
5α-reductase inhibitors		
Finasteride (1–5 mg/day, continuously)	Well tolerated No adverse effect on ovulation	Necessary to avoid pregnancy

[a]Available in several countries for use in women, as a combination oral contraceptive pill containing 2 mg cyproterone acetate and 35 μg of ethinylestradiol.

Table 2
Chief Cosmetic Procedures for Symptomatic
Management of Hirsutism

Shaving
Depilatory creams
Plucking
Waxing
Bleaching
Electrolysis
Laser photothermolysis

self-reported severity of facial hair and time spent on hair removal in a randomized controlled trial carried out in 88 women with PCOS *(9)*. A retrospective study assessed 242 patients with hirsutism who received diode laser treatments over 4 years *(10)*. After an average of two treatments (range 1–6), a sufficient reduction in terminal hairs was achieved for a mean period of 8 months, and the hair-plucking interval was raised from a mean of 3.7 days before treatment to 15.2 days after laser epilation. The procedure was well accepted by about 80% of the subjects. At least in the short term, ruby, alexandrite, and diode lasers or the intense pulsed light resulted in similar success rates. The long-pulsed diode and Nd:YAG wavelength-based laser systems are best suited to treat patients with darker skin.

2.2. Eflornithine

Eflornithine is an irreversible inhibitor of ornithine decarboxylase. This enzyme catalyzes the conversion of ornithine to polyamines, which are involved in the regulation of cell growth and differentiation in several tissues. The enzyme is modulated by androgens and takes part in the physiology of hair growth, regulating the proliferation of matrix cells in the hair follicle. Studies have indicated that blockade of this enzyme activity in hair follicles slows hair growth, and the drug has recently been licensed for topical treatment of facial hirsutism. Percutaneous absorption of the drug is negligible. In short-term clinical studies, eflornithine 11.5–15% cream was better than placebo in reducing hair growth in women with unwanted facial hair, as demonstrated by objective and subjective methods *(11)*. However, hair growth returned to pretreatment rates within a few weeks after stopping treatment. Mild irritation and folliculitis may affect the skin with treatment. Anecdotal evidence also suggests that eflornithine may also be used in reducing unwanted hair in other body sites. Currently there are no controlled studies comparing this drug with the systemic pharmacological tools used for hirsutism. However, it appears reasonable to suggest eflornithine as single-drug therapy in mild hirsutism to reduce the frequency for the need for hair removal, although other pharmacological strategies should be considered in more severe hirsutism.

2.3. Estrogen–Progestogen Combinations

In women who are hyperandrogenic, suppression of ovarian androgens by estrogen–progestogen combinations is widely used *(1)*. This strategy is effective in lowering circulating free androgens resulting from the decrease in luteinizing hormone and androgen secretions and via an estrogen-induced increase in sex hormone-binding globulin. In addition, there is some evidence of a mild reduction in adrenal androgen secretion, attributed to the effect of the progestin. Despite the fact that these drugs have been used for hirsutism treatment for decades, there have been few controlled studies assessing their efficacy on the cutaneous manifestation of androgen excess. In addition, there are pros and cons in using oral contraceptives for the treatment of hyperandrogenism in PCOS. These medications counteract the risk of endometrial cancer as a result of unopposed estrogen. Furthermore, they ensure contraception, which may be very useful in these women, who have in many cases reduced but unpredictable ovulation and often receive other medications for hirsutism that necessitate the avoidance of pregnancies. Alternatively, their efficacy in established hirsutism appears limited. In addition, their use may result in a worsening of the metabolic impairment of these patients, both in the lipid profile and in insulin sensitivity. This is a concern in women with PCOS who often have concomitant metabolic abnormalities, although the impact of these therapy-related changes on relevant clinical outcomes has not yet been assessed.

In several countries, although not in the United States, a contraceptive pill containing the antiandrogen cyproterone acetate is available and widely used by these women. Preliminary data also suggest that a combination oral contraceptive including the novel progestin drospirenone, a 17α-spironolactone derivative, might have some application in the treatment of hirsutism *(12)*.

2.4. Antiandrogens

2.4.1. Cyproterone Acetate

Cyproterone acetate is a steroidal drug derived from 17-hydroxyprogesterone. It is capable of interacting with several steroid receptors, demonstrating progestin, antiandrogen, and weak glucocorticoid activities. This drug is effective both in reducing gonadotropin-dependent androgen secretion, because of its progestin effect, and in counteracting androgen action, as a result of competitive binding with the androgen receptor. The progestin activity of cyproterone acetate requires, in women who have a uterus, its combination with estrogens. Therefore, studies concerning this drug have generally been conducted in women receiving combined therapy. For this reason cyproterone acetate should not be used when estrogens are not recommended.

Cyproterone acetate is an effective treatment for hirsutism and acne and is widely used throughout the world for this indication. However, it is not available in the United States. Doses of the drug used for treatment of hirsutism range widely, between 2 and 100 mg/day (2 mg are contained in the previously mentioned contraceptive pill).

Accumulation of the drug in adipose tissue prolongs its biological half-life. For this reason, to avoid menstrual abnormalities cyproterone acetate (25–100 mg/day) is usually administered during the first 10 days of a cyclic 21-day course of estrogen (usually 20–35 µg ethinylestradiol), with 7 days off both medications (reversed sequential regimen). Cyproterone acetate can also be added during the first 10 days to any contraceptive pill.

A recent Cochrane review assessed the evidence-based information available on the efficacy of this treatment for hirsutism *(13)*. It is noteworthy that, despite cyproterone acetate having been available for some decades, only nine small studies were suitable for inclusion in the analysis. The review concluded that there is limited but consistent evidence that cyproterone acetate is effective in the treatment of hirsutism. Moreover, comparing the clinical outcomes of different active drugs, no differences were found, but because of the limitations of such studies, differences cannot be excluded.

The evidence from dose-ranging studies is limited and still equivocal. A controlled study assessing the efficacy of different schedules of cyproterone acetate administration did not find differences between the oral contraceptive containing 2 mg/day of cyproterone acetate and the combination of an oral contraceptive plus either 20 or 100 mg of the drug in the first 10 days of the cycle *(14)*. However, several investigators report that differences between these doses may exist. Side effects are usually mild, particularly at low doses, and include metabolic alterations, weight gain, edema, loss of libido, mood changes, and headaches. Liver toxicity has been rarely reported in subjects treated with high doses.

2.4.2. Spironolactone

Spironolactone is an antimineralocorticoid drug, originally used to treat primary and secondary hyperaldosteronism. Reports of gynecomastia in men given this drug highlighted its antiandrogenic properties. Subsequent studies demonstrated that it is a competitive inhibitor of androgen binding to the androgen receptor *(15)*. In fact, this steroidal compound has several intrinsic hormonal activities.

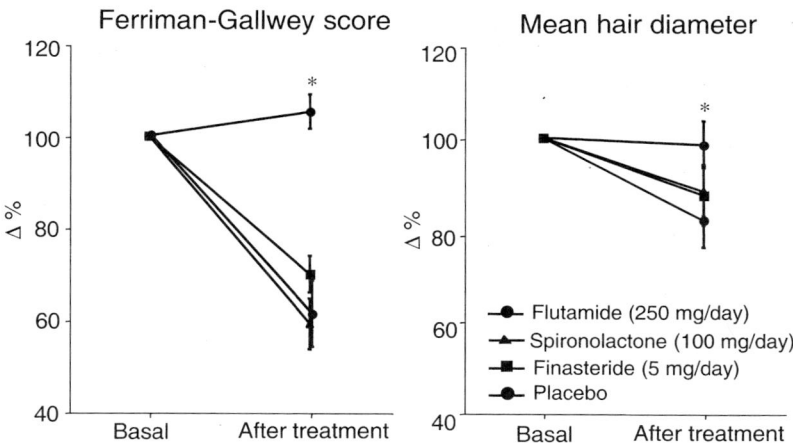

Fig. 2. Changes after therapy in hirsutism score (left) and hair shaft diameter (right) in women randomized to receive spironolactone, flutamide, finasteride, or placebo. (From ref. *17*.)

It is a weak progestin, and some data suggest that it may also be a selective estrogen receptor modulator, demonstrating weak estrogenic or antiestrogenic properties according to the hormonal milieu. In addition, spironolactone may interfere directly with steroidogenesis, at least at high doses. Spironolactone is used worldwide for the treatment of hirsutism, although it is not licensed for this use.

Spironolactone as a treatment for hirsutism was first assessed in 1978. Subsequently, this drug has been administered to hirsute women in a number of studies, mostly uncontrolled, in doses ranging from 50 to 400 mg/day *(16)*. In a short-term trial, Lobo and colleagues demonstrated a greater reduction in hair shaft diameters with 200 mg/day than with 100 mg/day *(17)*. Nevertheless, the large majority of investigators use daily doses of 100 mg because of the increased frequency of side effects at higher doses.

The largest study concerning this drug reported an improvement in about 85% of more than 300 subjects with hirsutism. However, this study was uncontrolled and short-term. Two small controlled studies assessing this drug as a single agent were not able to recognize statistically significant differences vs placebo in posttreatment hirsutism measures. More recently, in a 6-month randomized, double-blind, placebo-controlled comparative study including spironolactone, flutamide, or finasteride as active drugs *(18)*, we found a 12% reduction in hair shaft diameter and a 41% reduction in the hirsutism score in women receiving spironolactone. These changes were statistically significant as compared to placebo. Interestingly, the improvements observed with spironolactone were similar to those seen with flutamide or finasteride (Fig. 2). These conclusions were supported by the patients' self-evaluations and their use of less frequent cosmetic measures for hair removal.

Tolerability is generally acceptable, with the noticeable exception of menstrual irregularity, especially polymenorrhea, in many subjects given this drug alone. However, in many women side effects are transient, resolving within a few months. If the menstrual abnormality does not reverse, an oral contraceptive may be added or the spironolactone dose may be reduced. It should be kept in mind that women treated with antiandrogens must avoid pregnancies and that an estrogen–progestogen combination may be contraceptive and could have synergic effects on hirsutism.

Alternatively, spironolactone may sometimes improve amenorrhea in women with PCOS. In this regard, a recent study reported similar rates of improvement in menstruation with spironolactone or metformin therapy *(19)*. However, when using this drug, the frequency of menstruation cannot be used as a surrogate marker of improved ovulation. Increased diuresis, breast tenderness, and abdominal discomfort are other frequently reported mild side effects of spironolactone, at least with high doses of the drug and especially in the first period of treatment, whereas hypotension and clinically significant increase in serum potassium concentrations are rare. Nevertheless, the drug should not be given to women with renal insufficiency and should be given with caution to women treated with other drugs that can cause hyperkalemia.

2.4.3. Flutamide

Flutamide is a nonsteroidal drug, considered a "pure" antiandrogen in that it seems to act only at the androgen receptor as a competitive antagonist *(20)*. Some data suggest that flutamide might also reduce synthesis of androgens and/or increase their metabolism to inactive molecules, lowering androgen serum levels. This drug is licensed for the treatment of advanced prostatic carcinoma and was successfully evaluated in hyperandrogenic women *(21)*. Flutamide has been administered to hirsute women in doses ranging from 62.5 to 750 mg/day.

Some studies reported that this drug is more effective than spironolactone or finasteride in the treatment of hirsutism. However, differences were small, and all these medications gave similar results in a controlled comparative trial *(18)*. Anecdotal evidence suggests that flutamide is more effective than other drugs in treating androgen-dependent acne, although no controlled study has been specifically designed to assess this aspect. Some studies reported that this drug might have favorable effects on visceral fat and on the lipid profile in patients with PCOS *(22,23)*. These effects are of great interest in subjects who frequently show abdominal obesity, insulin resistance, and multiple metabolic abnormalities.

Flutamide does not interfere with ovulation and is generally well tolerated. The only common complaint of patients given flutamide is dry skin, attributable to reduced sebum production. Liver toxicity is an uncommon, but potentially severe, risk with this drug *(24)*. On the whole, from several points of view, flutamide is probably the best available antiandrogen drug. However, this drug should be used with caution for the treatment of hirsutism, and in these cases serum transaminases should be carefully monitored.

2.4.4. Finasteride

Finasteride, strictly considered, is not an antiandrogen, in that it does not act at the androgen receptor. It is a competitive inhibitor of the type 2 isoenzyme of 5α-reductase, the enzyme responsible for conversion of testosterone to the more potent dihydrotestosterone. This drug has been approved for treatment of benign prostatic hyperplasia and male pattern baldness.

Increased 5α-reductase activity in the skin is considered to be the major pathogenetic mechanism of "idiopathic" hirsutism (i.e., excessive hair growth occurring in the absence of increased serum androgens and with ovulatory cycles). Selective enzyme inhibition has been proposed as a rational medical approach to this condition. However, because 5α-reductase plays a key role in the androgenic regulation of hair growth, enzyme inhibition may be potentially effective in all forms of hirsutism, irrespective of specific pathogenic mechanisms. In the skin, the type 1 isoenzyme of 5α-reductase isoenzymes appears to be primarily expressed; the importance of the type 2 isoenzyme in hair growth is still unclear. However, the isoenzyme selectivity of finasteride is not absolute at therapeutic doses. Studies assessing finasteride efficacy in women with hirsutism consistently indicated positive results, without noticeable side effects. Patients with hirsutism given the drug only occasionally reported a decrease in libido. Finasteride has been used in a dose of 5 mg/day in most published trials, although similar effects on skin androgens were found with 1 mg. A recent uncontrolled study reported similar results with continuous 2.5 mg/day or intermittent 2.5 mg every 3 days of finasteride.

In the controlled comparative study mentioned previously *(18)*, the efficacy of finasteride in hirsute women was similar to that of spironolactone or flutamide. Consistent results were also obtained in other studies comparing finasteride, spironolactone, or flutamide efficacy. Alternatively, other studies showed less improvement of hirsutism with finasteride than with spironolactone or flutamide. Interestingly, preliminary data suggest that the addition of finasteride might increase the efficacy of spironolactone in women with hirsutism *(25)*.

2.5. Insulin-Sensitizing Drugs

In the past decade, several studies reported that insulin sensitization may be effective in improving several abnormalities of PCOS. In particular, metformin has been proposed as the first-line therapy for both reproductive and metabolic abnormalities in these women. Insulin sensitizers also lower serum testosterone, and a few controlled studies assessed the efficacy of these drugs on hirsutism. These studies, carried out with either metformin or troglitazone, showed limited effect of this approach on established hirsutism (26,27), suggesting that the treatment of hirsutism should not be a primary indication for using insulin sensitizers. However, it can be hypothesized that these drugs might be helpful in hirsute women with PCOS for the maintenance of the clinical improvement obtained with antiandrogen treatment.

3. CONCLUSIONS

Although few controlled studies have assessed the clinical efficacy of antiandrogen drugs in women who are hyperandrogenic, there is adequate evidence that this approach is effective in many subjects with hirsutism. Antiandrogen drugs appear to work better than oral contraceptives. However, considering that antiandrogens require contraception and that their effect on hair growth may be potentiated by combination with estrogen–progestin combination, these drugs are usually used in conjunction with an oral contraceptive. Alternatively, oral contraceptives may have unfavorable metabolic effects, which should be taken into account, especially in women who are insulin resistant and have PCOS.

The relative efficacy of cyproterone acetate, spironolactone, flutamide, and finasteride in the treatment of hirsutism is still debated. However, in most cases, differences, if any, do not appear marked, and thus safety should be the main consideration in choosing an antiandrogen medication. Given the extensive clinical experience, lack of severe side effects, and low cost of spironolactone and cyproterone acetate, these drugs should still be considered first-line antiandrogen drugs in the treatment of hirsutism. However, cyproterone acetate cannot be used when estrogens are not recommended. In terms of tolerability, finasteride, devoid of appreciable side effects, appears to be an effective alternative. Flutamide is probably the best available antiandrogen drug. However, when using flutamide the risk of potentially severe liver toxicity should be considered.

Most pharmacological treatments for hirsutism appear to be safe. However, long-term data are limited. As a general rule, the lowest effective dose should be used for maintenance therapy, and low doses of the drugs will probably prevent worsening hair growth once the desired improvement in hirsutism is reached. In women with PCOS, insulin sensitizers could be a useful alternative in maintaining the clinical improvement obtained with antiandrogens long term, although this assumption needs to be confirmed.

In general, treatment of hirsutism needs to be individualized. The patient's perception of symptoms and her reliability, reproductive goals, metabolic features, and safety concerns are major elements to be considered in the decision making.

4. FUTURE AVENUES OF INVESTIGATION

Future research should investigate the efficacy of type 1 5α-reductase inhibition in hirsute women and continue to search for pure antiandrogens devoid of liver toxicity. Further research should also investigate the effects of combined therapies, with drugs acting at different steps in androgen action. Finally, in women with PCOS, the potential role of antiandrogen therapy in reversing the pathogenic mechanisms of this condition, including the associated metabolic dysfunction, should be investigated.

KEY POINTS

- The treatment of hirsutism should be based on the dual approach of pharmacological therapy of hyperandrogenism and removal of terminal hairs already present.

- The efficacy of suppression of ovarian androgen secretion with oral contraceptives alone is limited.
- The most reliable medical therapy for hirsutism is the antiandrogens. However, these drugs are not licensed for use in women, except for—in many countries—cyproterone acetate in a combination oral contraceptive, and require avoidance of pregnancy.
- Because there are no clear differences in potency between the different antiandrogen drugs, safety should be the main consideration in the choice of the drug.
- There is generally a delay in the development of clinical benefits, and it can take a long time before the full benefits are observed. In addition, generally there is a regrowth of hair growth after treatment is stopped. Therefore, therapies should be maintained for long periods.
- Electrolysis and laser photothermolysis appear to be the most effective cosmetic procedures, although the effects of these methods should not be considered permanent. Cosmetic treatments are an important adjunct in women receiving medical therapy.

REFERENCES

1. Azziz R. The evaluation and management of hirsutism. Obstet Gynecol 2003;101:995–1007.
2. Barth JH. How robust is the methodology for trials of therapy in hirsute women? Clin Endocrinol 1996;45:379–380.
3. Azziz R, Ochoa TM, Bradley EL Jr, Potter HD, Boots LR. Leuprolide and estrogen versus oral contraceptive pills for the treatment of hirsutism: a prospective randomized study. J Clin Endocrinol Metab 1995;80:3406–3411.
4. Venturoli S, Marescalchi O, Colombo FM, et al. A prospective, randomized trial comparing low dose flutamide, finasteride, ketoconazole, and cyproterone acetate-estrogen regimens in the treatment of hirsutism. J Clin Endocrinol Metab 1999;84:1304–1310.
5. Moghetti P. Insulin resistance: What is its role in PCOS? Curr Opin Endocrinol Diabetes 2002;9:444–450.
6. Gambineri A, Patton L, De Iasio R, et al. Efficacy of octreotide-LAR in dieting women with abdominal obesity and polycystic ovary syndrome. J Clin Endocrinol Metab 2005;90:3854–3862.
7. Richards RN. Electrolysis for the treatment of hypertrichosis and hirsutism. Skin Therapy Lett 1999;4:3–4.
8. Sanchez LA, Perez M, Azziz R. Laser hair reduction in the hirsute patient: a critical assessment. Hum Reprod Update 2002;8:169–181.
9. Clayton WJ, Lipton M, Elford J, Rustin M, Sherr L. A randomized controlled trial of laser treatment among hirsute women with polycystic ovary syndrome. Br J Dermatol 2005;152:986–992.
10. Kopera D. Hair reduction: 48 months of experience with 800nm diode laser. J Cosmet Laser Ther 2003;5:146–149.
11. Balfour JA, McClellan K. Topical eflornithine. Am J Clin Dermatol 2001;2:197–201.
12. Guido M, Romualdi D, Giuliani M, et al. Drospirenone for the treatment of hirsute women with polycystic ovary syndrome: a clinical, endocrinological, metabolic pilot study. J Clin Endocrinol Metab 2004;89:2817–2823.
13. Van der Spuy ZM, le Roux PA. Cyproterone acetate for hirsutism. Cochrane Database Syst Rev 2003;CD001125.
14. Barth JH, Cherry CA, Wojnarowska F, Dawber RP. Cyproterone acetate for severe hirsutism: results of a double-blind dose-ranging study. Clin Endocrinol (Oxf) 1991;35:5–10.
15. McMullen GR, Van Herle AJ. Hirsutism and the effectiveness of spironolactone in its management. J Endocrinol Invest 1993;16:925–932.
16. Lee O, Farquhar C, Toomath R, Jepson R. Spironolactone versus placebo or in combination with steroids for hirsutism and/or acne. Cochrane Database Syst Rev 2000;CD000194.
17. Lobo RA, Shoupe D, Serafini P, Brinton D, Horton R. The effects of two doses of spironolactone on serum androgens and anagen hair in hirsute women. Fertil Steril 1985;43:200–205.
18. Moghetti P, Tosi F, Tosti A, et al. Comparison of spironolactone, flutamide and finasteride efficacy in the treatment of hirsutism: a randomized, double-blind, placebo-controlled trial. J Clin Endocrinol Metab 2000;85:89–94.
19. Ganie MA, Khurana ML, Eunice M, et al. Comparison of efficacy of spironolactone with metformin in the management of polycystic ovary syndrome: an open-labeled study. J Clin Endocrinol Metab 2004;89:2756–2762.
20. Simard J, Luthy I, Guay J, Belanger A, Labrie F. Characteristics of interaction of the antiandrogen flutamide with the androgen receptor in various target tissues. Mol Cell Endocrinol 1986;44:261–270.
21. Cusan L, Dupont A, Belanger A, Tremblay RR, Manhes G, Labrie F. Treatment of hirsutism with the pure antiandrogen flutamide. J Am Acad Dermatol 1990;23:462–469.
22. Gambineri A, Pelusi C, Genghini S, et al. Effect of flutamide and metformin administered alone or in combination in dieting obese women with polycystic ovary syndrome. Clin Endocrinol (Oxf) 2004;60:241–249.
23. Diamanti-Kandarakis E, Mitrakou A, Raptis S, Tolis G, Duleba AJ. The effect of a pure antiandrogen receptor blocker, flutamide, on the lipid profile in the polycystic ovary syndrome. J Clin Endocrinol Metab 1998;83:2699–2705.
24. Wysowski DK, Freiman JP, Tourtelot JB, Horton ML. Fatal and nonfatal hepatotoxicity associated with flutamide. Ann Intern Med 1993;118:860–864.
25. Kelestimur F, Everest H, Unluhizarci K, Bayram F, Sahin Y. A comparison between spironolactone and spironolactone plus finasteride in the treatment of hirsutism. Eur J Endocrinol 2004;150:351–354.

26. Harborne L, Fleming R, Lyall H, Sattar N, Norman J. Metformin or antiandrogen in the treatment of hirsutism in polycystic ovary syndrome. J Clin Endocrinol Metab 2003;88:4116–4123.

27. Azziz R, Ehrmann D, Legro RS, et al. PCOS/Troglitazone Study Group. Troglitazone improves ovulation and hirsutism in the polycystic ovary syndrome: a multicenter, double blind, placebo-controlled trial. J Clin Endocrinol Metab 2001;86:1626–1632.

35

Polycystic Ovary Syndrome and Ovulation Induction

Bulent O. Yildiz and Ricardo Azziz

SUMMARY

Polycystic ovary syndrome (PCOS) is the most common cause of anovulatory infertility. Although many options are available for ovulation induction in these patients, there is currently no evidence-based algorithm to guide the initial and subsequent choices of medical ovulation induction methods. In obese women with PCOS, mild to moderate weight loss results in improvement of ovulatory dysfunction and should be advocated at the onset of the evaluation. Clomiphene citrate is currently the first-line pharmacological therapy for ovulation induction, although metformin appears to also be a promising agent for first-line therapy, at least in some patients. Studies are currently ongoing to address which first-line agent (clomiphene or metformin) is of greater benefit and/or which subpopulations of patients with PCOS will benefit from either treatment alone or in combination. Alternatively, glucocorticoids do not result in consistent ovulation and have significant side effects, although they may have a role in the treatment of the patient who is clomiphene resistant. These patients may also benefit from the addition of metformin. In this setting, exogenous pulsatile gonadotropin-releasing hormone (GnRH) treatment has also been advocated, although overall it has low ovulation and pregnancy rates, with a high risk of miscarriage. The most commonly used medical agents for ovulation induction in clomiphene-resistant women with PCOS, or those who fail to conceive following treatment with first-line agents and their adjuvants, are parenteral gonadotropins. Various gonadotropin preparations and different protocols are available; however, the risk of multiple pregnancy and ovarian hyperstimulation is high with gonadotropin therapy in the patient with PCOS. As such, various investigators advocate proceeding first to laparoscopic ovarian drilling, possibly combined with retreatment with clomiphene and/or metformin or, in an effort to control more aggressively the number of multiples, to in vitro fertilization.

Key Words: Ovulation induction; clomiphene citrate; human menopausal gonadotropins; menotropins; glucocorticoids; metformin; polycystic ovary syndrome.

1. INTRODUCTION

Polycystic ovary syndrome (PCOS), characterized by hyperandrogenism, ovulatory dysfunction, and polycystic ovaries, is the most common endocrine disorder of reproductive-aged women, with an estimated prevalence of 6–7% (1). In clinical practice, women with PCOS typically present complaining of menstrual irregularity, hirsutism, acne, alopecia, or infertility. Overall, PCOS has been reported to be the leading cause of anovulatory infertility (2). In this chapter, we review the nonsurgical treatment options for the treatment of oligo-ovulatory infertility in PCOS.

From: *Contemporary Endocrinology: Androgen Excess Disorders in Women:*
Polycystic Ovary Syndrome and Other Disorders, Second Edition
Edited by: R. Azziz et al. © Humana Press Inc., Totowa, NJ

2. BACKGROUND

2.1. Lifestyle Modification in the Infertile Patient With PCOS

Obesity is a very common feature of women with PCOS, with an estimated prevalence of 35–63% among women with the disorder *(3)*. Ghrelin homeostasis and measures of hunger and satiety are significantly impaired in subjects with PCOS, although not affected by dietary composition *(4)*. Abdominal obesity, characterized by a waist-to-hip ratio of more than 0.8, has been reported in 63% of women with PCOS whether they are obese or not *(5,6)*. Obesity, particularly abdominal, is often associated with insulin resistance and hyperinsulinemia, which stimulates the biosynthesis of androgens and the decreased hepatic production of sex hormone-binding globulin (SHBG). Other factors such as increased estrogen production rate, increased activity of the opioid system and of the hypothalamic–pituitary–adrenal axis, and, possibly, high dietary lipid intake may be mechanisms by which obesity worsens the degree of hyperandrogenism and ovulatory function in PCOS *(7)*. Obesity is associated with a reduced chance of ovulation *(8)*. Moreover, overweight and obese women with PCOS are less likely to achieve pregnancy spontaneously or with medical assistance, are more likely to miscarry, have a higher prevalence of fetal abnormality, and suffer more pregnancy complications *(9)*.

Several studies of lifestyle modification in obese women with PCOS have demonstrated that weight loss resulted in a reduction in hyperandrogenism *(10)*, an improvement in ovulatory and conception rates *(5,11,12)*, and a decrease in the miscarriage rates *(11)*. Mild to moderate weight loss of only 2–5% of body weight often results in the restoration of regular vagina bleeding suggestive of normal ovulation *(13)*. These results are often associated with decreased androgen levels, improved insulin sensitivity, and reduction of abdominal fat, although the degree of reduction of hyperinsulinemia seems to be the key factor responsible for the beneficial effect of weight loss on ovulatory dysfunction *(13,14)*. Nevertheless, improvements in menstrual irregularity and ovulation rates with weight reduction do not occur universally in all obese women with PCOS *(13–15)*. A recent study evaluating the effects of caloric restriction and dietary composition in women with PCOS observed that approximately 50% of subjects did not respond to treatment with improved menstrual cyclicity (i.e., the nonresponders) *(16)*. In that study, both nonresponders and responders demonstrated similar degrees of abdominal fat reduction; the only significant difference between the two groups was a greater reduction in insulin resistance in responders *(16)*.

In obese women with PCOS, weight loss through lifestyle modification should be considered a primary intervention for the treatment of oligo-ovulatory infertility. The optimal means to achieve and maintain weight loss and the most reliable predictors of response to lifestyle modification have yet to be determined, although an improvement in insulin sensitivity and a reduction in circulating insulin levels appears to be important. Although some investigators have suggested that calorie restriction with a diet low in saturated fat and high in fiber from predominantly low glycemic-index carbohydrate foods be recommended *(17)*, the optimum type of diet for PCOS patients remains to be determined prospectively. Similarly, the relative impact of exercise regimen, behavior modification, and drug therapy on weight reduction in obese PCOS patients has yet to be defined.

2.2. Pharmacological Options for Ovulation Induction in PCOS

In women with PCOS who are not obese or overweight, who are obese but unable to lose weight, or who fail to ovulate despite weight loss, ovarian stimulation with pharmacological agents is the next step in the treatment of oligo-ovulatory infertility. Clomiphene citrate is currently the first-line pharmacological therapy for ovulation induction, although metformin appears to also be a promising agent for first-line therapy, at least in some patients. Studies are currently ongoing to address which first-line agent (clomiphene or metformin) is of greater benefit and/or which subpopulations of patients with PCOS will benefit from either treatment alone or in combination. Alternatively, glucocorticoids do not result in consistent ovulation and have significant side effects. Patients who are resistant to ovulation with clomiphene may benefit from the addition of metformin or may proceed to laparoscopic ovarian drilling or exogenous pulsatile GnRH treatment. However, the most commonly used medical agents for ovulation induction in clomiphene-resistant women with PCOS, or those who fail to conceive following treatment with first-line agents and their adjuvants, are parenteral gonadotropins.

2.2.1. Clomiphene Citrate

2.2.1.1. MECHANISM OF ACTION

The antiestrogenic compound clomiphene citrate currently represents the first-line medical treatment of ovulatory dysfunction in PCOS. This easy-to-use, convenient, inexpensive, and relatively safe drug has been used extensively since the first report of clomiphene-induced ovulation in 1961 (18) and its first approval for clinical use by the US Food and Drug Administration (FDA) in 1967.

Clomiphene is an oral synthetic triphenylethylene derivative with estrogen agonist/antagonist characteristics (19). The clinically available preparations contain an approximate 3:2 mixture of two stereoisomers, enclomiphene and zuclomiphene, which show distinctly different patterns of agonistic and antagonistic activity in vitro (20). Clomiphene is metabolized in the liver, and its biological half-life is reported to be 5 days (21). The drug is contraindicated in patients with liver disease, endometrial carcinoma, undiagnosed abnormal uterine bleeding, ovarian cysts not rrelated to PCOS, and during pregnancy.

Clomiphene competes with estradiol for binding to estrogen nuclear receptor protein in a variety of estrogen-responsive tissues, including the hypothalamus, pituitary, ovary, endometrium, and vaginal mucosa. Most of the clinical action of clomiphene in initiating ovulation is attributed to its antiestrogenic effect, whereby clomiphene occupies the estrogen receptor in the hypothalamus for prolonged periods of time, creating a hypo-estrogenic state and diminishing the normal ovarian–hypothalamic estrogen feedback (20). This effect increases the GnRH pulse amplitude, which in turn results in increased follicle-stimulating hormone (FSH) and luteinizing hormone (LH) secretion from the pituitary and subsequent ovarian follicular development (22).

2.2.1.2. CLINICAL USE

Clomiphene is available in the form of 50-mg tablets. Empirically, it is administered at a starting dose of 50–100 mg/day for 5 days, beginning on days 2–5 after either a spontaneous or progestogen-induced menses. Careful monitoring is important to achieve optimum results. The methods used to predict or confirm ovulation include basal body temperature charting, urinary LH surge testing, measurement of mid-luteal phase progesterone, and serial ultrasound monitoring. Ultrasonography, at least during the first treatment cycle, provides accurate monitoring of follicular development (23). Important prognostic indicators of a conception cycle include the number and size of preovulatory follicles and endometrial thickness. More than one preovulatory follicle can be expected in approximately one-third of clomiphene-treated cycles, even at the lowest dose required for an ovulatory response (24). Monitoring by ultrasonography enables identification of patients who do not develop a dominant follicle (i.e., generally measuring 16–18 mm or greater), and therefore require an increased dose in the next cycle, and those who overrespond to standard doses of clomiphene, requiring cycle cancelation and a lower dose in subsequent cycles (25). If ovulation fails to occur, the dose of clomiphene is increased in a stepwise fashion by 50-mg increments, up to 150–200 mg/day for 5 days, until either ovulation or the maximum dose is achieved. Overall, approximately 80% of all patients receiving clomiphene ovulate and 40% (one-half of those ovulating) conceive (26). Few patients ovulate on doses of 150 mg/day or more, and conceptions occur rarely at these higher doses (27,28). Almost all pregnancies occur within the first six ovulatory cycles, and continuing clomiphene therapy beyond this is of little benefit. In addition, we should note that FDA has approved the use of clomiphene only in a dose up to 100 mg/day for 5 days and for only three cycles.

2.2.1.3. Predictors of Response and Treating Failure to Ovulate With Clomiphene

Approximately 20% of women with PCOS do not ovulate on the maximum clomiphene dose (i.e., clomiphene-resistant patients). For these women, a number of therapeutic alternatives are available, including extending the duration of clomiphene administration, generally with the addition of midcycle human chorionic gonadotropin (hCG). An extended regimen of clomiphene consisting of 250 mg of clomiphene for 8 days followed by the administration of 10,000 IU of hCG 6 days later has been proposed *(29)*. However, treating 13 oligomenorrheic women who had previously failed to ovulate when treated with 250 mg of clomiphene for 5 days and hCG, these investigators were able to achieve only three pregnancies during 25 treatment cycles, a very low rate of success. Other adjuvants include the administration of midcycle hCG *(see* below), the addition of an insulin sensitizer *(see* Subsection 2.3.) or glucocorticoids *(see* Subsection 2.2.2.), performing laparoscopic ovarian drilling *(see* Chapter 37), or proceeding to pulsatile GnRH or gonadotropin *(see* Subsections 2.2.4. and 2.2.5.).

The administration of an ovulatory dose of hCG (2500–10,000 USP units im) when follicle maturity is confirmed sonographically is suggested in those patients who do not exhibit an LH surge despite ultrasonographic evidence of a mature follicle of appropriate size (>18–20 mm in diameter) and acceptable endometrial development (>8–10 mm in thickness). Ovulation typically occurs 34–46 hours after hCG administration *(30)*. However, this adjuvant therapy is rarely warranted because most treatment cycles in which ovulation does not occur following the administration of clomiphene are characterized by failure of development of a dominant follicle *(23)*.

Obesity is associated with a decreased response to clomiphene in women with PCOS *(10,31,32)*. A study by Murakawa et al. *(33)* reported that PCOS patients who do not respond to clomiphene had higher androgen levels and insulin resistance than those who respond to clomiphene. In this study, the receiver operating characteristic curve of the area under the curve of insulin provided the most appropriate cutoff point for the prediction of clomiphene resistance *(33)*. In a series of studies Imani and colleagues determined the factors predicting an ovulatory response and chance of conception in patients with World Health Organization group 2 ovulatory dysfunction (although not necessarily PCOS) treated with clomiphene. Patients who were less likely to ovulate were mostly those with higher body mass index (BMI), higher androgen levels, or a greater ovarian volume *(34)*. These data also suggest that decreased insulin sensitivity, hyperandrogenemia, and obesity, all associated with PCOS, are prominent factors involved in reducing the probability that the ovaries will respond to clomiphene. Although insulin and insulin-like growth factor binding protein (IGFBP)-1 levels predicted patients who were not responsive to clomiphene, this relationship disappeared when the free androgen index (FAI; measured by the testosterone-to-SHBG ratio), was entered into the model *(35)*, suggesting that insulin sensitivity is most likely associated with abnormal ovarian function through its action on androgens. Factors predicting the chances for a conception and live birth included the FAI, BMI, severity of menstrual disturbance (oligomenorrhea vs amenorrhea), and maternal age *(36,37)*.

2.2.1.4. Results and Side Effects

Overall, approximately 40–60% of patients will conceive following six cycles of ovulation induction with clomiphene *(27,28)*. Pregnancies achieved with clomiphene are usually singletons, with a multiple pregnancy incidence of about 7–10%, mostly twins *(38)*. Miscarriage and ectopic pregnancy rates are found to be comparable with those in general population *(25,39)*. There is no increased risk of congenital abnormalities with clomiphene use *(40,41)*.

Side effects include hot flushes, bloating, nausea and vomiting, breast tenderness, headache, intermenstrual spotting, and menorrhagia. Less commonly, visual symptoms may occur, warranting cessation of the therapy. One of the rare complications of clomiphene therapy is a mild form of ovarian hyperstimulation syndrome (OHSS). After an unsuccessful treatment cycle, there may be persistence of one or more functional ovarian cysts; however, there are no studies on the effect of residual ovarian cysts on the outcome of subsequent clomiphene-treated cycles.

The relationship between ovarian cancer and clomiphene use remains controversial. Infertility caused by ovulatory dysfunction may itself increase the risk for the development of ovarian cancer, and available data in the literature for a causal relationship between clomiphene and ovarian cancer are contradictory *(42)*. Regardless, patients should be counseled regarding this potential risk before beginning treatment with clomiphene.

Of those women who are clomiphene responders, approximately 40% fail to conceive despite having ovulatory cycles *(23,26)*. Several studies have suggested that the discrepancy between ovulation and pregnancy rates in these patients could be the result of negative effects of clomiphene on oocyte or granulosa cells or on endometrial receptivity and cervical mucus *(43,44)*. However, other investigators have not been able to confirm these findings *(24,45)*. Women with high LH levels are also reported to be less likely to have a successful pregnancy with clomiphene treatment *(46)*. Alternatively, more than one-third of patients failing to conceive after four ovulatory cycles of clomiphene citrate have significant intrapelvic pathology *(47)*, suggesting that a laparoscopy is indicated in women who fail to conceive on clomiphene despite verified ovulation. Women who fail to conceive past successful ovulation induction with clomiphene and have minimal or repairable intrapelvic pathology are candidates for gonadotropin ovulation induction *(see* Subsection 2.2.5.).

2.2.2. Glucocorticoids for Ovulation Induction

The role of adrenal hyperandrogenism in producing oligo-ovulation is unclear, because both dehydroepiandrosterone and dehydroepiandrosterone sulfate (DHEAS) are relatively weak androgens. Nevertheless, DHEAS circulates in a concentration 10,000 times that of testosterone, and thus may result in significant androgenicity. Furthermore, circulating DHEAS has been found to be the precursor for almost 50% of testosterone within follicular fluid in women being treated with menotropins *(48)*. Thus, it is possible that excess circulating DHEAS results in elevated levels of intrafollicular testosterone and a higher risk of follicular atresia. Consistent with this, a number of investigators have suggested a beneficial effect of glucocorticoid administration, alone or in combination with other agents, on ovulatory function in these patients *(49–51)*.

2.2.2.1. GLUCOCORTICOID-ONLY THERAPY

Jones et al. and Greenblatt first reported in 1953 the use of corticosteroid suppression in the treatment of ovulatory dysfunction *(52,53)*. Subsequently, a number of investigators have reported improved of menstrual regularity in 30–66% of oligo-ovulatory patients treated with glucocorticoids only *(54–57)*. However, it should be noted that these reports primarily noted changes in menstruation, without further documentation of ovulatory function.

We prospectively studied the impact of dexamethasone 0.5 mg/day for 4 months in 36 hirsute/oligo-ovulatory patients *(58)*. Ovulatory function was documented by basal body temperature charting and a day 22–24 progesterone level. With therapy, all patients demonstrated a significant decrease in all androgens (–40 to –60%), a 24% increase in SHBG, and no change in LH/FSH. Mean body weight increased by more than 4 kg (4.4%) during treatment. Of the 138 cycles monitored, 78% remained anovulatory, with no difference between the first and fourth cycle (25 vs 20%, respectively). Of the 36 patients studied, 50% did not demonstrate a single ovulatory cycle, and of the remaining, 28% had only one, 14% had two, and 8% had three ovulatory cycles. There were no significant differences in physical features, basal hormones, including DHEAS, adrenal response to adrenocorticotropin hormone stimulation, or hormonal levels at the end of treatment between those women ovulating and those not. These data did not support the use of continuous glucocorticoid suppression for ovulation induction in PCOS, regardless of basal DHEAS levels. Furthermore, this treatment was associated with significant side effects, notably weight gain.

2.2.2.2. GLUCOCORTICOIDS IN IMPROVING OVULATION INDUCTION WITH CLOMIPHENE CITRATE

Dexamethasone in doses ranging from 150 to 250 mg/day appears to enhance clomiphene ovulation induction in patients who are resistant to clomiphene alone. Pregnancy rates have ranged from 8 to 69% *(49,51,59,60)*. While some investigators used higher dosages for short intervals (e.g., 2 mg/day from days 5–14 of the cycle) *(59)*, other investigators have used a continuous lower-dose (0.25–0.5 mg/day) regimen *(49,51)*. Clomiphene-resistant women with PCOS who responded to the combination of dexamethasone and clomiphene generally had higher mean circulating DHEAS levels than nonresponders *(51)*, although not all investigators agree. In a randomized, double-blind, placebo-controlled clinical trial, Parsanezhad and colleagues studied 230 women with PCOS and normal DHEAS levels who failed to ovulate after a routine protocol of clomiphene citrate; the treatment group received 200 mg of clomiphene citrate from day 5 to day 9 and 2 mg of dexamethasone from day 5 to day 14 of the menstrual cycle, for a maximum of six cycles *(60)*. The control group received the same protocol of clomiphene citrate combined with placebo. Mean follicular diameters (18.4 ± 2.4 mm vs 13.8 ± 2.1 mm, respectively), ovulation rates (88 vs 20%, respectively), and cumulative pregnancy rates (40.5 vs 4.2%, respectively) were greater in the treatment group compared to controls.

Attempting to determine whether all anovulatory hyperandrogenic women receiving clomiphene would benefit from concomitant glucocorticoid suppression, Daly and colleagues carried out a prospective study in 50 anovulatory patients, randomly assigned to receive either 50 mg clomiphene citrate on days 5–9, alone or with 0.5 mg of dexamethasone *(50)*. Their results suggested a significantly higher rate of ovulation and conception in dexamethasone- and clomiphene-treated patients. Furthermore, the improvement in pregnancy rate was noted only for those patients with circulating DHEAS levels greater than 200 μg/dL. Other investigators, however, have not been able to correlate circulating DHEAS or 11-OHA4 levels, with ovulatory response to dexamethasone and clomiphene in PCOS *(61)*. Thus, it remains unclear whether the administration of dexamethasone to all women with PCOS undergoing clomiphene ovulation induction improves pregnancy rates, and whether this benefit can be predicted by basal circulating adrenal androgen levels.

2.2.2.3. GLUCOCORTICOIDS IN IMPROVING OVULATION INDUCTION WITH GONADOTROPINS

Few studies are available determining the value of glucocorticoid suppression in patients undergoing ovulation induction with human menopausal gonadotropin (hMG). In one observational uncontrolled study, 27 women with PCOS who had failed to conceive on hMG-only therapy were treated with dexamethasone 0.5 mg/day throughout the cycle along with hMG in the usual protocol *(62)*. Of these women, 74% conceived and 55% delivered at term. Prior to dexamethasone, 40% of cycles were ovulatory (but not conceived), whereas 66% were ovulatory after the addition of glucocorticoid. However, it is difficult to fully interpret these data because of the uncontrolled nature of the report. No prospective randomized and controlled studies are available. In a controlled trial, 290 patients were randomized to receive either 1 mg dexamethasone or placebo, in addition to a standard long protocol GnRH analog with gonadotropin stimulation regime *(63)*. A significantly lower cancelation rate for poor ovarian response was observed in the dexamethasone-treated group compared with controls (2.8% vs 12.4%, respectively), independent of the presence of polycystic and normal ovaries. Alternatively, there were no significant differences in the median fertilization rates, implantation rates, and pregnancy rate per cycle started. These results were similar to those of an earlier pilot study *(64)*. Overall, at least for in vitro fertilization, the addition of glucocorticoid suppression does not appear to significantly improve ovulatory response to gonadotropins.

2.2.3. Insulin-Sensitizing Agents

Insulin resistance with compensatory hyperinsulinemia is a prominent feature of PCOS diagnosed in both lean and obese patients *(see* Chapter 24). The exact mechanisms for abnormalities of insulin action in the syndrome have yet to be elucidated *(65)*. However, hyperinsulinemia has been show to increase ovarian androgen biosynthesis *(66)* and decrease hepatic synthesis of SHBG *(67)*, leading to increased bioavailability of free androgens. The increase in local ovarian androgen production mediated by hyperinsulinemia can also result in premature follicular atresia and anovulation *(68)*.

The strong association between PCOS and insulin resistance and the role of hyperinsulinemia in hyperandrogenism and disrupted folliculogenesis provide the rationale for the use of insulin sensitizers in the treatment of the syndrome. It seems logical that the therapeutic interventions directed at increasing insulin sensitivity, thereby decreasing hyperinsulinemia, would ameliorate the hyperandrogenism and improve ovulatory dysfunction in women with PCOS. Indeed, insulin sensitizers are becoming widely used in PCOS, both as single ovulatory agents and as an adjuvant to other ovulation-induction drugs, principally clomiphene. However, we should note that not all women with PCOS have documented insulin resistance, and no measure of insulin sensitivity is included in the current diagnostic criteria of PCOS. Moreover, there are no generally acknowledged guidelines or criteria for the clinical quantification of insulin resistance *(69)*. Nevertheless, up to 65% of patients with PCOS can actually be classified as having insulin resistance *(70–72)*.

A number of insulin-sensitizing agents with different mechanisms of action have been used in the setting of PCOS. These medications include biguanide oral hypoglycemic agents such as metformin and thiazolidinediones such as troglitazone, rosiglitazone, and pioglitazone.

2.2.3.1. METFORMIN

Metformin is the most extensively used insulin-sensitizing agent for ovulation induction in PCOS *(73)*. Metformin is a biguanide antihyperglycemic that has been used to treat type 2 diabetes mellitus since the early 1970s in Europe and was introduced (actually reintroduced) in the United States in 1994. The glucose-lowering effects of metformin are mainly a consequence of reduced hepatic glucose output (primarily through inhibition of gluconeogenesis and, to a lesser extent, glycogenolysis) and increased insulin-stimulated glucose uptake in skeletal muscle and adipocytes. By increasing insulin sensitivity, metformin reduces peripheral and hepatic insulin resistance, insulin secretion, and hyperinsulinemia *(74)*.

There has been increasing interest in the use of metformin for the treatment of PCOS since the first report by Velazquez et al. *(75)* in 1994. Recently, the effects of metformin for ovulation induction were assessed in a meta-analysis including 13 randomized controlled trials and 543 participants *(73)*. The dose range of the metformin used in these trials was 1500–1700 mg/day. There was a significant effect of metformin in achieving ovulation in women with PCOS, with odds ratios (OR) of 3.88 (95% confidence interval [CI] 2.25–6.69, $p < 0.00001$) for metformin vs placebo and 4.41 (95% CI 2.37–8.22, $p < 0.00001$) for metformin plus clomiphene vs clomiphene alone *(73)*. The overall ovulation rate achieved by metformin, or metformin and clomiphene, was reported as 57%. The data from this meta-analysis suggested that metformin acts relatively quickly to improve ovulation, with significant treatment effects reported after 2 months. We should note that there was a significant correlation between trial length and the proportion of patients ovulating on placebo, suggesting that the treatment effect of metformin will appear less with the longer trials becaus of the higher rate of spontaneous ovulation in the placebo arm *(73)*. Alternatively, and in contrast to its beneficial effect in clomiphene-resistant patients, metformin does not enhance ovulatory or pregnancy rates in gonadotropin ovulation-induction cycles *(76,77)* or in vitro fertilization *(78)*, although current data are limited.

It has been suggested that metformin could be used as a first-line agent for ovulation induction in women with PCOS, instead of clomiphene. Recently, Palomba et al. *(79)* reported the results of a randomized controlled trial comparing metformin with clomiphene as the first-line treatment for ovulation induction in non-obese PCOS patients (Table 1). One hundred non-obese PCOS patients were enrolled and randomized to metformin (850 mg twice daily) plus placebo or clomiphene (150 mg for 5 days from the third day of a progesterone withdrawal bleeding) plus placebo for 6 months. At the end of the study, the ovulation rate was not significantly different between the two treatment groups, whereas patients receiving metformin demonstrated a higher continuing pregnancy rate (15.1 vs 7.2%, respectively, $p = 0.009$) and a lower miscarriage rate (9.7 vs 37.5%, respectively, $p = 0.045$) compared to clomiphene. These promising results remain to be confirmed in larger studies. Additionally, it is important to note that ovulation and pregnancy rates are surrogate outcomes, and live birth

Table 1
Ovulation and Pregnancy Rates in PCOS Women Randomized to Treatment
With Metformin Cloridrate or Clomiphene Citrate (CC) During Each Cycle of Treatment

	Ovulation rate [No. ovulatory cycles/no. cycles (%)]			Pregnancy rate [No. pregnancies/no. cycles (%)]		
	Metformin	CC	p	Metformin	CC	p^a
Cycle						
1	19/45 (42.2)	39/47 (83.0)	<0.001	3/45 (6.7)	6/47 (12.8)	0.49
2	24/42 (57.1)	33/41 (80.5)	0.02	4/42 (9.5)	5/41 (12.2)	0.74
3	25/38 (65.8)	25/36 (69.4)	0.74	6/38 (15.8)	2/36 (5.6)	0.26
4	22/32 (68.8)	19/34 (55.9)	0.28	5/32 (15.6)	2/34 (5.9)	0.25
5	21/27 (77.8)	17/32 (53.1)	0.049	6/27 (22.2)	1/32 (3.1)	0.04
6	18/21 (85.7)	15/31 (48.4)	0.006	7/21 (33.3)	0/31 (0.0)	0.001

Data were analyzed using χ^2 text unless otherwise specified.
[a]Fisher's exact test.
PCOS, polycystic ovary syndrome.

rate should be the primary outcome in future studies. One such large multicenter randomized double-blind study is currently being conducted by The National Institute of Child Health and Human Development's Reproductive Medicine Network comparing clomiphene vs metformin vs clomiphene plus metformin for the induction of ovulation in women with PCOS; these data should be available by the end of 2006.

Observational studies suggested that metformin may reduce the risk of first-trimester miscarriage and may also result in a more predictable follicular response to ovulation induction, reducing the risk of ovarian hyperstimulation and multiple pregnancy *(80)*. However, these observations remain to be validated in large randomized controlled trials. Metformin is a category B pregnancy drug, and there is no evidence that this drug is teratogenic. Uncontrolled observational studies suggest that metformin therapy during pregnancy in women with PCOS appears to be safe *(81,82)*. However, larger longitudinal follow-up trials are required to determine the full effects of this drug on the neonate and child.

2.2.3.2. THIAZOLIDINEDIONES

The thiazolidinediones are insulin-sensitizing agents that improve insulin resistance by acting on the intranuclear hormone receptor peroxisome proliferator activated receptor-γ. Enhanced insulin sensitivity is noted particularly in the liver, adipose tissue, and muscle, reducing insulin resistance and improving pancreatic β-cell function.

Troglitazone was the first member of this category used in the treatment of PCOS *(83,84)*. In a large, multicenter, double-blindm placebo-controlled dose-determining study of 305 patients with PCOS, 60% of "expected" cycles were ovulatory on troglitazone 600 mg/day after 6 months of treatment, compared with a baseline rate of 30% in the placebo group *(84)* (Fig. 1). In addition to increasing the rates of ovulation, troglitazone has been shown to improve androgen levels and insulin sensitivity *(83,85)*. However, troglitazone was withdrawn from the market because of an increased, albeit minimal, risk of hepatotoxicity and fulminant liver failure.

The more recently introduced thiazolidinediones, rosiglitazone and pioglitazone, appear not to have the same degree of hepatic side effects observed for troglitazone *(86)*. However, only a few studies are available, with small numbers of participants, that address the effects of these new agents in the treatment of PCOS *(64,87–90)*. In an uncontrolled study of rosiglitazone and clomiphene vs clomiphene for ovulation induction in PCOS, the rates of improvement in menstrual pattern were 92 and 68%, respectively *(91)*. In a randomized, double-blind, placebo-controlled study of 25 clomiphene-resistant PCOS patients, ovulation rates of 33 and 77%, respectively, were observed after rosiglitazone alone vs rosiglitazone and clomiphene *(92)*. In another randomized, controlled double-

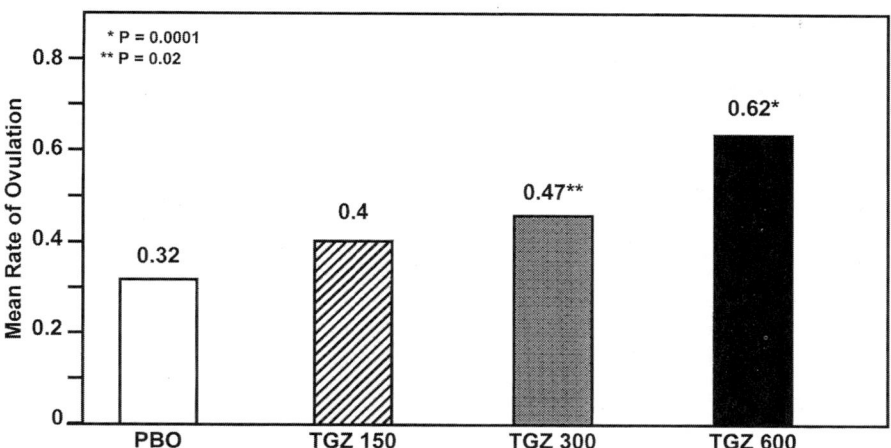

Fig. 1. The mean rate of ovulation in patients with polycystic ovary syndrome (PCOS) increased in a dose-related fashion with troglitazone (TGZ) treatment and was significantly different from placebo (PBO) for the patients randomized to treatment with TGZ 300 mg/day (TGZ 300) and 600 mg/day (TGZ 600). (From ref. *84*.)

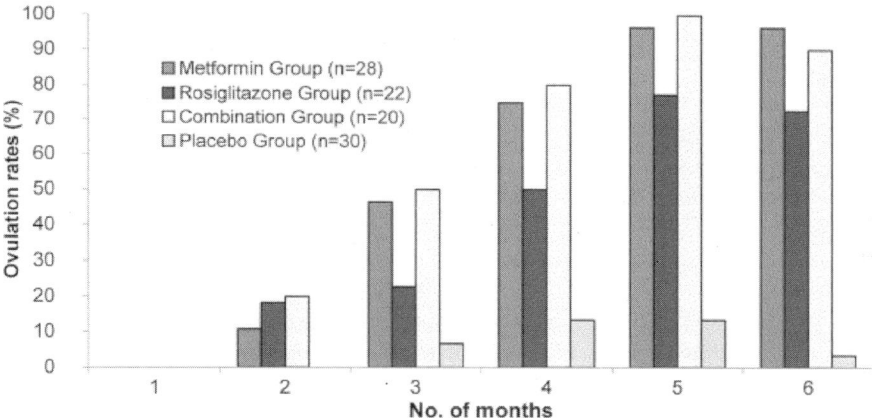

Fig. 2. Monthly ovulation rates in non-obese women with polycystic ovary syndrome and normal indices of insulin sensitivity after administration of insulin-sensitizing drugs or placebo for 6 months. Values are the number of women who ovulated during each month in a group divided by the total number of subjects in that group. $p < 0.001$ for differences among months using mixed-model repeated-measures logistic regression. (From ref. *93*.)

blind trial, Baillargeon and colleagues randomized 100 non-obese PCOS women with normal indices of insulin sensitivity to receive metformin (850 mg twice a day), rosiglitazone (4 mg twice a day), the combination of both drugs, or at least one placebo for 6 months *(93)*. The ovulatory response was higher after treatment with an insulin-sensitizing drug compared to placebo, and the ovulatory frequency was significantly greater with metformin than rosiglitazone, while the combination was not more effective (ovulations per subject in 6 months: metformin 3.3, rosiglitazone 2.4, combination 3.4, and placebo 0.4) (Fig. 2).

Overall, insulin-sensitizing agents appear to improve the ovulatory response of PCOS, regardless of weight. Several points remain to be addressed in further studies regarding use of these agents for ovulation induction in women with PCOS, including the optimum dose and duration of therapy, the pregnancy and neonatal outcome, and the variability in ovarian response. Current evidence suggests that insulin-sensitizing agents may be beneficial in a subset, but probably not all PCOS patients. Of particular importance is whether therapy with insulin-sensitizing agents should be limited to patients with documented insulin resistance. Needless to say, the greatest challenge remains to predict who will benefit and who will not.

2.2.4. Pulsatile GnRH

Exogenous pulsatile GnRH has been used in an attempt to simulate the physiological pattern of pulsatile gonadotropin secretion, with the expected outcome of monofollicular development. It was suggested that this approach was more "physiological," potentially avoiding the risks of OHSS and multiple gestation. However, the results in women with PCOS have not been promising *(94,95)*. Exogenous GnRH stimulation often caused brisk gonadotropin secretion that resulted in the maturation of more than a single dominant follicle with a high risk of multiple pregnancy. Even after pretreatment with a GnRH agonist, the rates of ovulation and pregnancy were low and miscarriage rates were as high as 45% *(95)*. Thus, pulsatile GnRH therapy is rarely used today for ovulation induction in women with PCOS *(96)*.

2.2.5. Gonadotropin Ovulation Induction

Exogenous gonadotropins have traditionally been used in PCOS patients who are resistant to ovulation induction with clomiphene and more recently those not responding to the addition of metformin or laparoscopic ovarian drilling. Gonadotropin preparations derived from hMG, a mixture that contains FSH, LH, and large quantities of urinary proteins, have been in use since the early 1960s *(97)*. Other gonadotropin preparations in use today include purified urinary FSH (uFSH) and recombinant FSH (rFSH). Highly purified uFSH contains a reduced amount of LH and very small amounts of urinary proteins. The lack of urinary proteins in this preparation reduces adverse reactions such as local allergy or hypersensitivity *(98)*. Preparations of rFSH were recently developed with a complete absence of LH and co-purified proteins, giving high specific bioactivity. These preparations share similar pharmacokinetic characteristics with purified uFSH *(99)*. Whereas hMG is administered intramuscularly, uFSH and rFSH can be given by subcutaneous injection, which can be self-administered and appear to be better tolerated by the patient. There are no data suggesting a higher pregnancy rate in PCOS with the use of one product compared with another *(100–102)*.

In general, ovulation induction with gonadotropins in clomiphene-resistant PCOS patients is less successful than in patients with hypogonadotrophic hypogonadism *(103)*. However, women with PCOS are more sensitive to gonadotropin stimulation compared with spontaneously cycling women. This increased sensitivity appears to result from a larger pool of small antral follicles available for recruitment, rather than on differences in the FSH threshold level *(104)*. Women with PCOS receiving gonadotropins for ovulation induction are particularly prone to a higher risk of overstimulation, multiple pregnancy, and OHSS rates *(105)*.

Inappropriately elevated serum LH levels are found in up to 70% of women with PCOS *(106)*. This persistent elevation in serum LH during the follicular phase has been correlated with decreased pregnancy and increased spontaneous abortion rates. These complications were also related to the higher LH levels commonly seen in women with PCOS *(46)*. Consequently, it has been suggested that purified uFSH would be more effective than hMG for ovulation induction in PCOS. However, a recent meta-analysis including 14 randomized, controlled trials did not find any differences in pregnancy rates between women with PCOS receiving uFSH or hMG *(107)*. The incidence of OHSS was lower, however, in women using uFSH compared with hMG (OR: 0.20; 95% CI 0.08–0.46). In this study no conclusions could be drawn regarding the relative effect of these preparations on miscarriage and multiple pregnancy rates because of insufficient reporting of these outcomes in the trials *(107)*.

The conventional regimen of ovulation induction with gonadotropins begins with a starting dose of 75–150 IU/day, with an increase of 75–150 IU/day every 3–5 days in those women with an inadequate response *(108)*. Regardless of the specific gonadotropin preparation used, this regimen generally induces multiple follicular development, resulting in high rates of multiple pregnancy and OHSS *(109)*. Administration of lower doses of gonadotropins in a stepwise fashion (low-dose step-up or step-down, or sequential step-up, step-down) has generally replaced the conventional regimen and is reported to be effective in significantly reducing risks. The low-dose step-up protocol *(110)* generally involves starting with a dose of 75 IU/day or less, which is increased after 14 days by 37.5 IU/day and every 7 days thereafter if no response is observed. The goal of the low-dose step-up protocol is to achieve the development of a single dominant follicle, rather than the development of multiple large follicles. This regimen is based on the concept of a follicular FSH threshold. The rationale is not to exceed the critical FSH concentration above which multiple follicles develop and result in increased risks of OHSS and multiple gestation *(111)*.

The step-down protocol *(112)* usually begins with a dose of 150–225 IU/day for 2 days, which is decreased to 75 IU/day for 7 days and increased to 150 IU/day after the seventh day if there is inadequate follicular response. More recently, a "sequential step-up, step-down" protocol has been used, in which the FSH dose is reduced by half when the leading follicle has reached 14 mm *(113)*. The purpose of the sequential step-up, step-down protocol is to mimic the normal menstrual cycle, in which the normal early follicular phase FSH elevation is followed by a decline until a small FSH peak accompanies the normal LH surge *(112)*.

Low-dose regimens appear to be a promising approach when using gonadotropins for ovulation induction in clomiphene-resistant women with PCOS. The mono-ovulatory cycle rate is approximately 70%. The cumulative pregnancy rate is approximately 40%, and can be as high as 20% per cycle, comparable with conventional regimens. However, a multiple pregnancy rate of 6% and a low rate of OHSS (<1%) with low-dose regimens make these protocols more attractive than conventional gonadotropin regimens *(114)*. A recent multicenter randomized study comparing the low-dose step-up and step-down protocols in clomiphene-resistant women with PCOS found that the low-dose step-up regimen of rFSH administration was as effective as, and safer than, the step-down regimen *(115)*. In this study, the rate of monofollicular development was significantly higher (68.2 vs 32%, $p < 0.0001$) and the rate of multifollicular development significantly lower (4.7 vs 36%, $p < 0.0001$) with the step-up protocol compared to the step-down regimen *(115)*.

3. CONCLUSIONS

There are many options for ovulation induction in anovulatory women with PCOS, although treatment should be individualized to the patient. Weight loss through lifestyle modification in obese women with PCOS has the beneficial effect of frequently restoring regular ovulation. When pharmacological agents are required, clomiphene citrate remains the first-line therapy for ovulation induction, pending studies evaluating its effectiveness to metformin. Approximately 20% of women with PCOS fail to ovulate on the maximum clomiphene dose, and therapeutic alternatives include extending the duration of clomiphene administration, generally with the addition of midcycle hCG; the addition of midcycle hCG, an insulin sensitizer, or glucocorticoids as adjuvants; or proceeding to gonadotropin or pulsatile GnRH therapy. The use of hCG as an adjuvant to clomiphene is generally not needed. Glucocorticoids are used with some success, particularly in clomiphene-resistant patients, but side effects are significant. Ovulatory response to insulin sensitizers is modest, but may not be much lower than that in response to clomiphene. Pulsatile GnRH has overall low rates of ovulation and pregnancy and high risk of miscarriage and is not generally used in PCOS. The most promising medical treatment option for the clomiphene-resistant women with PCOS is low-dose gonadotropin therapy, although it has an increased risk of multiple pregnancy and OHSS.

4. FUTURE AVENUES OF INVESTIGATION

Despite significant advances in the field of PCOS, induction of ovulation in women with PCOS remains a challenge for the clinical investigator and practicing physician. Future studies should address the alternative approaches to achieve and maintain weight loss in obese women with PCOS. It is also important to determine the predictors of response to lifestyle modification. The success rate of insulin-sensitizing agents for short-term reproductive outcomes is promising. Large, preferably multicenter randomized controlled studies investigating possible direct effects, optimal dose-finding studies, and treatment regimens under different conditions in different types of patients are warranted to optimize the efficacy of insulin-sensitizing agents and tailor treatment strategies to individual needs. Finally, the use of newer ovulatory agents, such as aromatase inhibitors, requires further study in large prospective randomized trials.

KEY POINTS

- Weight loss through lifestyle modification is the first intervention for the management of anovulatory infertility in obese women with PCOS.
- Clomiphene citrate, an easy-to-use, convenient, cheap, and relatively safe drug, is currently the first-line pharmacological therapy for ovulation induction in PCOS.
- Insulin sensitizers are potentially another first-line ovulation induction agent and are effective in improving ovulation rates in clomiphene-resistant patients.
- Glucocorticoids are used with some success, particularly in clomiphene-resistant patients, but side effects are significant.
- Pulsatile GnRH has overall low rates of ovulation and pregnancy and high risk of miscarriage and is not generally used in PCOS.
- Use of gonadotropins is a second-line treatment in PCOS patients who fail to respond to clomiphene citrate, preferably using a low-dose regimen to reduce the risk of OHSS and multiple births.

ACKNOWLEDGMENTS

Supported in part by an endowment from the Helping Hand of Los Angeles and by grants RO1-HD29364 and K24-D01346 from the National Institutes of Health.

REFERENCES

1. Azziz R, Woods KS, Reyna R, Key TJ, Knochenhauer ES, Yildiz BO. The prevalence and features of the polycystic ovary syndrome in an unselected population. J Clin Endocrinol Metab 2004;89:2745–2749.
2. Hull MG. Epidemiology of infertility and polycystic ovarian disease: endocrinological and demographic studies. Gynecol Endocrinol 1987;1:235–245.
3. Norman RJ, Davies MJ, Lord J, Moran LJ. The role of lifestyle modification in polycystic ovary syndrome. Trends Endocrinol Metab 2002;13:251–257.
4. Moran LJ, Noakes M, Clifton PM, et al. Ghrelin and measures of satiety are altered in polycystic ovary syndrome but not differentially affected by diet composition. Asia Pac J Clin Nutr 2003;12(Suppl):S52.
5. Pasquali R, Casimirri F, Venturoli S, et al. Body fat distribution has weight-independent effects on clinical, hormonal, and metabolic features of women with polycystic ovary syndrome. Metabolism 1994;43:706–713.
6. Kirchengast S, Huber J. Body composition characteristics and body fat distribution in lean women with polycystic ovary syndrome. Hum Reprod 2001;16:1255–1260.
7. Gambineri A, Pelusi C, Vicennati V, Pagotto U, Pasquali R. Obesity and the polycystic ovary syndrome. Int J Obes Relat Metab Disord 2002;26:883–896.
8. Grodstein F, Goldman MB, Cramer DW. Body mass index and ovulatory infertility. Epidemiology 1994;5:247–250.
9. Moran LJ, Norman RJ. The obese patient with infertility: a practical approach to diagnosis and treatment. Nutr Clin Care 2002;5:290–297.
10. Pasquali R, Casimirri F, Vicennati V. Weight control and its beneficial effect on fertility in women with obesity and polycystic ovary syndrome. Hum Reprod 1997;12(Suppl 1):82–87.
11. Clark AM, Thornley B, Tomlinson L, Galletley C, Norman RJ. Weight loss in obese infertile women results in improvement in reproductive outcome for all forms of fertility treatment. Hum Reprod 1998;13:1502–1505.
12. Clark AM, Ledger W, Galletly C, et al. Weight loss results in significant improvement in pregnancy and ovulation rates in anovulatory obese women. Hum Reprod 1995;10:2705–2712.
13. Huber-Buchholz MM, Carey DG, Norman RJ. Restoration of reproductive potential by lifestyle modification in obese polycystic ovary syndrome: role of insulin sensitivity and luteinizing hormone. J Clin Endocrinol Metab 1999;84:1470–1474.

14. Holte J, Bergh T, Berne C, Wide L, Lithell H. Restored insulin sensitivity but persistently increased early insulin secretion after weight loss in obese women with polycystic ovary syndrome. J Clin Endocrinol Metab 1995;80:2586–2593.
15. Pasquali R, Antenucci D, Casimirri F, et al. Clinical and hormonal characteristics of obese amenorrheic hyperandrogenic women before and after weight loss. J Clin Endocrinol Metab 1989;68:173–179.
16. Moran LJ, Noakes M, Clifton PM, Tomlinson L, Norman RJ. Dietary composition in restoring reproductive and metabolic physiology in overweight women with polycystic ovary syndrome. J Clin Endocrinol Metab 2003;88:812–819.
17. Marsh K, Brand-Miller J. The optimal diet for women with polycystic ovary syndrome? Br J Nutr 2005;94:154–165.
18. Greenblatt RB, Barfield WE, Jungck EC, Ray AW. Induction of ovulation with MRL/41: preliminary report. J Am Med Assoc 1961;178:101–104.
19. Clark JH, Markaverich BM. The agonistic and antagonistic effects of short acting estrogens: a review. Pharmacol Ther 1983;21:429–453.
20. Clark JH, Guthrie SC. Agonistic and antagonistic effects of clomiphene citrate and its isomers. Biol Reprod 1981;25:667–672.
21. McKenna KM, Pepperell RJ. Anti-oestrogens: their clinical physiology and use in reproductive medicine. Baillieres Clin Obstet Gynaecol 1988;2:545–565.
22. Adashi E. Ovulation induction: clomiphene citrate. In: Adashi EY Jr, Rosenwaks Z, eds. Reproductive Endocrinology, Surgery, and Technology, Vol. 1. Philadelphia: Lippincott-Raven, 1996:1181–1206.
23. Polson DW, Kiddy DS, Mason HD, Franks S. Induction of ovulation with clomiphene citrate in women with polycystic ovary syndrome: the difference between responders and nonresponders. Fertil Steril 1989;51:30–34.
24. Opsahl MS, Robins ED, O'Connor DM, Scott RT, Fritz MA. Characteristics of gonadotropin response, follicular development, and endometrial growth and maturation across consecutive cycles of clomiphene citrate treatment. Fertil Steril 1996;66:533–539.
25. Kousta E, White DM, Franks S. Modern use of clomiphene citrate in induction of ovulation. Hum Reprod Update 1997;3:359–365.
26. Garcia J, Jones GS, Wentz AC. The use of clomiphene citrate. Fertil Steril 1977;28:707–717.
27. Gysler M, March CM, Mishell DR Jr, Bailey EJ. A decade's experience with an individualized clomiphene treatment regimen including its effect on the postcoital test. Fertil Steril 1982;37:161–167.
28. Dickey RP, Taylor SN, Curole DN, Rye PH, Lu PY, Pyrzak R. Relationship of clomiphene dose and patient weight to successful treatment. Hum Reprod 1997;12:449–453.
29. Lobo RA, Granger LR, Davajan V, Mishell DR, Jr. An extended regimen of clomiphene citrate in women unresponsive to standard therapy. Fertil Steril 1982;37:762–766.
30. Andersen AG, Als-Nielsen B, Hornnes PJ, Franch Andersen L. Time interval from human chorionic gonadotrophin (HCG) injection to follicular rupture. Hum Reprod 1995;10:3202–3205.
31. Lobo RA, Gysler M, March CM, Goebelsmann U, Mishell DR Jr. Clinical and laboratory predictors of clomiphene response. Fertil Steril 1982;37:168–174.
32. Galtier-Dereure F, Pujol P, Dewailly D, Bringer J. Choice of stimulation in polycystic ovarian syndrome: the influence of obesity. Hum Reprod 1997;12(Suppl 1):88–96.
33. Murakawa H, Hasegawa I, Kurabayashi T, Tanaka K. Polycystic ovary syndrome. Insulin resistance and ovulatory responses to clomiphene citrate. J Reprod Med 1999;44:23–27.
34. Imani B, Eijkemans MJ, te Velde ER, Habbema JD, Fauser BC. Predictors of patients remaining anovulatory during clomiphene citrate induction of ovulation in normogonadotropic oligoamenorrheic infertility. J Clin Endocrinol Metab 1998;83:2361–2365.
35. Imani B, Eijkemans MJ, de Jong FH, et al. Free androgen index and leptin are the most prominent endocrine predictors of ovarian response during clomiphene citrate induction of ovulation in normogonadotropic oligoamenorrheic infertility. J Clin Endocrinol Metab 2000;85:676–682.
36. Imani B, Eijkemans MJ, te Velde ER, Habbema JD, Fauser BC. Predictors of chances to conceive in ovulatory patients during clomiphene citrate induction of ovulation in normogonadotropic oligoamenorrheic infertility. J Clin Endocrinol Metab 1999;84:1617–1622.
37. Imani B, Eijkemans MJ, te Velde ER, Habbema JD, Fauser BC. A nomogram to predict the probability of live birth after clomiphene citrate induction of ovulation in normogonadotropic oligoamenorrheic infertility. Fertil Steril 2002;77:91–97.
38. Levene MI, Wild J, Steer P. Higher multiple births and the modern management of infertility in Britain. The British Association of Perinatal Medicine. Br J Obstet Gynaecol 1992;99:607–613.
39. Hammond MG. Monitoring techniques for improved pregnancy rates during clomiphene ovulation induction. Fertil Steril 1984;42:499–509.
40. Shoham Z, Zosmer A, Insler V. Early miscarriage and fetal malformations after induction of ovulation (by clomiphene citrate and/or human menotropins), in vitro fertilization, and gamete intrafallopian transfer. Fertil Steril 1991;55:1–11.
41. Venn A, Lumley J. Clomiphene citrate and pregnancy outcome. Aust NZ J Obstet Gynaecol 1994;34:56–66.
42. Bristow RE, Karlan BY. Ovulation induction, infertility, and ovarian cancer risk. Fertil Steril 1996;66:499–507.

43. Hammond MG, Halme JK, Talbert LM. Factors affecting the pregnancy rate in clomiphene citrate induction of ovulation. Obstet Gynecol 1983;62:196–202.
44. Randall JM, Templeton A. Transvaginal sonographic assessment of follicular and endometrial growth in spontaneous and clomiphene citrate cycles. Fertil Steril 1991;56:208–212.
45. Check JH, Dietterich C, Lurie D. The effect of consecutive cycles of clomiphene citrate therapy on endometrial thickness and echo pattern. Obstet Gynecol 1995;86:341–345.
46. Homburg R, Armar NA, Eshel A, Adams J, Jacobs HS. Influence of serum luteinising hormone concentrations on ovulation, conception, and early pregnancy loss in polycystic ovary syndrome. BMJ 1988;297:1024–1026.
47. Capelo F-O, Kumer A, Steinkampf MP, Azziz R. Laparoscopic evaluation following failure to achieve pregnancy after clomiphene citrate ovulation induction. Fertil Steril 2003;80:1450–1453.
48. Haning RV Jr, Hackett RJ, Flood CA, Loughlin JS, Zhao QY, Longcope C. Plasma dehydroepiandrosterone sulfate serves as a prehormone for 48% of follicular fluid testosterone during treatment with menotropins. J Clin Endocrinol Metab 1993;76:1301–1307.
49. Diamant YZ, Evron S. Induction of ovulation by combined clomiphene citrate and dexamethasone treatment in clomiphene citrate nonresponders. Eur J Obstet Gynecol Reprod Biol 1981;11:335–340.
50. Daly DC, Walters CA, Soto-Albors CE, Tohan N, Riddick DH. A randomized study of dexamethasone in ovulation induction with clomiphene citrate. Fertil Steril 1984;41:844–848.
51. Lobo RA, Paul W, March CM, Granger L, Kletzky OA. Clomiphene and dexamethasone in women unresponsive to clomiphene alone. Obstet Gynecol 1982;60:497–501.
52. Jones GE, Howard JE, Langford H. The use of cortisone in follicular phase disturbances. Fertil Steril 1953;4:49–62.
53. Greenblatt RB. Cortisone in treatment of the hirsute woman. Am J Obstet Gynecol 1953;66:700–710.
54. Ettinger B, Goldfield EB, Burrill KC, Von Werder K, Forsham PH. Plasma testosterone stimulation-suppression dynamics in hirsute women. correlation with long-term therapy. Am J Med 1973;54:195–200.
55. Rodriguez-Rigau LJ, Smith KD, Tcholakian RK, Steinberger E. Effect of prednisone on plasma testosterone levels and on duration of phases of the menstrual cycle in hyperandrogenic women. Fertil Steril 1979;32:408–413.
56. Emans SJ, Grace E, Woods ER, Mansfield J, Crigler JF, Jr. Treatment with dexamethasone of androgen excess in adolescent patients. J Pediatr 1988;112:821–826.
57. Loughlin T, Cunningham S, Moore A, Culliton M, Smyth PP, McKenna TJ. Adrenal abnormalities in polycystic ovary syndrome. J Clin Endocrinol Metab 1986;62:142–147.
58. Azziz R, Black VY, Knochenhauer ES, Hines GA, Boots LR. Ovulation after glucocorticoid suppression of adrenal androgens in the polycystic ovary syndrome is not predicted by the basal dehydroepiandrosterone sulfate level. J Clin Endocrinol Metab 1999;84:946–950.
59. Stojanov S. [Ovulation induction with clomiphene and dexamethasone]. Zentralbl Gynakol 1984;106:107–110.
60. Parsanezhad ME, Alborzi S, Motazedian S, Omrani G. Use of dexamethasone and clomiphene citrate in the treatment of clomiphene citrate-resistant patients with polycystic ovary syndrome and normal dehydroepiandrosterone sulfate levels: a prospective, double-blind, placebo-controlled trial. Fertil Steril 2002;78:1001–1004.
61. Polson DW, Reed MJ, Scanlon MJ, Quiney N, Franks S. Androstenedione concentrations following dexamethasone suppression: correlation with clomiphene responsiveness in women with polycystic ovary syndrome. Gynecol Endocrinol 1988;2:257–264.
62. Evron S, Navot D, Laufer N, Diamant YZ. Induction of ovulation with combined human gonadotropins and dexamethasone in women with polycystic ovarian disease. Fertil Steril 1983;40:183–186.
63. Keay SD, Lenton EA, Cooke ID, Hull MG, Jenkins JM. Low-dose dexamethasone augments the ovarian response to exogenous gonadotrophins leading to a reduction in cycle cancellation rate in a standard IVF programme. Hum Reprod 2001;16:1861–1865.
64. Rein MS, Jackson KV, Sable DB, Thomas PP, Hornstein MD. Dexamethasone during ovulation induction for in-vitro fertilization: a pilot study. Hum Reprod 1996;11:253–255.
65. Dunaif A. Insulin resistance and the polycystic ovary syndrome: mechanism and implications for pathogenesis. Endocr Rev 1997;18:774–800.
66. Barbieri RL, Makris A, Ryan KJ. Insulin stimulates androgen accumulation in incubations of human ovarian stroma and theca. Obstet Gynecol 1984;64:73S–80S.
67. Nestler JE, Powers LP, Matt DW, et al. A direct effect of hyperinsulinemia on serum sex hormone-binding globulin levels in obese women with the polycystic ovary syndrome. J Clin Endocrinol Metab 1991;72:83–89.
68. Ben-Shlomo I, Homburg R, Shalev E. Hyperandrogenic anovulation (ôve polycystic ovary syndrome)—back to the ovary? Hum Reprod Update 1998;4:296–300.
69. Ovalle F, Azziz R. Insulin resistance, polycystic ovary syndrome, and type 2 diabetes mellitus. Fertil Steril 2002;77:1095–1105.
70. Dunaif A, Segal KR, Futterweit W, Dobrjansky A. Profound peripheral insulin resistance, independent of obesity, in polycystic ovary syndrome. Diabetes 1989;38:1165–1174.
71. Legro RS, Finegood D, Dunaif A. A fasting glucose to insulin ratio is a useful measure of insulin sensitivity in women with polycystic ovary syndrome. J Clin Endocrinol Metab 1998;83:2694–2698.

72. DeUgarte CM, Bartolucci AA, Azziz R. Prevalence of insulin resistance in the polycystic ovary syndrome using the homeostasis model assessment. Fertil Steril 2005;83:1454–1460.
73. Lord J, Flight I, Norman R. Insulin-sensitising drugs (metformin, troglitazone, rosiglitazone, pioglitazone, D-chiro-inositol) for polycystic ovary syndrome. Cochrane Database Syst Rev 2003;3:CD003053.
74. Kirpichnikov D, McFarlane SI, Sowers JR. Metformin: an update. Ann Intern Med 2002;137:25–33.
75. Velazquez EM, Mendoza S, Hamer T, Sosa F, Glueck CJ. Metformin therapy in polycystic ovary syndrome reduces hyperinsulinemia, insulin resistance, hyperandrogenemia, and systolic blood pressure, while facilitating normal menses and pregnancy. Metabolism 1994;43:647–654.
76. Yarali H, Yildiz BO, Demirol A, et al. Co-administration of metformin during rFSH treatment in patients with clomiphene citrate-resistant polycystic ovarian syndrome: a prospective randomized trial. Hum Reprod 2002;17:289–294.
77. Palomba S, Falbo A, Orio F Jr, et al. A randomized controlled trial evaluating metformin pre-treatment and co-administration in non-obese insulin-resistant women with polycystic ovary syndrome treated with controlled ovarian stimulation plus timed intercourse or intrauterine insemination. Hum Reprod 2005;20:2879–2886.
78. Kjotrod SB, von During V, Carlsen SM. Metformin treatment before IVF/ICSI in women with polycystic ovary syndrome; a prospective, randomized, double blind study. Hum Reprod 2004;19:1315–1322.
79. Palomba S, Orio F, Jr., Falbo A, et al. Prospective parallel randomized, double-blind, double-dummy controlled clinical trial comparing clomiphene citrate and metformin as the first-line treatment for ovulation induction in nonobese anovulatory women with polycystic ovary syndrome. J Clin Endocrinol Metab 2005;90:4068–4074.
80. Stadtmauer LA, Wong BC, Oehninger S. Should patients with polycystic ovary syndrome be treated with metformin? Benefits of insulin sensitizing drugs in polycystic ovary syndrome—beyond ovulation induction. Hum Reprod 2002;17:3016–3026.
81. Glueck CJ, Phillips H, Cameron D, Sieve-Smith L, Wang P. Continuing metformin throughout pregnancy in women with polycystic ovary syndrome appears to safely reduce first-trimester spontaneous abortion: a pilot study. Fertil Steril 2001;75:46–52.
82. Glueck CJ, Wang P, Goldenberg N, Sieve-Smith L. Pregnancy outcomes among women with polycystic ovary syndrome treated with metformin. Hum Reprod 2002;17:2858–2864.
83. Hasegawa I, Murakawa H, Suzuki M, Yamamoto Y, Kurabayashi T, Tanaka K. Effect of troglitazone on endocrine and ovulatory performance in women with insulin resistance-related polycystic ovary syndrome. Fertil Steril 1999;71:323–327.
84. Azziz R, Ehrmann D, Legro RS, et al. Troglitazone improves ovulation and hirsutism in the polycystic ovary syndrome: a multicenter, double blind, placebo-controlled trial. J Clin Endocrinol Metab 2001;86:1626–1632.
85. Dunaif A, Scott D, Finegood D, Quintana B, Whitcomb R. The insulin-sensitizing agent troglitazone improves metabolic and reproductive abnormalities in the polycystic ovary syndrome. J Clin Endocrinol Metab 1996;81:3299–3306.
86. Reasner CA. Where thiazolidinediones will fit. Diabetes Metab Res Rev 2002;18 Suppl 2:S30–35.
87. Romualdi D, Guido M, Ciampelli M, et al. Selective effects of pioglitazone on insulin and androgen abnormalities in normo- and hyperinsulinaemic obese patients with polycystic ovary syndrome. Hum Reprod 2003;18:1210–1218.
88. Cataldo NA, Abbasi F, McLaughlin TL, Lamendola C, Reaven GM. Improvement in insulin sensitivity followed by ovulation and pregnancy in a woman with polycystic ovary syndrome who was treated with rosiglitazone. Fertil Steril 2001;76:1057–1059.
89. Brettenthaler N, De Geyter C, Huber PR, Keller U. Effect of the insulin sensitizer pioglitazone on insulin resistance, hyperandrogenism, and ovulatory dysfunction in women with polycystic ovary syndrome. J Clin Endocrinol Metab 2004;89:3835–3840.
90. Cataldo NA, Abbasi F, McLaughlin TL, et al. Metabolic and ovarian effects of rosiglitazone treatment for 12 weeks in insulin-resistant women with polycystic ovary syndrome. Hum Reprod 2006;21:109–120.
91. Shobokshi A, Shaarawy M. Correction of insulin resistance and hyperandrogenism in polycystic ovary syndrome by combined rosiglitazone and clomiphene citrate therapy. J Soc Gynecol Investig 2003;10:99–104.
92. Ghazeeri G, Kutteh WH, Bryer-Ash M, Haas D, Ke RW. Effect of rosiglitazone on spontaneous and clomiphene citrate-induced ovulation in women with polycystic ovary syndrome. Fertil Steril 2003;79:562–566.
93. Baillargeon JP, Jakubowicz DJ, Iuorno MJ, Jakubowicz S, Nestler JE. Effects of metformin and rosiglitazone, alone and in combination, in nonobese women with polycystic ovary syndrome and normal indices of insulin sensitivity. Fertil Steril 2004;82:893–902.
94. Eshel A, Abdulwahid NA, Armar NA, Adams JM, Jacobs HS. Pulsatile luteinizing hormone-releasing hormone therapy in women with polycystic ovary syndrome. Fertil Steril 1988;49:956–960.
95. Filicori M, Flamigni C, Dellai P, et al. Treatment of anovulation with pulsatile gonadotropin-releasing hormone: prognostic factors and clinical results in 600 cycles. J Clin Endocrinol Metab 1994;79:1215–1220.
96. Shoham Z, Homburg R, Jacobs HS. Induction of ovulation with pulsatile GnRH. Baillieres Clin Obstet Gynaecol 1990;4:589–608.
97. Rabau E, David A, Serr DM, Mashiach S, Lunenfeld B. Human menopausal gonadotropins for anovulation and sterility. Results of 7 years of treatment. Am J Obstet Gynecol 1967;98:92–98.
98. Biffoni M, Battaglia A, Borrelli F, Cantelmo A, Galli G, Eshkol A. Allergenic potential of gonadotrophic preparations in experimental animals: relevance of purity. Hum Reprod 1994;9:1845–1848.

99. Lathi RB, Milki AA. Recombinant gonadotropins. Curr Womens Health Rep 2001;1:157–163.

100. Hughes E, Collins J, Vandekerckhove P. Ovulation induction with urinary follicle stimulating hormone versus human menopausal gonadotropin for clomiphene-resistant polycystic ovary syndrome. Cochrane Database Syst Rev 2000;CD000087.

101. Bayram N, van Wely M, van Der Veen F. Recombinant FSH versus urinary gonadotrophins or recombinant FSH for ovulation induction in subfertility associated with polycystic ovary syndrome. Cochrane Database Syst Rev 2001;CD002121.

102. Van Wely M, Bayram N, Van Der Veen F. Recombinant FSH in alternative doses or versus urinary gonadotrophins for ovulation induction in subfertility associated with polycystic ovary syndrome: a systematic review based on a Cochrane review. Hum Reprod 2003;18:1143–1149.

103. Balen AH, Braat DD, West C, Patel A, Jacobs HS. Cumulative conception and live birth rates after the treatment of anovulatory infertility: safety and efficacy of ovulation induction in 200 patients. Hum Reprod 1994;9:1563–1570.

104. Van Der Meer M, Hompes PG, De Boer JA, Schats R, Schoemaker J. Cohort size rather than follicle-stimulating hormone threshold level determines ovarian sensitivity in polycystic ovary syndrome. J Clin Endocrinol Metab 1998;83:423–446.

105. Farhi J, West C, Patel A, Jacobs HS. Treatment of anovulatory infertility: the problem of multiple pregnancy. Hum Reprod 1996;11:429–434.

106. Fox R, Corrigan E, Thomas PA, Hull MG. The diagnosis of polycystic ovaries in women with oligo-amenorrhoea: predictive power of endocrine tests. Clin Endocrinol (Oxf) 1991;34:127–131.

107. Nugent D, Vandekerckhove P, Hughes E, Arnot M, Lilford R. Gonadotrophin therapy for ovulation induction in subfertility associated with polycystic ovary syndrome. Cochrane Database Syst Rev 2000;CD000410.

108. Thompson CR, Hansen LM. Pergonal (menotropins): a summary of clinical experience in the induction of ovulation and pregnancy. Fertil Steril 1970;21:844–853.

109. Wang CF, Gemzell C. The use of human gonadotropins for the induction of ovulation in women with polycystic ovarian disease. Fertil Steril 1980;33:479–486.

110. Shoham Z, Patel A, Jacobs HS. Polycystic ovarian syndrome: safety and effectiveness of stepwise and low-dose administration of purified follicle-stimulating hormone. Fertil Steril 1991;55:1051–1056.

111. Brown JB. A new approach to capital payment in a competitive era. Health Aff (Millwood) 1986;5:47–62.

112. Fauser BC, Donderwinkel P, Schoot DC. The step-down principle in gonadotrophin treatment and the role of GnRH analogs. Baillieres Clin Obstet Gynaecol 1993;7:309–330.

113. Hugues JN, Cedrin-Durnerin I, Avril C, Bulwa S, Herve F, Uzan M. Sequential step-up and step-down dose regimen: an alternative method for ovulation induction with follicle-stimulating hormone in polycystic ovarian syndrome. Hum Reprod 1996;11:2581–2584.

114. Homburg R, Howles CM. Low-dose FSH therapy for anovulatory infertility associated with polycystic ovary syndrome: rationale, results, reflections and refinements. Hum Reprod Update 1999;5:493–499.

115. Christin-Maitre S, Hugues JN. A comparative randomized multicentric study comparing the step-up versus step-down protocol in polycystic ovary syndrome. Hum Reprod 2003;18:1626–1631.

36

In Vitro Fertilization for the Polycystic Ovary Syndrome

Hanadi Ba-Akdah, Ezgi Demirtas, Hananel Holzer, William Buckett, and Seang Lin Tan

SUMMARY

In vitro fertilization (IVF) is the final option for fertility treatment of all etiologies and must often be performed on women with polycystic ovaries or polycystic ovary syndrome (PCOS). Many women undergoing IVF are noted to have polycystic ovaries or PCOS. These women have ovaries that are more sensitive to ovarian stimulation, they develop more follicles, produce more oocytes, and are more prone to ovarian hyperstimulation syndrome (OHSS). This may be because of the increased vascularity of the ovaries as exemplified by increased ovarian stromal peak systolic blood flow velocity (V_{max}) and increased levels of serum and intrafollicular vascular endothelial growth factor. In order to minimize the risk of OHSS, it is therefore necessary to commence ovarian stimulation with a smaller dose of gonadotropins. An alternative strategy could be to perform in vitro maturation of oocytes.

Key Words: Polycystic ovaries, polycystic ovary syndrome (PCOS), in vitro fertilization (IVF), in vitro maturation (IVM), ovarian hyperstimulation syndrome (OHSS).

1. INTRODUCTION

In vitro fertilization (IVF) involves the collection of mature oocytes from the woman's ovaries in order to achieve fertilization outside the body followed by the transfer of embryo(s) into the uterus. The first live birth following IVF occurred in 1978 *(1)*, and since then more than 2 million IVF babies have been born worldwide. With its increasing success rates, IVF has become a life-transforming treatment and the final path for couples with long-standing female or male factor infertility *(2,3)*. Although IVF was initially developed to treat tubal factor infertility, it has since become a successful treatment for various forms of infertility including endometriosis, ovulatory dysfunction, cervical factor, luteal defects, immunological causes, male-factor and unexplained infertility *(4)*. A couple's chance of success with IVF is linked to the length of infertility, the woman's age, the cause of infertility, and the treating clinic. The presence of polycystic ovaries and the polycystic ovary syndrome (PCOS) have important implications for the prognosis and treatment of the patient undergoing IVF, and, in turn, IVF and in vitro maturation (IVM) are essential options for the treatment of the patient with PCOS.

2. BACKGROUND

2. 1. In Vitro Fertilization

In IVF and other assisted reproduction technologies, the ovaries are stimulated with gonadotropins, which contain follicle-stimulating hormone (FSH) with or without luteinizing hormone (LH).

From: *Contemporary Endocrinology: Androgen Excess Disorders in Women:*
Polycystic Ovary Syndrome and Other Disorders, Second Edition
Edited by: R. Azziz et al. © Humana Press Inc., Totowa, NJ

This leads to multiple follicular development, which results in increased numbers of oocytes being available for retrieval. Higher numbers of oocytes after fertilization generate higher numbers of embryos. This, in turn, allows for the selection of good-quality embryos for transfer and therefore higher pregnancy rates. Good-quality embryos may also be cryopreserved, allowing further attempts at pregnancy without the necessity of repeated ovarian stimulation and oocyte retrieval.

Initially, oocyte retrieval was performed by laparoscopy, but today the transvaginal ultrasound-guided method is used *(5)*. Different anesthesia and analgesia regimens can be administered. General or spinal anesthesia is not generally needed; paracervical block with bupivacaine and mild sedation in combination with opioid analgesics can be used for analgesia during most cases. Following retrieval, the mature oocytes are inseminated and fertilize spontaneously (in IVF) or undergo intracytoplasmic sperm injection (ICSI). Then the best quality embryo or embryos are transferred into the uterine cavity by using, preferably, soft transfer catheters.

In order to achieve optimal ovarian stimulation, it is necessary to determine the ideal stimulation regimen and dose of gonadotropin for the individual patient. This is because inadequate stimulation leads to few oocytes and poor outcome or even a complete absence of embryos to transfer, whereas excessive stimulation leads to an increased risk of developing ovarian hyperstimulation syndrome (OHSS) *(6)*, a potentially life-threatening condition.

There are a number of different ovarian-stimulation protocols; namely: (1) human menopausal gonadotropin (hMG) or FSH, with or without clomiphene citrate, which was typically used in the early 1980s; (2) the "long," "short," "ultrashort," and "microdose flare" protocols using gonadotropin-releasing hormone (GnRH) agonists and hMG or FSH, which have gained widespread acceptance since the "long" protocol was first described in the mid-1980s; and, more recently, (3) protocols using hMG or FSH followed by the addition of newer GnRH antagonists. The highest pregnancy and live-birth rates reported in all age groups and for all causes of infertility have been with the "long" protocol, which induces pituitary desensitization with GnRH agonist followed by ovarian stimulation with hMG or FSH and which is the most widely used protocol today *(7)*.

The results of IVF today exceed spontaneous pregnancy rates in healthy fertile couples in many IVF clinics, although the success declines quite significantly with age. As an example, at the McGill Reproductive Center in 2003 for all infertility indications, clinical pregnancy rates per cycle started were 56% for women younger than 35, 42.2% for those 35–37, 35.4% for those 38–40, and 12.3% for those older than 40 years, and live-birth rates per cycle started for the same age groups were 46, 33.3, 25.5, and 4.8%, respectively. Although there is variation between different centers, the results for all clinics in the United States in 2002 showed a clinical pregnancy rate of 34% and a live-birth rate of 28% per cycle *(8)*, which is still higher than the rate of spontaneous conception.

2.2. IVF in PCOS

PCOS is the most common cause of anovulatory infertility. As described in Chapters 35 and 37, the treatment of anovulation comprises lifestyle modification through diet, exercise, and weight loss, insulin sensitizers, oral ovulation induction agents, surgical treatments, and gonadotropins. Patients who have not conceived with other treatment approaches or couples who have additional infertility factors may require IVF. In vitro fertilization treatment outcomes for PCO/PCOS patients may even be better than for patients with normal ovaries. After three cycles of IVF treatment, the odds of achieving a pregnancy with polycystic ovaries were 69% higher than those of a woman with normal ovaries *(9)*.

PCOS has been shown to exist in 4–10% of the general population and can be found in even higher incidences in certain populations. Polycystic ovary is a morphological definition based on ultrasound without any manifestation of the syndrome and can be defined as the presence of 12 or more follicles in one ovary measuring 2–9 mm in diameter and/or increased ovarian volume (>10 mL) (Fig. 1). Isolated polycystic ovaries in ultrasound scans has been shown to occur in 16–23% of the normal population *(10–12)*; however, in an IVF population, the incidence may be as high as 33% *(13)*. Therefore, many women with polycystic ovaries or PCOS will need IVF for indications other than anovulation.

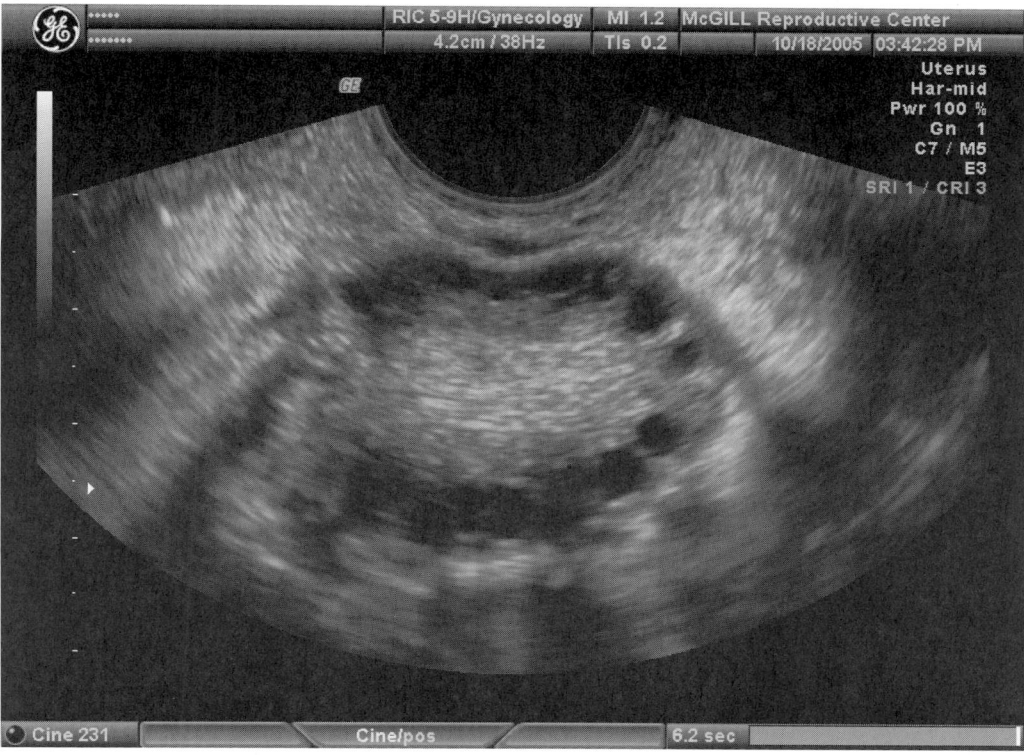

Fig. 1. Ultrasound appearance of a polycystic ovary. Polycystic ovaries is a morphological definition based on ultrasound without any manifestation of the syndrome and can be defined as the presence of 12 or more follicles in one ovary measuring 2–9 mm in diameter, and/or increased ovarian volume (>10 mL).

When treated with gonadotropins, patients with polycystic ovaries/PCOS have a narrow therapeutic window because a slight increase in dose may lead to an exaggerated response. When gonadotropin-treatment cycles for ovulation induction and/or intrauterine insemination result in hyperresponse with multiple follicle counts, this increases the rate of high-order multiple pregnancy. In order to avoid this, cycles need to be either canceled or converted to IVF. Controlled ovarian hyperstimulation in patients with polycystic ovaries or PCOS also carries the risk of OHSS.

2.3. Pretreatment Counseling in PCOS

2.3.1. Insulin Resistance

A large number of women with PCOS exhibit insulin resistance, which has been shown to play an important role in the pathophysiology of PCOS *(14)*. Patients with PCOS should be screened for type 2 diabetes mellitus or impaired glucose tolerance before treatment. In addition, insulin resistance has also been suggested to cause higher doses of gonadotropin consumption *(15)*. Some authors suggest that when metformin is added to the IVF treatment, although the gonadotropin requirement does not change, more oocytes are retrieved *(16)*, and that co-administration of metformin in IVF treatment decreases the total number of follicles *(17)* with higher clinical pregnancy rates *(18)*. However, other authors found that metformin had no effect on duration of stimulation, total FSH dose, number of follicles 10–16 mm and 17 mm or greater in diameter, serum levels of estradiol (E2) on the day of human chorionic gonadotropin (hCG) injection, number of retrieved oocytes, fertilization rate, number of transferred embryos, and total and clinical pregnancy rates *(19,20)*.

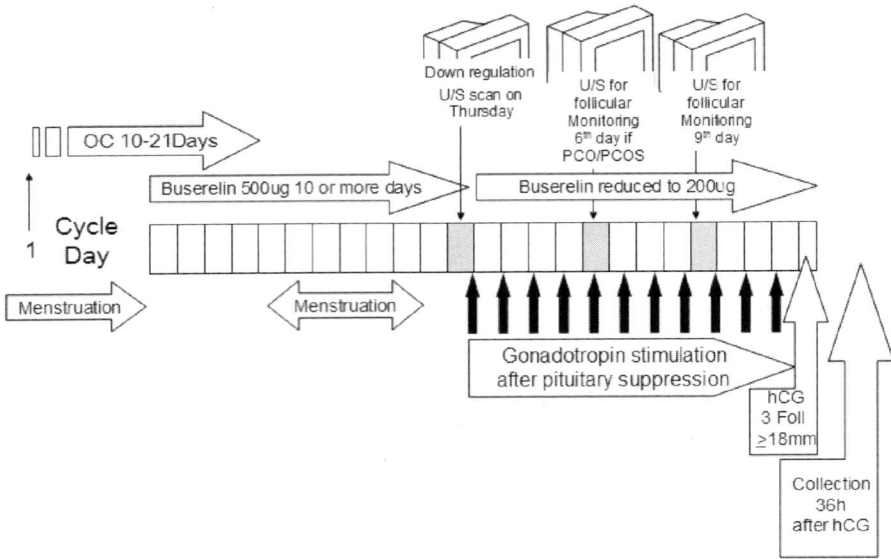

Fig. 2. The "long" gonadotropin-releasing hormone (GnRH) suppression protocol for in vitro fertilization (IVF), accompanied by pretreatment with combined oral contraceptive pills, has been widely accepted as the most effective stimulation protocol for polycystic ovary/polycystic ovary syndrome (PCO/PCOS) patients.

2.3.2. Weight Loss

Because obese women are more prone to insulin resistance, and obesity, independent of hyperinsulinemia, may increase gonadotropin requirement in IVF cycles *(21)*, weight loss should be considered in obese polycystic ovary/PCOS patients before starting treatment. In addition, weight loss in these patients also results in better obstetric outcomes such as lower cesarean section and gestational diabetes rates.

2.3.3. Endometrial Evaluation

Patients with irregular bleeding or thick endometrium detected in ultrasound examination may have endometrial hyperplasia because of unopposed estrogen. Such cases should be evaluated for these pathologies before starting treatment.

2.4. Ovarian Stimulation for IVF in Polycystic Ovary/PCOS

Ovarian stimulation with gonadotropins for IVF in polycystic ovaries and/or PCOS is a particular problem and is indeed a challenge for the physician and patient. The "long" GnRH agonist-suppression protocol accompanied by pretreatment with combined oral contraceptive pills has been widely accepted as the most effective stimulation protocol for polycystic ovary/PCOS patients (Fig. 2). Pituitary suppression takes longer for polycystic ovary/PCOS patients *(22)*. Although the optimal time to commence GnRH agonist is not clearly determined, commencement in the early follicular phase combined with oral contraceptive pills would avoid the risk of inadvertent administration during early pregnancy. The GnRH agonist protocol has been found to suppress elevated LH and androgen levels and prevent a premature LH surge, which appears to improve the pregnancy rate and reduce the miscarriage rate in PCOS patients undergoing IVF treatment *(23–26)*.

It is very important to select the appropriate dose of gonadotropins, particularly in polycystic ovary/PCOS patients, because excessive dosage of gonadotropins increases the risk of OHSS whereas low dosage leads to poor response. Recombinant FSH products without LH are better for ovarian stimulation in patients with polycystic ovary/PCOS because of the possible deleterious effects of an excessive LH level *(25,27,28)*.

Patients with polycystic ovaries respond the same as those with PCOS during ovarian stimulation. When comparing polycystic ovary-only women to those who had normal ovarian morphology on ultrasound examination, it was found that they produced more follicles, oocytes, and embryos, but the fertilization, cleavage, and miscarriage rates were not significantly different *(9)*.

Hyperresponse of not only PCOS patients but also those with solely polycystic ovary morphology may be associated with increased ovarian stromal blood flow. Ovarian blood flow has been shown to be increased in both polycystic and PCOS ovaries (Fig. 3) *(29)*. In one study women were divided into three groups: normal, polycystic ovary, and PCOS. Ultrasonographic and Doppler examinations were performed on each group on the second or third day of the menstrual cycle. Mean ovarian stromal peak systolic blood flow velocity (V_{max}) in the polycystic ovary and PCOS groups was found to be significantly higher than in the normal group (16.88 cm/second and 16.89 cm/second compared with 8.74 cm/second). In addition, mean ovarian stromal time averaged maximum velocity was found to be higher in the polycystic ovary and PCOS groups and normal ovaries (10.55 cm/sec and 10.89 cm/second compared with 5.44 cm/second). However, there was no significant difference between ultrasound polycystic ovaries and PCOS. The intraovarian and uterine artery vascular differences were likely to be a result of a primary disorder within the polycystic ovary and their different hormonal status *(30)*.

Ovarian volume, stromal volume, and stromal peak blood flow velocity are all significantly higher in the ovaries of women with polycystic ovaries and PCOS. It has been shown that there is no difference in the mean stromal echogenicity, although the stromal index is significantly greater in women with polycystic ovaries or PCOS *(31)*.

The reason for increased blood flow is probably the increased vascular endothelial growth factor (VEGF) levels in polycystic ovary/PCOS patients. A positive correlation was observed between the serum VEGF and E2 concentrations on the day of hCG and oocyte retrieval and between the serum VEGF concentration and Doppler blood flow velocities throughout the IVF cycle *(32)*. Serum VEGF seems to be a major capillary permeability factor in the development of OHSS ascites *(33)*. VEGF concentrations rise after hCG administration and are higher in women who develop OHSS. Moreover, ascitic fluid obtained from OHSS patients contains high levels of VEGF, and follicular fluid VEGF concentrations are higher in women with OHSS and those with polycystic ovary/PCOS. Within ovarian and uterine blood vessels, blood flow velocities are higher both in the early follicular phase and on the day of hCG administration in women with polycystic ovary/PCOS. According to these findings, it can be concluded that there is most probably a link between VEGF, polycystic ovary/PCOS, and OHSS.

Therefore, ovarian stimulation for polycystic ovary/PCOS in women undergoing IVF should be commenced with a smaller dose, ideally with a preparation of FSH only, and with increased ultrasound and endocrine monitoring.

2.5. Ovarian Hyperstimulation Syndrome

OHSS is a potentially serious iatrogenic complication of ovarian stimulation with gonadotropins. A controlled study comparing patients with and without polycystic ovaries undergoing IVF showed that 10.5% of the polycystic ovary patients developed moderate/severe OHSS compared with none of the controls *(13)*. OHSS is characterized by increased vascular permeability and transudation of protein-rich fluid from the vascular space into the peritoneal cavity. The incidence of severe OHSS ranges between 0.6 and 1.9% but may occur as frequently as 6% in women with polycystic ovary/PCOS *(34–36)*. Severe OHSS is characterized by ascites, plural effusion, and electrolyte imbalances. While there is no successful strategy to completely predict and prevent this potentially life-threatening complication, patients are commonly identified when the serum E2 concentration is high (10,000–15,000 pmol/L) and the number of ovarian follicles is greater than 20. Although follicular puncture

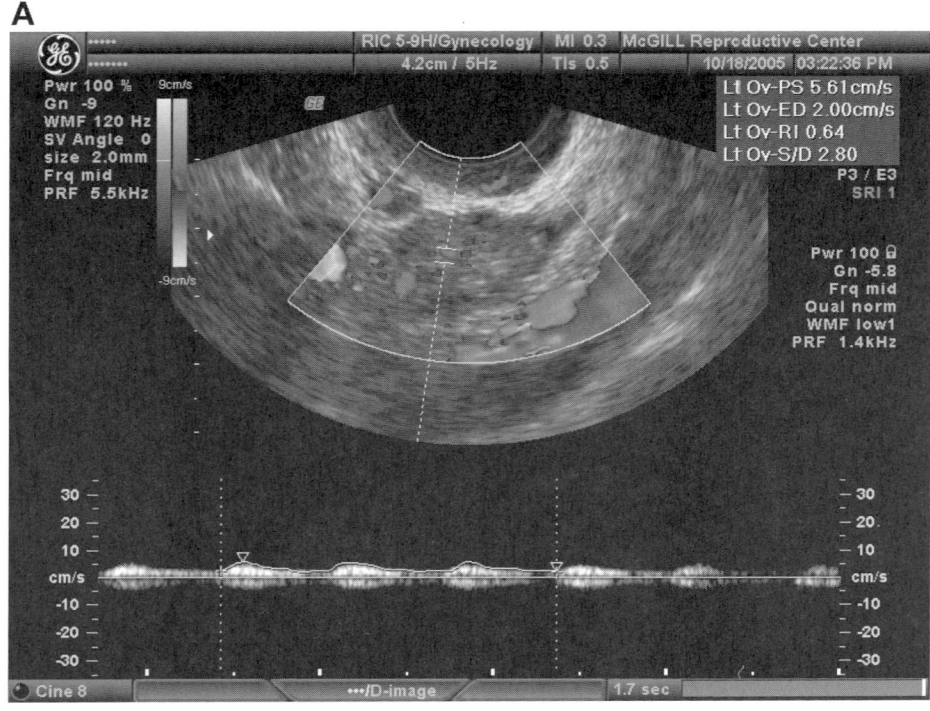

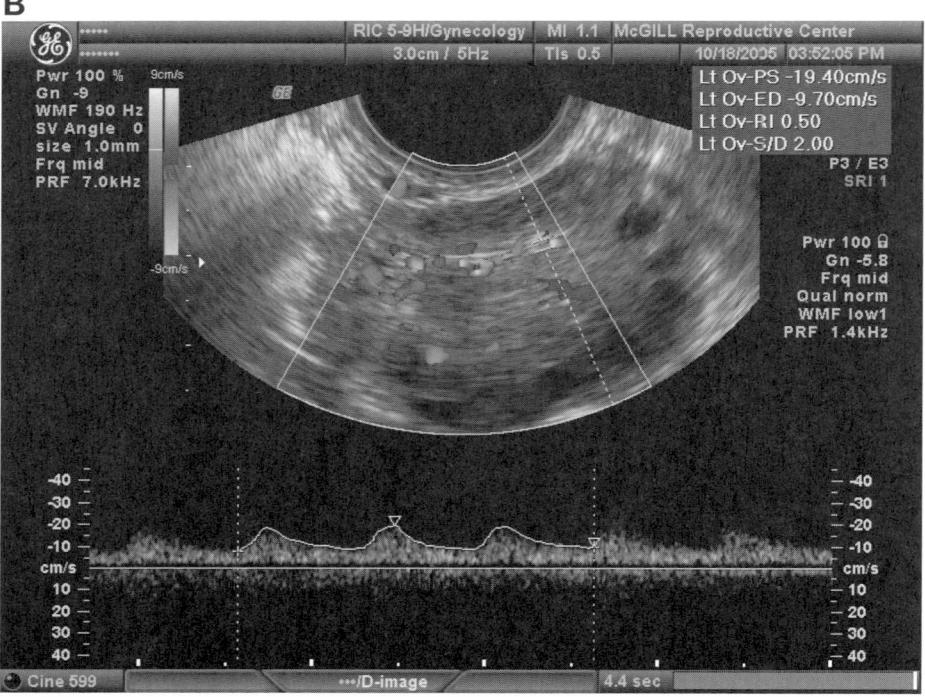

Fig. 3. Stromal blood flow in a normal ovary (**A**) and a polycystic ovary (**B**). Note that ovarian blood flow is increased in the polycystic ovary.

and aspiration of follicular fluid have been suggested to be protective against OHSS, it seems that follicular aspiration does not necessarily prevent OHSS.

The strategies to reduce the risk of OHSS include the following:

1. Starting stimulation with lower dose of FSH.
2. Close monitoring by ultrasound scans and serum E2 level measurements.
3. Coasting: When E2 levels exceed 10,000 pmol/L following gonadotropin injections, hCG can be with-held. Coasting does not seem to decrease pregnancy rates. In a series of 51 patients, Sher et al. *(37)* stopped further gonadotropin injections when E2 exceeded 22,000 pmol/L for 3–11 days and administered hCG when serum E2 levels decreased below 11,000 pmol/L; the pregnancy rate was 41%.
4. Withholding or decreasing the hCG dose for final follicular maturation, or in suitable cases substituting hCG with GnRH agonist *(38)*.
5. Intravascular volume expanders, hydroxyethyl starch, or human albumin can be administered during and after oocyte retrieval; however, it is not always helpful in preventing OHSS *(39)*.
6. Glucocorticoid use has been recommended, but it does not seem to have a significant effect in preventing OHSS *(40)*.
7. Decreasing the number of embryos transferred or freezing all the embryos to be transferred in a later frozen embryo transfer cycle.
8. Metformin pretreatment and/or concomitant metformin use has been suggested for PCOS patients. A study showed that metformin pretreatment in coasted cycles may help to decrease the maximum E2 concentrations and the number of days of coasting *(18)*. However, large prospective studies regarding OHSS risk with metformin pretreatment are not available yet.

None of the above-mentioned solutions avoids the risk of OHSS completely. The only reliable way to prevent OHSS is to avoid ovarian stimulation with FSH completely. In addition, other than severe OHSS risk, there are numerous disadvantages associated with gonadotropin stimulation, including high drug costs, the need for daily injections, frequent monitoring, and potential side effects, such as abdominal bloating, breast tenderness, mood swings, and nausea. IVM would avoid the risk of OHSS in women with polycystic ovary/PCOS who need IVF.

2.6. In Vitro Maturation

The recovery of immature oocytes (Fig. 4) followed by IVM and IVF is an attractive alternative to conventional IVF treatment in which controlled ovarian stimulation with gonadotropins is used to increase the number of available oocytes and embryos *(41)*. Significant progress has been made in improving implantation and pregnancy rates from in vitro matured oocytes (Fig. 4). The high numbers of antral follicles in patients with polycystic ovaries or PCOS make them prime candidates for IVM treatment, even if the appearance of polycystic ovaries in the scan is not associated with an ovulation disorder. Indeed, the main determinant of clinical success rates of IVM treatment is antral follicle count *(42)*. When hCG priming is used before oocyte retrieval, it has been found that immature oocytes retrieved from normal ovaries, polycystic ovaries, or women with PCOS have a similarly high maturation, fertilization, and cleavage potential *(43)*. However, although the implantation rate was lower, the live-birth rates were not significantly different, and, as expected, the OHSS rate was zero in the IVM group. These results suggested that IVM is a promising alternative to conventional IVF treatment for women with polycystic ovaries or a high antral follicle count that require assisted conception. The oocytes are retrieved without FSH pretreatment, 36 hours after a single 10,000 IU hCG administration *(44)*. Once immature oocytes have been retrieved, they are cultured in the IVM medium for 24–48 hours and ICSI is performed on the matured oocytes. Follicle size may be important for subsequent embryonic development, but the developmental competence of oocytes derived from small antral follicles is not adversely affected by the presence of a dominant follicle *(45,46)*. As of the date of writing, more than 1000 healthy infants have been born worldwide following immature oocyte retrieval and IVM. In general, the clinical pregnancy per cycle started and implantation rates per embryo transfer have reached 30–35% and 10–15%, respectively. Pregnancy rates after IVM correlate with the number of immature oocytes retrieved. IVM is can be routinely

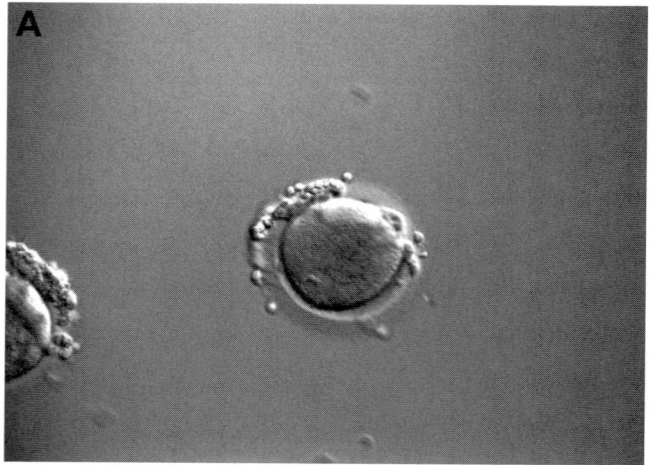

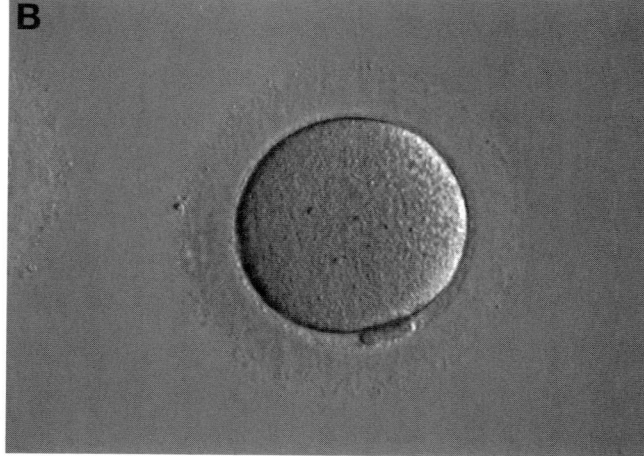

Fig. 4. Depicted is an immature oocyte (**A**) and an in vitro matured oocyte (**B**). The recovery of immature oocytes followed by in vitro maturation (IVM) and in vitro fertilization (IVF) is an attractive alternative to conventional IVF treatment in PCOS. (From R.C. Chian.)

offered to polycystic ovary/PCOS patients and also to patients previously treated by IVF who repeatedly produce virtually all poor-quality embryos for no obvious reason. Further research remains to be done to address the mechanism of oocyte maturation in order to further refine culture conditions and improve the implantation rate of oocytes matured in vitro.

3. CONCLUSION

The presence of polycystic ovaries and/or PCOS is important for determining the results of treatment with IVF because a longer time is required for pituitary suppression and there is higher sensitivity to gonadotropins during ovarian stimulation and a higher risk of OHSS. However, clinical pregnancy rates and live-birth rates after IVF treatment are generally higher in women with polycystic ovary/PCOS compared with women with morphologically normal ovaries. IVM is an alternative treatment option that avoids the risks of ovarian stimulation in this group. In some centers, IVM outcomes are comparable to overall IVF outcomes.

KEY POINTS

- IVF is an effective treatment for infertility of many causes.
- Ovarian stimulation is used to generate multiple oocytes and therefore multiple embryos.
- Polycystic ovaries and PCOS react to ovarian stimulation in the same manner. This may be because of increased blood flow and increased VEGF in polycystic ovaries. They are both at increased risk of OHSS
- Clinical pregnancy and live-birth rates after three-cycle IVF treatment are higher in patients with polycystic ovary/PCOS when compared with patients with morphologically normal ovaries.
- IVM avoids the risk of OHSS and is an effective alternative for patients with polycystic ovary/PCOS.

REFERENCES

1. Steptoe PC, Edwards RG. Birth after the reimplantation of a human embryo. Lancet 1978;2:366.
2. Schultz RM, Williams CJ. The science of ART. Science 2002;296:2188–2190.
3. Tan SL, Royston P, Campbell S, et al. Cumulative conception and livebirth rates after in vitro fertilisation. Lancet 1992;339:1390–1394.
4. Klonoff-Cohen, H. Female and male lifestyle habits and IVF: what is known and unknown. Hum Reprod Update 2005;11:179–203.
5. Tan SL, Bennett S, Parsons J. Surgical techniques for oocyte recovery and embryo transfer. British Medical Bulletin 1990;46:628–642.
6. Brinsden PR, Wada I, Tan SL, Balen A, Jacobs HS. Diagnosis, prevention and management of ovarian hyperstimulation syndrome. Br J Obstet Gynaecol 1995;102:767–772.
7. Tan SL, Maconochie N, Doyle P, et al. Cumulative conception and live-birth rates after in vitro fertilization with and without the use of long, short, and ultrashort regimens of the gonadotropin-releasing hormone agonist buserelin. Am J Obstet Gynecol 1994;171:513–520.
8. Wright VC, Schieve LA, Reynolds MA, Jeng G. Assisted reproductive technology surveillance—United States, 2002. MMWR Surveill Summ 2005;54:1–24.
9. Engmann L, Maconochie N, Sladkevicius P, Bekir J, Campbell S, Tan SL. The outcome of in-vitro fertilization treatment in women with sonographic evidence of polycystic ovarian morphology. Hum Reprod 1999;14:167–171.
10. Polson DW, Adams J, Wadsworth J, Franks S. Polycystic ovaries—a common finding in normal women. Lancet 1988;1:870–872.
11. Abdel Gadir A, Khatim MS, Mowafi RS, Alnaser HM, Muharib NS, Shaw RW. Implications of ultrasonically diagnosed polycystic ovaries. I. Correlations with basal hormonal profiles. Hum Reprod 1992;7:453–457.
12. Clayton RN, Ogden V, Hodgkinson J, et al. How common are polycystic ovaries in normal women and what is their significance for the fertility of the population? Clin Endocrinol (Oxf) 1992;37:127–134.
13. MacDougall MJ, Tan SL, Balen A, Jacobs HS. A controlled study comparing patients with and without polycystic ovaries undergoing in-vitro fertilization. Hum Reprod 1993;8:233–237.
14. Dunaif A, Graf M. Insulin administration alters gonadal steroid metabolism independent of changes in gonadotropin secretion in insulin-resistant women with the polycystic ovary syndrome. J Clin Invest 1989;83:23–29.
15. Dale O, Tanbo T, Haug H, Abyholm T. The impact of insulin resistance on the outcome of ovulation induction with low dose FSH in women with polycystic ovary syndrome. Hum Reprod 1998;13:567–570.
16. Fedorcsak P, Dale PO, Storeng R, Abyholm T, Tanbo T. The effect of metformin on ovarian stimulation and in vitro fertilization in insulin-resistant women with polycystic ovary syndrome: an open-label randomized cross-over trial. Gynecol Endocrinol 2003;17:207–214.
17. Stadtmauer LA, Toma SK, Riehl RM, Talbert LM. Metformin treatment of patients with polycystic ovary syndrome undergoing in vitro fertilization improves outcomes and is associated with modulation of the insulin-like growth factors. Fertil Steril 2001;75:505–509.
18. Stadtmauer LA, Toma SK, Riehl RM, Talbert LM. Impact of metformin therapy on ovarian stimulation and outcome in 'coasted' patients with polycystic ovary syndrome undergoing in-vitro fertilization. Reprod Biomed Online 2002;5:112–116.
19. Kjotrod SB, von During V, Carlsen SM. Metformin treatment before IVF/ICSI in women with polycystic ovary syndrome; a prospective, randomized, double blind study. Hum Reprod 2004;19:1315–1322.
20. Onalan G, Pabuccu R, Goktolga U, Ceyhan T, Bagis T, Cincik M. Metformin treatment in patients with polycystic ovary syndrome undergoing in vitro fertilization: a prospective randomized trial. Fertil Steril 2005;84:798–801.
21. Fedorcsak P, Dale PO, Storeng R, Tanbo T, Abyholm T. The impact of obesity and insulin resistance on the outcome of IVF or ICSI in women with polycystic ovarian syndrome. Hum Reprod 2001;16:1086–1091.
22. Biljan MM, Mahutte NG, Dean N, Hemmings R, Bissonnette F, Tan SL. Effects of pretreatment with an oral contraceptive on the time required to achieve pituitary suppression with gonadotropin-releasing hormone analogues and on subsequent implantation and pregnancy rates. Fertil Steril 1998;70:1063–1069.

23. Smitz J, Ron-El R, Tarlatzis BC. The use of gonadotrophin releasing hormone agonists for in vitro fertilization and other assisted procreation techniques: experience from three centres. Hum Reprod 1992;7(Suppl 1):49–66.

24. Balen AH, Tan SL, MacDougall J, Jacobs HS. Miscarriage rates following in-vitro fertilization are increased in women with polycystic ovaries and reduced by pituitary desensitization with buserelin. Hum Reprod 1993;8:959–964.

25. Homburg R, Armar NA, Eshel A, Adams J, Jacobs HS. Influence of serum luteinising hormone concentrations on ovulation, conception, and early pregnancy loss in polycystic ovary syndrome. BMJ 1988;297:1024–1026.

26. Homburg R, Berkowitz D, Levy T, Feldberg D, Ashkenazi J, Ben-Rafael Z. In vitro fertilization and embryo transfer for the treatment of infertility associated with polycystic ovary syndrome. Fertil Steril 1993;60:858–863.

27. Stanger JD, Yovich JL. Reduced in-vitro fertilization of human oocytes from patients with raised basal luteinizing hormone levels during the follicular phase. Br J Obstet Gynaecol 1985;92:385–393.

28. Howles CM, Macnamee MC, Edwards RG, Goswamy R, Steptoe PC. Effect of high tonic levels of luteinising hormone on outcome of in-vitro fertilisation. Lancet 1986;2:521–522.

29. Zaidi J, Campbell S, Pittrof R, et al. Ovarian stromal blood flow in women with polycystic ovaries—a possible new marker for diagnosis? Hum Reprod 1995;10:1992–1996.

30. Zaidi J, Jacobs H, Campbell S, Tan SL. Blood flow changes in the ovarian and uterine arteries in women with polycystic ovary syndrome who respond to clomiphene citrate: correlation with serum hormone concentrations. Ultrasound Obstet Gynecol 1998;12:188–196.

31. Buckett WM, Bouzayen R, Watkin KL, Tulandi T, Tan SL. Ovarian stromal echogenicity in women with normal and polycystic ovaries. Hum Reprod 1999;14:618–621.

32. Agrawal R, Conway G, Sladkevicius P, et al. Serum vascular endothelial growth factor and Doppler blood flow velocities in in vitro fertilization: relevance to ovarian hyperstimulation syndrome and polycystic ovaries. Fertil Steril 1998;70:651–658.

33. McClure N, Healy DL, Rogers PA, et al.. Vascular endothelial growth factor as capillary permeability agent in ovarian hyperstimulation syndrome. Lancet 1994;344:235–236.

34. Smitz J, Camus M, Devroey P, Erard P, Wisanto A, Van Steirteghem AC. Incidence of severe ovarian hyperstimulation syndrome after GnRH agonist/HMG superovulation for in-vitro fertilization. Hum Reprod 1990;5:933–937.

35. Forman RG, Frydman R, Egan D, Ross C, Barlow DH. Severe ovarian hyperstimulation syndrome using agonists of gonadotropin-releasing hormone for in vitro fertilization: a European series and a proposal for prevention. Fertil Steril 1990;53:502–509.

36. MacDougall MJ, Tan SL, Jacobs HS. In-vitro fertilization and the ovarian hyperstimulation syndrome. Hum Reprod 1992;7:597–600.

37. Sher G, Zouves C, Feinman M, Maassarani G. 'Prolonged coasting': an effective method for preventing severe ovarian hyperstimulation syndrome in patients undergoing in-vitro fertilization. Hum Reprod 1995;10:3107–3109.

38. Orvieto R. Can we eliminate severe ovarian hyperstimulation syndrome? Hum Reprod 2005;20:320–322.

39. Chen CD, Wu MY, Yang JH, Chen SU, Ho HN, Yang YS. Intravenous albumin does not prevent the development of severe ovarian hyperstimulation syndrome. Fertil Steril 1997;68:287–291.

40. Tan SL, Balen A, el Hussein E, Campbell S, Jacobs HS. The administration of glucocorticoids for the prevention of ovarian hyperstimulation syndrome in in vitro fertilization: a prospective randomized study. Fertil Steril 1992;58:378–383.

41. Chian RC, Buckett WM, Too LL, Tan SL. Pregnancies resulting from in vitro matured oocytes retrieved from patients with polycystic ovary syndrome after priming with human chorionic gonadotropin. Fertil Steril 1999;72:639–642.

42. Tan SL, Child TJ, Gulekli B. In vitro maturation and fertilization of oocytes from unstimulated ovaries: predicting the number of immature oocytes retrieved by early follicular phase ultrasonography. Am J Obstet Gynecol 2002;186: 684–689.

43. Child TJ, Abdul-Jalil AK, Gulekli B, Tan SL. In vitro maturation and fertilization of oocytes from unstimulated normal ovaries, polycystic ovaries, and women with polycystic ovary syndrome. Fertil Steril 2001;76:936–942.

44. Chian RC, Gulekli B, Buckett WM, Tan SL. Priming with human chorionic gonadotropin before retrieval of immature oocytes in women with infertility due to the polycystic ovary syndrome. N Engl J Med 1999;341:1624, 1626.

45. Chian RC, Buckett WM, Tan SL. In-vitro maturation of human oocytes. Reprod Biomed Online 2004;8:148–166.

46. Chian RC, Buckett WM, Abdul Jalil, et al. Natural-cycle in vitro fertilization combined with in vitro maturation of immature oocytes is a potential approach in infertility treatment. Fertil Steril 2004;82:1675–1678.

37

Surgical Management of the Polycystic Ovary Syndrome

Adam H. Balen

SUMMARY

Surgical approaches to ovulation induction have developed from the traditional wedge resection to modern minimal access techniques, usually employing laparoscopic ovarian diathermy (LOD) or laser. This provides a single treatment option with a good rate of unifollicular ovulation, thereby minimizing the need for extensive ultrasound monitoring because of a low risk of multiple pregnancy. There is no difference between the cumulative conception rates between LOD and gonadotropin therapy after 12 months, although the latter produces higher pregnancy rates at 6 months. Ovarian surgery carries minimal risks, namely those of general anaesthesia, pelvic adhesions, and ovarian damage—all of which can be avoided by careful operative technique.

Key Words: Laparoscopic ovarian diathermy; wedge resection; polycystic ovary syndrome; ovulation induction; clomiphene citrate resistance; gonadotropin therapy.

1. INTRODUCTION

The management of anovulatory infertility in polycystic ovary syndrome (PCOS) has traditionally involved the use of clomiphene citrate and then gonadotropin therapy or laparoscopic ovarian surgery in those who are clomiphene resistant. The principles of therapy are first to optimize health before commencing therapy (e.g., weight loss for those who are overweight) and then induce regular unifollicular ovulation, while minimizing the risks of ovarian hyperstimulation syndrome (OHSS) and multiple pregnancy. Weight loss improves the endocrine profile and the likelihood of ovulation and a healthy pregnancy.

From the 1930s to the early 1960s wedge resection of the ovary was the only treatment for PCOS. Wedge resection required a laparotomy—removal of up to 75% of each ovary—and often resulted in extensive pelvic adhesions. The modern-day, minimal-access alternative to gonadotropin therapy for clomiphene-resistant PCOS is laparoscopic ovarian surgery. Laparoscopic ovarian surgery has therefore replaced ovarian wedge resection as the surgical treatment for clomiphene resistance in women with PCOS. It is free of the risks of multiple pregnancy and ovarian hyperstimulation and does not require intensive ultrasound monitoring. Furthermore, ovarian diathermy is said to be as effective as routine gonadotropin therapy in the treatment of clomiphene-insensitive PCOS, although the evidence will be discussed in this chapter. In addition, laparoscopic ovarian surgery is a useful therapy for anovulatory women with PCOS who need a laparoscopic assessment of their pelvis or who live too far away from the hospital to be able to attend for the intensive monitoring required for gonadotropin therapy. Laparoscopic ovarian surgery by its effect on reducing serum luteinizing hormone (LH) concentrations is recommended for patients who persistently hypersecrete LH either during natural cycles or in response to clomiphene. Surgery, of course, carries its own risks and should be performed only by properly trained laparoscopic surgeons.

From: *Contemporary Endocrinology: Androgen Excess Disorders in Women:*
Polycystic Ovary Syndrome and Other Disorders, Second Edition
Edited by: R. Azziz et al. © Humana Press Inc., Totowa, NJ

2. BACKGROUND

2.1. History

Wedge resection of the ovaries was initially described by Stein and Leventhal *(1)* at the time that polycystic ovaries were diagnosed during a laparotomy. It was found that ovarian biopsies taken to make the diagnosis led to subsequent ovulation. The rationale was to "normalize" ovarian size and hence the endocrinopathy by removing between 50 and 75% of each ovary. A large review of 187 reports summarized data on 1079 ovarian wedge resections, with overall rates of ovulation of 80% and pregnancy of 62.5% (range 13.5–89.5%) *(2)*. Another 30 or so years later, Donesky and Adashi *(3)* were able to increase the summarized experience in the literature to 1766 treatments, with an average pregnancy rate of 58.8%. Because of the realization that significant postoperative adhesion formation occurred and that initial favorable reports of pregnancy rates were not sustained, wedge resection became less popular in the 1970s—the same time medical therapies for ovulation induction appeared more successful.

2.2. Methods and Dose

Commonly employed methods for laparoscopic surgery include monopolar electrocautery (diathermy) and laser. In the first reported series, ovarian diathermy resulted in ovulation in 90% and conception in 70% of the 62 women treated *(4)*. The outcome of 62 pregnancies was no different from the normal population *(5)*, and the miscarriage rate was 15%. A number of subsequent studies have produced similarly encouraging results, although the techniques used and the degree of ovarian damage varies considerably. Gjoannaess *(4)* cauterized each ovary at five to eight points for 5–6 seconds at each point with 300–400 W. Using a similar technique, Dabirashrafi et al. *(6)* reported mild to moderate adhesion formation in 20% of patients. Naether et al. treated 5–20 points per ovary with 400 W for approximately 1 second *(7)*. They found that the rate of adhesions was 19.3% and that this was reduced to 16.6% by peritoneal lavage with saline *(8)*. They also reported that the postdiathermy fall in serum testosterone concentration was proportional to the degree of ovarian damage, with up to 40 cauterization sites being used in some patients *(9)*.

The greater the amount of damage to the surface of the ovary, the greater the risk of peri-ovarian adhesion formation. This led Armar to develop a strategy of minimizing the number of diathermy points *(10)*. This is a logical technique in which the ovary is simply cauterized at four points and the lowest effective dose is used. The high pregnancy rate (86% of those with no other pelvic abnormality) indicates that the small number of diathermy points used leads to a low rate of significant adhesion formation *(11)*.

The difficulty when deciding how to perform laparoscopic ovarian diathermy (LOD) is knowing the dose response for a particular patient. We have shown in a small study that LOD using 40 W for 4 seconds in four places on one ovary can lead to bilateral ovarian activity and ovulation (our usual protocol involves the same on each ovary) *(12)*, our ovulation rate was 50% and conception rate 40% (some patients were sensitized to exogenous stimulation). It has been proposed that the degree of ovarian destruction should be determined by the size of the ovary *(13)*. Naether et al. reported their method of laparoscopic electrocautery of the ovarian surface, which causes greater destruction of the ovary than the method we use, as they apply 400 W at 5–20 sites on each ovary *(13)*. Despite such a large amount of ovarian destruction, in Naether's series of 206 patients 45.2% of those who conceived required additional ovarian stimulation (with an 8% multiple pregnancy rate) and the overall miscarriage rate was 20%. There is also no doubt that different patient populations are being treated, and we only recommend operation for women with irregular, anovulatory cycles who have not responded to antiestrogen therapy, whereas in Naether's series approximately 24% of the women operated on had regular cycles and 15% were ovulating before their operation *(13)*.

Amer and colleagues performed first a retrospective assessment of the effect of the amount of energy used on the outcome of LOD *(14)* and then proceeded to assess this prospectively *(15)*. They found in their analysis no difference between the use of 3–10 punctures but suggested that 2 punctures were too few *(14)*. There are many variables in the potential for response, including the anthropometric characteristics of the patients and ovarian morphology. In a prospective study using a modified Monte Carlo protocol and a standardized energy of 150 J/puncture, the rates of ovulation and pregnancy were, respectively, 67 and 67% with four punctures/ovary ($n = 12$), 44 and 56% with three punctures/ovary ($n = 9$), 33 and 17% with two punctures/ovary ($n = 6$), and 33 and 0% with one puncture/ovary ($n = 12$), indicating a clear dose response albeit in relatively small numbers *(15)*.

2.3. Risks: Adhesion Formation and Ovarian Failure

The risk of periovarian adhesion formation may be reduced by abdominal lavage and early second-look laparoscopy, with adhesiolysis if necessary *(16)*. Others have also used liberal peritoneal lavage to good effect *(10)*. Greenblatt and Casper *(17)* found no correlation between the degree of ovarian damage and subsequent adhesion formation, nor did they find benefit from the adhesion barrier Interceed® (Ethico, Inc), as assessed by second-look laparoscopy. In another interesting study, 40 women undergoing laser photocoagulation of the ovaries using an Nd-YAG laser set at 50 W at 20–25 points per ovary were randomized to a second-look laparoscopy and adhesiolysis *(18)*. Of those who underwent a second-look laparoscopy, adhesions described as minimal or mild were found in 68%, yet adhesiolysis did not appear to be necessary because the cumulative conception rate after 6 months was 47% compared with 55% in the expectantly managed group—this difference was not significant.

Laser treatment seems to be as efficacious as diathermy, and it has been suggested that it may result in less adhesion formation *(19,20)*, although the only study to compare the two techniques was nonrandomized, reported similar ovulation and pregnancy rates and did not examine adhesion formation *(20)*. Various types of laser have been used, from the CO_2 laser to the Nd:YAG and KTP lasers. As in other spheres of laparoscopic surgery, whether laser or diathermy is employed appears to depend upon the preference of the surgeon and the availability of the equipment.

An additional concern is the possibility of ovarian destruction leading to ovarian failure—an obvious disaster in women wishing to conceive. Cases of ovarian failure have been reported after both wedge resection and laparoscopic surgery. An unfortunate vogue has developed whereby women with polycystic ovaries who have overresponded to superovulation for in vitro fertilization (IVF) are subjected to ovarian diathermy as a way of reducing the likelihood of subsequent OHSS *(21)*. If one accepts that appropriately performed ovarian diathermy works by sensitizing the ovary to follicle-stimulating hormone (FSH) and ovarian diathermy certainly makes the clomiphene-resistant polycystic ovary sensitive to clomiphene *(22)*, then one could extrapolate that ovarian diathermy prior to superovulation for IVF should make the ovary more and not less likely to overstimulate. The amount of ovarian destruction that is required to reduce the chance of overstimulation is therefore likely to be considerable, and one should be very cautious before proceeding with such an approach because of concerns about permanent ovarian atrophy.

2.4. Endocrine Changes After Laparoscopic Ovarian Surgery

With restoration of ovarian activity after LOD, the serum concentrations of LH and testosterone fall. A fall in serum LH concentrations may both increase the chance of conception and reduce the risk of miscarriage *(11,23)*. Whether patients respond to LOD appears to depend on their pretreatment characteristics, with patients with high basal LH concentrations having a better clinical and endocrine response *(24,25)*. The Cochrane meta-analysis did not, however, find any difference in miscarriage rates compared with the use of gonadotropins (OR 0.61, 95% CI 0.17, 2.16) *(26)*. Although hyperinsulinemia plays a major role in the pathophysiology of anovulatory PCOS, the procedure of LOD does not appear to influence insulin sensitivity *(27)*. With respect to insulin resistance there has been much recent interest in insulin-lowering drugs such as metformin in enhancing reproductive function. A prospective randomized, controlled trial (RCT) in 120 clomiphene citrate-resistant women found no significant difference in rates of ovulation (approximately 55% in each group), yet those treated with metformin had significantly higher pregnancy rates (18.6 vs 13.4%, $p < 0.05$) and live-birth rates (82.1 vs 64.5%, $p < 0.05$) *(28)*.

2.5. Pregnancy Rates

Most early studies were of an observational nature and have also been reported in the context of large reviews *(2,3)*. An unfortunate feature of many of the papers that describe laparoscopic treatment wedge resection is the poor characterization of the patients such that many appear to have been ovulating prior to treatment. Furthermore, as the polycystic ovary becomes more sensitive to either endogenous or exogenous FSH after LOD, many practitioners have taken a pragmatic approach by commencing ovarian stimulation with either clomiphene or gonadotropins if ovulatory activity is not immediately induced *(30,31)*. The first RCT suggested that LOD was as effective as routine gonadotropin therapy in the treatment of clomiphene-insensitive PCOS *(29)*. In this study 88 patients were randomized prospectively to receive either human menopausal gonadotropin, FSH, or LOD. There were no differences in the rates of ovulation or pregnancy between the two groups, although those treated with LOD had fewer cycles with multiple follicular growth and a lower rate of miscarriage *(29)*. The largest RCT to date was the multicenter study performed in the the Netherlands in which 168 patients resistant to clomiphene were randomized to either LOD ($n = 83$) or ovulation induction with recombinant FSH (rFSH) ($n = 65$) *(32)*. The initial cumulative pregnancy rate after 6 months was 34% in the LOD arm vs 67% with rFSH. Those who did not ovulate in response to LOD were then given first clomiphene and then rFSH so that by 12 months the cumulative pregnancy rate was similar in each group at 67% *(32)*. Thus, those treated with LOD took longer to conceive, and 54% required additional medical ovulation-induction therapy.

It has been suggested that to demonstrate a 20% increase in pregnancy rate over 6 months from 50 to 70% with an 80% power, at least 235 patients would be required in each arm of a study to compare LOD with gonadotropin therapy. The current meta-analysis in the Cochrane database includes only 303 women *(26)*. The ongoing pregnancy rate following ovarian drilling compared with gonadotropins differed according to the length of follow-up. Overall, the pooled odds ratio for all studies was not statistically significant (OR 1.27, 95% CI 0.77, 1.98). Multiple pregnancy rates were reduced in the ovarian drilling arms of the four trials where there was a direct comparison with gonadotropin therapy (OR 0.16, 95% CI 0.03,0.98). There was no difference in miscarriage rates in the drilling group when compared with gonadotropin in these trials (OR 0.61, 95% CI 0.17, 2.16).

The duration of follow-up varied among the studies that were included in the meta-analysis. Furthermore, it is difficult to produce a temporal comparison because not all women receiving gonadotropin therapy are treated in consecutive months and so it is necessary to compare treatment cycles. The meta-analysis found that when comparing 6 months after ovarian drilling with six cycles of gonadotropin therapy, the ongoing cumulative pregnancy rate was higher among women who received gonadotropins (OR 0.48, 95% CI 0.28, 0.81). Thus, it was concluded that there is insufficient evidence of a difference in cumulative ongoing pregnancy rates between laparoscopic ovarian drilling after 6–12 months of follow-up and three to six cycles of ovulation induction with gonadotropins as a primary treatment for subfertile patients with anovulatory PCOS. The greatest advantage is that multiple pregnancy rates are considerably reduced.

3. CONCLUSIONS

Laparoscopic ovarian surgery is free of the risks of multiple pregnancy and ovarian hyperstimulation and does not require intensive ultrasound monitoring. Laparoscopic ovarian surgery is a useful therapy for an anovulatory woman with PCOS who fails to respond to clomiphene and who either persistently hypersecretes LH, needs a laparoscopic assessment of her pelvis, or lives too far away from the hospital to be able to undergo the intensive monitoring required of gonadotropin therapy. Surgery carries its own risks and must be performed only by fully trained laparoscopic surgeons.

Compared with medical ovulation induction, the additional advantage of laparoscopic diathermy is that it need only be performed once and intensive monitoring is not required because there is no danger of multiple ovulation or ovarian hyperstimulation. Furthermore, minimal ovarian damage only is required to achieve this effect. There is still uncertainty, however, about the right dose of diathermy to reliably stimulate the resumption of ovulatory cycles. The chance of achieving an ongoing pregnancy within 6 months is less than with carefully conducted ovulation induction with gonadotropins, but if adjuvant ovulation induction agents are used in those who do not initially respond, the 12-month pregnancy rates are similar.

4. FUTURE AVENUES OF INVESTIGATION

A number of questions remain to be answered, including what the predictors for a successful response to LOD are, what the appropriate diathermy dose for each individual is, and what the exact mechanism of action is.

KEY POINTS

- Laparoscopic ovarian surgery can achieve unifollicular ovulation in PCOS, with no risk of OHSS or high-order multiples and without requiring detailed ovulation monitoring.
- Gonadotropin preparations can be expensive but provide a quicker cumulative pregnancy rate than LOD.
- LOD is a single treatment using existing equipment.
- The risks of surgery are minimal and include the risk of laparoscopy, adhesion formation, and destruction of normal ovarian tissue.
- LOD may achieve a normalization of the patient's endocrinology, with a fall in LH and testosterone concentrations, albeit no change in insulin sensitivity.

REFERENCES

1. Stein IF, Leventhal ML. Amenorrhoea associated with bilateral polycystic ovaries. Am J Obstet Gynaecol 1935;29:181–191.
2. Goldzieher JW, Axelrod LR. Clinical and biochemical features of polycystic ovarian disease. Fertil Steril 1963;14:631–653.
3. Donesky BW, Adashi EY. Surgically induced ovulation in the polycystic ovary syndrome: wedge resection revisited in the age of laparoscopy. Fertil Steril 1995;63:439–463.
4. Gjoannaess H. Polycystic ovarian syndrome treated by ovarian electrocautery through the laparoscope. Fertil Steril 1984;41:20–25.
5. Gjoannaess H. The course and outcome of pregnancy after ovarian electrocautery with PCOS: the influence of body weight. Br J Obstet Gynaecol 1989;96:714–719.
6. Dabirashrafi H, Mohamad K, Behjatnia Y, et al. Adhesion formation after ovarian electrocauterization on patients with PCO syndrome. Fertil Steril.1991;55:1200–1201.
7. Naether OGJ, Fischer R, Weise HC, Geiger-Kotzler L, Delfs T, Rudolf K. Laparoscopic electrocoagulation of the ovarian surface in infertile patients with polycystic ovarian disease. Fertil Steril 1993;60:88–94.
8. Naether OGJ, Fischer R. Adhesion formation after laparoscopic electrocoagulation of the ovarian surface in polycystic ovary patients. Fertil Steril 1993;60:95–99.
9. Naether O, Weise HC, Fischer R. Treatment with electrocautery in sterility patients with polycystic ovarian disease. Geburtsh Frauenheilk 1991;51:920–924.
10. Armar NA, McGarrigle HHG, Honour JW, Holownia P, Jacobs HS, Lachelin GCL. Laparoscopic ovarian diathermy in the management of anovulatory infertility in women with polycystic ovaries: endocrine changes and clinical outcome. Fertil Steril 1990;53:45–49.
11. Armar NA, Lachelin GCL. Laparoscopic ovarian diathermy: an effective treatment for anti-oestrogen resistant anovulatory infertility in women with polycystic ovaries. Br J Obstet Gynaecol 1993;100:161–164.
12. Balen AH, Jacobs HS. A prospective study comparing unilateral and bilateral laparoscopic ovarian diathermy in women with the polycystic ovary syndrome. Fertil Steril 1994;62:921–925.
13. Naether lGJ, Baukloh V, Fischer R, Kowalczyk T. Long-term follow-up in 206 infertility patients with polycystic ovarian syndrome after laparoscopic electrocautery of the ovarian surface. Hum Reprod 1994;9:2342–2349.
14. Amer SAK, Li TC, Cooke ID. Laparoscopic ovarian diathermy in women with polycystic ovary syndrome: a retrospective study on the influence of the amount of energy used on the outcome. Hum Reprod 2002;17:1046–1051.
15. Amer SAK, Li TC, Cooke ID. A prospective dose-finding study of the amount of thermal energy required for laparoscopic ovarian. Hum Reprod 2003;18:1693–1698.
16. Naether OGJ. Significant reduction in of adnexal adhesions following laparoscopic electrocautery of the ovarian surface by lavage and artificial ascites. Gynaecol Endoscopy 1995;4:17–19.

17. Greenblatt E, Casper RF. Adhesion formation after laparoscopic ovarian cautery for POCS: lack of correlation with pregnancy rate. Fertil Steril 1993;60:766–769.

18. Gurgan T, Urman B, et al. The effect of short internal laparoscopic lysis of adhesions in pregnancy rates following ND:YAG laser photocoagulation of PCO. Obstet. Gynaecol 1992;80:45–47.

19. Daniell JF, Miller N. Polycystic ovaries treated by laparoscopic laser vaporization. Fertil Steril 1989;51:232–236.

20. Heylen SM, Puttemans PJ, Brosens LH. Polycystic ovarian disease treated by laparoscopic argon laser capsule drilling: comparison of vaporization versus perforation technique. Hum Reprod 1994;9:1038–1042.

21. Rimmington MR, Walker SM, Shaw RW. The use of laparoscopic ovarian electrocautery in preventing cancellation of in-vitro fertilization treatment cycles due to risk of ovarian hyperstimulation syndrome in women with polycystic ovaries. Hum Reprod 1997;7:1443–1447.

22. Farhi J, Soule S, Jacobs H. Effect of laparoscopic ovarian electrocautery on ovarian response and outcome of treatment with gonadotrophins in clomifene citrate resistant patients with PCOS. Fertil Steril 1995;64:930–935.

23. Balen AH, Tan SL, Jacobs HS. Hypersecretion of luteinising hormone—a significant cause of subfertility and miscarriage. Br J Obstet Gynaecol 1993;100:1082–1089.

24. Balen AH, Jacobs HS. A prospective study comparing unilateral and bilateral laparoscopic ovarian diathermy in women with the polycystic ovary syndrome. Fertil Steril 1994;62:921–925.

25. Abdel Gadir A, Alnaser HMI, Mowafi RS, Shaw RW. The response of patients with polycystic ovarian disease to human menopausal gonadotrophin therapy after ovarian electrocautery or a luteinizing hormone-releasing hormone agonist. Fertil Steril 1992;57:309–313.

26. Farquhar C, Vanderkerckhove P, Arnot M, Lilford R. Laparoscopic "drilling" by diathermy or laser for ovulation induction in anovulatory polycystic ovary syndrome (Cochrane Review). In The Cochrane Library, Issue 4, 1999. Oxford: Update Software.

27. Tulandi T, Saleh A, Morris D, Jacobs HS, Payne N, Tan SL. Effects of laparoscopic ovarian drilling on serum vascular endothelial growth factor and on insulin responses to the oral glucose tolerance test in women with polycystic ovary syndrome. Fertil Steril 2000;74:585–588.

28. Palomba S, Orio F, Nardo LG, et al. Metformin administration versus laparoscopic ovarian diathermy in clomifene citrate-resistant women with polycystic ovary syndrome: a prospective parallel randomized double-blind placebo-controlled trial. JCEM 2004;89:4801–4809.

29. Abdel Gadir A, Mowafi RS, Alnaser HMI, Alrashid AH, Alonezi OM, Shaw RW. Ovarian electrocautery versus human menopausal gonadotrophins and pure follicle stimulating hormone therapy in the treatment of patients with polycystic ovarian disease. Clin Endocrinol (Oxf) 1990;33:585–592.

30 Ostrzenski A. Endoscopic carbon dioxide laser ovarian wedge resection in resistant polycystic ovarian disease. Int J Fertil 1992;37:295–299.

31. Farhi J, Soule S, Jacobs H. Effect of laparoscopic ovarian electrocautery on ovarian response and outcome of treatment with gonadotrophins in clomifene citrate resistant patients with PCOS. Fertil Steril 1995;64:930–935.

32. Bayram N, van Wely M, Kaaijk EM, Bossuyt PMM, van der Veen F. Using an electrocautery strategy or recombinant FSH to induce ovulation in polycystic ovary syndrome: a randomised controlled trial. BMJ 2004;328:192–195.

38

Use of Insulin Sensitizers in the Polycystic Ovary Syndrome

Kai I. Cheang and John E. Nestler

SUMMARY

Insulin resistance is a central feature of the polycystic ovary syndrome (PCOS). Women with PCOS suffer both short- and long-term health consequences from insulin resistance and the compensatory hyperinsulinemia associated with PCOS. Short-term consequences include anovulation, hyperandrogenism, infertility, and early pregnancy loss. Over the long term, chronic hyperinsulinemia predisposes women with PCOS to increased risks of diabetes and cardiovascular events. Current data support the use of metformin for treatment of anovulation in women with PCOS, both in monotherapy and in combination with clomiphene in clomiphene-resistant patients. The effect of insulin sensitizers on hirsutism will also be reviewed. In addition, novel evidence, although limited, suggests that insulin sensitizers may play a role in preventing early pregnancy loss. Over the long term, insulin sensitizers may prevent cardiovascular and diabetes complications by ameliorating the metabolic syndrome associated with chronic insulin resistance in PCOS, pending further evidence. Several insulin-sensitizing agents are available commercially: metformin, rosiglitazone, and pioglitazone. Metformin has the most desirable pregnancy safety profile and is the agent that has been most frequently studied in PCOS. In conclusion, there is strong support for the use of insulin-sensitizing drugs, particularly metformin, in women with PCOS.

Key Words: Metformin; thiazolidinediones; insulin resistance; polycystic ovary syndrome.

1. INTRODUCTION

Current evidence suggests that insulin resistance is an important pathophysiological feature of the polycystic ovary syndrome (PCOS). An important implication of insulin resistance in PCOS is that insulin-sensitizing agents are a useful therapeutic approach in this disorder, as documented by multiple clinical trials. This chapter reviews the evidence supporting the use of insulin-sensitizing agents in ameliorating the short-term problems associated with PCOS, such as hyperandrogenism, anovulation, infertility, and early pregnancy loss. In addition, the role of insulin sensitizers in the prevention or management of long-term health issues in PCOS, such as cardiovascular disease and diabetes, will also be discussed.

2. BACKGROUND/SUMMARY

2.1. Insulin Resistance and PCOS

The role of insulin resistance in the pathogenesis of PCOS has been extensively reviewed in Chapter 24. Briefly, insulin resistance is an intrinsic and virtually universal feature of PCOS. Insulin resistance exists in PCOS independent of obesity, as evidenced by the presence of insulin resistance even in lean women with PCOS (1). In addition, about 50–80% of women with PCOS are obese.

From: Contemporary Endocrinology: Androgen Excess Disorders in Women:
Polycystic Ovary Syndrome and Other Disorders, Second Edition
Edited by: R. Azziz et al. © Humana Press Inc., Totowa, NJ

Hence, in PCOS obesity further induces an added burden of insulin resistance and hyperinsulinemia in addition to the component of insulin resistance that is intrinsic to the disorder *(2)*.

2.1.1. Effect of Hyperinsulinemia on Hyperandrogenism in PCOS

Several lines of evidence suggest that the compensatory hyperinsulinemia associated with insulin resistance is critical to the pathogenesis of hyperandrogenism in PCOS. In in vitro cultures of isolated human ovarian thecal cells, ovarian testosterone biosynthesis stimulated by insulin was fourfold greater in cells from women with PCOS than those of normal women, and in a dose–response study insulin stimulated thecal androgen production at physiological concentrations *(3)*. This stimulation of testosterone production was almost fully prevented by antibody blockade of the insulin receptors. This suggests that insulin stimulates ovarian thecal testosterone biosynthesis via activation of its homologous receptor.

Insulin also decreases circulating levels of sex hormone-binding globulin (SHBG), the primary circulating binding protein for testosterone *(4)*. Because of testosterone's high-affinity binding to SHBG, lower SHBG levels translate to increased levels of free and bioavailable testosterone to target tissues. In addition, insulin may also influence secretory dynamics of gonadotropins by the pituitary, resulting in further stimulation of ovarian androgen production *(5–7)*.

In vivo data also suggest that hyperinsulinemia is associated with an increased level of circulating testosterone in PCOS. When pancreatic insulin secretion was suppressed in women with PCOS by diazoxide, both serum total and free testosterone levels decreased *(8)*. However, diazoxide's suppression of insulin release in healthy normal women did not result in changes in serum testosterone levels *(9)*. Hence, hyperinsulinemia seems to play a role in hyperandrogenemia in PCOS, and women with PCOS may be more susceptible than normal women to stimulation of ovarian androgen production by insulin.

2.1.2. Effect of Hyperinsulinemia on Anovulation in PCOS

Numerous studies have demonstrated that hyperinsulinemia contributes to the chronic anovulation of PCOS. In the largest long-term study, 305 women with PCOS were randomized to the insulin sensitizer troglitazone (150, 300, or 600 mg daily) or placebo for 44 weeks *(10)*. Women receiving troglitazone at doses of 300 and 600 mg daily had a significantly higher ovulation rate (0.42 and 0.58) than those receiving placebo (0.32; $p < 0.05$ and 0.0001, respectively). This increase in ovulation rate was dose dependent, suggesting that the improvement in ovulatory function was at least in part accounted for by the improvement in insulin sensitivity.

2.2. Clinical Trials of Insulin Sensitizers in Short-Term Management of PCOS

The observations indicating that insulin resistance plays a pivotal role in the pathogenesis of PCOS led investigations to explore the use of insulin sensitizers in the disorder. Both metformin and the thiazolidinediones (troglitazone, pioglitazone, and rosiglitazone) have been assessed. Among these agents, metformin has been most widely studied in PCOS, and troglitazone is no longer commercially available because of reports of liver toxicity.

2.2.1. Effect of Metformin on Anovulation and Infertility

2.2.1.1. METFORMIN MONOTHERAPY IN PCOS

Numerous clinical trials have evaluated the effect of metformin on ovulation in women with PCOS. Nester et al. reported the first randomized, placebo-controlled trial evaluating the use of metformin to reestablish ovulation *(11)*. Sixty-one obese women with PCOS were randomized to metformin 500 mg three times daily or placebo for 35 days. If spontaneous ovulation did not occur, they were entered into the second phase of the study and were given clomiphene 50 mg daily for 5 days while continuing to take the metformin or placebo for another month. During the first 35 days in which metformin monotherapy was compared to placebo, 24% of the women (12 in 35) taking metformin ovulated spontaneously, whereas only 4% (1 in 26) in the placebo group had spontaneous ovulation.

Table 1
Ovulation Rate With Metformin Compared to Placebo or No Treatment

Study (ref.)	Proportion of women with spontaneous ovulation		Peto odds ratio (95% CI)
	Treatment	Control	
Fleming et al., 2002 *(14)*	37/45	30/47	2.51 (1.01–6.25)
Jakubowicz et al., 2001 *(15)*	8/28	0/28	9.89 (2.24– 43.61)
Nestler and Jakubowicz, 1996 *(7)*	5/11	1/13	6.89 (1.12–42.33)
Nestler et al., 1998 *(11)*	12/35	1/26	5.96 (1.74–20.38)
Ng et al., 2001 *(19)*	3/9	3/9	1.00 (0.15–6.72)
Vandermolen et al., 2001 *(22)*	1/12	1/15	1.26 (0.07– 21.72)
Yarali et al., 2002 *(23)*	6/16	1/16	5.88 (1.13–30.61)
Total (95% CI)	72/156	37/154	3.88 (1.13–30.61)

Adapted from ref. *12.*

In the second phase of the study, 19 of 21 women (90%) receiving metformin and clomiphene ovulated, compared with only 2 of 25 women (8%) receiving placebo plus clomiphene. Of note, more women receiving metformin monotherapy for 7 weeks ovulated than those who underwent a single cycle of induction with clomiphene at the 50 mg dose (12% vs 8%, respectively).

Other studies have corroborated the above findings. The Cochrane library has published a systematic review of the use of metformin in PCOS *(12)*. The review included only the most rigorously performed clinical trials for analysis. These included randomized controlled trials comparing insulin-sensitizing drugs to placebo or to an ovulation-induction agent such as clomiphene or gonadotropin. Thirteen trials were included for analysis *(7,11,13–24)*. Almost uniformly across studies, metformin monotherapy showed a beneficial effect on ovulation rates in PCOS (Table 1). Overall, the ovulation rate increased 3.9-fold when metformin was given as monotherapy ($p < 0.00001$; CI 2.3–6.7). Metformin treatment resulted in spontaneous ovulation in 46% of women, while the ovulation rate was 24% with placebo. Clinical pregnancy rate with metformin was also increased 2.8-fold ($p = 0.09$; CI 0.9–9.0), but fell short of statistical significance probably as a result of the small number of women included in these studies and the fact that pregnancy was not an outcome measure of these studies. Of note, these studies were short term in nature (most were only up to 3 months in duration) and therefore were biased against showing a positive effect with metformin. Despite a short intervention period, it was evident that metformin was beneficial in improving ovulation in women with PCOS.

Clomiphene citrate has been the initial therapeutic option for the management of anovulatory infertility for many years. Metformin was directly compared to clomiphene in a recent double-blind, placebo-controlled trial in non-obese women (average body mass index [BMI] = 27 kg/m^2) with PCOS *(24)*. After progesterone-induced withdrawal bleeds, 100 women with PCOS were randomized to metformin (850 mg twice daily) or clomiphene citrate (150 mg for 5 days on a monthly basis) for 6 months. Although there was no difference in ovulation rates between the metformin and clomiphene groups, the pregnancy rate was higher (15.1% vs 7.2%; $p = 0.009$) and spontaneous abortion rate lower (9.7% vs 37.5%; $p = 0.045$) in the metformin group. There was also a favorable trend for the live-birth rate with metformin compared to clomiphene (83.9% vs 56.3%; $p = 0.07$).

2.2.1.2. METFORMIN IN CLOMIPHENE-RESISTANT WOMEN

Although clomiphene citrate has been historically the primary agent for ovulation induction in PCOS, obese women with PCOS often have an inadequate response to clomiphene. This is likely due to a high degree of insulin resistance and concomitant hyperinsulinemia *(25)*. Obese patients with PCOS often require progressively increased doses and multiple courses of clomiphene for successful

Table 2
**Ovulation Rate With Metformin Combined With Ovulation-Induction Agent
Compared With Induction Agent Alone**

Study (ref.)	Treatment	Control	Peto odds ratio (95% CI)
	Proportion of women with ovulation		
Clomiphene resistant:			
Kocak et al., 2002 *(16)*	21/27	4/28	12.36 (4.32–35.39)
Malkawiand Qublan, 2002 *(17)*	11/16	3/12	5.41 (1.24–23.51)
Subtotal (95% CI)	32/43	7/40	9.34 (3.97–21.97)
Clomiphene sensitivity not defined:			
El-Biely and Habba, 2001 *(13)*	35/45	29/45	1.90 (0.77–4.70)
Subtotal (95% CI)	35/45	29/45	1.90 (0.77– 4.70)
Total (95% CI)	67/88	36/85	4.41 (2.37–8.22)

Adapted from ref. *12.*

ovulation induction, and the required dose of clomiphene directly correlates with the degree of obesity *(26)*. Metformin has been evaluated in women with a poor response to clomiphene.

The Cochrane review analyzed several studies evaluating combination therapy with metformin and clomiphene in PCOS (Table 2) *(12)*. Overall, ovulation occurred in 76% of women taking metformin plus clomiphene compared to 42% with clomiphene alone ($p < 0.00001$; CI 2.4–8.2). Importantly, the clinical pregnancy rate was increased significantly by 4.4-fold with the combination of metformin and clomiphene compared to clomiphene alone ($p = 0.0003$; CI 2.0–9.9).

Recently, metformin has been compared to laparoscopic ovarian diathermy in overweight (BMI 25–40 kg/m^2), clomiphene-resistant women with PCOS *(27)*. In a double-blind, randomized, placebo-controlled trial, women who underwent a mock laparoscopy and then took metformin 850 mg twice daily for 6 months were compared to those who underwent laparoscopic ovarian diathermy and then took placebo for 6 months. Although the ovulation rate did not differ between the metformin (54.8%) and diathermy (55.1%) groups, the metformin group had a significantly higher pregnancy rate (18.5 vs 13.4%), lower abortion rate (15.4 vs 29.0%), and, consequently, a higher live-birth rate (82.1 vs 13.4%).

2.2.2. Effect of Thiazolidinediones on Anovulation and Infertility

2.2.2.1. EFFECTS OF THIAZOLIDINEDIONE MONOTHERAPY

As described earlier, the largest long-term study evaluating the use of an insulin-sensitizing agent in PCOS has been conducted with troglitazone. In this study, troglitazone (150, 300, or 600 mg daily) was compared to placebo in 305 women with PCOS for 44 weeks *(10)*. Women receiving troglitazone daily at the 300- and 600-mg doses had significantly higher ovulation rates than those receiving placebo. This increase in ovulation rate was dose dependent, suggesting that improvement in insulin sensitivity ameliorated the anovulation in PCOS. However, because of liver toxicity, troglitazone is on longer commercially available.

The newer thiazolidinediones rosiglitazone and pioglitazone have subsequently been evaluated in women with PCOS. In a small study involving 12 obese women with PCOS, rosiglitazone 4 mg daily was administered for 4 months *(28)*. Rosiglitazone therapy restored regular ovulatory cycles in 11 of the 12 women studied. Serum androgen levels and insulin sensitivity also significantly improved. In another study in non-obese (BMI < 27 kg/m^2) women with PCOS and impaired glucose tolerance, rosiglitazone 2 or 4 mg daily improved ovulation rate in a dose-dependent fashion *(29)*. However, the above studies with rosiglitazone lacked a placebo-control arm and were not controlled.

Pioglitazone has also been studied in women with PCOS. In 40 women who were randomized to pioglitazone 30 mg daily vs placebo for 3 months, pioglitazone significantly improved the ovulation rate (41.2 vs 5.6%; $p < 0.02$) and also improved insulin sensitivity *(30)*.

2.2.2.2. THIAZOLIDINEDIONES AND CLOMIPHENE COMBINATION THERAPY

Shobokshi et al. evaluated the combination of rosiglitazone 4 mg daily with clomiphene 100 mg vs clomiphene alone in 50 women with PCOS *(31)*. The combination treatment had a significant reduction in the area under the insulin curve (AUC-insulin), which was not observed in the clomiphene monotherapy group. The study also reported that combination therapy with rosiglitazone and clomiphene led to regular menstrual cycles in 72% of the women, compared with 48% with clomiphene alone. In addition, improvement in menstrual pattern (among classifications of amenorrhea, oligomenorrhea, and regular menses) were reported in 68% of the women taking clomiphene alone vs 92% of the women taking the rosiglitazone–clomiphene combination (OR 0.185, CI 0.035–0.993). However, the study was only 12 weeks in duration, and hence the results on menstrual pattern should be interpreted with caution. In addition, this study was not performed in clomiphene-resistant women with PCOS.

2.2.3. Effect of Insulin Sensitizers on Hirsutism

Current data on the effect of insulin sensitizers on hirsutism are conflicting. In an open-label study of 39 women with PCOS, metformin 500 mg three times daily significantly decreased Ferriman–Gallwey hirsutism scores at the end of the 12-week study period *(32)*. However, in a more rigorous randomized, placebo-controlled, 44-week trial, metformin at the same dose did not significantly reduce clinical hirsutism scores *(17)*. In another randomized controlled 12-month study, metformin 500 mg three times daily was compared to Dianette (ethinyl estradiol 35 µg and cyproterone acetate 2 mg) in 52 women *(33)*. Both groups demonstrated significant reductions in the Ferriman–Galwey scores, with a significantly greater reduction in the metformin arm ($p < 0.01$). In addition, in patient self-assessment, women taking metformin scored their hirsutism as having improved significantly more than the contraceptive group ($p = 0.01$). In the long-term, placebo-controlled study using troglitazone, a significant decrease in the Ferriman–Gallwey score was observed with troglitazone 600 mg daily compared with placebo *(10)*. However, pioglitazone 30 mg daily did not confer a significant change in Ferriman–Galwey scores *(34)*. These disparate results of the effect of insulin sensitizers on hirsutism could be a result of the limited number of women studied in these trials or the fact that improvement in hirsutism may not be directly associated with improvement in insulin sensitivity. In addition, most clinical studies, with the exception of the one conducted by Harborne et al. *(33)*, were relatively short term in nature (<6 months) because hirsutism was not a primary outcome measure, and hence these studies were not optimally designed to evaluate the effect of insulin sensitizers on hirsutism.

2.2.4. Comparative Efficacy of Different Insulin-Sensitizing Agents

Few studies have directly compared the utility of the various insulin-sensitizing drugs in women with PCOS. Ortega-Gonzalez et al. studied 52 obese (BMI > 28 kg/m^2), insulin-resistant women with PCOS who had not been previously treated and randomized them to either pioglitazone 30 mg/day or metformin 850 mg three times daily for 6 months *(35)*. Both metformin and pioglitazone significantly improved AUC-insulin and fasting glucose-to-insulin ratio despite a significant increase in weight (from BMI of 32.2 ±1.0 kg /m^2 at baseline to 34.0 ±1.2 kg/m^2 at 6 months) in the pioglitazone group. At 6 months, fasting AUC-insulin was significantly lower in the women treated with pioglitazone compared with those treated with metformin. However, women in the metformin group also had a higher AUC-insulin at baseline, and hence whether there was a true difference in the reduction of AUC-insulin between pioglitazone and metformin is unknown. Ovulation rates were not evaluated in this study.

Baillargeon et al. evaluated 128 non-obese women (BMI < 27 kg/m^2) with normal glucose tolerance and normal indices of insulin sensitivity who were randomized to metformin 850 mg twice daily, rosiglitazone 4 mg twice daily, metformin–rosiglitazone combination, or placebo for 6 months *(36)*. Both the metformin and metformin–rosiglitazone combination groups, but not the rosiglitazone monotherapy group, had an improvement in fasting insulin and AUC-insulin during an oral glucose tolerance test. Body weight significantly increased by 1.1 kg with rosiglitazone monotherapy, but not in other groups. In addition, both the metformin monotherapy and the combination therapy groups had a significantly higher ovulation rate than the rosiglitazone monotherapy arm. The addition of rosiglitazone to metformin did not further improve the ovulation rate.

The above two studies seemed to yield disparate results. In obese, insulin-resistant women with PCOS, pioglitazone and metformin seemed to be equally efficacious in PCOS. However, in non-obese women with PCOS with normal indices of insulin sensitivity, metformin was more effective in improving ovulation than rosiglitazone. These differences may be a result of the different insulin-sensitizing agents being compared (rosiglitazone vs pioglitazone) or the different weight and insulin sensitivity characteristics of these women.

The use of pioglitazone in women with PCOS who had an inadequate response to metformin was evaluated in a small observational, open-label study *(37)*. In 13 women with PCOS who had suboptimal response (menses rate of 14%) to metformin 2.55 g/day and a 1500- to 2000-calorie diet, pioglitazone 45 mg per day was added. Compared with metformin alone at baseline, ovulation rate significantly improved with the addition of pioglitazone to 77% at 6 months ($p = 0.017$) and 73% at 9 months ($p = 0.032$). Although the results of this study on metformin treatment failure are encouraging, a randomized, placebo-controlled, double-blind study will be necessary to confirm the utility of the addition of a thiazolidinedione to metformin in women with PCOS who fail to respond to metformin.

2.2.5. Insulin Sensitizers and Pregnancy Outcomes

Although ovulatory dysfunction is an important etiological feature of the infertility of PCOS, ovulation is only one aspect of fertility. In addition to ovulatory problems, women with PCOS suffer a high rate of early pregnancy loss (EPL) during the first trimester (30–50% in PCOS vs 10–15% in normal women) *(38–42)*. It is possible that insulin resistance may contribute to EPL by adversely affecting the endometrial environment and/or endometrial function.

The effect of metformin on endometrial function has been studied using surrogate markers, such as circulating levels of glycodelin and insulin-like growth factor binding protein-1(IGFBP-1) *(43)*. Glycodelin is secreted by endometrial glands *(44,45)* to lessen the endometrial immune response against the developing embryo *(46,47)*. Decreased endometrial secretion of glycodelin has been associated with EPL *(48,49)*. IGFBP-1 modulates adhesion processes at the feto/maternal interface *(50,51)* and hence may be important in the peri-implantation period. Besides being synthesized by the endometrium, IGFBP-1 is primarily synthesized in the liver, and insulin is known to inhibit hepatic IGFBP-1 production. Importantly, serum levels of glycodelin and IGFBP-1 during the first trimester are decreased in women with PCOS compared with normal women *(52)*.

Hyperinsulinemia in PCOS may contribute to EPL by decreasing glycodelin and IGFBP-1 expression. The insulin sensitizer metformin has been shown to have a beneficial effect to increase glycodelin and IGFBP-1 levels in women with PCOS *(15)*. In one study, luteal phase glycodelin and IGFBP-1 levels were measured in PCOS women whose ovulations were induced by clomiphene at baseline. These women were then randomized to metformin 500 mg three times daily or placebo for 4 weeks. After the treatment period, luteal phase glycodelin and IGFBP-1 levels were again determined after a clomiphene-induced ovulation. The metformin group had a significant threefold increase in luteal-phase glycodelin (from 3434 ± 1299 to 10624 ± 1803 pmol/L; $p < 0.001$) and IGFBP-1 levels (from 1220 ± 136 to 4916 ± 596 pmol/L; $p < 0.001$). Glycodelin and IGFBP-1 levels did not change in the placebo group. In addition, uterine vascular penetration and blood flow of uterine spiral arteries increased in the metformin group (demonstrated by a decrease of the resistance index), whereas these parameters did not change in the placebo group. These data suggest that hyperinsulinemia in PCOS decreases production of glycodelin and IGFBP-1 and modulates undesirable uterine vascular changes, all of which contribute to an adverse endometrial environment for the establishment and continuation of pregnancy. Metformin reduces serum insulin and its unfavorable effects on luteal-phase endometrial function and therefore may improve the peri-implantation endometrial milieu in PCOS.

A recent retrospective study compared the pregnancy outcomes of women with PCOS who became pregnant but were not exposed to metformin with those who became pregnant while taking metformin and remained on metformin throughout pregnancy *(53)*. These were nondiabetic women with PCOS who were seen at an academic endocrinology clinic in Caracas within a 4.5-year period. Among these women, 65 became pregnant using metformin, while 31 women never exposed to metformin became pregnant. The EPL rate in the metformin group was 8.8% (6 of 68 pregnancies), while the EPL rate in women never exposed to metformin was 41.9% (13 of 31 pregnancies; $p <$ 0.001). In women with a prior history of miscarriage, the EPL rate in the metformin group was 11.1% (4 of 36 pregnancies) vs 58.3% (7 of 12 pregnancies) in women not taking metformin ($p = 0.002$). Metformin was not associated with adverse fetal outcomes except for a single infant born with achondrodysplasia.

The observations from this retrospective study suggest that improving insulin sensitivity is associated with a reduction in EPL rates in PCOS. However, because the women in this study took metformin throughout pregnancy, the optimal duration of metformin administration to achieve desirable pregnancy outcomes remains unknown. For example, metformin may need to be administered throughout the pregnancy, only during the first trimester, or only during conception to confer maximal protection, as suggested by two recent studies by Palomba et al. *(24,27)*. Importantly, the beneficial effect of metformin on EPL seen in this retrospective study needs to be confirmed by a randomized and controlled prospective trial.

Metformin is classified by the U.S. Food and Drug Administration (FDA) as a Category B drug for use in pregnancy, which means that either animal studies have not shown a teratogenic risk or that adverse effects shown in animal data were not confirmed in controlled studies in women in the first trimester. Current evidence of metformin's use in pregnancy does not suggest that it is teratogenic. Women with PCOS who conceived while taking metformin and continued taking metformin throughout pregnancy have been described in several studies. Glueck et al. described several cohort studies examining metformin's maternal and fetal effects in women with PCOS who conceived on metformin and continued metformin therapy throughout their pregnancies *(54)*. In these reports, metformin use during pregnancy was not associated with maternal lactic acidosis or with neonatal or maternal hypoglycemia. In 126 infants who were studied up to 18 months of life, metformin did not appear to affect birth length and weight, motor-social development, or growth *(55)*. In addition, metformin use throughout pregnancy may reduce the incidence of gestational diabetes *(54–56)*. However, it is important to note that these reports utilized historical controls that may limit interpretation of the data.

Despite several reports suggesting that metformin may lead to favorable pregnancy outcomes, there is still uncertainty regarding metformin use during pregnancy. Metformin has been shown to cross the placenta in women with PCOS, and fetal serum levels are comparable to maternal serum concentrations *(57)*. In addition, data in one retrospective study in diabetic women suggested an increased rate of pre-eclampsia in women treated with metformin compared to those treated with sulphonylureas or insulin during their pregnancies. However, the groups in this study were not matched due to the retrospective nature of the study *(58)*, and the women who took metformin were more obese and had more poorly controlled diabetes, both of which are known risk factors for pre-eclampsia. Moreover, the excess perinatal mortality in the metformin group in this study occurred in two subjects who had poorly controlled diabetes and were noncompliant, and therefore cannot be attributed to metformin treatment alone *(58)*.

The only prospective, randomized study assessing metformin's effects on maternal and fetal outcomes was conducted by Vanky et al. *(59)*. In this study 40 pregnant women with PCOS were randomized to metformin 850 mg twice daily or placebo. Randomization occurred at gestational age between 5 and 12 weeks. At conception, 11 of 22 women in the placebo group and 8 of 18 women in the metformin group were taking metformin. All women who were on metformin at conception stopped their metformin therapy for at least 2 days before inclusion in the study. Seven of the 22 women (32%) in the placebo group experienced severe pregnancy complications (such as preterm deliveries and severe pre-eclampsia) or postpartum complications (sepsis, acute respiratory distress syndrome, deep vein and pulmonary embolism). None in the metformin group had any severe pregnancy or postpartum complications.

In summary, preliminary data suggest that metformin may decrease spontaneous abortion, gestational diabetes, and pregnancy-related complications in women with PCOS. Importantly, current data do not suggest that metformin is teratogenic. However, questions remain regarding the optimal time to discontinue metformin (or whether metformin should be continued throughout the entire pregnancy) once pregnancy has been achieved.

The commercially available thiazolidinediones rosiglitazone and pioglitazone are both FDA Pregnancy Category C medications, which means that teratogenicity has been demonstrated in animal models, and thus they are unsuitable drugs for women desiring pregnancy.

2.2.6. When Should Insulin Sensitizers Be Used for Ovulation Induction?

Although current data suggest that metformin increases the frequency of spontaneous ovulation and improves the efficacy of ovulation induction and clinical pregnancy with clomiphene, large-scale studies with metformin have not been completed. Hence, the optimal therapy or combination of therapies to achieve pregnancy most efficiently is currently unknown. If pregnancy is desired as soon as possible, currently clomiphene is still the historical standard of care until larger-scale studies have been completed. If ovulation induction is not successful with clomiphene or if the establishment of pregnancy is difficult despite successful ovulation induction with clomiphene, current data suggest that the addition of metformin is the logical next step.

In the scenario where a woman with PCOS desires pregnancy but time to achieving pregnancy is not of essence, current evidence suggests it would be reasonable to use metformin as an initial therapy combined with lifestyle intervention. If metformin therapy is not successful in increasing ovulation, clomiphene should be added to metformin therapy. While waiting for the results of prospective randomized studies, the addition of metformin may be beneficial for the establishment and maintenance of pregnancy in women with PCOS who have a prior history of spontaneous abortions after conceiving with clomiphene.

In order to establish the optimal therapy or combination of therapies to achieve pregnancy most efficiently, the National Institutes of Health's Reproductive Medicine Network is conducting the Pregnancy in Polycystic Ovary Syndrome (PPCOS) study. The PPCOS study is a randomized, double-blind controlled trial comparing three regimens in women with PCOS desiring pregnancy: (a) monotherapy with extended-release metformin (metformin XR), (b) clomiphene citrate monotherapy, and (c) a combination of metformin XR and clomiphene citrate. In addition to the usual inclusion criteria for PCOS women (elevated serum testosterone and eight or fewer menstrual cycles per year), partners of these women with PCOS must have normal sperm counts, and the enrolled subjects must agree to have intercourse two to three times per week during the study. The primary endpoint is the number of live births. Thirteen U.S. centers are participating in the clinical trial, and enrollment for this study has been completed and totals more than 600 women. Results of this study are eagerly awaited and will define the best initial treatment for women with PCOS desiring pregnancy.

2.3. Insulin Sensitizers for the Chronic Management of PCOS

2.3.1. Long-Term Health Consequences of PCOS

Insulin resistance is associated with diabetes, hypertension, dyslipidemia, endothelial dysfunction, a procoagulant state, and cardiovascular disease (*see* Chapters 28 and 29). Recently, the National Cholesterol Education Program Adult Treatment Panel defined the metabolic syndrome as the presence of three of the five following risk factors: waist circumference greater than 88 cm in females; fasting serum glucose 110 mg/dL or more, fasting serum triglycerides greater than 150 mg/dL; serum high-density lipoprotein cholesterol less than 50 mg/dL; and blood pressure greater than 130/85 mmHg *(60)*. The metabolic syndrome has been found to be present in 43–46% of women with PCOS, a twofold higher prevalence compared with women in the general population of the same age *(61,62)*.

2.3.2. Chronic Therapy in PCOS with Insulin Sensitizers

Both metformin and the thiazolidinediones have been shown to reduce blood pressure and inflammatory markers. Obese women on metformin also tend to lose weight while on metformin in a dose-dependent manner *(63)*. Although the insulin sensitizers' effects on cardiovascular risk factors are favorable, currently there are no prospective randomized outcome trials examining the use of insulin sensitizers in the prevention of diabetes or cardiovascular events in women with PCOS.

2.3.2.1. INSULIN SENSITIZERS IN PREVENTING TYPE 2 DIABETES MELLITUS

Although not specifically conducted in women with PCOS, there are several outcome studies indicating that interventions to improve insulin sensitivity may decrease the incidence of diabetes in individuals at high risk. The Diabetes Prevention Project was a prospective study sponsored by the National Institutes of Health *(64)*. The study included 3234 patients at high risk of diabetes (history of gestational diabetes or presence of impaired glucose tolerance and a first-degree relative with diabetes). Subjects were randomized to standard management, intensive lifestyle intervention, or metformin. During the average follow-up period of 3 years, intensive lifestyle intervention reduced the incidence of newly developed diabetes by 58%. Subjects treated with metformin had a 31% reduction in the risk of developing diabetes. These results indicate that improvement in insulin sensitivity, through either intensive lifestyle modification or metformin, reduces the risk of developing diabetes in high-risk individuals.

The insulin sensitizer troglitazone has also been studied in the prevention of diabetes. Buchanan et al. studied 235 Hispanic women with gestational diabetes who were randomized to troglitazone or placebo *(65)*. During a median follow-up of 30 months, compared with the placebo group, subjects taking troglitazone reduced their rate of progression to diabetes by 56%. This protective effect persisted for 8 months even after troglitazone was discontinued. Protection from diabetes in the troglitazone group was also associated with preservation of pancreatic β-cell function. In the Diabetes Prevention Project, troglitazone was one of the initial arms of the study, but was later discontinued after a mean duration of treatment of 0.9 year following reports of hepatotoxicity *(66)*. During the active treatment period, the diabetes incidence rate was 3.0 cases/100 person-years in the troglitazone arm as compared with 12.0, 6.7, and 5.1 cases/100 person-years in the placebo, metformin, and intensive lifestyle-style groups, respectively. The difference in the reduction in risk was significant between troglitazone and placebo and between troglitazone and metformin. However, this protective effect against progression to diabetes did not persist after the discontinuation of troglitazone *(66)*.

2.3.2.2. INSULIN SENSITIZERS IN PREVENTING CARDIOVASCULAR EVENTS

The available evidence suggesting that insulin sensitizers may be cardioprotective is circumstantial, but nonetheless encouraging. In another insulin-resistant state, type 2 diabetes, the United Kingdom Prospective Diabetes Study showed that metformin may be cardioprotective *(67)*. In this large prospective trial in obese diabetic patients, metformin significantly decreased all-cause mortality and stroke endpoints.

The thiazolidinediones have been studied using surrogate cardiovascular endpoints. Specifically in PCOS, troglitazone has been shown to improve endothelial function in obese women with PCOS *(68)*. In another study involving women at high risk of developing diabetes, 266 nondiabetic Hispanic women with gestational diabetes were randomized to placebo or troglitazone. Carotid intima media thickness progression was 31% lower in troglitazone group ($p = 0.048$), and this reduction was independent of obesity, baseline lipids, and serum levels of glucose or insulin. This beneficial effect was present only in subjects who had an increase in insulin sensitivity on troglitazone during the first 3 months of study. This finding suggests that improving insulin sensitivity using a thiazolidinedione may reduce the progression of subclinical atherosclerosis *(69)*.

2.4. Choice of Insulin-Sensitizing Agent in PCOS

Currently, the commercially available insulin-sensitizing agents include metformin and the thiazolidinediones (rosiglitazone and pioglitazone). Most of the clinical studies in women with PCOS have been conducted with metformin. In addition, for several decades metformin has been used worldwide to treat diabetes, and thus its side effect profile has been well described. Adverse effects of metformin include gastrointestinal distress, such as diarrhea and nausea. More serious toxicity includes a documented, but rare, risk of lactic acidosis. Nearly all reports of lactic acidosis occurred in patients with renal insufficiency (plasma creatinine > 1.4 in women), hepatic dysfunction, heart failure, or other pulmonary and circulatory dysfunctions that can by themselves lead to hypoxia and lactic acidosis *(70,71)*. Careful attention to these contraindications will prevent most occurrences of lactic acidosis.

As discussed in a previous section, metformin is classified by FDA as Pregnancy Category B, and human data suggest that it is nonteratogenic. Although not studied specifically in PCOS, metformin has been shown to reduce the progression to diabetes in a large prospective randomized trial of individuals at high risk for developing type 2 diabetes *(64)*. It also reduces the stroke and all-cause mortality endpoints in obese diabetic subjects in a large prospective trial in which cardiovascular disease accounted for 62% of the total mortality *(67)*.

The thiazolidinediones (troglitazone, rosiglitazone, and pioglitazone) are also insulin-sensitizing agents. Troglitazone showed favorable improvements in insulin sensitivity and ovulation rate in the largest prospective randomized controlled trial in PCOS to date. However, because of hepatotoxicity, it was removed from the market by FDA. Current evidence also suggests that rosiglitazone and pioglitazone are effective in improving ovulation in PCOS. Because these are newer agents, postmarketing surveillance is still ongoing to assess their safety profiles. Weight gain and fluid retention have been reported with the thiazolidinediones. Although hepatotoxicity is still a concern, current data do not suggest that they are as hepatotoxic as troglitazone. Importantly, both rosiglitazone and pioglitazone are FDA Pregnancy Category C medications, which means that teratogenicity has been shown in animal models. Thus, these agents may not be suitable for women with PCOS desiring pregnancy.

3. CONCLUSIONS

Insulin resistance is a central feature of PCOS. Hyperinsulinemia contributes to short-term health consequences of PCOS such as anovulation, hirsutism, infertility, and EPL. Chronic hyperinsulinemia also predisposes women with PCOS to increased risks of diabetes and cardiovascular events. Available data indicate that metformin is an effective treatment for anovulation in PCOS and is a suitable pharmacotherapy for women with PCOS desiring pregnancy. There is also encouraging evidence suggesting that metformin and the thiazolidinediones may ameliorate the metabolic syndrome associated with insulin resistance and prevent long-term cardiovascular and diabetes complications. Based on these data, insulin sensitizers, particularly metformin, may represent optimal first-line therapy for both the short-term and long-term treatment of women with PCOS.

4. FUTURE AVENUES OF INVESTIGATION

Several unresolved questions exist concerning the use of metformin in PCOS, and these may serve as future investigative directions, including (a) the efficacy of metformin in large-scale trials in increasing clinical pregnancy, live-birth rates, and its effects in multiparity; (b) ovulation induction with combination therapy of metformin and gonadotropins; (c) prospective studies of metformin in decreasing EPL; (d) whether metformin enhances the quality of eggs retrieved for in vitro fertilization; (e) whether measurable biological factor(s) exist that can be used to predict response to insulin sensitizers; (f) the effects of metformin on birth defects, pre-eclampsia, and gestational diabetes in large-scale studies; (g) the development of a more potent and optimal insulin-sensitizing drug for PCOS; and (h) large prospective trials evaluating the chronic use of metformin in women with PCOS for long-term cardiovascular and diabetes risk reduction.

KEY POINTS

- Insulin resistance is a central feature of PCOS.
- Hyperinsulinemia contributes to anovulation, infertility, and EPL in women with PCOS.
- Chronic hyperinsulinemia may predispose women with PCOS to increased risks of diabetes and cardiovascular events.
- Current data indicate that the insulin sensitizer metformin is an effective treatment for anovulation in PCOS and is a suitable agent for use in women with PCOS desiring pregnancy.
- Although thiazolidinediones have also been studied in PCOS, data concerning their safety in pregnancy are not available.
- Beginning evidence also suggests that metformin and thiazolidinediones may ameliorate the long-term risks of cardiovascular and diabetic complications, pending prospective studies.
- Based on these data, insulin sensitizers, particularly metformin, may represent optimal first-line therapy for both the short-term and long-term treatment of women with PCOS.

REFERENCES

1. Dunaif A, Segal KR, Futterweit W, Dobrjansky A. Profound peripheral insulin resistance, independent of obesity, in polycystic ovary syndrome. Diabetes 1989;38(9):1165–1174.
2. Campbell PJ, Gerich JE. Impact of obesity on insulin action in volunteers with normal glucose tolerance: demonstration of a threshold for the adverse effect of obesity. J Clin Endocrinol Metab 1990;70(4):1114–1118.
3. Nestler JE, Jakubowicz DJ, Falcon de Vargasm A., Brik C, Quinterro N, Medina F. Insulin stimulates testosterone biosynthesis by human thecal cells from women with polycystic ovary syndrome by activating its own receptor and using inositolglycan mediators as the signal transduction system. J Clin Endocrinol Metab 1998;83:2001–2005.
4. Nestler JE, Powers LP, Matt DW, et al. A direct effect of hyperinsulinemia on serum sex hormone-binding globulin levels in obese women with the polycystic ovary syndrome. J Clin Endocrinol Metab 1991;72(1):83–89.
5. Adashi EY, Hsueh AJ, Yen SS. Insulin enhancement of luteinizing hormone and follicle-stimulating hormone release by cultured pituitary cells. Endocrinology 1981;108(4):1441–1449.
6. Berga SL, Guzick DS, Winters SJ. Increased luteinizing hormone and alpha-subunit secretion in women with hyperandrogenic anovulation. J Clin Endocrinol Metab 1993;77(4):895–901.
7. Nestler JE, Jakubowicz DJ. Decreases in ovarian cytochrome P450c17α activity and serum free testosterone after reduction of insulin secretion in polycystic ovary syndrome. N Engl J Med 1996;335(9):617–623.
8. Nestler JE, Barlascini CO, Matt DW, et al. Suppression of serum insulin by diazoxide reduces serum testosterone levels in obese women with polycystic ovary syndrome. J Clin Endocrinol Metab 1989;68(6):1027–1032.
9. Nestler JE, Singh R, Matt DW, Clore JN, Blackard WG. Suppression of serum insulin level by diazoxide does not alter serum testosterone or sex hormone-binding globulin levels in healthy, nonobese women. Am J Obstet Gynecol 1990;163(4 Pt 1):1243–1246.
10. Azziz R, Ehrmann D, Legro RS, et al. Troglitazone improves ovulation and hirsutism in the polycystic ovary syndrome: a multicenter, double blind, placebo-controlled trial. J Clin Endocrinol Metab 2001;86(4):1626–1632.
11. Nestler JE, Jakubowicz DJ, Evans WS, Pasquali R. Effects of metformin on spontaneous and clomiphene-induced ovulation in the polycystic ovary syndrome. N Engl J Med 1998;338(26):1876–1880.

12. Lord JM, Flight IHK, Norman RJ. Metformin in polycystic ovary syndrome: systematic review and meta-analysis. BMJ 2003;327(7421):951–0.
13. El-Biely MM, Habba M. The use of metformin to augment the induction of ovulation in obese infertile patients with polycystic ovary syndrome. Middle East Fertil Soc J 2001;6:43–49.
14. Fleming R, Hopkinson ZE, Wallace AM, Greer IA, Sattar N. Ovarian function and metabolic factors in women with oligomenorrhea treated with metformin in a randomized double blind placebo-controlled trial. J Clin Endocrinol Metab 2002;87(2):569–574.
15. Jakubowicz DJ, Seppala M, Jakubowicz S, et al. Insulin reduction with metformin increases luteal phase serum glycodelin and insulin-like growth factor-binding protein 1 concentrations and enhances uterine vascularity and blood flow in the polycystic ovary syndrome. J Clin Endocrinol Metab 2001;86(3):1126–1133.
16. Kocak M, Caliskan E, Simsir C, Haberal A. Metformin therapy improves ovulatory rates, cervical scores, and pregnancy rates in clomiphene citrate-resistant women with polycystic ovary syndrome. Fertil Steril 2002;77(1):101–106.
17. Malkawi HY, Qublan HS. The effect of metformin plus clomiphene citrate on ovulation and pregnancy rates in clomiphene-resistant women with polycystic ovary syndrome. Saudi Med J 2002;23(6):663–666.
18. Moghetti P, Castello R, Negri C, et al. Metformin effects on clinical features, endocrine and metabolic profiles, and insulin sensitivity in polycystic ovary syndrome: a randomized, double-blind, placebo-controlled 6-month trial, followed by open, long-term clinical evaluation. J Clin Endocrinol Metab 2000;85(1):139–146.
19. Ng EHY, Wat NMS, Ho PC. Effects of metformin on ovulation rate, hormonal and metabolic profiles in women with clomiphene-resistant polycystic ovaries: a randomized, double-blinded placebo-controlled trial. Hum Reprod 2001;16(8):1625–1631.
20. Pasquali R, Gambineri A, Biscotti D, et al. Effect of long-term treatment with metformin added to hypocaloric diet on body composition, fat distribution, and androgen and insulin levels in abdominally obese women with and without the polycystic ovary syndrome. J Clin Endocrinol Metab 2000;85(8):2767–2774.
21. Sturrock ND, Lannon B, Fay TN. Metformin does not enhance ovulation induction in clomiphene resistant polycystic ovary syndrome in clinical practice. Br J Clin Pharmacol 2002;53(5):469–473.
22. Vandermolen DT, Ratts VS, Evans WS, Stovall DW, Kauma SW, Nestler JE. Metformin increases the ovulatory rate and pregnancy rate from clomiphene citrate in patients with polycystic ovary syndrome who are resistant to clomiphene citrate alone. Fertility and Sterility 2001;75(2):310–315.
23. Yarali H, Yildiz BO, Demirol A, et al. Co-administration of metformin during rFSH treatment in patients with clomiphene citrate-resistant polycystic ovarian syndrome: a prospective randomized trial. Hum Reprod 2002;17(2):289–294.
24. Palomba S, Orio F Jr, Falbo A, et al. Prospective parallel randomized, double-blind, double-dummy controlled clinical trial comparing clomiphene citrate and metformin as the first-line treatment for ovulation induction in nonobese anovulatory women with polycystic ovary syndrome. J Clin Endocrinol Metab 2005;90(7):4068–4074.
25. Shepard MK, Balmaceda JP, Leija CG. Relationship of weight to sucessful induction of ovulation with clomiphene citrate. Fertil Steril 1979;32:641–645.
26. Lobo RA, Gysler.M., March CM, Goebelsmann U, Mishell DR. Clinical and laboratory predictors of clomiphene response. Fertil Steril 1982;37:168–174.
27. Palomba S, Orio F Jr, Nardo LG, et al. Metformin administration versus laparoscopic ovarian diathermy in clomiphene citrate-resistant women with polycystic ovary syndrome: a prospective parallel randomized double-blind placebo-controlled trial. J Clin Endocrinol Metab 2004;89(10):4801–4809.
28. Sepilian V, Nagamani M. Effects of rosiglitazone in obese women with polycystic ovary syndrome and severe insulin resistance. J Clin Endocrinol Metab 2005;90(1):60–65.
29. Dereli D, Dereli T, Bayraktar F, Ozgen AG, Yilmaz C. Endocrine and metabolic effects of rosiglitazone in non-obese women with polycystic ovary disease. Endocr J 2005;52(3):299–308.
30. Brettenthaler N, De Geyter C, Huber PR, Keller U. Effect of the insulin sensitizer pioglitazone on insulin resistance, hyperandrogenism, and ovulatory dysfunction in women with polycystic ovary syndrome. J Clin Endocrinol Metab 2004;89(8):3835–3840.
31. Shobokshi A, Shaarawy M. Correction of insulin resistance and hyperandrogenism in polycystic ovary syndrome by combined rosiglitazone and clomiphene citrate therapy. J Soc Gynecol Invest 2003;10(2):99–104.
32. Kolodziejczyk B, Duleba AJ, Spaczynski RZ, Pawelczyk L. Metformin therapy decreases hyperandrogenism and hyperinsulinemia in women with polycystic ovary syndrome. Fertil Steril 2000;73(6):1149–1154.
33. Harborne L, Fleming R, Lyall H, Sattar N, Norman J. Metformin or antiandrogen in the treatment of hirsutism in polycystic ovary syndrome. J Clin Endocrinol Metab 2003;88(9):4116–4123.
34. Glintborg D, Stoving RK, Hagen C, et al. Pioglitazone treatment increases spontaneous growth hormone secretion and stimulated growth hormone levels in polycystic ovary syndrome. J Clin Endocrinol Metab 2005;90:5605–5612 .
35. Ortega-Gonzalez C, Luna S, Hernandez L, et al. Responses of serum androgen and insulin resistance to metformin and pioglirazone in obese insulin-resistant women with polycystic ovary syndrome. J Clin Endocrinol Metab 2005;90:1360–1365.
36. Baillargeon JP, Jakubowicz DJ, Iuorno MJ, Jakubowicz S, Nestler JE. Effects of metformin and rosiglitazone, alone

and in combination, in nonobese women with polycystic ovary syndrome and normal indices of insulin sensitivity. Fertil Steril 2004;82(4):893–902.

37. Glueck CJ, Moreira A, Goldenberg N, Sieve L, Wang P. Pioglitazone and metformin in obese women with polycystic ovary syndrome not optimally responsive to metformin. Hum Reprod 2003;18(8):1618–1625.

38. Balen AH, Tan SL, MacDougall J, Jacobs HS. Miscarriage rates following in-vitro fertilization are increased in women with polycystic ovaries and reduced by pituitary desensitization with buserelin. Hum Reprod 1993;8(6):959–964.

39. Homburg R, Armar NA, Eshel A, Adams J, Jacobs HS. Influence of serum luteinising hormone concentrations on ovulation, conception, and early pregnancy loss in polycystic ovary syndrome. BMJ 1988;297(6655):1024–1026.

40. Regan L, Owen EJ, Jacobs HS. Hypersecretion of luteinising hormone, infertility, and miscarriage. Lancet 1990;336(8724):1141–1144.

41. Sagle M, Bishop K, Ridley N, et al. Recurrent early miscarriage and polycystic ovaries. BMJ 1988;297(6655):1027–1028.

42. Watson H, Kiddy DS, Hamilton-Fairley D, et al. Hypersecretion of luteinizing hormone and ovarian steroids in women with recurrent early miscarriage. Hum Reprod 1993;8(6):829–833.

43. Jakubowicz DJ, Essah PA, Seppala M, et al. Reduced serum glycodelin and insulin-like growth factor-binding protein-1 in women with polycystic ovary syndrome during first trimester of pregnancy. J Clin Endocrinol Metab 2004;89(2):833–839.

44. Julkunen M, Koistinen R, Sjoberg J, Rutanen EM, Wahlstrom T, Seppala M. Secretory endometrium synthesizes placental protein 14. Endocrinology 1986;118(5):1782–1786.

45. Julkunen M, Koistinen R, Suikkari AM, Seppala M, Janne OA. Identification by hybridization histochemistry of human endometrial cells expressing mRNAs encoding a uterine beta-lactoglobulin homologue and insulin-like growth factor-binding protein-1. Mol Endocrinol 1990;4(5):700–707.

46. Bolton AE, Pockley AG, Clough KJ, et al. Identification of placental protein 14 as an immunosuppressive factor in human reproduction. Lancet 1987;1(8533):593–595.

47. Okamoto N, Uchida A, Takakura K, et al. Suppression by human placental protein 14 of natural killer cell activity. Am J Reprod Immunol 1991;26(4):137–142.

48. Dalton CF, Laird SM, Serle E, et al. The measurement of CA 125 and placental protein 14 in uterine flushings in women with recurrent miscarriage; relation to endometrial morphology. Hum Reprod 1995;10(10):2680–2684.

49. Tulppala M, Julkunen M, Tiitinen A, Stenman UH, Seppala M. Habitual abortion is accompanied by low serum levels of placental protein 14 in the luteal phase of the fertile cycle. Fertil Steril 1995;63(4):792–795.

50. Giudice LC, Mark SP, Irwin JC. Paracrine actions of insulin-like growth factors and IGF binding protein-1 in non-pregnant human endometrium and at the decidual-trophoblast interface. J Reprod Immunol 1998;39(1–2):133–148.

51. Jones JI, Gockerman A, Busby WH Jr, Wright G, Clemmons DR. Insulin-like growth factor binding protein 1 stimulates cell migration and binds to the α5β1 integrin by means of its Arg-Gly-Asp sequence. PNAS 1993;90(22):10553–10557.

52. Jakubowicz DJ, Seppala M, Jakubowicz S, et al. Insulin reduction with metformin increases luteal phase serum glycodelin and insulin-like growth factor-binding protein 1 concentrations and enhances uterine vascularity and blood flow in the polycystic ovary syndrome. J Clin Endocrinol Metab 2001;86(3):1126–1133.

53. Jakubowicz DJ, Iuorno MJ, Jakubowicz S, Roberts KA, Nestler JE. Effects of metformin on early pregnancy loss in the polycystic ovary syndrome. J Clin Endocrinol Metab 2002;87(2):524–529.

54. Glueck CJ, Wang P, Goldenberg N, Sieve-Smith L. Pregnancy outcomes among women with polycystic ovary syndrome treated with metformin. Hum Reprod 2002;17(11):2858–2864.

55. Glueck CJ, Goldenberg N, Pranikoff J, Loftspring M, Sieve L, Wang P. Height, weight, and motor-social development during the first 18 months of life in 126 infants born to 109 mothers with polycystic ovary syndrome who conceived on and continued metformin through pregnancy. Hum Reprod 2004;19(6):1323–1330.

56. Glueck CJ, Wang P, Fontaine R, Tracy T, Sieve-Smith L. Metformin to restore normal menses in oligo-amenorrheic teenage girls with polycystic ovary syndrome (PCOS). J Adolesc Health 2001;29(3):160–169.

57. Vanky E, Zahlsen K, Spigset O, Carlsen SM. Placental passage of metformin in women with polycystic ovary syndrome. Fertil Steril 2005;83(5):1575–1578.

58. Hellmuth E, Damm P, Molsted-Pedersen L. Oral hypoglycaemic agents in 118 diabetic pregnancies. Diabet Med 2000;17(7):507–511.

59. Vanky E, Salvesen KA, Heimstad R, Fougner KJ, Romundstad P, Carlsen SM. Metformin reduces pregnancy complications without affecting androgen levels in pregnant polycystic ovary syndrome women: results of a randomized study. Hum Reprod 2004;19(8):1734–1740.

60. Third Report of the National Cholesterol Education Program (NCEP) Expert Panel on Detection, Evaluation, and Treatment of High Blood Cholesterol in Adults (Adult Treatment Panel III) final report. Circulation 2002;106(25):3143–3421.

61. Apridonidze T, Essah PA, Iuorno MJ, Nestler JE. Prevalence and characteristics of the metabolic syndrome in women with polycystic ovary syndrome. J Clin Endocrinol Metab 2005;90(4):1929–1935.

62. Glueck CJ, Papanna R, Wang P, Goldenberg N, Sieve-Smith L. Incidence and treatment of metabolic syndrome in newly referred women with confirmed polycystic ovarian syndrome. Metabolism 2003;52(7):908–915.

63. Harborne LR, Sattar N, Norman JE, Fleming R. Metformin and weight loss in obese women with polycystic ovary syndrome: comparison of doses. J Clin Endocrinol Metab 2005;90(8):4593–4598.

64. Diabetes Prevention Program Research Group. Reduction in the incidence of type 2 diabetes with lifestyle intervention or metformin. N Engl J Med 2002;346(6):393–403.

65. Buchanan TA, Xiang AH, Peters RK, et al. Preservation of pancreatic beta-cell function and prevention of type 2 diabetes by pharmacological treatment of insulin resistance in high-risk hispanic women. Diabetes 2002;51(9):2796–2803.

66. Knowler WC, Hamman RF, Edelstein SL, et al. Prevention of type 2 diabetes with troglitazone in the Diabetes Prevention Program. Diabetes 2005;54(4):1150–1156.

67. UK Prospective Diabetes Study (UKPDS) Group. Effect of intensive blood-glucose control with metformin on complications in overweight patients with type 2 diabetes (UKPDS 34). UK Prospective Diabetes Study (UKPDS) Group. Lancet 1998;352(9131):854–865.

68. Paradisi G, Steinberg HO, Hempfling A, et al. Polycystic ovary syndrome is associated with endothelial dysfunction. Circulation 2001;103(10):1410–1415.

69. Xiang AH, Peters RK, Kjos SL, et al. Effect of thiazolidinedione treatment on progression of subclinical atherosclerosis in premenopausal women at high risk for type 2 diabetes. J Clin Endocrinol Metab 2005;90(4):1986–1991.

70. Brown JB, Pedula K, Barzilay J, Herson MK, Latare P. Lactic acidosis rates in type 2 diabetes. Diabetes Care 1998;21(10):1659–1663.

71. Jones GC, Macklin JP, Alexander WD. Contraindications to the use of metformin. BMJ 2003;326(7379):4–5.

39

Weight Reduction and Lifestyle Modifiction in the Treatment of Androgen Excess

Manny Noakes, Lisa J. Moran, Grant D. Brinkworth, and Robert J. Norman

SUMMARY

Polycystic ovary syndrome (POCS) is a common disorder in women resulting in anovulation and reproductive dysfunction. Androgen excess and hyperinsulinemia are key contributing features that are exacerbated by obesity. Both weight reduction and increases in physical activity are highly effective in increasing insulin sensitivity but underutilized in clinical practice as evidence on effective strategies to achieve this are lacking. The use of metformin is effective in restoring ovulatory function through improvement in insulin sensitivity, possibly also mediated by weight loss. Orlistat and sibutramine have also been shown to assist in weight loss and thereby restore reproductive function. In the morbidly obese with PCOS, bariatric surgery has been shown to achieve substantial weight loss and to be effective in improving fertility. The literature on effective diet and exercise programs for PCOS has been sparse. Studies examining the role of meal replacements as well as structured dietary patterns with higher protein composition and/or lower glycemic-index carbohydrates show promise. Further studies on novel lifestyle programs are needed compared to standard care and pharmacotherapy to guide clinical practice.

Key Words: Polycystic ovary syndrome; weight loss; insulin resistance; lifestyle programs; metformin.

1. INTRODUCTION

The prevalence of the different pathological conditions causing clinically evident androgen excess is largely dominated by polycystic ovary syndrome (PCOS), which represents 82% of this population *(1)*. There is also a strong link between obesity and conditions of androgen excess such as PCOS, with half being overweight or obese. The role of weight reduction and lifestyle modification in restoring reproductive function is examined in this chapter.

2. BACKGROUND

Various studies have reported different levels of obesity, with the highest percentages being in the Western world, including the United States, the United Kingdom, and Australia. Obesity, but also associated menstrual disturbances, hirsutism, acne, and male pattern of alopecia, in women with androgen excess all contribute to the poor quality of life of these patients. Furthermore, obesity is also associated with a number of health problems, including infertility, miscarriage, adverse pregnancy outcomes, type 2 diabetes mellitus, hyperlipidemia, heart disease, osteoarthritis, and other serious health outcomes. PCOS is a heterogeneous clinical syndrome characterized by hyperandrogenism with chronic anovulation in women without specific adrenal or pituitary gland

From: *Contemporary Endocrinology: Androgen Excess Disorders in Women:*
Polycystic Ovary Syndrome and Other Disorders, Second Edition
Edited by: R. Azziz et al. © Humana Press Inc., Totowa, NJ

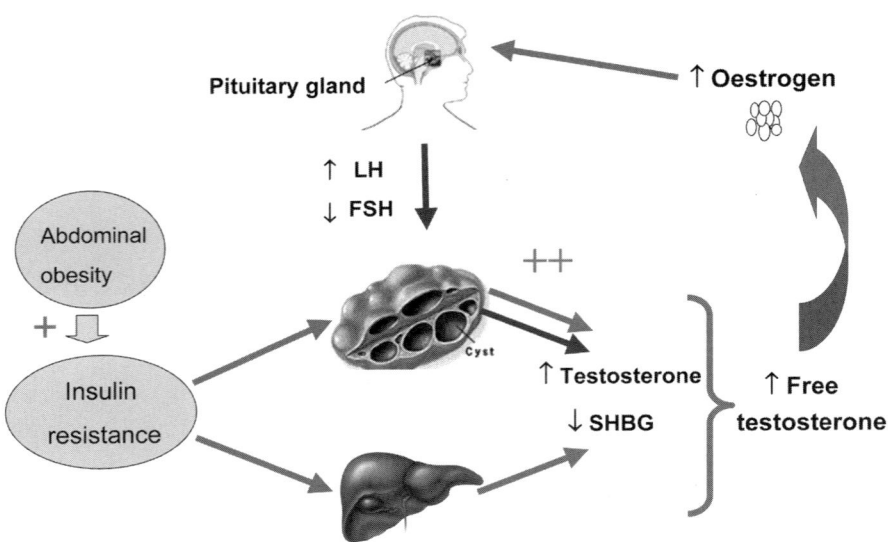

Fig. 1. Schematic representation of how insulin resistance may contribute to the androgen excess of polycystic ovary syndrome (PCOS). SHBG, sex hormone-binding globulin; LH, luteinizing hormone; FSH, follicle-stimulating hormone.

disease. The age of onset is frequently perimenarchal, and, in some cases, premature adrenarche may be a precursor to the development of the syndrome.

2.1. Mechanism Linking Obesity and PCOS

The mechanisms linking obesity and PCOS are unclear but may be related to insulin resistance and hyperandrogenism, both of which are commonly documented in lean and obese women with this condition. Insulin resistance is a common but not universal feature of PCOS, although women with insulin resistance appear to be more clinically affected *(2)*.

Insulin is a physiological hormone regulating ovarian function, specifically ovarian steroidogenesis and androgen blood transport and/or activity in the target tissues *(3)*. The high insulin levels associated with insulin resistance stimulate the ovary to make excessive amounts of androgens. Additionally, high insulin levels decrease levels of sex hormone-binding globulin (SHBG), increasing the bioavailability of androgens. Moreover, high insulin levels may also work at the level of the brain, causing increased luteinizing hormone (LH) secretion, which in turn stimulates greater ovarian androgen production.

Because insulin resistance is strongly influenced by obesity in non-PCOS subjects, it was initially debated whether insulin resistance and hyperinsulinemia are a primary metabolic disturbance of PCOS or a symptom of the obesity commonly observed in PCOS. Hyperandrogenemia and insulin resistance appear to be independent features of PCOS, with hyperinsulinemia enhancing the expression of hyperandrogenemia by increasing bioavailability of androgens *(4)* (Fig. 1). Obese women with PCOS show decreased insulin sensitivity and hyperinsulinemia to an extent greater than can be explained by obesity alone *(4,5)*. A synergistic interaction appears to exist with the degree of insulin resistance and hyperinsulinemia in lean PCOS women augmented in those who have obesity. Not all women with PCOS exhibit hyperinsulinemia and insulin resistance, and discrepant results may be explained by the heterogeneity and complex etiology of the syndrome. Silfen et al. *(6)* found a more pronounced alteration in the hypothalamic–pituitary–adrenal axis in non-obese adolescents with PCOS, including higher levels of LH, SHBG, androstenedione, dehydroepiandrosterone sulfate

(DHEAS), dihydrotestosterone, and free insulin-like growth factor-1, but a more marked insulin resistance in obese adolescents with PCOS.

Hyperandrogenism is postulated to result from either or both increased adrenal and ovarian androgen production. The predominance of abdominal obesity, insulin resistance, glucose intolerance, hypertension, and other conditions suggests that the metabolic syndrome may be more common in PCOS, and indeed this has been reported by a number of investigators (7). It is therefore obvious that for adequate treatment of patients with androgen excess, there needs to be a reduction in hyperinsulinemia and insulin resistance as well as in obesity *per se*. Both weight reduction and increases in physical activity are highly effective in increasing insulin sensitivity. These interrelated and yet separate issues are the focus of the following discussion.

2.2. Metformin, Weight Loss, and PCOS

The initial use of metformin in the treatment of PCOS has now been widely accepted to be a valuable and inexpensive therapeutic modality. Recent systematic reviews have indicated that metformin is highly effective in inducing ovulation and increasing pregnancy rates (8,9). The numbers needed to treat for ovulation are around four patients, and there is an improvement in serum insulin levels and a reduction in free testosterone in response to metformin. The drug appears to be safe in early pregnancy with respect to congenital abnormalities, although it is controversial whether miscarriage rates are reduced. The predictors of success of metformin have not been established, although there is some evidence that patients who are substantially overweight do not respond as well. Side effects of metformin include nausea, vomiting, diarrhea, and other forms of gastrointestinal intolerance, and patients need to be warned about the interaction between metformin and alcohol. The ovulatory response to clomiphene can be increased in obese women with PCOS by decreasing insulin secretion with metformin (10). The role of metformin in weight loss is controversial with a number of studies showing some weight loss and others showing no change at all. The National Institute of Child Health and Human Development's Reproductive Medicine Network has begun a randomized, double-blind trial of clomiphene vs metformin vs clomiphene plus metformin for the induction of ovulation in patients with PCOS seeking pregnancy, with live-birth rate as the primary outcome (11). This trial should definitively answer the question of the relative efficacy of metformin, clomiphene, and combination therapy in the treatment of infertile women with PCOS.

One of the best studies relating to metformin use in androgen excess was that by Pasquali et al. (12). They randomized patients with PCOS or obesity alone who were weight index-matched to a lifestyle-modification program including diet plus metformin or placebo. After 6 months the frequency of menstrual cycles was better with metformin, and metformin was superior to placebo alone in the loss of weight, reduction in waist circumference and visceral fat, and reduction in testosterone. There was no differential benefit of metformin on fasting serum glucose, insulin, or SHBG. Patients with androgen excess responded better than matched patients with obesity to metformin with respect to reduction of visceral fat and testosterone, but not in terms of weight loss, weight circumference, fasting serum glucose, or insulin.

This suggests that the addition of metformin to lifestyle modification may be of some benefit to all obese subjects in terms of weight loss and reduction of central fat, particularly so for PCOS patients with regard to central fat. Whether this is a direct effect of metformin alone or indirect through the contribution of gastrointestinal side effects remains to be verified. Weight loss (3.9 kg after 8 months) is a feature of protracted metformin therapy (1.5 g/day) in obese women with PCOS, with greater weight reduction potentially achievable with higher doses (1.50 or 2.55 g/day) (13). No dose-response effect in weight loss was noted in the morbidly obese with PCOS, and there was no dose-response effect with reductions in androstenedione, which was significant with both doses. Systematic reviews of metformin's effect on weight loss do not suggest that this trend can be generalized beyond PCOS. Given the relatively fewer side effects of metformin, there would appear to be no harm in using metformin as part of a weight loss program in patients with PCOS. At present, however, metformin should probably be stopped as soon as pregnancy is established. Alcohol should be avoided, and the very rare occurrence of lactic acidosis should be borne in mind.

It is likely that weight reduction plays the most significant role in restoration of ovulation in obese women with PCOS and that this may be the primary mechanism for the effects noted with metformin, independent of its insulin-sensitizing effects *(14)*. More recently, drugs that target weight loss have been compared with metformin therapy in women with PCOS. Treatment with orlistat for 3 months led to a 4.7% reduction in body weight, a more significant loss than was observed in metformin-treated women (1.0%). Serum testosterone levels declined significantly in both groups with no change in levels of SHBG. This agent has potential as an adjunct to the treatment of PCOS *(15)*.

The effects of metformin (1700 mg/day) vs acarbose (300 mg/day) on insulin resistance, hormone profiles, and ovulation rates in patients with clomiphene citrate-resistant PCOS have recently been assessed *(16)*. The ratio of LH to follicle-stimulating hormone (FSH) and total testosterone concentrations decreased and ovulation rates increased in both groups. Reduction in weight and body mass index (BMI) was only significant in the acarbose group.

In a placebo-controlled, double-blind study, Tang et al. *(17)* randomized 143 oligo-/amenorrheic obese women with PCOS to metformin (850 mg) or placebo twice daily for 6 months. All received diet and lifestyle advice from a dietitian. Both groups showed significant improvements in menstrual frequency and weight loss, with no significant differences between the groups. Logistic regression analysis showed that only percentage weight loss correlated with an improvement in menses. Women who received metformin achieved a significant reduction in waist circumference and free androgen index (FAI). Because this is one of the largest such studies conducted, it suggests that metformin does not improve weight loss or menstrual frequency in obese patients with PCOS receiving weight-management advice and that weight loss alone through lifestyle changes, if this can be achieved, can significantly improve menstrual frequency.

The benefits of weight reduction in women with androgen excess and PCOS are well documented. In overweight women with hyperandrogenism and PCOS, weight loss decreases abdominal fat, hyperandrogenism, and insulin resistance and improves lipid profiles, menstrual cyclicity, and fertility and risk factors for diabetes and cardiovascular disease *(18–21)*. An important point is that a minimal amount of weight loss is sufficient to improve the presentation of PCOS. A moderate weight loss of 5% initial body weight on an energy-restricted diet results in reduction in insulin concentrations, reciprocal changes in SHBG, and improved menstrual cyclicity and fertility in the majority of patients *(21)*. In this study a reduction in hirsutism was also noted in 40% of women.

Moran et al. have *(20)* also confirmed that a 7.5% weight loss in overweight women with PCOS increases SHBG, decreases the FAI and testosterone, and improves menstrual cyclicity, and that this can occur as soon as within the first 4 weeks of energy restriction and be sustained through weight loss. Energy restriction and weight loss is thus a desirable outcome in overweight women with PCOS for short- and long-term improvements in reproductive and metabolic health.

Replacing protein for carbohydrate within the context of an energy-restricted diet was not associated with significant differences in reproductive outcomes, although postprandial glucose response was 3.5-fold lower on the higher protein dietary pattern. No changes in hirsutism were noted *(20)*.

High concentrations of serum LH in the follicular phase are associated with PCOS and with decreased reproductive function *(22)*. Tonic hypersecretion of LH appears to induce premature oocyte maturation, causing problems with fertilization and miscarriage. Van Dam et al. *(23)* noted that 7 days of calorie restriction on a very-low-calorie diet (VLCD) (471 kcal/day) paradoxically increased basal and pulsatile LH secretion, despite reductions in plasma glucose, insulin, leptin, and testosterone concentrations, which decreased by 18, 75, 50, and 23%, respectively. Serum estrone, estradiol, SHBG, and androstenedione concentrations remained unchanged.

Van Dam et al. *(24)* examined predictors of improvements in ovarian function in response to a 6-month weight loss, noting that responders exhibited a significant decline of circulating estradiol concentrations and a concurrent increase in LH secretion in response to 7 days of acute energy restriction using a VLCD, whereas neither parameter changed significantly in nonresponders. The authors suggest that this hormonal response to energy restriction may be a marker or precedent for follicle maturation and ovulation in these patients. It is concluded that in PCOS women receiving long-term downregulation and stimulation with recombinant FSH, insulin resistance is related to neither hormone levels nor the outcome of in vitro fertilization. Obesity, independent of insulin resistance, is associated with relative gonadotropin resistance *(25)*.

2.4. Dietary Composition and Weight Management

Although low-fat, high-carbohydrate diets have been the mainstream approach for weight management, they appear to be no more effective than other dietary patterns that restrict kilojoules *(26)*. Furthermore, high-carbohydrate dietary patterns may worsen the metabolic profile if weight loss is not achieved. Modifying the type of dietary carbohydrate or glycemic index (GI) has been highly controversial *(27,28)*. GI is proposed to both improve the cardiovascular risk profile and aid in weight loss *(29,30)*, although education on GI has not shown an improvement in weight loss at 10 weeks *(31)* or 1 year *(32)*. Surprisingly, there appear to be no studies on the utility of using GI as a strategy for weight management in women PCOS. Increasing the amount of dietary protein at the expense of carbohydrate has been shown to reduce abdominal fat in insulin-resistant subjects *(33,34)* and has been shown to be more effective in improving weight loss after 6 months and 1 year *(35,36)*. However, in two small studies in PCOS subjects, no differential effect of protein was observed *(20,37)*, although these dietary studies were not *ad libitum*. Improvements in postprandial glucose response were noted on the higher protein pattern. A proposed mechanism whereby protein is thought to exert effects on weight management is through increased satiety *(38)*. This could be advantageous, as it has been shown that satiety appears impaired in women with PCOS *(39)*.

PCOS is frequently associated with morbid obesity, in which conventional lifestyle modification may present a challenge. Sustained and marked weight loss has been achieved by bariatric surgery *(40)*. A weight loss of 41 kg after 12 months was paralleled by a decrease in the hirsutism score and free testosterone, androstendione, and DHEAS and the restoration of regular menstrual cycles and/or ovulation in all patients. There is some indication that weight loss studies in women with PCOS have increased drop-out rates: 26–38% over 1–4 months *(20,37)* and 8–9% over 4 months in non-PCOS subjects *(33,38)*. This may be a result of the increased difficulty of energy restriction consequent to lower satiety *(41)*.

The use of alternative dietary compositions provides patients with an increased range of dietary options and may be more successful for optimally improving the metabolic profile and achieving and sustaining a reduced weight. However, continued follow-up is required to sustain weight loss, as this is likely to outweigh the effects of dietary composition *(42)*.

Meal replacements have been shown to be an effective weight loss strategy in overweight women with PCOS. Moran et al. *(43)* placed 34 overweight women with PCOS on a weight-loss intervention (two meal replacements, low-fat snacks, and evening meal daily) for 8 weeks and reviewed them every 2 weeks (Fig. 2). The intervention resulted in a 5.6-kg (6%) reduction in weight and a 6-cm reduction in waist circumference. There was no change in SHBG, but a significant reduction in serum testosterone (0.3 ± 0.7 nmol/L) and FAI (3.1 ±4. 6nmol/L, 16.8%). This change in FAI occurred from weeks 0 to 2 and corresponded with a weight loss of 2.4 ± 1.0 kg (2.5%), with no further changes in FAI occurring from weeks 2 to 8. These data suggest that reproductive function can be restored very quickly by acute energy restriction. The implication is that only short-term energy restriction may be required to improve reproductive function and that weight loss is needed to sustain it, but this requires further investigation.

2.5. Physical Activity, Insulin Sensitivity, and Hormonal Control

Despite the considerable evidence that physical activity provides an effective medium for improving insulin sensitivity in a wide range of insulin-resistant states, the potential benefits of exercise for

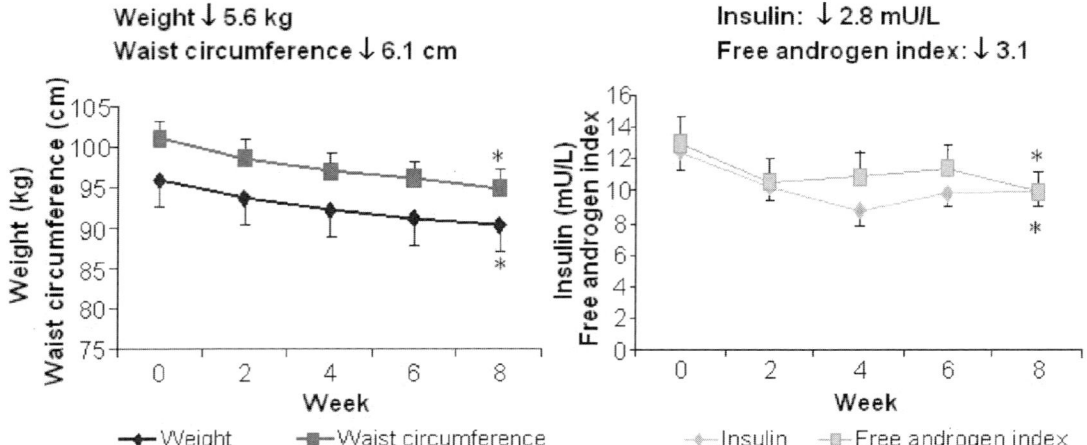

Fig. 2. Effects of short-term weight loss using meal replacements two times daily plus snacks plus low-fat meal; dietary intake/day 4904 kJ, composed of 21% fat, 24% protein, 53% carbohydrate—on insulin and free androgen index in polycystic ovary syndrome (PCOS). *$p < 0.01$ from baseline. (Adapted from ref. *43.*)

improving hormonal and metabolic abnormalities in individuals with hyperandrogenism is not well understood. To date, only two controlled studies have examined the direct effects of physical exercise in women with excessive androgen levels *(44,45)*. In 2002 Randeva et al. *(44)* showed no effects of a 6-month exercise program on either fasting insulin or FAI in young overweight and obese women with PCOS. However, baseline levels of insulin resistance were relatively low in their subjects, and this study only evaluated the effects of light aerobic exercise, which could have limited the changes observed. Jaatinen *(45)* challenged nine oligomenorrheic women with PCOS (BMI 19.5–46.0 kg/m^2) and eight control women with regular menstrual cycles (BMI 20.0–53.5 kg/m^2) with a bicycle ergometer test. The exercise-induced increase in circulating growth hormone levels was significantly greater in controls than in PCOS patients. There was also a negative correlation between the growth hormone response and BMI. The increases in the concentrations of adrenaline, noradrenaline, 3,4-dihydroxyphenylglycol, glucose, and insulin:C-peptide ratios during the bicycle ergometer test were correlated negatively to BMI. The authors concluded that obesity is an important determinant of the hormonal responses to physical exercise.

Further research is required to examine hormonal adaptations to exercise training and to evaluate the role of differing forms and duration of exercise for optimizing management of metabolic and hormonal abnormalities in populations with excess androgen levels. Equally important is the assessment of compliance to exercise regimes and strategies to optimize this. The possible interactions between insulin and testosterone in response to exercise therapies also warrant further study. There is little doubt that exercise provides a means of increasing energy expenditure and plays a pivotal role in management and that a dose-response effect exists, such that more exercise is associated with greater weight loss and maintenance.

3. CONCLUSION

Both weight reduction and increases in physical activity are highly effective in increasing insulin sensitivity and restoring reproductive function in PCOS. The use of metformin is effective in restoring ovulatory function through improvement in insulin sensitivity, possibly also mediated by weight loss. Orlistat and sibutramine have also been shown to assist in weight loss and thereby restore reproductive function. In the morbidly obese with PCOS, bariatric surgery has been shown achieve substantial weight loss and improve fertility.

The literature on effective diet and exercise programs for PCOS has been sparse compared with studies evaluating the role of pharmaceutical agents. Hence, lifestyle management has been underutilized in clinical management. Studies examining the role of meal replacements as well as structured dietary patterns with higher protein composition and/or lower glycemic index carbohydrates show promise, but further studies on novel lifestyle programs are needed compared to standard care and pharmacotherapy to guide clinical practice.

4. FUTURE AVENUES OF INVESTIGATION

The role of diet composition in PCOS needs further study. In particular, the utility of the gastrointestinal system in assisting energy restriction in PCOS needs to be assessed, as does the effect of high-protein and low-carbohydrate diets on satiety in this group. More data on the type and intensity of physical activity required to improve reproductive performance is needed. Whether this is modified by a concomitant energy-restricted diet will guide the development of well-defined lifestyle-intervention programs. These programs need to be tailored to PCOS patients and evaluated against and in addition to metformin. Understanding the motivation of responders and nonresponders to lifestyle intervention, both behavioral and metabolic, will also assist in tailoring clinical management of PCOS.

KEY POINTS

- PCOS is a significant cause of infertility and is related to excess androgen production and insulin resistance compounded by obesity.
- Weight loss and physical activity can significantly improve the reproductive milieu, with improvements noted within 2 weeks of introduction.
- Metformin is commonly used to improve insulin sensitivity mediated in part by a small weight loss.
- Structured dietary patterns with higher protein composition and/or lower glycemic index carbohydrates show promise, as does the use of meal replacements.
- Lifestyle programs incorporating weight management and physical activity can be as effective as pharmacotherapy, but have been underutilized in clinical management.

REFERENCES

1. Azziz R, Sanchez LA, Knochenhauer ES, et al. Androgen excess in women: experience with over 1000 consecutive patients. J Clin Endocrinol Metab 2004;89:453–462.
2. DeUgarte CM, Bartolucci AA, Azziz R. Prevalence of insulin resistance in the polycystic ovary syndrome using the homeostasis model assessment. Fertil Steril 2005;83:1454–1460.
3. Pasquali R, Gambineri A. Role of changes in dietary habits in polycystic ovary syndrome. Reprod Biomed Online 2004;8:431–439.
4. Rajkhowa M, Talbot JA, Jones PW, et al. Prevalence of an immunological LH beta-subunit variant in a UK population of healthy women and women with polycystic ovary syndrome. Clin Endocrinol (Oxf) 1995;43:297–303.
5. Acien P, Quereda F, Matallin P, et al. Insulin, androgens, and obesity in women with and without polycystic ovary syndrome: a heterogeneous group of disorders. Fertil Steril 1999;72:32–40.
6. Silfen ME, Manibo AM, Ferin M, McMahon DJ, Levine LS, Oberfield SE. Elevated free IGF-I levels in prepubertal Hispanic girls with premature adrenarche: relationship with hyperandrogenism and insulin sensitivity. J Clin Endocrinol Metab 2002;87:398–403.
7. Apridonidze T, Essah PA, Iuorno MJ, Nestler JE. Prevalence and characteristics of the metabolic syndrome in women with polycystic ovary syndrome. J Clin Endocrinol Metab 2005;90:1929–1935.
8. Lord JM, Flight IH, Norman RJ. Metformin in polycystic ovary syndrome: systematic review and meta-analysis. BMJ 2003;327:951–953.
9. Lord JM, Flight IH, Norman RJ. Insulin-sensitising drugs (metformin, troglitazone, rosiglitazone, pioglitazone, D-chiro-inositol) for polycystic ovary syndrome. Cochrane Database Syst Rev 2003;CD003053.
10. Nestler JE, Jakubowicz DJ, Evans WS, Pasquali R. Effects of metformin on spontaneous and clomiphene-induced ovulation in the polycystic ovary syndrome. N Engl J Med 1998;338:1876–1880.
11. Myers ER, Silva SG, Hafley G, Kunselman AR, Nestler JE, Legro RS. Estimating live birth rates after ovulation induction in polycystic ovary syndrome: sample size calculations for the pregnancy in polycystic ovary syndrome trial. Contemp Clin Trials 2005;26:271–280.

12. Pasquali R, Gambineri A, Biscotti D, et al. Effect of long-term treatment with metformin added to hypocaloric diet on body composition, fat distribution, and androgen and insulin levels in abdominally obese women with and without the polycystic ovary syndrome. J Clin Endocrinol Metab 2000;85:2767–2774.

13. Harborne LR, Sattar N, Norman JE, Fleming R. Metformin and weight loss in obese women with polycystic ovary syndrome: comparison of doses. J Clin Endocrinol Metab 2005;90:4593–4598.

14. Hoeger KM, Kochman L, Wixom N, Craig K, Miller RK, Guzick DS. A randomized, 48-week, placebo-controlled trial of intensive lifestyle modification and/or metformin therapy in overweight women with polycystic ovary syndrome: a pilot study. Fertil Steril 2004;82:421–429.

15. Jayagopal V, Kilpatrick ES, Holding S, Jennings PE, Atkin SL. Orlistat is as beneficial as metformin in the treatment of polycystic ovarian syndrome. Obstet Gynecol Surv 2005;60:442–443.

16. Sonmez AS, Yasar L, Savan K, et al. Comparison of the effects of acarbose and metformin use on ovulation rates in clomiphene citrate-resistant polycystic ovary syndrome. Hum Reprod 2005;20:175–179.

17. Tang T, Glanville J, Hayden CJ, White D, Barth JH, Balen AH. Combined lifestyle modification and metformin in obese patients with polycystic ovary syndrome. A randomized, placebo-controlled, double-blind multicentre study. Hum Reprod 2006;21(1):80–89.

18. Clark AM, Ledger W, Galletly C, et al. Weight loss results in significant improvement in pregnancy and ovulation rates in anovulatory obese women. Hum Reprod 1995;10:2705–2712.

19. Huber-Buchholz MM, Carey DG, Norman RJ. Restoration of reproductive potential by lifestyle modification in obese polycystic ovary syndrome: role of insulin sensitivity and luteinizing hormone. J Clin Endocrinol Metab 1999;84:1470–1474.

20. Moran LJ, Noakes M, Clifton PM, Tomlinson L, Norman RJ. Dietary composition in restoring reproductive and metabolic physiology in overweight women with polycystic ovary syndrome. J Clin Endocrinol Metab 2003;88:812–819.

21. Kiddy DS, Hamilton-Fairley D, Bush A, et al. Improvement in endocrine and ovarian function during dietary treatment of obese women with polycystic ovary syndrome. Clin Endocrinol (Oxf) 1992;36:105–111.

22. Homburg R. Adverse effects of luteinizing hormone on fertility: fact or fantasy. Baillieres Clin Obstet Gynaecol 1998;12:555–563.

23. van Dam EW, Roelfsema F, Veldhuis JD, et al. Increase in daily LH secretion in response to short-term calorie restriction in obese women with PCOS. Am J Physiol Endocrinol.Metab 2002;282:E865–E872.

24. van Dam EW, Roelfsema F, Veldhuis JD, et al. Retention of estradiol negative feedback relationship to LH predicts ovulation in response to caloric restriction and weight loss in obese patients with polycystic ovary syndrome. Am J Physiol Endocrinol Metab 2004;286:E615–E620.

25. Fedorcsak P, Dale PO, Storeng R, Tanbo T, Abyholm T. The impact of obesity and insulin resistance on the outcome of IVF or ICSI in women with polycystic ovarian syndrome. Hum Reprod 2001;16:1086–1091.

26. Pirozzo S, Summerbell C, Cameron C, Glasziou P. Should we recommend low-fat diets for obesity? Obes Rev 2003;4:83–90.

27. Raben A. Should obese patients be counselled to follow a low-glycaemic index diet? No. Obes Rev 2002;3:245–256.

28. Pawlak DB, Ebbeling CB, Ludwig DS. Should obese patients be counselled to follow a low-glycaemic index diet? Yes. Obes Rev 2002;3:235–243.

29. Brand-Miller JC. Glycemic index in relation to coronary disease. Asia Pac J Clin Nutr 2004;13:S3.

30. Brand-Miller JC, Holt SH, Pawlak DB, McMillan J. Glycemic index and obesity. Am J Clin Nutr 2002;76:281S–285S.

31. Sloth B, Krog-Mikkelsen I, Flint A, et al. No difference in body weight decrease between a low-glycemic-index and a high-glycemic-index diet but reduced LDL cholesterol after 10-wk ad libitum intake of the low-glycemic-index diet. Am J Clin Nutr 2004;80:337–347.

32. Carels RA, Darby LA, Douglass OM, Cacciapaglia HM, Rydin S. Education on the glycemic index of foods fails to improve treatment outcomes in a behavioral weight loss program. Eat Behav 2005;6:145–150.

33. Farnsworth E, Luscombe ND, Noakes M, Wittert G, Argyiou E, Clifton PM. Effect of a high-protein, energy-restricted diet on body composition, glycemic control, and lipid concentrations in overweight and obese hyperinsulinemic men and women. Am J Clin Nutr 2003;78:31–39.

34. Parker B, Noakes M, Luscombe N, Clifton P. Effect of a high-protein, high-monounsaturated fat weight loss diet on glycemic control and lipid levels in type 2 diabetes. Diabetes Care 2002;25:425–430.

35. Due A, Toubro S, Skov AR, Astrup A. Effect of normal-fat diets, either medium or high in protein, on body weight in overweight subjects: a randomised 1-year trial. Int J Obes Relat Metab Disord 2004;28:1283–1290.

36. Skov AR, Toubro S, Ronn B, Holm L, Astrup A. Randomized trial on protein vs carbohydrate in ad libitum fat reduced diet for the treatment of obesity. Int J Obes Relat Metab Disord 1999;23:528–536.

37. Stamets K, Taylor DS, Kunselman A, Demers LM, Pelkman CL, Legro RS. A randomized trial of the effects of two types of short-term hypocaloric diets on weight loss in women with polycystic ovary syndrome. Fertil Steril 2004;81:630–637.

38. Luscombe-Marsh ND, Noakes M, Wittert GA, Keogh JB, Foster P, Clifton PM. Carbohydrate-restricted diets high in either monounsaturated fat or protein are equally effective at promoting fat loss and improving blood lipids. Am J Clin Nutr 2005;81:762–772.

39. Moran LJ, Noakes M, Clifton PM, et al. Ghrelin and measures of satiety are altered in polycystic ovary syndrome but not differentially affected by diet composition. J Clin Endocrinol Metab 2004;89:3337–3344.
40. Escobar-Morreale HF, Botella-Carretero JI, Alvarez-Blasco F, Sancho J, San Millan JL. The polycystic ovary syndrome associated with morbid obesity may resolve after weight loss induced by bariatric surgery. J Clin Endocrinol Metab 2005;90(12):6364–6369.
41. Hirschberg AL, Naessen S, Stridsberg M, Bystrom B, Holtet J. Impaired cholecystokinin secretion and disturbed appetite regulation in women with polycystic ovary syndrome. Gynecol Endocrinol 2004;19(2):79–87.
42. Noakes M, Clifton P. Weight loss, diet composition and cardiovascular risk. Curr Opin Lipidol 2004;15:31–35.
43. Moran LJ, Noakes M, Clifton PM, Wittert G, Norman RJ. Short term energy restriction (using meal replacements) improves reproductive parameters in polycystic ovary syndrome. Asia Pac J Clin Nutr 2004;13:S88.
44. Randeva HS, Lewandowski KC, Drzewoski J, et al. Exercise decreases plasma total homocysteine in overweight young women with polycystic ovary syndrome. J Clin Endocrinol Metab 2002;87:4496–4501.
45. Jaatinen TA, Anttila L, Erkkola R, et al. Hormonal responses to physical exercise in patients with polycystic ovarian syndrome. Fertil Steril 1993;60:262–267.

INDEX